# Advances in Microwave Chemistry

# New Directions in Organic and Biological Chemistry

Series Editor

*Philip Page*

For more information about this series, please visit: https://www.crcpress.com/New-Direct
ions-in-Organic--Biological-Chemistry/book-series/CRCNDOBCHE

# Advances in Microwave Chemistry

Edited by
Bimal Krishna Banik
Debasish Bandyopadhyay

**CRC Press**
Taylor & Francis Group
Boca Raton London New York

CRC Press is an imprint of the
Taylor & Francis Group, an **informa** business

CRC Press
Taylor & Francis Group
6000 Broken Sound Parkway NW, Suite 300
Boca Raton, FL 33487-2742

First issued in paperback 2021

ISBN-13: 978-1-03-209416-8 (pbk)
ISBN-13: 978-0-8153-7519-7 (hbk)

# Contents

# Preface

Modern technology that facilitates the development of new science in a number of ways is the topic of current interest. Highly expeditious preparation and testing of new molecules have been demonstrated in recent years. For example, microwave-assisted reactions have been used extensively for the synthesis of privileged molecules with known and unknown structures. In many experiments, the superiority of microwave-assisted chemistry over the conventional heating method has been observed and proved. A few authors have attempted to explain the cause of the acceleration rates, stereoselectivity of the products, cleaner synthesis, increased yields, and the environmentally benign nature of these reactions that are associated with domestic and automated microwave-induced reactions. It has been argued that the reactants absorb energy, thereby permitting them to activate at the molecular level and helping to form a non-equilibrium situation, allowing molecules to transfer heat into other components. The cause of acceleration of numerous reactions by the microwave-induced method has expanded this field. However, both thermal and non-thermal microwave effects are proposed for rapid reactions that are observed under microwave-induced methods compared to conventional heating. The synthesis of molecules under microwave irradiation is one of the most attractive areas of current research in chemistry, both academic and industrial. In many instances, the molecules synthesized by microwave-induced reactions have demonstrated reasonable pharmacological activities.

Microwave-induced methods are not restricted to chemistry. Their relevance has expanded to other areas of science: medicine, biotechnology, biology, and engineering. Significant progress has been made in these areas using microwave technology. Some of the methods described herein are new and, therefore, the scope in these areas is unlimited.

The diversity of microwave-assisted chemistry as presented in this book is exceptionally high. The chapters are not organized based upon a specific subject matter. Rather, they are numbered in the order of acceptance date. Authors from Argentina, Brazil, China, India, and the United States have contributed chapters in this book.

Each of the twelve chapters presented in this book aimed to provide the contemporary aspects of advances in microwave-mediated science.

In Chapter 1, Jimenez, Zanin, Ferreira, Araújo, and Porto described recent applications of microwave radiation in biocatalysis. The synergism of the microwave radiation microbial cells, and enzymes is claimed to produce molecules that are of interest to chemistry, biology, and natural products, as well as the biodegradation of recalcitrant xenobiotic compounds. In Chapter 2, Alvarez, Lorenzetti, Acebal, Lista, and Domini demonstrated microwave-assisted sample preparation. In Chapter 3, Gomez, Aguinaga, LLmas, Garrido, Acebal, and Domini reported green microwave-mediated processes for degradation and microextraction. In Chapter 4, Wang and Zhong advanced the preparation and application of functional rare earth (RE)-based micro/nanomaterials using the microwave. In Chapter 5, Batra, Panwar, Pratap, and Nath described microwave-accelerated synthesis and functionalization of numerous oxygen-containing six-membered heterocyclic compounds: benzopyrans, 1, 3-dioxanes, 4H-pyrans, spiropyrans, chromones, xanthenes, and coumarins as well as some fused six-membered oxygen heterocycles. In Chapter 6, Xu investigated the origin of selectivities in the microwave-mediated organic reactions. Diverse selectivities, including chemoselectivity, regioselectivity, diastereoselectivity, and enantioselectivity in organic reactions were discussed under the microwave-assisted and conventional heating conditions. In Chapter 7, Brahmachari reported and updated microwave-induced Hirao and Kabachnik-Fields phosphorus-carbon bond forming reactions. Various examples of unusual carbon-phosphorous compounds were shown for the first time. In Chapter 8, Seelaboyina, Kumar, and Singh studied the microwave-induced synthesis of absorber layer materials for thin film photovoltaic application. Recent progress in the microwave-assisted synthesis of size and composition defined CIS, CIGS, and CZTS nanopowder for the absorber layer in TFPV application is reported. In Chapter 9, Shukla, Purohit, and Chakraborti studied microwave-assisted transition metal-catalyzed synthesis of pharmaceutically important heterocycles. Examples were provided where different heterocyclic scaffolds are obtained in the presence of a transition metal-based catalyst under microwave heating compared to that obtained by conventional heating. In Chapter 10, Bandyopadhyay and Banik discussed the microwave-induced synthesis of versatile important lactams. In Chapter 11, Sahoo, Banik, and Panda demonstrated the use of microwave technology in synthesizing various organic compounds following an eco-friendly route. In Chapter 12, Sahoo, Banik, and Panda investigated a green chemistry approach for the synthesis of drugs and drug candidates.

After reading this book, one may become convinced that the microwave has no boundary in science: it can be applied to almost all types of discoveries. In conclusion, this book provides an exploratory, as well as a concise and critical understanding of the significances in microwave research. Therefore, this book will be useful for scientists particularly for chemists, biologists, biotechnologists, and engineers who are working in diverse and specific new research areas.

We take the opportunity to thank the authors for their highly significant and exceptional contributions. Thanks are also due to Ms. Hilary Lafoe and Taylor & Francis who have realized that such a book on microwave chemistry will be extremely useful for researchers around the world.

**Bimal Krishna Banik**
**Debasish Bandyopadhyay**

# About the Editors

**Dr. Bimal Krishna Banik** is the Vice President of Research and Education Development of Community Health Systems of South Texas. He was a Tenured Full Professor and First President's Endowed Professor at the University of Texas-Pan American and an Assistant Professor of University of Texas M. D. Anderson Cancer Center for many years. He was awarded a Bachelor of Science Honors Degree in Chemistry from Itachuna Bejoy Narayan College and a Master of Science Degree in Chemistry from Burdwan University. He obtained his Ph.D. degree based upon his thesis work performed at the Indian Association for the Cultivation of Science, Jadavpur. Dr. Banik was a Postdoctoral Fellow at Case Western Reserve University (Ohio) and Stevens Institute of Technology (New Jersey). He is a Fellow (FRSC) and Chartered Chemist (CChem) of the Royal Society of Chemistry. Dr. Banik has been involved in organic, medicinal chemistry, and biomedical research for many years. As Principal Investigator, he has been awarded $7.25 million USD grants from National Institutes of Health, National Cancer Institute, Kleberg Foundation, University of Texas M. D. Anderson Cancer Center, University of Texas Health Science Center and University of Texas-Pan American (UTPA).

**Dr. Debasish Bandyopadhyay** studied chemistry at the Chandernagore Government College and the University of Burdwan (India). He received his Ph.D. in 2004 from the University of Calcutta (India), the oldest university in South-East Asia. He performed his first postdoctoral research at the same university with Professor Asima Chatterjee and Professor Julie Banerji. In 2007, he joined the University of Texas-Pan American as a NIH/NCI postdoctoral fellow with Professor Bimal Krishna Banik in the Department of Chemistry. In 2011, he was appointed as Assistant Professor of Research in the same department. His research foci include, but are not limited to, the development of greener methodologies to synthesize novel pharmacophores; design, in silico validation and synthesis of anticancer compounds; extraction, purification, structure elucidation, and chemical modification of plant natural products targeting pharmacologically active molecules. He has authored 67 international patent/book chapter/journal articles. He is currently engaged as Associate Editor with two internationally reputed journals and Editorial Board Member of eight international journals.

# List of Contributor

**Carolina C. Acebal**
INQUISUR, Departamento de Química
Universidad Nacional del Sur
Bahía Blanca, Argentina

**Maite V. Aguinaga**
INQUISUR, Departamento de Química
Universidad Nacional del Sur
Bahía Blanca, Argentina

**Mónica B. Alvarez**
INQUISUR, Departamento
    de Química
Universidad Nacional del Sur
Bahía Blanca, Argentina

**Yara J. K. Araújo**
Institute of Chemistry of São Carlos
University of São Paulo
São Paulo, Brazil

**Debasish Bandyopadhyay**
Department of Chemistry
University of Texas
Edinburg, Texas

**Bimal Krishna Banik**
Community Health Systems of South
    Texas
Edinburg, Texas

**Neha Batra**
Department of Chemistry
University of Delhi
Delhi, India

**Goutam Brahmachari**
Department of Chemistry
Visva-Bharati University
Santiniketan, India

**Asit K. Chakraborti**
Department of Medicinal Chemistry
National Institute of Pharmaceutical
    Education and Research (NIPER)
S.A.S Nagar, India

**Claudia E. Domini**
INQUISUR, Departamento de Química
Universidad Nacional del Sur
Bahía Blanca, Argentina

**Irlon M. Ferreira**
Institute of Chemistry of São Carlos
University of São Paulo
São Paulo, Brazil

**Mariano Garrido**
INQUISUR, Departamento de Química
Universidad Nacional del Sur
Bahía Blanca, Argentina

**Natalia A. Gomez**
INQUISUR, Departamento de Química
Universidad Nacional del Sur
Bahía Blanca, Argentina

**David E. Q. Jimenez**
Department of Chemistry
Federal University of Amapá
Amapá, Brazil

**Manoj Kumar**
Centre for Nanotechnology
Bharat Heavy Electricals Limited
    (BHEL) Corporate R & D,
Vikasnagar, India

**Adriana G. Lista**
INQUISUR, Departamento de Química
Universidad Nacional del Sur
Bahía Blanca, Argentina

**Natalia Llamas**
INQUISUR, Departamento de Química
Universidad Nacional del Sur
Bahía Blanca, Argentina

**Anabela Lorenzetti**
INQUISUR, Departamento de Química
Universidad Nacional del Sur
Bahía Blanca, Argentina

**Mahendra Nath**
Department of Chemistry
University of Delhi
Delhi, India

**Jnyanaranjan Panda**
Department of Pharmaceutical
    Chemistry
Roland Institute of Pharmaceutical
    Sciences
Berhampur, India

**Rahul Panwar**
Department of Chemistry
University of Delhi
Delhi, India

**André L. M. Porto**
Institute of Chemistry of São Carlos
University of São Paulo
São Paulo, Brazil

**Ramendra Pratap**
Department of Chemistry
University of Delhi
Delhi, India

**Priyank Purohit**
Department of Medicinal Chemistry
National Institute of
    Pharmaceutical Education
    and Research (NIPER)
S.A.S Nagar, India

**Biswa Mohan Sahoo**
Department of Pharmacy
Vikas Group of Institutions
Andhra Pradesh, India

**Raghunandan Seelaboyina**
Centre for Nanotechnology
Bharat Heavy Electricals Limited
    (BHEL) Corporate R & D,
Vikasnagar, India

**Dipti Shukla**
Department of Medicinal Chemistry
National Institute of Pharmaceutical
    Education and Research (NIPER)
S.A.S Nagar, India

**Kulvir Singh**
Centre for Nanotechnology
Bharat Heavy Electricals Limited
    (BHEL) Corporate R & D,
Vikasnagar, India

**Lei Wang**
College of Chemistry and Chemical
    Engineering
Jiangxi Normal University
Nanchang, China

**Jiaxi Xu**
Department of Organic Chemistry
Beijing University of Chemical
    Technology
Beijing, China

**Lucas Lima Zanin**
Institute of Chemistry of São Carlos
University of São Paulo
São Paulo, Brazil

**Shengliang Zhong**
College of Chemistry and Chemical
    Engineering
Jiangxi Normal University
Nanchang, China

# 1 Microwave Radiation in Biocatalysis

*David E. Q. Jimenez, Lucas Lima Zanin,*
*Irlon M. Ferreira, Yara J. K. Araújo,*
*and André L. M. Porto*

## CONTENTS

## 1.1 INTRODUCTION

Although microwave ovens manufactured for homes have been used since the 1970s, the first report that these energy sources were appropriately being used to accelerate organic reactions was in 1986. In their pioneering studies, Gedye [1] and Guiguere [2] used the domestic microwave as a tool for conducting organic reactions. In these studies, the authors described the results obtained in esterification reactions and cycloaddition with a domestic microwave apparatus [3].

The risk associated with the flammability of organic solvents and the lack of available systems to control temperature and pressure were the main reasons for using microwave reactors developed especially for organic synthesis. Today, this device is safe and allows the synthetic organic chemist control over all reaction parameters (temperature, pressure and power), thus achieving greater reproducibility and safety in the experiments [3, 4].

In the last decade, microwave radiation has been used to simplify and improve the reaction conditions of many classic organic reactions. Reactions carried out under microwave radiation are generally faster and cleaner and have better yields than reactions performed under conventional heating in similar conditions [5, 6]. The microwave methods provide an efficient and safe technology, according to the principles of "Green Chemistry" [7], because this technique enables solvent-free reactions to be performed, decreasing the number of competing side reactions, increasing the yield and reducing the reaction time [8–10].

More recently, microwave radiation became an important tool for performing bio-catalytic reactions. The potential of this technique has been exploited, particularly in the resolution of racemates to obtain enantiomerically pure compounds using immobilized lipases [11–14].

The organic synthesis presents a great contribution to obtain molecules with biological activities. Thus, the development of methodologies that apply the principles of Green Chemistry in the synthesis of new selective products with chemo-, regio- and enantio-selective and environmentally benign characteristics is required. Therefore, the use of microwave radiation in synthetic protocols has been very advantageous because the reactions are performed in a very short time in the absence of organic solvents and with a low consumption of energy [15].

### 1.1.1  PRINCIPLES OF MICROWAVE RADIATION

Microwaves are electromagnetic waves like energy carriers; these are located in the region of the electromagnetic spectrum between infrared light and radio waves in the frequency range between 300 and 300,000 MHz (Figure 1.1) [3, 7].

Domestic microwave ovens operate at a frequency of 2450 MHz (wavelength of 12.24 cm) to avoid interference with frequency telecommunications and mobile phones. According to the Federal Communications Commission (www.fcc.gov), only four frequencies are reserved for Industrial, Scientific and Medical (ISM) purposes: $915 \pm 25$, $2450 \pm 13$, $5800 \pm 75$ and $22,125 \pm 125$ MHz, with the most commonly used frequencies being 915 and 2450 MHz.

Microwave ovens that can process a frequency change of 0.9 to 18 GHz have been developed for the transformation of materials [16–18].

The microwave is a type of non-ionizing radiation capable of causing molecular motion in dipolar polarization and ionic conduction, but not changes in the molecular structure of molecules [11]. Since the energy of a microwave photon in this

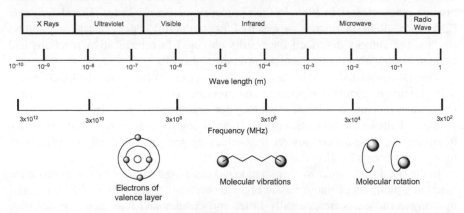

**FIGURE 1.1**  Illustration of electromagnetic spectrum. (Adapted from Young DD, Nichols J, Kelly RM, Deiters A (2008) Microwave activation of enzymatic catalysis. *Journal of the American Chemical Society* 130: 10048–10049.)

frequency region is 0.037 kcal.mol$^{-1}$, very low energy is needed to break a chemical bond, which is generally of the order of 80–120 kcal.mol$^{-1}$ [19].

The conventional heating process is fundamentally different to microwave heating upon radiation. In conventional heating an external power source reaches the walls of the flask and heat is transferred from the flask surface into the solution (reagents and solvents); through the driving process, such heating can often cause a convection current in solution. In contrast, with heating under microwave radiation, the energy is transferred directly to the substances by molecular interaction with the ions dissolved in the solution and/or the solvent; thus localized overheating of the substance absorbs the microwave (Figure 1.2) [16, 17]. This type of heating will depend on the ability of that particular material, reagent or solvent to absorb the microwave energy and convert it into heat [19].

Microwave heating, also called dielectric heating, converts electromagnetic energy into heat by two main mechanisms: dipolar polarization and ionic conduction [3]. The mechanism of dipolar polarization associated with the alignment of the molecules has permanent or induced dipoles with the applied electric field. When the field is removed the molecules return to a disordered state, and the energy that was absorbed into this guidance, these dipoles, is dissipated as heat, causing the molecules to be quickly heated in the system (Figure 1.3) [18, 20].

The ionic conduction process consists of an electromagnetic migration of ions when an electric field is applied in the solution. The friction generated between these ions in solution, due to the migratory movement, causes heating of the solution through friction losses. These losses depend on the size, charge and conductivity of dissolved ions and their interaction with the solvent. The applied electric field alternates quickly, about 5 billion times per second to the applied frequency 2450 MHz [20, 21]. This oscillation causes considerable intermolecular friction resulting in the generation of heat. A schematic representation is shown in Figure 1.4 for the driving ionic mechanism [20].

When the power microwave focuses on the material (or substance), there are three possibilities regarding the penetration of an electromagnetic wave: reflection, transparency and absorption (Figure 1.5). Materials such as metals reflect the microwave without being affected by it due to the high dielectric loss factor ($\varepsilon''$) and intensity of penetration of the electromagnetic wave close to zero. When the microwave

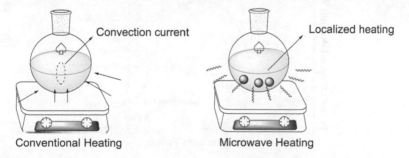

**FIGURE 1.2** Illustration of conventional and microwave heating processes. (From Kingston HM, Jassie LB. *Introduction to microwave sample preparation: theory and practice.* American Chemical Society, ACS professional reference book, Washington DC, 1998. With permission.)

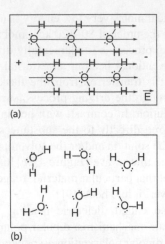

(a)

(b)

**FIGURE 1.3**  Illustration for molecular behavior in the presence of an electromagnetic field. (a) Polarized molecules aligned with the electromagnetic field. (b) Thermally induced disorder when the electromagnetic field changes occur. E=electric field. (From Souza GB, Nogueira ARA, Rassini JB. *Circular Técnica da Embrapa,* 33: 1–9, 2002. With permission.)

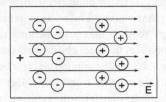

**FIGURE 1.4**  Illustration of ionic conduction in the presence of an electromagnetic field. E=electric field. (From Sanseverino AM. "Micro-ondas em síntese orgânica." *Química Nova* 25: 660–667, 2002. With permission.)

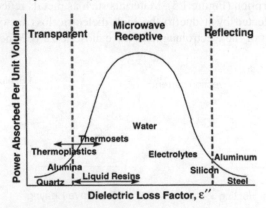

**FIGURE 1.5**  Relationship between the dielectric loss factor and ability to absorb microwave power for some materials. (From Thostenson ET, Chou TW. "Microwave processing: fundamentals and applications." *Composites Part A. Applied Science and Manufacturing* 30: 1055–1071, 1999. With permission.) (License Number: 3614930551491.)

passes through the material without causing any effect, this material is transparent to microwave. When a material absorbs the microwave, there is an interaction of the electromagnetic wave with the electric dipole of the molecule at the molecular level [17]. Physico-chemical properties such as concentration, heat capacity, molecular structure and dielectric constant affect the absorption of energy [20].

The heat amount of a material under microwave radiation will depend on its dielectric properties, i.e., the ability of this specific substance to convert electromagnetic energy into heat at a given temperature and frequency. This is determined by the dissipation factor (tan δ) which is measured by the ratio of the dielectric loss factor (ε″) and the dielectric constant (ε′) of the substance (tan δ = ε″/ε′). The dielectric loss factor measures the efficiency of electromagnetic energy conversion into heat. Thus, the higher the dissipation factor, the greater the substances will be heated under microwave [18, 19]. Polar molecules and ionic solutions can strongly absorb microwave energy to present a permanent dipole moment, and nonpolar solvents, such as hexane, do not heat under microwave radiation [18]. Another important factor is the dielectric constant, which is a measure of molecular polarity.

When polar solvents are used in a microwave over a high dissipation factor, they can still be rapidly overheated to temperatures above the boiling point; in this case it is necessary to use the microwave reactor vials sealed with the "closed vessel", enhancing the efficiency and speed of the reaction. This rapid increase in the temperature can be further enhanced for extreme solvents using dissipation factors such as ionic liquid [22]. Obviously, a particular solvent with a low dissipation factor may heat up significantly in the presence of polar compounds, such as salts or ionic liquid.

Table 1.1 presents the physical parameters, such as dielectric constant and dissipation factor, for some solvents [18].

The chemical reactions carried out in the presence of microwave radiation have several advantages including: reduction of competitive reactions, shorter reaction time, increased productivity and better reproducibility [23]. These effects can be rationalized by three principles: thermal effects, specific effects of microwave and non-thermal effects of microwave. The thermal effects can be understood as a direct result of the high reaction temperatures, which can be obtained when the reactants and polar solvents are irradiated by microwave. The decrease in reaction time at elevated temperatures may be evidenced by the implementation of the Arrhenius Law [24].

The specific effects of microwave can be defined as acceleration rates, which cannot be achieved by conventional heating, for example, overheating of solvents at atmospheric pressure, the selective heating of the reactants and the formation of "hot spots" (high-temperature isolated points in the reaction medium). Mainly, although these effects are produced exclusively in the presence of microwave radiation, observed results are due to the increase in reaction temperature [8, 25].

The non-thermal effects of microwave result from a direct interaction between the electric field and the species present in the reaction medium [9, 26]. The presence of an electric field affects the dipolar orientation in some types of reactions by increasing the factor of the pre-exponential Arrhenius equation and decreasing the ΔG activation of the transition state. The first is based on the increased probability of intermolecular shocks due to subsequent changes in the orientation of polar

**TABLE 1.1**
**Data of Dielectric Constant ($\varepsilon'$), Dielectric Loss ($\varepsilon''$), tan δ and Boiling Point (p.f.) for Some Organic Solvents**

| Solvent | $\varepsilon'$ | $\varepsilon''$ | tan δ | b.p. (°C) | b.p.* |
|---|---|---|---|---|---|
| Hexane | 1.89 | 0.038 | 0.020 | 69 | N.a. |
| Toluene | 2.4 | 0.096 | 0.040 | 111 | N.a. |
| Dichloromethane | 9.1 | 0.382 | 0.042 | 40 | N.a. |
| THF | 7.4 | 0.348 | 0.047 | 66 | N.a. |
| Acetone | 20.7 | 1.118 | 0.054 | 56 | 164 |
| Ethyl acetate | 6.0 | 0.354 | 0.054 | 77 | N.a. |
| Acetonitrile | 37.5 | 2.325 | 0.062 | 82 | 194 |
| Nitromethane | 36.0 | 2.034 | 0.064 | 101 | N.a. |
| Chloroform | 4.8 | 0.437 | 0.091 | 61 | N.a. |
| Chlorobenzene | 2.6 | 0.263 | 0.101 | 132 | N.a. |
| Water | 78.3 | 9.889 | 0.123 | 100 | N.a. |
| 1,2-Dichloroethane | 10.4 | 1.321 | 0.127 | 83 | N.a. |
| Methanol | 32.6 | 21.483 | 0.659 | 65 | 151 |
| 2-Propanol | 19.9 | 14.622 | 0.799 | 82 | 145 |
| Ethanol | 24.3 | 22.866 | 0.941 | 78 | 164 |

N.a. = not appropriate to use in closed vessel; b.p. = boiling point; b.p.* = point temperature in closed vessel.

Source:　Eskilsson CS, Bjorklund E, 2000. "Analytical-scale microwave-assisted extraction." *Journal of Chromatography* A 902: 227-250.

molecules as a function of the electric field oscillation. The second suggests that the mechanisms of formation of polar charged species in the transition state must be favored by microwave radiation due to the interaction with the electric field generated [24, 25].

## 1.1.2 INFLUENCE OF MICROWAVE RADIATION ON ENZYMES

The effects of microwave radiation in terms of the rate and selectivity of the enzymatic reactions have not been fully clarified; however, there are some studies that suggest explanations for the observed effects.

Yadav and Lathi carried out a study of transesterification reactions of methyl acetate with different alcohols using Novozyme 435® lipase as a catalyst under microwave radiation. The authors found full conversion and the reaction rate was higher under microwave radiation than conventional heating. However, after checking the kinetic model based on the initial rates, they found no change in the ping-pong Bi-Bi mechanism using lipase. The authors suggest that the enzyme seems to become slightly different when heated under microwave radiation [27].

Rejasse et al. analyzed a lot of biocatalytic reactions in the presence of microwave radiation and concluded that in aqueous media the properties of irradiated enzymes are identical to those obtained under conventional heating. In non-aqueous media activity, selectivity and stability of enzymes can be improved under microwave heating, which occurs through absorption of electromagnetic energy polar species [28].

Some conformational changes can occur with enzymes in the presence of microwave. In a study with β-glucosidase from *Pyrococcus furiosus* in the presence of microwave radiation, Young et al. (2008) showed an increased molecular mobility of the enzyme structure by rapidly induced dipole alignment of the peptide bond with the oscillating electric field. This structural change can lead to an increased activity or even inactivation of the enzyme compared with conventional heating [11].

An interesting observation of the effect of electromagnetic fields on biological structures was reported by Laurence et al. (2000) that showed that the energy of the microwave was sufficient to induce a conformational change in some proteins. The water molecules that are around the proteins form hydrogen bonds with the hydrophilic regions of the protein and are essential to maintaining their structure of the same. Upon increasing temperature, these bonds are disrupted and alter the conformation of the protein [29].

The influence of non-thermal effects on the reactivity and selectivity of reactions was described by Loupy et al. (2001). They showed that in polar mechanisms the transition state is more polar than the ground state, resulting in an increase of the reactivity by decreasing the activation energy. This study suggests that stabilization of the transition state can occur in two types of mechanisms: (a) neutral species reactions leading to a transition state dipole; (b) loaded nucleophilic reactions leading to the dissociated ion pairs in the transition state [30].

The research of Loupy et al. (2001) showed that the transition state for biocatalytic reactions under microwave radiation involves the formation of tetrahedral enzyme-substrate complexes that have dipole characteristics and the reaction occurs quickly (Figure 1.6) [30, 31]. Figure 1.7 shows the relative stabilization of a more polar transition state when compared to the ground state [32].

Finally, although the influence of microwave on the mechanism of enzymatic reactions has been studied for more than two decades, there is still no consensus on the real effects of electromagnetic radiation on the structure and activity of proteins.

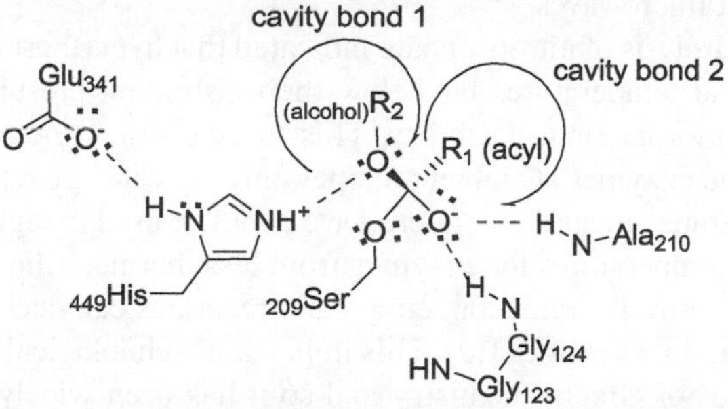

**FIGURE 1.6** Possible tetrahedral enzyme-substrate complex presenting dipolar characteristics for ester hydrolysis catalyzed using *Candida rugosa* lipase. (Adapted from Cygler M, Grochulski P, Kazlauskas JK, Schrag JD, Bouthillier F, Rubin B, Serreqi AN, Gupta AK, 1994. A structural basis for the chiral preferences of lipases. *Journal of American Chemical Society* 116: 3180-3186.)

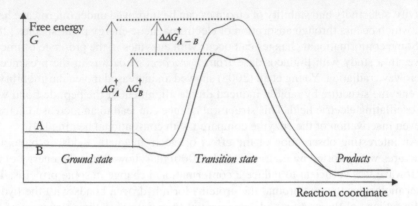

**FIGURE 1.7** Relative stabilization of more polar transition state when compared to ground state. (From Fransson L. *Molecular modeling: understanding and prediction of enzyme selectivity. Licentiate thesis.* Royal Institute of Technology School of Biotechnology, AlbaNova University Center SE-106 91 Stockholm, Sweden, 2009. With permission.)

## 1.2 APPLICATION OF MICROWAVE RADIATION IN BIOCATALYSIS

Microwave radiation has been used for increasing the yield and selectivity in biocatalytic reactions and decreasing the reaction times. Lipases are used as versatile biocatalysts in biological and chemical process, such as modification fats and lipids, kinetic resolutions of alcohols and amines, hydrolysis, and esterification and interesterification reactions [33, 34].

### 1.2.1 USE OF ISOLATED ENZYMES IN BIOCATALYSIS UNDER MICROWAVE RADIATION AND CONVENTIONAL HEATING

Enzymes can be used in synergism between biocatalytic processes with microwave radiation. In general, immobilized enzymes and thermophilic enzymes are certainly more stable under microwaves.

Enzymatic hydrolysis of nitrophenoate indicated that hyperthermophilic enzymes can be activated at temperatures far below their optimum, presumably by microwave-induced conformational flexibility. This finding offers the prospect of using hyperthermophilic enzymes at ambient temperatures to catalyze reactions with thermally labile substrates. In addition, microwave could be used to regulate biocatalytic rates at very low temperatures for enzymes from less thermophilic sources [11].

When using microwave radiation, enzymatic reactions can occur in minutes and with high yields and selectivities [12]. This important technological tool is consistent with the principles of Green Chemistry and so it has been widely used in organic reactions and more recently in reactions catalyzed by enzymes [10, 22, 35].

Ribeiro et al. (2012) researched the Enzymatic Kinetic Resolution (EKR) of (±)-mandelonitrile by *Candida antarctica* lipase (CALB) under microwave radiation and conventional heating in toluene. The ethyl (*S*)-mandelonitrile was obtained with an enantiomeric excess of 92% and isolated yield of 35% at 8 h of reaction under

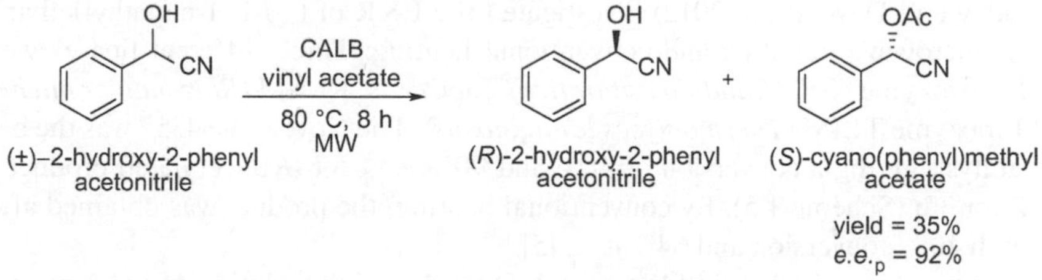

SCHEME 1.1 EKR of (±)-mandelonitrile under MW.

microwave radiation (Scheme 1.1). The transesterification of (±)-mandelonitrile conducted under conventional heating yielded the (R)-mandelonitrile with 51% enantiomeric excess and the (S)-mandelonitrile with 98% enantiomeric excess after 184 h [13].

In another study, Ribeiro et al. (2013) carried out enzymatic resolution on a number of fluoroaromatic compounds using CALB in toluene and vinyl acetate as an acylating agent. The reactions showed shorter reaction times (4–14 h), higher enantiomeric excesses for the (S)- or (R)-acetates and better yields under microwave radiation in comparison with conventional heating (48–144 h) (Scheme 1.2) [36].

Yu et al. (2007) conducted the kinetic resolution of (±)-2-octanol catalyzed by CALB using vinyl acetate under microwave radiation at 60°C for 2 h to yield the (S)-2-octanol with 50.5% conversion and enantiomeric excess of 99% (Scheme 1.3). Reaction under microwave radiation afforded 50% conversion of acetylated (R)-enantiomer (3 h) and conventional heating was required for 12 h at 40°C [37]. In addition, Yu et al. (2012) investigated the enzymatic resolution of (±)-2-octanol under microwave radiation in the ionic liquid. In this study, the reaction under microwave radiation afforded better results than in solvent-free systems under conventional heating [22].

Souza et al. (2009) reported the kinetic resolution of (±)-1-phenylethanol with vinyl acetate as acyl donor and cyclohexane as solvent under microwave heating and conventional heating. In this study, the authors concluded about the existence of non-thermal microwave effects in the kinetic resolution of a secondary alcohol with five immobilized lipases. This study showed that lipases do not differ when heating in an oil bath or microwave radiation using (±)-1-phenylethanol as substrate (Scheme 1.4) [38].

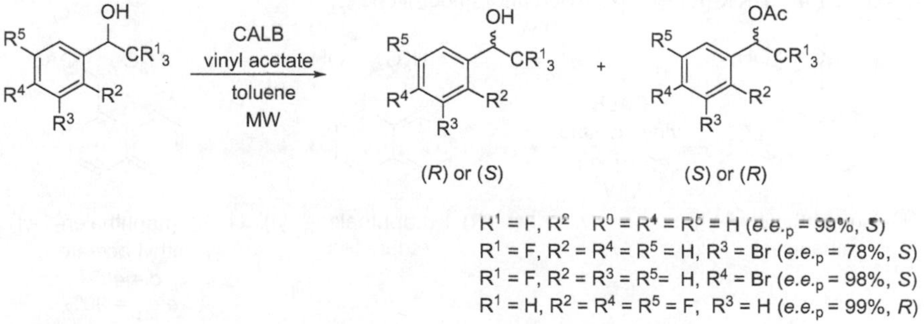

$R^1 = F, R^2$  $R^0 = R^4 = R^5 = H$ $(e.e._p = 99\%, S)$
$R^1 = F, R^2 = R^4 = R^5 = H, R^3 = Br$ $(e.e._p = 78\%, S)$
$R^1 = F, R^2 = R^3 = R^5 = H, R^4 = Br$ $(e.e._p = 98\%, S)$
$R^1 = H, R^2 = R^4 = R^5 = F, R^3 = H$ $(e.e._p = 99\%, R)$

SCHEME 1.2 EKR of fluoroaromatic compounds under MW.

Yadav and Devendran (2012) investigated the EKR of (±)-1-(1-naphthyl)ethanol under microwave radiation and conventional heating. Three different lipases were used: Novozyme 435® (*Candida antarctica*), Lipozyme RMIM (*Rhizomucor miehei*) and Lipozyme TLIM (*Thermomyces lanuginosus*). The Novozyme 435® was the best biocatalyst, giving a conversion of 48% and 90% $e.e._p$ for (*R*)-acetylated product at 60°C for 3 h (Scheme 1.5). By conventional heating, the product was obtained after 5 h with 40% conversion and 64% $e.e._p$ [5].

Devendran and Yadav (2014) researched the kinetic resolution of (±)-1-phenyl-2-propyn-1-ol through transesterification reaction with acyl acetate to evaluation of synergism between microwave radiation and enzymatic catalysis. Lipases from different microbial origins were employed in this study (Novozyme 435®, Lipozyme RM-IM and Lipozyme TL IM). The *Candida antarctica* lipase B, immobilized on acrylic resin was the best catalyst in *n*-hexane with vinyl acetate as acyl donor. Under optimum conditions, maximum conversion (48.78%) and high enantiomeric excess (93.25%) were obtained in 2 h at 60°C under MW in synergism with CALB (Scheme 1.6) [39].

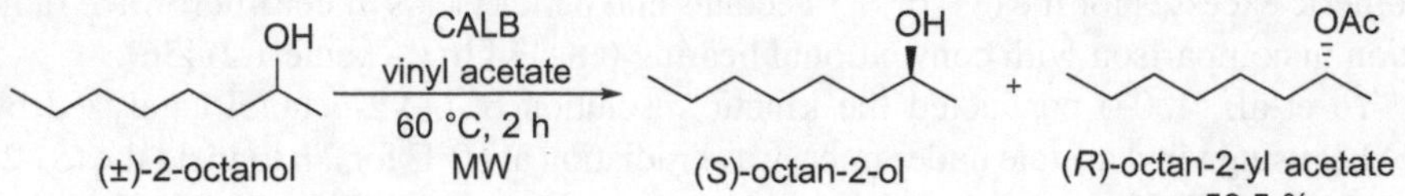

**SCHEME 1.3**    EKR of (±)-2-octanol under MW.

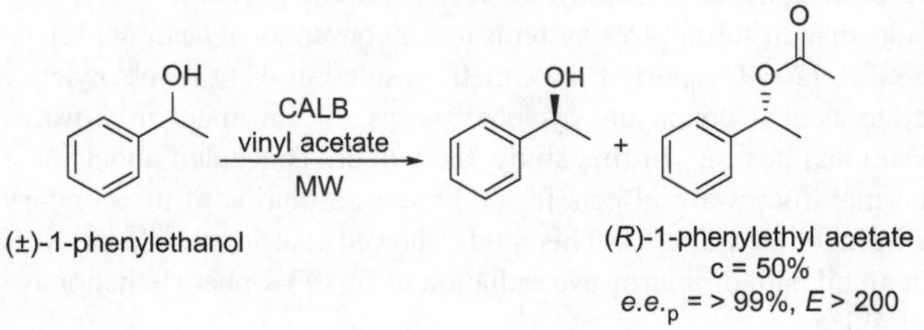

**SCHEME 1.4**    EKR of (±)-1-phenylethanol under MW.

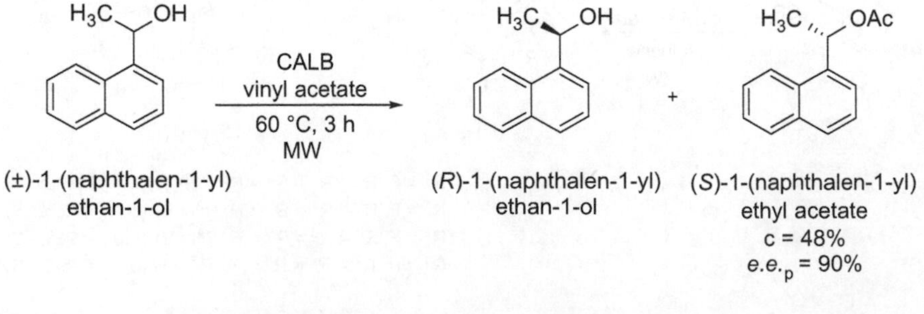

**SCHEME 1.5**    EKR of (±)-1-(1-naphthyl) ethanol under MW.

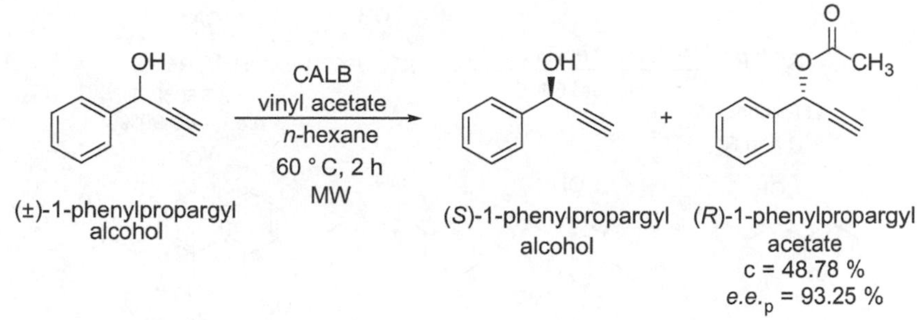

**SCHEME 1.6** EKR of (±)-1-phenylpropargyl alcohol under MW.

Rouillard et al. (2014) reported the use of lipases under microwave radiation on the kinetic resolution of homochiral (±)-(Z)-cyclooct-5-ene-1,2-diol and (±)-(Z)-2-acetoxycyclooct-4-enyl acetate. In order to best achieve the kinetic resolution, different parameters were studied including the type of lipase, temperature and the impact of microwave power compared to conventional heating for kinetic resolution of (±)-(Z)-cyclooct-5-ene-1,2-diol. Optimization of the reaction parameters led to highly enriched mono and diacetylated products in a clean, efficient and safe way (Scheme 1.7) [40].

EKR of secondary alcohols was performed using CALB in toluene at 60°C under microwave radiation and conventional heating. The conversions under microwave radiation were obtained between 30–50% and 43–99% $e.e._p$. However, conventional heating obtained conversions of 6–36% and 6–99% $e.e._p$ (Scheme 1.8) [23].

The regioselective esterification of isoquercitrin and floridzine with different saturated (oleic, stearic, linoleic, linolenic, eicosapentaenoic and docosahexaenoic acid), monounsaturated and polyunsaturated fatty acids was performed under three different reaction conditions: (i) by conventional heating; ii) by microwave radiation in acetone; and iii) by microwave radiation in the absence of solvent. When lipase Novozyme 435® was used as a biocatalyst at temperatures of 45–60°C, the reaction times varied from 18 to 24 h, giving yields of 81–97% for conventional heating. The reaction under microwave radiation in acetone gave the monoacetylated isoquercitrin in 120 s with 98% yield (stearic acid) and the floridzine acetylated in 160 s with 98% yield (linolenic acid) (Scheme 1.9). The most effective method for the reaction without solvent yielded 98% for acetylated isoquercitrin and 85% for acetylated floridzine with reaction times of 75 and 105 s, respectively [6]. The enzymatic acetylation

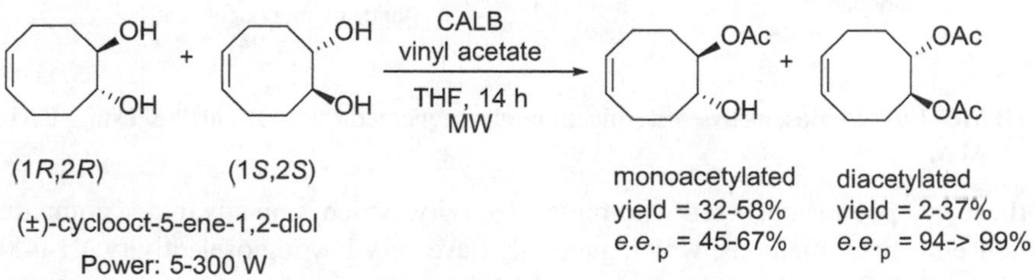

**SCHEME 1.7** Chemoenzymatic acetylation of (±)-diols using immobilized CALB under MW.

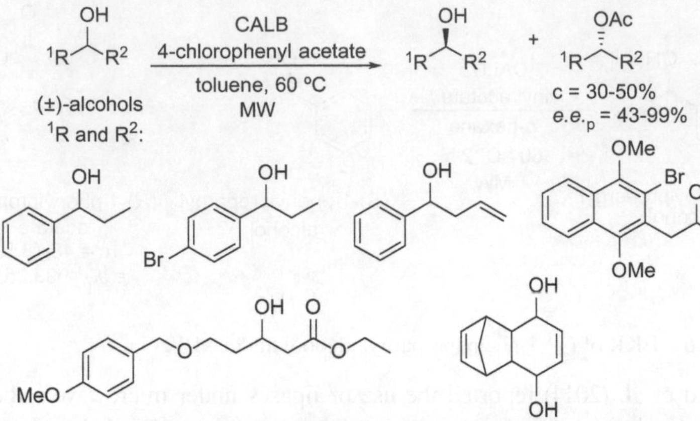

**SCHEME 1.8**   EKR of (±)-secondary alcohols under MW.

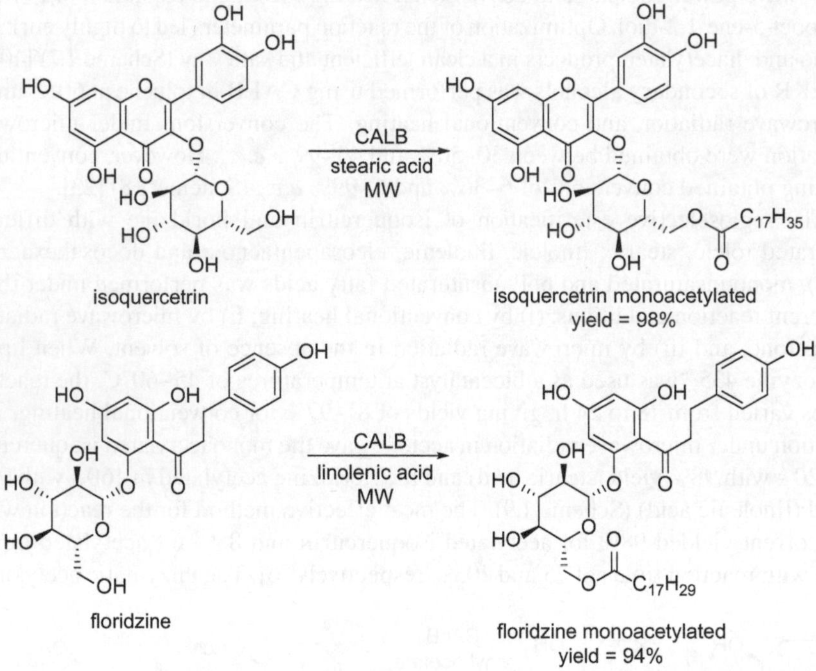

**SCHEME 1.9**   Regioselective esterification of isoquercetin and floridzine using CALB under MW.

of these compounds occurred with regioselectivity, which is an advantage compared to non-enzymatic methods, which generally have very low regioselectivity [41–45].

Yadav and Devendran (2012) researched the synthesis of isoamyl myristate in a solvent-free reaction using CALB under microwave radiation through the esterification of myristic acid and isoamyl alcohol. In 60 min, the reaction gave 96% conversion

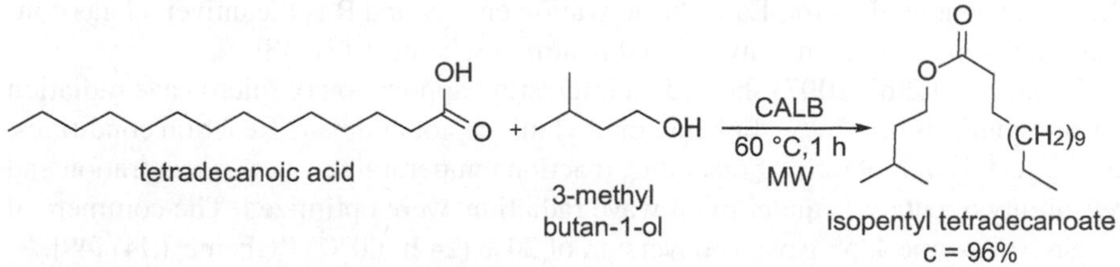

**SCHEME 1.10** Biocatalytic synthesis of isoamyl myristate under MW.

at 60°C, while conventional heating at 60°C gave 56% conversion in the same time. This reaction was performed in the absence of solvent with CALB, the best biocatalyst among the evaluated enzymes [Lipozyme RM IM (*Rhizomucor miehei*) and Lipozyme TL IM (*Thermomyces lanuginosus*)]. Isoamyl myristate is a compound used in the food, cosmetic and pharmaceutical industries (Scheme 1.10) [5].

Chemoenzymatic esterification of 4-chloro-2-methylphenoxyacetic acid (MCPA) was investigated by Shinde and Yadav (2014) using immobilized enzymes (Novozyme 435®, Lipozyme TL IM, Lipozyme RM IM and Lipase AYS Amano) under microwave radiation. The Novozyme 435® was the best catalyst, yielding 83% of the MCPA ester using *n*-butanol as an acylating agent and 1,4-dioxane as a solvent (60°C, 6 h) (Scheme 1.11) [46].

Yang et al. (2013) reported the enantioselective esterification of Ibuprofen in organic solvent was catalyzed by recombinant APE 1547, a thermophilic esterase from the archaeon *Aeropyrum pernix* K1. The reaction was performed under microwave radiation and in the optimum conditions, the enzyme activity from APE 1547 was 4.16 $\mu$mol$^{-1}$ mg$^{-1}$ h$^{-1}$ and the enantioselectivity ($E$) was 52.9. Compared with conventional heating, the enzyme activity and the enantioselectivity were increased about 21.9-fold and 1.4-fold, respectively. This study showed that APE 1547 can maintain 95% of its activity even after being used five times, suggesting that the enzyme is stable in low power under MW (Scheme 1.12) [47].

The enzymatic esterification between *n*-caprylic acid and *n*-pentanol catalyzed by lipozyme RM IM (lipase from *Mucor miehei* immobilized on an anionic resin) under microwave radiation and conventional heating was investigated by Huang et al. (2005). The reaction under MW reduced the apparent activation energy of the enzymatic reaction according to the Arrhenius equation ($k = Ae^{-Ea/RT}$), where A is

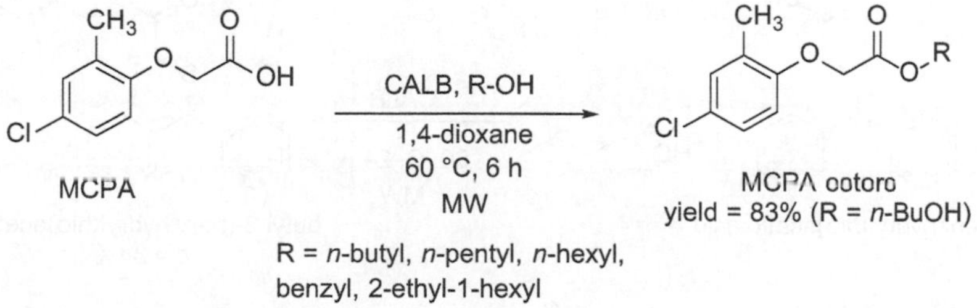

R = *n*-butyl, *n*-pentyl, *n*-hexyl, benzyl, 2-ethyl-1-hexyl

**SCHEME 1.11** Esterification of MCPA under MW.

the pre-exponential factor, Ea is the activation energy and R is the universal gas constant in comparison with conventional heating (Scheme 1.13) [48].

Yadav and Lathi (2007) showed an interesting application of microwave radiation for the synthesis of *n*-butyl diphenyl methyl mercaptobutanoate. Reaction conditions, such as catalyst, mol ratio of reactants, reaction temperature, water concentration and reusability of catalysts under microwave radiation, were optimized. The commercial lipase Novozyme 435® gave a conversion of 34% (24 h, 60°C) (Scheme 1.14) [49].

Risso et al. (2012) employed the lipase B from *Candida antarctica* (Novozyme 435®) with non-activated acyl donors under microwave radiation. The synergism of CALB and microwave radiation presented excellent conditions in solvent-free and afforded high yields (up to 96%) in the transesterification of β-ketoesters (Scheme 1.15) [8].

β-Ketoesters are important synthons since they can be easily transformed into chiral building blocks via enzymatic transformation. Yadav and Lathi (2007) studied

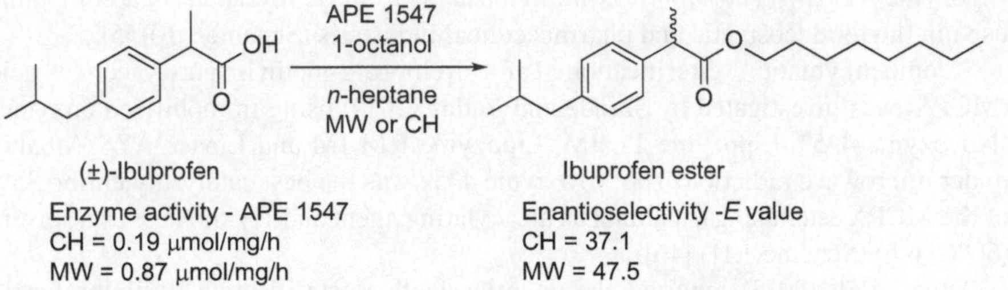
APE 1547
1-octanol
OH
O
n-heptane
MW or CH
O
O
(±)-Ibuprofen
Ibuprofen ester
Enzyme activity - APE 1547
CH = 0.19 μmol/mg/h
MW = 0.87 μmol/mg/h
Enantioselectivity -E value
CH = 37.1
MW = 47.5

**SCHEME 1.12** Esterification of (±)-Ibuprofen by recombinant lipase from *A. pernix* under MW and CH.

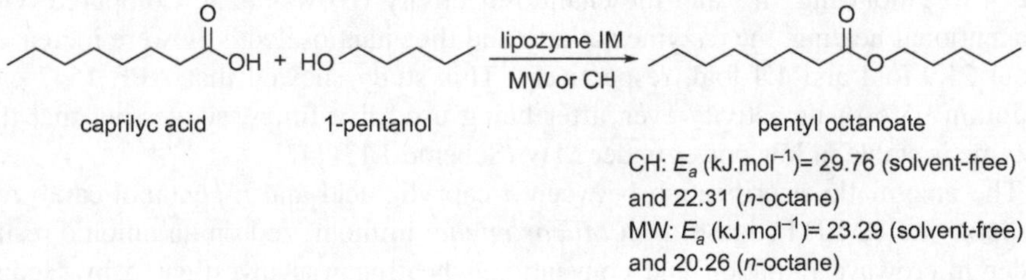
O
OH + HO
lipozyme IM
MW or CH
O
O
caprilyc acid
1-pentanol
pentyl octanoate
CH: Ea (kJ.mol−1)= 29.76 (solvent-free)
and 22.31 (n-octane)
MW: Ea (kJ.mol−1)= 23.29 (solvent-free)
and 20.26 (n-octane)

**SCHEME 1.13** Esterification of caprylic acid using lipozyme RM IM under MW and CH.

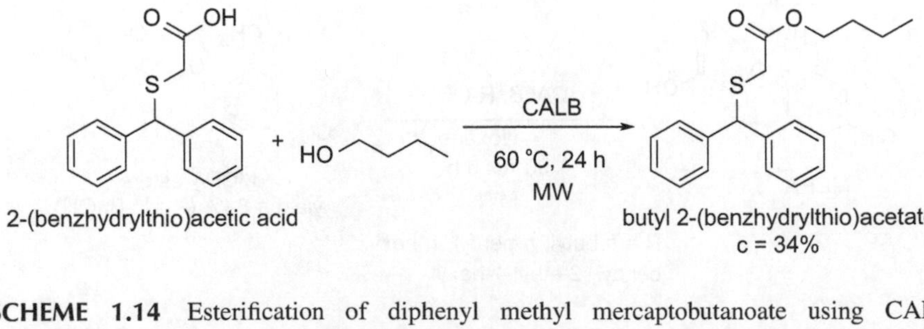
O
OH
S
+
HO
CALB
60 °C, 24 h
MW
O
O
S
2-(benzhydrylthio)acetic acid
butyl 2-(benzhydrylthio)acetate
c = 34%

**SCHEME 1.14** Esterification of diphenyl methyl mercaptobutanoate using CALB under MW.

the transesterification of methyl acetoacetate with various alcohols (1-propanol, 1-butanol, 1-pentanol, 1-hexanol, 1-octanol, 1-decanol, 2-propanol, 2-butanol and 2-pentanol) in the presence of immobilized lipases (Novozyme 435®, Lipozyme RM IM and Lipozyme TL IM) under microwave radiation [49]. This study found that the overall conversion as well as the rate of reaction was higher under microwave than under conventional heating. The reaction under microwave radiation afforded 74% yield (60 min, 50°C) and under conventional heating obtained 57% yield using CALB and 1-butanol (Scheme 1.16) [49].

Yadav and Borkar (2009) described the use of *Candida antarctica* lipase B immobilized on polyacrylic resin to synthesize citronellol acetate under microwave radiation in comparison with conventional heating (Scheme 1.17). The effects of various parameters affecting the conversion in transesterification reaction were studied. In the reaction under microwave radiation, there was an increase in lipase activity due to the enhanced collision of molecules, which can in turn be attributed to an increase in the entropy of the system. Microwave radiation leads to an increase in the affinity of the substrate toward the active site of lipase. The frequency factor was increased by 190 fold in the presence of microwaves over conventional heating [14].

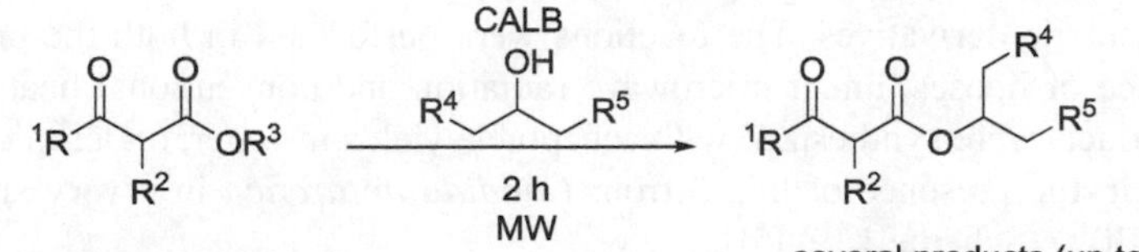

$R^1$ = Me, $R^2$ = H, $R^3$ = Et
$R^1$ = $R^2$ = Me, $R^3$ = Et
$R^1$ = Me, $R^2$ = H, $R^3$ = *t*-Bu
$R^1$ = $R^3$ = Et, $R^2$ = H
$R^1$ = iso-Pr, Me, $R^2$ = H, $R^3$ = Et

several products (up to 96%)
$R^1$ = Me, $R^2$ = H, $R^5$ = Et, $R^4$ = H
$R^1$ = $R^2$ = Me, $R^5$ = Et, $R^4$ = H
$R^1$ = Me, $R^2$ = H, $R^4$ = $R^5$ = Et
$R^1$ = $R^4$ = Et, $R^2$ = H
$R^1$ = iso-Pr, Me, $R^2$ = H, $R^4$ = $R^5$ = CH$_3$

**SCHEME 1.15**   Transesterification of esters using CALB under MW.

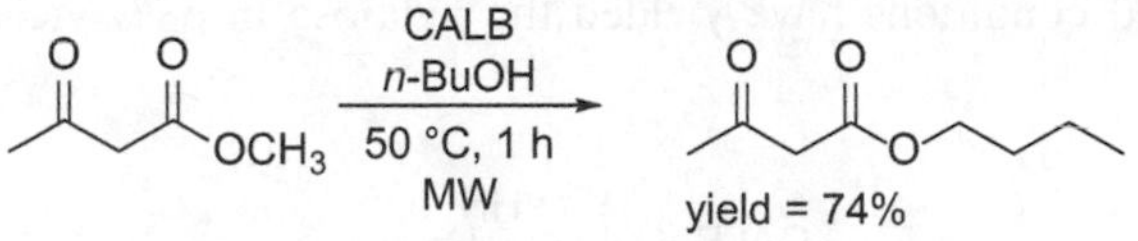

**SCHEME 1.16**   Transesterification of methyl acetoacetate using CALB under MW.

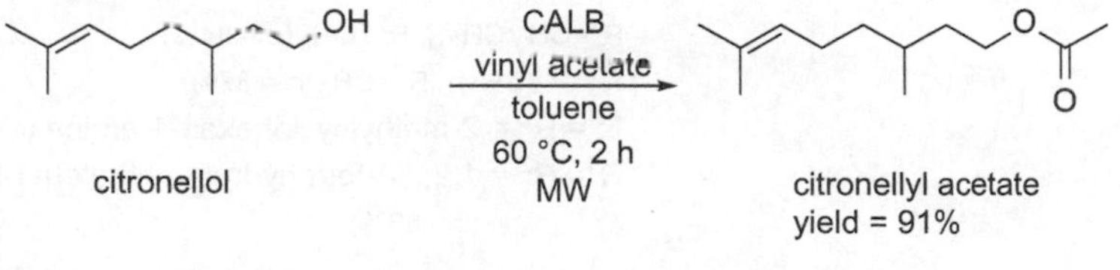

**SCHEME 1.17**   Esterification of citronellol using CALB under MW.

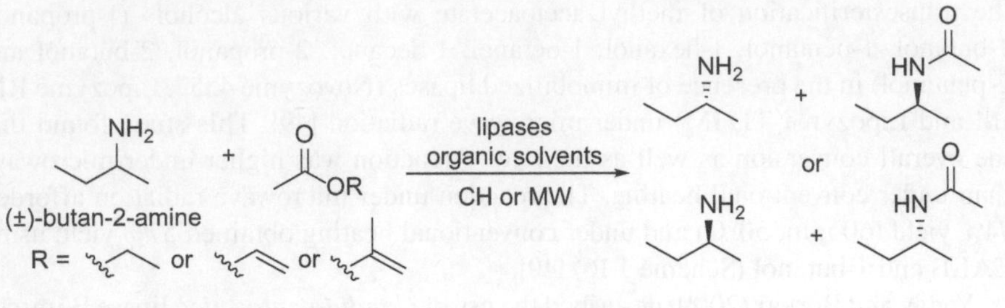

**SCHEME 1.18** EKR of (±)-*sec*-butylamine under MW and CH.

Pilissão et al. (2012) used lipases CAL-B, PSL, PSL-C, PSL-D and *A. niger* lipase, free or immobilized in the acylation of (±)-*sec*-butylamine, with different acyl donors and various organic solvents under microwave radiation or conventional heating. For free lipase from *A. niger* the conversion was three times higher under microwave radiation when compared to conventional heating at 35°C. In this case, with free lipase from *A. niger* the (*R*)-amide was obtained with a conversion of 21%, resulting in 99% *e.e.*$_p$ and (*E*=200), at 1 min of reaction under MW radiation (Scheme 1.18) [50].

Araújo and Porto (2014) investigated the aza-Michael addition reactions to synthesize propanenitrile derivatives. The reactions were performed in both the presence and absence of lipases, under microwave radiation and conventional heating. aza-Michael adducts were synthesized with acceptable yields in water, under microwave radiation in the presence of lipase from *Candida antarctica* in a very short period of time (30 s) (Scheme 1.19) [51].

Ribeiro et al. (2014) investigated the aza-Michael reactions between 1-phenyl-methanamine and α, β-unsaturated cyclohexenones using lipase from *Candida antarctica*. The reactions were performed in different organic solvents ($CH_2Cl_2$, hexane, MeOH, toluene, THF) in mild conditions under microwave radiation and conventional heating. The reactions in the presence of CALB yielded the Michael adducts and imines (Scheme 1.20) [52].

Yu et al. (2012) reported the use of ionic liquid under microwave radiation in the isomerization of xylose to xylulose using immobilized xylose isomerase. Reactions under optimized conditions have yielded the xylulose in 68% yield at 3 h. These

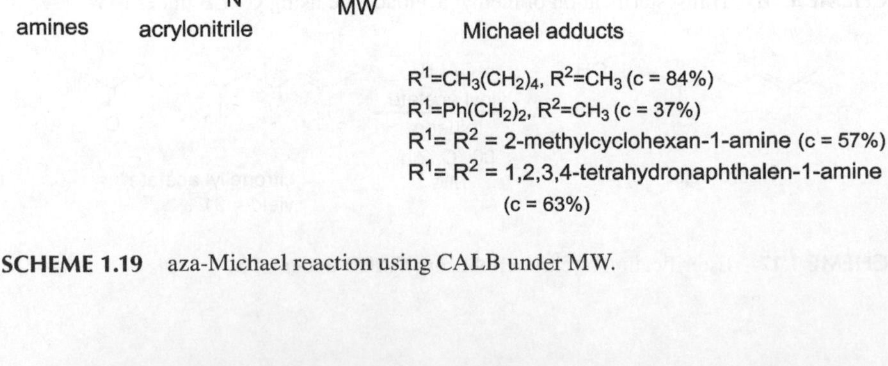

$R^1$=$CH_3(CH_2)_4$, $R^2$=$CH_3$ (c = 84%)
$R^1$=Ph(CH_2)_2, $R^2$=$CH_3$ (c = 37%)
$R^1$= $R^2$ = 2-methylcyclohexan-1-amine (c = 57%)
$R^1$= $R^2$ = 1,2,3,4-tetrahydronaphthalen-1-amine
(c = 63%)

**SCHEME 1.19** aza-Michael reaction using CALB under MW.

results indicated that the use of microwave radiation and ionic liquid was an efficient method for enzymatic isomerization of xylose to xylulose (Scheme 1.21) [53].

Ren et al. (2014) investigated the treatment of 2,4-dichlorophenol (2,4-DCP) using 2,4-DCP hydroxylase under MW and ionic liquid. 2,4-Dichlorophenol is a serious global contaminant because of its carcinogenicity, toxicity and persistence. A highly active 2,4-DCP hydroxylase removed the 2,4-DCP under microwave radiation and ionic liquid [EMIM][PF$_6$] as an additive. The reaction was a fast, efficient and environmentally benign method for the dechlorination of 2,4-DCP to produce a catechol derivative (Scheme 1.22) [54].

Young et al. (2008) employed a β-glucosidase (CelB) from the hyperthermophilic archaeon, *Pyrococcus furiosus*, for biocatalytic hydrolysis of *p*-nitrophenyl-α-D-galactopyranoside under microwave radiation. *P. furiosus* CelB cleaves exoglycosidic linkages in natural (e.g., cellobiose) and synthetic substrates; no significant enzymatic activity ($<10^{-11}$ mol min$^{-1}$ μg$^{-1}$) was detected in the absence of microwave radiation. However, a greater than 4 orders of magnitude increase in enzymatic

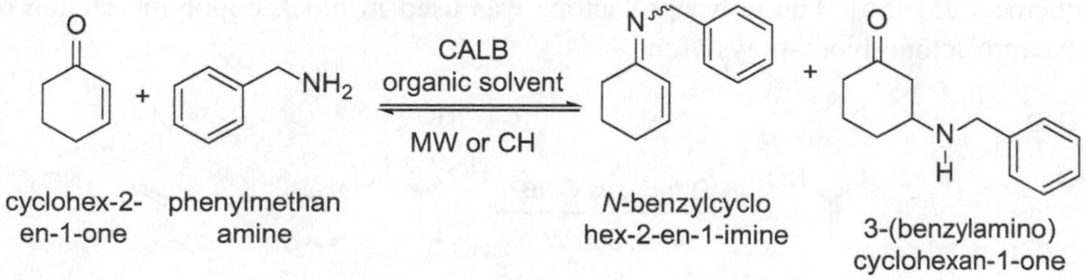

cyclohex-2-en-1-one    phenylmethanamine        *N*-benzylcyclohex-2-en-1-imine     3-(benzylamino)cyclohexan-1-one

**SCHEME 1.20**    aza-Michael reaction of 1-phenylmethanamine and cyclohexenone using CALB under MW and CH.

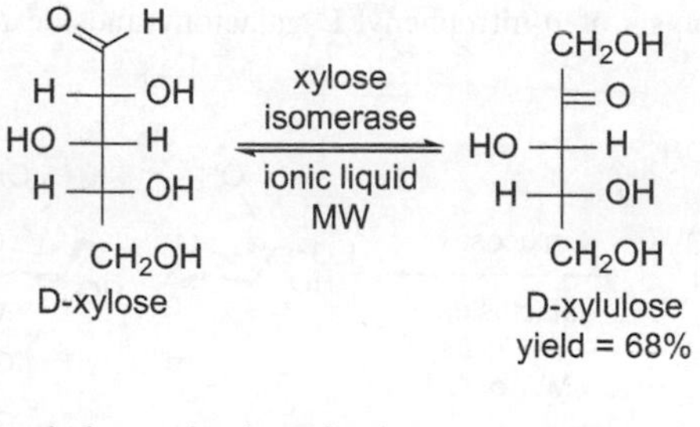

D-xylose             D-xylulose
yield = 68%

**SCHEME 1.21**    Enzymatic isomerization of xylose under MW.

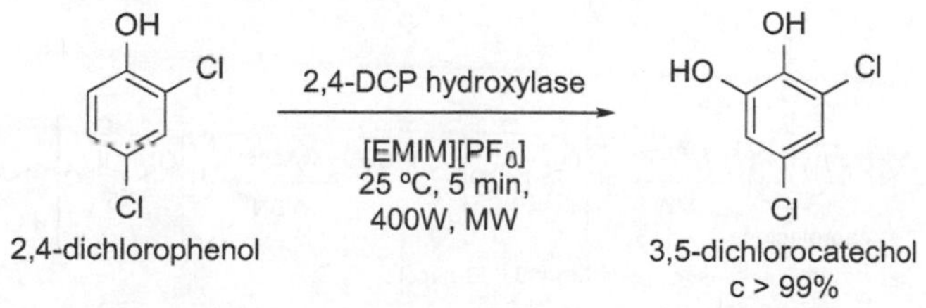

2,4-dichlorophenol            3,5-dichlorocatechol
c > 99%

**SCHEME 1.22**    2,4-DCP dechlorination using 2,4-DCP hydroxylase under MW.

activity of *P. furiosus* CelB was achieved under microwave radiation at 300 W $(2.3 \times 10^{-6} \text{ mol min}^{-1} \mu\text{g}^{-1})$ (Scheme 1.23) [11].

Maugard et al. (2003) synthesized galacto-oligosaccharide (GOS) from immobilized lactose and free β-galactosidase from *Kluyveromyces lactis* (Lactozym 3000 L HP-G) under microwave radiation and conventional heating. Immobilization of the β-galactosidase on to Duolite A-568 with the addition of phosphate buffer as solvent (pH 6.5) under microwave radiation resulted in an increase in the production of GOS (40°C, 480 min, 12 W) (Scheme 1.24) [55].

Lipase-catalyzed polymerization was explored as an alternative for Ru-catalyzed ring opening metathesis polymerization in the synthesis of polyesters bearing large functional moieties as part of the main chain. Kerep et al. (2007) showed the application in the chemoenzymatic synthesis of polymers under microwave radiation of ε-caprolactone that was polymerized with 2-mercaptoethanol as the initiator, in an oil bath (60°C) using CALB. The polymerization in equal conditions led to higher yields and less formation of side products. The resulting polyester with a terminal –SH moiety had a Mn (number average molar mass) of 3.600 g mol$^{-1}$ (Scheme 1.25) [56]. The polycaprolactone was used in block copolymerization of polycaprolactone-block-polystyrene.

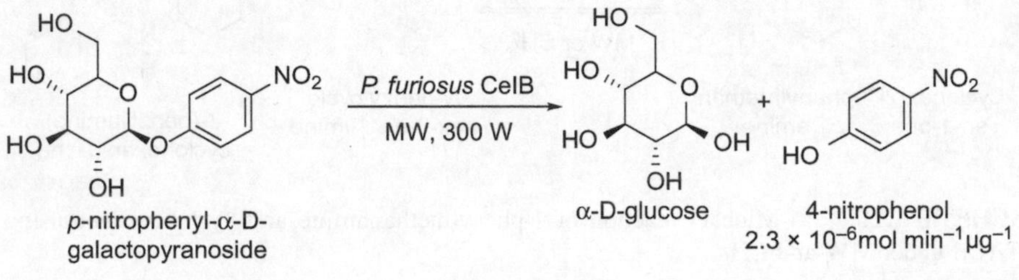

**SCHEME 1.23** Hydrolysis of *p*-nitrophenyl-D-galactopyranoside using *P. furiosus* CelB under MW.

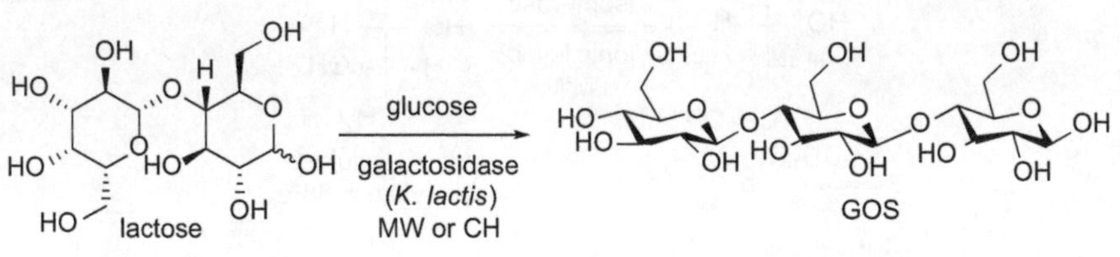

**SCHEME 1.24** Synthesis of galacto-oligosaccharides under MW and CH.

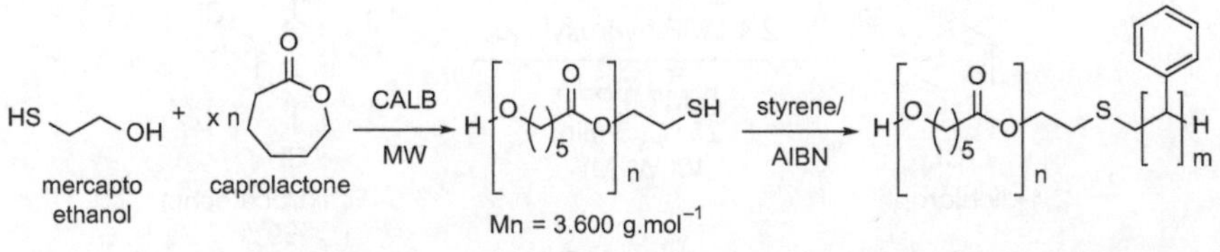

**SCHEME 1.25** Chemoenzymatic synthesis of polycaprolactone-block-polystyrene using CALB under MW.

## 1.2.2 BIOCATALYTIC REACTIONS USING WHOLE CELLS OF MICROORGANISMS UNDER MW

Recently, we investigated the first study of reduction of 2,2,2-trifluoroacetophenone to obtain the corresponding S-alcohol with the marine-derived fungus *Mucor racemosus* CBMAI 847 under microwave radiation and conventional heating [57].

The reactions were performed in different concentrations of 2,2,2-trifluoroacetophenone (2.9, 5.7, 8.5 and 14 mmol L$^{-1}$) at pH 8, at 40°C for 6 h (Table 1.2). Conventional heating gave 100% bioreduction and 74% enantiomeric excess in the concentrations of 2.9 and 5.7 mmol L$^{-1}$, respectively. Microwave radiation obtained 64% conversion and 91% *e.e.* for the S-alcohol at 5.7 mmol L$^{-1}$. This was the first study of biocatalytic reduction with filamentous fungus under microwave radiation (Table 1.2) [57].

The thermophilic microorganisms are defined as those whose optimum growth temperature is above 37°C, these being isolated from terrestrial or marine environments [58–60]. Thermophilic bacteria have an optimum growth in temperatures ranging from 30 to 65°C and most investigated thermophilic bacteria

---

**TABLE 1.2**

**Bioreduction of 2,2,2-Trifluoro-1-Phenylethanone with Whole Cells of Marine-Derived Fungus *M. racemosus* CBMAI 847**

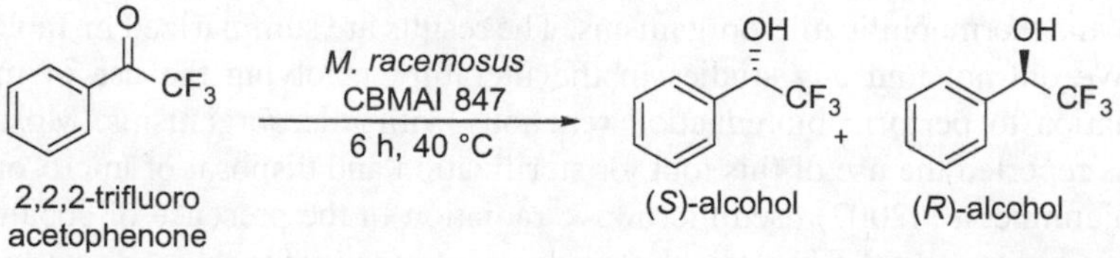

2,2,2-trifluoro acetophenone → *M. racemosus* CBMAI 847, 6 h, 40 °C → (S)-alcohol + (R)-alcohol

| Entry | c (mmol/L) ketone | pH | c (%) alcohol | e.e. (%) alcohol | ac alcohol |
|-------|-------------------|-----|---------------|------------------|------------|
| | | Conventional Heating (40°C, 130 rpm) | | | |
| 1 | 2.9 | 8 | 100 | 74 | S |
| 2 | 5.7 | 8 | 100 | 74 | S |
| 3 | 8.5 | 8 | 56 | 90 | S |
| 4 | 14.0 | 8 | 39 | 95 | S |
| 5 | 14.0 | 5 | 12 | 99 | S |
| | | Microwave Radiation (40°C, 200 W) | | | |
| 6 | 2.9 | 8 | 38 | 73 | S |
| 7 | 5.7 | 8 | 64 | 91 | S |
| 8 | 8.5 | 8 | 51 | 92 | S |
| 9 | 14.0 | 8 | 28 | 96 | S |
| 10 | 14.0 | 5 | 42 | 56 | S |

c=initial concentration; c=conversion of (S)-2,2,2-trifluoro-1-phenylethanol determined by GC-FID; *e.e.*=enantiomeric excess; ac=absolute configuration.

belonging to the genus *Bacillus* were isolated from thermophilic and mesophilic environments [61, 62].

In the literature, many reported biotransformation reactions using bacteria, but few described the use of thermophilic microorganisms in biocatalytic reactions. Ishihara et al. (2000) reported the bioreduction of aliphatic and aromatic-ketoesters using the thermophilic bacterium *Streptomyces thermocyaneoviolaceus* IFO14271 as a biocatalyst at different temperatures (37, 45 and 55°C, 20 h) in an orbital shaking. At the temperatures of 37 and 45°C, high conversions were obtained (98%) and the enantiomeric excess varied between 18% and 99%. At 55°C the conversion was between 31 and 99% and the enantiomeric excess was 15–96% for products [63].

Nishii et al. (1989) reported the bioreduction of 2,2,6-trimethyl-1,4-cyclohexanedione to 4-hydroxy-2,2,6-trimethyl cyclohexanone in 75% conversion after 20 h incubation at 50°C using the thermophilic bacterium *Bacillus stearothermophilus* isolated from soil. Four enantiomers were obtained from 4-hydroxy-2,2,6-trimethyl cyclohexanone [*cis*-(4R,6S), *cis*-(4S,6R), *trans*-(4R,6R) and *trans*-(4S,6S)] in the proportions of 68:25:5:2 [64].

Recently, bioreduction reactions of prochiral ketones using strains of thermophilic bacteria (genus *Bacillus*), isolated from composted organic material, has been a subject of study in our research group. These reactions were carried out under microwave radiation and conventional heating. The experiments were performed at 45°C and under orbital shaker reaction time varied according to the substrate used (96–144 h). In this work it was possible to obtain 100% conversion and enantiomeric excess >99%. After 6 h, microwave radiation gave 30% conversion and 62% *e.e.* This study has opened new frontiers for biocatalytic reactions conducted under microwave radiation and thermophilic microorganisms. The results are summarized in Table 1.3 [65].

We did not find any studies in the literature involving the use of microwave radiation to perform bioreduction reactions with microorganisms. Most publications reported the use of this tool for sterilization and disposal of micro-organisms. Benjamin et al. (2007) usedmicrowave radiation in the presence of cobalt and iron salts to eliminate the bacteria *Enterococcus faecalis*, *Staphylococcus aureus* and *Escherichia coli* in 1–2 min [66]. The rapid reduction in bacteria during microwave treatment in the presence of metal ions can be explained by the increased penetration of these ions in bacteria causing cell death [66].

Almajhdi et al. (2009) also carried out the inactivation of pathogenic and non-pathogenic micro-organisms in wastewater sewage; after 90 s under microwave radiation 80% of the microbial population were killed [67].

The bacteria (genus *Bacillus*) used in biocatalytic reduction reactions in our studies were subjected to microwave radiation. Subsequently, the bacteria were inoculated in a solid medium and, surprisingly, the bacteria grown in the culture medium survived [65].

## 1.3  CONCLUSIONS AND PERSPECTIVES

The use of microwave radiation presents new perspectives for organic synthesis, in particular by biocatalytic methods. Enzymes have been used for catalyzing esterification and transesterification reactions under microwave radiation, and although it

## TABLE 1.3
### Reduction of Fluoroacetophenones by Thermophilic Bacteria Under Conventional Heating and Microwave Radiation

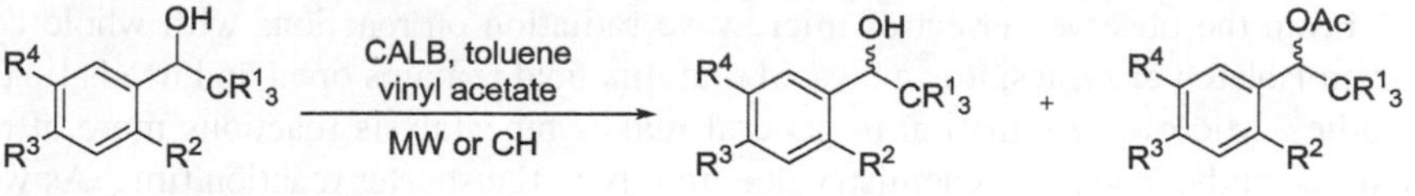

$$R^4 \underset{R^3}{\overset{OH}{\underset{R^2}{\bigcirc}}} CR^1_3 \xrightarrow[\text{MW or CH}]{\substack{\text{CALB, toluene} \\ \text{vinyl acetate}}} R^4 \underset{R^3}{\overset{OH}{\underset{R^2}{\bigcirc}}} CR^1_3 \;(R)\text{ or }(S) \;+\; R^4 \underset{R^3}{\overset{OAc}{\underset{R^2}{\bigcirc}}} CR^1_3 \;(S)\text{ or }(R)$$

$R^1 = F, R^2 = R^3 = R^4 = H$ (A)
$R^1 = H, R^2 = R^3 = R^4 = F$ (B)
$R^1 = H, R^2 = Br, R^3 =, R^4 = H$ (C)
$R^1 = H, R^2 = CF_3, R^3 = R^4 = F$ (**D**)

| Entry | Biocatalyst | ketones | t | Conventional Heating c | Conventional Heating e.e. [ac] | Microwave Irradiation* c | Microwave Irradiation* e.e. [ac] |
|---|---|---|---|---|---|---|---|
| 1 | *Bacillus* sp. FPZSP005 | A | 96 | >99 | 82 (R) | 30 | 62 (R) |
| 2 | | B | 96 | 87 | >99 (S) | (–) | (–) |
| 3 | | C | (–) | (–) | (–) | (–) | (–) |
| 4 | | D | (–) | (–) | (–) | (–) | (–) |
| 5 | *Bacillus subtilis* FPZSP088 | A | 144 | >99 | 24 (R) | 42 | 22 (R) |
| 6 | | B | 144 | >99 | >99 (S) | (–) | (–) |
| 7 | | C | (–) | (–) | (–) | (–) | (–) |
| 8 | | D | (–) | (–) | (–) | (–) | (–) |
| 9 | *Bacillus licheniformis* FPZSP055 | A | 144 | > 99 | 34 (R) | 26 | 5(R) |
| 10 | | B | 120 | > 99 | >99 (S) | (–) | (–) |
| 11 | | C | (–) | (–) | (–) | (–) | (–) |
| 12 | | D | (–) | (–) | (–) | (–) | (–) |
| 13 | BC-FPZSP051 | A | 144 | 92 | 10 (R) | 34 | 26 (R) |
| 14 | | B | 144 | 85 | >99 (S) | (–) | (–) |
| 15 | | C | (–) | (–) | (–) | (–) | (–) |
| 16 | | D | (–) | (–) | (–) | (–) | (–) |

\* Reactions were performed under microwave radiation for 6 h; (–): product not formed; $c$ (%): conversion of alcohol determined by GC–FID analysis; *e.e.* (%): enantiomeric excess of alcohol; ac: absolute configuration; $t$: time (h).

has great potential, its use is still inconspicuous as a synthetic methodology, particularly in the production of enantiomerically pure compounds.

There is a myth about the use of microbial cells under microwave radiation to promote chemical reactions. Few examples are described in the literature because of the destructive effects that microwave radiation can have on living organisms, especially when subjected to extreme conditions: high power, high temperature and long reaction time.

However, over short time periods, radiation exposure is often sufficient to catalyze several chemical reactions under microwave radiation without causing the death of micro-organisms, especially thermophiles.

From the observed effect of microwave radiation on reactions with whole cells or on isolated enzymes, it can be said that this field remains open and needs further studies. Microwave radiation in general makes biocatalysis reactions more attractive to synthetic organic chemistry due mainly to the shorter reaction time. As well, adapting this type of reaction to the biodegradation of xenobiotic compounds is an interesting area that also lacks further studies.

## ACKNOWLEDGMENTS

DEQ thanks CAPES for the scholarship. IMF and IJKA thank CNPq for the scholarships. The authors thank Conselho Nacional de Desenvolvimento Científico e Tecnológico (CNPq) and Fundação de Amparo a Pesquisa do Estado de São Paulo (FAPESP) for the financial support. The authors thank Núcleo de Pesquisa em Ciência e Tecnologia de BioRecursos (CITECBio) of the University of São Paulo for the financial and structural support.

## REFERENCES

1. Gedye R, Smith F, Westaway K, Ali H, Baldisera L, Laberge L, Rousell J (1986) The use of microwave ovens for rapid organic synthesis. *Tetrahedron Letters* 27: 279–282. doi:10.1016/S0040-4039(00)83996-9.
2. Giguere RJ, Bray TL, Duncan SM (1986) Application of commercial microwave ovens to organic synthesis. *Tetrahedron Letters* 27: 4945–4948. doi:10.1016/S0040-4039(00)85103-5.
3. Larhed M, Hallberg A (2001) Microwave-assisted high-speed chemistry: A new technique in drug discovery. *Drugs Discovery Today* 6: 406–416. doi:10.1016/S1359-6446(01)01735-4.
4. De Souza ROMA, Miranda LSM (2011) Microwave assisted organic synthesis: A history of success in Brazil. *Química Nova* 3: 497–506. doi:10.1590/S0100-40422011000300023.
5. Yadav GD, Devedran S (2012) Lipase catalyzed kinetic resolution of (±)-1-(1-naphthyl) ethanol under microwave irradiation. *Journal of Molecular Catalysis B: Enzymatic* 81: 58–65. doi:10.1016/j.molcatb.2012.05.007.
6. Ziaullah HPVR (2013) An efficient microwave-assisted enzyme-catalyzed regioselective synthesis of long chain acylated derivatives of flavonoid glycosides. *Tetrahedron Letters* 54: 1933–1937. doi:10.1016/j.tetlet.2013.01.103.
7. Bassyouni FA, Abu-Bakr SM, Rehim MA (2012) Evolution of microwave irradiation and its application in green chemistry and biosciences. *Research on Chemical Intermediates* 38: 283–322. doi:10.1007/s11164-011-0348-1.
8. Risso M, Mazzini M, Kroger S, Saenz-Mendez P, Seoane G, Gamenara D (2012) Microwave-assisted solvent-free lipase catalyzed transesterification of β-ketoesters. *Green Chemistry Letters Review* 5: 539–543. doi:10.1080/17518253.2012.672596.
9. Wan HD, Sun SY, Hu XY, Xia YM (2012) Non-thermal effect of microwave irradiation in non-aqueous enzymatic esterification. *Biotechnology and Applied Biochemistry* 166: 1454–1462. doi:10.1007/s12010-012-9539-5.

10. Yu D, Wang C, Yin Y, Zhang A, Gao G, Fang X (2011) A synergistic effect of microwave irradiation and ionic liquids on enzyme-catalyzed biodiesel production. *Green Chemistry* 13: 1869–1875. doi:10.1039/C1GC15114B.
11. Young DD, Nichols J, Kelly RM, Deiters A (2008) Microwave activation of enzymatic catalysis. *Journal of the American Chemical Society* 130: 10048–10049. doi:10.1021/ja802404g.
12. Yadav GD, Pawar SV (2012) Synergism between microwave irradiation and enzyme catalysis in transesterification of ethyl-3-phenylpropanoate with *n*-butanol. *Bioresource Technology* 109: 1–6. doi:10.1016/j.biortech.2012.01.030.
13. Ribeiro SS, De Oliveira JR, Porto ALM (2012) Lipase-catalyzed kinetic resolution of (±)-mandelonitrile under conventional condition and microwave irradiation. *Journal of the Brazilian Chemical Society* 23: 1395–1399. doi:10.1590/S0103-50532012000700025.
14. Yadav GD, Borkar IV (2009) Kinetic and mechanistic investigation of microwave-assisted lipase catalyzed synthesis of citronellyl acetate. *Journal of Industrial and Engineering Chemistry* 48: 7915–7922. doi:10.1021/ie800591c.
15. Dallinger D, Kappe CO (2008) Microwave-assisted synthesis in water as solvent. *Chemical Reviews* 107: 2563–2591. doi:10.1021/cr0509410.
16. Kingston HM, Jassie LB (1988) *Introduction to Microwave Sample Preparation: Theory and Practice.* American Chemical Society , Washington, DC .
17. Thostenson ET, Chou TW (1999) Microwave processing: Fundamentals and applications. *Composites Part A. Applied Science and Manufacturing* 30: 1055–1071. doi:10.1016/S1359-835X(99)00020-2.
18. Eskilsson CS, Bjorklund E (2000) Analytical-scale microwave-assisted extraction. *Journal of Chromatography A* 902: 227–250. doi:10.1016/S0021-9673(00)00921-3.
19. Kappe CO, Stadler A (2006) Chapter 2. Microwave theory. In: *Microwaves in Organic and Medicinal Chemistry.* Wiley-VCH Verlag GmbH & Co., Weinheim, 9–28.
20. Sanseverino AM (2002) Micro-ondas em síntese orgânica. *Química Nova* 25: 660–667. doi:10.1590/S0100-40422002000400022.
21. Souza GB, Nogueira ARA, Rassini JB (2002) Determinação de matéria seca e umidade em solos e plantas com forno de microondas doméstico. *Circular Técnica da Embrapa* 33: 1–9. ISSN:1516-411X.
22. Yu D, Ma D, Wang Z, Wang Y, Pan Y, Fang X (2012) Microwave-assisted enzymatic resolution of (*R/S*)-2-octanol in ionic liquid. *Process Biochemistry* 47: 479–484. doi:10.1016/j.procbio.2011.12.007.
23. Bachu P, Gibson JS, Sperry J, Brimble MA (2007) The influence of microwave irradiation on lipase-catalyzed kinetic resolution of racemic secondary alcohols. *Tetrahedron: Asymmetry* 18: 1618–1624. doi:10.1016/j.tetasy.2007.06.035.
24. Lukasiewicz M, Kowalski S (2012) Low power microwave-assisted enzymatic esterification of starch. *Starch-Starke* 64: 188–197. doi:10.1002/star.201100095.
25. Kappe CO (2006) The use of microwave radiation in organic synthesis. From laboratory curiosity to standard practice in twenty years. *Chimia International Journal for Chemistry* 60: 308–312. ISSN:0009-4293.
26. Corrêa AG, Bueno MA (2009) Reações químicas ativadas por ultrassom e irradiação de micro-ondas. In: *Química Verde: Fundamentos e Aplicações.* Edufscar, São Carlos. ISBN:978-85-7600-150-8.
27. Yadav GD, Lathi PS (2004) Synergism between microwave and enzyme catalysis in intensification of reactions and selectivities: Transesterification of methyl acetoacetate with alcohols. *Journal of Molecular Catalysis A: Chemical* 223: 51–56. doi:10.1016/j.molcata.2003.09.050.
28. Rejasse B, Lamare S, Legoy MD, Besson T (2007) Influence of microwave radiation on enzymatic properties: Applications in enzyme chemistry. *Journal of Enzyme Inhibition and Medicinal Chemistry* 22: 519–527. doi:10.1080/14756360701424959.

29. Laurence JA, French PW, Lindner RA, Mckenzie DR (2000) Biological effects of electromagnetic fields-mechanisms for the effects of pulsed microwave radiation on protein conformation. *Journal of Theoretical Biology* 206: 291–298. doi:10.1006/jtbi.2000.2123.

30. Loupy A, Perreux L, Liagre M, Burle K, Moneuse M (2001) Reactivity and selectivity under microwaves in organic chemistry. Relation with medium effects and reaction mechanisms. *Pure and Applied Chemistry* 73: 161–166. doi:10.1351/pac200173010161.

31. Cygler M, Grochulski P, Kazlauskas JK, Schrag JD, Bouthillier F, Rubin B, Serreqi AN, Gupta AK (1994) A structural basis for the chiral preferences of lipases. *Journal of American Chemical Society* 116: 3180–3186. doi:10.1021/ja00087a002.

32. Fransson L (2009) Molecular modeling: Understanding and prediction of enzyme selectivity. Licentiate thesis. Royal Institute of Technology School of Biotechnology, AlbaNova University Center, Stockholm, Sweden. ISSN: 1654–2312.

33. Lill JR, Ingle ES, Liu P, Pham V, Sandoval WN (2007) Microwave-assisted proteomics. *Mass Spectrometry Reviews* 26: 657–671. doi:10.1002/mas.20140.

34. Bradoo S, Rathi P, Saxena RK, Gupta R (2002) Microwave-assisted rapid characterization of lipase selectivities. *Journal of Biochemical and Biophysical Methods* 51: 115–120. doi:10.1016/S0165-022X(02)00005-2.

35. Zhu S, Wu Y, Yu Z, Zhang X, Li H, Gao M (2006) The effect of microwave radiation on enzymatic hydrolysis of rice straw. *Bioresource Technology* 97: 1964–1968. doi:10.1016/j.biortech.2005.08.008.

36. Ribeiro SS, Raminelli C, Porto ALM (2013) Enzymatic resolution by CALB of organofluorine compounds under conventional condition and microwave radiation. *Journal of Fluorine Chemistry* 154: 53–59. doi:10.1016/j.jfluchem.2013.06.014.

37. Yu D, Wang Z, Chen P, Jin L, Cheng Y, Zhou J, Cao S (2007) Microwave-assisted resolution of (R/S)-2-octanol by enzymatic transesterification. *Journal of Molecular Catalysis B: Enzymatic* 48: 51–57. doi:10.1016/j.molcatb.2007.06.009.

38. De Souza ROMA, Antunes OAC, Wolfgang K, Kappe CO (2009). Kinetic resolution of rac-1-phenylethanol with immobilized lipases: A critical comparison of microwave and conventional heating protocols. *Journal of Organic Chemistry* 74: 6157–6162. doi:10.1021/jo9010443.

39. Devendran S, Yadav GD (2014) Microwave assisted enzymatic kinetic resolution of (±)-1-phenyl-2-propyn-1-ol in nonaqueous media. *Biomedicinal Research International* 2014: 1–9. doi:10.1155/2014/482678.

40. Rouillard H, Deau E, Domon L, Cherouvrier JR, Graber M, Thiery V (2014) Microwave-assisted kinetic resolution of homochiral (Z)-cyclooct-5-ene-1, 2-diol and (Z)-2-acetoxycyclooct-4-enyl acetate using lipases. *Molecules* 19: 9215–9227. doi:10.3390/molecules19079215.

41. Hilt P, Schieber A, Yildirim C, Arnold G, Klaiber I, Conrad J, Beifuss U, Carle R (2003) Detection of phloridzin in strawberries (*Fragaria × ananassa* duch.) by HPLC-PDA-MS/MS and NMR spectroscopy. *Journal of Agricultural and Food Chemistry* 51: 2896–2899. doi:10.1021/jf021115k.

42. Salem J, Humeau C, Chevalot I, Harscoat-Schiavo C, Vanderesse R, Blanchard F, Fick M (2010) Effect of acyl donor chain length on isoquercitrin acylation and biological activities of corresponding esters. *Process Biochemistry* 45: 382–389. doi:10.1016/j.procbio.2009.10.012.

43. Aursand M, Grasdalen H (1992) Interpretation of the $^{13}$C-NMR spectra of omega-3 fatty acids and lipid extracted from the white muscle of Atlantic salmon (*Salmo salar*). *Chemistry and Physics of Lipids* 62: 239–251. doi: 0009-3084/92/$05.00.

44. Enaud E, Humeau C, Piffaut B, Girardin M (2004) Enzymatic synthesis of new aromatic esters of phloridzin. *Journal of Molecular Catalysis B: Enzymatic* 27: 1–6. doi:10.1016/j.molcatb.2003.08.002.

45. Chebil L, Anthoni J, Humeau C, Gerardin C, Engasser JM, Ghoul M (2007) Enzymatic acylation of flavonoids: Effect of the nature of the substrate, origin of lipase, and operating conditions on conversion yield and regioselectivity. *Journal of Agricultural and Food Chemistry* 55: 9496–9502. doi:10.1021/jf071943j.

46. Shinde SD, Yadav GD (2014) Process intensification of immobilized lipase catalysis by microwave radiation in the synthesis of 4-chloro-2-methylphenoxyacetic acid (MCPA) esters. *Biochemical Engineering Journal* 90: 96–102. doi:10.1016/j.bej.2014.05.015.

47. Yang Z, Niu X, Fang X, Chen G, Zhang H, Yue H, Wang L, Zhao D, Wang Z (2013) Enantioselective esterification of Ibuprofen under microwave radiation. *Molecules* 18: 5472–5481. doi:10.3390/molecules18055472.

48. Huang W, Xia Y-M, Gao Y-J, Fang Y, Wang Y, Fang Y (2005) Enzymatic esterification between *n*-alcohol homologs and *n*-caprylic acid in non-aqueous medium under microwave radiation. *Journal of Molecular Catalysis B: Enzymatic* 35: 113–116. doi:10.1016/j.molcatb.2005.06.004.

49. Yadav GD, Lathi PS (2007) Microwave assisted enzyme catalysis for synthesis of *n*-butyl dipheyl methyl mercapto acetate in non-aqueous media. *Clean Technologies and Environmental Policy* 9: 281–287. doi:10.1007/s10098-006-0082-3.

50. Pilissão C, De Oliveira CP, Nascimento MG (2012) The influence of conventional heating and microwave radiation on the resolution of (*R/S*)-*sec*-butylamine catalyzed by free or immobilized lipases. *Journal of the Brazilian Chemical Society* 23: 1688–1697. doi:10.1590/S0103-50532012005000033.

51. Araujo YJK, Porto ALM (2014) aza-Michael addition of primary amines by lipases and microwave irradiation: A green protocol for the synthesis of propanenitrile derivatives. *Current Microwave Chemistry* 1: 87–93. doi:10.2174/2213335601666140610201546.

52. Ribeiro SS, Uliana MP, Brocksom TJ, Porto ALM (2014) Analysis by GC-MS of an aza-Michael reaction catalyzed by CALB on an orbital shaker and under microwave radiation. *Global Journal of Science Frontier Research* 14: 7–21. Code:100505,250106.

53. Yu D, Wang Y, Wang C, Ma D, Fang X (2012) Combination use of microwave radiation and ionic liquid in enzymatic isomerization of xylose to xylulose. *Journal of Molecular Catalysis B: Enzymatic* 79: 8–14. doi:10.1016/j.molcatb.2012.04.005.

54. Ren H, Zhan Y, Fang X, Yu D (2014) Enhanced catalytic activity and thermal stability of 2,4-dichlorophenol hydroxylase by using microwave radiation and imidazolium ionic liquid for 2,4-dichlorophenol removal. *RSC Advances* 4: 62631–62638. doi:10.1039/C4RA10637G.

55. Maugard T, Gaunt D, Legoy MD, Besson T (2003) Microwave-assisted synthesis of galacto-oligosaccharides from lactose with immobilized β-galactosidase from *Kluyveromyces lactis*. *Biotechnology Letters* 25: 623–629. doi:10.1023/A:1023060030558.

56. Kerep P, Ritter H (2007) Chemoenzymatic synthesis of polycaprolactone-block-polystyrene via macromolecular chain transfer reagents. *Macromolecular Rapid Communications* 28: 759–766. doi:10.1002/marc.200600701.

57. Ribeiro SS, Sette LD, Porto ALM (2016) First bioreduction of 2,2,2-trifluoro-1-phenyl-ethanone by whole hyphae of marine fungus *Mucor racemosus* CBMAI 847 under microwave radiation. *Current Microwave Chemistry* 3: 92–96 (accepted 2015, published 2016). doi:10.2174/2213335602666150616222320

58. Maheshwari R, Bharadwaj G, Bhat MK (2000) Thermophilic fungi: Their physiology and enzymes. *Microbiology and Molecular Biology Reviews* 64: 461–488. doi:10.1128/MMBR.64.3.461-488.2000.

59. Trincone A (2013) Biocatalytic processes using marine biocatalysts: Ten cases in point. *Current Organic Chemistry* 17: 1058–1066. doi: 1875-5348/13 $58.00+.00.

60. Bhalla A, Bansal N, Kumar S, Bischoff KM, Sani RK (2013) Improved lignocellulose conversion to biofuels with thermophilic bacteria and thermostable enzymes. *Bioresource Technology* 128: 751–759. doi:10.1016/j.biortech.2012.10.145.

61. Royter M, Schmidt M, Elend C, benreich H, Schafer Y, Bornscheuer UT, Antranikian G (2009) Thermostable lipases from the extreme thermophilic anaerobic bacteria *Thermoanaerobacter thermohydrosulfuricus* SOL1 and *Caldanaerobacter subterraneus* subsp. *tengcongensis*. *Extremophiles* 13: 769–783. doi: 10.1007/s00792-009-0265-z.

62. Martins CAM, Lelis MLL (2009) Produção de poligalacturonase, pelo termofílico *Bacillus* sp. e algumas de suas propriedades. *Ciência e Tecnologia de Alimentos* 29: 135–141. doi:10.1590/S0101-20612009000100021.

63. Ishihara K, Yamaguchi H, Hamada H, Nakajima N, Nakamura K (2000). Stereocontrolled reduction of α-keto esters with thermophilic actinomycete, *Streptomyces thermocyaneoviolaceus* IFO 14271. *Journal of Molecular Catalysis B: Enzymatic* 10: 429–434. doi:10.1016/S1381-1177(99)00115-0.

64. Nishii K, Sode K, Karube I (1989) Microbial conversion of dihydroox oisophorone (DOIP) to 4-hydroxy-2,2,6-trimethylcyclohexanone (4-HTMCH) by thermophilic bacteria. *Journal of Biotechnology* 9: 117–128. doi:10.1016/0168-1656(89)90081-3.

65. Ribeiro SS, De Vasconcellos SP, Ramos PL, Da Cruz JB, Porto ALM (2016) Application of conventional heating and microwave radiation for the biocatalytic reduction of fluoroacetophenones by thermophilic bacteria. *Current Microwave Chemistry* 3: 9–13 (accepted 2015, published 2016). doi:10.2174/2213335602666150205102756.

66. Benjamin E, Reznik A, Benjamin E, Williams A (2007) Mathematical models of cobalt and iron ions catalyzed microwave bacterial deactivation. *International Journal of Environmental Research and Public Health* 4: 203–210. ISSN: 1661-7827.

67. Almajhdi F, Albrithen H, Alhadlaq H, Farrag M A, Abdel-Meggeed A (2009) Microorganisms inactivation by microwaves radiation in Riyadh Sewage treatment water plant. *World Applied Sciences Journal* 6: 600–607. ISSN: 1818-4952.

# 2 Tracking Microwave-Assisted Sample Preparation Through the Last Years

*Mónica B. Alvarez, Anabela Lorenzetti,*
*Carolina C. Acebal, Adriana G. Lista,*
*and Claudia E. Domini*

## CONTENTS

## 2.1 INTRODUCTION

Microwave (MW) radiation is an electromagnetic energy constituted by a non-ionizing radiation that is not affected by the propagation medium. It lies between radio waves and infrared radiation, with wavelengths ranging from near one meter to about one millimeter, or equivalently with frequencies between 300 MHz (0.3 GHz) and 300 GHz [1]. Microwaves interact with matter through both electric and magnetic components of electromagnetic radiation.

MW technology has been applied in different fields of science. Traditionally it has been used for telecommunication/communication purposes, such as radar, television, and satellite applications [2] but it is also used for different sensing and imaging applications besides its use as a heating source to enhance the kinetics of diverse chemical processes.

Figure 2.1 shows a summary of MW applications. As can be seen, Chemistry is one of the fields where this kind of energy is widely used due to its versatility

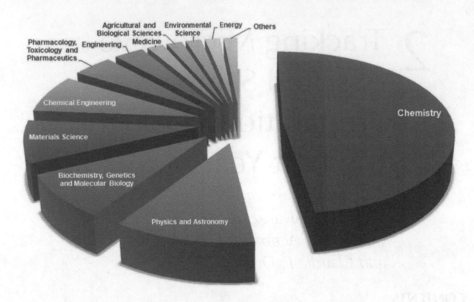

**FIGURE 2.1**   Microwave energy applications.

and simplicity. MW presents remarkable characteristics such as very fast heating and better homogeneity in temperature [3]. Also, MW has a series of advantages when is compared with the applications of other traditional technologies, i.e., better recovery of volatiles compounds, lower pollution levels, lesser volumes of the required reagents, more reproducible procedures, and better operating environment [4, 5]. All of these advantages are due to the interaction of dielectric materials with MW radiation. The heating involves primarily two mechanisms: dipolar polarization and ionic conduction. The first one is caused by dipole–dipole interactions between the electromagnetic field and polar molecules that generate heat [6–8]. The ionic conduction affects charged particles in a sample, usually ions. In this case, the MW irradiation moves the ions back and forth through the sample (colliding with each other) and generates heat. At the same time, the probability of molecular encounters is increased by accelerating the molecular/ionic movement which leads to a shift in reaction rates from days or hours to minutes or seconds (when used in synthesis or chemical reactions) [3, 9].

The MW energy is applied in organic chemistry to achieve best yields in the synthesis of different products with shorter times and lower costs. The implementation of electromagnetic energy to sample preparation by the analytical chemists came several years after the appearance of the first article published on the topic at the beginning of the 70s by Samra et al. [10]. The use of MW as a tool in analytical chemistry is nowadays one of the most important applications in sample preparation. This step of the analytical process is still the most cumbersome of chemical analysis.

Despite the heating effects of high-frequency fields on some materials being recognized even in the 19th century, the first compact microwave oven was only available in the 1950s [11]. It is assumed that one of the most important reasons for this delay was that the mechanism of energy transfer using a microwave field is very different from that of the three well-established modes of heat transfer, that is, conduction, radiation, and convection [12].

Today, MW sample treatment has been established as a standard method for preparation of samples for elemental analysis. Thus, in 2007, the Environmental Protection Agency (EPA) of the United States of America established the method 3051A for microwave-assisted acid extraction/digestion of sediments, sludges, soils, and oils for various elements, which are determined by flame atomic absorption spectrophotometry (FLAA), graphite furnace atomic absorption spectrophotometry (GFAA), inductively coupled plasma optical emission spectrometry (ICP-OES), and inductively coupled plasma mass spectrometry (ICP-MS).

Both, microwave-assisted extraction (MAE) and microwave-assisted digestion (MAD) have been proved to be quick and efficient methods for sample pretreatment. In recent years, many approaches of MAE were reported, such as dynamic microwave-assisted extraction [13–19] and solvent-free microwave-assisted extraction [20–22]. On the other hand, the MAD procedures have become widely used for the pretreatment of different matrices, since they make it possible to simultaneously process large numbers of samples, and reduce contamination and losses of volatile analytes. Figure 2.2 shows the bar chart with the number of publications about microwave-assisted sample preparation between 2004 and 2018.

The instrumentation for the MW sample treatment may be implemented on closed or open vessel devices. The closed devices make it possible (i) to reach high temperatures that decrease the time needed for the extraction process, due to the high

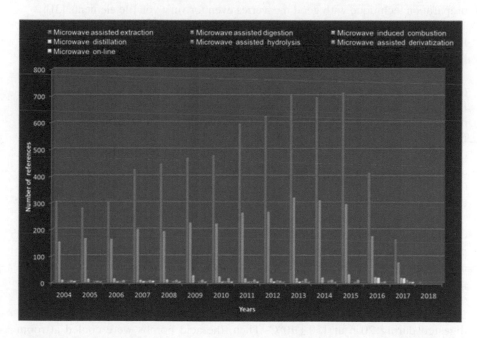

**FIGURE 2.2** Number of references in MW sample preparation (Data from Scopus, 2017).

pressures inside the vessel, (ii) to diminish the losses of volatile substances, and, finally, (iii) to use a lower amount of organic solvent [23].

In addition, the MW sample-preparation procedures can be performed in a continuous way or can be coupled on-line to other analytical steps. This can be accomplished by means of a partial or total automation of the analytical process, reducing the analysis time and analyte loss and increasing personal security. These improvements are due to the synergy effect of the two methodologies, the MW radiations and the continuous flow methods.

In this chapter, the application of MW energy in the sample preparation is discussed in detail, taking into account extraction, digestion, and some other procedures using this type of radiation. Moreover, remarks about the continuous flow sample treatment procedure using microwave radiation are included.

## 2.2  MICROWAVE-ASSISTED DIGESTION (MAD)

The MAD technique has been widely used for sample decomposition or dissolution of organic and inorganic materials with complex matrices such as soils, sediments, food, environmental, and biological samples among others [1, 24–26]. Currently, numerous publications deal with MAD where significant parameters are studied such as sample mass, acid type, volume of reagents, oxidant mixture concentrations, temperature, pressure, and heating time [27–30].

The use of MAD has advantages compared to conventional heating methods in the number of samples that could be processed each run, in the risk reduction of losses and contamination. In addition, MAD has become a rapid and reproducible sample preparation technique with good recoveries even for very volatile elements [31].

With the intention of discussing the application of MAD in the digestion of different matrices, various proposed methods were selected and are described in the following sections.

Soil samples were prepared for multi-element analysis using inductively coupled plasma sector field mass spectrometry (ICP-SFMS) analysis by Engstrom et al. [32]. In this work, the authors compared two sample preparation procedures: $HNO_3$ leaching and pseudo-total digestion with $HNO_3$, HCl, and HF in a microwave oven. The first digestion procedure was carried out with 0.5 g of sample and 5 mL of concentrated $HNO_3$ and the digestion conditions were: $2 \times 30$ min at 600 W and then 10 min at 800 W. The second procedure was performed using the same MW parameters and 0.2 g of sample in a mixture of $HNO_3$ (3 mL), HCl (2 mL), and HF (1 mL). By using this acid mixture, quantitative recoveries for many elements were obtained.

Tuncel and collaborators proposed the digestion of lichen samples by three different procedures [33] in order to determine trace metals. Inductively coupled plasma atomic emission spectroscopy (ICP-OES) technique was applied for the determination of major, minor, and trace elements. Before the digestion, the samples were stored in the refrigerator at 4°C until analysis. The first method is based on an acid bomb digestion in which 0.2 g of lichens was mixed with 8 mL of $HNO_3$, 2 mL of $H_2O_2$, and 0.5 mL of HF. The bombs were closed and the samples were digested during 20 h at $150 \pm 10$°C. Then, the acid bombs were cooled at room temperature and Teflon containers kept them at $140 \pm 10$°C to obtain 0.5–1 mL.

After that, the hot plate was removed and the sample was diluted to 50 mL with deionized water. The second procedure was an open vessel digestion. Here, 0.2 g of lichen samples were mixed with 15 mL of $HNO_3$ and 4 mL of $H_2O_2$ in 150 mL Teflon beakers. The digestion was carried out on the hot plate for 12–14 h at $140 \pm 10°C$. When the process was finished, an aliquot was evaporated to obtain a final volume of 1–2 and then 1 mL of HF was added. This sample was refluxed for 6–8 h. After this, the aliquots were evaporated and then 5 mL of $HNO_3$ was added in order to remove the excess HF. Again, evaporation to dryness was carried out. The sample was removed from the hot plate when all the HF was eliminated, cooled, and diluted to 50 mL with deionized water. The third digestion was carried out as a microwave digestion using 0.2 g of a lichen sample and a mixture of 8 mL of $HNO_3$, 2 mL of $H_2O_2$, and 0.5 mL of HF in tetrafluoromethaxil digestion vessels, which are placed in an Ethos 900 Milestone Microwave Digestion equipment. The optimized conditions were 2 min at 250 W, 2 min at 0 W, 6 min at 250 W, 5 min at 400 W, and 5 min at 600 W. As it was necessary to remove the reagent excess, the containers were placed on the hot plate at $140 \pm 10°C$. When the liquid volume in the vessel was reduced to 0.5–1 mL of the sample, it was cooled and diluted to 50 mL with deionized water.

The authors concluded that good accuracy was obtained in the determination of many elements when open-vessel and acid bomb digestion systems were used. They recommended the microwave digestion system because high analyte recoveries, with a significant reduction of the digestion procedure time, are achieved.

During the last year, several research groups continued to use microwave digestion with diverse acid mixtures. The mixture $H_2O_2/HNO_3$ was involved in different matrices decomposition, such as dietary supplements, tree leaves, fish oil, lipoid matrices, berries, medicinal plants, and traditional fermented foods [34–39]. Kaya and Türkoğlu studied the bioaccumulation of metals in various fish species by treatment with $HNO_3$ while, Barros et al. carried out the determination of molybdenum ultratraces in plants using the same acid [40, 41]. Other authors reported the determination of metals in dietary supplements using $HNO_3$ and HCl; however, for other matrices, such as mastic gum and particulate matter, the sample was treated with $HNO_3$, $H_2O_2$, and HF [42].

A MAD with an alkaline solution (tetramethylammonium hydroxide, TMAH), the addition of $Cu^{2+}$, aqueous-phase derivatization of methylmercury with sodium tetrapropylborate, and subsequent extraction with n-heptane was developed by Chen and co-workers [43]. The authors presented a treatment for fish samples in order to determine methylmercury. For that purpose, a focused microwave digestion was carried out where 0.5 g of the sample was mixed with 5 mL of tetramethylammonium hydroxide (TMAH) and placed in a microwave oven during 2.5 min at 45 W. Later, this solution was mixed with 1 mL of 40 mM $Cu^{2+}$ solution and the pH was adjusted at 4.0 with an acetate buffer. Then, 1 mL of 1% sodium tetrapropylborate/potassium hydroxide solution was added and the solution was mixed and remained at room temperature for 10 min. At last, 2 mL of n-heptane was added, shaken for 10 min and the supernatant was analyzed by GC-MS. The digestion procedure took place in a focused microwave digester Microdigest 3.6 (2.45 GHz, maximum power 300 W), Prolabo, France). A rigorous study of the influence of all extraction variables was done and under the optimal conditions, good recoveries were obtained for tuna, marlin, and shark samples.

Another proposal of alkaline MAD procedure – using tetramethylammonium hydroxide (TMAH) – was reported as a sample pretreatment for total iodine and bromine determinations in seaweed for human consumption by Romarís–Hortas and colleagues [44]. Temperature, TMAH volume, ramp time, and hold time were the studied variables for the MAD procedure. ICP-MS using tellurium and yttrium as internal standards were employed for iodine and bromine determinations, respectively. The accuracy of the method was assessed by analyzing a NIES-09 certified reference material. Working with the same matrix, Domínguez-González et al. [45] carried out a microwave acid digestion to assess total trace element content and non-dialyzable fractions, using 8 mL of 69% (w/v) $HNO_3$ and 2 mL of 33% (w/v) $H_2O_2$ at 1000 W for 0.2 g of seaweed sample or 5 g of residue. The initial and final temperatures were 25–90°C for 10 min, 90–90°C hold 5 min, 90–120°C for 10 min, 120–190°C for 10 min, and 190–190°C hold 5 min. As the previous method, the accuracy was assessed by analyzing a NIES-09 certified reference material (Sargasso seaweed). In the same research group, Moreda-Piñeiro and co-workers [46] developed a MAD procedure to replicate an in vitro digestion method (dialyzation) for assessing the bioavailability ratios of trace and minor elements (Al, Cd, Co, Cr, Cu, Fe, Mn, Ni, V, and Zn) from fish, shellfish, and seaweed. Microwave-assisted acid digestion on 0.5 g of crushed samples was achieved by using 69% $HNO_3$ (4 mL) and 33% (m/v) $H_2O_2$ (2 mL). ICP-MS was used for metal determination in the extracts. Trueness of the procedure for total metal determination was evaluated by analyzing different certified reference materials (DORM-2, BCR 279, and NIES-09). Also, this group of researchers [47] determined trace Cd and Pb in fresh fish samples by ETAAS. Sample preparation was carried out by means of a microwave-assisted acid digestion procedure using $HNO_3$ and $H_2O_2$. A four stages temperature program was applied. Thus, 2.5 min was necessary to reach 90°C from room temperature, followed by a temperature ramp of 8.3°C $min^{-1}$ to 140°C during 6.0 min and from 140°C to 200°C for 5 min. Then it was maintained at 200°C for 10 min. Cd and Pb were satisfactorily determined at levels lower than the acceptable levels in fresh marine products (European Union legislation). LODs obtained were 0.21 and 0.67 ng $g^{-1}$ for Cd and Pb respectively. Besides, Moreda-Piñeiro and colleagues have determined essential and toxic metals also in seeds and nuts using MAD. The most relevant data of the proposed procedure are shown in Table 2.1.

Rudolph et al. [48] proposed a microwave digestion sample preparation followed by an open vessel treatment for determining platinum at background levels in human autopsy tissues. Inductively coupled plasma sector field mass spectrometry (ICP-SFMS) was used. The microwave digestion assistance provided high temperature (240°C) and high-pressure (30 bar). The sample mass (human tissue 0.5 g) was digested with an acid mixture of 4 mL $HNO_3$ and 1 mL $H_2O_2$. Then, the samples were evaporated at 120°C to a residue and redissolved with 500 μL $HNO_3$. Then, 2 mL HCl and 0.1 mL HF were added and the samples were digested again following these steps: 1 min, 250 W at 220°C; 2 min, 0 W at 220°C; 1 h, 400 W at 220°C. The samples were evaporated again to approximately 200 μL. Then, 2 mL aqua regia was added and re-evaporated to 200 μL. The same procedure was performed using 1 mL of

**TABLE 2.1**
**Operational Conditions for MAD Procedure of Different Matrices**

| Analyte | Matrix | Experimental Conditions | | | Detection Technique | Ref. |
|---|---|---|---|---|---|---|
| | | Sample Mass | Solvent | Preselected Program | | |
| Li, Be, Al, V, Cr, Mn, Co, Ni, Cu, Zn, Ga, As, Se, Rb, Sr, Ag, Cd, Cs, Ba, Tl, Pb, Th, and U | Freshwater sponges | 0.05 g dry sponge | 0.5 mL 70% $HNO_3$ 0.2 mL 30% $H_2O_2$ | MW oven (MW power 1000 W) 10 s, at each power of MW irradiation from 10 to 50% with a step of 10%. dry residue 1 mL of 10% HF MW irradiation 10 to 20%. | ICP-MS | [76] |
| 242Pu, 243Am, 253U, 232Th | Swipe samples | 2.00 g | 10.0 mL $HNO_3$ | predigested for a night up to 80°C, 15 min, held 2 min; up to 120°C in 20 min and held at that temperature for 10 min; up to 150°C, 15 min, held 10 min. The applied microwave power was 300 and 600 W, respectively | ICP-SFMS | [77] |
| Ca, Cd, Co, Cr, Cu, Fe, K, Mg, Mn, Na, Ni, Pb, V, and Zn | Paris polyphylla samples | 0.30 g | 2.0 mL $HNO_3$–$H_2O_2$ (2:1, v/v) | 2 min at 250 W 2 min at 0 W 6 min at 250 W 5 min at 400 W 8 min at 550 W | FASS – GF AAS | [78] |
| Al, As, B, Ba, Ca, Co, Cr, Cu, Fe, K, Li, Mg, Mn, Na, Ni, Pb, Sr, and Zn | Olive tree (Olea europaea L.) leaves | 0.25 g | 7.0 mL $HNO_3$ 1.0 mL 35% $H_2O_2$, | 180°C, 10 min 15 min, 180°C, 1,200 W, 400 psi | ICP-MS | [79] |

*(Continued)*

**TABLE 2.1 (CONTINUED)**
**Operational Conditions for MAD Procedure of Different Matrices**

| Analyte | Matrix | Experimental Conditions | | | Detection Technique | Ref. |
|---|---|---|---|---|---|---|
| | | Sample Mass | Solvent | Preselected Program | | |
| Total mercury | Ground rice | 0.50 g | 3.0 mL $H_2O$ 3.0 mL $HNO_3$ | Power 400 W 100% power operation Ramp time 20 min Maximum temperature 100°C Hold time 20 min Cool down time 60 min | ICP-MS | [80] |
| Hg, Pd, Os | Pharmaceutical substances | 0.25 g | 2.5 mL aqua regia | 1 min, 250 W 1 min, 0 W 4 min, 250 W 4 min, 400 W 4 min, 600 W | ICP-MS | [81] |
| Al, Ba, Ca, Cu, Fe, K, Mg, Mn, Mo, and Zn | Biodiesel | 0.20 g | 8.0 mL 7 mol $L^1$ $HNO_3$ | 20, 180°C, 750 W | ICP-OES | [82] |
| Arsenic | Shrimp paste sample | 0.10 g | 2.0 mL $HNO_3$ 1.0 mL $H_2O_2$ | 1400 W, 5 min <30 min | AAS | [83] |
| Ag, Al, Ba, Ca, Cd, Co, Cr, Cu, Fe, K, Mg, Mn, Na, Ni, Pb, Sr, and Zn | Apples | 0.20 g dried apple | 5.0 mL $HNO_3$, (50:50 v/v). | Closed MW 4 min, 700 W, $T_1$ 70°C, $T_2$ 100°C, 100 bar 12 min, 700 W, $T_1$ 130°C, $T_2$ 70°C, 100 bar 7 min, 1000 W, $T_1$ 180°C, $T_2$ 70°C, 120 bar 7 min, 1000 W, $T_1$ 200°C, $T_2$ 70°C, 120 bar 15 min, 800 W, $T_1$ 200°C, $T_2$ 70°C, 120 bar 30 min, 0 W, $T_1$ 20°C, $T_2$ 20°C, 10 bar | ICP-OES | [84] |

*(Continued)*

**TABLE 2.1 (CONTINUED)**
**Operational Conditions for MAD Procedure of Different Matrices**

| Analyte | Matrix | Experimental Conditions | | | Detection Technique | Ref. |
|---|---|---|---|---|---|---|
| | | Sample Mass | Solvent | Preselected Program | | |
| | | 1.00 g fresh apple | 5.0 mL $HNO_3$ | Open MW 3 min $T_1$, 0°C $T_2$, 110°C; 10 min $T_1$, 110°C $T_2$, 110°C; 2 min $T_1$, 110°C $T_2$, 25°C | | |
| Pb | Lipstick | 0.50 g raw material; 0.15 g of lipstick | 7.5 mL of a mixture of $HNO_3$–$H_2O$–$H_2O_2$ in 5:1.5:1 ratios | 0 W to 1400 W, 25 min, 1400 W, 30 min, 0 W, 15 min. With this program, the sample was digested at 200°C | GF AAS | [85] |
| As, Cd, Pb, and Hg, Fe, Na, Ca, P, and Zn | Botanical extracts | 1.00 g | $HNO_3$ | 1200 W, 15 min | ICP-OES | [86] |
| Pt | Whole blood, plasma, residue (pellet fraction), plasma-ultrafiltrate, and protein-residue fraction | 0.20 g | $HNO_3$, $H_2O_2$ | 2 min, 250 W; 2 min 0 W; 6 min, 250 W; 5 min, 400 W; 5 min, 650 W | ICP-SFMS | [87] |
| As, Zn | Leaves of *Tropaeolum majus* (protein extract) | 3.00 g | $HNO_3$, $H_2O_2$ | 3 min, 85°C, 700 W; 5 min, 145°C, 1000 W; 18 min, 210°C, 1000 W | AAS | [88] |

*(Continued)*

**TABLE 2.1 (CONTINUED)**
**Operational Conditions for MAD Procedure of Different Matrices**

| Analyte | Matrix | Experimental Conditions | | | Detection Technique | Ref. |
|---|---|---|---|---|---|---|
| | | Sample Mass | Solvent | Preselected Program | | |
| Cu, Cr, Pb, Mn, and Cd | Tomatoes, onions, and peppers | 0.50 g | HNO₃, H₂O₂ | 2 min, 250 W<br>2 min, 0 W<br>6 min, 250 W<br>5 min, 400 W<br>5 min, 600 W | GFAAS | [89] |
| 20 essential and non-essential elements (Ag, Al, As, Ca, Cd, Co, Cr, Cu, Fe, K, Li, Mg, Mn, Mo, Ni, Pb, Se, Sn, V, and Zn) | Liver and muscle samples of harbor porpoises (*Phocoena phocoena*) | 0.25–0.30 g | HNO₃ (65%) HCl (30%) | The temperature was risen stepwise to 80°C, 140°C, and 180°C within 25 min<br>The temperature was maintained at 180°C for 10 min | CC-ICP-MS | [90] |
| Al, As, Ba, Cd, Cr, Cu, Fe, Mn, Ni, Pb, and Zn, and for Ba, Cu, and Mn | Raft mussels | 0.50 g | HNO₃ (70%) H₂O₂ (33% (m/v)) | 1000 W, 300°C | ICP-OES | [91] |
| Trace elements | Sulfide ore samples | 0.20 g | HNO₃-HCl-HF | 1200 W, ramp from room temperature to 200°C in 15 min<br>holding at 200°C for 30 min, then cool to room temperature | ICP-OES | [92] |

*(Continued)*

**TABLE 2.1 (CONTINUED)**

**Operational Conditions for MAD Procedure of Different Matrices**

| Analyte | Matrix | Sample Mass | Solvent | Preselected Program | Detection Technique | Ref. |
|---|---|---|---|---|---|---|
| | | | | **Experimental Conditions** | | |
| Brominated flame retardants | Polymeric matrices, flame retardant paints and enamels | 0.05 g | 25% TMAH | 5 min, increased up to 800 W, 120°C constant rate and then it was kept for 30 min at 800 W and 120°C | FI–ICP-MS | [93] |
| Trace elements | Red wine | 2.00 (mL) | $HNO_3$ | 3 min to reach 85°C<br>12 min to reach 145°C<br>6 min to reach 180°C<br>15 min at 180°C | ICP-OES<br>– ICP-MS | [94] |
| 72 elements | Vendace and whitefish caviar | 1.00 g | $HNO_3$ | 1380 kPa, 600 W, 60 min | ICP-SFMS | [95] |
| Co | Pharmaceutical products (B12 vitamin powder, B12 ampoules, Centrum, Spectrum ABC, and Optima Forte) | 0.50 g | 6.0 mL $HNO_3$<br>1.0 mL $H_2O_2$ | 2 min at 250 W<br>2 min at 0 W<br>6 min at 250 W<br>5 min at 400 W<br>5 min at 650 W | FAAS<br>– GFAAS<br>– ICP-AES<br>– AdSV[a] | [96] |
| Se, Zn, Fe, and Cu | Chicken meat and feed | 0.60 g | 5.0 mL $HNO_3$<br>2.0 mL $H_2O_2$ | 10°C min⁻¹ up to 120°C and holding, 5 min,<br>10°C min⁻¹ up to 150°C and holding for 5 min, again at 10°C min⁻¹ up to 180 1°C and a 5 min hold, and finally at 10°C min⁻¹ up to 200°C with holding for 10 min. | ICP-AES<br>– ICP-MS | [97] |

*(Continued)*

**TABLE 2.1 (CONTINUED)**

**Operational Conditions for MAD Procedure of Different Matrices**

| Analyte | Matrix | Experimental Conditions | | | Detection Technique | Ref. |
|---|---|---|---|---|---|---|
| | | Sample Mass | Solvent | Preselected Program | | |
| Pt and Rh | Dust and plant samples | 100–500 | 5.0 mL HNO$_3$ <br> 3.0 mL H$_2$O$_2$ <br> 0.1 mL HF <br> 3.5 mL HCl <br><br> 2.0–5.0 mL 4% (w/v) H$_3$BO$_3$ | 5 min, 60 W <br> 12 min, 632 W <br> 15 min, 190 W <br> 25 min, 632 W <br> 40 min, 474 W <br> 15 min, 632 W <br> 5 min, 221 W | ICP-MS | [98] |
| Total Se | Plasma | | 10.0 mL HNO$_3$ 0.5% sample:acid ratio of 1 mL:1 mL | 5 min, 250 W <br> 5 min, 400 W <br> 5 min, 600 W | ICP-MS | [99] |
| 63 Elements | German White Wines | 2.50 mL | 2.5 mL 14 mol L$^{-1}$ HNO$_3$ | 25 min, 1000 W (T$_{max}$ of 170–190°C, P$_{max}$ of 30 bar) | ICP-MS | [100] |
| As | Different Types of Nuts | 20.0 g | 100 mL chloroform/methanol (2:1) | Shaken vigorously | IC-ICP-MS[b] | [101] |
| | | 1.00 g subsamples | 10.0 mL HNO$_3$ (50% v/v) | 25% of maximum power (950 W), ramp time: 5 min, hold time: 2 min <br> 45% of power, ramp time: 5 min and hold time: 2 min | | |
| | | 1.00 g oil | 9:1 mixture of 50% (v/v) HNO$_3$-30% (v/v) H$_2$O$_2$ | 55%, ramp time: 5 min, hold time: 2 min <br> 65%, ramp time: 5 min and hold time: 2 min microwave oven using the same program | | |

(Continued)

**TABLE 2.1 (CONTINUED)**

**Operational Conditions for MAD Procedure of Different Matrices**

| Analyte | Matrix | Experimental Conditions | | | Detection Technique | Ref. |
|---|---|---|---|---|---|---|
| | | Sample Mass | Solvent | Preselected Program | | |
| Ru | Ru(OH)Cl$_3$ preparation | 1.00–7.00 mg | 3.0–5.0 mL 0.1 mol L$^{-1}$ KOH or 5% NaHCO$_3$ until the pH of the solution reached 9.0<br>5.0 mL HCl<br>1.0–2.0 mL HCOOH | domestic commercial 350 W, p$_{max}$=7.8 atm, 15 min<br>600 W, p$_{max}$=1.7 atm, 3 min<br>300 W, p$_{max}$=2.4 atm, 6 min | Coulometric determination | [102] |
| Al, As, Cd, Co, Cr, Cu, Fe, Hg, Mg, Mn, Ni, Pb, Se, V, Zn, Ca, Na, and P | Hair | 100 mg | Stirred in acetone for 10 min,<br>Dried to constant weight at 50°C, 2.5 mL 65% HNO$_3$ and 5 mL Milli-Q water | 1000 W, 10 min (80°C,80 bar);<br>000 W, 10 min (130°C,80 bar);<br>1000 W, 10 min (180°C,80 bar);<br>1000 W, 15 min (200°C,100 bar), 50 min (cooling) | ICP-OES (Ca, Na, P)<br>ICP-MS | [103] |
| As, Cd, Cu, Ni, Pb, and Zn | Artichoke leaves and fruits | 0.50 g | 10.0 mL HNO$_3$ | 400–1800 W, 15 min ramp time until 200°C, 10 min hold at 200°C, and 500 psi pressure (reference method total digestion) | ETAAS | [104] |
| Ca, Fe, K, Mg, Na, B, Ba, Cd, Cu, Mn, Mo, Pb, Sr, and Zn | Milk powder | 500 mg | 8.0 mL concentrated HNO$_3$ 50% H$_2$O$_2$ | 10 min of ramp and hold for 20 min at 250°C. Cooling down (65°C) | ICP-OES (Ca, Fe, K, Mg, Na) ICP-MS (B, Ba, Cd, Cu, Mn, Mo, Pb, Sr, Zn) | [105] |

*(Continued)*

**TABLE 2.1 (CONTINUED)**
**Operational Conditions for MAD Procedure of Different Matrices**

| Analyte | Matrix | Experimental Conditions | | | Detection Technique | Ref. |
|---|---|---|---|---|---|---|
| | | Sample Mass | Solvent | Preselected Program | | |
| Al, Fe, Hg; Ba, Ca, Cd, Co, Cu, K, Li, Mg, Mn, Mo, P, Pb, Se, Sr, Tl, Zn; As, Cr, and Ni | Nut and seed walnuts, Brazil nuts, macadamia nuts, pecans, hazelnuts, chestnuts, cashews, peanuts, pistachios, and seeds (almond, pine, pumpkin, and sunflower) | 0.30–25.00 g (non-dialyzable fractions) | 4.0 mL ultrapure $HNO_3$ 2.0 mL $H_2O_2$, 3.0 mL ultrapure water 4.0 mL ultrapure $HNO_3$, and 2.0 mL $H_2O_2$ (non-dialyzable fractions) but microwave irradiation at 200°C (fourth stage) was maintained for 25 min | Four stage microwave irradiation (1000 W): 90°C, 2.5 min 90°C – 140°C, 6 min 140 C – 200 C, 5.0 min; 200°C, 10 min. Cooling (50 min MW irradiation at 200°C (fourth stage) was maintained for 25 min (non-dialyzable fractions) | ICP-MS | [106] |
| N-glycopeptides | Human/animal serum or tissue | 100 µg | 60 µL (10 µL of 50 mmol $L^{-1}$ ammonium bicarbonate and 50 µL of 10 mmol $L^{-1}$ dithiothreitol) | 400 mL water bath (MW oven for 15 min, at a power setting 45–47°C) – Microwave incubation | MALDI-TOF MS[c] | [107] |
| Al, As, Ba, Co, Cr, Cu, Mn, Mo, Ni, Pb, Se, Sr, V, and Zn | Nut and babassu coconut | 150–250 mg | 6.0 mL $HNO_3$ | 1300 W over 10 min; 1300 W, 25 min Cooling | ICP-MS | [108] |
| Perchlorate (trace level) | biological samples (seafood) | 3.00 g | 25.0 mL 100 mmol $L^{-1}$ $HNO_3$ | Irradiation power 250 W 100°C, 10 min | TS-FF-AAS[d] | [109] |

*(Continued)*

**TABLE 2.1 (CONTINUED)**

**Operational Conditions for MAD Procedure of Different Matrices**

| Analyte | Matrix | Sample Mass | Solvent | Preselected Program | Detection Technique | Ref. |
|---|---|---|---|---|---|---|
| | | | | **Experimental Conditions** | | |
| Cd, Hg, and Pb | Cosmetic powder | 0.20 g | 6.0 mL $HNO_3$ (overnight) 2.0 mL HF (~50%), 2 mL $H_2O_2$ (~30%) (2 h of equilibration) | 200°C, 10 min, MW irradiation 20 min at 200°C | ID ICP-MS[e] | [110] |
| Al, P, Ti, V, Cr, Fe, Ni, Cu, Ga, Ge, As, Y, Ag, Cd, Sb, La, Ce, Eu, Gd, Tb, W, Au, Hg, and Pb | Fluorescence lamp shredder | 0.10 mg | $HClO_4$– $HNO_3$– HF aqua regia | DIN ISO 11466, German Standardization Organization (Deutsches Institut für Normung), Soil quality-Extraction of Trace Elements Soluble in Aqua Regia. 1997 | ICP-OES – ICP-MS | [111] |
| Cd | Sunflower (Heliánthus ánnuus) seeds | 0.20 g | $HNO_3$ – $H_2O_2$ (1:2) | 800 W, 30 min | ASV[f] | [112] |
| Be, Sc, Ti, V, Cr, Mn, Co, Ni, Cu, Zn, Ga, Ge, As, Sr, Y, Zr, Cs, Ba, La, Ce, Sm, Hf, Pb, Th, and U | Coal | 0.10 g | 12.0 mL $H_2O_2$ (3 mol $L^{-1}$) 3.0 mL $HNO_3$ (7 mol $L^{-1}$) | 200°C,10 min | ICP-MS | [113] |

*(Continued)*

**TABLE 2.1 (CONTINUED)**
**Operational Conditions for MAD Procedure of Different Matrices**

| | | Experimental Conditions | | | | |
|---|---|---|---|---|---|---|
| Analyte | Matrix | Sample Mass | Solvent | Preselected Program | Detection Technique | Ref. |
| Si, Al, B, Ca, Cu, Fe, K, Mg, Mn, Mo, S, and Zn | Plant (leaves of sugar cane, corn, soy, and alfalfa) | 0.10 g | 5.0 mL $HNO_3$ (1.0 mol $L^{-1}$), 5.0 mL $H_2O_2$ (30% v/v). NaOH (1.5 mol $L^{-1}$) Dilution: 750 μL of $HNO_3$ (14 mol $L^{-1}$) | 1305 W<br>5 min to reach 120°C<br>5 min at 120°C<br>5 min to reach 160°C<br>5 min at 160°C<br>3 min to reach 230°C<br>5 min at 230°C.<br>5 min to reach 150°C<br>5 min at 150°C<br>5 min to reach 230°C<br>10 min at 230°C | ICP OES | [114] |
| S | Coal | 0.05 g | $H_2O_2$ 3 mol $L^{-1}$ | 150°C, 5 min, | ICP OES | [115] |
| Ca, Cu, Mg, Zn, crude proteins | Beef, pork, and chicken | 250 mg | 4.0 mL 7 mol $L^{-1}$ $HNO_3$<br>2.0 mL 30% (m/v) $H_2O_2$ | 2 min, up to 80°C<br>3 min, hold at 80°C<br>4 min, up to 120°C<br>5 min, up to 180°C<br>5 min, up to 210°C<br>5 min, cooling down | ICP OES | [116] |
| Ti, Al, Mg, Ca, Fe, Zn, Cu, and Mn | Human oral mucosa | 0.15 g | 0.5 mL $HNO_3$<br>0.5 mL $H_2O_2$ | Ramp time 15 min<br>Hold time 25 min, 1100 W, 150°C | LA-ICP-MS[g] | [117] |

*(Continued)*

**TABLE 2.1 (CONTINUED)**
**Operational Conditions for MAD Procedure of Different Matrices**

| Analyte | Matrix | Experimental Conditions | | | Detection Technique | Ref. |
|---|---|---|---|---|---|---|
| | | Sample Mass | Solvent | Preselected Program | | |
| Lanthanides, actinides and transition metals | Mastic gum | 0.25 g | 2.0 mL HCl<br>4.0 mL HNO$_3$<br>3.5 mL HF | 80% (1600 W), 400 psi, rump time 15 min, hold time 15 min, 230°C | ICP-MS | [118] |
| Al, V, Cr, Mn, Fe, Co, Ni, Cu, Zn, As, Se, Mo, Cd, Pb, and U | Tropical marine fishes | 0.50 g | 6.0 mL (65%) HNO$_3$<br>2.0 mL H$_2$O$_2$ | 1500 W, 100°C, 3 min (ramp), 3 min (hold)<br>1500 W, 150°C, 7 min (ramp), 3 min (hold)<br>1500 W, 170°C, 5 min (ramp), 3 min (hold)<br>1500 W, 190°C, 5 min (ramp), 10 min (hold) | ICP-MS | [119] |
| Al and Na | Aluminosilicates | 0.10–0.50 g | 4.0 mL 65% (v/v) HNO$_3$<br>3.0–4.0 mL 40% (v/v) HF | 400 W, 5 min<br>790 W, 8 min<br>320 W, 4 min<br>0 W, 3 min | ICP-OES | [120] |
| 237Np | Soils | 0.6 g of dry ashed | 4.0 mL HNO$_3$<br>4.0 mL HF<br>10.0 mL of a 7.5 mol L$^{-1}$ HNO$_3$ + 2% boric acid mixture | 7 min, 30% of maximum power (900 W)<br>4 min, 10%<br>ice bath for 30 min (cooling)<br>evaporate to dryness with hotplate and heat lamp<br>5 min, 30% of power | ICP-MS | [121] |

*(Continued)*

## TABLE 2.1 (CONTINUED)
## Operational Conditions for MAD Procedure of Different Matrices

| Analyte | Matrix | Sample Mass | Solvent | Preselected Program | Detection Technique | Ref. |
|---------|--------|-------------|---------|---------------------|---------------------|------|
| | | | **Experimental Conditions** | | | |
| Trace elements (Boron and phosphorous) | Silicon | $500 \pm 5$ mg | Samples with particle size of <2 mm: 28 mL of an acid mixture of 5.65 mol $L^{-1}$ HF/0.85 mol $L^{-1}$ $HNO_3$ Samples with particle size of >2 mm: 18 mL of an acid mixture of 8.6 mol $L^{-1}$ HF/1.35 mol $L^{-1}$ $HNO_3$ | MLS Ethos Start, 100 mL PTFE vessels, max 120°C 50°C, 3 min (ramp), 20 min (hold) 80°C, 3 min (ramp), 20 min (hold) 120°C, 5 min (ramp), 30 min (hold) | ICP-OES | [122] |

a   Adsorptive stripping voltammetry (**AdSV**)

b   Ion Chromatography–Inductively Coupled Plasma Mass Spectrometry (**IC-ICP-MS**)

c   Laser desorption/ionization-time of flight-mass spectrometry (**MALDI-TOF MS**)

d   Thermospray flame furnace atomic absorption spectrometry (**TS-FF-AAS**)

e   Double isotope dilution inductively coupled plasma/mass spectrometry (**ID ICP-MS**)

f   Anodic stripping voltammetry (**ASV**)

g   Laser ablation inductively coupled plasma mass spectrometry (**LA-ICP-MS**)

HCl. The samples were stored and, before ICP-MS analysis, were diluted with 0.2 M HCl. In this way, platinum was determined at very low concentration levels.

Asfaw and Wibetoe [49] presented another interesting approach since they developed a new method to determine hydride (Se) and non-hydride-forming (Ca, Mg, K, P, S, and Zn) elements in various beverages. To obtain accurate results a simple pretreatment was needed. For this purpose, three different modes of sample preparation were compared, i.e., dilution only, partial decomposition (aqua regia treatment), and complete decomposition by using a MAD. The last procedure was performed using 0.5–2 g of samples, 7 mL concentrated $HNO_3$ and 1 mL $H_2O_2$. For beer samples 6–8 g was weighted and 8 mL of concentrated $HNO_3$ was used. The parameters of the digestion were 50 bar pressure, temperature ramp from 25 to 200°C for 10 min, kept for 20 min at 200°C and then reduced to 25°C in 20 min.

Welna et al. [50] assessed diverse sample treatments prior to the determination of total inorganic arsenic and selenium in slim instant coffees using hydride generation inductively coupled plasma optical emission spectrometry (HG-ICP-OES). Various sample preparation procedures, including a MAD using a $HNO_3/H_2O_2$ mixture, were examined and compared. The corresponding As and Se hydrides were generated in the reaction of an acidified sample solution with the $NaBH_4$ reductant in the presence of antifoam A. The MW-assisted sample decomposition was carried out using a closed vessel microwave-assisted digestion system (Milestone, Italy, high-pressure MW digestion system MLS-1200 MEGA, equipped with a rotor MDR 300/10) at a maximum power of 600 W for 45 min using concentrated $HNO_3$ and $H_2O_2$ (6:1 mL). Szymczycha-Madeja et al. [51] evaluated Ba, Ca, Cd, Cr, Cu, Fe, Mg, Mn, Ni, P, Pb, Sr, and Zn total content in slim instant coffees. Total sample digestion with $HNO_3$ and $H_2O_2$ mixture was performed using a closed vessel microwave at 600 W for 45 min. The same authors compared different sample treatments prior to the multielement analysis by ICP-OES of slim teas [52].

A new procedure for determining low lead levels in bone tissues has been developed by Grotti and co-workers [53]. The researchers used a wet acid digestion in a pressurized microwave-heated system for sample preparation. For that, 0.1–0.3 g freeze-dried sub-samples were mixed with 5 mL of concentrated nitric acid in a microwave digestion system MDS-2000. The procedure was carried out using 150 psi for 40 min, 100% power. The determination was performed using the ICP-OES.

The determination of elemental composition of cyanobacteria *Spirulina platensis* and *Nostoc commune* by different techniques (ICP–OES, FAAS, and ETAAS) was studied [54]. In this work, an inorganic biomass mineralization with an autoclave heater and microwave digestion were carried out. For the MAD procedure, the parameters were 10–100 mg sample, $HNO_3$, and $H_2O_2$ (4 mL each), 10 min at 10% and 20 min at a 100% output power. The authors stated that the procedure selection for the mineralization of biological samples without volatiles loss is especially significant. Using microwave digestion, the time was reduced from 4 h to 20 min, but in this way, not all the samples could be mineralized. So, the mineralization procedure was selected depending on the task. The conclusions of the paper indicate that both mineralization methods are appropriate to determine many elements in the biomass of cyanobacteria *Spirulina platensis* and *Nostoc commune*, and some elements (Se, Mo, and Zn) in cell fractions.

Hearn and colleagues [55] have demonstrated that the use of ICP-MS in combination with microwave digestion for the measurement of low levels of sulfur in gas oils is equally adequate as the traditional method used for this purpose. The procedure for microwave digestion was carried out in the Multiwave 2000 microwave oven (Anton Paar, Austria) with these operational parameters: 0.1 g of the sample, 5 mL $HNO_3$, 0.8 mL $H_2O_2$, maximum temperature 200°C, and maximum pressure 75 bar. The microwave was set as follows: ramp from 50 to 1000 W over 15 min and hold at maximum temperature for 30 min. After digestion, the samples were diluted with deionized water to a minimum final weight of 20 g and were introduced in the ICP-MS instrument. The authors concluded that the two methodologies used for the sample treatment, HPA and microwave digestion, were robust and combined with ICP-MS constituted a simpler and faster procedure than the traditional method.

As mentioned previously, MAD procedure accelerates the process, leading to a shorter preparation time. In this sense, in order to reduce the time of analysis consumed by the standard Kjeldahl method, an open focused microwave digestion system was developed to determine Kjeldahl nitrogen in wastewaters of different origins [56]. The process consisted of two steps: (1) decomposition of organic matter with sulfuric acid and (2) oxidation with hydrogen peroxide. In the first step, 50 mL of the sample and 10 mL of 96% (w/w) $H_2SO_4$ were digested at a fixed time. In the second step, 20 mL of 30% (v/v) $H_2O_2$ was added and a higher temperature was used to complete the digestion. The most influencing variables in microwave digestion were identified and optimized using an experimental design. The equipment to carry out the treatment was especially designed and a focused microwave heating system at 800 W was included [57]. Then, the method was validated by the application to real samples such as industrial wastewater samples of different origins. A comparison between the results provided by the proposed method and the results obtained by analyzing the same samples with the standard method demonstrated that the new method is more effective taking into account the recoveries and the process acceleration.

Time analysis reduction was also demonstrated by Priego-Capote and Luque de Castro [58] that have developed a method to determine Cr (III) and Cr (VI) in glass material, using three different sample preparation procedures: EPA method 3060A, ultrasound-assisted leaching, and MAD. The sample treatment was carried out using 2 g of the sample, 10 mL of $Na_2CO_3/NaOH$, microwave energy at 80%, and 120 s in 8 cycles with a delay time of 30 s. The authors mentioned that the determination of Cr (VI) in solid samples required an additional effort and for this reason, auxiliary energies such as microwaves and ultrasound have been used to assist the treatment of these kinds of samples. They compared the MAD procedure with EPA 3060A method based on alkaline digestion for the determination of Cr (VI). The results shown that MAD provided similar results – 99.6 and 98.3% for Cr (VI) and Cr (III), respectively – compared to the EPA 3060A method and the authors concluded that these results can be ascribed to the fact that both procedures are similar with the only difference being the type of energy that assists the digestion process. However, the time for MAD is 20 min and for EPA method is 60 min, demonstrating the process acceleration by using MAD.

Ashoka and co-workers [59] have developed a microwave digestion method for the determination of forty trace elements (Li, B, Na, Mg, Al, K, Ca, V, Cr, Mn, Fe,

Co, Ni, Cu, Zn, As, Se, Rb, Sr, Y, Ag, Cd, Cs, Ba, La, Ce, Pr, Nd, Sm, Eu, Gd, Y, Ho, Er, Tm, Yb, Lu, Pb, Th, U) in fish tissue. The proposed method was compared with another five digestion methods in order to show that the time of analysis could be drastically reduced using MAD. The measurements were made using an ICP-MS spectrometer with an octopole collision cell. All of the proposed methods employed $HNO_3$ in conjunction with other reagents to digest three certified marine biological samples (DOLT-3, DORM-3, IAEA-407) and a fish bone homogenate. For the microwave method, a domestic microwave oven (Toshiba model ER-694ETN, 650 W) was employed. The samples were placed in a screw cap polypropylene sample tube with 3 mL of concentrated $HNO_3$, 2 mL $H_2O_2$, and 3 mL Milli-Q water. The sample tube was encased in an airtight plastic container to prevent corrosion of the components of the microwave oven by acid fumes. Then, the sample tube was placed into a domestic microwave oven and the digestion was carried out in three interleaved cycles applying 30% power during 10 min, and cooling for another 10 min. In addition, a digestion with $HNO_3$ and $H_2O_2$ in a closed high-pressure microwave system was performed, but it was found that this method was not more accurate (except for Cd and Ni) than the method which used the domestic microwave oven. Overall, the microwave method was chosen to be the closest approximation to an ideal method because very few steps are involved, the equipment is readily available and the digestion can be completed within an hour. The effectiveness of the digestion methods was evaluated by comparison with certified values.

MAD (EPA method, 10 mL 65% $HNO_3$, or 10 mL of aqua regia) was also tested on sample pretreatment procedures for the ICP-OES determination of Al, B, Ca, Cu, Fe, K, Mg, Mn, Na, P, and Zn from pine needles by Väisänen and co-workers [60]. A digestion procedure (1000 W for 5.5 min, 175°C and a second step using 1000 W for 10 min, 175°C) with a standard reference material (SRM) 1575a or real sample (0.25 g) was compared with ultrasound-assisted digestion and dry ashing procedure. The microwave digestion gave slightly higher element concentrations than ultrasound-assisted digestion. B, Cu, and Na were not accurately determined by the microwave digestion method. The sample matrix of pine needles had a considerable influence on the atomization and excitation conditions of the plasma. Therefore, all the ICP-OES measurements were carried out in robust plasma conditions, tested by measuring the Mg (II) 280.270 nm/Mg (I) 285.213 nm line intensity ratios.

A procedure to prevent analyte losses in some samples was optimized by Reis and Almeida [61]. Cd, Cr, Cu, Hg, Ni, Pb, V, and Zn in animal material, specifically in fish, lobster, and mussel tissues was determined with a two-step MAD procedure in a Parr reactor bomb combined with a commercial microwave oven. Certified reference materials Dorm-2, Tort-2, and SRM 2976 were used to develop, evaluate, and validate the analytical procedure. Metal analysis was performed by atomic absorption spectrometry. A good correlation between the certified and the measured concentrations was achieved. The microwave digestion procedure was fast and accurate for the metals in lobster hepatopancreas and mussel tissue, but was not suitable for fish muscle, regardless of matrix resemblance.

Torres et al. [62] have proposed an easy and reliable MAD method to determine total and inorganic mercury in biological certified reference material (CRM) by cold vapor atomic absorption spectrometry (CV AAS). Inorganic mercury was determined after treatment with TMAH and total mercury was determined after MAD ($HNO_3/H_2O_2$) by CV AAS. The proposed method assured that all mercury species have been oxidized to $Hg^{2+}$. It was adequate for fractionation analysis of mercury in biological samples of pig kidney, lobster hepatopancreas, dogfish liver, and mussel tissue and did not require any chromatographic technique. Also, Wu et al. [63] developed a MAD method (150°C for 10 min) based on the oxidation process ($H_2O_2$ 1 mL) for assessing speciation analysis of ultratrace inorganic mercury and total mercury in water and biological samples by CV AAS and ICP-MS detection. Validation data were provided based on the analysis of water samples, as well as a CRM human hair sample (GBW 070601).

Gonzalez et al. [64] developed a different proposal of the MAD method using diluted nitric acid solutions (instead of concentrated solutions) as an alternative procedure for digesting organic samples. The extraction efficiency was evaluated by the determination of residual carbon content (RCC) and element recoveries (Ca, Fe, K, Mg, Na, P, Zn) by ICP-OES. 0.20 g sample mass was weighed and 2 mL $HNO_3$ solution and 1 mL of 30% (w/v) $H_2O_2$ were employed. Five steps were carried out: 2 min at 250 W; 2 min at 0 W; 4 min at 650 W; 5 min at 850 W; and 5 min at 1000 W. A 2 mol $L^{-1}$ $HNO_3$ solution was adequate for the digestion of most of the samples.

Niemelä and co-workers [65] studied a microwave-assisted aqua regia digestion of catalyst samples prior to the determination of Pt, Pd, Rh, and Pb by ICP-OES. Good recoveries were obtained for digestion at 160°C temperature in aqua regia, while higher digestion temperatures were used to quantitatively extract elements (Al, Ce, Mg, and Zn) associated with impurities in the catalysts and support materials of the catalysts. The addition of a small amount of HF on the aqua regia showed no major differences in metal recoveries when catalyst samples were digested.

Tyutyunnik and co-workers [66] have developed different microwave digestion procedures under elevated pressure, using closed systems with partial removal of the gas phase. These methods were suitable for sample preparation for mercury and other toxic elements (As, Cd, Pb, Se) determination in organic matrices (natural high-color waters, soils, bottom sediments, aquatic organism tissue). Under optimal oxidative and temperature-time conditions, the partial removal of the gas phase does not lead to volatile elements losses when using sample portions less than 2 g. The accuracy of determination was confirmed by the results of analysis of certified reference materials of water and plant materials and by the standard addition method. The selected conditions of preparation of sludge samples have ensured the determination of mercury by the CV AAS in drinking, natural, and sewage waters with a detection limit of 0.07 µg $L^{-1}$.

Dantas et al. [67] proposed the combined use of infrared radiation (commercially available infrared lamps) and microwave radiation (IR-MW). This procedure increases the mass of organic samples which can be digested while using small volumes of nitric acid (4 mL of 65% w/w $HNO_3$ for 1.0 g of sample approximately). Al, Ca, Cu, Fe, K, Mg, Mn, Na, P, and Zn in human-feed samples were measured by ICP OES. The recovery results agreed with those obtained from conventional decomposition by microwave radiation (closed system). The results obtained using the proposed procedure IR-MW for standard reference material (whole milk powder, NIST 8435) ranged between 85 and 100% for the studied elements. The use of a cavity microwave oven contributed to lower LODs in comparison to conventional closed microwave oven method.

A MAD procedure for organic matter decomposition in wine was optimized [68], previous to the detection of total Fe, Cu, Pb, and Cd using potentiometric sensors. The wine oxidation procedure involved the use of concentrated $HNO_3$ (4 mL) and 30% $H_2O_2$ (2 mL). Microwave digestion (Ethos Microsynth System oven, Milestone Microwave Laboratory Systems) was carried out following the sequence: 3 min for heating 85°C at 800 W; 12 min for heating 145°C; 6 min for heating 180°C; and 15 min at 180°C. After the end of the program, the digested wine was left to cool and its volume was adjusted (15 mL). Digestion of organic matter was assessed using IR spectroscopy and total phenolic content. The potentiometric sensors with chalcogenide glass and plasticized PVC membranes were able to simultaneously detect Fe (III), Cu, Pb, and Cd in digested wines. In addition, two analytical approaches for sample pretreatment to reduce the organic matrix in wines were developed by Durante et al. [69]. A MAD protocol and a low-temperature mineralization procedure were investigated to develop an accurate and precise analytical method for the determination of an 87Sr/86Sr isotopic ratio, which is an indicator used for geographical origin and authenticity of food. The two procedures led to comparable results.

Barbosa et al. [70] reported a new MAD procedure for the determination of essential and non-essential trace elements in soybeans and their products (i.e., soy extract, textured soy protein, transgenic soybeans, and whole soy flour). The effects related to the concentration of $HNO_3$ (2.1–14.5 mol $L^{-1}$) and the use of $H_2O_2$ on the efficiency of decomposition was evaluated. Recoveries were higher than 90% for all analytes. Using this procedure, the difficulties involved in soybean digestion due to the high content of proteins and fats were overcome.

Kronewitter et al. [71] described an improvement in sample preparation and analysis to broaden electrospray ionization-mass spectrometry (ESI-MS) glycan characterization including polysialylated N-glycans. Sample preparation improvements involved an acidified microwave-accelerated PNGase F N-glycan release to promote lactonization, and sodium borohydride reduction. Both procedures were optimized to improve quantitative yields and preserve the glycoforms detected. The microwave-assisted acidic digestion was a secondary objective to begin the acid lactonization stabilization of polysialic acid residues. These advances enabled the first report and direct measurement of families of polysialylated glycans in human serum using MS.

Yafa and Farmer [72] presented four acid digestion methods ($HNO_3$, $HNO_3$/HCl, aqua regia, and $HNO_3$/HF) that were compared using a Canadian fen peat (OGS 1878P-6) because a standard reference wet method for the determination of inorganic elements in peats was not available. The technique used for the determination of different metals was ICP-OES. Two microwave-assisted methods, $HNO_3$ and $HNO_3$/HF, representing "acid-extractable" and total digestion respectively, were developed. The studied elements were Al, Co, Cr, Cu, Fe, Mn, P, Pb, S, Ti, V, and Zn. The first method procedure consists of 0.25 g of a sample being ashed at 450°C for 4 h prior to digestion in 10 mL concentrated $HNO_3$ using Teflon microwave digestion vessels. The program used was maximum power 1200 W, 100%, ramp 30 min, hold 20 min, 150 psi, 205°C. Upon cooling, the solution was filtered to remove any remaining solid material. The solution was diluted up to 25 mL with 2% (v/v) $HNO_3$ prior to analysis by ICP-OES. The second extraction procedure was a modified version of USEPA 3051 a method that consisted of the use of $HNO_3/HCl$ as digestion solvent. The procedure was similar to the procedure described previously using 9 mL of $HNO_3$ and 1 mL of HCl instead of $HNO_3$ only. In the third procedure, the standard aqua regia extraction according to ISO Standard 11466 was used, in the absence of a microwave oven. The last method was carried out in order to achieve total dissolution of peat material. Again, the first procedure was used but now adding 9 mL of concentrated $HNO_3$ and 0.5 mL of concentrated HF.

Błazewicz et al. [73] reported a sample pretreatment method for total iodine determination in human serum and urine. Both alkaline (25% tetramethylammonium hydroxide, 6 min) and acidic (65% $HNO_3$, 7 min) digestion procedures were carried out in a closed system with the assistance of microwaves. The simplest combination of reagents and the appropriate parameters for the digestion were optimized to find a better method of analysis. An ion chromatography method with pulsed amperometric detection applied to evaluate a variety of parameters affecting the determination of total iodine in biological fluids at low levels. The best recoveries were 96.8 and 98.8% for urine and serum samples respectively using 25% tetramethylammonium hydroxide solution within 6 min of effective digestion time.

Balarama Krishna et al. [74] developed a simple and efficient MAD method for the rapid determination of chloride and fluoride in nuclear-grade boron carbide powders using a closed microwave-vessel (G30 Anton Parr, resisting pressure up to 30 bar and temperature up to 300°C) and a domestic microwave oven (Videocon, Mumbai, India) programmable for time and microwave power. Quantitative recovery of chloride and fluoride (98–101%) were obtained by irradiating ~0.5 g of the sample in 10 mL of 10% $HNO_3$ (v/v) for about 30 s at 480 W. Both anions were determined by an ion-selective electrode (ISE) using matrix-matched standards. The LOD values for chloride and fluoride in conjunction with the proposed MAD were found to be 1.9 $\mu$g g$^{-1}$ and 1.2 $\mu$g g$^{-1}$ respectively with a relative standard deviation (RSD) of less than 10%.

Leme and co-workers [75] proposed a method to determine inorganic constituents (Se, Mg, Ca, Al, P, Mn, Fe, Cu, Zn, Ba, and Pb) in 60 honey samples from different regions of Brazil. Mineralization with 2 mol L$^{-1}$ $HNO_3$ at 1000 W using a microwave digestion system with PFA closed vessels was the most suitable method for determination of the analytes. Microwave-assisted acid digestion was optimized by means of a central composite design (13 experiments). The $HNO_3$ concentration (2–14 mol L$^{-1}$) and microwave power (500–1000 W) were evaluated. Final acidity and residual carbon were the monitored and optimized responses. Simultaneous optimization of the responses was performed using the global desirability function. In this method, 4 mL of $HNO_3$ in different concentrations was added to 0.250 g of honey. This mixture was predigested by keeping it overnight. Then, it was added to 1 mL of $H_2O_2$ and 3 mL of $H_2O$. The heating program of the microwave consisted of a step of 5 min (ramp) and a step of 10 min (hold), for all conditions proposed in the central composite design. After digestion, the vessels were submitted to a cooling program of 10 min. Every digested sample was dissolved up to 15 mL with deionized water. The accuracies were evaluated using two certified reference materials (tomato leaves—NIST 1573a for Cr and peach leaves—NIST 1547 for the other elements) and the recoveries ranged from 82 to 115%.

Other works of interest using MAD are summarized in Table 2.1. Information regarding the analyte, the matrix, the procedure used and the determination technique applied is included.

## 2.3   MICROWAVE-INDUCED COMBUSTION (MIC)

In recent years, the MIC technique which combines the advantages of combustion and microwave-assisted digestion, in a unique system, has been developed [123]. The high amounts of concentrated acids (i.e., sulfuric acid) necessary to attain high temperatures at atmospheric pressure may increase blank values, which is one of the major disadvantages associated with wet digestion procedures [124]. MIC considers the combustion of organic samples in closed quartz vessels pressurized in the presence of an oxygen excess with an ignition step carried out by microwave radiation. The process removes the organic matrix with a high efficiency of sample oxidation and facilitates the digestion of relatively high sample masses in a few minutes [125]. The combustion products are absorbed in a suitable solution into the same reaction vessel. In the cases in which is necessary to extract the remaining oxidized portions of the samples, a complete sample digestion by reflux can be carried out. Focused microwave-induced combustion (FMIC) avoids the use of concentrated acids, significantly reducing laboratory wastes and possible interferences in the determination step. The MIC device is represented in Figure 2.3.

A digestion procedure based on sample combustion ignited by microwave radiation was proposed by de Moraes Flores et al. [126] for determining cadmium and

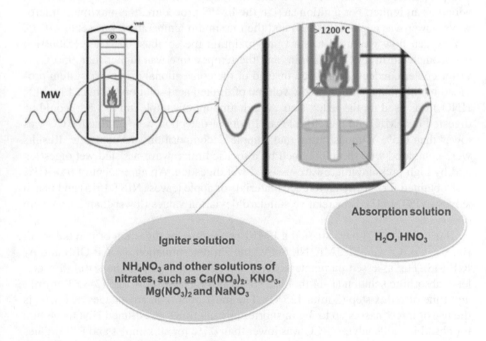

**FIGURE 2.3**   MIC digestion device.

copper by ETAAS in organic samples such as bovine liver, pig kidney, and skim milk. A Multiwave 3000 microwave sample preparation system equipped with up to eight high-pressure quartz vessels was used in this study. A home-produced holder was used to put the samples inside the quartz vessels. Samples between 50 and 250 mg were weighed and they were placed in the quartz vessels which contained 6 mL of $HNO_3$ as an absorbing solution. The holder containing the sample was located into the quartz vessel, and 50 μL of a $NH_4NO_3$ solution was immediately added. The microwave energy program used for the combustion procedure was: (1) 1400 W for 20 s, (2) 0 W for 2 min, (3) 1400 W for 8 min (optional step), and (4) 0 W for 20 min for cooling if step 3 was applied. After finishing the digestion, each vessel was carefully opened to release the pressure. The resultant solution was diluted with water and transferred into a 25 mL polypropylene vessel. After each run, holders were soaked in concentrated $HNO_3$ for 10 min and rinsed with water. The determination of Cd and Cu by AAS was carried out to demonstrate the applicability of the combustion procedure. The proposed combustion method using microwave oven allowed the digestion of bovine liver and skim milk samples in a faster way than conventional wet digestion using a microwave for heating. The digestion can be performed in only a few minutes and with a minimum use of solvents.

Mesko et al. [124] presented another procedure for sample digestion based on FMIC. Here, a commercially-focused microwave oven (Star System 2, 800 W, CEM, Matthews, NC) with a lab-made quartz sample holder and a modified glass vessel (180 mL of capacity) was applied for the assessment of botanical sample digestion prior to the determination of Al, Ba, Ca, Fe, Mg, Mn, Sr, and Zn by ICP-OES. Sample pellets were placed on the quartz holder and 6 mol $L^{-1}$ $NH_4NO_3$ (50 μL) was added as an igniter. For ignition step in the FMIC procedure, the maximum microwave power was applied (800 W) and the maximum temperature was set at 430°C. An oxygen flow was used to start and maintain the combustion. The combustion was completed in less than 2 min, and the temperature was higher than 950°C. A water-cooler condenser was used instead of the conventional air-cooler system provided by the manufacturer. A low volume of diluted acid solution (10 mL, 4 mol $L^{-1}$ $HNO_3$) was used for the reflux step. A high amount of sample (up to 1.5 g) could be digested by FMIC and therefore lower limits of detection were attained. RCC was lower than 0.5%, demonstrating that sample decomposition was complete. Results were compared with those obtained by a focused microwave-assisted wet digestion and by high pressure microwave-assisted wet digestion. An agreement of 95–103% was obtained for certified reference materials of apple leaves (NIST 1515) and peach leaves (NIST 1547). Low relative standard deviation values (lower than 3.8%) were observed.

Barin et al. [127] have studied a FMIC procedure for digestion of plants for Al, Ba, Ca, Co, Cr, Cu, Mg, Mn, Ni, Sr, V, and Zn determination by ICP-OES and by ICP-MS. The assessed parameters were, among others, sample mass (0.1–3 g pellets), absorbing solution (1–14 mol $L^{-1}$ $HNO_3$), oxygen gas flow-rate (2–15 L $min^{-1}$), and time of reflux step (5 min, 125°C). The most important advantages of FMIC is the use of large masses up to 3 g in shorter heating times and diluted $HNO_3$ solution for absorbing all analytes. RCC was lower than 0.7% for all samples and RSD values ranged from 1.5 to 14.1% using the selected experimental conditions for FMIC.

Accuracy was checked using certified reference materials (NIST 1515 apple leaves and NIST 1547 peach leaves). Determined values for all metals were in agreement with certified values at a 95% confidence level. Detection limits using FMIC were about 3 and 6 times lower than those obtained by FMAWD and MAWD, respectively ($0.02$–$0.15$ $\mu g\ g^{-1}$ for ICP OES and $0.001$–$0.01$ $\mu g\ g^{-1}$ for ICP-MS).

Flores and co-workers [128] have developed an effective approach for fluoropolymers digestion using MIC in closed quartz vessels pressurized with oxygen for the determination of Ag, Ca, Cd, Co, Cr, Cu, Fe, K, Mg, Mn, and Ni impurities. The samples submitted to study were polytetrafluorethylene (PTFE), polytetrafluoroethylene with an additional modifier, perfluoropropylvinylether (PTFE-TFM) and fluorinated ethylene propylene (FEP). The sample preparation system was equipped with high-pressure quartz vessels (80 mL internal volume) and the maximum pressure and temperature were 80 bar and 280°C, respectively. For the combustion step, the microwave power was set at 1400 W for 10 min. A quartz device was employed as the sample holder. Sample masses were between 100 and 400 mg. Dilute $HNO_3$ (5 mol $L^{-1}$, 6 mL) selected as the absorbing medium, was also used to reflux the sample for 5 min after the combustion. The method effectiveness was demonstrated by the RCC in the digests (lower than 1%). Trace elements determination was performed by ICP-OES and MS, and the neutron activation analysis (NAA) was used for validation purposes obtaining a good agreement for all the studied analytes (higher than 98%). Results were also compared to those obtained using microwave-assisted acid extraction in high-pressure closed systems. Taking into account the inertness of the fluoropolymers, a suitable sample throughput was obtained using MIC (up to 8 samples each 30 min).

Pereira et al. [129] proposed a sample preparation method based on FMIC for milk powder digestion prior to determination of Ba, Ca, Co, Cr, Cu, Fe, Mg, Mn, Sr, V, and Zn by ICP-OES. Milk powder digestion was carried out using a commercial microwave oven with focused microwave radiation with two cavities (Star System 2, 800 W, CEM, Matthews, NC, USA) equipped with glass vessels of 180 mL of capacity and a lab-made water-cooler condenser instead of the conventional air-cooler system provided by the manufacturer. Sample digestion was carried out by combustion under oxygen flow rate (2 to 15 L $min^{-1}$) using $NH_4NO_3$ (50 $\mu L$ of 6 mol $L^{-1}$ solution) as an ignition aid. Pressed samples (pellets up to 3 g) were placed on a quartz holder positioned inside the glass vessel and diluted $HNO_3$ (2 mol $L^{-1}$) was used as the absorbing solution. The maximum microwave power was 800 W and the temperature applied was 430°C. The FMIC system operates at atmospheric pressure using a glass vessel especially designed to allow oxygen flow inside. Agreement between results obtained with certified reference materials of milk powder after FMIC compared with the results obtained by microwave-assisted wet digestion (MAWD) and MIC was higher than 95% for all analytes.

Muller and colleagues [130] have assessed a sample preparation method for nuts based on the MIC in closed vessels. The determination of As, Cd, and Pb was carried out by ICP-MS while flow injection cold vapor generation coupled with ICP-MS (FI-CVG-ICP-MS) was used for Hg. High-fat content samples (packaged in polyethylene films) were combusted using 20 bar of oxygen and 50 $\mu L$ of 6 mol $L^{-1}$ ammonium nitrate as an aid for ignition. Hazelnuts, almonds, cashew nuts, Brazil nuts, and walnuts (sample masses up to 500 mg) were combusted using 7 mol $L^{-1}$ $HNO_3$ as the absorbing solution. A certified reference material was used to evaluate the accuracy of the procedure (better than 96% for all analytes). The results were compared with the results obtained with conventional microwave-assisted acid digestion (MW-AD) in pressurized vessels. MIC showed lower RCC (less than 1.5%) and lower limits of detection (LODs). The LOD obtained using MIC was 3, 2, and 6 ng $g^{-1}$ for As, Cd, and Pb, respectively. The LOD for Hg by FI-CVG-ICP-MS was 7 ng $g^{-1}$.

The determination of As, Cd, Hg, and Pb in tricyclic active pharmaceutical ingredients (APIs) after digestion using both combustion and wet digestion methods was performed by Barin et al. by ICP-MS [131]. The United States Pharmacopeia (35th edition) recommended the dry ashing method for carbamazepine digestion (muffle furnace, 500–600°C). Nevertheless, significant losses of analytes were observed in recovery tests ($38.0 \pm 8.9$, $99.5 \pm 7.1$, and $89.4 \pm 6.3\%$ of recovery for As, Cd, and Pb, respectively), besides Hg was completely lost by volatilization during digestion. Solid residues were observed for carbamazepine and amitriptyline hydrochloride using high-pressure asher (HPA) digestion systems under drastic digestion conditions (280°C, 120 min). The decomposition of all samples was only feasible by reducing sample mass, using high temperature and long digestion time (0.08 g, 320°C, 180 min). Moreover, digestion of carbamazepine, amitriptyline hydrochloride, and imipramine hydrochloride by microwave-assisted wet digestion in closed vessels was not efficient (yellow-orange solid residue was observed for all substances). The tricyclic APIs digestion by MIC method was adequate for the digestion of 0.5 g of all substances with high efficiency (RCC lower than 1%) using diluted $HNO_3$ as an absorbing solution (7 mol $L^{-1}$). For this purpose, a Multiwave 3000 microwave sample preparation system (Anton Paar, Graz, Austria) equipped with eight high-pressure quartz vessels (XQ-80) was employed at 1400 W for 10 min. Recovery values for the assessed elements were between 94 and 103%. MIC allowed the compatibility with ICP-MS due to low RCC and acidity of the digests obtained.

Another approach based on the MIC for fish (*Engraulis anchoita*) digestion and later determination of As, Cd, Co, Cr, Cu, Fe, Mn, Mo, Ni, Se, and Zn by ICP-MS was developed by Maciel et al. [132]. A 5 min reflux step was applied in order to improve absorption and analyte recovery. An absorbing solution of 5 mol $L^{-1}$ $HNO_3$ gave appropriate results. In addition, MAD in closed vessels using concentrated $HNO_3$ was assessed to compare results. Both sample preparation methods were suitable but MIC was recommended because higher digestion efficiency and better detection limits were provided. Furthermore, diluted $HNO_3$ can also be used as an absorbing solution. Accuracy was assessed by the analysis of certified reference materials (DORM-2 and TORT-2) after MIC digestion and ICP-MS determination (more than 94%agreement with CRM) (Table 2.2).

## 2.4 MICROWAVE-ASSISTED EXTRACTION (MAE)

The MAE procedure is most interest for the analytical chemists. Microwave-assisted Soxhlet extraction (MASE) can be included in this instrumental methodology. Microwave heating involves two mechanisms, ionic conductance and dipolar rotation. In this way, the heating production is more efficient and polar solvents can

**TABLE 2.2**
**Application of MIC to Different Matrices**

| Analyte | Matrix | Experimental Conditions | | | Detection Technique | Ref. |
|---|---|---|---|---|---|---|
| | | Sample Mass | Solvent | Preselected Program | | |
| Al, Fe, Mn, Sr, and Zn | Carbon black-containing elastomers (30%) | 50–500 mg | $NH_4NO_3$ (6 mol $L^{-1}$, 50.0 μL) as ignition aid $HNO_3$ (4 mol $L^{-1}$, 6.0 mL) as absorbing solution 20 bar $O_2$ | 1400 W, 60 s (sample ignition) 1400 W, 5 min (optional reflux step) 0 W, 20 min (cooling step, if optional step was applied) | ICP OES | [133] |
| Cu and Zn | Biological materials | 50–500 mg | $NH_4NO_3$ (50% m/v, 50.0 μL) as ignition aid | 1400 W, 60 s (sample ignition) 1400 W, 5 min (optional reflux step) 0 W, 20 min (cooling step, if optional step was applied) | FAAS | [134] |
| Bromide, chloride, fluoride, and iodide | Coal | 50–500 mg | $NH_4NO_3$ (6 mol $L^{-1}$) as ignition aid $(NH_4)_2CO_3$ (50 mmol $L^{-1}$) as absorbing solution, 5 min of reflux 2 MPa $O_2$ | 1400 W, 60 s (sample ignition) 1400 W, 5 min (optional reflux step) 0 W, 20 min (cooling step, if optional step was applied) | $IC^a$ | [135] |
| As, Hg, Cd, Pb, Cr, Cu, Mn, Mo, Ni, Pd, Pt, V, Os, Rh, Ru, and Ir | Pharmaceutical products (aspirin) | 400–700 mg | $NH_4NO_3$ (1.5 mol $L^{-1}$) as igniter 20% (v/v) $HNO_3$ (5.0 mL) as absorbing solution, 20 bar high purity $O_2$ | 1400 W (initial high power microwave irradiation) Reflux cycle pressure: 80 bar, temperature reached: 220°C | ICP-MS | [136] |

*(Continued)*

**TABLE 2.2 (CONTINUED)**
**Application of MIC to Different Matrices**

| Analyte | Matrix | Sample Mass | Solvent | Preselected Program | Detection Technique | Ref. |
|---|---|---|---|---|---|---|
| | | | **Experimental Conditions** | | | |
| Total arsenic | Mussel tissue, fish tissue, and whole shrimp | 500 mg | $NH_4NO_3$ (6 mol $L^{-1}$, 40.0 mL) as igniter $HNO_3$ (0.1 mol $L^{-1}$, 6.0 mL) as absorbing solution, 5 min of reflux | 1400 W, 1 min 1400 W, 5 min (optional reflux step) 0 W, 20 min (cooling step) | ICP-MS, FI-HG AAS, FI-HG-ICP-MS | [137] |
| Nitrates | Whole milk powder pellets | 100 mg milk 50 μL Nitrates solution | $NH_4NO_3$ (7 mol $L^{-1}$, 50.0 mL) and aqueous solutions of $Ca(NO_3)_2$, $KNO_3$, $Mg(NO_3)_2$, $NaNO_3$ and $NH_4NO_3$ as igniters, $H_2O$ (6.0 mL) as absorbing solution, 20 bar $O_2$ | 1400 W, 60 s | Combustion was observed | [138] |
| As, Cd, Pb, and Hg | Pellets of coal | 500 mg | $NH_4NO_3$ (6 mol $L^{-1}$, 50.0 μL) as igniter $HNO_3$ (5.0 mol $L^{-1}$ 6.0 mL) as absorbing solution, 20 bar $O_2$ | 1400 W, 1 min 1400 W, 5 min (optional reflux step) 0 W, 20 min (cooling) | CV[b] ICP-MS ICP-MS | [139] |
| As, Cd, Pb, and Hg | Soil | 300 mg | $NH_4NO_3$ (6 mol $L^{-1}$, 50.0 μl) as igniter 6.0 mL absorbing solution $HNO_3$ (2 mol $L^{-1}$) + HCl (2 mol $L^{-1}$) 20 bar $O_2$ (initial pressure) | 1400 W, 5 min (ignition and reflux step) 0 W, 20 min (cooling) | ICP-MS, ICP-OES (As, Cd, Pb) CVG-ICP-MS (Hg) | [140] |

(*Continued*)

**TABLE 2.2 (CONTINUED)**
**Application of MIC to Different Matrices**

| Analyte | Matrix | Experimental Conditions | | | Detection Technique | Ref. |
| --- | --- | --- | --- | --- | --- | --- |
| | | Sample Mass | Solvent | Preselected Program | | |
| Bromine and iodine | Honey | 500 mg | 6.0 mL of the alkaline solution (0.11 mol L⁻¹ TMAH solution, 50 or 100 mmol L⁻¹ NH₄OH solution) 400 mg microcrystalline cellulose, PE film (combustion aid) | 1400 W, 50 s 0 W, 3 min 1400 W, 5 min 0 W, 20 min (cooling) | ICP-MS | [141] |
| Halogens and sulfur | High polyimide | 600 mg | NH₄OH (50 mmol L⁻¹) as absorbing solution, 20 bar of O₂ | 1400 W, 5 min 0 W, 20 min (cooling) | IC¹ conductivity detection ICP-MS | [142] |
| Bromine and iodine | Milk | 500 mg | 6.0 mL of absorbing solution (25 mmoL⁻¹ NH₄OH) | 1400 W, 5 min 0 W, 20 min (cooling) | ICP-MS | [143] |
| Sulfur | Coals | ≈0.50 g | 5.0 mL 70% HNO₃ in water+4.0 mL 30% H₂O₂ | 1400 W, 1.5 min 60 min, 1400 W | ID –SF-ICPMSᶜ | [144] |
| Cl and F | Portland cement | 100 mg | 6.0 mL of absorbing solution Ammonium nitrate solution (50.0 μL, 6 mol L⁻¹) 20 bar of oxygen. 300 mg of cellulose | 1400 W, 1, 5, 10, or 15 min (volatilization+reflux step) 0 W, 20 min (cooling) | IC¹ | [145] |

*(Continued)*

**TABLE 2.2 (CONTINUED)**
**Application of MIC to Different Matrices**

| Analyte | Matrix | Experimental Conditions | | | Detection Technique | Ref. |
|---|---|---|---|---|---|---|
| | | Sample Mass | Solvent | Preselected Program | | |
| Ag, As, Cd, Ga, Hg, In, Pb, Bi, Sb, and Sn | Flexible graphite | 100 mg | 50.0 µL (6 mol L⁻¹ NH₄NO₃) 4 mol L⁻¹ HNO₃ absorbing solutions (Ag, As, Cd, Ga, Hg, In, Pb, and Zn) and inversed aqua regia (Bi, Sb, and Sn), cellulose pellet (300 mg) | 1400 W, 10 min (ignition and reflux steps) 0 W, 20 min (for cooling) | ICP-OES and CVG-AAS | [146] |
| Bromine and iodine | Milk powder for adult and infant nutrition | 700 mg | 50.0 µL of ammonium nitrate solution (6 mol L⁻¹), 6.0 mL of absorbing solution, 20 bar of oxygen | 1400 W, 5 min (optional reflux step) 0 W, 20 min (cooling step) | ICP-MS and ICP-OES | [147] |
| Trace, residual catalyst metal | Single-wall carbon nanotubes | 25 mg | 6.0 mL HNO₃ and HCl acids (3:1) (absorbing solution) 50.0 µL of ammonium nitrate solution (6 mol L⁻¹), 20 bar of oxygen | 1400 W, 15 min 0 W, 20 min (cooling step) | ICP-MS | [148] |
| Chlorine | Pet food | 600 mg | 20 bar of O₂ (NH₄)₂CO₃ or ultrapure water (as absorbing mediums) | 5 min, 1400 W 20 min, 0 W for cooling | ISEᵈ | [149] |

*(Continued)*

## TABLE 2.2 (CONTINUED)
## Application of MIC to Different Matrices

| Analyte | Matrix | Experimental Conditions | | | Detection Technique | Ref. |
|---|---|---|---|---|---|---|
| | | Sample Mass | Solvent | Preselected Program | | |
| Br and I | Edible flours | 100–1000 mg | $NH_4NO_3$ (6 mol $L^{-1}$, 50.0 µL) 25 mmol $L^{-1}$ $NH_4OH$ as absorbing solution | 1400 W, 5 min (reflux step) 0 W, 20 min (cooling) | ICP MS | [150] |
| L-T$_4$ | Pharmaceutical products | 400 mg | $NH_4NO_3$ (6 mol $L^{-1}$, 50.0 µL) 20 bar $O_2$ | 1400 W, 5 min (reflux step) 0 W, 20 min (cooling) | ISE for I[1] | [151] |
| Br and I | Whole eggs | 350 mg | 6.0 mL $NH_4OH$ solution (50 mmol $L^{-1}$) | 1400 W, 5 min (reflux step) 0 W, 20 min (cooling step) | ICP MS | [152] |
| Cd, Co, Cr, Cu, Mn, Ni, Pb, and V | Diesel | 100 ± 5 mg | 50.0 µL of 6 mol $L^{-1}$ $NH_4NO_3$ | 1400 W, 5 min (reflux step) 0 W, 20 min (cooling step) | ICP MS | [153] |
| F, Cl, Br, and I | Soil | 100 mg | 50.0 µL $NH_4NO_3$ (6 mol $L^{-1}$) 6.0 mL $NH_4OH$ (100 mmol $L^{-1}$) | 1400 W, 5 min (reflux step) 0 W, 20 min (cooling step) | ICP MS | [154] |

a Ion chromatography
b Cold vapor generation
c Isotope dilution analysis and sector-field inductively coupled plasma mass spectrometry
d Ion selective electrode

absorb and convert much more MW energy. In conventional heating, a finite period of time is needed to heat the vessel before the heat is transferred to the solution, while microwaves heat the solution directly. Besides, better control of the experimental conditions and shorter analysis times prevent problems of samples degradation. It is interesting to note that water can be used as a solvent for MAE. Compared with methanol, acetone, hexane, and dichloromethane it is shown that water in MAE is as efficient as organic solvents, and is inexpensive, safe, and environmentally friendly [155]. The high sample throughput and the relatively small extraction times required make this technique quite attractive and together with sonication, it is the most used extraction methodology.

On the other hand, focused MASE is a variant of MAE which contemplates the advantages of traditional Soxhlet method and tries to overcome its main drawbacks [156].

Here, there are comments about the most relevant papers including this procedure for the sample treatment and Table 2.3 contains the principal applications of this instrumental methodology in other papers of interest.

Karthikeyan and collaborators [157] used a closed vessel microwave digestion system for both digestion (MAD) and extraction (MAE) experiments for determining the water-soluble fraction of trace elements in urban dust. The vessels were sonicated (15 min, 10% $HNO_3$) and soaked in 2% $HNO_3$ acid overnight to prevent contamination. Finally, these vessels were rinsed with ultrapure water at least three times. For the determination of total metals, half of the filtered air samples were placed in Teflon vessels, and 4 mL of $HNO_3$, 2 mL of $H_2O_2$, and 0.2 mL of HF were added. The protocol for the digestion was: 5 min at 250 W and 95°C, 5 min at 400 W and 120°C, and 2 min at 600 W and 130°C. When room temperature was achieved, the sample digests were transferred to polyethylene (high density) (HDPE) sample vials and diluted to 50 mL with ultrapure water. Then, they were filtered and stored at 4°C until analysis. Different filter substrates were used. The other half of filtered air samples were used for the extraction of water-soluble metals. The microwave extraction conditions were: 15 mL of water and 100 W for 5 min. The extracts were filtered when they were at room temperature. Then, 20 µL (1:1) of ultra-pure $HNO_3$ was added and the extracts were stored at 4°C until analysis. The analysis of the samples was performed with a Perkin-Elmer Elan 6100 ICP-MS. The authors concluded that the two new extraction methods are very simple, fast, reliable, and quality-assured and it was possible to analyze many air particulate samples in order to monitor air pollution.

An easy method for the potentiometric determination of cadmium, lead, and copper in milk and fermented milk products was developed by Suturović et al. [158] using a microwave-assisted acid extraction sample pretreatment. Various samples of pasteurized milk and fermented milk products with 2.8% milk fat (declared by producers) were studied. 5.00 g of the samples were diluted with double distilled water (3-fold for the milk and 5-fold for the milk products samples), acidified with HCl and then stirred intensively (4000 min⁻¹) for 5 min. Metal extraction (1.00 g samples) was carried out with 3 mL of diluted $HNO_3$ (1:1), rather than HCl. The partial chemical digestion of the sample guaranteed an efficient trace metal leaching from insoluble sample fractions. The microwave acid extraction was carried out during 40 s at microwave energy of 160 W, and, additionally, 20 s at 320 W, using the microwave

**TABLE 2.3**

**Application of MAE to Different Matrices**

| Analyte | Matrix | Experimental Conditions | | | Detection Technique | Ref. |
|---|---|---|---|---|---|---|
| | | Sample Mass | Solvent | Preselected Program | | |
| V, Cr, Cu, Co, Se, Sr, Sn, Sb, Ba, Bi, Pb, Cd, As, Ni, Mn, Fe, Mg, and Zn | Leaves, flowers, and the infusion from *Salvia fruticosa* | 0.125 g | 5.0 mL 65% $HNO_3$ | • 1,600 W 165°C 2 min; • 1,600 W 175°C 8 min | FAAS ICP MS | [181] |
| N-nornuciferine, O-nornuciferine, and nuciferine | Lotus leaf | 0.50 g of dried sample | 10.0 mL 1.0 mol $L^{-1}$ [$C_6$MIM]Br | • 280 W, 2 min, solid–liquid ratio 1:30 (gmL$^{-1}$). | HPLC-UV | [182] |
| Ibuprofen, naproxen, ketoprofen, and diclofenac | Sewage sludge | 0.50 g freeze-dried | 50.0 mL water adjusted pH=6,0 (0.1 mol $L^{-1}$ HCl) | • Preheating (700 W, 5 min, 60°C) • 600 W, 5 min, up to 100°C • 700 W, 100°C, 30 min. | GC-MS | [183] |
| Chlorogenic acid | Tobacco residues | 0.50 g residue powder | 20.0 mL acetone- water (3:7, v/v) | • 300 W, 1 min | CZE[a] | [184] |
| Bergenin | *Ardisia crenata* sims and *Rodgersia sambucifolia* hemsl | 2.00 g | 60% (v/v) aqueous methanol | • liquid/solid ratio 10/1 (mL/g), 60°C, 15 min | HSCCC[b] | [185] |
| Sulfonamides | Soil | 2.00 g | 20.0 mL Triton X-114 (5.0%, v/v) | • 76°C, 13 min | HPLC-UV | [186] |
| Polycyclic Aromatic Hydrocarbons (FAHs) | Solid | 100 mg of SRM 1649a | 40.0 mL acetone/hexane (2:3, v/v) | • 600 W, 110°C 6 min • Holding 110°C, 14 min | TD–GC/MS[c] | [187] |
| Diethofencarb and pyrimethanil | Apple pulp and peel | 0.50 g spiked apple | 5.0 mL acetonitrile | • Focused Microwave oven • 80°C for 30 min | HPLC-DAD | [188] |

*(Continued)*

**TABLE 2.3 (CONTINUED)**
**Application of MAE to Different Matrices**

| Analyte | Matrix | Sample Mass | Experimental Conditions | | Detection Technique | Ref. |
|---|---|---|---|---|---|---|
| | | | Solvent | Preselected Program | | |
| Alpinetin, pinocembrin chalcone, and 1,7-diphenyl-4,6-heptadien-3-one Pinocembrin and cardamomin | *Alpinia katsumadai hayata* (*A. katsumadai*) | 2.00 g shattered crude | 10.0 mL methanol | • 40°C, 3 min, 300 W | HSCCC^b | [189] |
| Biphenyl cyclooctene lignans | *Schisandra chinensis* Baill | 25.00 g | 0.25 mol L$^{-1}$ [C$_{12}$mim] Br | • 385 W, 40 min, solid–liquid ratio 1:12 | HPLC | [190] |
| Cu, Mo, and Sb | Airborne particulate matter | 200 mg | 12.0 mL of aqua regia | • 175°C, 40 bar, P$_{90\%}$, Ramp (min) 1, 10 min | HR-CS GF AAS^d | [191] |
| Cd and Pb | Fresh fish | 250 mg lyophilized | 6.0 mL HNO$_3$ (65% v/v) 2.0 mL H$_2$O$_2$ (30% v/v) | • 200°C, 40 bar, 90% of power, ramp of 15 min 40 min | ETAAS | [192] |
| Cr, Mo, and Se | Infant formulas | 1.00 g | 7.0 mL HNO$_3$ 1.0 mL H$_2$O$_2$ | • 190°C, 15 min ramp to temperature, 1000 W, 30 min | ICP-OES – ICP-MS | [193] |
| V, Ni, Co, Mn, and Mo | Petroleum source rocks | 0.20 g | 10.0 mL 0.04 mol L$^{-1}$ Polyoxyethylene(23) dodecyl ether 0.04 mol L$^{-1}$ polyoxyethylene (10) dodecyl ether | • 200°C, 195 psi, maximum microwave irradiation power | ICP-OES | [194] |

*(Continued)*

**TABLE 2.3 (CONTINUED)**
**Application of MAE to Different Matrices**

| Analyte | Matrix | Experimental Conditions | | | Detection Technique | Ref. |
|---|---|---|---|---|---|---|
| | | Sample Mass | Solvent | Preselected Program | | |
| Phenolic compounds | *P. calliantha* H. Andr | 1.00 g | liquid-solid ratio 20:1, ionic liquid 0.5 mol L$^{-1}$ [C$_4$MIM]BF$_4$ | • 700 W, −0.07 MPa, 40°C, 15 min | HPLC | [195] |
| As(iii), As(v), DMA, and MMA | Chicken feed | 200 mg | 8.0 mL 2% v/v HNO$_3$ | • 650 W, 100°C for 30 min | HPLC-ICP-MS | [196] |
| 3-Amino-1,4- dimethyl-5 H-piridol[4,3-b]indole, 3-Amino-1-methyl-5 H- piridol[4,3-b]-ndole, 2-amino-1-methy-6-phenylimidazo-[4,5-b] pyridine, 2-amino-9 H-pyrido-[2,3-b] indole, 2-amino-3-methyl-9 H-pyrido-[2,3- b] indole 2-Amino-1,6-dimethylimidazo [4,5-b]-pyridine) | Cooked beefburguers | 3.00 g | 9.0 mL n-heptane (in three steps of 3 mL each one) | • 1600 W in less than 2 min <br> • 800 W, 3 min, ramp of 90, 100, and 110°C | LC–FD-DAD | [197] |

(*Continued*)

**TABLE 2.3 (CONTINUED)**
**Application of MAE to Different Matrices**

| Analyte | Matrix | Experimental Conditions | | | Detection Technique | Ref. |
|---|---|---|---|---|---|---|
| | | Sample Mass | Solvent | Preselected Program | | |
| Ba, Ca, Cd, Cr, Co, Cu, Fe, K, Mg, Mn, Na, Ni, Pb, and Zn | Honeybees | 250 mg | 6.0 mL HNO$_3$ | • Predigestion: room temperature, 1 h, 1.0 mL of H$_2$O$_2$ (30%, w w$^{-1}$) <br> • Increased up to 90°C, 4 min 500 W. <br> • 90°C for 2 min. <br> • Increased up to 180°C, 6 min <br> • 180°C, 10 min | ICP OES | [198] |
| Yeast lipids | *Saccharomyces cerevisiae* | Freeze-dried cells (~10 mg) | 7.0 mL chloroform– methanol (2:1, v/v) | • Ramped to 60°C from room temperature, 800 W, 6 min <br> • Kept constant for 10 min | HPLC-CAD | [199] |
| Total Mercury | Fish | 0.20 g of ground and dried fish | 6.0 mL HNO$_3$ | • 0–1000 W (10 min of ramp) <br> • 1000 W for 10 min <br> • 0 W for 20 min <br> • During sample digestion, the temperature reached about 230°C | FI-CVG-ICPMS | [200] |

*(Continued)*

**TABLE 2.3 (CONTINUED)**
**Application of MAE to Different Matrices**

| Analyte | Matrix | Experimental Conditions | | | Detection Technique | Ref. |
|---|---|---|---|---|---|---|
| | | Sample Mass | Solvent | Preselected Program | | |
| As, Sb, and Pb | Glass beads | 0.10±0.05 g | 3.0 mL of $HNO_3$, 1.0 mL of HF | • 180±5°C, 5 min<br>• hold at this temperature for 10 min | ICP-OES | [201] |
| Alprazolam, estazolam, lorazepam, clorazepam, diazepam, and tetrazepam | Human hair | 45 mg | 980 µL borate buffer (pH=9.5), 3.0 mL ethyl acetate | • 10 min at 75°C (ramping in 5 min), 1600 W | UHPLC-TOF-MS | [202] |
| Lead | Lipstick | 300 mg | 5.0 mL $HNO_3$, 2.0 mL HF | • 10 min to reach 180°C, This temperature was then maintained for 30 min | GFAAS | [203] |
| iAs(III) and iAs(V) | Biological samples | 100 mg BCR414 | 10.0 mL $H_2O$ | • 200 W, 2 min<br>• 800 W, 10 min<br>• 0 W, 30 min | FI-HG-MF-AAS | [204] |
| Lanthanides | Plant materials | 0.50 g | 4.0 mL $HNO_3$, 0.5 mL 30% $H_2O_2$ | • 500 W, 10 min<br>• 1000 W, 15 min | IC UV | [205] |
| Cocaine, benzoylecgonine, cocaethylene, morphine, 6-monoacethylmorphine, and codeine | Human hair | 0.05 g | MeOH, Dichloromethane | • 9 min, 60°C | HPLC-UV | [206] |
| Chlorogenic acid | Honeysuckle | 1.00 g | 70% ethanol | • Domestic MO<br>• 400 W, 4 min | nano-LC-ESI/MSe | [207] |
| Ethyl glucuronide | Hair | 0.01 g | n-hexan/water (1:1) | • 110°C, 11 min | GC-MS | [208] |

(Continued)

**TABLE 2.3 (CONTINUED)**
**Application of MAE to Different Matrices**

| Analyte | Matrix | Sample Mass | Solvent | Preselected Program | Detection Technique | Ref. |
|---|---|---|---|---|---|---|
| | | | **Experimental Conditions** | | | |
| Total mercury (Hg), inorganic mercury (Hg$^{2+}$), and methylmercury (CH$_3$Hg$^+$) | Fish tissues | 0.50 g | HCl, NaCl | • 60°C, 10 min, 3 min ramping time was used to reach the desired temperature of 60°C | ICP–MS for total Hg, LC–ICP–MS for mercury species | [209] |
| Phenolic alkaloids | Medicinal plant *Nelumbo nucifera* Gaertn | 1.00 g | ionic liquids | • Domestic MO • 280 W, 1.5 min | HPLC | [210] |
| Polybrominated diphenyl ethers | Biological matrices | 0.50 g | 30.0 mL dichloromethane | • 30 W, 10 min | GC | [211] |
| Chlorogenic acid, geniposidic acid, neochlorogenic ethyl ester, Isochlorogenic ethyl ester | Herbal Chinese medicine (Eucommia ulmodies Oliver) | 0.50 g | 60% ethanol | • 2 min, 3 atm | HPLC and LC–MS | [212] |
| Active pharmaceutical ingredients | Release tablet formulations, IR-1 and IR-2, and two controlled release tablet formulations, CR-1 and CR-2 | 0.389–0.558 g | IR-1:0.1 N HCl/ACN, 80/20 (v/v); IR-2: 20 mmol L$^{-1}$ phosphate buffer, pH 3/MeOH, 45/55 (v/v); CR-1: 0.1 N phosphate buffer, pH 6/MeOH, 55/45 (v/v); CR-2: ACN/MeOH, 50/50 (v/v) | • 45°C in 1–3 min (IR tablets) or to 65°C in 3–5 min (CR tablets); 10 min hold (IR tablets) or 15 min hold (CR tablets); and 10 min cool down for all formulations | HPLC-UV | [213] |

*(Continued)*

**TABLE 2.3 (CONTINUED)**
**Application of MAE to Different Matrices**

| Analyte | Matrix | Experimental Conditions | | | Detection Technique | Ref. |
|---|---|---|---|---|---|---|
| | | Sample Mass | Solvent | Preselected Program | | |
| Total mercury, $Hg^{2+}$ (as Hg) and $CH_3Hg$ (as Hg) | Tuna Fish Tissue Certified Reference Material (ERM-CE464) | 0.50 g | 5% (w/v) tetramethylammonium hydroxide (TMAH) in methanol | • 100°C, 10 min with magnetic stirring • 2 min ramping time was used to reach 100°C | ICP-MS and HPLC-ICP-MS | [214] |
| Organophosphorus fire retardants and plasticizers | Wastewater | 2.00 (mL) | | • 2 min, 100 W | GC-ICP-MS/ GC-TOF-MS | [215] |
| Active pharmaceutical ingredient (API) | Immediate released (IR) tablet formulation | 0.05 g | 10.0 mL $H_2O$ followed by 40.0 mL ACN | • 40°C, 70 min | HPLC-UV | [216] |
| Active pharmaceutical ingredients (APIs) | Immediate release tablets (model compound) | 0.005 g | acetonitrile | • 300 W, 50°C, 5 min | HPLC-UV | [217] |
| Polybrominated biphenyls (PBBs) and polybrominated diphenyl ethers (PBDEs) | Aquaculture feeds and products | 1.00 g | hexane/dichloromethane (1:1) | • 15 min, 85°C | GC-MS/MS | [218] |
| Polycyclic aromatic hydrocarbons (PAHs) | Wastewater from the scrubber of a pilot-scale fluidized bed incinerator | 20.0 (mL) | water | • Modified version of the domestic with a temperature-control cooling system • 145 W for 30 min | MA-HS-SPME GC/ FID | [219] |
| Organophosphate flame retardants and plasticizers | Dust | 0.50 g | acetone | • 130°C, 30 min | GC-NPDf | [220] |

*(Continued)*

## TABLE 2.3 (CONTINUED)
## Application of MAE to Different Matrices

| Analyte | Matrix | Sample Mass | Solvent | Preselected Program | Detection Technique | Ref. |
|---|---|---|---|---|---|---|
| | | | **Experimental Conditions** | | | |
| Isoflavonoids (genistein, daidzein, formononetin, and biochanin A) | Yellow soybeans | 0.10 g | HCl 12 mol $L^{-1}$ | • 600 W, 1 min | HPLC–ESI-MS/MS | [221] |
| Caffeine, 17β-estradiol, ibuprofen, ketoprofen, musk ketone, naproxen, and triclosan | Soil and sediments | 3.00 g | methanol/methylene chloride | • Temperature ramped from room temperature to 115°C in 8 min, with the final temperature held for 15 min, 800 W | GC-MS | [222] |
| Saponins | Herb slices of *Dioscorea nipponica* | 10.00 g | 100.0 mL MeOH | • Domestic microwave oven with a magnetic stirrer, water condenser, temperature measurement and time controlling<br>• 40 min, 80°C, 700 W | HPLC ELSD[g] | [223] |
| Artemisinin | Aerial parts of *Artemisia annua* L. plant | 10.00 g | 100.0 mL petroleum ether–acetone (4:1 v/v) | • Domestic microwave oven, with the addition of a magnetic stirrer, water condenser, temperature measurement and time controlling<br>• 40 min, 50°C, 700 W | HPLC-ELSD[g] | [224] |

*(Continued)*

**TABLE 2.3 (CONTINUED)**
**Application of MAE to Different Matrices**

| Analyte | Matrix | Experimental Conditions | | | Detection Technique | Ref. |
|---|---|---|---|---|---|---|
| | | Sample Mass | Solvent | Preselected Program | | |
| Solanesol | Tobacco | 10.00 g | 100.0 mL (hexane and ethanol, 1:3, v/v), | • 770 W, 30 min. | HPLC-ELSDg | [225] |
| As and Hg | Rice flour | 0.50 g | 15.0 mL of 1% HNO$_3$ | • 55°C, 5 min and held for 10 min, • 75°C, 5 min and held for 10 min • 95°C, 5 min and held for 30 min | HPLC-ICP-MS | [226] |
| iAs | Fish – seafood | 0.30 g | 7.3 mL of water, 0.7 mL HCl (36% w/w) | • 400 W, 5 min (ramp), 30 min (Hold), 1 min (Fan) • 0 W, 0 min (ramp), 20 min (Hold), 3 min (Fan) (cooling) | HG AASh | [227] |
| Sb | Soils – sediments –volcanic ash | 50.00 mg | 6.0 mL HNO$_3$, 0.5 mL HBF$_4$ | • Room °C (initial temperature –IT–), 200°C (final temperature –FT), 10 min (time slope –TS–), 0 min (Time hold- TH) • 200°C (IT), 200°C (FT), 0 min (TS), 20 min (TH) • 200°C (IT), 0°C (FT), 0 min (TS), 0 min (TH) | HPLC-HG-AFS | [228] |

*(Continued)*

**TABLE 2.3 (CONTINUED)**
**Application of MAE to Different Matrices**

| Analyte | Matrix | Experimental Conditions | | | Detection Technique | Ref. |
|---|---|---|---|---|---|---|
| | | Sample Mass | Solvent | Preselected Program | | |
| Sulfated polysaccharides | Green algae *Ulva* spp. – *Monostroma latissimum* | 1.00 g | 20.0 mL distilled water | • 4 min (come-up time), 10 min (extraction time) 100°C – 180°C | HPLC | [229] |
| Pb, Cd, and Zn | Water – Beverages (ice tea, mixed fruit juice) – foods (corn, potato chips, cumin, cheese, egg, canned tuna fish, canned corn, bean stew, chicken shawarma, black tea, green tea, spinach, rice, and meat) | 1.0 mL 1.00 g | 6.0 mL $HNO_3$ (65%), 2.0 mL $H_2O_2$ (30%) | • 6 min, 250 W • 6 min, 400 W • 6 min, 550 W • 6 min, 250 W • Ventilation (8 min) | AAS | [230] |
| Bioactive carbohydrates | Artichoke (*Cynara scolymus* L.) | 0.10 g dry sample 0.30 g dry sample | 10.0 mL of ultra-pure water | • 50°C, 3 min (inositol) • 120°C, 3 min (inulin) | GC-FID | [231] |
| Oleuropein | Olive leaves | 10.00 mg | 10.0 mL methanol and water (80:20, v:v) | • 80°C, 8 min • 80°C, 6 min (remained) | Square wave voltammetry | [232] |
| Cr | Edible animal oils | 0.10 g | 5.0 mL 0.4% v/v HF, 2% Triton X-100 | • 90°C, 45 min, a ramp time of 10 min | HPLC-ICP-MS | [233] |
| Lipids | Microalgae | 1.00 g | 5.00 g [BMIM][$HSO_4$] | • 800 W, 120°C and hold for 10–60 min | GC-MS | [234] |

*(Continued)*

**TABLE 2.3 (CONTINUED)**
**Application of MAE to Different Matrices**

| | | | Experimental Conditions | | | Detection | |
|---|---|---|---|---|---|---|---|
| Analyte | Matrix | Sample Mass | Solvent | Preselected Program | | Technique | Ref. |
| Cd | Omega-3 dietary supplements | 1.00 g | 6.0 mL HNO$_3$, 4.0 mL of H$_2$O$_2$ (30% v/v) | • 80°C, 5 min<br>• 80°C, 5 min<br>• 120°C, 7 min<br>• 120°C, 5 min<br>• 200°C, 11 min<br>• 200°C, 13 min | | TS-FF-AAS$^i$ | [235] |
| Benzotriazole, benzothiazole, and benzenesulfonamide | Soil | 3.00 g | 6.0 mL methanol | • pre-stirring (5 min)<br>• 200 W, 120°C, 10 min | | HPLC-UV | [236] |
| Endocrine disrupting chemicals | Soil, sediment, sewage sludge | 0.50–2.00 g | 10.0 mL methanol | • 350 W, 3 min | | GC | [237] |
| Polyphenols | Eclipta prostrata | 1.00 g<br>Ratio of liquid/solid 30 mL/g | 50% ethanol | • 400 W, 70°C, 2 min | | HPLC–DAD–ESI–MS/MS | [238] |
| Methyl β-Cyclodextrin-Complexed Curcumin | Turmeric rhizome oleoresin | 2.00 g | Particle size of 0.30–0.60 mm 10.0 mL water. | • 700 W, 3 min, 60°C | | HPLC | [239] |
| As(III), As(V), dimethyl arsenic acid, monomethyl arsenic acid, p-arsanilic acid, and roxarsone | Chicken tissue | 100 mg | 5.0 mL 22% v/v methanol, 90 mmol/L (NH$_4$)$_2$HPO$_4$, 0.07% v/v trifluoroacetic acid (pH 10.0 ammonium hydroxide solution) | • 71°C, 10 min, 11 min (holding) | | HPLC-MS | [240] |

*(Continued)*

**TABLE 2.3 (CONTINUED)**
**Application of MAE to Different Matrices**

| Analyte | Matrix | Experimental Conditions | | | Detection Technique | Ref. |
|---|---|---|---|---|---|---|
| | | Sample Mass | Solvent | Preselected Program | | |
| Pb | Geological samples | 0.50 g | 1.5% (m/v) NaBH₄, 3.0% (m/v) K₃[Fe(CN)₆] 0.3 mol L⁻¹ (HCl, HNO₃or CH₃COOH) | • 350 W, 15 min | HG-ICP-OESⁱ | [241] |
| Phenolic compounds | Almond Skin Byproducts (*Prunus amygdalus*) | 4.00 g | 60.0 mL of 70% (v/v) ethanol | • 100 W, 60 s | HPLC-DAD-ESI-MS/MS | [242] |
| Hg²⁺, CH₃Hg⁺, C₂H₅Hg⁺, and C₆H₅Hg⁺ | River water | — | 5.0 mL 6 mol L⁻¹ HCl | • 55°C, 15 min, 500 W | DGT-MAE-LC-CV-AFS | [243] |
| Phenolic compounds | Rice grains | 2.50g | 100% MeOH, 10:1 (ratio solvent-to-sample) | • 1000 W, 185°C, 20 min | HPLC | [244] |
| Total phenolics content | *Myrtus communis* | | 42% etanol, 32 (ratio liquid/solid mL/g) | • 500 W, 1.04 min | Folin–Ciocalteu method | [245] |
| Chiral pharmaceuticals and illicit drugs | Wastewater and sludge | 1.00 g | 20.0 mL methanol:water (1:1) | • 120°C, 30 min | LC-MS/MS | [246] |
| Fluoroquinolones | Compost | 0.30 g | 10.0 mL, 40% (w/v) Mg(NO₃)₂·6 H₂O, 4% (v/v) NH₃ | • 15 min pre-stirring • 200 W, 135°C, 15 min | UPLC – MS | [247] |

*(Continued)*

**TABLE 2.3 (CONTINUED)**

**Application of MAE to Different Matrices**

| Analyte | Matrix | Experimental Conditions | | | Detection Technique | Ref. |
|---|---|---|---|---|---|---|
| | | Sample Mass | Solvent | Preselected Program | | |
| Phyto-pharmaceutic compounds | Insect | <0.05 g dry mass | | • 1000 W, ramp to 110°C, 10 min<br>• 110°C, 10 min (hold)<br>• 30 min cool down | LC-MS/MS | [248] |
| Carotenoids | Shrimp matrices | 100 g dry sample | 20:1 mL g⁻¹<br>Hexane:acetone:ethanol 2:1:1 (v/v/v) as extraction solvent | • 7 min, 30 W | UV-Vis | [249] |
| Saponins and antioxidant capacity | Xao Tam Phan (*Paramignya trimera*) root | | 100 mL g⁻¹ | • 40 min, 360 W | UV-Vis | [250] |
| Betalains | Dragon fruit peel powder | 20.00 g | Distilled water | • 35°C, 8 min, 100 W | UV-Vis | [251] |
| Antioxidant phytochemicals | *Berberis asiatica* Roxb. Ex DC. leaves | | 1:45 g mL⁻¹ sample to solvent ratio<br>MeOH 60% | • 5 min, 500 W | HPLC DAD | [252] |
| Polyphenolics | *Berberis jaeschkeana* CK. Schneid. fruits | | 80% MeOH 0.1 mol L⁻¹ HCl<br>1.40 g mL⁻¹ sample to solvent ratio | • 180°C, 5 min, 670 W | UV-Vis | [253] |
| Pectin and naringin | Pomelo peels | 1.00 g | 20.0 mL 10 mmolL⁻¹ [HO₃S(CH₂)₄mim] HSO₄ aqueous solution 26 mL g⁻¹ solvent to sample ratio | • 15 min, 331 W | UV=Vis | [254] |

*(Continued)*

**TABLE 2.3 (CONTINUED)**
**Application of MAE to Different Matrices**

| Analyte | Matrix | Sample Mass | Experimental Conditions | | Detection Technique | Ref. |
|---|---|---|---|---|---|---|
| | | | Solvent | Preselected Program | | |
| Fluoroquinolones | Fish | 1.00 g | 10.0 mL MeOH | • 8 min, 150 W | UHPLC FD^k | [255] |
| Polyphenolic compounds | Ocimum basilicum | 5.00 g | 50.0 mL 50% EtOH | • 15 min, 442 W | UV-Vis | [256] |
| As(III) and As(V) | Leafy vegetables | 0.50 g | 20.0 mL 1% $HNO_3$ | • 90°C, 90 min | HPLC-ICP-MS | [257] |
| Polyphenols | Brewer's spent grain | 1.00 g | 0.75% NaOH 1:20 (w/v) solid to liquid ratio | • 100°C, 15 min | HPLC-DAD | [258] |
| Sulfur | Coal | 0.10 g | 12.0 mL 10% (m/v) NaOH-$H_2O_2$ | • 150°C, 5 min | ICP-OS | [259] |
| Essential oils | Citrus fruits | 500 g | 400.0 mL distilled water | • 100°C, 60–80°C, 600 lowered to 500 W | DRIFT^l | [260] |
| Flavonoids | Crotalaria sessiliflora IL. | 0.40 g Particle size 80 mesh | EtOH 32% $(NH_4)_2SO_4$ 22% 50:1 solvent to matrial ratio | • 80°C, 8 min | HPLC | [261] |
| PAHs | Ambient air | Glass fiber filters | 20.0 mL hexane: acetone 1:1 | • 100°C (rate of 10°C $min^{-1}$), 10 min | CG-MS | [262] |
| | surface soil | 2.00 g | 20.0 mL hexane: acetone 1:1 | | | |
| | wheat grain | 10.00 g | 20.0 mL ACN | | | |

*(Continued)*

**TABLE 2.3 (CONTINUED)**
**Application of MAE to Different Matrices**

| Analyte | Matrix | Experimental Conditions | | | Detection Technique | Ref. |
|---|---|---|---|---|---|---|
| | | Sample Mass | Solvent | Preselected Program | | |
| Phenolic compounds | Pitaya fruit | White-fleshed red pitaya | 1/149.95 g mL$^{-1}$ solid solvent ratio | • 72.27°C, 39.39 min | HPLC-DAD-ESI-MS | [263] |
| | | Yellow pitaya | 1/148.96 g mL$^{-1}$ solid solvent ratio | • 72.56°C, 5.02 min | | |

a  Capillary zone electrophoresis
b  High-speed counter-current chromatography
c  Thermal desorption–gas chromatography–mass spectrometry
d  High-resolution continuum source graphite furnace atomic absorption spectrometry
e  Nano-liquid chromatography-electrospray ionization mass spectrometry
f  Gas Chromatography with nitrogen–phosphorus detection
g  HPLC with evaporative light scattering detection
h  Hydride generation atomic absorption spectrometry
i  Thermospray flame furnace atomic absorption spectrometry
j  Hydride generation inductively coupled plasma optical emission spectrometry.
k  Ultra-high performance liquid chromatographic with fluorescence detection
l  Diffuse Reflectance Infrared Fourier Transform

digestion bomb (Parr, USA) and the household microwave oven (AVM 606, Philips Whirlpool). The authors constructed the automatic system for potentiometric and chronopotentiometric stripping analysis. The method accuracy was confirmed by analysis of standard reference material (SRM 1577b).

Mizanur Rahman and Kingston [159] who have developed a method to extract and determine inorganic mercury and methylmercury in soils and sediments carried out another attempt that illustrates the usefulness of MAE. The variables taking into account in the extraction procedure were $HNO_3$ concentration, sample amount, extraction temperature, and irradiation time. Approximately 1.0 g of homogenized soil or sediment and 10 mL of 4 mol $L^{-1}$ $HNO_3$ were placed in the microwave extraction vessels. Then, the vessels were irradiated at 100°C for 10 min. A 2 min ramping time was used to reach the desired temperature of 100°C. The MAE method was validated by using different sets of reference soil samples.

A MAE system was applied to determine tributyltin chloride (TBT), triphenyltin chloride (TPhT), tetraphenyltin (TrPhT), triethyltin chloride (TET), and tetraethyltin (TrET) in flour samples by HPLC [160]. This system has the capacity to handle a maximum of 10 reaction vessels at the same time. About 0.5 g of the sample was put into high-pressure PTFE tubes with a mixture of 4.5 mL of hexane–acetic acid (80/20, v/v). The samples were irradiated at 100°C for 3 min. Then, the extract was centrifuged at 12,000 rpm for 10 min to remove residues. An aliquot of the supernatant was diluted to 8 mL using hexane. The sample solution was filtered through a 0.45 μm membrane and injected into the HPLC–UV system. The proposed extraction system provided higher extraction efficiency compared with the conventional methods. A hexane replaced the methanol and water system as the extract-solvent and the application of MAE allows the simplification of the determination step.

Regueiro et al. [161] presented a microwave-assisted solvent extraction method for the determination of polybrominated diphenyl ethers (PBDEs) in domestic dust samples using GC–MS/MS. The extraction of the analytes was performed using an Ethos E Microwave Solvent Extraction Labstation, equipped with 12 pressurized 100 mL vessels. For selecting the optimal condition, an experimental design was used (mixed level fraction factorial design). The obtained results were 4 mL of a 10% NaOH aqueous solution, an extraction temperature of 80°C and no addition of sodium chloride. Therefore, samples (0.8 g) were extracted using a 10% NaOH (aq)/n-hexane 1:2 mixture at 80°C for 15 min. To evaluate the validation of the method a recovery study using spiked dust samples was done. The results were ranged from 92 to 114% for all compounds. The relative standard deviation of three replicates at a two-concentration level was calculated and the results were 11% and 15% at the low and higher level respectively. In this way, the authors evaluated the precision of the method. One of the most important conclusions in this work is that an efficient extraction could be reached using a combination of NaOH phase and n-hexane. Besides, it is possible to reduce the chromatographic background. It is important to point out that the manipulation of the samples was also reduced.

Two different experimental designs for microwave-assisted solvent extraction for the determination of POPs in marine sediment were proposed by Carro et al. [162]. In the first MASE, the studied variables were microwave power, extraction time and temperature, amount of sample, solvent volume, and sample moisture. A two-level full design $2^6$ is applied, so 64 experiments were needed. The analysis of the obtained results demonstrated that the amount of sample affected the extraction of α-HCH, pp'-DDE, dieldrin, pp'-DDD, and endrin and the solvent volume affected the extraction of pp'-DDE. The other variables (microwave power, extraction time and temperature, and sample moisture) and the interaction between factors did not affect the procedure. Based on the results of this first procedure, the authors proposed a second extraction procedure with only 8 experiments and 1 center point. The optimal variables were fixed: microwave power at a level of 20%, extraction temperature at 120°C, and solvent volume at 50 mL.

The best results were obtained when a low level of sample amount and higher extraction time were used. According to these two studies, under these conditions, the extraction of POPs was satisfactorily carried out. The MASE procedure was compared with Soxhlet one, and the obtained recoveries were similar.

Another attractive MAE approach for the extraction of various organophosphates in flame retardants and plasticizers in sediment was presented by García-López et al. [163]. Analyte determination was carried out by GC–ICP-MS. The sample was weighted (0.5 g) and extracted at 150°C in two sequential steps of 15 min, using 5 mL of acetone in the first extraction and 5 mL of acetonitrile in the second one. Both supernatants were combined, centrifuged, and evaporated under a nitrogen stream to 0.5 mL, approximately. An experimental factorial design type $3^1 \times 2^2$ was employed for the optimization of the extraction parameters (temperature, time, and acetone volume). The MAE method was compared with the conventional Soxhlet method where extractions were performed with a much higher volume of solvent (90 mL of acetone) for longer times (16 h against 30 min with MAE).

An interesting environmentally friendly MAE attempt was proposed by Varga et al. [164] using water as an extractant. They carried out studies on the Danube River in Budapest (Hungary) by collecting water and sediment samples simultaneously to elucidate the potential risk of special acidic pharmaceuticals (ibuprofen, naproxen, ketoprofen, and diclofenac) on the water supply used for the production of drinking water by bank filtration. The sample preparation procedure, in the case of water samples, included solid phase extraction (SPE). The MAE procedure followed by dispersive matrix extraction (DME) for pre-cleaning as well as SPE for enrichment was used in the case of sediment samples. 5 g of the dried sample was weighed and 50 mL of distilled water was used as an extractant. The extraction was performed by using a Milestone Start E Microwave Extraction System (Milestone, Italy). The quantification was carried out using GC-MS. The calculated recoveries were 97–99% (±7%) for the water and 95–103% (±12%) for the sediment samples. In the river water, ketoprofen concentration was always below the limit of quantification level and ibuprofen, naproxen, and diclofenac could be quantified in the range of 8–50, 2–30, 7–90 ngL$^{-1}$. In sediments, only naproxen and diclofenac were found in the range of 2–20 and 5–38 ngg$^{-1}$, respectively. The applied procedure for the investigation of sediments resulted in good recoveries and reproducibility using water as an extractant.

Mazzarino et al. [165] described a microwave-assisted liquid/liquid extraction (LLE) method to assess various prohibited substances – by the World Anti-doping Agency – from human urine samples. The LLE was carried out using a temperature-controlled single beam microwave oven with an extraction unit and closed vessels. For the most thermolabile components such as triamcinolone, prednisolone, chlorothiazide, chlorthalidone, epi-trembolone, and oxandrolone, the optimal power was 600 W (70°C), while 1020 W (150°C) was considered for the other compounds. Total extraction time was diminished between 30 and 60 s. The method effectiveness was evaluated by GC-MS (anabolic steroids, beta2-agonists, and narcotics) and by LC-MS/MS (diuretics, glucocorticoids, and beta-blockers). The authors have demonstrated that using microwave irradiation is feasible to speed up the LLE step of diuretics, narcotics, glucocorticoids, beta-blockers, and beta2-agonists, all included in the list of prohibited substances.

Xu and co-workers [166] have developed a new MAE and in situ clean-up method for the determination of six fluoroquinolone antibiotics (FQs: fleroxacin, levofloxacin, ciprofloxacin, lomefloxacin, enrofloxacin, and sparfloxacin) in chicken breast muscle. FQs were extracted from chicken breast muscle (5 g thawed at room temperature) into an aqueous solution of acetonitrile (ACN), meanwhile lipids were removed by hexane, simplifying the sample preparation process and reducing the operation errors. The microwave irradiation method included the extraction and clean-up procedures in one single step. The mixture of ACN containing 0.3% v/v phosphoric acid/water pH 3.0 (70:30, v/v) was used as the extraction solution and hexane was used as the clean-up solution. The extract was analyzed by LC-MS. The RSD values of intra- and inter-day obtained were in the range of 1.0–10.4 and 3.8–13.6%, respectively. Recoveries of FQs ranging from 66.0–97.2% were obtained in the three fortified levels of chicken breast muscle (20, 100, and 500 ng/g). The LODs were in the range of 2.7–6.7 ngg$^{-1}$.

Hu and colleagues [167] developed a new sample preparation method for auxin analysis in plant samples. A vacuum microwave-assisted extraction (VMAE) followed by molecularly imprinted clean-up procedure was proposed. Mainly, the method was based on two steps. In the first step, conventional solvent extraction was replaced by VMAE for extraction of auxins from plant tissues providing efficient extraction of 3-indole acetic acid (IAA) with an important reduction in extraction time. Auxin degradation was prevented by creating a reduced oxygen environment under vacuum conditions. In the second step, the raw extract of VMAE was further subjected to a clean-up procedure by magnetic molecularly imprinted polymer (MIP) beads, which had high molecular recognition ability for the two target auxins in plants, IAA and 3-indole-butyric acid (IBA). Selective enrichment and removal of interfering substances were achieved by dealing with a magnetic separation procedure. The proposed sample preparation method was coupled with HPLC and fluorescence detection for determination of IAA and IBA in peas and rice. The detection limits obtained for IAA and IBA were 0.47 and 1.6 ng mL$^{-1}$ and the relative standard deviation was 2.3% and 2.1%, respectively. The content of IAA was determined in pea and rice seeds, and pea embryo and roots. The recoveries ranged from 70.0–85.6%.

Wang et al. [168] proposed a sample pretreatment method for steroid hormones determination in seven kinds of fish tissues by coupling dynamic microwave-assisted extraction (DMAE) with salting-out liquid-liquid extraction (SLLE). The procedure can continually provide an extraction vessel with a fresh extraction solvent, and analytes could be transferred out of the extraction vessel as soon as they were extracted by a homogeneous LLE. A suitable salting-out agent was added to obtain phase separation of organic solvent from the bulk aqueous phase. In this work, steroid hormones were sequentially extracted from fish tissue (3.00 g) with acetonitrile and water (5 mL each). Then, the extract was separated, assisted by microwave energy, into an acetonitrile phase and an aqueous phase with ammonium acetate (a mass-spectrometry friendly salt). The acetonitrile phase containing the target analytes was concentrated and determined by LC-MS/MS. The steroid hormones recoveries for the spiked samples were in the range of $75.3 \pm 4.9\%$ to $95.4 \pm 6.2\%$. The proposed method reduced the consumption of the organic solvent, shortened the sample preparation time, and improved the sample throughput in comparison with the conventional procedure.

An innovative one-step sample preparation method combining MAE and solvent bar microextraction (MAE-SBME) was developed by Guo and Lee [169] for the determination of polycyclic aromatic hydrocarbons (PAHs) in environmental soil samples. SBME was carried out simultaneously with MAE and the extract from the SBME was straightforwardly analyzed by GC-MS without the requirement of an independent clean-up and/or preconcentration step. A CEM (Matthews, NC, USA) MES-1000 microwave extraction system equipped with a solvent detector was used to carry out MAE. 1.0 g of soil sample was transferred into a Teflon-lined extraction vessel (3 cm inner diameter) with 10 mL of water. An acceptor solvent, which was introduced into a sealed hollow fiber segment, the "solvent bar", was placed in the sample. Then the temperature was raised from room temperature to the selected temperature. MAE was performed at the desired temperature for an adequate time. After the extraction, the vessel was cooled to room temperature and the "solvent bar" was removed. The analyte-enriched acceptor solvent was carefully withdrawn into a microsyringe and an aliquot of the extract (1 µL) was directly injected into the GC–MS system for analysis. Since water was used as the extraction solvent in MAE, and only some microliters (less than 10 µL) of organic solvent were used in SBME, the procedure was environmentally friendly.

Balarama Krishna and co-workers [170] developed a method based on the conjunction of MAE and an oxidative pyrolysis (OP) using a homemade quartz chamber. This procedure constitutes a new sample preparation proposal for the determination of metallic impurities in boron carbide powders. A known amount of boron carbide powder (100–400 mg) was weighed into a quartz boat, inserted in the pyrolysis chamber and heated (~500°C, Bunsen burner) in the presence of an oxygen stream until the burning process was completed. Once the quartz boat was cooled to room temperature, MAE was applied with an extractant mixture of 10 mL of 30% $HNO_3 + 10\%$ HCl + 5% HF (microwave power: 1200 W, ramp: 10 min, temperature: $200 \pm 5°C$, hold: 10 min). Then, the supernatant was separated from the undissolved residues by centrifugation for 4 min at 5000 rpm. W, Si, Pb, Ni, Fe, Ca, Mg, Ti, and Al were determined by ICP-OES. Cross-validation of obtained data by the proposed OP-MAE method was conducted by sodium carbonate fusion. The detection limits obtained were significantly lower than those of the fusion method.

A new, efficient, rapid, cheap, and environmentally friendly method has been developed by Ares et al. [171] to determine glucosinolates in broccoli leaves using

liquid chromatography (LC) coupled with diode array (DAD) and electrospray ionization mass spectrometry (ESI-MS) detection. A microwave extraction procedure
using heated water was proposed and optimized (Box–Behnken design) based on the
deactivation of myrosinase. 50 mg of freeze-dried or dried ground broccoli leaf powder was put into a beaker and water (23 mL) was added. The resulting mixture was
heated for 3.5 min using a 70 W MW power. Low limits of detection and quantification were obtained, ranging from 10 to 72 µg g$^{-1}$ with DAD and 0.01 to 0.23 µg g$^{-1}$
with ESI-MS, and the resulting recovery values ranged from 87 to 106% in all cases.

Chang et al. [172] developed a method based on MAE and LC-MS for simultaneous analysis of drugs and their metabolites in hair: methamphetamine, amphet-
amine, methylenedioxymethamphetamine, methylenedioxyamphetamine, ketamine,
norketamine, dehydronorketamine, 6-acetylmorphine, morphine, and codeine. Hair
(10 mg) was incubated for 3 min with methanol-trifluoroacetic acid during MAE
at 700 W using MARS 5 microwave-accelerated reaction system. The total sample
preparation and analysis time was 50 min. Additional clean-up of the sample was not
required. Similar or greater extraction yields were obtained with the MAE method
(3 min) in comparison with the traditional overnight incubation process. Sample
preparation by MAE was a consistent procedure for the analytes extraction from
hair. Recovery values were higher than 90% and the SD for each compound was less
than 6%. Precision and accuracy for each analyte were within 15%.

Stan and co-workers [173] evaluated different sample preparations for extracting
ascorbic acid (AA) from parsley, dill, and celery. A new and original pulse micro-
wave-extraction procedure for AA extraction was carried out using a homemade
device, characterized by repetitive pulses of microwave power (900 W) at a rate of 1
pulse per second with controllable duty cycle (40%). Operation time and temperature
could also be controlled. The HPLC method was used for detecting AA. Ascorbate
from the plant extracts obtained was also quantified. Parsley contains the highest
amount of ascorbic acid (264 mg AA/100 g of fresh plant), followed by dill (121 mg
AA/100 g of fresh plant), and celery (103 mg AA/100 g of fresh plant).

Carrasco and Vassileva [174] developed and validated a method for MeHg determination in marine biota samples by aqueous-phase ethylation and GC coupled with
pyrolysis-atomic fluorescence spectrometry (Py-AFS). Various MAE procedures
were evaluated to extract MeHg from Scallop soft tissue CRM IAEA-452 in an effective way. Leaching with 5 mol L$^{-1}$ HCl at 60°C for 10 min, with 25% (w/w) KOH in
methanol at 70°C for 8 min, and with aqueous 25% (w/v) TMAH at 70°C for 8 min
was assessed. The extraction procedures were also tested with fish protein Dorm-3
CRM in order to validate the analytical procedure. Matrix effects observed when
using 25% (w/w) KOH in methanol or 25% (w/v) aqueous TMAH could be successfully overcome using standard addition method or by the introduction of a dilution
step. The MAE procedure with 5 M HCl gave adequate extraction efficiency (94%).
Recently, the same authors have evaluated a MAE method for mercury speciation
analysis in marine sediment samples using derivatization with 0.5% (v/v) 2-mercap-
toethanol in 5% (v/v) methanol [175]. This procedure was followed by separation and
detection steps using purge and trap and hyphenated GC-Py-AFS. A full validation
approach in-line with ISO 17025 and Eurachem guidelines was followed using an
estuarine sediment certified reference material (CRM IAEA-405).

Wu et al. [176] developed a novel method by coupling DMAE and microwave-accelerated solvent elution for the extraction of organophosphorus pesticides (OPPs) in fresh vegetables. Extraction, separation, enrichment, and elution procedures were completed in a simple step, simplifying the pretreatment of the sample. Assessment and optimization of experimental parameters involved in extraction efficiency (microwave output power, extraction solvent, extraction time, amount of sorbent, elution microwave power, elution solvent, and elution solvent flow rate) were conducted. The obtained recoveries were between 71.5 and 105.2% and relative standard deviations were lower than 11.6% under optimized conditions. The analytes were determined by GC-MS.

In 1879, Franz von Soxhlet invented the extraction technique that takes his name. Afterward, Luque de Castro et al. [177, 178] developed a new extraction technique called focused microwave-assisted Soxhlet extraction (FMASE). The FMASE technique maintains the main features of the conventional Soxhlet extraction technique namely, sample fresh solvent contact, no filtration required after extraction, easy manipulation and adds the acceleration of the extraction process by focusing the microwave radiation on the sample compartment as an outstanding advantage. In this way, microwave radiation acts on both the sample and solvent allowing the extraction of the most strongly retained analytes and shortening the extraction time. Thereby, the completeness of the analyte extraction is not always guaranteed with conventional methods. Furthermore, environmental pollution is minimized due to the small amount of solvent released into the atmosphere and low degradation of thermolabile analytes [179]. Following the principles of this mode, Zhou and co-workers [180] developed a novel one-step selective extraction technique; called microwave-accelerated selective Soxhlet extraction (MA-SSE). The effectiveness of the MA-SSE procedure was verified by the determination of organophosphorus (OPP) and carbamate pesticide residues in ginseng using GC/MS. A Soxhlet extraction system containing a glass filter with 10–15 μm pore size was designed as an extractor and used to hold the entire sample (1.0 g) and sorbent (Florisil). Both the target analytes and the interfering components were extracted from the sample into the extraction solvent by microwave irradiation during the procedure of MA-SSE (75°C, 900 W microwave power, 15 min, 1000 rpm stirring speed). After the solvent flowed through the sorbent, the interfering components were adsorbed by the sorbent, and the target analytes remaining in the solvent were collected in the extraction bottle. No clean-up or filtration was required after extraction. Low detection limits (0.050–0.50 μg kg$^{-1}$) were obtained under the optimized conditions. The recoveries were in the range of 72.0–110.1% with relative standard deviations less than 7.1%.

## 2.5 MICROWAVE DISTILLATION (MD)

Two solvent-free sample preparation techniques of MD and solid-phase microextraction (SPME) were combined and developed for essential oil compounds determination in traditional Chinese medicine (TCM) by Deng and co-workers [264] using a homemade MD–SPME equipment. The principal experimental variables, such as the fiber coating of SPME, irradiation time and microwave power were tested using 2 g of sample. The polydimethylsiloxane/divinylbenzene (PDMS-DVB) fiber was selected and the other optimum values were 400 W for microwave power and 3 min of irradiation time. In this way, essential oil compounds (49 analytes) could be determined using GC-MS. The proposed method was compared with steam distillation (SD), which is the conventional technique demonstrated that it was possible to reduce drastically the extraction time from 6 h to 3 min and organic solvents were not used.

An improved microwave steam distillation (MSD) method for the isolation of essential oils from lavender flowers prior to GC-MS analysis was developed by Sahraoui et al. [265]. A cartridge containing 20 g of dry lavender flowers was placed in a multimode microwave reactor (Milestone "Dry Dist", Bergamo, Italy) with a 2.45 GHz frequency and a maximum delivered power of 1000 W variable in 10 W increments. A microwave irradiation power of 200 W was chosen in order to ensure a quick extraction and no losses of volatile compounds. For comparison, the same glassware and same operating conditions have been used for conventional SD, which is one of the reference methods in essential oil isolation. Both processes had comparable yields, but extraction could be performed in 6 min applying MSD method against 30 min for SD which demonstrated that MSD highly accelerated the isolation process, without causing changes in the volatile oil composition. The proposed method was evaluated in terms of time, energy, and environmental impact, obtaining satisfactory results. Thus, the authors ensure that the MSD method could be appropriate for the routine quality control analysis of essential oils from aromatic herbs, spices, or flowers.

Jiao et al. [266] developed an ionic liquids-assisted microwave distillation procedure coupled with headspace single-drop microextraction (ILAMD-HS-SDME) followed by GC-MS for the analysis of essential oil in *Fructus forsythiae*. Ionic liquids (ILs) were used as the absorption medium of microwave irradiation and simultaneously as the destruction agent of plant cell walls. Four ILs including 1-butyl-3-methylimidazoliumbromide ([C$_4$mim]Br), 1-butyl-3-methylimidazolium chloride ([C$_4$mim]Cl), 1-allyl-3-methylimidazolium chloride ([Amim]Cl), and 1-ethyl-3-methylimidazolium acetate ([C$_2$mim]OAc) were evaluated. [C$_2$mim]OAc was the selected IL and n-heptadecane (2.0 µL) was the suspended solvent for the extraction and concentration of the essential oil. The microwave extractor (MAS-II model, Sineo Chemical Equipment Corp., Shanghai, China) was equipped with a time controller, an infrared temperature sensor, an electromagnetic stirrer, and a circulating water-cooling system. The temperature was monitored by the infrared temperature sensor and controlled by feedback to the microwave power regulator. The reaction flask (50 mL) was placed in the microwave resonance cavity and connected to a microsyringe as well as a cooling system through a hole at the top of the microwave oven. The optimized operational parameters for the extraction process were irradiation power (300 W), sample mass (0.7 g), mass ratio of ILs to sample (2:4), temperature (78°C), and time (3.4 min). In comparison to previous reports, the proposed method was faster and required a smaller sample quantity and could equally examine all EO constituents with no significant differences. In the same way, the authors also developed a miniaturized sample preparation technique based on ionic-liquid-assisted microwave distillation coupled with headspace single-drop microextraction for the extraction of essential oil from dried Dryopteris fragrans [267]. In this work, 1-ethyl-3-methylimidazolium acetate was the optimal ionic liquid as the destruction agent of plant cell walls and microwave absorption was medium. As in the previous work, n-heptadecane (2.0 μL) was adopted as the suspended microdrop solvent in the headspace for the extraction and concentration of essential oil. An irradiation power of 300 W, sample mass of 0.9 g, mass ratio of ionic liquids to a sample of 2:8, extraction temperature of 79°C, and extraction time of 3.6 min were the optimal parameters of the proposed method. Compared with other methods found in the literature, the proposed technique could equally monitor all the essential oil components in a simple way, more rapid and with a much lower amount of the sample with no significant differences.

## 2.6　MICROWAVE-ASSISTED HYDROLYSIS (MAH)

Damm et al. [268] developed a protein hydrolysis method assisted by microwave energy. They used silicon carbide-based microtiter platforms, which strongly absorbed microwaves. The total amino acid in proteins and peptides can be hydrolyzed and analyzed using the same vial; therefore, errors caused by sample transfer could be minimized. The plates with 20 boreholes had adequate dimensions for holding standard screw-capped HPLC/GC vials, and allowed to heat up 4 heating platforms at the same time (80 vials), performing various microwave-assisted acid hydrolysis simultaneously under regulated conditions. Thus, the total time (necessary for protein hydrolysis and the following evaporation step) required for larger volumes of acid was notably reduced. The optimization of the reaction temperature was performed (Monowave 300) applying standard 10 mL Pyrex vessels equipped with a stir bar, and adding 25 μL bovine serum albumin stock solution (0.5 mg, 20 mg mL$^{-1}$), 100 μL internal standard solution (100 nmol mL$^{-1}$), and 2 mL HCl containing 0.1% (w/v) of phenol. Runs were performed at different temperatures (controlled by fiber optic) ranging from 140 to 200°C for 10 min holding time. An optimization of the hydrolysis parameters has established that 5 min irradiation at 160°C with 6 N HCl led to comparable results for total and individual amino acid recoveries, while the conventional method required 24 h heating at 110°C. Proteins and synthetic peptides hydrolysis were carried out with 25 μg of the sample and 100 μL of 6 N HCl in a low-volume vial.

## 2.7　MICROWAVE-ASSISTED DERIVATIZATION (MADe)

The aim of the work presented by Silva and co-workers [269] was to use a microwave oven for accelerating the pretreatment of fatty acids in terms of transmethylation and saponification and then to develop a fast HPLC derivatization procedure. The reactions were carried out in a domestic microwave oven at 160 W (20% of total exit power), using closed polypropylene vials. For saponification, 50 μL of 20 mg mL$^{-1}$ samples and 50 μL of margaric acid as internal standard were placed in a polypropylene vial with 50 μL of KOH 2% in ethanol and reacted for 3 min. Then, 100 μL of acetic acid 10% in water and 100 μL of toluene were added and vortex mixed for 15 s. After phase separation, 50 μL of the organic phase was withdrawn to another vial and dried under a stream of nitrogen at room temperature. For phenacyl derivatization, the procedure of Wood and Lee [270] was adapted. 100 μL of phenacyl chloride and triethylamine (both 20 mg mL$^{-1}$ in acetone) were added, vortex mixed for 15 s, and reacted for 2 min. The next step was to destroy the excess of reagent, using 50 µL of acetic acid, and submitting the vial to the microwave oven for 1 min. 150 µL mobile phase was added to the vial, for dilution and injection. The quantitative data of all fatty acids in these samples are in agreement with the published composition data.

Jurado-Sánchez et al. [271] described the determination of monocarboxylic, dicarboxylic, and tricarboxylic acids (35 compounds) in water applying MADe as an alternative heating approach for the rapid silylation of carboxylic acids. Target analytes were retained immediately and the sample matrix was sent to waste. An air stream (flow rate 3 mL min$^{-1}$) was used to remove any residual water from the system and to transport the eluent (200 µL of methanol containing 2 mg L$^{-1}$ of triphenylphosphate (TPP) as internal standard). After evaporation of the extract to approximately 10 µL, the analytes were spiked with 60 µL of the derivatizing reagent and placed in a household microwave oven and derivatized for 3 min. The evaluated reagents were BF 3/1-butanol; acetyl chloride/1-butanol; isobutyl chloroformate/ 1-butanol; trimethylphenylammonium hydroxide, N,O-bis-(trimethylsilyl)acet-amide, N,O-bis-(trimethylsilyl) trifluoroacetamide and trimethylchlorosilane. The mixture of 1% trimethylchlorosilane in N,O-bis-(trimethylsilyl)trifluoroacetamide produced the best reaction yield and stability of the derivatives. Detection limits were in the range of 0.6–15 ng L$^{-1}$, precision values varied between 4.0 and 6.0% (as a within-day relative standard deviation) and recoveries ranged from 93–101% for all the target analytes.

Sternbauer et al. [272] have developed an efficient methodology for the analysis by GC/MS of sorbitol-based additives for polyolefin-based materials. A two-step approach for sample preparation applying MAE and MADe by silylation was proposed to improve the GC-suitability of the analytes. Tetrahydrofuran (THF) was the best extraction solvent. The reproducibility and recovery values in real samples ranged from approximately 2.0 to 6.0%, and from 95.8% up to 104.2%, respectively. In addition, a single-step procedure comprising simultaneous extraction and derivatization lead to reduction and acceleration of sample preparation steps without losing the analytical performance.

Table 2.4 includes microwave-assisted techniques different from MAD, MIC, and MAE.

## 2.8 MICROWAVE ON-LINE SAMPLE PREPARATION

The different microwave techniques that have been already described could be automated with the inherent advantages of flow methods, such as reproducibility in sample handling, reducing the analysis time, solvent consumption, consequent waste generation, and sample contamination. Furthermore, microwave sample preparation could be on-line coupled to other analytical steps, resulting in a fully automated procedure. Different attractive approaches have been developed and some of them are commented in the following sections.

Criado et al. [300] proposed a simple fully continuous flow system for screening of PAHs in soil samples. The extraction system was constructed by making a vent

**TABLE 2.4**
**Other Microwave-Assisted Techniques**

| Analyte | Matrix | Experimental Conditions | | | Techniques | Ref. |
| | | Sample Mass | Solvent | Preselected Program | | |
| --- | --- | --- | --- | --- | --- | --- |
| Polysaccharide-based plasma | Human urine | 50.0 µL | 0.6 mL HCl (3 mol $L^{-1}$, pH lower than 2) 50 µL (internal standard ISTD: Glucose-$^{13}$C6, 1 mg/mL | chemical hydrolysis 1200 W, 100°C, 2 min derivatization 1020 W, T 100 C, 5 min | Chemical hydrolysis and derivatization followed by GC–MS | [273] |
| Glycols and GHB | Urine or plasma | 50.0 µL | 10.0 µL of internal standard (1,3-propyleneglycol, 0.5 g/L) 50.0 µL acetonitrile | 450 W, 5 min | Derivatization followed by GC–MS analysis | [274] |
| Amino acid | Wheat (*Triticum durum* Desf.) flour. Grain other cereal crops | 200 mg | 10.0 mL 6 mol $L^{-1}$ HCl, | 1200 W 150°C, 3 h | Acid hydrolysis followed by automated amino acid analyzer | [275] |
| Phenolic compounds | Ber (*Ziziphus mauritiana L.*) fruits | (500 ± 1 mg) | 10.0 mL base hydrolysis solution (0.372 g of EDTA, 1.00 g ascorbic acid in 2 mol $L^{-1}$ NaOH) in 35.0 mL | 1000 W (50% power), 10 min, 56°C | Hydrolysis followed by HPLC–DAD | [276] |
| Ellagitannins | Lyophilized strawberry | 125 mg | 5.0 mL formic acid/water (80:20, v/v) | 30 min, 100°C | Hydrolysis | [277] |
| Fatty acids profile | Herbal medicine | 3.00 g ground perilla seeds | 40.0 mL n-hexane 0.30 g KOH 4.0 mL methanol $V_f$ 100.0 mL | 435 W, 45.0°C, 12 min | Derivatization followed by GC–MS | [278] |
| Essential oils | *Perilla frutescens* (Chinese medicine) | 1.0 g | — | Domestic MO 230 W, 2 min | Microwave distillation followed by GC–MS | [279] |

*(Continued)*

**TABLE 2.4 (CONTINUED)**
**Other Microwave-Assisted Techniques**

| Analyte | Matrix | Experimental Conditions | | | Techniques | Ref. |
| | | Sample Mass | Solvent | Preselected Program | | |
|---|---|---|---|---|---|---|
| Chlorophenols | Water samples | 10.0 mL | 1-octanol | stirring speed 500 rpm 167 W, 10 min | Microwave-assisted headspace liquid-phase microextraction followed by GC-ECD | [280] |
| Volatile compounds | Chinese herb (*Artemisia capillaris*) | 2.0 g | — | Domestic MO (700 W) 400 W, 4 min | Microwave distillation followed by MD–HS-SDME GC–MS | [281] |
| PAHs | Sewage sludge and soil | 0.20 g | n-hexane/saturated methanolic KOH | 1000 W, 260°C, 35 bar, 129°C, 17 min | Microwave-assisted saponification–extraction followed by HPLC-DAD-Fluorescence | [282] |
| Glucocorticoids residues | Liver tissue | 5.00 g | sodium acetate buffer pH 5.2 | Domestic MO (700 W) Incubation time 2 min, 300 W | Microwave-assisted enzymatic hydrolysis | [283] |
| Aminoacids | Yeast | 1.00 mg | 6 mol L$^{-1}$ HCl | Vapor Phase Hydrolysis Microwave System 200 W, 15 min (155°C) | MW-assisted vapor phase acid hydrolysis | [284] |

*(Continued)*

**TABLE 2.4 (CONTINUED)**
**Other Microwave-Assisted Techniques**

| Analyte | Matrix | Sample Mass | Solvent | Preselected Program | Techniques | Ref. |
|---|---|---|---|---|---|---|
| | | | **Experimental Conditions** | | | |
| Seleno-amino acids (Se-Met, Se-MetSeCys, Se-Cys) | Extra virgin olive oil | 5.00 g | 15% HCl (v/v) | Milestone Start D microwave system 900 W, 5.5 min | Microwave-assisted hydrolysis | [285] |
| Mineral oil | Cereal-based products (pasta, bread, biscuits, cakes) | 5.00 g | 10.0 mL saturated methanolic KOH (20.0 mL for high-fat content) 10.0 mL n-hexane | Microwave extractor 20 min, 120°C | Microwave-assisted saponification | [286] |
| Total choline and carnitine | Infant Formula and Adult/Pediatric Nutritional Formula | 1.00 g | 5.0 mL water 2.5 mL 70% (w/w) $HNO_3$ | 120°C, 10 min ramp 1000 W, 40 min | Microwave-assisted acid hydrolysis | [287] |
| Milk and non-milk proteins | Dried milk powder | 30 mg | 10.0 mL 6 mol $L^{-1}$ HCl | 200 W, 15 min | Microwave-assisted hydrolysis | [288] |
| Polymeric carbohydrates | Astragalus membranaceus (traditional Chinese medicine) | 0.50 g | 15.0 mL 2 mol $L^{-1}$ TFA | 800 W with a ramp time of 3 min 800 W, 10 min | Microwave-assisted acidic hydrolysis | [289] |
| Amino acids | Human and bovine hemoglobin | 40.0 μL | 0.5 μL 500 mmol $L^{-1}$ DTT 40.0 mL 6 mol $L^{-1}$ HCl | 200 W or 300 W, 30–90 s 150°C, 147psi | Microwave-assisted acid hydrolysis | [290] |
| Xylooligosaccharides | Sugarcane bagasse | 1.00 g | H2SO4 0.1–0.3 mol $L^{-1}$ | 700 W, 20–40 min | Microwave-assisted acid hydrolysis | [291] |

*(Continued)*

## TABLE 2.4 (CONTINUED)
## Other Microwave-Assisted Techniques

| Analyte | Matrix | Experimental Conditions | | | Techniques | Ref. |
|---|---|---|---|---|---|---|
| | | Sample Mass | Solvent | Preselected Program | | |
| Protein sequence | BSA, human plasma proteins, bovine alpha casein | 40.0 μg | 120.0 μL 12 mmol L⁻¹ DTT 40.0 μL TFA | Domestic MO (1200 W) 1200 W, 8–10 min | Microwave-assisted acid hydrolysis | [292] |
| Free and total myo-inositol | Soy-based infant formula | 1.00 g | 30.0 mL purified water 10.0 mL concentrated HCl | 800 W, 10 min 800 W, 60 min | Microwave-assisted acid hydrolysis | [293] |
| Levulinic acid and furfural | Cellulose | 100 mg | 5.0 mL aqueous solution HCl 1.37 mol L⁻¹ | 200°C, 3.32 min, 400 W | Microwave-assisted acid hydrolysis | [294] |
| | Xylan | | 5.0 mL aqueous solution HCl 0.36 mol L⁻¹ | 195°C, 1 min, 400 W | | |
| Cellulose nanofibers | Wheat Straw | 1.50 g | 30.0 mL of 2% NaOH aqueous solution | 140 ± 2°C within 3 min and maintained for 20 min, 1200 W | Microwave-assisted alkaline hydrolysis | [295] |
| Cellulose and hemicellulose | Tropical Plant Wastes | 0.05 g | 1/60 substrate/solvent 0.4 mol L⁻¹ H₂SO₄ | 140°C, 30 seg | Microwave-assisted acid hydrolysis | [296] |
| Essential oil | Rindera lanata var canescens | 165.00 g | 50 mL of water | 40 min, 110°C, 600 W | Microwave-assisted distillation followed by CG-MS | [297] |
| Essential oil, proanthocyanidins, and polysaccharides | Cinnamomi cortex | 30.00 g | 18.0 mL/g of liquid-solid ratio Ethyl ether as extractant solvent | 38 min, 374 W | Microwave-assisted distillation followed by CG-MS | [298] |
| Essential Oil | Lavender | | 5 mg Fe₃O₄ | 15 min, 600 W | Microwave-assisted distillation followed by HS-SPME-CG-MS | [299] |

hole to a household microwave oven (2450 MHz, 800 W), where the PTFE tubing was inserted to carry out the aspiration of the sample. To accomplish the extraction, the soil samples were directly weighed in PTFE bottles, which were perforated on the top, and the extraction solvent (acetonitrile) was added. The extraction was performed at a power of 425 W for 10 min, and after this time, the organic extract was aspirated by starting the peristaltic pump, passed through a cooler, and filtered. The filtered extract solution was transferred to a vial in order to acidify and dilute it before passing through a RP-C18 sorbent column located in the loop of a low-pressure injection valve. The PAHs were retained and then eluted into a liquid chromatography (LC) column for separation. The flow and the LC systems were interfaced employing a six-port high-pressure injection valve.

The extraction efficiency was tested by comparison of the signals obtained by applying the proposed MAE method to standard solutions and those provided by the same concentration spiked into the soils.

In a similar way, Serrano et al. [301] proposed a novel, fast and inexpensive continuous extraction method for the determination of aliphatic hydrocarbons (AHs) in soil and sediment samples. Here, the continuous MAE system was combined with LLE, for clean-up purposes. The extraction was performed in a vial using a mixture of n-hexane-water in a ratio 1:0.02 at a power of 425 W for 5 min. After extraction, the organic extract was pumped for 2 min at a 1 ml min$^{-1}$ flow-rate, cooled by immersion in an ice beaker and filtered. The filtered organic solution was merged with a water stream (flow-rate, 1 mL min$^{-1}$) in order to clean it and introduced in a membrane phase separator. The extract was manually injected into a GC-MS system to separation and determination of the AHs. The continuous MAE system was compared with the Soxhlet method obtaining a good correlation in the results of the extraction of AHs from C14 to C27, and better recovery values for the most volatile hydrocarbons (C9–C13).

The microwave extraction also could be done by passing the extraction solvent through the sample in a continuous way, DMAE, where the extraction could be performed either in a closed recirculating system or in a system where the sample is continuously extracted with fresh solvent [302]. DMAE simplifies the coupling of the extraction step with other steps of the analytical process, and thereby facilitates the automation of the whole analytical process.

Taking advantage of this, the determination of lignans and naphthoquinones in Wuweizi and Zicao from different growing areas were investigated [303]. The extraction was performed on a modified household microwave oven (800 W) and a polytetrafluoroethylene coil (250 cm × 3 mm i.d.) was used as extraction coil. The sample (100 mg) was first mixed with 10 mL of extraction solvent (80% methanol for Wuweizi, 80% ethanol for Zicao) and the resulting suspension was introduced into a sample loop by a peristaltic pump. Then, the pump was stopped, a microinfusion pump was activated and the suspension was delivered into the extraction coil located in the microwave-assisted system. The extraction was performed for 9 min at 240 W and 12 min at 300 W for Zicao and Wuweizi respectively. The suspension was filtrated with a filter placed at the end of the coil and injected into a second sample loop of the HPLC-DAD system. No significant difference was found by comparing the obtained results of the on-line DMAE and off-line DMAE, ultrasound-assisted extraction (UAE), and Soxhlet extraction (SE). Furthermore, the extraction time and the solvent consumption of the on-line DMAE were less than those of UAE (30 min, 25 mL) and SE (180 or 240 min, 100 mL). Therefore, the authors affirm that the proposed method was quicker and more effective than the others that were tested.

The extraction of safflower yellow from Flos Carthami samples was accomplished by developing an automated DMAE system [304]. To perform the on-line extraction, a microwave resonance cavity was constructed in the laboratory and was placed between the microwave source (maximum microwave power of 100 W) and the flow system. The power was selected using two tuning screws that were located in the cavity. The sample was weighted, placed in a PTFE extraction column held by two glass fiber plugs and positioned in the microwave resonance cavity. The extraction column was filled with 60% aqueous methanol solvent using a high-pressure pump, and the sample was irradiated at 60 W. After extraction, the extract solution merged with water and passed through the absorption flow cell where the detection at 401 nm was done. The extraction yield of safflower yellow and precision obtained by on- and off-line DMAE and pressurized microwave-assisted extraction (PMAE) were comparable.

Morales-Muñoz et al. [305] presented a closed DMAE approach for the continuous quantitative determination of Cr (VI) in spiked and natural sediment and soil samples. For this purpose, an automated flow system was developed using a focused microwave digestor that was on-line coupled to a derivatization step using 1,5-diphenylcarbazide (DCP) and subsequent photometric detection at 540 nm. The sample was located into a laboratory-made chamber of Teflon (7 cm × 7.5 mm i.d.) and filled with ammonium buffer solution (extraction solvent) propelled by a peristaltic pump. The chamber with the sample was placed in the microwave vessel and the extraction was performed by irradiation at 300 W. The time of extraction was set taking into account the sample matrix. During microwave irradiation, the direction of the extraction solvent flow was changed through the sample cell in an iterative way to overcome the compactness of the sample and therefore, overpressure in the system. In addition, the contact between the sample and the extraction solvent was favored. After the extraction, the system was opened by switching the position of a selection valve and the extract was directed to the derivatization step where the reaction with DCP took place. If lower detection limits are required, the extract was conducted to a preconcentration column packed with a strong anion-exchange resin and eluted with 0.5 M ammonium buffer solution before derivatization. The proposed method was compared with the reference method (EPA method 3060/7196) in terms of efficiency and precision, obtaining good agreement between results. Moreover, the time of extraction was significantly shorter (10–14 min) than the reference method (1 h).

Chen et al. [306] also proposed the employment of a DMAE recirculating system coupled on-line with HPLC-DAD through a flow injection interface for the determination of two components (andrographolide and dehydroandrographolide) in medicinal herbs. In this case, the extraction was performed using a mixture of methanol and water (60:40, v/v) as extractant, and the sample was irradiated at 80 W for 6 min. The proposed method was demonstrated to obtain a higher extraction yield in a shorter time, requiring minor quantities of sample and solvent (5.0 mL), compared with ultrasonic extraction used in the Chinese pharmacopeia.

An automatic DMAE method for the clean-up and extraction of sulfonamide (SA) antibiotic residues from chicken breast muscle samples was developed by Wang et al. [307]. The authors designed an on-line DMAE-SPE system by coupling a household microwave oven to a SPE manifold equipped with a vacuum pump. For that purpose, the extraction vessels had an inlet, in order to perform the introduction of the solvent, and an outlet to conduct the eluate directly to the SPE column. Briefly, the samples were dried and dispersed with anhydrous sodium sulfate and placed in extraction vessels. Then, the extraction solvent (acetonitrile) was introduced and the samples were irradiated at 900 W. Subsequently, the extract was directed to the SPE cartridge to perform the clean-up and preconcentration of the SAs. The analytes were determined off-line by LC-MS. The DMAE variables such as the power of the microwave and the volume and flow rate of the extraction solvent were optimized using a Box–Behnken design. Inter- and intra-day precision were evaluated by extracting chicken samples spiked at three concentration levels, obtaining RSD values between 4.9 and 7.4%. In addition, acceptable recovery values in the range of 82.6–93.2% were obtained. The proposed on-line DMAE-SPE system allows the pretreatment of complex samples as chicken breast muscle in a short time (up to 20 samples in 6 min) which lead to a significantly high sample throughput.

DMAE technique was also coupled on-line to the single-drop microextraction (SDME) technique to perform the extraction, clean-up, and enrichment of organophosphorus pesticides in tea samples [308]. The microwave extraction device was similar to the one developed by Wang et al. [307]. The tea samples were dispersed (acidic alumina) and placed into the glass extraction vessel (5.0 cm × 0.6 cm i.d.) between two cotton plugs. A 25% ethanol solution, used as extractant, was propelled by a microinfusion pump and passed through the sample at 1.0 mL min$^{-1}$ flow rate in the direction of a home-made extraction chamber. A 5 µL drop of carbon tetrachloride was injected into the chamber and remained in the bottom part due to its high density. Then, the sample was irradiated for 10 min at 230 W to perform the extraction and enrichment of the analytes. After this time, the pump was stopped; the drop was retracted into the microsyringe and injected into the GC-MS. The proposed method was compared with other common extraction methods found in the literature in terms of volume of the extraction solvent, time of extraction, and recoveries, concluding that it was suitable for the determination of the selected analytes in tea samples. The authors also proposed the use of this device to the dynamic extraction of the analytes from fresh vegetables [309]. In this case, the sample was first dispersed with quartz sand and a 3% NaCl solution was used as an extraction solvent. The analytes were extracted and preconcentrated (toluene drop) by applying a power of 250 W for 10 min. Likewise, they have developed a microwave processing system by coupling the DMAE technique to a microwave-accelerated solvent elution (MASe) to simplify the extraction and enrichment of organophosphorus pesticides from fresh vegetable samples [178]. To accomplish this goal, the extraction vessel was connected to an active carbon fiber packed column, and both were located inside a household microwave, with an inlet and an outlet holes. First, 2.0 g of the sample was dispersed (quartz sand) and placed into the extraction vessel between glass fiber plugs. N-hexane and ethyl acetate were chosen as extraction and elution solvents, respectively. To achieve the extraction, n-hexane was propelled into the extraction vessel and the column by a peristaltic pump through a six-way valve, in a recirculation mode. Once the extraction vessel was filled with the extractant, it was irradiated for 5 min at 300 W. After that, 12 mL of the elution solvent passed through the column in the opposite direction and the irradiation was performed at 250 W.

Gao et al. have proposed an on-line DMAE method to perform the determination of lipophilic compounds in a widely used traditional Chinese medicine [310]. For that purpose, a PTFE extraction coil (120 cm×3 mm i.d.) was located inside a household microwave oven. The sample was mixed with a 2.0 mol L$^{-1}$ [C6MIM]Cl ionic liquid used as extractant, and stirred for 15 min. Then, the sample suspension was pumped into the system in order to fill a 1.0 mL injection loop. Subsequently, the sample was injected and passed through the PTFE coil at a 1.60 mL min$^{-1}$ flow rate while it was irradiated at 180 W. After that, the sample was filtrated and directly introduced into the loop of the HPLC instrument.

Likewise, the FMASE technique was automated to perform the determination of linear alkylbenzene sulfonates (LAS) in sediments [311]. The FMASE extractor consists of a sample vessel, situated inside a household microwave connected to a distillation flask that is above an electrical heater. To perform the procedure in an automatic way, the distillation flask was modified in order to perform the aspiration of the extract using a peristaltic pump. In this way, the device designed by the authors enables the use of water as an extractant. The sample was contained in a cellulose extraction cartridge and was placed inside the sample vessel. As the conventional Soxhlet extraction, the on-line FMASE involves evaporation of the extraction solvent, condensation, and drips back down into the sample. Once the cartridge of the sample was filled with the extractant, the sample was irradiated for 200 s at 200 W. The extractor was coupled on-line to an FI system to perform the preconcentration of the analytes, and the subsequent derivatization with methyl orange. The derivatization product was measured at 465 nm. In addition, individual determination of LAS could be carried out after extraction by using a HPLC method with fluorescence detection.

On-line microwave sample preparation could be also employed to perform reactions, as hydrolysis, oxidation, or reduction, or to the digestion of samples.

An on-line method for the assisted microwave derivatization of aliphatic and Chávez et al. [312] developed polyethoxylated alcohols with phenyl isocyanate or benzyl chloride reagents. The solution containing the alcohols and the reagents (phenyl-isocyanate and benzoyl chloride) in acetonitrile medium was inserted into acetonitrile carrier stream and driven to a spiral Teflon coil reactor placed into a household microwave (450 W). The sample was irradiated at 400 W for 60 s and the derivatization reaction took place. After derivatization process, the reactions products were loaded in the loop of an injection valve and injected into a HPLC system with photodiode array detection for separation. Compared to the conventional conditions, the proposed method presented high sensitivity, and analysis time has been drastically reduced (10 times), as has the number of reagents used.

A fast, simple, and relatively safe continuous approach for the determination of total creatinine in meat and chicken dehydrated broths was developed [313, 314]. For this purpose, a simple continuous system composed by a single line and Teflon reaction coil was designed to introduce the samples into the focused microwave oven.

The samples were previously mixed with a 1.5 mol L$^{-1}$ HCl solution and stirred for 10 min at 70–80°C. Then, the acidic sample solution was propelled into the focused microwave oven at a fixed flow rate and was irradiated (105 W) to carry out the hydrolysis reaction to turn creatine into creatinine. After the treatment, the sample solution was collected and creatinine was determined by the classical Jaffé reaction and the products were determined spectrometrically at 500 nm. The microwave-assisted hydrolysis method was validated by comparison with the official AOAC method obtaining good agreement between the results. The proposed method reduces the analysis time from 6 h (AOAC method) to less than 15 min per sample, with the inherent advantages of an on-line procedure.

To determine the total amount of phosphorus in wastewater samples, Almeida et al. [315] have developed a multi-syringe flow injection (MSFIA) system to perform the in-line digestion of the samples and the spectrophotometric determination based on the reaction of phosphorus with molybdenum blue. A domestic microwave oven (MWO, Becken MWB 1000, maximum power of 700 W) with two vent holes, in the upper and in the bottom part of the body was employed. For the digestion, a lab-made PTFE digestion vessel (0.73 mL of capacity) was located inside MWO and a glass spiral was situated in the exit aperture of the vessel in order to release the pressure due to bubble formation. In this way, the presence of de-bubblers or cooling coils originated in the traditional long tubular reactors was avoided. The other aperture of the digestion vessel was connected to a solenoid valve the MSFIA system through a confluence connector. In this way, the sample and digestion solution (a sulfuric acid and potassium persulfate mixture) were propelled to the digestion vessel where sample digestion took place (30 s at 595 W). After digestion, the sample was conducted to the MSFIA system to carry out the reaction and the spectrometric detection.

An interesting sequential injection approach was developed by Egorov et al. [316] for the rapid at-line/on-line analysis to determine the total 99Tc content of aged nuclear waste samples, a chemically and radiologically complex matrix. Radiochemical sample preparation using MAD, separation in an anion-exchange column, and detection using a solid scintillator detector were integrated and fully automated in the proposed prototype. Concerning microwave-assisted treatment, acidification by the addition of HNO$_3$ and oxidation using peroxydisulfate were performed in the digestion vessel in order to achieve the oxidation of 99Tc to 99Tc (VII). The vessel was located in a lab-made vessel holder, which replaced the standard commercial digestion vessel, in the center of a microwave cavity. To perform the procedure in an automated way, the digestion vessel was modified and two lines were introduced; one line for the delivery of the reagents and the aspiration of the oxidation product, and a second line for the release of the gas products that were originated during the irradiation. A certain volume of the sample (0.286 mL for analysis of feed samples or 0.495 mL for analysis of blended samples with lower Tc activities) was aspirated by a piston pump and dispensed into the digestion vessel. Then, 2.8 mL of HNO$_3$ solution was loaded in a loop and dispensed into the vessel. After that, 2.5 mL of air was aspirated and delivered to the vessel in order to perform the agitation during the microwave irradiation. In order to allow a moderate boiling of the reaction mixture, the sample with the HNO$_3$ was irradiated for 30 s with a microwave power of 10% of the full power, using a ramp power ramp period of 10 s. A second oxidative digestion cycle was performed by delivering 0.7 mL of a $Na_2S_2O_8$ solution to the reaction vessel and applying a microwave power of 5% during 90 s under continuous sample agitation. After microwave treatment, the digested sample was aspirated by one of the piston pumps and delivered to the column were separation was carried out. After digestion, the vessel was properly washed with a $HNO_3$ solution, twice. The efficiency of the automated microwave-assisted acidification and oxidation procedure was validated in the analysis of nuclear waste matrixes, with satisfactory results. The authors affirm that the flow-based system with MAD of complex matrices such as nuclear waste in an open-vessel format allows the execution of different operations in a completely fully automated way. Moreover, the outgassing during sample treatment operations was possible.

Quaresma et al. have designed a flow injection analysis (FIA) system coupled with a focused microwave oven to perform the on-line dissolution of silicate rocks samples to quantify iron [317]. To perform the dissolution in an automated way, a PTFE reactor coil (300 cm length and 0.8 mm i.d.) was placed on the microwave cavity of the focused microwave oven. Sample slurry was prepared by weighing 50 mg of rock and mixing with 200 mL of a mixture of HF, HCl, and $HNO_3$. Then, the slurry was pumped towards the reaction coil and irradiated for 210 s at a power of 90 W. After this time, the digested sample was pumped in order to fill a 500 μL loop and was injected in a water carrier stream to be introduced in the nebulizer of a FAAS. In order to improve the precision of FAAS signals, a de-bubbler device was connected at the exit of the focused microwave oven. A sample throughput of 10 $h^{-1}$ was obtained, taking into account the complete analytical procedure. The accuracy of the proposed method was tested by the analysis of seven certified rock samples with respect to their $Fe_2O_3$ content.

In the last year, Marques et al. [318] assessed a flow digestion method using a microwave system for metal determination in juice and milk samples. In this case, the sample preparation was performed at high-pressure using a mixture of diluted $HNO_3$ and HCl for juice samples and only $HNO_3$ for milk samples.

An interesting in-line flow injection approach was employed by Silva et al. [319] for the quantitation of the accessible fraction of Mg, V, Cr, Mn, Co, Ni, Cu, Zn, Mo, Sb, and Pb in crushed rock (hematite) samples. Here, a mini-column packed with the sample was submerged in the bottom part of the cavity of a focused microwave digestion system, filled with water at 90°C. The sample size was chosen taking into account the reproducibility of the signals and backpressure in the flow system. The mini-column consisted of a fluorinated ethylene propylene (FEP) tube packed with 200 mg of hematite sample, and placed between a by-pass valve, which directed the flow to either to the mini-column or to the nebulizer of an ICP-MS instrument. Leaching was carried out by sequentially passing doubly-deionized water and $HNO_3$ solutions at increasing concentrations (1%, 10%, and 30%) through the sample at a flow rate of 0.8 mL $min^{-1}$ (residence time, 32 s) for up to 10 min each, and the progressive release of elements was continuously monitored. In comparison to the batch mode, the use of the focused microwave oven allows the increment of the release of the majority of the elements, and the reduction of the time of analysis from 2 h or more per sample to 15 min.

## 2.9 CONCLUDING REMARKS

As can be seen from the previously reviewed works, microwave techniques have been applied to many fields of analytical chemistry. The reception of the use of microwave radiation in the sample preparation step is based on its unique ability to induce a variety of physical and chemical phenomena. In this way, the different microwave techniques play an increasingly significant role in the development of new and existing technologies in analytical chemistry. Microwave-assisted digestion and extraction are the most developed procedures by researchers, as it is shown in this review. It is clear that the wide use of these techniques is due to the advantages that they possess: greater recoveries of analytes compared to conventional methods with an important decrease of analysis time, reduction of organic solvents and a significant contribution to the care of the environment and analyst. Furthermore, when these techniques are automated the inherent advantages of the flow methodologies are added.

## ACKNOWLEDGMENTS

The authors deeply acknowledge Universidad Nacional del Sur (UNS). A. Lorenzetti, C.C. Acebal, and C.E. Domini acknowledges Consejo Nacional de Investigaciones Científicas y Técnicas (CONICET).

## REFERENCES

1. Kingston HM, Haswell SJ (1997) *Microwave-enhanced Chemistry: Fundamentals, Sample Preparation, and Applications.* American Chemical Society. Washington, DC.
2. Oghbaei M, Mirzaee O (2010) Microwave versus conventional sintering: A review of fundamentals, advantages and applications. *J. Alloys Compd.* 494: 175–189. doi:10.1016/j.jallcom.2010.01.068
3. Dallinger D, Irfan M, Suljanovic A, Kappe CO (2010) An investigation of wall effects in microwave-assisted ring-closing metathesis and cyclotrimerization reactions.*J. Org. Chem.* 75: 5278–5288. doi:10.1021/jo1011703.
4. Agazzi A, Pirola C (2000) Fundamentals, methods and future trends of environmental microwave sample preparation. *Microchem. J.* 67: 337–341. doi:10.1016/S0026-265X(00)00085-0
5. Jones DA, Lelyveld TP, Mavrofidis SD, Kingman SW, Miles NJ (2002) Microwave heating applications in environmental engineering—A review. *Resour. Conserv. Recyl.* 34: 75–90. doi:10.1016/S0921-3449(01)00088-X.
6. Kappe CO (2008) Microwave dielectric heating in synthetic organic chemistry. *Chem. Soc. Rev.* 37: 1127–1139. doi:10.1039/B803001B.
7. Sajjadi B, Abdul Aziz AR, Ibrahim S (2014) Investigation, modelling and reviewing the effective parameters in microwave-assisted transesterification. *Renew. Sust. Energ. Rev.* 37: 762–777. doi:10.1016/j.rser.2014.05.021.
8. Perreux L, Loupy A (2001) A tentative rationalization of microwave effects in organic synthesis according to the reaction medium, and mechanistic considerations. *Tetrahedron* 57: 9199–9223. doi:10.1016/S0040-4020(01)00905-X.
9. Komorowska-Durka M, Dimitrakis Bogdał GD, Stankiewicz AI, Stefanidis GD (2015) A concise review on microwave-assisted polycondensation reactions and curing of polycondensation polymers with focus on the effect of process conditions. *Chem. Eng. J.* 264: 633–644. doi:10.1016/j.cej.2014.11.087.

10. Samra AA, Moms JS, Koirtyohann SR (1975) Wet ashing of some biological samples in a microwave oven. *Anal. Chem.* 47: 1475–1477 doi:10.1021/ac60358a013.

11. Saltiel, C. and Datta, A. K. (1999). Heat and mass transfer in microwave processing. *Advances in heat transfer*, 33(1): 1–94.

12. Zlotorzynski A (1995) The application of microwave radiation to analytical and environmental chemistry. *Crit. Rev. Anal. Chem.* 25: 43–76.doi:10.1080/10408349508050557.

13. Ericsson M, Colmsjö A (2003) Dynamic microwave-assisted extraction coupled online with solid-phase extraction and large-volume injection gas chromatography: determination of organophosphate esters in air samples. *Anal. Chem.* 75: 1713–1719. doi:10.1021/ac026287v.

14. Wang H, Li GJ, Zhang YQ, Chen HY, Zhao Q, Song WT, Xu Y, Jin HY, Ding L (2012) Determination of triazine herbicides in cereals using dynamic microwave-assisted extraction with solidification of floating organic drop followed by high-performance liquid chromatography. *J. Chromatogr. A* 1233: 36–43. doi:10.1016/j.chroma.2012.02.034.

15. Chen C, Shao Y, Tao Y, Wen H (2015) Optimization of dynamic microwave-assisted extraction of Armillaria polysaccharides using RSM, and their biological activity. *LWT- Food Sci. Technol. Int.* 64: 1263–1269. doi:10.1016/j.lwt.2015.07.009

16. Wang H, Zhao Q, Song WT, Xu Y, Zhang XP, Zeng QL, Chen HY, Ding L, Ren NQ (2011) High-throughput dynamic microwave-assisted extraction on-line coupled with solid-phase extraction for analysis of nicotine in mushroom. *Talanta* 85: 743–748. doi:10.1016/j.talanta.2011.04.058.

17. Zhang Y, Li Y, Liu Z, Zhong L, Chi R, Yu, J (2015) Dynamic microwave-assisted extraction of total ginsenosides from ginseng fibrous roots.*Wuhan Univ. J. Nat. Sci.* 20: 247–254.doi:10.1007/s11859–015-1089-6.

18. Chen LG, Zeng QL, Wang H, Su R, Xu Y, Zhang XP, Yu AM, Zhang HQ, Ding L (2009) On-line coupling of dynamic microwave-assisted extraction to solid-phase extraction for the determination of sulfonamide antibiotics in soil. *Anal. Chim. Acta* 648: 200–206. doi:10.1016/j.aca.2009.07.010.

19. Li N, Wu L, Nian L, Song Y, Lei L, Yang X, Wang K, Wang Z, Zhang L, Zhang H, Yu A, Zhang Z (2015) Dynamic microwave assisted extraction coupled with dispersive micro-solid-phase extraction of herbicides in soybeans. *Talanta* 142: 43–50. doi:10.1016/j.talanta.2015.04.038.

20. Wang ZM, Ding L, Li TC, Zhou X, Wang L, Zhang HQ, Liu L, Li Y, Liu ZH, Wang HJ, Zeng H, He H (2006) Improved solvent-free microwave extraction of essential oil from dried *Cuminum cyminum* L. and *Zanthoxylum bungeanum* Maxim. *J. Chromatogr. A* 1102: 11–17. doi:10.1016/j.chroma.2005.10.032.

21. Lucchesi ME, Chemat F, Smadja J (2004) Solvent-free microwave extraction of essential oil from aromatic herbs: comparison with conventional hydro-distillation. *J. Chromatogr. A* 1043: 323–327. doi:10.1016/j.chroma.2004.05.083.

22. Qi XL, Li TT, Wei ZF, Guo N, Luo M, Wang W, Zu YG, Fu YJ, Peng X (2014) Solvent-free microwave extraction of essential oil from pigeon pea leaves [*Cajanus cajan* (L.) Millsp.] and evaluation of its antimicrobial activity. *Ind. Crop. Prod.* 58: 322–328. doi:10.1016/j.indcrop.2014.04.038.

23. Renoe BW (1994) Microwave assisted extraction. *Am. Lab.* 34: 34–40.

24. Matusiewicz H (2014) Systems for microwave-assisted wet digestion. In Flores EMM (ed). *Microwave-assisted Sample Preparation for Trace Element Analysis*. Elsevier, Amsterdam, pp 77–98.

25. Kingston HM, Jassie LB (1988) *Introduction to Microwave Sample Preparation. Theory and Practice*. American Chemical Society, Washington, DC.

26. Matusiewicz H, Sturgeon RE (1989) Present status of microwave sample dissolution and decomposition for elemental analysis. *Prog. Anal. Spectrosc.*12: 21–39.

27. Hristozov D, Domini CE, Kmetov V, Stefanova V, Georgieva D, Canals A (2004) Direct ultrasound-assisted extraction of heavy metals from sewage sludge samples for ICP-OES analysis. *Anal. Chim. Acta* 516: 187–196. doi:10.1016/j.aca.2004.04.026.

28. Domini CE, Hidalgo M, Marken F, Canals A (2006) Comparison of three optimized digestion methods for rapid determination of chemical oxygen demand: Closed microwaves, open microwaves and ultrasound irradiation. *Anal. Chim. Acta* 561: 210–217. doi:10.1016/j.aca.2006.01.022.

29. Navarro P, Raposo JC, Arana G, Etxebarria N (2006) Optimisation of microwave assisted digestion of sediments and determination of Sn and Hg. *Anal. Chim. Acta* 566: 37–44. doi:10.1016/j.aca.2006.02.056.

30. Sant'Ana FW, Santelli RE, Cassella AR, Cassella RJ (2007) Optimization of an open-focused microwave oven digestion procedure for determination of metals in diesel oil by inductively coupled plasma optical emission spectrometry. *J. Hazard. Mater.* 149: 67–74. doi:10.1016/j.jhazmat.2007.03.045.

31. Soylak M, Tuzen M, Souza AS, Korn MdGA, Ferreira SLC (2007) Optimization of microwave assisted digestion procedure for the determination of zinc, copper and nickel in tea samples employing flame atomic absorption spectrometry. *J. Hazard. Mater.* 149: 264–268. doi:10.1016/j.jhazmat.2007.03.072.

32. Engstrom E, Stenberg A, Baxter DC, Malinovsky D, Makinen I, Pönni S, Rodushkin I(2004) Effects of sample preparation and calibration strategy on accuracy and precision in the multi-elemental analysis of soil by sector-field ICP. *J. Anal. At. Spectrom.* 19: 858–866. doi:10.1039/B315283A.

33. Tuncel SG, Yenisoy-Karakas S, Dogangün A (2004) Determination of metal concentrations in lichen samples by inductively coupled plasma atomic emission spectroscopy technique after applying different digestion procedures. *Talanta* 63: 273–277. doi:10.1016/j.talanta.2003.10.055.

34. Udousoro I, Ikem A, Akinbo OT (2017) Content and daily intake of essential and potentially toxic elements from dietary supplements marketed in Nigeria. *J. Food Composit. Anal.* 62: 23–34. doi:10.1016/j.jfca.2017.04.017.

35. Bilo F, Borgese L, Dalipi R, Zacco A, Federici S, Masperi M, Leonesio P, Bontempi E, Depero L. (2017) Elemental analysis of tree leaves by total reflection X-ray fluorescence: New approaches for air quality monitoring. *Chemosphere* 178: 504–512. doi:10.1016/j.chemosphere.2017.03.090.

36. Schneider M, Pereira ÉR, de Quadros DPC, Welz B, Carasek E, de Andrade JB, del Campo Menoyo J, Feldmann J (2017) Investigation of chemical modifiers for the determination of cadmium and chromium in fish oil and lipoid matrices using HR-CS GF AAS and a simple 'dilute-and-shoot' approach. *Microchem. J.* 133: 175–181. doi:10.1016/j.microc.2017.03.038.

37. Llorent-Martínez EJ, Spínola V, Castilho PC (2017) Evaluation of the inorganic content of six underused wild berries from Portugal: Potential new sources of essential minerals. *J. Food Composit. Anal.* 59: 153–160. doi:10.1016/j.jfca.2017.02.016.

38. Santos ADF, Jr., Matos RA, Andrade EMJ, Dos Santos WNL, Magalhães HIF, Costa FDN, Korn MDGA (2017) Multielement determination of Macro and micro contents in medicinal plants and phytomedicines from Brazil by ICP OES. *J. Brazil. Chem. Soc.* 28 (2): 376–384.

39. Hwang J, Kim JC, Moon H, Yang JY, Kim M (2017) Determination of sodium contents in traditional fermented foods in Korea. *J. Food Composit. Anal.* 56: 110–114. doi:10.1016/j.jfca.2016.11.013.

40. Barros JAVA, Virgilio A, Schiavo D, Nóbrega JA (2017) Determination of ultra-trace levels of Mo in plants by inductively coupled plasma tandem mass spectrometry (ICP-MS/MS). Microchem. J. 133: 567–571. doi:10.1016%2fj.microc.2017.04.037&partnerID=40&md5=b2fafd6ccb197177689a674cfc9fbe2.

41. Kaya G, Türkoğlu S (2017)Analysis of certain fatty acids and toxic metal bioaccumulation in various tissues of three fish species that are consumed by Turkish people. *Environ. Sci. Pollut. Res.* 24 (10): 9495–9505. doi:10.1007/s11356-017-8632-2.

42. Gómez D, Nakazawa T, Furuta N, Smichowski P (2017) Multielemental chemical characterisation of fine urban aerosols collected in Buenos Aires and Tokyo by plasmabased techniques. *Microchem. J.* 133: 346-351. doi:10.1016/j.microc.2017.03.041.

43. Chen SS, Chou SS, Hwang DF (2004) Determination of methylmercury in fish using focused microwave digestion following by $Cu^{2+}$ addition, sodium tetrapropylborate derivatization, n-heptane extraction, and gas chromatography–mass spectrometry. *J. Chromatogr. A* 1024: 209–215. doi:10.1016/j.chroma.2003.10.015.

44. Romarís–Hortas V, García-Sartal C, Barciela-Alonso MC, Domínguez-González R, Moreda-Piñeiro A, Bermejo-Barrera P (2011) Bioavailability study using an in-vitro method of iodine and bromine in edible seaweed. *Food Chem.* 124(15): 1747–1752. doi:10.1016/j.foodchem.2010.07.117.

45. Domínguez-González R, Romarís-Hortas V, García-Sartal C, Moreda-Piñeiro A, Barciela-Alonso MC, Bermejo-Barrera P (2010) Evaluation of an in vitro method to estimate trace elements bioavailability in edible seaweeds. *Talanta* 82: 1668–1673. doi:10.1016/j.talanta.2010.07.043.

46. Moreda-Piñeiro J, Moreda-Piñeiro A, Romarís-Hortas V, Domínguez-González R, Alonso-Rodríguez E, López-Mahía P, Muniategui-Lorenzo S, Prada-Rodríguez D, Bermejo-Barrera P (2012) Trace metals in marine foodstuff: Bioavailability estimation and effect of major food constituents. *Food Chem.* 134: 339–345. doi:10.1016/j.foodchem.2012.02.165.

47. Barciela-Alonso MC, Plata-García V, Rouco-López A, Moreda-Piñeiro A, Bermejo-Barrera P (2014) Ionic imprinted polymer based solid phase extraction for cadmium and lead pre-concentration/determination in seafood. *Microchem. J.* 114: 106–110. doi:10.1016/j.microc.2013.12.008.

48. Rudolph E, Hann S, Stingeder G, Reiter C (2005) Ultra-trace analysis of platinum in human tissue samples. *Anal. Bioanal. Chem* 382: 1500–1506. doi:10.1007/s00216-005-3370-6.

49. Asfaw A, Wibetoe G (2005) Simultaneous determination of hydride (Se) and nonhydride-forming (Ca, Mg, K, P, S and Zn) elements in various beverages (beer, coffee and milk), with minimum sample preparation, by ICP–AES and use of a dualmode sample-introduction system. *Anal. Bioanal. Chem.* 382: 173–179. doi:10.1007/s00216-005-3188-2.

50. Welna M, Szymczycha-Madeja A, Pohl P (2014) Improvement of determination of trace amounts of arsenic and selenium in Slim Coffee Products by HG-ICP-OES. *Food Anal. Methods* 7(5): 1016–1023. doi:10.1007/s12161-013-9707-4.

51. Szymczycha-Madeja A, Welna M, Pohl P (2014) Fast method of elements determination in slim coffees by ICP-OES. *Food Chem.* 146: 220–225. doi:10.1016/j.foodchem.2013.09.054.

52. Szymczycha-Madeja A, Welna M, Pohl P (2014) Simple and fast sample preparation procedure prior to multi-element analysis of slim teas by ICP-OES. *Food Anal. Methods* 7(10): 2051–2063. doi:10.1007/s12161-014-9850-6.

53. Grotti M, Abelmoschi ML, Dalla Riva S, Soggia F, Frache R (2005) Determination of lead in bone tissues by axially viewed inductively coupled plasma multichannelbased emission spectrometry. *Anal. Bioanal. Chem.* 381: 1395–1400. doi:10.1007/s00216-005-3057-z.

54. Sedykh EM, Lyabusheva OA, Tambiev AKh, Bannykh LN (2005) Determination of the elemental composition of cyanobacteria cells and cell fractions by atomic emission and atomic absorption spectrometry. *J. Anal. Chem.* 60: 29–33. doi:1061-9348/05/6001.

55. Hearn R, Berglund M, Ostermann M, Pusticek N, Taylor P (2005) A comparison of high accuracy isotope dilution techniques for the measurement of low level sulfur in gas oils. *Anal. Chim. Acta* 532: 55–60. doi:10.1016/j.aca.2004.10.070.
56. Ramón R, Del Valle M, Valero, F (2005) Use of a Focused Microwave system for the determination of Kjeldahl nitrogen in industrial wastewaters. *Anal. Lett.* 38: 2415–2430. doi:10.1080/00032710500318106.
57. Ramon R, Valero F, Del Valle M (2003) Rapid determination of chemical oxygen demand using a focused microwave heating system featuring temperature control. *Anal. Chim. Acta* 491: 99–109. doi:10.1016/S0003-2670(03)00795-5.
58. Priego-Capote F, Luque de Castro MD (2006) Speciation of chromium by in-capillary derivatization and electrophoretically mediated microanalysis. *J. Chromatogr. A* 1113: 244–250. doi: 10.1016/j.chroma.2006.01.122.
59. Ashoka S, Peake BM, Bremner G, Hageman KJ, Reid MR (2009) Comparison of digestion methods for ICP-MS determination of trace elements in fish tissues. *Anal. Chim. Acta* 653: 191–199 doi:10.1016/j.aca.2009.09.025.
60. Väisänen A, Laatikainen P, Ilander A, Renvall S (2008) Determination of mineral and trace element concentrations in pine needles by ICP-OES: Evaluation of different sample pre-treatment methods. *Int. J. Environ. Anal. Chem.* 88: 1005-1016. doi:10.1080/03067310802308483.
61. Reis PA, Almeida CMR (2008) Matrix importance in animal material pre-treatment for metal determination. *Food Chem.* 107: 1294–1299. doi:10.1016/j.foodchem.2007.09.002.
62. Torres DP, Frescura VLA, Curtius AJ (2009) Simple mercury fractionation in biological samples by CV AAS following microwave-assisted acid digestion or TMAH pre-treatment. *Microchem. J.* 93: 206–210. doi:10.1016/j.microc.2009.07.003.
63. Wu Y, Lee YI, Wu L, Hou X (2012) Simple mercury speciation analysis by CVG-ICP-MS following TMAH pre-treatment and microwave-assisted digestion. *Microchem. J.* 103: 105–109. doi:10.1016/j.microc.2012.01.011.
64. Gonzalez MH, Souza GB, Oliveira RV, Forato LA, Nóbrega JA, Nogueira ARA (2009) Microwave-assisted digestion procedures for biological samples with diluted nitric acid: Identification of reaction products. *Talanta* 79 396–401. doi:10.1016/j.talanta.2009.04.001.
65. Niemelä M, Pitkäaho S, Ojala S, Keiski RL, Perämäki P (2012) Microwave-assisted aqua regia digestion for determining platinum, palladium, rhodium and lead in catalyst materials. *Microchem. J.* 101: 75–79. doi:10.1016/j.microc.2011.11.001.
66. Tyutyunnik OA, Getsina ML, Toropchenova ES, Kubrakova IV (2013) Microwave preparation of natural samples to the determination of mercury and other toxic elements by atomic absorption spectrometry. *J. Anal. Chem.* 68: 377–385. doi:10.1134/S1061934813050158.
67. Dantas ANS, Matos WO, Gouveia ST, Lopes GS (2013) The combination of infrared and microwave radiation to quantify trace elements in organic samples by ICP OES. *Talanta* 107: 292–296. doi:10.1016/j.talanta.2013.01.047
68. Simões Da Costa AM, Delgadillo I, Rudnitskaya A (2014) Detection of copper, lead, cadmium and iron in wine using electronic tongue sensor system. *Talanta* 129: 63–71. doi:10.1016/j.talanta.2014.04.030.
69. Durante C, Baschieri C, Bertacchini L, Bertelli D, Cocchi M, Marchetti A, Manzini D, Papotti G, Sighinolfi S (2015) An analytical approach to Sr isotope ratio determination in Lambrusco wines for geographical traceability purposes. *Food Chem.* 173: 557–563. doi:10.1016/j.foodchem.2014.10.086.
70. Barbosa JTP, Santos CMM, Peralva VN, Flores EMM, Korn M, Nóbrega JA, Korn MGA (2015) Microwave-assisted diluted acid digestion for trace elements analysis of edible soybean products. *Food Chem.* 175: 212–217. doi:10.1016/j.foodchem.2014.11.092.

71. Kronewitter SR, Marginean I, Cox JT, Zhao R, Hagler CD, Shukla AK, Carlson TS, Adkins JN, Camp II DG, Moore RJ, Rodland KD, Smith RD (2014) Polysialylated N-glycans identified in human serum through combined developments in sample preparation, separations, and electrospray ionization-mass spectrometry. *Anal. Chem.* 86(17): 8700–8710. doi:10.1021/ac501839b.

72. Yafa C, Farmer JG (2006) A comparative study of acid-extractable and total digestion methods for the determination of inorganic elements in peat material by inductively coupled plasma-optical spectrometry. *Anal. Chim. Acta* 557: 296–303. doi:10.1016/j.aca.2005.10.043.

73. Błazewicz A, Klatka M, Dolliver W, Kocjan R (2014) Determination of total iodine in serum and urine samples by ion chromatography with pulsed amperometric detection – Studies on analyte loss, optimization of sample preparation procedures, and validation of analytical method. *J. Chromatogr. B: Anal. Technol. Biomed.* Life Sci. 962: 141–146. doi:10.1016/j.jchromb.2014.05.022.

74. Balarama Krishna MV, Rao SV, Balaji Rao Y, Shenoy NS, Karunasagar D (2014) Development of a microwave-assisted digestion method for the rapid determination of chloride and fluoride in nuclear-grade boron carbide powders. *Anal. Methods* 6(1): 261–268. doi:10.1039/c3ay41313f.

75. Leme ABP, Bianchi SR, Carneiro RL, Nogueira ARA (2014) Optimization of Sample Preparation in the Determination of Minerals and Trace Elements in Honey by ICP-MS. *Food Anal. Methods* 7(5): 1009–1015. doi:10.1007/s12161-013-9706-5.

76. Saibatalova EV, Kulikova NN, Suturin AN, Paradina LF, Pakhomova NN, Vodneva EN, Semiturkina NA (2010) Influence of sample preparation on the determination of the elemental composition of fresh water sponges by inductively coupled plasma mass spectrometry. *J. Anal. Chem.* 65: 674–681. doi:10.1134/S1061934810070038.

77. Széles E, Varga Z, Stefánka Z (2010) Sample preparation method for analysis of swipe samples by inductively coupled plasma mass spectrometry. *J. Anal. Atom. Spectrom.* 25: 1014–1018. doi:10.1039/b926332b.

78. Wang H, Liu, Y (2010) Evaluation of trace and toxic element concentrations in Paris polyphylla from China with empirical and chemometric approaches. *Food Chem.* 121(3): 887–892. doi:10.1016/j.foodchem.2010.01.012.

79. Turan D, Kocahakimoglu C, Kavcar P, Gaygisiz H, Atatanir L, Turgut C, Sofuoglu SC (2011) The use of olive tree (Olea europaea L.) leaves as a bioindicator for environmental pollution in the Province of Aydin, Turkey. *Environ. Sci. Pollut. R.* 18: 355–364. doi:10.1007/s11356-010-0378-z.

80. Drennan-Harris LR, Wongwilawan S, Tyson JF (2013) Trace determination of total mercury in rice by conventional inductively coupled plasma mass spectrometry.*J. Anal. Atom. Spectrom.* 28 259–265. doi:10.1039/c2ja30278k.

81. Vanhaecke F, Van Hoecke K, Catry C (2012) Optimization of sample preparation and a quadrupole ICP-MS measurement protocol for the determination of elemental impurities in pharmaceutical substances in compliance with USP guidelines. *J. Anal. Atom. Spectrom.* 27: 1909–1919. doi:10.1039/c2ja30128h.

82. de Andrade FP, Nascentes CC, Costa LM (2012) Exploratory analysis for elemental characterization of biomass residues from biodiesel production by inductively coupled plasma optical emission spectrometry. *Anal. Lett.* 45: 2835–2844. doi:10.1080/0003271 9.2012.702175.

83. Ngah CWZCW, Yahya MA (2012) Optimisation of digestion method for determination of arsenic in shrimp paste sample using atomic absorption spectrometry. *Food Chem.* 134: 2406–2410. doi:10.1016/j.foodchem.2012.04.032.

84. Juranović Cindrić I, Krizman I, Zeiner M, Kampić S, Medunić G, Stingeder G. (2012) ICP-AES determination of minor- and major elements in apples after microwave assisted digestion. *Food Chem*, 135: 2675–2680. doi:10.1016/j.foodchem.2012.07.051.

85. Lemaire R, Bianco DD, Garnier L, Beltramo JL (2013) Determination of lead in lipstick by direct solid sampling high-resolution continuum source graphite furnace atomic absorption spectrometry: Comparison of two digestion methods. *Anal. Lett.* 46: 2265–2278. doi:10.1080/00032719.2013.787430.
86. Castro J, Spraul JC, Marcus RK (2009) Metals analysis of botanical products in various matrices using a single microwave digestion and inductively coupled plasma optical emission spectrometry (ICP-OES) method. *Anal. Method* 1:188–194. doi:10.1039/B9AY00080A.
87. Falta T, Koellensperger G, Standler A, Buchberger W, Mader RM, Hann S (2009) Quantification of cisplatin, carboplatin and oxaliplatin in spiked human plasma samples by ICP-SFMS and hydrophilic interaction liquid chromatography (HILIC) combined with ICP-MS detection. *J. Anal. At. Spectrom* 24(10):1336–1342. doi:10.1039/b907011g.
88. Schmidt A, Ahlswede J, Störr B (2009) Sample preparation strategies for one- and two-dimensional gel electrophoretic separation of plant proteins and the influence on arsenic and zinc bindings. *J. Chromatogr.* B877: 3097–3104. doi:10.1016/j.jchromb.2009.07.034.
89. Bakkali K, Martos NR, Souhail B, Ballesteros E (2009) Characterization of trace metals in vegetables by graphite furnace atomic absorption spectrometry after closed vessel microwave digestion. *Food Chem* 116: 590–594. doi:10.1016/j.foodchem.2009.03.010.
90. Fahrenholtz S, Griesel S, Pröfrock D, Kakuschke A (2009) Essential and non-essential elements in tissues of harbour porpoises (*Phocoena phocoena*) stranded on the coasts of the north and baltic seas between 2004–2006. *J. Environ. Monitor.* 11:1107–1113. doi:10.1039/b821504a
91. Santiago-Rivas S, Moreda-Piñeiro A, Barciela-Alonso MDC, Bermejo-Barrera P (2009) Characterization of raft mussels according to total trace elements and trace elements bound to metallothionein-like proteins. *J. Environ. Monitor.* 11: 1389–1396. doi:10.1039/b905942n
92. Al-Harahsheh M, Kingman S, Somerfield C, Ababneh F (2009) Microwave-assisted total digestion of sulphide ores for multi-element analysis. *Anal. Chim. Acta* 638: 101–105. doi:10.1016/j.aca.2009.02.030
93. Vázquez AS, Costa-Fernandez JM, Encinar JR, Pereiro R, Sanz-Medel A. (2008) Bromine determination in polymers by inductively coupled plasma-mass spectrometry and its potential for fast first screening of brominated flame retardants in polymers and paintings. *Anal. Chim. Acta* 623: 140–145. doi:10.1016/j.aca.2008.06.029.
94. Gonzálvez A, Armenta S, Pastor A, De La Guardia M. (2008) Searching the most appropriate sample pretreatment for the elemental analysis of wines by inductively coupled plasma-based techniques. *J. Agric. Food Chem.* 56: 4943–54. doi:10.1021/jf800286y
95. Rodushkin I, Bergman T, Douglas G, Engström E, Sörlin D, Baxter DC (2007) Authentication of kalix (N.E. Sweden) vendace caviar using inductively coupled plasma-based analytical techniques: Evaluation of different approaches. *Anal. Chim. Acta* 583: 310–318. doi:10.1016/j.aca.2006.10.038.
96. Stoica A, Peltea M, Baiulescu G, Ionica M (2004) Determination of cobalt in pharmaceutical products. *J. Pharm. Biomed. Anal.* 36: 653–656. doi:10.1016/j.jpba.2004.07.030.
97. Bou R, Guardiola F, Padró A, Pelfort E, Codony R (2004) Validation of mineralisation procedures for the determination of selenium, zinc, iron and copper in chicken meat and feed samples by ICP-AES and ICP-MS. *J. Anal. At. Spectrom.*19: 1361–1369. doi:10.1039/B404558K.
98. Niemelä M, Perämäki P, Piispanen J, Poikolainen J (2004) Determination of platinum and rhodium in dust and plant samples using microwave-assisted sample digestion and ICP-MS. *Anal. Chim. Acta* 521: 137–142. doi:10.1016/j.aca.2004.05.075.

99. Featherstone AM, Townsend AT, Jacobson GA, Peterson GM (2004) Comparison of methods for the determination of total selenium in plasma by magnetic sector inductively coupled plasma mass spectrometry. *Anal. Chim. Acta* 512: 319–327. doi:10.1016/j.aca.2004.02.058.

100. Castiñeira Gómez MDM, Brandt R, Jakubowski N, Andersson JT (2004) Changes of the metal composition in German white wines through the winemaking process. A study of 63 elements by inductively coupled plasma-mass spectrometry. *J. Agric. Food Chem.* 52: 2953–2961. doi:10.1021/jf035119g.

101. Kannamkumarath SS, Wróbel K, Wróbel K, Caruso JA (2004) Speciation of arsenic in different types of nuts by ion chromatography – inductively coupled plasma Amass spectrometry. *J. Agric. Food Chem.* 52:1458–1463. doi:10.1021/jf0351801.

102. Ezerskaya NA, Toropchenova ES, Pachgin DB, Kiseleva IN (2004) Controlled-potential coulometric determination of ruthenium in Ru(OH)Cl$_3$ with microwave sample preparation. *J. Anal. Chem.* 59: 296–298. doi:1061–9348/04/5903.

103. Drobyshev EJ, Solovyev ND, Ivanenko NB, Kombarova MY, Ganeev AA (2017) Trace element biomonitoring in hair of school children from a polluted area by sector field inductively coupled plasma mass spectrometry. *J. Trace Elem. Med Biol.* 39: 14–20. doi:10.1016/j.jtemb.2016.07.004.

104. Machado I, Dol I, Rodríguez-Arce E, Cesio MV, Pistón M (2016) Comparison of different sample treatments for the determination of As, Cd, Cu, Ni, Pb and Zn in globe artichoke (Cynara cardunculus L.). *Microchem. J.* 128: 128–133. doi:10.1016/j.microc.2016.04.016.

105. Muller EI, Souza JP, Muller CC, Muller ALH, Mello PA, Bizzi CA (2016) Microwave-assisted wet digestion with H$_2$O$_2$ at high temperature and pressure using single reaction chamber for elemental determination in milk powder by ICP-OES and ICP-MS. *Talanta* 156–157:232–238. doi:10.1016/j.talanta.2016.05.019.

106. Moreda-Piñeiro J, Herbello-Hermelo P, Domínguez-González R, Bermejo-Barrera P, Moreda-Piñeiro A (2016) Bioavailability assessment of essential and toxic metals in edible nuts and seeds. *Food Chem.* 205: 146–154. doi:10.1016/j.foodchem.2016.03.006.

107. Sharma A, Tapadia K (2016) Green tea-synthesized magnetic nanoparticles accelerate the microwave digestion of proteins analyzed by MALDI-TOF-MS. *J. Iran. Chem. Soc.* 13: 1723–1732. doi:10.1007/s13738-016-0889-8.

108. Lopes GS, Silva FLF, Grinberg AP, Sturgeon RE (2016) An evaluation of the use of formic acid for extraction of trace elements from Brazil nut and babassu coconut and its suitability for multi-element determination by ICP-MS. *J. Braz. Chem. Soc.* 27: 1229–1235. doi:10.5935/0103-5053.20160018.

109. Nsubuga H, Basheer C, Bushra MM, Essa MH, Omar MH, Shemsi AM (2016) Microwave-assisted digestion followed by parallel electromembrane extraction for trace level perchlorate detection in biological samples. *J. Chromatogr. B*1012: 1–7. doi:10.1016/j.jchromb.2016.01.014.

110. Kim SH, Lim Y, Hwang E, Yim YH (2016) Development of an ID ICP-MS reference method for the determination of Cd, Hg and Pb in a cosmetic powder certified reference material. *Anal. Methods* 8: 796–804. doi:10.1039/C5AY02040A

111. Hobohm J, Kuchta K, Krüger O, van Wasen S, Adam C (2016) Optimized elemental analysis of fluorescence lamp shredder waste. *Talanta* 147: 615–620. doi:10.1016/j.talanta.2015.09.068.

112. Zubakina EA, Solovyev ND, Savinkova ES, Slesar NI (2016) Sample preparation for cadmium quantification in sunflower (Heliánthus ánnuus) seeds using anodic stripping voltammetry. *Anal. Methods* 8: 326–332. doi:10.1039/C5AY02275D.

113. Mketo N, Nomngongo PN, Ngila JC (2016) An innovative microwave-assisted digestion method with diluted hydrogen peroxide for rapid extraction of trace elements in coal samples followed by inductively coupled plasma-mass spectrometry. *Microchem. J.* 124: 201–208 doi:10.1016/j.microc.2015.08.010

114. Barros JAVA, de Souza PF, Schiavo D, Nobrega JA (2016) Microwave-assisted diges-
     tion using diluted acid and base solutions for plant analysis by ICP OES. *J. Anal. At.
     Spectrom.* 31: 337–343. doi:10.1039/C5JA00294J.
115. Mketo N, Nomngongo PN, Ngila JC (2015) Development of a novel and green micro-
     wave-assisted hydrogen peroxide digestion method for total sulphur quantitative extrac-
     tion in coal samples prior to inductively coupled plasma-optical emission spectroscopy
     and ion-chromatography determination. *RSC Adv.* 5: 38931–38938. doi:10.1039/
     C5RA03040D.
116. Menezes EA, Oliveira AF, França CJ, Souza GB, Nogueira ARA (2018) Bioaccessibility
     of Ca, Cu, Fe, Mg, Zn, and crude protein in beef, pork and chicken after thermal pro-
     cessing. *Food Chem.* 240: 75–83. doi:10.1016/j.foodchem.2017.07.090.
117. Sajnóg A, Hanć A, Koczorowski R, Barałkiewicz D (2017) New procedure of quan-
     titative mapping of Ti and Al released from dental implant and Mg, Ca, Fe, Zn, Cu,
     Mn as physiological elements in oral mucosa by LA-ICP-MS. *Talanta* 175: 370–381.
     doi:10.1016/j.talanta.2017.07.058.
118. Rousis NI, Thomaidis NS (2017) Reduction of interferences in the determination of
     lanthanides, actinides and transition metals by an octopole collision/reaction cell
     inductively coupled plasma mass spectrometer – Application to the analysis of Chios
     mastic. *Talanta* 175: 69–76. doi:10.1016/j.talanta.2017.07.034.
119. Li J, Sun C, Zheng L, Jiang F, Wang S, Zhuang Z, Wang X (2017) Determination of trace
     metals and analysis of arsenic species in tropical marine fishes from Spratly islands.
     *Mar. Pollut. Bull.* 122 (1–2): 464–469. doi:10.1016/j.marpolbul.2017.06.017.
120. Ramos FS, Almeida RK, Júnior CAL, Arrudab MAZ, Pastore HO (2017) A
     Straightforward Method for Determination of Al and Na in Aluminosilicates
     Using ICP OES. *J. Braz. Chem. Soc.* 28(8): 1557–1563. doi:10.21577/0103-5053.201
     70008.
121. Snow MS, Morrison SS, Clark SB, Olson JE, Watrous MG (2017) 237 Np analytical
     method using 239 Np tracers and application to a contaminated nuclear disposal facil-
     ity. *J. Environ. Radioact.* 172: 89–95. doi:10.1016/j.jenvrad.2017.02.018.
122. Rietig A, Acker J (2017) Development and validation of a new method for the precise
     and accurate determination of trace elements in silicon by ICP-OES in high silicon
     matrices. *J. Anal. At. Spectrom.* 32(2): 322–333. doi:10.1039/c6ja00241b.
123. Pereira JSF, Moraes DP, Antes FG, Diehl LO, Santos MFP, Guimarães RCL, Fonseca
     TCO, Dressler VL, Flores EMM (2010) Determination of metals and metalloids in light
     and heavy crude oil by ICP-MS after digestion by microwave-induced combustion.
     *Microchem. J.* 96: 4–11. doi:10.1016/j.microc.2009.12.016.
124. Mesko MF, Pereira JSF, Moraes DP, Barin JS, Mello PA, Paniz JNG, Nóbrega JA, Korn
     MGA, FEMM (2010) Focused microwave-induced combustion: A new technique for
     sample digestion. *Anal. Chem.* 82: 2155–2160. doi:10.1021/ac902976j.
125. Fu Y-P, Lin C-H, HsuC-S (2005) Preparation of ultrafine CeO$_2$ powders by microwave-
     induced combustion and precipitation. *J. Alloys Compd.* s 391: 110–114. doi:10.1016/
     j.jallcom.2004.07.079.
126. Flores EMM, Smanioto Barin J, Gottfried Paniz JN, Medeiros JA, Knapp G (2004)
     Microwave-assisted sample combustion: A technique for sample preparation in trace
     element determination. *Anal. Chem.* 76: 3525–3529. doi:10.1021/ac0497712.
127. Barin JS, Pereira JSF, Mello PA, Knorr CL, Moraes DP, Mesko MF, Nóbrega JA, Korn
     MGA, Flores EMM (2012) Focused microwave-induced combustion for digestion of
     botanical samples and metals determination by ICP OES and ICP-MS. *Talanta* 94:
     308–314. doi:10.1016/j.talanta.2012.03.048.
128. Flores EMM, Muller EI, Duarte FA, Grinberg P, Sturgeon RE (2013) Determination
     of trace elements in fluoropolymers after microwave-induced combustion. *Anal. Chem.*
     85: 374–380. doi:10.1021/ac3029213.

129. Pereira JSF, Pereira LSF, Schmidt L, Moreira CM, Barin JS, Flores EMM (2013) Metals determination in milk powder samples for adult and infant nutrition after focused-microwave induced combustion. *Microchem. J.* 109: 29–35. doi:10.1016/j.microc.2012.05.010.

130. Muller ALH, Muller CC, Lyra F, Mello PA, Mesko MF, Muller EI, Flores EMM (2013) Determination of toxic elements in nuts by inductively coupled plasma mass spectrometry after microwave-induced combustion. *Food Anal. Method* 6: 258–264. doi:10.1007/s12161-012-9381-y.

131. Barin JS, Tischer B, Picoloto RS, Antes FG, Da Silva FEB, Paula FR, Flores EMM (2014) Determination of toxic elements in tricyclic active pharmaceutical ingredients by ICP-MS: A critical study of digestion methods. *J. Anal. At. Spectrom.* 29: 352–358. doi:10.1039/c3ja50334h.

132. Maciel JV, Knorr CL, Flores EMM, Müller EI, Mesko MF, Primel EG, Duarte FA (2014) Feasibility of microwave-induced combustion for trace element determination in Engraulis anchoita by ICP-MS. *Food Chem.* 145: 927–931. doi:10.1016/j.foodchem.2013.08.119.

133. Moraes DP., Mesko MF., Mello PA., Paniz JNG., Dressler VL., Knapp G, Flores EMM (2007) Application of microwave induced combustion in closed vessels for carbon black-containing elastomers decomposition. *Spectrochim. Acta, Part B* 62: 1065–1071. doi:10.1016/j.sab.2007.03.011.

134. Mesko MF, Moraes DP, Barin JS, Dressler VL, Knapp G, Flores EMM(2006) Digestion of biological materials using the microwave-assisted sample combustion technique. *Microchem. J.* 82: 183–188. doi:10.1016/j.microc.2006.01.004.

135. Flores EMM, Mesko MF, Moraes DP, Pereira JSF, Mello PA, Barin JS, Knapp G (2008) Determination of halogens in coal after digestion using the microwave-induced combustion technique. *Anal. Chem.*80: 1865–1870. doi:10.1021/ac8000836.

136. Nam KH, Isensee R, Infantino G, Putyera K, Wang X (2011) Microwave-induced combustion for ICP-MS: A generic approach traces elemental analysis of pharmaceutical products. *Spectroscopy* 26: 1–7.

137. Duarte FA, Pereira JSF, Barin JS, Mesko MF, Dressler VL, Flores EMM, Knapp G (2009) Seafood digestion by microwave-induced combustion for total arsenic determination by atomic spectrometry techniques with hydride generation. *J. Anal. At. Spectrom.* 24: 224–227. doi:10.1039/B810952D.

138. Pereira LSF, Bizzi CA, Schmidt L, Mesko MF, Barin JS, Flores EMM (2015) Evaluation of nitrates as igniters for microwave-induced combustion: understanding the mechanism of ignition. *RSC Adv.* 5: 9532–9538. doi:10.1039/C4RA12554A.

139. Antes FG, Duarte FA, Mesko MF, Nunes MAG, Pereira VA, Müller EI, Dressler VL, Flores EMM (2010) Determination of toxic elements in coal by ICP-MS after digestion using microwave-induced combustion. *Talanta* 83: 364–369. doi:10.1016/j.talanta.2010.09.030.

140. Picoloto RS, Wiltsche H, Knapp G, Mello PA, Barin JS, Flores EMM (2013) Determination of inorganic pollutants in soil after volatilization using microwave-induced combustion. *Spectrochim. Acta, Part B* 86: 123–130. doi:10.1016/j.sab.2013.01.010.

141. Costa VC, Picoloto RS, Hartwig CA, Mello PA, Flores EMM, Mesko MF (2015) Feasibility of ultra-trace determination of bromine and iodine in honey by ICP-MS using high sample mass in microwave-induced combustion. *Anal. Bioanal. Chem.* 407: 7957–7964. doi:10.1007/s00216–015-8967-9.

142. Krzyzaniak SR, Santos RF, Dalla Nora FM, Cruz SM, Flores EM, Mello PA (2016) Determination of halogens and sulfur in high-purity polyimide by IC after digestion by MIC. *Talanta* 158: 193–197. doi:10.1016/j.talanta.2016.05.032.

143. da Silva SV, Picoloto RS, Flores EMM, Wagner R, dos Santos Richards NSP, Barin JS (2016) Evaluation of bromine and iodine content of milk whey proteins combining digestion by microwave-induced combustion and ICP-MS determination. *Food Chem.* 190: 364–367. doi:10.1016/j.foodchem.2015.05.087.

144. Christopher SJ, Vetter TW (2016) Application of microwave-induced combustion and isotope dilution strategies for quantification of sulfur in coals via sector-field inductively coupled plasma mass spectrometry. *Anal. Chem.* 88: 4635–4643. doi:10.1021/acs.analchem.5b03981.

145. Pereira RM, Costa VC, Hartwig CA, Picoloto RS, Flores EMM, Duarte FA, Mesko MF (2016) Feasibility of halogen determination in noncombustible inorganic matrices by ion chromatography after a novel volatilization method using microwave-induced combustion. *Talanta* 147: 76–81. doi:10.1016/j.talanta.2015.09.031.

146. Enders MSP, de Souza JP, Balestrin P, de Azevedo Mello P, Duarte FA, Muller EI (2016) Microwave-induced combustion of high purity nuclear flexible graphite for the determination of potentially embrittling elements using atomic spectrometric techniques. *Microchem. J.* 124: 321–325. doi:10.1016/j.microc.2015.09.015.

147. Picoloto RS, Doned M, Flores ELM, Mesko MF, Flores EMM, Mello PA (2015) Simultaneous determination of bromine and iodine in milk powder for adult and infant nutrition by plasma based techniques after digestion using microwave-induced combustion. *Spectrochim. Acta, Part B* 07: 86–92. doi:10.1016/j.sab.2015.02.007.

148. Grinberg P, Sturgeon RE, Diehl LO, Bizzi CA, Flores EMM (2015) Comparison of sample digestion techniques for the determination of trace and residual catalyst metal content in single-wall carbon nanotubes by inductively coupled plasma mass spectrometry. *Spectrochim. Acta, Part B* 105: 89–94. doi:10.1016/j.sab.2014.09.009.

149. Crizel MG, Hartwig CA, Novo DLR, Toralles IG, Schmidt L., Muller EI, Mesko MF (2015) A new method for chlorine determination in commercial pet food after decomposition by microwave-induced combustion. *Anal. Methods* 7: 4315–4320. doi:N10.1039/C5AY00649J

150. Silva JS, Diehl LO, Frohlich AC, Costa VC, Mesko MF, Duarte FA, Flores EMM (2017) Determination of bromine and iodine in edible flours by inductively coupled plasma mass spectrometry after microwave-induced combustion. *Microchem. J.* 133: 246–250.

151. Mesko MF, Teotonio AC, Oliveira DTT, Novo DLR, Costa VC (2017) A feasible method for indirect quantification of LT 4 in drugs by iodine determination. *Talanta* 166: 223–227.

152. Toralles IG, Coelho GS, Costa VC, Cruz SM, Flores EMM, Mesko MF (2017) A fast and feasible method for Br and I determination in whole egg powder and its fractions by ICP-MS. *Food Chem.* 221: 877–883.

153. Dalla Nora FM, Cruz SM, Giesbrecht CK, Knapp G, Wiltsche H, Bizzi CA, Barin JS, Flores EMM (2017) A new approach for the digestion of diesel oil by microwave-induced combustion and determination of inorganic impurities by ICP-MS. *J. Anal. At. Spectrom.* 32(2): 408–414.

154. Pereira LSF, Pedrotti MF, Enders MSP, Albers CN, Pereira JSF, Flores EMM (2016) Multitechnique determination of halogens in soil after selective volatilization using microwave-induced combustion. *Anal. Chem.* 89(1): 980–987.

155. Morales-Muñoz S, Luque-García JL, Luque de Castro MD (2006) Pure and modified water assisted by auxiliary energies: An environmental friendly extractant for sample preparation. *Anal. Chim. Acta* 557: 278–286. doi:10.1016/j.aca.2005.10.013.

156. Prado-Rosales RC, Herrera MC, Luque-García JL, Luque de Castro MD (2004) Study of the feasibility of focused microwave-assisted Soxhlet extraction of N-methylcarbamates from soil. *J. Chromatogr. A* 953: 133–140. doi:10.1016/S0021-9673(02)00118-8.

157. Karthikeyan S, Joshi UM, Balasubramanian R (2006) Microwave assisted sample preparation for determining water-soluble fraction of trace elements in urban airborne particulate matter: Evaluation of bioavailability. *Anal. Chim. Acta* 576: 23–30. doi:10.1016/j.aca.2006.05.051.

158. Suturović Z, Kravić S, Milanović S, Crossed D, Signurović A, Brezo T (2014) Determination of heavy metals in milk and fermented milk products by potentiometric stripping analysis with constant inverse current in the analytical step. *Food Chem.* 155: 120–125. doi:10.1016/j.foodchem.2014.01.030.

159. Mizanur Rahman GM, Kingston HM (2005) Development of a microwave-assisted extraction method and isotopic validation of mercury species in soils and sediments. *J. Anal. At. Spectrom.* 20:183–191. doi:10.1039/b404581e.

160. Wang X, Ding L, Zhang H, Cheng J, Yu A, Zhang H, Liu L, Liu Z, Li Y (2006) Development of an analytical method for organotin compounds in fortified flour samples using microwave-assisted extraction and normal-phase HPLC with UV detection. *J. Chromat. B* 843: 268–274. doi:10.1016/j.jchromb.2006.06.013.

161. Regueiro J, Llompart M, Garcia-Jares C, Cela R (2006) Determination of polybrominated diphenyl ethers in domestic dust by microwave-assisted solvent extraction and gas chromatography–tandem mass spectrometry. *J. Chromat. A* 1137: 1–7. doi:10.1016/j.chroma.2006.09.080.

162. Carro N, García I, Ignacio M, Moreira A (2006) Microwave-assisted solvent extraction and gas chromatography ion trap mass spectrometry procedure for the determination of persistent organochlorine pesticides (POPs) in marine sediment. *Anal. Bioanal. Chem.* 385: 901–909. doi:10.1007/s00216-006-0485-3.

163. García-López M, Rodríguez I, Cela R, Kroening KK Caruso JA (2009) Determination of organophosphate flame retardants and plasticizers in sediment samples using microwave-assisted extraction and gas chromatography with inductively coupled plasma mass spectrometry. *Talanta* 79: 824–829. doi:10.1016/j.talanta.2009.05.006.

164. Varga M, Dobor J, Helenkár A, Jurecska L, Yao J, Záray G (2010) Investigation of acidic pharmaceuticals in river water and sediment by microwave-assisted extraction and gas chromatography-mass spectrometry. *Microchem. J.* 95: 353–358. doi:10.1016/j.microc.2010.02.010.

165. Mazzarino M, Riggi S, de la Torre X, Botrè F (2010) Speeding up the process urine sample pre-treatment: Some perspectives on the use of microwave assisted extraction in the anti-doping field. *Talanta* 81: 1264–1272. doi:10.1016/j.talanta.2010.02.019.

166. Xu H, Chen L, Sun L, Sun X, Du X, Wang J, Wang T, Zeng Q, Wang H, Xu Y, Zhang X, Ding L (2011) Microwave-assisted extraction and in situ clean-up for the determination of fluoroquinolone antibiotics in chicken breast muscle by LC-MS/MS. *J. Sep. Sci.* 34: 142–149. doi:10.1002/jssc.201000365.

167. Hu Y, Li Y, Zhang Y, Li G, Chen Y (2011) Development of sample preparation method for auxin analysis in plants by vacuum microwave-assisted extraction combined with molecularly imprinted clean-up procedure. *Anal. Bioanal. Chem.* 399: 3367–3374. doi:10.1007/s00216-010-4257-8.

168. Wang H, Zhou X, Zhang Y, Chen H, Li G, Xu Y, Zhao, Q, Song W, Jin H, Ding L (2012) Dynamic microwave-assisted extraction coupled with salting-out liquid-liquid extraction for determination of steroid hormones in fish tissues. *J. Agric. Food. Chem.* 60: 10343–10351. doi:10.1021/jf303124c.

169. Guo L, Lee HK (2013) Microwave assisted extraction combined with solvent bar microextraction for one-step solvent-minimized extraction, cleanup and preconcentration of polycyclic aromatic hydrocarbons in soil samples. *J. Chromatogr. A* 1286: 9–15. doi:10.1016/j.chroma.2013.02.067.

170. Balarama Krishna MV, Venkateswarlu G, Thangavel S, Karunasagar D (2013) Oxidative pyrolysis combined with the microwave-assisted extraction method for the multi-elemental analysis of boron carbide powders by inductively coupled plasma optical emission spectrometry (ICP-OES). *Anal. Methods* 5: 1515–1523. doi:10.1039/c3ay26480g.

171. Ares AM, Nozal MJ, Bernal JL, Bernal J (2014) Optimized extraction, separation and quantification of twelve intact glucosinolates in broccoli leaves. *Food Chem.* 152: 66–74. doi:10.1016/j.foodchem.2013.11.125.

172. Chang YJ, Chao MR, Chen SC, Chen CH, Chang YZ (2014) A high-throughput method based on microwave-assisted extraction and liquid chromatography-tandem mass spectrometry for simultaneous analysis of amphetamines, ketamine, opiates, and their metabolites in hair. *Anal. Bioanal. Chem.* 406: 2445–2455. doi:10.1007/s00216-014-7669-z.

173. Stan M, Soran ML, Marutoiu C (2014) Extraction and HPLC determination of the ascorbic acid content of three indigenous spice plants. *J. Anal. Chem.* 69: 998–1002. doi:10.1134/S106193481410013X.

174. Carrasco L, Vassileva E (2014) Determination of methylmercury in marine biota samples: Method validation. *Talanta* 122: 106–114. doi:10.1016/j.talanta.2014.01.027.

175. Carrasco L, Vassileva E (2015) Determination of methylmercury in marine sediment samples: Method validation and occurrence data. *Anal. Chim. Acta* 853: 167–178. doi:10.1016/j.aca.2014.10.026.

176. Wu L, Song Y, Hu M, Xu X, Zhang H, Yu A, Ma Q, Wang Z (2015) Integrated microwave processing system for the extraction of organophosphorus pesticides in fresh vegetables. *Talanta* 134: 366–372. doi:10.1016/j.talanta.2014.11.035.

177. García-Ayuso LE, Sánchez M, Fernández De Alba A, Luque de Castro MD (1998) Focused microwave-assisted soxhlet: an advantageous tool for sample extraction. *Anal. Chem.* 70: 2626–2431. doi:10.1021/ac9711044.

178. Priego-López E, Velasco J, Dobarganes MC, Ramis-Ramos G, Luque de Castro MD (2003) Focused microwave-assisted Soxhlet extraction: an expeditive approach for the isolation of lipids from sausage products. *Food Chem.* 83: 143–149. doi:10.1016/S0308-8146(03)00220-6.

179. Luque-García JL, Luque de Castro MD (2004) Focused microwave-assisted Soxhlet extraction: devices and applications. *Talanta* 64: 571–577. doi:10.1016/j.talanta.2004.03.054.

180. Zhou T, Xiao X, Li G (2012) Microwave accelerated selective Soxhlet extraction for the determination of organophosphorus and carbamate pesticides in ginseng with gas chromatography/mass spectrometry. *Anal. Chem.* 84: 5816–5822. doi:10.1021/ac301274r.

181. Pasias IN, Farmaki EG, Thomaidis NS, Piperaki EA (2010) Elemental content and total antioxidant activity of Salvia fruticosa. *Food Anal. Method* 3: 195–204. doi:10.1007/s12161-009-9122-z.

182. Ma W, Lu Y, Hu R, Chen J, Zhang Z, Pan Y (2010) Application of ionic liquids based microwave-assisted extraction of three alkaloids N-nornuciferine, O-nornuciferine, and nuciferine from lotus leaf. *Talanta* 80: 1292-1297. doi:10.1016/j.talanta.2009.09.027.

183. Dobor J, Varga M, Yao J, Chen H, Palkó G, Záray G (2010) A new sample preparation method for determination of acidic drugs in sewage sludge applying microwave assisted solvent extraction followed by gas chromatography-mass spectrometry. *Microchem. J.* 94: 36–41. doi:10.1016/j.microc.2009.08.007.

184. Li Z, Huang D, Tang Z, Deng C, Zhang X (2010) Fast determination of chlorogenic acid in tobacco residues using microwave-assisted extraction and capillary zone electrophoresis technique. *Talanta* 82: 1181–1185. doi:10.1016/j.talanta.2010.06.037.

185. Deng J, Xiao X, Tong X, Li G (2010) Preparation of bergenin from Ardisia crenata sims and Rodgersia sambucifolia hemsl based on microwave-assisted extraction/high-speed counter-current chromatography. *Sep. Purif. Technol.* 74: 155–159. doi:10.1016/j.seppur.2010.05.018.

186. Chen L, Zhao Q, Xu Y, Sun L, Zeng Q, Xu H, Wang H, Zhang X, Yu A, Zhang H, Ding L (2010) A green method using micellar system for determination of sulfonamides in soil. *Talanta* 82: 1186–1192. doi:10.1016/j.talanta.2010.06.031.

187. Yamaguchi C, Lee WY (2010) A cost effective, sensitive, and environmentally friendly sample preparation method for determination of polycyclic aromatic hydrocarbons in solid samples. *J. Chromatogr. A*, 1217: 6816–6823. doi:10.1016/j.chroma.2010.08.055.

188. Zhou Y, Han L, Cheng J, Guo F, Zhi X, Hu H, Chen G (2011) Dispersive liquid-liquid microextraction based on the solidification of a floating organic droplet for simultaneous analysis of diethofencarb and pyrimethanil in apple pulp and peel. *Anal. Bioanal. Chem.* 399: 1901–1906. doi:10.1007/s00216–010-4567-x.

189. Xiao X, Si X, Tong X, Li G (2011) Preparation of flavonoids and diarylheptanoid from *Alpinia katsumadai* hayata by microwave-assisted extraction and high-speed counter-current chromatography. *Sep. Purif. Technol.* 81: 265–269. doi:10.1016/j.seppur.2011.07.013.

190. Ma CH, Liu TT, Yang L, Zu YG., Chen X, Zhang L, Zhang Y, Zhao C (2011) Ionic liquid-based microwave-assisted extraction of essential oil and biphenyl cyclooctene lignans from *Schisandra chinensis* Baill fruits. *J. Chromatogr. A* 1218: 8573–8580. doi:10.1016/j.chroma.2011.09.075

191. Castilho INB, Welz B, Vale MGR, De Andrade JB, Smichowski P, Shaltout AA, Colares L, Carasek E (2012) Comparison of three different sample preparation procedures for the determination of traffic-related elements in airborne particulate matter collected on glass fiber filters. *Talanta* 88: 689–695. doi:10.1016/j.talanta.2011.11.066.

192. Zmozinski AV, Passos LD, Damin ICF, Espírito Santo MAB, Vale MGR, Silva MM (2013) Determination of cadmium and lead in fresh fish samples by direct sampling electrothermal atomic absorption spectrometry. *Anal. Method* 5: 6416–6424. doi:10.1039/c3ay40923f.

193. Khan N, Jeong IS, Hwang IM, Kim JS, Choi SH, Nho EY, Choi JY, Kwak BM, Ahn JH, Yoon T, Kim KS (2013) Method validation for simultaneous determination of chromium, molybdenum and selenium in infant formulas by ICP-OES and ICP-MS. *Food Chem.* 141: 3566–3570. doi:10.1016/j.foodchem.2013.06.034.

194. Akinlua A, Torto N, McCrindle RI (2013) A new approach to sample preparation for the determination of trace metals in petroleum source rocks. *Anal. Method* 5: 4929–4934. doi:10.1039/c3ay40278a.

195. Zhang DY, Yao XH, Duan MH, Luo M, Wang W, Fu YJ, Zu YG, Efferth T (2013) An effective negative pressure cavitation-microwave assisted extraction for determination of phenolic compounds in *P. calliantha* H. Andr. *Analyst* 138: 4631–4641. doi:10.1039/c3an36534d.

196. Amaral CDB, Dionísio AGG, Santos MC, Donati GL, Nóbrega JA, Nogueira ARA (2013) Evaluation of sample preparation procedures and krypton as an interference standard probe for arsenic speciation by HPLC-ICP-QMS. *J. Anal. Atom. Spectrom.* 28: 1303–1310. doi:10.1039/c3ja50099c.

197. Mesa LBA, Padro JM, Reta M (2013) Analysis of non-polar heterocyclic aromatic amines in beefburguers by using microwave-assisted extraction and dispersive liquid-ionic liquid microextraction. *Food Chem.* 141: 1694–1701. doi:10.1016/j.foodchem.2013.04.076.

198. Korn MGA, Guida MAB, Barbosa JTP, Torres EA, Fernandes AP, Santos JCC, Dantas KDGF, Nóbrega JA (2013) Evaluation of sample preparation procedures for trace element determination in Brazilian propolis by inductively coupled plasma optical emission spectrometry and their discrimination according to geographic region. *Food Anal. Method* 6: 872–880. doi:10.1007/s12161-012-9497-0.

199. Khoomrung S, Chumnanpuen P, Jansa-Ard S, Staìshlman M, Nookaew I, Borén J, Nielsen J (2013) Rapid quantification of yeast lipid using microwave-assisted total lipid extraction and HPLC-CAD. *Anal. Chem.* 85: 4912–4919. doi:10.1021/ac3032405.

200. Duarte FA, Soares BM, Vieira AA, Pereira ER, MacIel JV, Caldas SS, Primel EG (2013) Assessment of modified matrix solid-phase dispersion as sample preparation for the determination of $CH_3Hg^+$ and $Hg^{2+}$ in fish. *Anal. Chem.* 85: 5015–5022. doi:10.1021/ac4002436.

201. dos Santos ÉJ, Herrmann AB, Prado SK, Fantin EB, dos Santos VW, de Oliveira AVM, Curtius AJ (2013) Determination of toxic elements in glass beads used for pavement marking by ICP OES. *Microchem. J.* 108: 233–238. doi:10.1016/j.microc.2012.11.003.

202. Wietecha-Posłuszny R, Woźniakiewicz M, Garbacik A, Chesy P, Kościelniak P (2013) Application of microwave irradiation to fast and efficient isolation of benzodiazepines from human hair. *J. Chromatogr. A* 1278: 22–28. doi:10.1016/j.chroma.2013.01.005.

203. Soares AR, Nascentes CC (2013) Development of a simple method for the determination of lead in lipstick using alkaline solubilization and graphite furnace atomic absorption spectrometry. *Talanta* 105: 272–277. doi:10.1016/j.talanta.2012.09.021.

204. Lehmann EL, Fostier AH, Arruda MAZ (2013) Hydride generation using a metallic atomizer after microwave-assisted extraction for inorganic arsenic speciation in biological samples. *Talanta* 104: 187–192. doi:10.1016/j.talanta.2012.11.009.

205. Bulska E, Danko B, Dybczyński RS, Krata A, Kulisa K, Samczyński Z, Wojciechowski M (2012) Inductively coupled plasma mass spectrometry in comparison with neutron activation and ion chromatography with UV/VIS detection for the determination of lanthanides in plant materials. *Talanta* 97: 303–311. doi:10.1016/j.talanta.2012.04.035.

206. Fernández P, Lago M, Lorenzo RA, Carro AM, Bermejo AM, Tabernero MJ (2009) Optimization of a rapid microwave-assisted extraction method for the simultaneous determination of opiates, cocaine and their metabolites in human hair. *J. Chromatogr. B* 877: 1743–1450. doi:10.1016/j.jchromb.2009.04.035.

207. Hu F, Deng C, Liu Y, Zhang X (2009) Quantitative determination of chlorogenic acid in honeysuckle using microwave-assisted extraction followed by nano-LC-ESI mass spectrometry. *Talanta* 77: 1299–1303. doi:10.1016/j.talanta.2008.09.003.

208. Álvarez I, Bermejo AM, Tabernero MJ, Fernández P, Cabarcos P, López P (2009) Microwave-assisted extraction: A simpler and faster method for the determination of ethyl glucuronide in hair by gas chromatography-mass spectrometry. *Anal. Bioanal. Chem.* 393: 1345–1350. doi:10.1007/s00216-008-2546-2.

209. Reyes LH, Rahman GMM, Kingston HMS (2009) Robust microwave-assisted extraction protocol for determination of total mercury and methylmercury in fish tissues. *Anal. Chim. Acta* 631: 121–128. doi:10.1016/j.aca.2008.10.044.

210. Lu Y, Ma W, Hu R, Dai X, Pan Y (2008) Ionic liquid-based microwave-assisted extraction of phenolic alkaloids from the medicinal plant nelumbo nucifera gaertn. *J. Chromatogr. A* 1208: 42–46. doi:10.1016/j.chroma.2008.08.070.

211. Tapie N, Budzinski H, Le Ménach K (2008) Fast and efficient extraction methods for the analysis of polychlorinated biphenyls and polybrominated diphenyl ethers in biological matrices. *Anal. Bioanal. Chem.* 391: 2169–2177. doi:10.1007/s00216-008-2148-z.

212. Tong L, Wang Y, Xiong J, Cui Y, YigangZhou, Yi L (2008) Selection and fingerprints of the control substances for plant drug eucommia ulmodies oliver by HPLC and LC-MS. *Talanta* 76: 80–84. doi:10.1016/j.talanta.2008.02.012.

213. Nickerson B, Arikpo WB, Berry MR, Bobin VJ, Houck TL, Mansour HL, Warzeka J (2008) Leveraging elevated temperature and particle size reduction to extract API from various tablet formulations. *J. Pharm. Biomed. Anal.* 47. 268–278. doi:10.1016/j.jpba.2008.01.006.

214. Kingston HM, Reyes LH, Mizanur Rahman GM, Fahrenholz T (2008) Comparison of methods with respect to efficiencies, recoveries, and quantitation of mercury species interconversions in food demonstrated using tuna fish. *Anal. Bioanal. Chem.* 390: 2123–2132. doi:10.1007/s00216-008-1966-3.

215. Ellis J, Shah M, Kubachka KM, Caruso JA (2007) Determination of organophosphorus fire retardants and plasticizers in wastewater samples using MAE-SPME with GC-ICPMS and GC-TOFMS detection. *J. Environ. Monitor.* 9: 1329–1336. doi:10.1039/B710667J.

216. Lee C, Gallo J, Arikpo W, Bobin V (2007) Comparison of extraction techniques for spray dried dispersion tablet formulations. *J. Pharm. Biomed. Anal.* 45: 565–571. doi:10.1016/j.jpba.2007.08.011.

217. Hoang TH, Sharma R, Susanto D, Di Maso M, Kwong E (2007) Microwave-assisted extraction of active pharmaceutical ingredient from solid dosage forms. *J. Chromatogr. A* 1156: 149–153. doi:10.1016/j.chroma.2007.02.060.

218. Carro AM, Lorenzo RA, Fernández F, Phan-Tan-Luu R, Cela R (2007) Microwave-assisted extraction followed by headspace solid-phase microextraction and gas chromatography with mass spectrometry detection (MAE-HSSPME-GC-MS/MS) for determination of polybrominated compounds in aquaculture samples. *Anal. Bioanal. Chem.* 388: 1021–1029. doi:10.1007/s00216-007-1220-4.

219. Wei M, Jen J (2007) Determination of polycyclic aromatic hydrocarbons in aqueous samples by microwave assisted headspace solid-phase microextraction and gas chromatography/flame ionization detection. *Talanta* 72: 1269–1274. doi:10.1016/j.talanta.2007.01.017.

220. García M, Rodríguez I, Cela R (2007) Microwave-assisted extraction of organophosphate flame retardants and plasticizers from indoor dust samples. *J. Chromatogr. A* 1152: 280–286. doi:10.1016/j.chroma.2006.11.046.

221. Careri M, Corradini C, Elviri L, Mangia A (2007) Optimization of a rapid microwave assisted extraction method for the liquid chromatography-electrospray-tandem mass spectrometry determination of isoflavonoid aglycones in soybeans. *J. Chromatogr. A* 1152: 274–279. doi:10.1016/j.chroma.2007.03.112.

222. Rice SL, Mitra S (2007) Microwave-assisted solvent extraction of solid matrices and subsequent detection of pharmaceuticals and personal care products (PPCPs) using gas chromatography-mass spectrometry. *Anal. Chim. Acta* 589: 125–132. doi:10.1016/j.aca.2007.02.051.

223. Liu C, Zhou H, Yan Q (2007) Fingerprint analysis of *Dioscorea nipponica* by high-performance liquid chromatography with evaporative light scattering detection. *Anal. Chim. Acta* 582: 61–68. doi:10.1016/j.aca.2006.08.057.

224. Liu C, Zhou H, Zhao Y (2007) An effective method for fast determination of artemisinin in *Artemisia annua* L. by high performance liquid chromatography with evaporative light scattering detection. *Anal. Chim. Acta* 581: 298–302. doi:10.1016/j.aca.2006.08.038.

225. Zhou HY, Liu CZ (2006) Rapid determination of solanesol in tobacco by high-performance liquid chromatography with evaporative light scattering detection following microwave-assisted extraction. *J. Chromatogr. B* 835: 119–122. doi:10.1016/j.jchromb.2006.02.055.

226. Fang Y, Pan Y, Li P, Xue M, Pei F, Yang W, Ma N, Hu Q (2016) Simultaneous determination of arsenic and mercury species in rice by ion-pairing reversed phase chromatography with inductively coupled plasma mass spectrometry. *Food Chem.* 213: 609–615. doi:10.1016/j.foodchem.2016.07.003.

227. Oliveira A, Gonzalez MH, Queiroz HM, Cadore S (2016) Fractionation of inorganic arsenic by adjusting hydrogen ion concentration. *Food Chem.* 213: 76–82. doi:10.1016/j.foodchem.2016.06.055.

228. Quiroz W, Astudillo F, Bravo M, Cereceda-Balic F, Vidal V, Palomo-Marín MR, Rueda-Holgado F, Pinilla-Gil E (2016) Antimony speciation in soils, sediments and volcanic ashes by microwave extraction and HPLC-HG-AFS detection. *Microchem. J.* 129: 111–116. doi:10.1016/j.microc.2016.06.016.

229. Tsubaki S, Oono K, Hiraoka M, Onda A, Mitani T (2016) Microwave-assisted hydrothermal extraction of sulfated polysaccharides from Ulva spp. and *Monostroma latissimum. Food Chem.* 210: 311–316. doi:10.1016/j.foodchem.2016.04.121.

230. Tuzen M, SahinerS, Hazer B (2016) Solid phase extraction of lead, cadmium and zinc on biodegradable polyhydroxybutyrate diethanol amine (PHB-DEA) polymer and their determination in water and food samples. *Food Chem.* 210: 115–120. doi:10.1016/j.foodchem.2016.04.079.

231. Ruiz-Aceituno L, García-Sarrió MJ, Alonso-Rodriguez B, Ramos L, Sanz ML (2016) Extraction of bioactive carbohydrates from artichoke (Cynara scolymus L.) external bracts using microwave assisted extraction and pressurized liquid extraction. *Food Chem.* 196: 1156–1162 doi:10.1016/j.foodchem.2015.10.046.

232. Cittan M, Koçak S, Çelik A, Dost K (2016) Determination of oleuropein using multiwalled carbon nanotube modified glassy carbon electrode by adsorptive stripping square wave voltammetry. *Talanta* 159: 148–154 doi:10.1016/j.talanta.2016.06.021.

233. Lin YA, Jiang SJ, Sahayam AC, Huang YL (2016) Speciation of chromium in edible animal oils after microwave extraction and liquid chromatography inductively coupled plasma mass spectrometry. *Microchem. J.* 128: 274–278. doi:10.1016/j.microc.2016.05.001.

234. Pan J, Muppaneni T, Sun Y, Reddy HK, Fu J, Lu X, Deng S (2016) Microwave-assisted extraction of lipids from microalgae using an ionic liquid solvent [BMIM][HSO$_4$]. *Fuel* 178: 49–55. doi:10.1016/j.fuel.2016.03.037.

235. Corazza MZ, Tarley CRT (2016) Development and feasibility of emulsion breaking method for the extraction of cadmium from omega-3 dietary supplements and determination by flow injection TS-FF-AAS. *Microchem. J.* 127: 145–151. doi:10.1016/j.microc.2016.02.021.

236. Speltini A, Sturini M, Maraschi F, Porta A, Profumo A (2016) Fast low-pressurized microwave-assisted extraction of benzotriazole, benzothiazole and benezenesulfonamide compounds from soil samples. *Talanta* 147: 322–327. doi:10.1016/j.talanta.2015.09.074.

237. Azzouz A, Ballesteros E (2016) Determination of 13 endocrine disrupting chemicals in environmental solid samples using microwave-assisted solvent extraction and continuous solid-phase extraction followed by gas chromatography–mass spectrometry. *Anal. Bioanal. Chem* 408: 231–241. doi:10.1007/s00216–015-9096-1.

238. Fang X, Wang J, Hao J, Li X, Guo N (2015) Simultaneous extraction, identification and quantification of phenolic compounds in Eclipta prostrate using microwave-assisted extraction combined with HPLC–DAD–ESI–MS/MS. *Food Chem.* 188: 527–536. doi:10.1016/j.foodchem.2015.05.037.

239. Hadi BJ, Sanagi MM, Aboul-Enein HY, Ibrahim WAW, Jamil S, Múazu MA (2015) Microwave-assisted extraction of methyl β-cyclodextrin-complexed curcumin from turmeric rhizome oleoresin. *Food Anal. Methods* 8: 2447–2456. doi:10.1007/s12161-015-0137-3.

240. Zhang W, Hu Y, Cheng H (2015) Optimization of microwave-assisted extraction for six inorganic and organic arsenic species in chicken tissues using response surface methodology. *J. Sep. Sci.* 38: 3063–3070. doi:10.1002/jssc.201500065.

241. Welna M, Borkowska-Burnecka J, Popko M (2015) Ultrasound- and microwave-assisted extractions followed by hydride generation inductively coupled plasma optical emission spectrometry for lead determination in geological samples. *Talanta* 144: 953–959. doi:10.1016/j.talanta.2015.07.058.

242. Valdés A, Vidal L, Beltrán A, Canals A, Garrigós MC (2015) Microwave-assisted extraction of phenolic compounds from almond skin byproducts (Prunus amygdalus): A multivariate analysis approach. *J. Agric. Food Chem.* 63: 5395–5402. doi:10.1021/acs.jafc.5b01011.

243. Pelcová P, Dočekalová H, Kleckerová A (2015) Determination of mercury species by the diffusive gradient in thin film technique and liquid chromatography – atomic fluorescence spectrometry after microwave extraction. *Anal. Chim. Acta* 866: 21–26. doi:10.1016/j.aca.2015.01.043.

244. Setyaningsih W, Saputro IE,Palma M, Barroso CG (2015) Optimisation and validation of the microwave-assisted extraction of phenolic compounds from rice grains. *Food Chem.* 169: 141–149. doi:10.1016/j.foodchem.2014.07.128.

245. Dahmoune F, Nayak B, Moussi K, Remini H, Madani K (2015) Optimization of microwave-assisted extraction of polyphenols from Myrtus communis L. leaves. *Food Chem.* 166: 585–595. doi:10.1016/j.foodchem.2014.06.066.

246. Evans SE, Davies P, Lubben A, Kasprzyk-Hordern B (2015) Determination of chiral pharmaceuticals and illicit drugs in wastewater and sludge using microwave assisted extraction, solid-phase extraction and chiral liquid chromatography coupled with tandem mass spectrometry. *Anal. Chim. Acta* 882: 112–126. doi:10.1016/j.aca.2015.03.039.

247. Speltini A, Sturini M, Maraschi F, Viti S, Sbarbada D, Profumo A (2015) Fluoroquinolone residues in compost by green enhanced microwave-assisted extraction followed by ultra performance liquid chromatography tandem mass spectrometry. *J. Chromatogr. A* 1410: 44–50. doi:10.1016/j.chroma.2015.07.093.

248. Haroune L, Cassoulet R, Lafontaine MP, Bélisle M, Garant D, Pelletier F, Cabana H, Bellenger JP (2015) Liquid chromatography-tandem mass spectrometry determination for multiclass pesticides from insect samples by microwave-assisted solvent extraction followed by a salt-out effect and micro-dispersion purification. *Anal. Chim. Acta* 891: 160–170. doi:10.1016/j.aca.2015.07.031.

249. Tsiaka T, Zoumpoulakis P, Sinanoglouc VJ, Makris C, Heropoulos GA, Calokerinos AC (2015) Response surface methodology toward the optimization of high-energy carotenoid extraction from *Aristeus antennatus* shrimp. *Anal. Chim. Acta* 877: 100–110. doi:10.1016/j.aca.2015.03.051.

250. Nguyen VT, Vuong QV, Bowyer MC, Van Altena IA, Scarlett CJ (2017) Microwave assisted extraction for saponins and antioxidant capacity from xao tam phan (Paramignya trimera) root. *J. Food Process. Preserv.* 41(2): e12851.

251. Thirugnanasambandham K, Sivakumar V (2017) Microwave assisted extraction process of betalain from dragon fruit and its antioxidant activities. *J. Saudi Soc. Agric. Sci.* 16(1): 41–48.

252. Belwal T, Bhatt ID, Rawal RS, Pande V (2017) Microwave-assisted extraction (MAE) conditions using polynomial design for improving antioxidant phytochemicals in Berberis asiatica Roxb. ex DC. leaves. *Ind. Crops Prod.* 95: 393–403.

253. Belwal T, Giri L, Bhatt ID, Rawal RS, Pande V (2017) An improved method for extraction of nutraceutically important polyphenolics from Berberis jaeschkeana CK Schneid. fruits. *Food Chem.* 230: 657–666.

254. Liu Z, Qiao L, Gu H, Yang F, Yang L (2017) Development of Brönsted acidic ionic liquid based microwave assisted method for simultaneous extraction of pectin and naringin from pomelo peels. *Sep. Purif. Technol.* 172: 326–337.

255. Aufartová J, Brabcová I, Torres-Padrón ME, Solich P, Sosa-Ferrera Z, Santana-Rodríguez JJ (2017) Determination of fluoroquinolones in fishes using microwave-assisted extraction combined with ultra-high performance liquid chromatography and fluorescence detection. *J. Food Composit. Anal.* 56: 140–146.

256. Filip S, Pavlić B, Vidović S, Vladić J, Zeković Z (2017) Optimization of microwave-assisted extraction of polyphenolic compounds from Ocimum basilicum by response surface methodology. *Food Anal. Methods* 10(7): 2270–2280.

257. Ma L, Yang Z, Kong Q, Wang L (2017) Extraction and determination of arsenic species in leafy vegetables: Method development and application. *Food Chem.* 217: 524–530.
258. Stefanello FS, dos Santos CO, Bochi VC, Fruet APB, Soquetta MB, Dörr AC, Nörnberg JL (2018) Analysis of polyphenols in brewer's spent grain and its comparison with corn silage and cereal brans commonly used for animal nutrition. *Food Chem.* 239: 385–401.
259. Mketo N, Nomngongo PN, Ngila JC (2017) Rapid total sulphur reduction in coal samples using various dilute alkaline leaching reagents under microwave heating: preventing sulphur emissions during coal processing. *Environ. Sci. Pollut. Res.* 24(24): 1–7.
260. Ciriminna R, Fidalgo A, Delisi R, Carnaroglio D, Grillo G, Cravotto G, Tamburino A, Ilharco LM, Pagliaro M (2017) High quality essential oils extracted by an eco-friendly process from different citrus fruits and fruit regions. *ACS Sustainable Chem. Eng.* 5 (6): 5578–5587.
261. Xie X, Zhu D, Zhang W, Huai W, Wang K, Huang X, Zhou L, Fan H (2017) Microwave-assisted aqueous two-phase extraction coupled with high performance liquid chromatography for simultaneous extraction and determination of four flavonoids in *Crotalaria sessiliflora* L. *Ind. Crops Prod.* 95: 632–642.
262. Liu W, Wang Y, Chen Y, Tao S, Liu W (2017) Polycyclic aromatic hydrocarbons in ambient air, surface soil and wheat grain near a large steel-smelting manufacturer in northern China. *J. Environ. Sci.* 57: 93–103.
263. Ferreres F, Grosso C, Gil-Izquierdo A, Valentão P, Mota AT, Andrade PB (2017) Optimization of the recovery of high-value compounds from pitaya fruit by-products using microwave-assisted extraction. *Food Chem.* 230: 463–474.
264. Deng H, Xu X, Yao N, Li N, Zhang X (2006) Rapid determination of essential oil compounds in *Artemisia Selengensis* Turcz by gas chromatography-mass spectrometry with microwave distillation and simultaneous solid-phase microextraction. *Anal. Chim. Acta* 556: 289–294. doi:10.1016/j.aca.2005.09.038.
265. Sahraoui N, Vian MA, Bornard I, Boutekedjiret C, Chemat F (2008) Improved microwave steam distillation apparatus for isolation of essential oils Comparison with conventional steam distillation. *J. Chromatogr. A* 1210: 229–233. doi:10.1016/j.chroma.2008.09.078.
266. Jiao J, Ma DH, Gai QY, Wang W, Luo M, Fu YJ, Ma W (2013) Rapid analysis of Fructus forsythiae essential oil by ionic liquids-assisted microwave distillation coupled with headspace single drop microextraction followed by gas chromatography-mass spectrometry. *Anal. Chim. Acta* 804: 143–150. doi:10.1016/j.aca.2013.10.035.
267. Jiao J, Gai QY, Wang W, Luo M, Zhao CJ, Fu YJ, Ma W (2013) Ionic-liquid-assisted microwave distillation coupled with headspace single-drop microextraction followed by GC-MS for the rapid analysis of essential oil in Dryopteris fragrans. *J. Sep. Sci.* 36: 3799–3806. doi:10.1002/jssc.201300906.
268. Damm M, Holzer M, Radspieler G, Marsche G, Kappe CO (2010) Microwave-assisted high-throughput acid hydrolysis in silicon carbide microtiter platforms-A rapid and low volume sample preparation technique for total amino acid analysis in proteins and peptides. *J. Chromatogr. A* 1217: 7826–7832. doi:10.1016/j.chroma.2010.10.062.
269. Silva FO, Ferraz V (2006) Double use of microwaves in fatty acid preparation for elaidic acid determination as phenacyl ester using high-performance liquid chromatography in Brazilian fat products. *Talanta* 68: 643–645. doi:10.1016/j.talanta.2005.05.002.
270. Wood R, Lee T (1983) High-performance liquid chromatography of fatty acids: quantitative analysis of saturated, monoenoic, polyenoic and geometrical isomers. *J. Chromatogr. A* 254: 237–246. doi:10.1016/S0021-9673(01)88338-2.
271. Jurado-Sánchez B, Ballesteros E, Gallego M (2012) Determination of carboxylic acids in water by gas chromatography-mass spectrometry after continuous extraction and derivatisation. *Talanta* 93: 224–232. doi:10.1016/j.talanta.2012.02.022.

272. Sternbauer L, Dieplinger J, Buchberger W, Marosits E (2014) Determination of nucleating agents in plastic materials by GC/MS after microwave-assisted extraction with in situ microwave-assisted derivatization. *Talanta* 128: 63–68. doi:10.1016/j.talanta.2014.04.022.

273. Mazzarino M, De Angelis F, Di Cicco T, de la Torre X, Botrè F (2010) Microwave irradiation for a fast gas chromatography-mass spectrometric analysis of polysaccharide-based plasma volume expanders in human urine. *J. Chromatogr. B* 878: 3024–3032. doi:10.1016/j.jchromb.2010.08.046.

274. Meyer MR, Weber AA, Maurer HH (2011) A validated GC-MS procedure for fast, simple, and cost-effective quantification of glycols and GHB in human plasma and their identification in urine and plasma developed for emergency toxicology. *Anal. Bioanal. Chem.* 400: 411–414. doi:10.1007/s00216-011-4760-6.

275. Kabaha K, Taralp A, Cakmak I, Ozturk L (2011) Accelerated hydrolysis method to estimate the amino acid content of wheat (Triticum durum Desf.) flour using microwave irradiation. *J. Agric. Food Chem.* 59: 2958–2965. doi:10.1021/jf103678c.

276. Memon AA, Memon N, Bhanger MI, Luthria DL (2013) Assay of phenolic compounds from four species of ber (*Ziziphus mauritiana* L.) fruits: Comparison of three base hydrolysis procedure for quantification of total phenolic acids. *Food Chem.* 139: 496–502. doi:10.1016/j.foodchem.2013.01.065.

277. Theocharis G, Andlauer W (2013) Innovative microwave-assisted hydrolysis of ellagitannins and quantification as ellagic acid equivalents. *Food Chem.* 138: 2430–2434. doi:10.1016/j.foodchem.2012.12.015.

278. Liu RL, Zhang J, Mou ZL, Hao SL, Zhang ZQ (2012) Microwave-assisted one-step extraction-derivatization for rapid analysis of fatty acids profile in herbal medicine by gas chromatography-mass spectrometry. *Analyst* 137: 5135–5143. doi:10.1039/c2an36178.

279. Ye Q, Zheng D (2009) Rapid analysis of the essential oil components of dried *Perilla frutescens* (L.) by magnetic nanoparticle-assisted microwave distillation and simultaneous headspace solid-phase microextraction followed by gas chromatography-mass spectrometry. *Anal. Method* 1: 39–44. doi:10.1039/B9AY00035F.

280. Shi Y, Chen M, Muniraj S, Jen J (2008) Microwave-assisted headspace controlled temperature liquid-phase microextraction of chlorophenols from aqueous samples for gas chromatography-electron capture detection. *J. Chromatogr. A* 1207: 130–135. doi:10.1016/j.chroma.2008.07.096.

281. Deng C, Mao Y, Hu F, Zhang X (2007) Development of gas chromatography-mass spectrometry following microwave distillation and simultaneous headspace single-drop microextraction for fast determination of volatile fraction in Chinese herb. *J. Chromatogr. A* 1152: 193–198. doi:10.1016/j.chroma.2006.08.074.

282. Pena MT, Pensado L, Casais MC, Mejuto MC, Cela R (2007) Sample preparation of sewage sludge and soil samples for the determination of polycyclic aromatic hydrocarbons based on one-pot microwave-assisted saponification and extraction. *Anal. Bioanal. Chem.* 387: 2559–2567. doi:10.1007/s00216-006-1110-1.

283. Feng J, Liu X, Li Y, Duan G (2016) Microwave-assisted enzymatic hydrolysis followed by extraction with restricted access nanocomposites for rapid analysis of glucocorticoids residues in liver tissue. *Talanta* 159: 155–162. doi:10.1016/j.talanta.2016.06.013.

284. Aviram L, Mc Cooeye M, Mester Z (2016) Determination of underivatized amino acids in microsamples of a yeast nutritional supplement by LC-MS following microwave assisted acid hydrolysis. *Anal. Methods* 8: 4497–4503. doi:10.1039/c6ay0407e.

285. Torres S, Gil R, Silva MF, Pacheco P (2016) Determination of seleno-amino acids bound to proteins in extra virgin olive oils. *Food Chem.* 197: 400–405. doi:10.1016/j.foodchem.2015.10.008.

286. Moret S, Scolaro M, Barp L, Purcaro G, Conte L.S. (2016) Microwave assisted saponification (MAS) followed by on-line liquid chromatography (LC)-gas chromatography (GC) for high-throughput and high-sensitivity determination of mineral oil in different cereal-based foodstuffs. *Food Chem.* 196: 50–57. doi:10.1016/j.foodchem.2015.09.032.

287. Ellingson D, Shippar J, Gilmore, J (2016) Determination of free and total choline and carnitine in infant formula and adult/pediatric nutritional formula by liquid chromatography/tandem mass spectrometry (LC/MS/MS): Single-laboratory validation, first action 2015.10. *J. AOAC Int.* 99: 204–209. doi:10.5740/jaoacint.15–0144.

288. Lu W, Lv X, Gao B, Shi H, Yu L (2015) Differentiating milk and non-milk proteins by UPLC amino acid fingerprints combined with chemometric data analysis techniques. *J. Agric. Food Chem.* 63: 3996–4002. doi:10.1021/acs.jafc.5b00702.

289. Lv GP, Hu DJ, Cheong KL, Li ZY, Qing XM, Zhao J, Li SP (2015) Decoding glycome of *Astragalus membranaceus* based on pressurized liquid extraction, microwave-assisted hydrolysis and chromatographic analysis. *J. Chromatogr. A* 1409: 19–29. doi:10.1016/j.chroma.2015.07.058.

290. Chen L, Wang N, Li L (2014) Development of microwave-assisted acid hydrolysis of proteins using a commercial microwave reactor and its combination with LC-MS for protein full-sequence analysis. *Talanta* 129: 290–295. doi:10.1016/j.talanta.2014.05.042.

291. Bian J, Peng P, Peng F, Xiao X, Xu F, Sun R (2014) Microwave-assisted acid hydrolysis to produce xylooligosaccharides from sugarcane bagasse hemicelluloses. *Food Chem.* 156: 7–13. doi:10.1016/j.foodchem.2014.01.112.

292. Sun D, Wang N, Li L (2014) In-gel microwave-assisted acid hydrolysis of proteins combined with liquid chromatography tandem mass spectrometry for mapping protein sequences. *Anal. Chem.* 86: 600–607. doi:10.1021/ac402802a.

293. Ellingson D, Pritchard T, Foy P, King K, Mitchell B, Austad J, Winters D, Sullivan D (2013) Determination of free and total myo-inositol in infant formula and adult/pediatric nutritional formula by high-performance anion exchange chromatography with pulsed amperometric detection, including a novel total extraction using microwave-assisted acid hydrolysis and enzymatic treatment: First action 2012.12. *J. AOAC Int.* 96: 1068–1072. doi:10.5740/jaoacint.13-128.

294. Sweygers N, Dewil R, Appels L (2017) Production of levulinic acid and furfural by microwave-assisted hydrolysis from model compounds: Effect of temperature, acid concentration and reaction time. *Waste Biomass Valorization* 9(3): 343–355.

295. Liu Q, Lu Y, Aguedo M, Jacquet N, Ouyang C, He W, Yan C. Bai W, Guo R, Goffin D, Richel A, Song J (2017) Isolation of high-purity cellulose nanofibers from wheat straw through the combined environmentally friendly methods of steam explosion, microwave-assisted hydrolysis, and microfluidization. *ACS Sustainable Chem. Eng.* 5(7): 6183–6191.

296. Su TC, Fang Z (2017) One-pot microwave-assisted hydrolysis of cellulose and hemicellulose in selected tropical plant wastes by NaOH-freeze pretreatment. *ACS Sustainable Chem. Eng.* 5(6): 5166–5174.

297. Yücel TB, Karaoğlu ŞA, Yaylı N (2017) Antimicrobial activity and composition of Rindera lanata (LAM.) Bunge var. canescens (ADC) Kosn. Essential oil obtained by hydrodistillation and microwave assisted distillation. *Rec. Nat. Prod.* 11(3): 328–333

298. Chen F, Du X, Zu Y, Yang L, Wang F (2016) Microwave-assisted method for distillation and dual extraction in obtaining essential oil, proanthocyanidins and polysaccharides by one-pot process from Cinnamomi Cortex. *Sep. Purif. Technol.* 164: 1–11.

299. Fu J, Zhao J, Zhu Y, Tang J (2017) Rapid analysis of the essential oil components in dried lavender by magnetic microsphere-assisted microwave distillation coupled with HS-SPME followed by GC-MS. *Food Anal. Methods* 10(7): 2373–2382.

300. Criado A, Cárdenas S, Gallego M, Valcárcel M (2004) Direct automatic screening of soils for polycyclic aromatic hydrocarbons based on microwave-assisted extraction/fluorescence detection and on-line liquid chromatographic confirmation. *J. Chromatogr. A* 1050: 111–118. doi:10.1016/j.chroma.2004.07.070.

301. Serrano A, Gallego M (2006) Continuous microwave-assisted extraction coupled on-line with liquid–liquid extraction: Determination of aliphatic hydrocarbons in soil and sediments. *J. Chromatogr. A* 1104: 323–330. doi:10.1016/j.chroma.2005.12.017.

302. Chen L, Song D, Tian Y, Ding L, Yu A, Zhang H (2008) Application of on-line microwave sample-preparation techniques. *Trends Anal. Chem.* 27: 151–159. doi:10.1016/j.trac.2008.01.003.

303. Gao S, You J, Wang Y, Zhang R, Zhang H (2012) On-line continuous sampling dynamic microwave-assisted extraction coupled with high performance liquid chromatographic separation for the determination of lignans in Wuweizi and naphthoquinones in Zicao. *J. Chromatogr. B.* 887–888: 35–42. doi:10.1016/j.jchromb.2012.01.005.

304. Chen L, Ding L, Zhang H, Li J, Wang Y, Wang X, Qu C, Zhang H (2006) Dynamic microwave-assisted extraction coupled with on-line spectrophotometric determination of safflower yellow in Flos Carthami. *Anal. Chim. Acta* 580: 75–82. doi:10.1016/j.aca.2006.07.040.

305. Morales-Muñoz S, Luque-García JL. Luque de Castro MD (2004) A continuous approach for the determination of Cr(VI) in sediment and soil based on the coupling of microwave-assisted water extraction, preconcentration, derivatization and photometric detection. *Anal. Chim. Acta* 515: 343–348. doi:10.1016/j.aca.2004.03.092.

306. Chen L, Jin H, Ding L, Zhang H, Wang X, Wang Z, Li J, Qu C, Wang Y, Zhang H (2007) On-line coupling of dynamic microwave-assisted extraction with high-performance liquid chromatography for determination of andrographolide and dehydroandrographolide in Andrographis paniculata Nees. *J. Chromatogr. A* 1140: 71–77. doi:10.1016/j.chroma.2006.11.070.

307. Wang H, Xu Y, Song W, Zhao Q, Zhang X, Zeng Q, Chen H, Ding L, Ren N (2011) Automatic sample preparation of sulfonamide antibiotic residues in chicken breast muscle by using dynamic microwave-assisted extraction coupled with solid-phase extraction. *J. Sep. Sci.* 34: 2489–2497. doi:10.1002/jssc.201100310.

308. Wu L, Hu M, Li Z, Song Y, Zhang H, Yu A, Ma Q, Wang Z (2015) Dynamic microwave-assisted extraction online coupled with single drop microextraction of organophosphorus pesticides in tea samples. *J. Chromatogr. A*, 1407: 42–45. doi:10.1016/j.chroma.2015.06.062.

309. Wu L, Hu M, Li Z, Song Y, Yu C, Zhang H, Yu A, Ma Q, Wang Z (2016) Dynamic microwave-assisted extraction combined with continuous-flow microextraction for determination of pesticides in vegetables. *Food Chem.* 192: 596–602. doi:10.1016/j.foodchem.2015.07.055.

310. Gao S, Yu W, Yang X, Liu Z, Jia Y, Zhang H (2012) On-line ionic liquid-based dynamic microwave-assisted extraction-high performance liquid chromatography for the determination of lipophilic constituents in root of *Salvia miltiorrhiza. Bunge. J. Sep. Sci.* 2012: 1–9. doi:10.1002/jssc.201200267.

311. Morales-Muñoz S, Luque-García JL, Luque de Castro MD (2004) Screening method for linear alkylbenzene sulfonates in sediments based on water Soxhlet extraction assisted by focused microwaves with on-line preconcentration/derivatization/detection. *J. Chromatogr. A* 1026: 41–46. doi:10.1016/j.chroma.2003.11.047.

312. Chávez G, Bravo B, Piña N, Arias M, Vivas E, Ysambertt F, Márquez N, Cáceres A (2004) Determination of aliphatic alcohols after on-line microwave-assisted derivatization by liquid chromatography-photodiode array detection. *Talanta* 64: 1323–1328. doi:10.1016/j.talanta.2004.05.055.

313. Acebal C, Centurión ME, Lista A, Fernandez B (2005) A new and fast continuous method for the pre-treatment of dehydrated broth for total creatinine determination. *Food Chem.* 93: 493–496. doi:10.1016/j.foodchem.2004.10.027.

314. Acebal C, Centurión ME, Lista A, Fernandez B (2006) Automated method for the total creatinine determination in dehydrated broths. *Anal. Lett.* 39: 387–394. doi:10.1080/00032710500477175.

315. Almeida MI, Segundo MA, Costa Lima JL, Rangel A (2004) Multi-syringe flow injection system with in-line microwave digestion for the determination of phosphorus. *Talanta* 64: 1283–1289. doi:10.1016/j.talanta.2004.04.006.

316. Egorov O, O'Hara M, Grate J (2012) Automated radioanalytical system incorporating microwave-assisted sample preparation, chemical separation, and online radiometric detection for the monitoring of total 99Tc in nuclear waste processing streams. *Anal. Chem.* 84: 3090–3098. doi:dx.doi.org/10.1021/ac300418b.

317. Quaresma MC, Cassella R, de la Guardia M, Santelli R (2004) Rapid on-line sample dissolution assisted by focused microwave radiation for silicate analysis employing flame atomic absorption spectrometry: iron determination. *Talanta* 62: 807–811. doi:10.1016/j.talanta.2003.10.002.

318. Marques TL, Wiltsche H, Nóbrega JA, Winkler M, Knapp G (2017) Performance evaluation of a high-pressure microwave-assisted flow digestion system for juice and milk sample preparation. *Anal. Bioanal. Chem.* 409(18): 4449–4458.

319. Silva M, Kyser K, Beauchemin D (2007) Enhanced flow injection leaching of rocks by focused microwave heating with in-line monitoring of released elements by inductively coupled plasma mass spectrometry. *Anal. Chim. Acta* 584: 447–454. doi:10.1016/j.aca.2006.11.043.

# 3 An Insight into Green Microwave-Assisted Techniques

## *Degradation and Microextraction*

Natalia A. Gomez, Maite V. Aguinaga,
Natalia LLamas, Mariano Garrido,
Carolina Acebal, and Claudia Domini

## CONTENTS

## 3.1   INTRODUCTION

The microwave (MW) comprises of wavelegnths in the region of the electromagnetic spectrum × between 0.1 and 100 cm (frequency range of 300 GHz to 300 MHz, respectively) [1]. This kind of energy has been successfully applied as a potent tool in chemistry due to its rapid and efficient selective heating properties [2].

MW irradiation combines both thermal and non-thermal effects [3]. The heating involves two processes: dipolar polarization and ionic conduction. Under MW irradiation, polar molecules orientate according to the rapid changes of the alternating electric field. Thus, they suffer rotation, friction and collisions, resulting in the generation of heat. Moreover, the electric conductor materials are polarized under the alternating electric field, so the ionic conduction takes place. [4]. In this manner, MW radiation is able to rapidly and simultaneously heat the bulk of the material and can accelerate the rates, enhance the yields and promote certain pathways in different chemical reactions [5–8].

On the other hand, some authors report the so-called non-thermal effect of MW radiation, which appears when some molecules are polarized under the electromagnetic field and are aligned, leading to the possible breakage of hydrogen bonds [9]. Other authors link the non-thermal effects (also called not purely thermal or specific MW effects) with the "hot spots" theory [3, 10–12]. Under MW irradiation, some materials, known as electromagnetic wave absorbers (conductive polymer, carbon, SiC, ferrite, $TiO_2$, magnetic metal and metal alloys) [13], can strongly absorb MW with the consequent generation of active "hot spots" on the surface of the materials. These

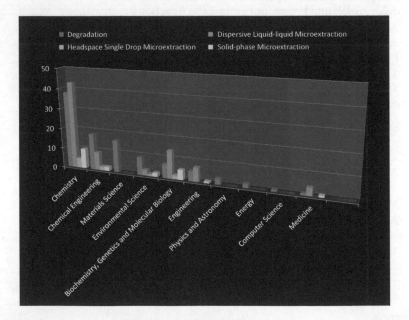

**FIGURE 3.1**   Number of publications in MW-assisted degradation and microextraction (Data from Scopus, 2017).

hotspots, which can reach temperatures around 1400 K, can accelerate the movement of electrons and increase the oxidative capacity in degradation reactions [10, 14].

All these characteristics have made MW applications increase in the last three decades whether in domestic, industrial or medical fields [15].

This chapter presents a review divided into two parts: the first section discusses the applications reported on using MW irradiation, particularly those involving their environmental use for improving the degradation efficiency of pollutants in water and wastewaters [16–21]. The second section is focused on the different types of microextraction assisted by MW [22–25]. Figure 3.1 offers a summary of the number of publications found in the literature devoted to MW-assisted degradation and microextraction.

## 3.2 DEGRADATION PROCESSES OF POLLUTANTS IN WATER AND WASTEWATER

Industrial and municipal wastewaters may contain several organic compounds, such as surfactants, pesticides, pharmaceuticals, dyes and endocrine disrupting chemicals, which can cause serious environmental problems [26]. Among the remediation technologies used to mineralize or degrade these pollutants into others with less harmfull effects, MW is a promising strategy either alone (Table 3.1) or combined with different advanced oxidation processes (AOPs) [2].

AOPs refer to chemical procedures that are able to generate highly reactive free radicals, especially the hydroxyl and hydroperoxyl radicals ($\cdot OH$, $\cdot HOO$) and the superoxide radical anion ($\cdot O_2^-$), which can degrade many persistent organic pollutants. Once generated, these free radicals can react with organic compounds by electron transfer, radical addition and/or hydrogen abstraction [26]. The AOPs have certain advantages such as high reaction rates, elevated oxidation potential and non-selective oxidation, which means that multiple pollutants can be simultaneously treated [26–28]. Figure 3.2 shows a scheme with the most common AOPs used to degrade different organic compounds [26].

## TABLE 3.1
### Microwave-assisted Degradations

| Analite | Microwave Power Temperature | Reaction Time (min) | Microwave | Ref |
|---|---|---|---|---|
| Phytosterol | 1000 W | 1–30 | Domestic multimodal MW oven (Panasonic NN-CT756, Domestic MW, Bracknell, U.K.) | [31] |
| Nature rubber | 115°C | 15 | Programmable microwave drying device (Guangzhou DiWei industrial microwave device Co.,LTD, China; 2.45 GHz) | [32] |
| Phycocyanobilin | 150°C | 90 | Microwave digestion system (Multiwave GO, Anton Paar GmbH, Graz, Austria) | [33] |

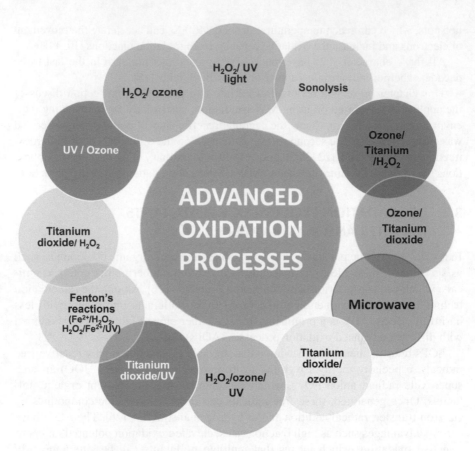

**FIGURE 3.2**   Scheme advanced oxidation processes.

Although there are many great advantages of AOPs, one disadvantage is that they use costly chemicals (e.g. $O_3$, $H_2O_2$ and ferrous iron) or expensive continuous UV radiation. Thus, many investigations are developing towards the economical improvement of their efficiency, specifically for their use at large-scale operations. [29, 30]. In this sense, the additional application of MW energy could significantly enhance the efficiency of AOPs. In the following chapter, we made a revision of the more relevant contributions found in the literature that combines the different AOPs and MW in different areas.

### 3.2.1   MW and Heterogeneous Catalysis

MW has been combined with different heterogeneous catalytic processes to degrade several contaminants. Lai et al. [34] have applied nickel oxides and MW irradiation to degrade 4-chlorophenol (4-CP) in air equilibrated solutions. They achieved the total removal of 4-CP at pH 7 in about 20 min. In another work [35], the same researchers used a mix-valenced nickel oxide for the degradation of 4-CP. They found that 4-CP could be degraded in 5 min into innocuous compounds (like $CO_2$,

$H_2O$ and mineral acids) at pH 7 and 70°C. They also employed nickel oxides and MW at low temperature for the complete degradation of 4-CP and 4-nitrophenol (PNP) [36, 37]. Hsu and coworkers [38] proposed a method for the degradation of chlorobenzene (CB) using platinum, palladium, rhodium and MW. They achieved 74–76% removal of CB without adding hydrogen gas, a strong base or strong acid at 69 to 78°C. Lee and Jou [39] also degraded CB with MW irradiation using zerovalent iron particles. They found a CB removal of 76.5% when freshly prepared iron nanoparticles were used.

Yuan et al. have combined MW with granular active carbon (GAC) for the treatment of p-nitrophenol (PNP) in wastewater [40]. The reported method reached a removal rate of around 99%. On the other hand, Bo and coworkers have used a similar catalyst based on copper supported with GAC, which is also used for the oxidation of PNP. Although the removal was about 92%, the authors reported a high catalytic capacity in comparison with GAC due to the presence of Cu [21]. In another work [41], the same research group proposed the use of platinum support in GAC for the degradation of PNP and pentachlorophenol. Degradation of PNP is also achieved using a catalyst of $Mn_2O_3$ supported with GAC and MW energy [42]. The authors have demonstrated that PNP degradation is low under MW irradiation, but the presence of $Mn_2O_3$ supported with GAC significantly enhance the removal of this pollutant. Zhang et al. [43] have developed a method for the degradation of congo red dye in the presence of activated carbon powder. The degradation ratio was about 98% in 2.5 min of irradiation time. Veksha et al. [44] used activated carbon for the degradation of methyl orange with MW and hydrogen peroxide. They attributed the removal of methyl orange to either adsorption on activated carbon and/or degradation by $H_2O_2$. The removal was around 92%.

GAC and MW were also successfully used for the degradation of the same dye. The authors report a synergistic effect of MW irradiation and granular-activated carbon that contributes to a high-degradation efficiency [45]. On the other hand, Zhang and coworkers reported that the combination of activated carbon and MW is not efficient enough to completely mineralize Bisphenol A (BPA) since the residues found in the degraded solution still interacted with human serum albumin, changing the inherent structure and the function of this protein [46]. Recently, a new catalytic degradation technology using MW-induced carbon nanotubes was proposed and applied in the treatment of different organic pollutants in an aqueous solution such as methyl orange, methyl parathion, sodium dodecyl benzene sulfonate, BPA and methylene blue (MB). The authors claimed that the combination of CNTs and MW led to faster absorption kinetics, more selective adsorption and easier regeneration than activated carbon [47].

Pang and Lei [10] developed a method for the degradation of PNP using peroxymonosulfate, manganese ferrite ($MnFe_2O_4$) and MW. After the reaction, the catalyst was collected by a magnet. More than 97% of PNP was decomposed within 2 min under the optimized conditions. Other works aimed at degrading PNP in soils used $MgFe_2O_4$ as a catalyst. [48]. The authors reported degradation efficiencies of up to 55% in dry soils, but these values increased to 100% when the water content was about 30%. Thus, the presence of water is crucial for obtaining high removal efficiencies, which confirms the role of ·OH in the PNP degradation under MW radiation. In another work, Qiu et al. have also investigated the degradation of PNP using an $\alpha$-Bi$_2$O$_3$ as a catalyst [49]. The calculated degradation efficiency of PNP reached up to 99% after 7 min of MW irradiation.

Chen and coworkers [50] proposed the degradation of PNP in soils under MW irradiation using copper-doped attapulgite (Cu/ATP) and copper silicon carbide (Cu/SiC) nanocomposites. Cu/SiC showed to have a better MW absorbing performance and thermal conductivity than Cu/ATP, and the MW activation made it possible to efficiently degrade the PNP. Hu et al. have proposed a MW treatment of water and groundwater to remove atrazine using zeolites and other mesoporous materials. Adsorption of atrazine takes place either in zeolites (CBV-720 and 4A), transition metal–exchanged dealuminated Y zeolites, mesoporous silica MCM-41, quartz sand or diatomite, and the degradation is produced by MW irradiation [51–54]. Zeolites (particularly ZSM-5) were also used in combination with MW for the removal of N-nitrosodimethylamine (NDMA) and its precursor, dimethylamine (DMA), in drinking water [54]. Microporous mineral sorption coupled with MW-induced degradation is able to simultaneously remove NDMA and DMA and achieve a complete mineralization of these pollutants.

Moreover, several studies have used heterogeneous catalysis and MW for the degradation of tetracyclines, which are persistent in the environment due to their poor absorption by animals and human beings and low degradation rate. Lv and coworkers have proposed the use of MW together with birnessite (manganese oxide) for achieving high percentages of degradation for tetracycline [55, 56]. Also, Wang and coworkers development a method for the degradation of MB in wastewater using manganese oxide under MW irradiation. The removal efficiency of MB reached 99.5% in 10 min. Oxidative capacity of manganese oxide in this study was directly related with the concentration of H$^+$ [57].

Other studies have developed combinations of MW and different catalysts for the degradation of crystal violet (CV) due to the potentially harmful effects of this dye on aquatic life and human health. Yin and coworkers have used mixtures of copper and cerium oxides supported with activated carbon and the results showed a high catalytic performance and degradation rate with short irradiation time [20]. Also, different nanosized magnetic catalysts based on barium, calcium and copper ferrites were used. MW-induced holes could be responsible for the efficient degradation of CV (more than 90%) [58–61]. Another catalyst used to degrade CV consisted of nanosized nickel oxide, which seemed to be a higher valence state nickel dioxide with a hydroxyl group and active oxygen in its structure. About 97% CV (81% total organic carbon (TOC) removal) was degraded in 5 min [62].

Table 3.2 shows the recent researches that use microwave combined with different heterogeneous catalysts.

## 3.2.2    MW AND FENTON OR FENTIN-LIKE REACTIONS

Fenton/Fenton-like processes are well-known reaction systems used for wastewater treatment, which involve transition metal ions (such as Fe, Cu or Mn among others) that act as catalysts and H$_2$O$_2$ that serves as an oxidant. While the Fenton process involves Fe ions, other metals give similar results and these reactions are usually

**TABLE 3.2**
**Overview of Recent Research Related to MW and Heterogeneous Catalysis**

| Analite | Catalyst | Oxidant | pH | Microwave Power Temperature (W) | Reaction Time (min) | Microwave | Ref. |
|---|---|---|---|---|---|---|---|
| 4-nitrophenol | Manganese ferrite | Peroxymonosulfate | 5–9 | 500 | 2 | Modified domestic microwave oven (2.45 GHz) | [10] |
| Crystal violet | CuO@AC and CeO$_2$-CuO@AC | | | 400 | 5 | Microwave generator (CEM Mars-5, USA) | [20] |
| 4-nitrophenol | Carbon-supported copper | | | 400 | 16 | Modified domestic microwave oven (2.45 GHz) | [21] |
| 4-chlorophenol | Nickel oxide | O$_2$ | 7 | 40°C | 25 | Thermostated static microwave apparatus (CEM. Discover, USA; 2.45 GHz; 300 W) | [34] |
| 4-chlorophenol | Nickel oxide | O$_2$ | 7 | 70°C | 3 | Thermostatic microwave apparatus (CEM, Discover, USA; 2.45 GHz; 300 W) | [35] |
| 4-chlorophenol | Nickel oxide | O$_2$ | 7 | 40°C | 20 | Thermostated static microwave apparatus (CEM. Discover, USA; 2.45 GHz; 300 W) | [36] |
| 4-nitrophenol | Nickel oxide | | 0 | 40°C | 15 | Thermostated static microwave apparatus (CEM. Discover, USA; 2.45 GHz; 300 W) | [37] |
| Chlorobenzene | Discarded Pt-Pd-Rh catalyst | | | 250 W | 5 | | [38] |
| Chlorobenzene | Zerovalent Iron Particles (ZVI) | | | 250 W | 2.5 | Modified household MW oven | [39] |

*(Continued)*

**TABLE 3.2 (CONTINUED)**
**Overview of Recent Research Related to MW and Heterogeneous Catalysis**

| Analite | Catalyst | Oxidant | pH | Microwave Power Temperature (W) | Reaction Time (min) | Microwave | Ref. |
|---|---|---|---|---|---|---|---|
| 4-nitrophenol | Granular active carbon | | | 500 W | 8 | Modified domestic microwave oven (2.45 GHz) | [40] |
| 4-nitrophenol and pentachlorophenol | Granular activated carbon-supported platinum | | | 400 W | 16 | Modified domestic microwave oven (2.45 GHz) | [41] |
| 4-nitrophenol | $Mn_2O_3$/AC | | | 400 W | 5 | Microwave generator (CEM Mars-5, USA) | [42] |
| Congo red | AC | | 8 | 800 W | 1.5 | Controllable microwave oven (G8023ESL–V8, Guangdong Galanz Company, China; 2.45 GHz) | [43] |
| Methyl orange | Biochar and AC | $H_2O_2$ | 2.5 | 750 W | 10 | Microwave oven (Ethos Synth, USA; 2.45 GHz) | [44] |
| Methyl orange | GAC | | | 539 W | 8 | Modified microwave apparatus (MM721AC8-PW, Meide household electrical appliances group Co., Ltd., China; 2.45 GHz; 700 W) | [45] |
| Bisphenol A | AC | | | 800 W | 5 | Controllable MW oven (G8023ESL–V8, Guangdong Galanz Company, China) | [46] |
| Methyl orange, methyl parathion, sodium dodecyl benzene sulfonate, bisphenol A and methylene blue | Carbon nanotubes | | 6 | 450 W | 7 | Controllable MW oven (WD750B, Galanz Company, China) | [47] |

*(Continued)*

**TABLE 3.2 (CONTINUED)**
**Overview of Recent Research Related to MW and Heterogeneous Catalysis**

| Analite | Catalyst | Oxidant | pH | Microwave Power Temperature (W) | Reaction Time (min) | Microwave | Ref. |
|---|---|---|---|---|---|---|---|
| 4-nitrophenol | $MgFe_2O_4$ | | | 750 W | 10 | Controllable MW oven (WD750B, Guangdong Galanz Company, China) | [48] |
| 4-nitrophenol | $\alpha$-$Bi_2O_3$ | | 6 | 600 W | 5 | Microwave apparatus (Shanghai Yiyao Technology Co., Ltd. China; 2.45 GHz) | [49] |
| 4-nitrophenol | Copper-doped attapulgite (Cu/ATP) and silicon carbide (Cu/SiC) | | 7 | 750 W | 5 | Controllable MW oven (WD750B, Guangdong Galanz Company, China) | [50] |
| Atrazine | Copper and iron-exchanged zeolites | | | 700 W | 10 | MARS system (CEM, U.S.) | [53] |
| Tetracycline | Fe-doped birnessite | | | 500 W | 30 | | [55] |
| Tetracycline | Manganese dioxides (birnessite) | | 1 | 400 W | 30 | Microwave reactor (Galanz, P70D20TPC6) | [56] |
| Methylene blue | Manganese oxide | | 1 | 700 W | 10 | Tangshan nano microwave thermal instrument manufacturing Co., Ltd. | [57] |
| Crystal violet | Nano-$BaFe_2O_4$ | | 5.4 | 500 W | 10 | Controllable XH100B MW oven (Beijing XiangHu Ltd., China) | [58] |
| Crystal violet | $CaFe_2O_4$ | | 5.8 | 700 W | 10 | Temperature-controllable microwave oven (XH100B, Beijing XiangHu Ltd. China) | [59] |

*(Continued)*

**TABLE 3.2 (CONTINUED)**
**Overview of Recent Research Related to MW and Heterogeneous Catalysis**

| Analite | Catalyst | Oxidant | pH | Microwave Power Temperature (W) | Reaction Time (min) | Microwave | Ref. |
|---|---|---|---|---|---|---|---|
| Crystal violet | $CuFe_2O_4$ | | 9 | | 30 | Modified domestic microwave oven (Huiyan Microwave System Engineering, Nanjing, Jiangsu, China). | [60] |
| Reactive yellow 3 | $CuFe_2O_4/PAC$ | $H_2O_2$ | 3–11 | 800 W | 5 | Microwave oven reactor modified by Huiyan Microwave System Engineering20 (Nanjing, Jiangsu, China) | [61] |
| Crystal violet | Nano-nickel dioxide | | 9 | 750 W | 5 | MW oven (Midea Company; 900 W), | [62] |
| Tetracycline | Manganese dioxides (birnessite) | | 3 | 600 W | 30 | Microwave reactor (Galanz, P70D20TPC6) | [63] |
| Corn starch and crystalline celluloses | Polyoxometalate (POM) clusters | | | 220°C | 13–14 | START D multimode microwave oven (Milestone Inc., Shelton, CT, USA; 2.45 GHz; 1 kW) | [64] |
| Aniline and nitrophenol | $Fe_2O_3$ and CuO | $H_2O_2$ | 3 | 800 W | 50 | Microwave reactor (Nanjing Yongan Architecture Company; 2.45 GHz; 1000 W). | [65] |
| Brilliant green | $CoFe_2O_4$ | | 6–10 | 600 W | 2 | NJL07-3 model MW apparatus (Jiequan Equipment Ltd., China) | [66] |
| Malachite green | $ZnFe_2O_4$ | | 6–10 | 500 W | 2 | NJL07-3 mcdel MW apparatus (Jiequan ecuipment Ltd. China) | [67] |
| Remazol golden yellow | $CuOn–La_2O_3/-Al_2O_3$ | $ClO_2$ | 7 | 400 W | 1.5 | Reconstructive commercial microwave oven | [68] |

*(Continued)*

**TABLE 3.2 (CONTINUED)**
**Overview of Recent Research Related to MW and Heterogeneous Catalysis**

| Analite | Catalyst | Oxidant | pH | Microwave Power (W)/Temperature (W) | Reaction Time (min) | Microwave | Ref. |
|---|---|---|---|---|---|---|---|
| Apocynol | Co(salen)/SBA-15 | $H_2O_2$ | | 300 W | 40 | CEM discover microwave reactor | [69] |
| Phenol | $CuO_x/Al_2O_3$ | $ClO_2$ | 9 | 50 W | 5 | Reconstructive commercial microwave oven (2.45 GHz) | [70] |
| Pentachlorophencl | AC | $O_2$ | | 800 W | 60 | Modified MW furnace (Whirlpool Model T120, China; 1000 W) | [71] |
| Acid orange 7 | Polyaniline | $O_2$ | | 30°C | 10 | Ladd Research Microwave oven model LBP-250, USA | [72] |
| Brilliant green | $NiFe_2O_4$ | | 7 | 500 W | 2 | Microwave apparatus (NJL07-3, Jiequan equipment Ltd, China; 2.45 GHz; 900 W) | [73] |
| Sodium dodecyl benzene sulfonate | Activated carbon powder/ferreous sulfate | | 6 | 750 W | 3.8 | Controllable microwave oven (WD750B, Guangdong Galanz Company, China; 2.45 GHz) | [74] |
| Acid orange 7 | AC | Persulfate | | 800 W | 3 | Modified domestic MW furnace | [75] |
| Carbofuran | GAC/ZVI | $H_2O_2$ | 10 | 750 W/80°C | 5 | Modified-MW reactor (2.45 GHz; 750 W) | [76] |

called Fenton-like processes. Typically, Fenton and Fenton-like processes consist of the formation of hydroxyl radicals (·OH) through the reaction between transition metal ions and $H_2O_2$, in acidic medium. The generated ·OH are extremely reactive and non-selective and can react with a large range of organic pollutants leading to their degradation or even mineralization [77–80]. Several studies have reported the combination of Fenton and Fenton-like processes and MW, which enhance the oxidation capacity of the system. In this sense, the localized MW superheating seems to be more efficient for the generation of ·OH from $H_2O_2$ [81]. The selection below concerns several articles that combine MW and Fenton processes. Xi and Shi [82] oxidized thiocyanate ($SCN^-$) by a MW-enhanced Fenton method in industrial wastewater. The increase of the MW power swells the number of ·OH and, therefore, the removal of $SCN^-$ that can reach values close to 95%. Also, the degradation of PNP was studied in a continuous flow system [77]. The method with MW irradiation was more efficient than conventional Fenton, with a degradation ratio of about 93%. The same group of researchers have proved a combination of Cu(II) and Fe(II) for the removal of 3-nitroaniline in water [79]. They state there is a synergic effect of copper in the Fenton process. 92% of the 3-nitroaniline was removed under the optimal conditions used. In another study, Wang and coworkers [77] presented a similar continuous pilot-scale system for the oxidation of PNP with a Fenton oxidation process. They found the amount of ·OH generated when they applied a MW/Fenton process was 2.8 times higher than that in a classical Fenton reaction. The PNP removal reaches values of around 93%. The PNP was also degraded by the catalytic action of $CuO/Al_2O_3$, the addition of $H_2O_2$ and the application of MW energy in a Fenton-like reaction [83]. Degradation of PNP was hardly observed when only MW was applied (about 15%), whereas the combination with the catalyst resulted in a degradation of around 90%.

On the other hand, Lin et al. [84] removed Cu–EDTA and Ni–EDTA complexes by Fenton oxidation under MW irradiation with hydroxide precipitation. EDTA complexes are very soluble and can reduce the efficiency of metal removal by conventional chemical precipitation. In addition, Li and coworkers [80] developed a method for the degradation of BPA in wastewater using a Fenton process enhanced by microwave and $Mn^{2+}$ ions. They reported 99.7% removal efficiency and 53.1% TOC under irradiation in a domestic MW. Among the different degradation processes, the authors conclude that a better efficiency was obtained with a MW-Mn Fenton process.

Other authors have investigated the effect of MW on Fenton-type reactions using solid reagents. In this sense, Cravotto and coworkers have studied the mechanism of Fenton reactions in the degradation of 2,4-dichlorophenoxyacetic acid, a widely-used herbicide [85]. The authors reported a very efficient degradation of the chlorophenol derivative under solvent-free conditions with solid Fenton-like reagents.

Cai et al. [12] proposed the degradation of the azo dye Orange G using a microwave-enhanced Fenton-like reaction with $CuFeO_2$ (delafossite). The degradation using the solid Fenton-type reagent reached almost 100% removal within 30 min. Also, Xu and coworkers [86] proposed a catalyst based on ferrihydrite supported with activated carbon for the removal of methyl orange in a Fenton-like reaction under MW irradiation. The removal efficiency of the system with ferrihydrite/activated carbon, $H_2O_2$ and MW reached 99.1%. The optimal degradation occurred at a pH range of 2.0–4.0 and the degradation efficiency was reduced as the pH increased.

Xu and coworkers [87] used GAC, $H_2O_2$ and $Fe^{2+}$-EDTA complex under MW irradiation for the treatment of old-age landfill leachate. A decrease of chemical oxygen demand (COD) and a significant increase of the ratio of 5-day biochemical oxygen demand ($BOD_5$) to chemical oxygen demand ($BOD_5/COD$) was obtained under the optimal conditions of the experiment.

Table 3.3 summarizes the works that improve the Fenton or Fenton-like methods with the use of microwaves.

### 3.2.3 MW AND DEGRADATION IN AQUEOUS SOLUTIONS

Several articles were found in the literature that investigate the effect of MW irradiation on the degradation of pollutants using different types of oxidants in aqueous solution. Most of the studies involve hydrogen peroxide alone or together with other oxidation agents. Ouyang and coworkers degraded lignin in the effluents of pulp and paper mill using $H_2O_2$ and MW irradiation [91]. These kinds of effluents have high COD and TOC mainly due to their large amount of lignin and its derivatives. The authors found that MW irradiation is able to achieve homogeneity in the temperature that favours the digestion of lignin, which accelerates the process in comparison with conventional heating. Under the optimal conditions, the reduction of TOC and discoloration of the effluent were accomplished and the lignin was converted into carbon dioxide, water, carbonates and little amounts of aliphatic compounds of low molecular weight. Oncu and Balcioglu [92] proposed a treatment of biological waste sludge using $H_2O_2$, persulfate and MW. The degradation was applied to two antibiotics, oxytetracycline and ciprofloxacin. The degree of antibiotic degradation strongly depends on the amounts of the oxidants. On the other hand, when only MW irradiation was used for 30 min, lower degradation of the antibiotics was obtained. Thus, the combination of the oxidants and MW irradiation has a synergic effect on the antibiotics' degradation. Hydrogen peroxide was also combined with mineral acids such as nitric and sulfuric. Chang et al. have used $H_2O_2$, $HNO_3$ and MW for the degradation of nonylphenol polyethoxylates in wastewater, a common endocrine-disrupting compound that is hazardous to the bio-environment [93]. They inform that over 95% of this pollutant in the sludge can be degraded in 3 min. The same authors have degraded several organic compounds using a combination of $H_2O_2$, $HNO_3$, $H_2SO_4$ and MW [94]. In 60 min of treatment, the maximum degradation efficiencies of polychlorinated dibenzo-p-dioxins/dibenzofurans, nonylphenol and BPA were up to 85.1%, 99.5% and 99.3%, respectively. However, the presence of chlorine in the organic pollutants makes them more difficult to destroy than nonchloride-containing aromatic compounds.

Another oxidant used in combination with MW is persulfate. Peng and coworkers [95] have treated soils with a high concentration of phenanthrene using persulfate activated by MW irradiation. Whereas only minor removal of phenanthrene was observed with MW alone after 5 min at 80°C, the degradation of phenanthrene reached 96% after 5 min and more than 99% after 30 min when persulfate is present. The authors also found the existence of a non-thermal effect in the MW method when

**TABLE 3.3**
**Summary of Published Works Combining MW Irradiation with Fenton or Fenton-like Reactions**

| Analite | Catalyst | Oxidant | pH | Microwave Power Temperature | Reaction Time (min) | Microwave | Ref |
|---|---|---|---|---|---|---|---|
| Orange G | $CuFeO_2$ | $H_2O_2$ | 5 | 420 W | 15 | Thermostated static MW apparatus (CEM. Discover, USA; 2.45 GHz; 1000 W) | [12] |
| Pharmaceutical wastewater | $Fe_2(SO_4)_3$ | $H_2O_2$ | 4.42 | 300 W | 6 | Ordinary family microwave oven (SANYO800) | [18] |
| 4-nitrophenol | Fe | $H_2O_2$ | 3.3 | 850 W | 12 | Custom-made pilot-scale microwave oven (NJY2-1; 2.45 GHz; 2000 W) | [77] |
| Bisphenol A | Mn | $H_2O_2$ | 4 | 300 W | 6 | Domestic microwave oven (Cooplex-E, PreeKem Scientific Instruments Co., Ltd, China) | [80] |
| 2,4-dibromophenol | $FeSO_4$ | $H_2O_2$ | 1.7–2 | 750 W/80°C | 360 | Modified domestic MW oven | [81] |
| KSCN | $FeSO_4$ | $H_2O_2$ | 3 | 900 W | 7 | MW reactor (University of Science and Technology Liaoning; 2.45 GHz; 900 W) | [82] |
| 4-nitrophenol | $CuO/Al_2O_3$ | $H_2O_2$ | 6 | 100 W | 6 | Chemical reaction microwave oven (EXCEL, 2.45 GHz) | [83] |
| Cu-EDTA and Ni-EDTA | $FeSO_4$ | $H_2O_2$ | 2.5 | 240 W | 20 | Microwave chemical reactor (MCR-3, Henan, China) | [84] |
| Methyl orange | Ferrihydrite/AC | $H_2O_2$ | 4 | 700 W | 4 | House microwave apparatus (Meide Household Electrical Appliances Group Co., Ltd, China) | [86] |

*(Continued)*

**TABLE 3.3 (CONTINUED)**
**Summary of Published Works Combining MW Irradiation with Fenton or Fenton-like Reactions**

| Analite | Catalyst | Oxidant | pH | Microwave Power Temperature | Reaction Time (min) | Microwave | Ref |
|---|---|---|---|---|---|---|---|
| Landfill leachate | Granular active carbon, $Fe^{2+}$-EDTA | $H_2O_2$ | 7.9 | 300 W | 8 | Commercial MW oven (Guangdong Galanz Group Co., Ltd., Foshan City, China; 2.45 GHz; 750 W) | [87] |
| Methylene blue | $FeSO_4$ | $H_2O_2$ | 3 | 700 W | 3 | WP700 domestic microwave oven (Galanz, China; 2.45 GHz; 700 W) | [88] |
| 4-chloronaphthol 2,4-dichlorophenoxyacetic acid, p-nonylphenol and 2,4-dibromophenol | | Urea with hydrogen peroxide or sodium percarbonate | | 700 W | 15 | Professional MW multimode oven (Milestone MycroSYNTH) | [89] |
| Methyl orange | $Fe_3O_4$ @ MIL-100 (Fe)-$OSO_3H$ | $H_2O_2$ | 5 | 500 W | 6 | Microwave oven (WD750B, Guangdong Galanz Company, China; 2.45 GHz; 700 W) | [90] |

compared with degradation in a water bath, which accelerated the decomposition of persulfate and contributed to the oxidation of the soil-sorbed organic contaminants. Persulfate and MW were also used by Chou et al. [96], who proposed a method for treating landfill leachate. Higher removal of TOC, color and UV absorption at 254 nm were achieved. For their part, Kim and Ahn [97] also irradiated landfill leachate with MW for the activation of persulfate solution and removal of organic material. They found that the degradation was pH-dependent and showed better results in acidic conditions.

On the other hand, Qi et al. have investigated a new method for removing BPA by the activation of peroxymonosulfate by MW irradiation [98]. The authors reported that peroxymonosulfate is more efficient than $H_2O_2$ in the degradation of BPA because it generates both $\cdot OH$ and $\cdot SO_4$ radicals. They also found that the degradation was enhanced at higher pH values. Moreover, when the method was applied to real water samples, the degradation increased, probably due to the effect of chlorine ions, which have the ability to activate peroxymonosulfate.

The works to degrade contaminants using homogeneous catalysts or oxidants are presented in Table 3.4.

### 3.2.4 MW AND PHOTOCHEMICAL PROCESSES

The AOPs involving photochemical reactions are perhaps the most efficient oxidation processes. They typically use UV radiation as an external source of energy and are combined with hydrogen peroxide, Fenton's reactions, ozone, catalyst, etc. [112].

Photodegradation reactions are produced by the breakdown of chemical bonds when the energy of incident photons is absorbed. This energy is transferred to electrons in the molecule, which changes its configuration (i.e., promotes the electrons from a ground state to an excited state). This fact also implies changes in the structure of the molecule, such as an increase in the interatomic distances or alterations in the molecular symmetry [113]. These excited-state molecules are more reactive than the corresponding ground states. They can deliver the excitation to other molecules around them and, as a result, provide chemical transformations [114]. The degradation by UV radiation can be produced by three mechanisms: the direct photoionization, the energy transfer from the excited molecule triplet state to molecular oxygen and the charge or electron transferred from an excited molecule singlet or triplet state to molecular oxygen. However, some investigations suggest that the last two are the dominant pathways in the process of degradation [115].

Photodegradation can also be mediated by the presence of a semiconductor material. In this photocatalytic degradation, the electrons in the valence band of the semiconductor could be excited to the conduction band. In this way, photo-induced electrons ($e^-$) and positive holes ($h^+$) are produced [2]. The holes can react with surface adsorbed $H_2O$ to produce $\cdot OH$, whereas the electrons can lead to $O_2^{-}\cdot$ when reacting with molecular oxygen. These reactive species are responsible for the degradation of several recalcitrant organic pollutants [2, 114].

UV-Vis radiation could also be used to favour the production of hydroxyl radicals in the so-called Fenton reactions. In this kind of homogeneous catalytic reaction, the $Fe^{2+}$ ions induce the formation of hydroxyl radicals from $H_2O_2$. Under UV radiation, the

**TABLE 3.4**
**Summary of Recently Reported on MW and Homogeneous Catalysis**

| Analite | Catalyst | Oxidant | pH | Microwave Power Temperature | Reaction Time (min) | Microwave | Ref |
|---|---|---|---|---|---|---|---|
| Lignin | | $H_2O_2$ | 5.5 | 300 W | 120 | Ethos-1 microwave digestion system (Milestone Inc., Sorisole, Italy) | [91] |
| Oxytetracycline (OTC) and ciprofloxacin (CIP) | | $H_2O_2$ or $S_2O_8^{2-}$ | | 120–160°C | 15 | Bench-scale microwave irradiation system (Berghof, Speedwave MWS-3; 2.45 GHz) | [92] |
| Nonylphenol | | $HNO_3$ and $H_2O_2$ | | 150°C | 3 | Upgraded Microwave Digestion System (MDS, MARS-Xpress/230/60, CEM Corporation, USA) | [93] |
| Nonylphenol, bisphenol A and polychlorinated dibenzo-p-dioxirs/dibenzofurans | $HNO_3/H_2SO_4$ | $H_2O_2$ | | 177°C | 60 | Upgraded Microwave Digestion System (MDS, MARSXpress/230/60, CEM Corporation) | [94] |
| Phenanthrene | | Persulfate | | 150 W/80°C | 5 | Microwave synthesizer (Model XH-MC-1, Xianghu Science and Technology Development Co., LTD., Beijing, China) | [95] |
| Landfill leachate | | Persulfate | 7 | 550 W/85°C | 30 | Milestone Terminal 320 (Milestone, Ethos Touch Control, USA) | [96] |
| Landfill leachate | | Persulfate | 3 | 1600 W/90°C | 10 | Microwave-Accelerated Reaction System (MARS, CEM Corporation, Matthews, North Carolina) | [97] |

*(Continued)*

**TABLE 3.4 (CONTINUED)**
**Summary of Recently Reported on MW and Homogeneous Catalysis**

| Analite | Catalyst | Oxidant | pH | Microwave Power Temperature | Reaction Time (min) | Microwave | Ref |
|---|---|---|---|---|---|---|---|
| Bisphenol A | | Peroxymonosulfate | 2.45 | 500 W/80°C | 60 | Commercial MW reactor (XH-100A, Beijing Xianghu Science and Technology Development Co., Ltd; 1000 W) | [98] |
| 4-nitrophenol | Fe/EDTA | | 3 | 400 W | 12 | Modified household microwave oven (2.45 GHz; 750 W) | [99] |
| Pentachlorophenol | Nanoscale Fe$^0$ or AC | | | 700 W | 0.5 | Household microwave oven (2.45 GHz; 750 W) | [100] |
| Alizarin red | poly(1-naphthylamine)/ZnO nanohybrids | | | 30°C | 40 | Ladd Research Microwave oven (model LBP-250, USA) | [101] |
| Hyaluronic acid | Ascorbic acid | $H_2O_2$ | 4 | 60°C | 30 | | [102] |
| RhodamineB | | $H_2O_2$ | 12 | 700 W | 10 | Microwave oven (Midea.Co.WuHan; 2.45 GHz; 700 W) | [103] |
| Sewage sludge | | $H_2O_2$ | 6–7 | >80°C | 7–10 | Closed vessel microwave-accelerated reaction system (MARS-5; CEM Corporation; 2.45 GHz; 1200 W, 260°C and 33 bars) | [104] |
| Phenol | | $H_2O_2$ | | 668 W | 1 | | [105] |

*(Continued)*

## TABLE 3.4 (CONTINUED)
## Summary of Recently Reported on MW and Homogeneous Catalysis

| Analite | Catalyst | Oxidant | pH | Microwave Power/Temperature | Reaction Time (min) | Microwave | Ref |
|---|---|---|---|---|---|---|---|
| 1,5-naphthalenedisulfonic acid | | $H_2O_2$ | 3 | 300 or 1200 W | 20 | Discover (CEM, Matthews, NC, USA) and an Ethos (Milestone, Sorisole, BG, Italy) | [106] |
| Carbofuran | | $H_2O_2$ | 6 | 750 W | 0.5 | Modified-MW system (2.45 GHz; 1000 W) | [107] |
| Pentachlorophenol | | $H_2O_2$ | 11 | 600 W | 60 | Modified domestic microwave oven | [108] |
| Nitrobenzene | | $H_2O_2$ | | 300 W/50°C | 30 | Modified domestic microwave oven (2.45 GHz) | [109] |
| Malachite green | | | | 900 W | | Midea microwave oven (2.45 GHz; 900 W) | [110] |
| Perfluorocarboxylic acids | | Persulfate | 2 | 70 W/90°C | 240 | Microwave digestion system (Milestone, Ethos Touch Control, USA) | [111] |

Fe(III) generated in the reaction is photoreduced to Fe(II), which can react again with $H_2O_2$, establishing a cycle with the additional production of hydroxyl radicals [116].

The increase in temperature can improve the efficiency of the photodegradation. In this perspective, several authors have reported better efficiency of the photochemical reactions when the non-thermal effect of MW is combined with UV radiation [117, 118]. Other authors have also demonstrated that MW enhances the formation of ·OH during photochemical processes [119]. These advantages have been exploited for the successful treatment of numerous pollutants in wastewaters (see Table 3.5).

At the beginning of the 2000s, Serpone and coworkers started to use MW irradiation to carry out the degradation of several pollutants. Different strategies were used for the degradation of the cationic dye rhodamine-B (RhB). For instance, a combination of $TiO_2$ dispersions and MW was applied with the aid of UV radiation through an optical fiber cable [120, 121]. Since the metal electrodes of the traditional mercury lamps are destroyed under MW field, they proposed a new device involving an electrodeless double-quartz cylindrical plasma photoreactor (DQCPP) containing Hg gas and very small proportions of Ne. In this way, the lamp is powered by MW irradiation [122]. This device was also tested in combination with heterogeneous catalysis via $TiO_2$ solutions with better results in terms of degradation percentage [123]. The same approaches were used to perform the degradation of carboxylic acids, aldehydes, alkoxycarbonyl, MB, BPA and other phenolic substrates [11, 124–126]. Also, the research group investigated the photocatalytic degradation of the systemic herbicide 2,4-dichlorophenoxyacetic acid (2,4-D) using similar strategies and studied the non-thermal effects of MW [127] and the influence of different new discharge electrodeless lamps [128]. Later, they designed an ultraviolet vacuum quartz lamp with a tungsten trigger, which requires lower MW power for the autoignition of the electrodeless lamp. This new device was successfully applied to the degradation of BPA and 2,4-D [129]. They also proved that by using multiple little lamps—named MW discharge granulated electrodeless lamp (MDGEL)—instead of a single MW discharged electrodeless lamp (MDEL), it is possible to achieve a higher photodegradation efficiency due to a greater irradiation surface area [130].

Other authors also used heterogeneous photocatalysis applied to the degradation of different pollutants. Ki and coworkers [118] developed a method for the removal of 4-CP from wastewaters using MW and ultraviolet radiation with a TiO2 photocatalytic system. They also found that the addition of hydrogen peroxide, injection of oxygen and lowering of the pH level improve the performance of the photodegradation. Other researchers investigate the photocatalysis of atrazine, an herbicide used in corn and other crops, by a combination of $TiO_2$ nanotubes and MW [131]. They reported that the photodegradation is much faster than other photocatalytic methods and atrazine is completely degraded after 5 min with a mineralization of about 98.5% in 20 min. A similar approach was proposed by Karthikeyan and Gopalakrishnan [132], who degraded several phenols. The MW-UV method in the presence of $TiO_2$ has demonstrated to be faster than other degradation methods for all the model phenols tested and is efficient for TOC removal. Yang and coworkers have synthesized a novel catalyst based on $TiO_2$ modified with fluorine and silicon for the photocatalysis of pentachlorophenol with the aid of MW [133]. The positive results are mainly due

**TABLE 3.5**

**Overview Recently Reported on MW and UV Irradiation**

| Analite | Catalyst | Oxidant | pH | Microwave Power Temperature | Reaction Time (min) | Microwave | Ref |
|---|---|---|---|---|---|---|---|
| Bisphenol A | TiO$_2$ | | | 45 W | 60 | Shikoku Keisoku Reactor (SMW-087) system – super high-pressure 150-W mercury lamp | [11] |
| Atrazine | TiO$_2$/MWCNTs | | | | 5 | Microwave device (MG08S, Huiyan Microwave System Engineering Co., Ltd.) microwave discharge electrodeless lamp (Foshan Full Sun Lighting Co., Ltd.) | [73] |
| PAH | Tween 80+citric acid | | | 650 W | 34 | MW controllable oven (1 kW) – Rayonet photochemical reactor (model RPR-100, Southern N.E. Ultraviolet Co.) | [115] |
| 4-chlorophenol | TiO$_2$ | H$_2$O$_2$ | 7 | 400 W | 20 | Microwave generator (2.45 GHz) – microwave discharge electrodeless lamp | [118] |
| 4-chlorophenol, bisphenol A and methylene blue | TiO$_2$ | | | | 60 | Microwave generator (Panasonic M5801; 700 W) | [125] |
| Atrazine | TiO$_2$ | | 8.1 | | 20 | Microwave discharge electrodeless lamp system | [131] |
| Atrazine | | | | | 2 | Panasonic NN7856BK; 2.45 GHz; 1500 W) – mercury discharge lamp electrode and Hg-EDL (UMEX GmbH Dresden, Hg:254 nm, Germany) | [134] |
| Tartrazine | | H$_2$O$_2$ | 2.6 | 224.2 W | 24 | Domestic MW oven – ctrode-less discharge lamps (Fusion UV Systems Inc.-D model) | [136] |
| Atrazine | | H$_2$O$_2$ | 5–7 | 30 W | 20 | Home-made reactor with MW and UV (electrodeless lamp) | [137] |

*(Continued)*

**TABLE 3.5 (CONTINUED)**
**Overview Recently Reported on MW and UV Irradiation**

| Analite | Catalyst | Oxidant | pH | Microwave Power Temperature | Reaction Time (min) | Microwave | Ref |
|---|---|---|---|---|---|---|---|
| Chlorfenvinphos and cypermethrin | $Fe^{2+}$ | $HNO_3$ and $H_2O_2$ | <5 | 140°C | 4 | A focused-microwave oven (Star System 6, CEM, Matthews, NC, USA; 950 W) – Immersed electrodeless Cd discharge lamps (Florian and Knapp) | [138] |
| Ciprofloxacin | $FeSO_4$ | $O_2/H_2O_2$ | 3 | | 180 | Microwave discharge electrodeless lamp photoelectro-Fenton reactor | [139] |
| 4-Chloro-2-nitrophenol | $TiO_2$ | | 6 | 150 W | 100 | Low-pressure mercury vapor lamps | [140] |
| 4-Chloro-2-nitrophenol | $TiO_2$ | $H_2O_2$ | 6 | 136 W | 30 | Microwave oven (Murphy Richard – Make, Model-MWO 20 MS; 800 W) – UV lamps of power 4 W (Philips TUV 4 W/G4T5) and 8 W (Philips TUV 8 W/G8T5) | [141] |
| 2,4-dichlorophenoxyacetic acid | $TiO_2$ | $O_3$ | | 400 W | 180 | Microwave generator (2.45 GHz; 1 kW) – Yumex Korea 200 W mercury lamp | [142] |
| Chlorodifluoromethane | $TiO_2$ | | | 400 W | 500 mL/min | Microwave irradiation equipment (Korea microwave instrument Co., Ltd.; 2.45 GHz; 1 kW) – microwave discharge electrodeless mercury lamp | [143] |
| Malachite green | | $H_2O_2$ | 5–7 | | 5 | Microwave oven (BM-3010BB1, Sanyo) coupled with microwave generator (2.45 GHz; 0.9 kW) – Cylindrical electrodeless discharge lamps (Nanhai Company, China; Hg and Ar) | [144] |
| Alizarin green | $TiO_2$ | | | 700 W | 90 | Modified domestic microwave oven (Galanz Electric Co. Ltd.; 2.45 GHz) – microwave discharged electrodeless lamp (Hg and Ar) | [145] |

*(Continued)*

**TABLE 3.5 (CONTINUED)**
**Overview Recently Reported on MW and UV Irradiation**

| Analite | Catalyst | Oxidant | pH | Microwave Power Temperature | Reaction Time (min) | Microwave | Ref |
|---|---|---|---|---|---|---|---|
| Direct red-81 and bromothymol blue | $TiO_2$ | | 4–6 | 300–700 W | 105 | Cylindrical single-mode cavity CMPR 250 microwave reactor (Wavemat Processing System, MI; 2.45 GHz; 1250 W) – Low-pressure black-light fluorescent lamps | [146] |
| Reactive Brilliant Red X-3B | $TiO_2$ | | 6.48 | 700 W/28°C | 40 | Domestic microwave oven (Haier Co. Ltd.; 2.45 GHz; 700 W) – Microwave electrodeless UV lamp (Hg) | [147] |
| Crystal violet | $TiO_2$ | | | | 3 | Apparatus cavity (BM-3010BB1, Sanyo) with external microwave generator (2.45 GHz; 800 W) – Cylindrical electrodeless discharge lamp (Nanhai Co, China; Hg and Ar) | [148] |
| Rhodamine B | $TiO_2$ | $H_2O_2$ | | 400 W | 180 | Microwave generator (Korea microwave instrument Co. Ltd; 2.45 GHz; 1 kW) – UV light (Yumex Korea 200 W mercury lamp) | [149] |
| Pentachlorophenol | $TiO_2$ nanotubes | | 10.32 | | 12 | Microwave oven (Midea Company; 2.45 GHz; 900 W) - electrodeless discharge lamp (Hg) | [150] |
| Rhodamine B | $TiO_2$ | | 4 | 850 W | 75 | Domestic microwave oven (LG Co.; 2.45 GHz; 850 W) - microwave electrodeless UV lamp (Hg and Ar) | [151] |
| Nitrobenzene | $TiO_2$ | $H_2O_2$ | 7 | 500 W | 100 | Microwave generator (2.45 GHz; 1 W) and microwave cavity – electrodeless microwave lamp | [152] |

*(Continued)*

**TABLE 3.5 (CONTINUED)**
**Overview Recently Reported on MW and UV Irradiation**

| Analite | Catalyst | Oxidant | pH | Microwave Power Temperature | Reaction Time (min) | Microwave | Ref |
|---|---|---|---|---|---|---|---|
| Guaiacol | | $H_2O_2$ | 6.35 | 500 W/98°C | 10 | Microwave reactor (WBFY-205, Gongyi, China) – electrodeless discharge lamp (U-shaped Pyrex, Hg and Ar; Shanghai Jiguang Special Illumination Instrument Factory) | [153] |
| Bromothymol blue | Nano TiO$_2$ | O$_3$ or H$_2$O$_2$ | | 600 W | | Microwave generator (2.45 GHz; 1 kW) - microwave discharge electrodeless lamp (UV-C) | [154] |

to the strong capacity of absorption and the efficient hydroxyl radicals formation on the surface of the catalyst under MW and UV irradiation.

On the other hand, Moreira and coworkers have studied the photolytic degradation of atrazine using a UV (electrodeless discharge lamp) and MW [134, 135]. The application of MW-assisted photolytic processes showed high efficiency of atrazine degradation.

Parolin et al. have developed a method for the degradation of tartrazine using UV (electrodeless discharge lamp), MW and $H_2O_2$ [136]. Chen et al., for their part, used $H_2O_2$ with UV/MW irradiation for the degradation of atrazine [137] with good degradation rates.

## 3.3 LIQUID PHASE MICROWAVE-ASSISTED MICROEXTRACTION

Different approaches based on liquid-phase microextraction assisted by microwaves can be found in the literature with interesting strategies to accomplish the extraction of the target analytes from complex matrices. Among them, microwave-assisted dispersive liquid-liquid microextraction (MA-DLLME) is the most used technique for the extraction of different organic compounds. Microwave radiation has been also used to assist techniques such as microwave-assisted headspace single drop microextraction (MA-HS-SDME) and microwave-assisted dispersive liquid-liquid microextraction based on solidification of the floating organic droplet (MA-DLLME-SFOD) with satisfactory results over the conventional applications.

At the end of this section, other approaches using microwaves for liquid extraction are described, and the summary of the extraction conditions are shown in Table 3.6.

### 3.3.1 MICROWAVE-ASSISTED DISPERSIVE LIQUID-LIQUID MICROEXTRACTION

Dispersive liquid-liquid microextraction (DLLME) technique was developed by Rezaee et al. in 2006, and since then many applications have been proposed [155]. In the DLLME technique, an appropriate volume of a mixture of the extraction solvent, which is immiscible with water, and the disperser solvent, which is miscible in both water and the extraction solvent, is rapidly injected into an aqueous sample solution containing the target analytes. As a result, the extraction solvent is dispersed in the sample, forming a cloudy solution consisting of fine microdroplets. The analytes are extracted and separated from the bulk solution by centrifugation.

The density of the extraction solvent should be higher than water so that separation can occur. Traditionally, halogenated solvents, such as carbon tetrachloride, chloroform and tetrachloroethylene, are used as extractants. On the other hand, methanol, ethanol, acetone and acetonitrile are most frequently used as disperser solvents due to their solubility with both the aqueous phase and the extraction solvent [156].

Recently, ionic liquids (ILs) have been proposed as extraction solvents due to their unique physicochemical properties [157]. Their miscibility in water and organic solvents can be controlled by selecting the cation/anion combination. In addition, their physical state can be manipulated by controlling the temperature during extraction. Moreover, ILs present low toxicity and biodegradability, which make them a greener alternative compared with the conventional solvents.

**TABLE 3.6**
**Liquid Phase Microwave-assisted Microextraction Methods**

| Type of Microextraction | Analyte | Sample | Extraction Solvent | Disperser | Ionic Strength | Type of Microwave | Microwave Power (W) | Irradiation Time or Extraction Time/Temperature | Sample pH | Ref. |
|---|---|---|---|---|---|---|---|---|---|---|
| MA-DLLME | Bisphenol A, Tetrabromo-bisphenol A | Milk (10 mL) | [C$_8$MIM][PF$_6$] (0.150 mL) | n.i | 5% NaCl | MWO | 700 | 5.0 min | 3.0 | [158] |
| MA-DLLME | Anthraqui-nones | *Rheum Palmatum* L. (0.010 g of sample powder) | [C$_8$MIM][BF$_4$] (140 uL) | n.i | n.i | Not informed | 180 | 60 s | 2.0 | [159] |
| MA-DLLME | Sulfonamides | Animal oil (2.0 g) | [C$_3$MIM][BF$_4$] (0.200 mL, pH 11.0) | n.i | n.i | MW reaction system | 1600 | 5 min, 100°C | Not adjusted | [160] |
| MA-DLLME | Sulfonamides | River water (10 mL), honey (1 g), milk (10 mL) and animal plasma (10 mL) | [C$_6$MIM][PF$_6$] (100 uL) | Methanol, (0.75 mL) | 3% | Household MWO | 240 | 90 s | 3.5 | [161] |

*(Continued)*

**TABLE 3.6 (CONTINUED)**
**Liquid Phase Microwave-assisted Microextraction Methods**

| Type of Microextraction | Analyte | Sample | Extraction Solvent | Disperser | Ionic Strength | Type of Microwave | Microwave Power (W) | Irradiation Time or Extraction Time/Temperature | Sample pH | Ref. |
|---|---|---|---|---|---|---|---|---|---|---|
| MA-DLLME | Aminoglycosides | Milk (5.0 mL) | $[C_6MIM][PF_6]$ (60 µL) | Triton X-100 (100 µL) | Not added | Household MWO | 180 | 60 s | 8.0 | [162] |
| MA-DLLME | Formaldehyde | Beverages (2.5 mL) | $[C_6MIM][PF_6]$ (70 uL) | Acetonitrile (0.4 mL) | n.i | Household MWO | 120 | 90 s | 3 | [163] |
| MA-DLLME | Plasticizers | Water (3.0 mL) | $[BMIM][PF_6]$ (0.100 mL) | Methanol (0.4 mL) | Not added | CEM MW synthesis system | 40 | 3 min, 50°C | n.i | [164] |
| MA-DLLME | Pyrethroid pesticides | Honey (1.0 g), almond milk (1.0 g), assorted fruits (10.0 g), water | $[N_{888}][Tf_2N]$ (52 µL) | Methanol (208 µL) | n.i | CEM Discover MW | 200 | 60 s | 5.0 | [165] |
| MA-DLLME | Triazine herbicides | Juice samples (5.0 mL) | $[C_6MIM][PF_6]$ (60 µL) | [C2MIM][BF4] (80 µL) | n.i | Modified household MWO | 180 | 90 s | not adjusted | [166] |
| MA-DLLME | Triazine herbicides | Water (5 mL) | $[C_4MIM][BF_4]$ (40 uL) | LiNTf$_2$ (500 µL) | n.i | Discover SP MWO | 30 | 90 s, 50°C | n.i | [167] |

*(Continued)*

**TABLE 3.6 (CONTINUED)**
**Liquid Phase Microwave-assisted Microextraction Methods**

| Type of Microextraction | Analyte | Sample | Extraction Solvent | Disperser | Ionic Strength | Type of Microwave | Microwave Power (W) | Irradiation Time or Extraction Time/ Temperature | Sample pH | Ref. |
|---|---|---|---|---|---|---|---|---|---|---|
| MA-DLLME | Phenylurea and triazine herbicides | Milk (4 mL) | [C$_6$MIM][PF$_6$] (55 uL) | n.i | 7% NaCl | Household MWO | 240 | 7 min | n.i | [168] |
| MA-HS-SDME | δ-3-carene, α-pinene, camphene | Prangos uloptera (100 g) | n-heptadecane (2 μL) | n.i | n.i | Domestic MWO | 300 | 2 min | n.i | [170] |
| MA-HS-SDME | Chloroben-zenes | Water (30 mL) | [C$_6$MIM][PF$_6$] (5 uL) | n.i | n.i | Domestic MWO | 200 | 20 min | n.i | [171] |
| MA-DLLME-SFOD | Sulfonamides | Environmental water (2 mL) | [C$_2$MIM][PF$_6$] (0.16 g) | n.i | Na$_2$SO$_4$, 5% | Modified household MWO | 240 | 90 s | 4.0 | [173] |
| MA-DLLME-SFOD | Sudan dyes | Red wines (4.0 mL) | [C$_{12}$MIM]Br | n.i | 5% NaCl | Modified household MWO | 180 | 90 s | not adjusted | [174] |
| MA-DLLME-SFOD | Triazine herbicides | Honey (2.0 g) | 1-dodecanol (70 μL) | n.i | Not added | Modified household MWO | 300 | 40 s | 5.0 | [175] |

*(Continued)*

**TABLE 3.6 (CONTINUED)**

**Liquid Phase Microwave-assisted Microextraction Methods**

| Type of Microextraction | Analyte | Sample | Extraction Solvent | Disperser | Ionic Strength | Type of Microwave | Microwave Power (W) | Irradiation Time or Extraction Time/Temperature | Sample pH | Ref. |
|---|---|---|---|---|---|---|---|---|---|---|
| MAE-HF-L/SME | Pharmaceutical, personal care products | Fish (10 g) | 1:1 (v/v) 1-octanol/toluene (quantity not informed) | n.i | 25% NaCl | A MW synthesis/extraction work station (MAS-II) | 100 | n.i | 7.0 | [176] |
| MA-HS-LPME | Trihalomethanes, haloketones | Fish tissue, green alga (3 g) | Toluene (500 µL) | n.i | Not informed | Laboratory MW extraction system | n.i | n.i | n.i | [177] |
| MA-HS-LPME | Chlorophenols | Aqueous samples (10 mL) | 1-octanol (quantity not informed) | n.i | n.i | NE-V32A inverter system | 167 | 10 min | 1 | [178] |
| MA-HS-LPME | Dichlorodiphenyltrichloroethane | Aqueous samples (10 mL) | 1-octanol (4 uL) | n.i | n.i | Domestic MWO | 249 | 6.5 min | 6 | [179] |

n.i., not informed.

The use of alternative energies, such as microwaves, has been proposed to improve the extraction efficiency of the classic DLLME technique and, at the same time, reduce the extraction time. Microwave energy facilitates the dispersion of the extraction solvent in the aqueous solution and thus ensures the proper formation of the microdroplets. In addition, an increase in the temperature of the extraction is observed, enhancing the kinetics of the partition process. In addition to the mentioned advantages, it is possible to perform the extraction of the analytes from the sample matrix or processes of derivatization of the studied analytes in a single step.

Nevertheless, there are very few articles where the microextraction assisted by microwaves is performed in one step. Generally, microwave extractions or derivatizations are carried out separately from the DLLME procedure. First, the target compounds are extracted from the matrix, and then the microextraction is performed in order to preconcentrate the analytes and, thus, reach lower detection limits.

The microwave-assisted DLLME approaches that can be found in the literature are described below.

Kang et al. have proposed the extraction of BPA and tetrabromobisphenol-A from commercial milk samples [158]. Both analytes can be found in different packaging materials contaminating the foodstuffs. To perform the extraction, a domestic microwave oven was modified in order to add a water condenser at the top. 10.0 mL of milk sample was placed in an erlenmeyer glass flask and 150 μL of 1-octyl-3-methylimidazolium hexafluorophosphate ([C$_8$MIM][PF$_6$]) was added. Since the analytes are weakly acidic compounds, the pH was optimized between 1.5 and 6.0, taking into account their acidic constants, and pH 3.0 was selected as the optimum level. The ionic strength was also studied in order to verify if the addition of salt can improve the extraction efficiency by decreasing the solubility of analytes and, thus, enhancing the partitioning in the extraction phase. It was proved that 5% of NaCl increased the recoveries of the analytes. Hence, the mixture of the sample, IL and NaCl, adjusted to the optimum pH, was irradiated at the maximum microwave power (700 W) for 5 min. After extraction, the sample was cooled in an ice bath and centrifuged and the extract with the analytes was analyzed by HPLC with a diode array detector (DAD).

The method is rapid, simple and, as an important advantage, a previous treatment of the milk sample for eliminating proteins and fat was not required.

MA-DLLME was also applied to the extraction of anthraquinones from the roots of *Rheum palmatum* L., a Chinese herbal medicine [159]. In this method, the sample was directly used in the solid state. Thus, when the sample was irradiated, a solid-liquid extraction was first performed to extract the analytes from the sample matrix. Once the analytes were in the solution, the liquid-liquid extraction took place. To perform the extraction, 0.010 g of the sample was mixed with 3.0 mL of aqueous solution (pH 9.0) and 140 μL of 1-octyl-3-methylimidazolium tetrafluoroborate [C$_8$MIM][BF$_4$]. The sample was irradiated by microwave at 180 W for 60 s. After that, a small amount of ammonium hexafluorophosphate (NH$_4$PF$_6$) (1.0 mL) was used as an ion-pairing agent to obtain a cloudy solution due to the formation of [C$_8$MIM][PF$_6$] that was kept in an ice-water bath for 5 min. Then, the sample was centrifuged and, with the IL phase, which was deposited at the bottom of the tube, the solid sample and the aqueous phase were separated. The IL phase containing the extracted analytes was diluted with acetonitrile and injected in the HPLC-DAD.

The proposed method was compared with ultrasound-assisted extraction (UAE) and heating reflux extraction (HRE) in terms of sample amount, extraction solvent, volume of solvent and extraction time. As a result, much less sample and extraction solvent were consumed and shorter extraction time was needed, which made the proposed method an interesting alternative.

MA-DLLME using IL as extractant (IL-based MA-DLLME) was also applied for the determination of six sulfonamides in animal oils [160]. The sample was heated at 55°C to obtain a liquid sample. Then, 2 g were placed in an extraction vessel and 10.0 mL of NaOH containing 200 μL of 1-butyl-3-methylimidazolium tetrafluoroborate ([$C_4MIM$][$BF_4$]) were added. The pH of the sample was adjusted to 11.0, since at higher pH values the saponification and solidification of the animal oil can occur.

The sample was then placed in a microwave reaction system (MARS, CEM Co., Matthews, USA) and irradiated for 5 min at a 1600 W, reaching a final temperature of 100°C. After the extraction, the sample was centrifuged, turned into solid and removed from the aqueous phase. $NH_4PF_6$ was added as ion-pairing agent in a 1:5 ratio ([$C_4MIM$][$BF_4$]: $NH_4PF_6$). The addition of the $NH_4PF_6$ solution significantly increased the peak areas of the selected analytes. After centrifugation, the hydrophobic IL with the extracted analytes was taken, diluted with acetonitrile and analyzed by HPLC.

The authors affirmed that the proposed method was suitable for the extraction of sulfonamides from food with a high content of fat. Moreover, the method was simple to operate and a little quantity of solvent was consumed.

Another interesting approach for extracting sulfonamides from river water, honey, milk and animal plasma was proposed by Xu et al. [161]. For this purpose, the sample solution was located in a centrifuge tube, along with methanol (disperser), fluorescamine solution (derivatization reagent) and 1-hexyl-3-methylimidazolium hexafluorophosphate ([$C_6MIM$][$PF_6$]) (extraction solvent), and it was irradiated under a microwave at 240 W for 90 s. It was important that the sample solution was acidic (pH 3.5) to ensure the proper derivatization of the sulfonamides. The derivatives of the analytes were transferred to the IL phase, and the obtained cloudy solution was centrifuged for 10 min at 0°C. The IL phase was dissolved in acetonitrile and then filtrated to finally determine the analytes by a HPLC system equipped with a fluorescence detector. The proposed method was compared with other methods that include many steps with a complex extraction procedure and longer process times. The IL-based MA-DLLME was proved to be a convenient method, since its procedure involves just two short steps, including preparation of sample, extraction, derivatization and preconcentration. In addition, the authors affirmed that the extraction is efficiently improved by the use of microwave irradiation.

A MA-DLLME method with the addition of surfactant was developed to perform a one-step derivatization and extraction of aminoglycosides in milk samples [162]. The derivatization is performed in order to turn the analytes into fluorescent compounds. On the other hand, the surfactant accelerated the dispersion of the extraction solvent into the aqueous sample and decreased the time of analysis. Thus, a mixture of 100 μL of a 0.20 mmol $L^{-1}$ Triton X-100 solution, 60 μL of [$C_6MIM$][$PF_6$] ionic liquid, used as the extraction solvent, and 50 μL of 2.5 mmol $L^{-1}$ of 9-fluorenylmethyl chloroformate solution, used as derivatization reagent, was injected into 5.00 mL of the sample. Then, the sample was irradiated in a modified household microwave oven at 180 W for 60 s to achieve the derivatization and extraction of the analytes. The IL phase was separated from the bulk solution, diluted with acetonitrile and injected into a HPLC instrument. The variables of the extraction were exhaustively optimized, taking into account the highest extraction recoveries.

Xu et al. described a new method for the quantitative determination of form-aldehyde in beverages [163]. The sample was located in a glass centrifuge tube along with acetonitrile (disperser), $[C_6MIM][PF_6]$ (IL, extraction solvent) and 2,4-dinitrophenylhydrazine (DNPH) (derivatization reagent). The mixture was shaken and immediately placed in a modified domestic microwave oven to be irradi-ated under a microwave power of 120 W for 90 s. Formaldehyde was simultaneously derivatized with DNPH, extracted and preconcentrated into the droplet of IL. The solution was centrifuged and the IL phase was dissolved in acetonitrile and then filtrated in order to analyze the compounds by HPLC, with a monitoring wavelength for the formaldehyde-DNPH derivative set at 352 nm. The proposed method was compared with other methods reported in the literature such as colorimetric phase extraction and flow injection analysis with spectrophotometric detection among oth-ers. It was verified that when the microwave-assisted derivatization and IL-based DLLME was applied, the limit of detection was lower as the microwave irradiation improved the analytical performance. Moreover, the extraction, derivatization and preconcentration can be performed in a single step, reducing the number of organic solvents and the analysis time.

Wang et al. proposed an IL-based MA-DLLME method for the determination of plasticizers in mineral water, soda water and carbonated beverages [164]. For the microextraction, 1-butyl-3-methylimidazolium hexafluorophosphate ([BMIM] $[PF_6]$) was selected as extraction solvent due to its hydrophobic characteristics, which allowed a stratification after centrifugation and the achievement of largest peak areas. Thus, 3 mL of the spiked sample was mixed with 100 mL $[BMIM][PF_6]$ and 0.4 mL of methanol, used as a dispersing agent. In this case, the addition of salt did not improve the extraction efficiency. A CEM microwave synthesis system apparatus was used to irradiate the cloudy solution at a microwave power of 40 W for 3 min, reaching a temperature of 50°C. After that, the mixture was centrifuged and the extract was separated for HPLC analysis with UV detection. The optimiza-tion of the extraction variables was performed by the response surface methodology in order to evaluate the most relevant parameters affecting the extraction efficiency in a simultaneous way. The authors affirmed that the proposed method was reliable for the determination of phthalic acid ester plasticizing agents in water samples with recovery percentages between 85 and 105%.

Environmentally relevant pyrethroid pesticides (cypermethrine, permethrin and allerthrine) were successfully extracted from tap water (10.0 mL), honey (1.0 g), almond milk (1.0 g) and assorted fruits (10.0 g) by MA-DLLME [165]. 260 µL of a mixture of IL and methanol, used as extracting and dispersing solvents respectively, in a 24: 80 v/v (IL/MeOH) ratio, was injected in 10.0 mL of the sample solution and subjected to microwave radiation at 200 W for 60 s under continuous stirring. The pH value was adjusted to 5.0 since the analytes demonstrated to be pH-depen-dent. After extraction, the samples were centrifuged to separate the aqueous and IL phases, and the IL phase was removed for HPLC analysis. For selecting the optimal variables, the recovered HPLC peak areas were checked. Particularly, to optimize the IL most suitable for extraction, the effect of the container wall material (glass vs. polypropylene) was studied. Trioctylmethylammonium bis(trifluoromethylsulfonyl) imide ($[N_{8881}][Tf_2N]$) was chosen as IL, taking into account the recovery efficiencies for the three analytes when glass containers were used. The pyrethroid stability under microwave irradiation was also studied, proving that the extraction can be performed without degradation of the analytes.

Comparison of conventional DLLME and MA-DLLME under the optimal conditions was performed. For allerthrine and permethrine, slightly higher recovery values can be obtained when MA-DLLME was used. For cypermethrine, conventional DLLME had better results. In spite of that, the extraction using MA DLLME can be performed efficiently in 60 s while DLLME required 5 min.

Triazine herbicides were extracted from fruit juice samples using IL based MA-DLLME for the subsequent analysis by HPLC [166]. To accomplish the extraction, two different ILs, 60 µL of $[C_6MIM][PF_6]$ (hydrophobic) and 80 µL of 1-Ethyl-3-methylimidazolium tetrafluoroborate ($[C_2MIM][BF_4]$, hydrophilic), used as the extraction and dispersion solvents, respectively, were added to 5.0 mL of the sample, without pH adjustment. The sample was placed in a modified household microwave oven and irradiated under a microwave power of 180 W for 90 s. The microwave energy increased the surface contact between both ILs and, as a consequence, the extraction efficiency increased. After extraction, $[NH_4][PF_6]$ was added to precipitate residual $[C_6MIM]^+$ and $[C_4MIM]^+$ cations. Next, the sample was centrifuged and the IL phase was dissolved in acetonitrile before HPLC analysis. The authors compared the proposed method with other DLLME methods reported in the literature in terms of sample volume, type and volume of solvent extraction, type and volume of dispersive solvent and percentages of recovery, concluding that similar results can be obtained with the different methods. The principal advantage of this approach was the use of ILs as extraction and dispersion solvents, which reduced the volume of organic solvents used with the benefits of environmental protection.

Another proposal for determining triazine herbicides in water samples was presented by Wu et al. [167]. In this approach, a microwave tube was filled with the sample and 40 µL of $[C_4MIM][BF_4]$ was added. After mild shaking, 500 µL 0.2 g mL$^{-1}$ of lithium bis[(trifluoromethane)sulfonyl]imide (LiNTf$_2$) solution was injected into the sample and the tube was immediately placed into a Discover SP microwave extraction apparatus and irradiated under a microwave power of 30 W at a temperature of 50°C for 90 s. Then, the suspension was centrifuged at 5000 rpm for 6 min, and the IL phase (at the bottom of the tube) was analyzed by HPLC-UV. This method was compared with other methods, and it could be verified that the limits of detection, recoveries and the relative standard deviation (RSD) were acceptable. In addition, it can be considered eco-friendly as there was no volatile organic solvent used. Moreover, the sample volume used was smaller and the time of analysis was shorter.

The extraction of phenylurea and triazine herbicides from milk samples using IL based MA-DLLME was performed by Wu et al. [168]. For the extraction of the analytes a modified domestic microwave oven was used. The milk sample together with NaCl and the IL, $[C_6MIM][PF_6]$, were located into a polytetrafluoroethylene

(PTFE) tube, and the mixture was sonicated in an ultrasonic bath to disperse the IL. Then, the suspension was intermittently irradiated under a microwave power of 240 W for 16 min. During the extraction, it was necessary to maintain a shaking between two irradiation processes. The formed suspension was centrifuged and then filtered. The compounds were finally analyzed by HPLC, with a monitoring wavelength of 228 nm for propazine, prometryne, terbutryn and trietazine, and 245 nm for isoproturon, monolinuron and linuron. It was verified that the proposed method provides good recoveries and precision, and the proposed procedure was able to efficiently extract the analytes from milk samples, consuming very little organic solvents in the process.

### 3.3.2 Microwave-Assisted Headspace Single Drop Microextraction (MA-HS-SDME)

In headspace single drop microextraction (HS-SDME) technique, the volatile compounds are evaporated, extracted and preconcentrated in a microdrop of water-immiscible solvent, generated by a syringe at the headspace of the sample. After the extraction, the microdrop is retracted into the syringe with the target analytes and prepared for the corresponding analysis, generally by gas chromatography (GC) [169].

HS-SDME is an advantageous alternative to conventional liquid-liquid extraction due to the low amount of solvent used and, thus, the reduction of the waste generation. In addition, the memory effects are eliminated due to a new solvent microdrop used for each determination

Microwave irradiation can assist this technique by heating the sample solution and, thus, accelerating the sample-to-headspace step. To achieve the MA-HS-SDME, a modification of the microwave oven should be done in order to incorporate a condenser and syringe with the extraction solvent on the outside part of the apparatus.

Two interesting approaches that applied this technique are described below.

MA-HS-SDME technique has successfully been applied for the isolation, extraction and concentration of essential oils from *Prangos uloptera*, a perennial herb used as medicine [170]. For the microextraction, a microwave oven was modified by assembling a Claisen adaptor at the top of the oven, and a condenser and a microsyringe were assembled at the upper joints. The sample (3.0 g) was located in a round flask inside the microwave oven and 1 mL of water was added. On the other hand, 3 µL of a mixture of the extracting solvent (n-heptadecane) and an internal standard (n-hexadecane), in a ratio of 1:200 v/v, was suspended on the top of the flask to generate the microdrop. The sample was irradiated at a microwave power of 300 W for 2 min and the microdrop was exposed for 4 min. By this way, the volatile compounds of the sample were transferred to the headspace by microwave heating, and the analytes in the headspace were extracted and concentrated into the suspended microdrop. After the extraction, the microdrop was retracted into the microsyringe and injected into the GC–MS instrument. The principal variables of the proposed method were optimized using the peak area of the four major compounds.

The MA-HS-SDME method was compared with the conventional hydrodistillation method obtaining a good correlation between results.

Another interesting approach was proposed by Vidal et al. in which a single step and *in situ* sample pretreatment for the quantitative determination of chlorobenzenes in water is involved [171]. For this purpose, the sample was irradiated with a modified household microwave oven coupled to a headspace single-drop microextraction system, under a power of 200 W that was set to be made by cycles of 30 s of irradiation followed by 60 s off, so the headspace temperature would not increase. Once the irradiation time was over, the analytes were extracted in a 5 μL microdrop of $[C_6MIM][PF_6]$ located in a microsyringe that was exposed to the headspace of the sample, situated outside the microwave oven. After extraction, the IL drop was injected into a HPLC-DAD system for the analysis of the compounds.

### 3.3.3 MICROWAVE-ASSISTED DISPERSIVE LIQUID-LIQUID MICROEXTRACTION BASED ON SOLIDIFICATION OF THE FLOATING ORGANIC DROPLET (MA-DLLME-SFOD)

As was previously mentioned, in the classical DLLME technique the density of the solvents used for extraction should be higher than water in order to remain at the bottom of the extraction tube after centrifugation. In general, the solvents that meet this requirement are highly toxic halogenated solvents. On the contrary, if the extraction solvent has a lower density than water, it will float on the surface of the aqueous solution, complicating the separation between the two phases. Thus, a simple and effective strategy is to select solvents with low melting and freezing points so that the solvent can be easily melted and solidified to be dispersed and separated from the bulk solution, respectively [172].

In the DLLME-SFOD approach, the solvent must be in the liquid state, the addition of a disperser is required and the dispersion is performed by manual or mechanical shaking. Microwave irradiation can accelerate this process by melting the solvent and dispersing it into the sample solution, and thus making the entire procedure simpler and faster compared to the conventional one.

The articles found in the literature that use microwave assistance for this technique are described below.

Song et al. presented an interesting approach to MA-DLLME-SFOD, to perform the extraction of sulfonamides from environmental water samples [173]. Here, the solvent selected for the extraction was an IL with a melting point between 40 and 100°C and solid at room temperature. So, the IL can be melted and dispersed in the sample solution under microwave irradiation. After the extraction, the IL with the extracted analytes was solidified at low temperature and then collected to perform the analysis. As an advantage, a dispersing solvent was not required.

Briefly, 2.0 mL of the sample at pH 4.0 was placed in a glass centrifuge tube and 0.1 g of $Na_2SO_4$ and 0.16 g of 1-(2-aminoethyl)-3-methylimidazolium hexafluorophosphate ($[C_2MIM][PF_6]$) were added. The mixture was irradiated in a microwave oven at 240 W for 90 s to melt and disperse the IL into the sample solution and achieve the extraction of the analytes. After that, the mixture was placed in an ice bath for 5 min to solidify the IL, and centrifuged in order to to separate the IL phase from the sample solution. Then, acetonitrile was used to dissolve the IL, and the analytes were separated and determined by HPLC with an UV detector.

The novel proposed method was compared with other methods for the extraction of sulfonamides reported in the literature. It presented lower extraction solvent volume and shorter extraction time. Moreover, the method was simple and environmentally friendly, since only the IL was used for the extraction.

The authors have also applied this strategy to the extraction of Sudan dyes (Sudan I-IV) in red wines [174]. Here, the analytes were extracted with 0.06 g of 1-dodecyl-3-methylimidazolium bromide ([$C_{12}$MIM]Br), which was completely dissolved in the sample solution. 0.20 g of NaCl was added in order to adjust the ionic strength and improved the extraction efficiency by promoting the analytes' partition into the IL phase. Then, the mixture was placed in a modified household microwave oven at 180 W for 90 s. To achieve the solidification of the extractant with the analytes, 0.0886 g of [$NH_4$][$PF_6$] was added and the sample was placed in an ice bath for 5 min. The sample was centrifuged and the extractant was removed and dissolved in acetonitrile for the analysis by HPLC-UV. As in the previous work, the comparison with other reported methods (liquid-liquid microextraction, magnetic solid-phase extraction among others) was performed to highlight the advantages of this approach (lower quantity of sample and organic solvent and shorter extraction time).

A similar approach was proposed by Hu et al. to extract triazines from honey samples [175]. Here, 1-Dodecanol was selected as an extracting solvent due to its low density and solubility in water and a melting point near room temperature. The extraction was performed using 2.0 g of honey dissolved in 10.0 mL of water, previously filtered. Since the extraction efficiency increased when the analytes where in their neutral form, the pH value of the sample was studied between 2.0 and 8.0, and 5.0 was selected as optimum. The addition of salt increased the viscosity of the sample solution, and the extraction efficiency decreased. Thus, the addition of salt was not recommended. 70 µL of 1-Dodecanol was added to the sample, and the mixture was irradiated at 300 W for 40 s to achieve the dispersion of the solvent by increasing its solubility and hence, accelerating the extraction. Then, the sample was centrifuged to separate the extractant from the solution. The tube was cooled in an ice bath for 5 min and the solvent drop was solidified, so it can be easily collected. The drop was transferred to a vial, diluted with methanol and analyzed by HPLC.

### 3.3.4 Other Interesting Microwave-Assisted Microextraction Approaches

Different approaches that take advantage of microwave irradiation to perform a liquid phase microextraction are presented below.

Zhang et al. presented an interesting proposal by designing a special device to perform a microwave-assisted extraction coupled to liquid/solid phase microextraction with the modification of a hollow fiber. The proposed MAE-HF-L/SME method was applied for the extraction of 54 trace pharmaceutical and personal care products from fish samples [176]. The designed device consisted of a modified hollow fiber tube that was cut in an appropriate length, cleaned and sealed at one of its ends. A SPME fiber, synthesized in the laboratory, and a 1:1 1-octanol/toluene (v/v) solution were placed into the hollow fiber tube and after that, the other end was sealed. Then, the hollow fiber tube was stuck to the top of a centrifuge tube that contained a NaCl solution with the target analytes in a known concentration. The salt solution reduced the solubility of the analytes in the aqueous phase and, thus, facilitating the adsorption onto the fiber. Next, the device was placed in a microwave synthesis/extraction workstation (MAS-II) and irradiated at a microwave power of 100 W for 12 min. After the extraction, the SPME fiber and the organic phase were sonicated in methanol to desorb the analytes. The determination was performed by LC–HRMS.

The selected analytes were successfully extracted and preconcentrated from a complex sample matrix by the proposed method with satisfactory validation parameters for almost all the compounds.

Alsharaa et al. proposed an original and novel approach to perform microwave-assisted headspace liquid-phase microextraction (MA-HS-LPME) without modifications of the microwave oven [177]. To accomplish this, the authors designed a porous polypropylene membrane envelope that contained the solvent for extraction, which had to be compatible with the porous membrane in order to dilate its pores, allowing the permeation of the volatile analytes. Thus, the sample was placed at the bottom of the extraction vessel, in an appropriate solution, and the envelope was placed at a certain height supported by a PTFE ring. The vessel was then placed in a commercial microwave extraction system with programmable temperature and pressure, and the sample was irradiated. A magnetic bar was introduced in the sample solution for stirring the sample during the extraction. The authors applied this approach to the extraction of six disinfection by-products in biota samples, mainly fish tissue and green algae. For this method, 3 g of the sample was digested in 15 mL of 100 mmol L$^{-1}$ of HNO$_3$ solution and the volatile analytes were extracted with 500 μL of toluene when the mixture was irradiated with a power equivalent to a temperature of 80°C for 12 min. The solvent in the polypropylene envelope was placed at a 7 cm depth in order to obtain good extraction efficiencies.

Therefore the digestion of the sample and extraction of the target analytes can be accomplished in a simple way, directly in a commercial microwave apparatus with no modification needed. The proposed method was compared with microwave digestion of the samples followed by headspace liquid-phase microextraction (HS-LPME). The conventional method not only required a longer time for analysis but also had lower recovery values.

A HS-LPME was also proposed for the determination of chlorophenols (CPs) in aqueous samples with complex matrices by using microwave irradiation in order to accelerate the evaporation of CPs into the headspace where the microextraction was going to take place [178].

The LPME device consisted of a polypropylene hollow fiber placed on the needle tip of a microsyringe and filled with 1-octanol and 2,4,6-tribromophenol, used as an internal standard. The hollow fiber was placed in a headspace sampling chamber with a cooling water jacket designed to control the temperature of the hollow fiber environment, and therefore to increase the partition coefficient of the analytes. The LPME device was adapted to be placed at the outside top of the microwave oven. On the other hand, the sample solution was placed in a round bottom flask inside the microwave oven. The flask was connected to the LPME device by a condenser. Thus, for the extraction, 10.0 mL of the sample solution was adjusted at pH 1.0 in order to ensure the neutral form of the analytes and, therefore, their volatility. The extraction and the release of the analytes were achieved when the sample was irradiated with a microwave power of 167 W at 45°C. After extraction, the extracted solvent in the hollow fiber was retracted into the microsyringe and analyzed by GC-electron capture detection (ECD).

In order to evaluate the performance of the proposed method, standard CP solutions were added in the sample matrix and subjected to the complete process. It could be verified that the accuracy and precision of the method are acceptable for environmental analysis. This method was compared with other methods where the technique did not include the cooling jacket to control the temperature. The extraction efficiency was much lower than that with the cooling system, and the reproducibility was much less since there was a continuous escape of vapors.

A similar MA-HS-CT-LPME approach to extract dichlorodiphenyltrichloroethane (DDT) and its main metabolites from environmental aqueous samples was proposed by Kumar and Jen [179]. The authors used the same apparatus configuration as Li et al. [24]. For the extraction, 10 mL of the sample was irradiated at a power of 249 W for 6.5 min. The LPME probe was filled with 4.0 μL of 1-octanol and placed on the center level of the cloud vapor zone where the temperature was controlled at 34°C to achieve the highest extraction efficiency.

The optimized method was compared in terms of the quantity of sample and solvent used, detection limit and sampling time with the MA-HS-SPME method, previously reported by Li et al. [180], and liquid–liquid extraction method, based on the extraction of DDT and its main metabolites from a certified reference material containing organochlorine pesticides. The authors affirmed that the performance obtained for the new approach is similar to the conventional and recently proposed extraction methods reported in the literature.

## 3.4   MICROWAVE-ASSISTED SOLID PHASE MICROEXTRACTION

Solid-phase microextraction (SPME) is a widely used sample preparation technique in which the analytes that are dissolved in the sample solution or in the headspace are in contact with a fiber coating with the sorbent. Once the target analytes are extracted, the technique can be coupled to a suitable instrument for desorption and determination of the compounds. The equilibrium of the analytes between the two phases is established by the distribution constant that depends on the temperature, nature of sorbents, ionic strength, pH of samples and organic-solvent content [181]. The application of SPME as an extraction technique has been significantly extended in recent years due to its outstanding characteristics, such as simplicity, solvent-free nature, extraction efficiency and easy coupling with separation techniques, mainly GC.

The most outstanding associated techniques and applications are described below, and the summary of the microwave conditions can be found in Table 3.7.

### 3.4.1   MICROWAVE-ASSISTED HEADSPACE SOLID PHASE MICROEXTRACTION (MA-HS-SPME)

The principles of HS-SPME extraction are similar to those already described for HS-SDME, but, in this case, the volatile analytes are extracted and preconcentrated in a coating fiber placed in the headspace above the sample.

**TABLE 3.7**
**Microwave-assisted Solid Phase Microextraction Methods**

| Type of Microextraction | Analyte | Sample | Type of Fiber/Microdrop Solvent | Ionic Strength | Type of Microwave | Microwave Power (W) | Irradiation Time/Temperature | Sample pH | Ref |
|---|---|---|---|---|---|---|---|---|---|
| MA-HS-SPME | Volatile oil | Citrus aurantium L. (2.5 g) | Nanoporous silica functionalized with amino propyl-triethoxysilane | n.i | MWO (model of GE614ST/GE614W, Samsung Korea, 900 W) | 450 | Not informed | n.i | [182] |
| MA-HS-SPME | Galaxolide, tonalide, musk xylene, musk ketone | Fish (2.0 g) | PDMS-DVB | 4 g NaCl | (CEM, Matthews, NC, USA), 80 W | 80 | 5 min | 2.0 | [183] |
| MA-HS-SPME | Phenol | Cigarette pad (95 mg) | CW/DVB | n.i | MWO (LWMC-205, Lingjiang Science and Technology Co. Ltd., China). | 800 | 60 s | n.i | [184] |
| MA-HS-SPME | PAHs | Environmental Water | PDMS/DVB | n.i | Domestic MWO | 209 | 40 min | n.i | [185] |
| MA-HS-SPME | PAHs | Wastewater (20 mL) | PDMS/DVB (65 µm) | n.i | Domestic NN-L520 inverter system | 145 | 30 min | n.i | [186] |
| MA-HS-SPME | Alkylphenols | WWTP-effluent, river water (20 mL) | PDMS/DVB (65 µm) | 2 g | CEM Mars Xpress MW system | 80 | 5 min | n.i | [187] |

(Continued)

**TABLE 3.7 (CONTINUED)**
**Microwave-assisted Solid Phase Microextraction Methods**

| Type of Microextraction | Analyte | Sample | Type of Fiber/ Microdrop Solvent | Ionic Strength | Type of Microwave | Microwave Power (W) | Irradiation Time/ Temperature | Sample pH | Ref |
|---|---|---|---|---|---|---|---|---|---|
| MA-HS-SPME | Organophosphate esters | Surface water, WWTP-effluent (20 mL) | PDMS/DVB (65 µm) | 3 g | CEM Mars Xpress MW system | 140 | 5 min | 3 | [188] |
| MA-HS-SPME | Pyrethroid residuals | Underground water (20 mL) | PDMS (100 µm) | n.i | Domestic NN-L520 inverter system | 157 | 10 min | 4 | [189] |
| MA-HS-SPME | Semi-volatile organic compounds | River water (20 mL) | PDMS/DVB (65 µm) | n.i | CEM Discover System | 30 | 30 min | n.i | [190] |
| MA-HS-SPME | Synthetic polycyclic musks | Oyster | PDMS/DVB (65 µm) | 3 g | CEM Mars Xpress MW system | 80 | 5 min | 1 | [191] |
| MA-HS-SPME | Synthetic polycyclic musks | Sewage sludge and sediments | PDMS/DVB (65 µm) | 3 g | CEM Mars Xpress MW system | 80 | 5 min | 1 | [192] |
| MA-HS-SPME | Synthetic polycyclic musks | River water | PDMS/DVB (65 µm) | 20% | CEM Mars Xpress MW system | 180 | Less than 4 min | | [193] |
| ASE-µ-SPE | Parabens | Human ovarian cancer tissues (5 g) | HayeSepA, HayeSepB | n.i | A MARS microwave extraction system | 600 | 20 min, 90°C | n.i | [195] |

*(Continued)*

**TABLE 3.7 (CONTINUED)**
**Microwave-assisted Solid Phase Microextraction Methods**

| Type of Microextraction | Analyte | Sample | Type of Fiber/ Microdrop Solvent | Ionic Strength | Type of Microwave | Microwave Power (W) | Irradiation Time/ Temperature | Sample pH | Ref |
|---|---|---|---|---|---|---|---|---|---|
| UMSE-HS-SPME | Essential oil | Angelica dahurica (0.1 g) | DVB/CAR/PDMS | n.i | Modified MWO | 600 | 10 min | n.i | [196] |
| UMSE-HS-SPME | Volatile components | Tobacco (0.5 g) | DVB/CAR/PDMS | n.i | Ultrasound-microwave apparatus (XH-300UA, Beijing, | 600 | 80°C | n.i | [197] |

WWTP, wastewater treatment plants; n.i., not informed.

MA-HS-SPME is one of the most applied solid-phase microextraction techniques assisted by microwave. Here, microwaves have the same role as in the liquid phase extraction, that is, the heating of the sample solution to achieve the volatilization of the analytes and, thus, accelerate the process.

As in the MA-HS-SDME technique, the microwave apparatus is modified to include a cooling system with circulating water in the sampling zone at a set temperature in order to improve the partition coefficient between the SPME fiber and the headspace zone.

The works combining the microwave irradiation with the HS-SPME technique are described below.

Gholivand et al. proposed the preparation of a nanocomposite silica fiber functionalized with amino propyl triethoxysilane for extracting volatile compounds from leaves of *Citrus aurantium* L. by MA-HS-SPME [182]. To perform the microextraction, a domestic microwave oven was modified and a Claisen adaptor was assembled at the top in order to place a condenser and a microsyringe with the SPME fiber. To prevent microwave leakage, an aluminum foil was inserted at both sides of the interface part. The sample (2.5 g) was placed into a round flask and 1.0 mL of water was added. The microwaves heated the sample at a power of 450 W and the volatile compounds were extracted in the SPME fiber that was exposed in the sample headspace for 3.5 min. To optimize the main variables of the extraction, a simplex method was used and the best conditions were chosen for monitoring the relative areas of the five main peaks in a GC-MS instrument.

The proposed method was compared with the conventional hydrodistillation method by a regression line approach, showing similar results. In addition, the extraction was simple and rapid, and a lower amount of samples were required. It is important to highlight that the extraction was performed without using organic solvents.

Another interesting approach using MA-HS-SPME was proposed by Wu et al. [183]. Here, synthetic polycyclic and nitro-aromatic musks were extracted from fish samples using a polydimethylsiloxane-divinylbenzene (PDMS-DVB) fiber. 2.0 g of the sample was mixed with 4.0 mL of methanol to prepare a slurry. The slurry was mixed with 15.0 mL of deionized water (water-headspace ratio 1:1), and placed in a SPME vial. It is well-known that the pH and the addition of salts can enhance the extraction efficiency by increasing the partitioning between the headspace and the sample. Thus, both variables were optimized, the sample was adjusted to a pH value of 2.0 and 4.0 g of NaCl was added. Then, the vial was sealed and placed in a CEM Mars Xpress microwave system. This apparatus had a hole at the top that permitted the insertion of the SPME device needle into the headspace over the sample. To perform the extraction, the sample was irradiated at 80 W for 5 min and the analytes were vaporized into the headspace and extracted with the fiber. The SPME device was inserted into the GC instrument to perform the separation and quantification of the analytes.

The proposed extraction procedure obtained satisfactory recoveries, showing to be a good alternative extraction method for the determination of lipophilic and semi-volatile organic compounds from fish samples. In addition, the method is simple, rapid and environmentally friendly since the extraction was solvent-free and only 4.0 mL of methanol was used to prepare the slurry.

The same technique was applied to the extraction of phenol in cigarette pad samples [184]. Here, 380 mg of the sample and 200 µL of phenol standard solution, prepared in anhydrous ethanol, were placed in a vial and irradiated at 800 W for 60 s. Then, the sample was left to equilibrate and a fiber coated with carbowax/divinylbenzene (CW/DVB) was used to extract the analytes in the headspace over the sample for 20 min. After that time, the coated fiber was inserted into GC-MS system for thermal desorption (250°C, 5 min). The validation parameters and the positive results obtained in the analysis of phenols in real samples demonstrated that the proposed approach was suitable for quantifying the selected analytes in a simple way.

Hsieh et al. designed a microwave device to develop a MA-HS-SPME method to determine polycyclic aromatic hydrocarbons (PAHs) in water samples [185]. For that purpose, a domestic microwave oven was modified in order to include a temperature controlled cooling system outside and at the top of the oven. The analytes were adsorbed onto a PDMS/DVB fiber that was located in a SPME commercial device. The SPME fiber was placed inside the cooling system in order to decrease the temperature and promote the sorption of the analytes onto the fiber. The temperature of the circulating water was optimized and 20°C was selected. Also, a microwave stirrer was located inside the oven to continuously stir the sample during extraction. Leakage of radiation was controlled by affixing an aluminum foil at the interface between the microwave oven and the cooling system, and a leak detector was also used. To perform the extraction, 20 mL of the water sample, spiked with 5 µg L$^{-1}$ of 16 PAHs, was placed in a flask and irradiated at 209 W for 40 min. These conditions were selected taking into account the direct relation between the sorption of the lower molecular weight PAHs and the temperature reached in the cooling system. The analytes were evaporated and adsorbed onto the fiber directly from the headspace during the same time. After the extraction, the fiber was removed and analyzed immediately by GC-MS. The analytical parameters and the recovery percentages were calculated for all the analytes, obtaining satisfactory results.

The authors also studied the effect of the dissolved organic matter (DOM) on the PAHs extraction, observing that the DOM did not significantly affect PAHs' extraction efficiency. Thus, they concluded that the proposed method can be applied to determine both freely dissolved and DOM-associated PAHs in environmental waters.

Another approach for the quantitative determination of PAHs in aqueous samples was proposed by Wei et al. [186]. In order to promote the vaporization of the PAHs from a water sample into headspace for SPME sampling, the sample was irradiated with a modified household NN-L520 inverter system, equipped with a temperature-control cooling system, under a microwave power of 145 W for 30 min. As adsorption of PAHs onto the fiber was not favored by high temperature, it was necessary to keep the temperature at 20°C around the sampling area with a circulating water system. The fibers used in the SPME device were 1 cm in length and coated with different thickness and materials (75 um CAR/PDMS, 65 um CW/DVB, 65 um DVB/PDMS, 85 um PA, 50 um DVB/CAR/PDMS and 7, 30 100 um PDMS). After adsorption of the analytes, the fiber was immediately placed into a GC injection-port where the analytes were desorbed at 290°C for 5 minutes. The analysis was finally made by GC with flame ionization detection. In order to study the method's performance, spiked water with different concentrations of standard PAHs was used, and then the same treatment procedure was applied to a real sample. It could be verified that the proposed method had an acceptable accuracy and precision in environmental samples with complicated matrices. This method was also compared with other SPME methods and, although all the SPME methods were fast, low-cost and used very little amounts of organic solvents, the sampling times were the longest. The proposed method had a shorter sampling time, taking only 30 min to complete the sample pretreatment.

Wu et al. presented a procedure for the determination of alkylphenols in aqueous samples by the derivatization of the analytes and MA-HS-SPME [187]. The method consisted of mixing an aliquot of the water sample, containing an acetic anhydride, with $KHCO_3$ and placing it into a bottle where NaCl was added. Once the acetylation was over, the bottle was located in a CEM Mars Xpress microwave system and the SPME needle was inserted in the bottle, where the PDM S-DVB fiber (65 um) was exposed to the headspace over the water. The sample was irradiated for 5 min under a power of 80 W and then the fiber was immediately placed into a GC injection-port, where the analytes were desorbed at 250°C for 3 min. In order to evaluate the method, spiked deionized water samples were used, and intra- and inter-day parameters were calculated. It could be verified that the proposed method is precise and provides a wide linear range and good detection limits at the ng $L^{-1}$ level. Moreover, this method can be considered eco-friendly, as it can be performed in a single step and is solvent-free.

A similar procedure was described by Tsao et al. [188] for the quantitative determination of organophosphate esters in aqueous samples. The difference lay in the microwave power, since in this method the sample was irradiated under a power of 140 W for 5 min. The method's performance was also studied by the analysis of spiked water samples, and it was concluded that this method was suitable for the determination of organic compounds in the analyzed samples, being simple, effective and green.

The MA-HS-SPME technique was also applied for the quantitative determination of pyrethroid residuals in aqueous samples [189]. For this purpose, an aliquot of the sample was mixed with buffer solution (pH 4.0) in a glass flask and then it was placed in a microwave oven (a modified version of the domestic NN-L520 inverter system) equipped with a cooling system connected to a water circulating machine. The flask was connected to the HS-SPME system and the sample was irradiated for 10 min under an irradiation power of 157 W. The analytes were vaporized into the headspace and adsorbed onto a PDMS fiber (100 μm). The sampling area had a controlled temperature of 30°C, so the partition coefficient of the analytes between the SPME fiber and headspace would increase. After adsorption, the fiber was immediately injected in a GC injector and heated to 290°C for 3 min for desorption of the pyretroids. The analytes were separated and measured by an electron capture detector held at 300°C. The proposed method was compared with the SPE and LLE/SPE methods. The extraction efficiency of the MA-HS-SPME method was much better than that of the SPE method and similar to that of LLE/SPE method. The precision of the proposed method was worse than the LLE/SPE method, but it was still acceptable for environmental analysis. The MA-HS-SPME method had shorter times of sample analysis, did not need the use of organic solvents and was simple, sensitive and cheap.

The extraction of semi-volatile compounds from aqueous samples was also performed by this pretreatment technique [190]. In order to perform the analysis, an aliquot of the sample was located in a vial, and a 65 μm PDMS-DVB SPME fiber was exposed to the headspace over the water sample. Then the vial was placed in a CEM Discover System and irradiated under a microwave power of 30 W for 30 min at 70°C. After extraction, the fiber was injected into the GC injector port (depth: 4.0 cm) and the analytes were desorbed at 250°C for 3 min. The proposed method was compared with the US Environmental Protection Agency's (EPA) methods, and the detection limit achieved by MA-HS-SPME turned out to be much lower. It was concluded that the developed method provided acceptable accuracy, precision, wide-range linearity and high sensitivity.

Wu et al. described a procedure for the quantitative determination of synthetic polycyclic musks in oyster samples that is simple, rapid and solvent-free [191]. For this purpose, oyster tissue was mixed with deionized water and NaCl in a vial, adjusting the pH to 1.0 by concentrated hydrochloric acid. Then, the vial was placed in a CEM Mars Xpress microwave System, the SPME needle was inserted into the sample vial and a 65 μm PDMS-DVB fiber was exposed to the headspace over the sample, which was irradiated under a microwave power of 80 W for 5 min. After extraction, the fiber was injected inside the the injector port (depth: 4.0 cm) of a GC-MS spectrometer and the analytes were desorbed at 270°C for 2 min. The proposed method was evaluated by the use of spiked oyster samples that were subjected to the same analytical procedure. Precision, accuracy and repeatability were evaluated by intra- and inter-day analysis. It could be verified that this method provided acceptable precision, accuracy and low detection limits. Moreover, the extraction procedure was little affected by the sample matrix, and was a green method as no organic solvents were used. The authors also applied the same method for the determination of synthetic polycyclic musks in sewage sludge and sediments [192], and Yu-Chen Wang et al. also described a similar method for the determination of polycyclic musks in water samples [193]. The only difference in this method was the microwave irradiation power, as the sample was irradiated under 180 W for less than 4 min.

### 3.4.2 Microwave-Assisted Solvent Extraction Combined with Micro-Solid Phase Extraction (MASE-μ-SPE)

Micro-solid phase extraction (μSPE) technique is characterized by the development of a small porous membrane bag that contains a suitable adsorbent. The bag can be placed in the sample headspace or inside the sample solution. The membrane can act as a filter and, therefore, it is possible to extract analytes from samples that are contaminated or contain large particles or suspended matter [194]

An interesting MASE-μ-SPE approach that combined the simultaneous use of microwave-assisted solvent extraction (MASE) and micro-solid phase extraction (μSPE) was proposed by Sajid et al. to determine parabens in cancer tissues [195]. The authors designed a μSPE device (2.0 cm×0.5 cm) by packing 25 mg of two polar polymer-based sorbents (divenylbenzene-ethyleneglycoldimethylacrylate, HayeSepA and divinylbenzene-polyethyleneimine, HayeSepB) materials in a polypropylene envelope. The extraction was performed by placing 5.0 g of the tissue and 10.0 mL of NaOH 5 mol L$^{-1}$ solution in a microwave vessel of a MARS microwave extraction system (CEM, Matthews, NC, USA). Next, two μSPE devices were introduced and the sample was stirred and irradiated at 600 W for 20 min, reaching a temperature of 90°C. After extraction, the devices were removed from the solution, and the analytes were desorbed in 200 μL of acetonitrile by using an ultrasound bath. Then, the extract was analyzed by HPLC with a UV detector.

The authors compared the extraction efficiency of the proposed method with the conventional MASE-SPE method using Oasis HLB® as a sorbent in terms of recovery percentages. The authors observed slightly higher recovery values when the MASE-μSPE was applied, and they attributed that to an adequate selection of the packing material. The main advantage of this method was that a sample clean-up was not needed as fatty compounds and other interferences were eliminated by the porous polypropylene membrane of the envelope.

### 3.4.3 Ultrasound-Microwave-Assisted Extraction Headspace Solid Phase Microextraction (UMSE-HS-SPME)

The articles described below combined the advantages of using microwaves with ultrasonic radiation to achieve higher extraction efficiencies. The first approach was developed by Feng et al., who presented, for the first time, a novel sample preparation technique by coupling ultrasound and microwave-assisted extraction to head-space solid-phase microextraction (UMSE-HS-SPME) to determine essential oils in dry roots of *Angelica dahurica*, a traditional Chinese medicine [196]. The UMHE-HS-SPME apparatus was similar to those employed for a MA-HS-SPME technique [170] but, in this case, a flask with three necks was used: one for an ultrasound probe, another for a temperature sensor and the last was assembled to the Claisen adaptor. The sample was placed in the flask with 10.0 mL of distilled water, and was sonicated and irradiated for 10 min at a power of 400 W and 600 W, respectively. The volatile compounds were extracted in a DVB/CAR/PDMS SPME fiber. Then, the fiber was inserted into the GC-MS instrument equipped with a trap analyzer in order to thermally desorb the analytes. The extraction variables were optimized, taking into account the peak area of the main representative compounds and the sum of the peak area.

The UMHE-HS-SPME method was compared with the MAE–HS-SPME procedure under the same experimental conditions. While microwave energy can lead to a fast heating of the mixture of sample and solvent, the ultrasound field can increase the pressure of the system and, thus, facilitate mass transfer. The net result is a better efficiency in the extraction compared with the MAE-HS-SPME procedure.

The UMHE-HS-SPME method was also compared with the steam distillation (SD) method. In general, the same compounds were identified by the two methods, except for the volatile compounds with a low boiling point that could be extracted only by SD. On the other hand, the compounds that were more oxygenated and had higher boiling points were not or little extracted by SD, while they were extracted by the proposed new method.

Conclusively, the authors demonstrated that the combination of both energies could release the essential oils in a short time, in a rapid and simple way. In addition, the proposed method presented the advantage of performing the isolation, extraction and concentration of the analytes in a single step without using organic solvents.

Yang et al. proposed the extraction of the volatile compounds from tobacco samples by using the same approach [197]. Here, 0.5 g of the powdered sample and 40 mL of distilled water were placed in a round flask and irradiated at 600 W to reach a temperature of 110°C along the experiment. The sample was then sonicated at an ultrasonic power of 100 W for 10 min. The volatile components of tobacco were trapped in a DVB/CAR/PDMS fiber, and subsequently analyzed by CG-MS.

The comparison with MA-HS-SPME and HS-SPME methods was performed. Unlike HS-SPME, the volatile compounds identified by UMSE-HS-SPME and MAE-HS-SPME were similar, but much better extraction efficiencies were obtained with the synergist approach due to the extra energy given by the sonication. Moreover, the proposed method was faster, efficient and environmentally friendly since no organic solvent was needed for the extraction.

The articles summarized in this chapter describe the use of MW irradiation for enhancing the degradation of different organic pollutants and the yields in microextraction methods. Both degradations and microextractions are more simple, environmentally friendly and efficient when MW is involved in these processes. In the case of microextraction, low volume of reagents and small amounts of solvent are consumed. Moreover, the time of analysis is shorter since the derivatization and extraction can be performed simultaneously. In addition, lower limits of detection can be achieved with high extraction efficiency. On the other hand, among the main advantages found when MW is combined with different AOPs for degradation of recalcitrant organic compounds, we can highlight the higher efficiency, significant shortening in process times and procedures with relatively low cost.

# REFERENCES

1. Mo, J., Zhang, Y., Xu, Q., Lamson, J. J., Zhao, R. (2009). Photocatalytic purification of volatile organic compounds in indoor air: A literature review. *Atmospheric Environment, 43*(14), 2229–2246.
2. Remya, N., Lin, J. G. (2011). Current status of microwave application in wastewater treatment—A review. *Chemical Engineering Journal, 166*(3), 797–813.
3. de la Hoz, A., Diaz-Ortiz, A., Moreno, A. (2005). Microwaves in organic synthesis. Thermal and non-thermal microwave effects. *Chemical Society Reviews, 34*(2), 164–178.
4. Dong, Y., Han, Z., Liu, C., Du, F. (2010). Preparation and photocatalytic performance of Fe(III)-amidoximated PAN fiber complex for oxidative degradation of azo dye under visible light irradiation. *Science of the Total Environment, 408*(10), 2245–2253.
5. Lee, Y. C., Lo, S. L., Chiueh, P. T., Liou, Y. H., Chen, M. L. (2010). Microwave-hydrothermal decomposition of perfluorooctanoic acid in water by iron-activated persulfate oxidation. *Water Research, 44*(3), 886–892.
6. Hu, E., Cheng, H. (2012). Impact of surface chemistry on microwave-induced degradation of atrazine in mineral micropores. *Environmental Science and Technology, 47*(1), 533–541.
7. Haque, K. E. (1999). Microwave energy for mineral treatment processes—A brief review. *International Journal of Mineral Processing, 57*(1), 1–24.
8. Tsintzou, G. P., Antonakou, E. V., Achilias, D. S. (2012). Environmentally friendly chemical recycling of poly (bisphenol-A carbonate) through phase transfer-catalysed alkaline hydrolysis under microwave irradiation. *Journal of Hazardous Materials, 241,* 137–145.

9. Eskicioglu, C., Terzian, N., Kennedy, K. J., Droste, R. L., Hamoda, M. (2007). Athermal microwave effects for enhancing digestibility of waste activated sludge. *Water Research*, *41*(11), 2457–2466.

10. Pang, Y., Lei, H. (2016). Degradation of p-nitrophenol through microwave-assisted heterogeneous activation of peroxymonosulfate by manganese ferrite. *Chemical Engineering Journal*, *287*, 585–592.

11. Horikoshi, S., Kajitani, M., Serpone, N. (2007). The microwave-/photo-assisted degradation of bisphenol-A in aqueous $TiO_2$ dispersions revisited: Re-assessment of the microwave non-thermal effect. *Journal of Photochemistry and Photobiology A: Chemistry*, *188*(1), 1–4.

12. Cai, M. Q., Zhu, Y. Z., Wei, Z. S., Hu, J. Q., Pan, S. D., Xiao, R. Y., Dong, C. Y., Jin, M. C. (2017). Rapid decolorization of dye Orange G by microwave enhanced Fenton-like reaction with delafossite-type $CuFeO_2$. *Science of The Total Environment*, *580*, 966–973.

13. Chen, X. G., Ye, Y., Cheng, J. P. (2011). Recent progress in electromagnetic wave absorbers. *Journal of Inorganic Materials*, 5, 001.

14. Horikoshi, S., Serpone, N. (2014). On the influence of the microwaves' thermal and non-thermal effects in titania photoassisted reactions. *Catalysis Today*, *224*, 225–235.

15. Ai, Z., Yang, P., Lu, X. (2005). Degradation of 4-chlorophenol by a microwave assisted photocatalysis method. *Journal of Hazardous Materials*, *124*(1), 147–152.

16. Noradoun, C., Engelmann, M. D., McLaughlin, M., Hutcheson, R., Breen, K., Paszczynski, A., Cheng, I. F. (2003). Destruction of chlorinated phenols by dioxygen activation under aqueous room temperature and pressure conditions. *Industrial and Engineering Chemistry Research*, *42*(21), 5024–5030.

17. Lee, C., Keenan, C. R., Sedlak, D. L. (2008). Polyoxometalate-enhanced oxidation of organic compounds by nanoparticulate zero-valent iron and ferrous ion in the presence of oxygen. *Environmental Science and Technology*, *42*(13), 4921–4926.

18. Yang, Y., Wang, P., Shi, S., Liu, Y. (2009). Microwave enhanced Fenton-like process for the treatment of high concentration pharmaceutical wastewater. *Journal of Hazardous Materials*, *168*(1), 238–245.

19. Bo, L., Quan, X., Chen, S., Zhao, H., Zhao, Y. (2006). Degradation of p-nitrophenol in aqueous solution by microwave assisted oxidation process through a granular activated carbon fixed bed. *Water Research*, *40*(16), 3061–3068.

20. Yin, J., Cai, J., Yin, C., Gao, L., Zhou, J. (2016). Degradation performance of crystal violet over CuO@AC and $CeO_2$-CuO@AC catalysts using microwave catalytic oxidation degradation method. *Journal of Environmental Chemical Engineering*, *4*(1), 958–964.

21. Bo, L. L., Zhang, Y. B., Quan, X., Zhao, B. (2008). Microwave assisted catalytic oxidation of p-nitrophenol in aqueous solution using carbon-supported copper catalyst. *Journal of Hazardous Materials*, *153*(3), 1201–1206.

22. Xu, X., Su, R., Zhao, X., Liu, Z., Zhang, Y., Li, D., Li, X., Zhang, H., Wang, Z. (2011). Ionic liquid-based microwave-assisted dispersive liquid–liquid microextraction and derivatization of sulfonamides in river water, honey, milk, and animal plasma. *Analytica Chimica Acta*, *707*(1), 92–99.

23. Su, R., Li, D., Wu, L., Han, J., Lian, W., Wang, K., Yang, H. (2017). Determination of triazine herbicides in juice samples by microwave-assisted ionic liquid/ionic liquid dispersive liquid–liquid microextraction coupled with high-performance liquid chromatography. *Journal of Separation Science*, *40*(14), 2950–2958.

24. Li, X., Chen, G., Liu, J., Liu, Y., Zhao, X., Cao, Z., Xia, L., Li, G., Sun, Z., Zhang, S., You, J., Wang, H. (2017). A rapid, accurate and sensitive method with the new stable isotopic tags based on microwave-assisted dispersive liquid-liquid microextraction and its application to the determination of hydroxyl UV filters in environmental water samples. *Talanta*, *167*, 242–252.

25. Motevalli, K., Yaghoubi, Z. (2017). Microwave-assisted liquid phase microextraction followed with flame atomic absorption spectrometry for trace determination of zinc in food samples. *Journal of the Chilean Chemical Society*, *62*(1), 3417–3420.
26. Stasinakis, A. S. (2008). Use of selected advanced oxidation processes (AOPs) for wastewater treatment—A mini review. *Global NEST Journal*, *10*(3), 376–385.
27. Goldstein, S., Aschengrau, D., Diamant, Y., Rabani, J. (2007). Photolysis of aqueous $H_2O_2$: Quantum yield and applications for polychromatic UV actinometry in photoreactors. *Environmental Science and Technology*, *41*(21), 7486–7490.
28. Horikoshi, S., Sakamoto, S., Serpone, N. (2013). Formation and efficacy of $Ti_2$/AC composites prepared under microwave irradiation in the photoinduced transformation of the 2-propanol VOC pollutant in air. *Applied Catalysis B: Environmental*, *140*, 646–651.
29. Gogate, P. R., Pandit, A. B. (2004). A review of imperative technologies for wastewater treatment I: Oxidation technologies at ambient conditions. *Advances in Environmental Research*, *8*(3), 501–551.
30. Xie, F., Xu, Y., Xia, K., Jia, C., Zhang, P. (2016). Alternate pulses of ultrasound and electricity enhanced electrochemical process for p-nitrophenol degradation. *Ultrasonics Sonochemistry*, *28*, 199–206.
31. Leal-Castañeda, E. J., Inchingolo, R., Cardenia, V., Hernandez-Becerra, J. A., Romani, S., Rodriguez-Estrada, M. T., Galindo, H. S. G. (2015). Effect of microwave heating on phytosterol oxidation. *Journal of Agricultural and Food Chemistry*, *63*(22), 5539–5547.
32. Zhang, F. Q., Huang, M. F., Chen, M., Wang, Y. Z., Liao, J. H. (2012). Thermal-oxidation degradation of microwave drying nature rubber. *Advanced Materials Research*, *399*, 1600–1603. Trans Tech Publications.
33. Roda-Serrat, M. C., Christensen, K. V., El-Houri, R. B., Fretté, X., Christensen, L. P. (2018). Fast cleavage of phycocyanobilin from phycocyanin for use in food colouring. *Food Chemistry*, *240*, 655–661.
34. Lai, T. L., Wang, W. F., Shu, Y. Y., Liu, Y. T., Wang, C. B. (2007). Evaluation of microwave-enhanced catalytic degradation of 4-chlorophenol over nickel oxides. *Journal of Molecular Catalysis A: Chemical*, *273*(1), 303–309.
35. Lai, T. L., Lee, C. C., Huang, G. L., Shu, Y. Y., Wang, C. B. (2008). Microwave-enhanced catalytic degradation of 4-chlorophenol over nickel oxides. *Applied Catalysis B: Environmental*, *78*(1), 151–157.
36. Lai, T. L., Liu, J. Y., Yong, K. F., Shu, Y. Y., Wang, C. B. (2008). Microwave-enhanced catalytic degradation of 4-chlorophenol over nickel oxides under low temperature. *Journal of Hazardous Materials*, *157*(2), 496–502.
37. Lai, T. L., Yong, K. F., Yu, J. W., Chen, J. H., Shu, Y. Y., Wang, C. B. (2011). High efficiency degradation of 4-nitrophenol by microwave-enhanced catalytic method. *Journal of Hazardous Materials*, *185*(1), 366–372.
38. Hsu, C., Jou, C. (2013). Catalytic degradation of chlorobenzene in aqueous solution with microwave and waste metals. *Sustainable Environment Research*, *23*, 49–52.
39. Lee, C. L., Jou, C. J. G. (2012). Degradation of chlorobenzene with microwave-aided zerovalent iron particles. *Environmental Engineering Science*, *29*(6), 432–435.
40. Yuan, D., Fu, D. Y., Luo, Z. W. (2013). Study on microwave combined with granular active carbon for treatment of p-nitrophenol wastewater. *Advanced Materials Research*, *602*, 2287–2290.
41. Bo, L., Quan, X., Wang, X., Chen, S. (2008). Preparation and characteristics of carbon-supported platinum catalyst and its application in the removal of phenolic pollutants in aqueous solution by microwave-assisted catalytic oxidation. *Journal of Hazardous Materials*, *157*(1), 179–186.
42. Yin, C., Cai, J., Gao, L., Yin, J., Zhou, J. (2016). Highly efficient degradation of 4-nitrophenol over the catalyst of $Mn_2O_3$/AC by microwave catalytic oxidation degradation method. *Journal of Hazardous Materials*, *305*, 15–20.

43. Zhang, Z., Shan, Y., Wang, J., Ling, H., Zang, S., Gao, W., Zhao, Z., Zhang, H. (2007). Investigation on the rapid degradation of congo red catalyzed by activated carbon powder under microwave irradiation. *Journal of Hazardous Materials, 147*(1), 325–333.

44. Veksha, A., Pandya, P., Hill, J. M. (2015). The removal of methyl orange from aqueous solution by biochar and activated carbon under microwave irradiation and in the presence of hydrogen peroxide. *Journal of Environmental Chemical Engineering, 3*(3), 1452–1458.

45. Xu, D., Cheng, F., Zhang, Y., Song, Z. (2014). Degradation of methyl orange in aqueous solution by microwave irradiation in the presence of granular-activated carbon. *Water, Air, and Soil Pollution, 225*(6), 1983.

46. Zhang, Z., Xu, D., Tie, M., Li, F., Chen, Z., Wang, J., Gao, W., Ji, X., Xu, Y. (2011). Spectroscopic study on interaction between bisphenol A or its degraded solution under microwave irradiation in the presence of activated carbon and human serum albumin. *Journal of Luminescence, 131*(7), 1386–1392.

47. Chen, J., Xue, S., Song, Y., Shen, M., Zhang, Z., Yuan, T., Tian, F., Dionysiou, D. D. (2016). Microwave-induced carbon nanotubes catalytic degradation of organic pollutants in aqueous solution. *Journal of Hazardous Materials, 310*, 226–234.

48. Zhou, H., Hu, L., Wan, J., Yang, R., Yu, X., Li, H., Chen, J., Wang, L., Lu, X. (2016). Microwave-enhanced catalytic degradation of p-nitrophenol in soil using $MgFe_2O_4$. *Chemical Engineering Journal, 284*, 54–60.

49. Qiu, Y., Zhou, J., Cai, J., Xu, W., You, Z., Yin, C. (2016). Highly efficient microwave catalytic oxidation degradation of p-nitrophenol over microwave catalyst of pristine α-$Bi_2O_3$. *Chemical Engineering Journal, 306*, 667–675.

50. Chen, J., Pan, H., Hou, H., Li, H., Yang, J., Wang, L. (2017). High efficient catalytic degradation of PNP over Cu-bearing catalysts with microwave irradiation. *Chemical Engineering Journal, 323*, 444–454.

51. Hu, E., Hu, Y., Cheng, H. (2015). Performance of a novel microwave-based treatment technology for atrazine removal and destruction: Sorbent reusability and chemical stability, and effect of water matrices. *Journal of Hazardous Materials, 299*, 444–452.

52. Hu, E., Cheng, H., Hu, Y. (2012). Microwave-induced degradation of atrazine sorbed in mineral micropores. *Environmental Science and Technology, 46*(9), 5067–5076.

53. Hu, E., Cheng, H. (2014). Catalytic effect of transition metals on microwave-induced degradation of atrazine in mineral micropores. *Water Research, 57*, 8–19.

54. He, Y., Cheng, H. (2016). Degradation of N-nitrosodimethylamine (NDMA) and its precursor dimethylamine (DMA) in mineral micropores induced by microwave irradiation. *Water Research, 94*, 305–314.

55. Gu, W., Lv, G., Liao, L., Yang, C., Liu, H., Nebendahl, I., Li, Z. (2017). Fabrication of Fe-doped birnessite with tunable electron spin magnetic moments for the degradation of tetracycline under microwave irradiation. *Journal of Hazardous Materials, 338*, 428–436.

56. Liu, M., Lv, G., Mei, L., Wang, X., Xing, X., Liao, L. (2014). Degradation of tetracycline by birnessite under microwave irradiation. *Advances in Materials Science and Engineering, 2014*, 409086.

57. Wang, X., Lv, G., Liao, L., Wang, G. (2015). Manganese oxide – An excellent microwave absorbent for the oxidation of methylene blue. *RSC Advances, 5*(68), 55595–55601.

58. Liu, X., Zhang, T., Xu, D., Zhang, L. (2016). Microwave-assisted catalytic degradation of crystal violet with barium ferrite nanomaterial. *Industrial and Engineering Chemistry Research, 55*(46), 11869–11877.

59. Shi, W., Li, Q., An, S., Zhang, T., Zhang, L. (2016). Magnetic nanosized calcium ferrite particles for efficient degradation of crystal violet using a microwave-induced catalytic method: Insight into the degradation pathway. *Journal of Chemical Technology and Biotechnology, 91*(2), 367–374.

60. Chen, H., Yang, S., Chang, J., Yu, K., Li, D., Sun, C., Li, A. (2012). Efficient degradation of crystal violet in magnetic $CuFe_2O_4$ aqueous solution coupled with microwave radiation. *Chemosphere*, *89*(2), 185–189.

61. Xiao, J., Fang, X., Yang, S., He, H., Sun, C. (2015). Microwave-assisted heterogeneous catalytic oxidation of high-concentration Reactive yellow 3 with $CuFe_2O_4$/PAC. *Journal of Chemical Technology and Biotechnology*, *90*(10), 1861–1868.

62. He, H., Yang, S., Yu, K., Ju, Y., Sun, C., Wang, L. (2010). Microwave induced catalytic degradation of crystal violet in nano-nickel dioxide suspensions. *Journal of Hazardous Materials*, *173*(1), 393–400.

63. Lv, G., Xing, X., Liao, L., An, P., Yin, H., Mei, L., Li, Z. (2017). Synthesis of birnessite with adjustable electron spin magnetic moments for the degradation of tetracycline under microwave induction. *Chemical Engineering Journal*, *326*, 329–338.

64. Tsubaki, S., Oono, K., Ueda, T., Onda, A., Yanagisawa, K., Mitani, T., Azuma, J. I. (2013). Microwave-assisted hydrolysis of polysaccharides over polyoxometalate clusters. *Bioresource Technology*, *144*, 67–73.

65. Li, X., Xu, F., Wang, J., Zhang, C., Chen, Y., Zhu, S., Shen, S. (2010). Preparation of Fe-Cu catalysts and treatment of a wastewater mixture by microwave-assisted UV catalytic oxidation processes. *Environmental Technology*, *31*(4), 433–443.

66. Zhang, L., Su, M., Guo, X. (2008). Studies on the treatment of brilliant green solution by combination microwave induced oxidation with $CoFe_2O_4$. *Separation and Purification Technology*, *62*(2), 458–463.

67. Zhang, L., Su, M., Liu, N., Zhou, X., Kang, P. (2009). Degradation of malachite green solution using combined microwave and $ZnFe_2O_4$ powder. *Water Science and Technology*, *60*(10), 2563–2569.

68. Bi, X., Wang, P., Jiao, C., Cao, H. (2009). Degradation of remazol golden yellow dye wastewater in microwave enhanced $ClO_2$ catalytic oxidation process. *Journal of Hazardous Materials*, *168*(2), 895–900.

69. Badamali, S. K., Luque, R., Clark, J. H., Breeden, S. W. (2009). Microwave assisted oxidation of a lignin model phenolic monomer using Co(salen)/SBA-15. *Catalysis Communications*, *10*(6), 1010–1013.

70. Bi, X. Y., Peng, W., Jiang, H., Xu, H. Y., Shi, S. J., Huang, J. L. (2007). Treatment of phenol wastewater by microwave induced $ClO_2$ $CuO_x$/$Al_2O_3$ catalytic oxidation process. *Journal of Environmental Sciences*, *19*(12), 1510–1515.

71. Quan, X., Zhang, Y., Chen, S., Zhao, Y., Yang, F. (2007). Generation of hydroxyl radical in aqueous solution by microwave energy using activated carbon as catalyst and its potential in removal of persistent organic substances. *Journal of Molecular Catalysis A: Chemical*, *263*(1), 216–222.

72. Riaz, U., Ashraf, S. M., Aqib, M. (2014). Microwave-assisted degradation of acid orange using a conjugated polymer, polyaniline, as catalyst. *Arabian Journal of Chemistry*, *7*(1), 79–86.

73. Zhang, L., Liu, X., Guo, X., Su, M., Xu, T., Song, X. (2011). Investigation on the degradation of brilliant green induced oxidation by $NiFe_2O_4$ under microwave irradiation. *Chemical Engineering Journal*, *173*(3), 737–742.

74. Zhang, Z., Deng, Y., Shen, M., Han, W., Chen, Z., Xu, D., Ji, X. (2009). Investigation on rapid degradation of sodium dodecyl benzene sulfonate (SDBS) under microwave irradiation in the presence of modified activated carbon powder with ferrous sulfate. *Desalination*, *249*(3), 1022–1029.

75. Shiying, Y., Ping, W., Xin, Y., Guang, W. E. I., Zhang, W., Liang, S. H. A. N. (2009). A novel advanced oxidation process to degrade organic pollutants in wastewater: Microwave-activated persulfate oxidation. *Journal of Environmental Sciences*, *21*(9), 1175–1180.

76. Remya, N., Lin, J. G. (2011). Microwave-assisted carbofuran degradation in the presence of GAC, ZVI and $H_2O_2$: Influence of reaction temperature and pH. *Separation and Purification Technology*, *76*(3), 244–252.

77. Wang, N., Zheng, T., Jiang, J., Lung, W. S., Miao, X., Wang, P. (2014). Pilot-scale treatment of p-nitrophenol wastewater by microwave-enhanced Fenton oxidation process: Effects of system parameters and kinetics study. *Chemical Engineering Journal*, *239*, 351–359

78. Homem, V., Alves, A., Santos, L. (2013). Microwave-assisted Fenton's oxidation of amoxicillin. *Chemical Engineering Journal*, *220*, 35–44.

79. Wang, N., Zheng, T., Jiang, J., Wang, P. (2015). Cu (II)–Fe (II)–$H_2O_2$ oxidative removal of 3-nitroaniline in water under microwave irradiation. *Chemical Engineering Journal*, *260*, 386–392.

80. Li, S., Zhang, G., Wang, P., Zheng, H., Zheng, Y. (2016). Microwave-enhanced Mn-Fenton process for the removal of BPA in water. *Chemical Engineering Journal*, *294*, 371–379.

81. Cravotto, G., Di Carlo, S., Curini, M., Tumiatti, V., Roggero, C. (2007). A new flow reactor for the treatment of polluted water with microwave and ultrasound. *Journal of Chemical Technology and Biotechnology*, *82*(2), 205–208.

82. Xi, B., Shi, Q. (2013). Removement of thiocyanate from industrial wastewater by microwave-Fenton oxidation method. *Journal of Environmental Sciences*, *25*, S201–S204.

83. Pan, W., Zhang, G., Zheng, T., Wang, P. (2015). Degradation of p-nitrophenol using $CuO/Al_2O_3$ as a Fenton-like catalyst under microwave irradiation. *RSC Advances*, *5*(34), 27043–27051.

84. Lin, Q., Pan, H., Yao, K., Pan, Y., Long, W. (2015). Competitive removal of Cu–EDTA and Ni–EDTA via microwave-enhanced Fenton oxidation with hydroxide precipitation. *Water Science and Technology*, *72*(7), 1184–1190.

85. Cravotto, G., Binello, A., Di Carlo, S., Orio, L., Wu, Z. L., Ondruschka, B. (2010). Oxidative degradation of chlorophenol derivatives promoted by microwaves or power ultrasound: A mechanism investigation. *Environmental Science and Pollution Research*, *17*(3), 674–687.

86. Xu, D., Zhang, Y., Cheng, F., Dai, P. (2016). Efficient removal of dye from an aqueous phase using activated carbon supported ferrihydrite as heterogeneous Fenton-like catalyst under assistance of microwave irradiation. *Journal of the Taiwan Institute of Chemical Engineers*, *60*, 376–382.

87. Xu, X. C., Zhang, H. T., Dong, Z. Y., Fan, Y. F. (2013). Pretreatment of old-age landfill leachate by microwave-assisted catalytic oxidation in the presence of activated carbon. *Environmental Technology*, *34*(20), 2853–2858.

88. Liu, S. T., Huang, J., Ye, Y., Zhang, A. B., Pan, L., Chen, X. G. (2013). Microwave enhanced Fenton process for the removal of methylene blue from aqueous solution. *Chemical Engineering Journal*, *215*, 586–590.

89. Cravotto, G., Di Carlo, S., Ondruschka, B., Tumiatti, V., Roggero, C. M. (2007). Decontamination of soil containing POPs by the combined action of solid Fenton-like reagents and microwaves. *Chemosphere*, *69*(8), 1326–1329.

90. Moradi, S. E., Dadfarnia, S., Haji, A. M., Shabani, S. (2017). Microwave-enhanced Fenton-like degradation by surface-modified metal-organic frameworks as a promising method for removal of dye from aqueous samples. *Turkish Journal of Chemistry*, *41*, 426–439.

91. Ouyang, X., Huang, X., Ruan, T., Qiu, X. (2015). Microwave-assisted oxidative digestion of lignin with hydrogen peroxide for TOC and color removal. *Water Science and Technology*, *71*(3), 390–396.

92. Oncu, N. B., Balcioglu, I. A. (2013). Microwave-assisted chemical oxidation of biological waste sludge: Simultaneous micropollutant degradation and sludge solubilization. *Bioresource Technology*, *146*, 126–134.

93. Chang, Y. M., Tsai, K. S., Tseng, C. H., Chen, J. H., Kao, C. M., Lin, K. L. (2015). Rapid nonylphenol degradation in wastewater sludge using microwave peroxide oxidation with nitric acid. *Environmental Progress and Sustainable Energy, 34*(2), 520–525.

94. Chang, Y. M., Tsai, K. S., Wang, L. P., Kao, J. C., Lin, K. L., Chou, C. M. (2015). Endocrine-disrupting chemical degradation in hazardous waste using microwave peroxide oxidation and acid. *Environmental Engineering Science, 32*(11), 907–911.

95. Peng, L., Deng, D., Ye, F. (2016). Efficient oxidation of high levels of soil-sorbed phenanthrene by microwave-activated persulfate: Implication for *in situ* subsurface remediation engineering. *Journal of Soils and Sediments, 16*(1), 28–37.

96. Chou, Y. C., Lo, S. L., Kuo, J., Yeh, C. J. (2015). Microwave-enhanced persulfate oxidation to treat mature landfill leachate. *Journal of Hazardous Materials, 284*, 83–91.

97. Kim, Y. B., Ahn, J. H. (2015). Microwave-assisted decomposition of landfill leachate with persulfate. *Journal of Environmental Engineering, 142*(3), 04015084

98. Qi, C., Liu, X., Lin, C., Zhang, H., Li, X., Ma, J. (2017). Activation of peroxymonosulfate by microwave irradiation for degradation of organic contaminants. *Chemical Engineering Journal, 315*, 201–209.

99. Liu, B., Li, S., Zhao, Y., Wu, W., Zhang, X., Gu, X., Li, Y., Yang, S. (2010). Enhanced degradation of 4-nitrophenol by microwave assisted Fe/EDTA process. *Journal of Hazardous Materials, 176*(1), 213–219.

100. Lee, H. Y., Lee, C. L., Jou, C. J. G. (2010). Comparison degradation of pentachlorophenol using microwave-induced nanoscale $Fe^0$ and activated carbon. *Water, Air, and Soil Pollution, 211*(1–4), 17–24.

101. Riaz, U., Ashraf, S. M., Budhiraja, V., Aleem, S., Kashyap, J. (2016). Comparative studies of the photocatalytic and microwave-assisted degradation of alizarin red using ZnO/poly (1-naphthylamine) nanohybrids. *Journal of Molecular Liquids, 216*, 259–267.

102. Chen, S., Chen, H., Gao, R., Li, L., Yang, X., Wu, Y., Hu, X. (2015). Degradation of hyaluronic acid derived from tilapia eyeballs by a combinatorial method of microwave, hydrogen peroxide, and ascorbic acid. *Polymer Degradation and Stability, 112*, 117–121.

103. Yuan, N. N., Hong, J. (2012). The research on rhodamine-B degradation in $MW/H_2O_2$ system under alkaline environment. *Applied Mechanics and Materials, 105*, 1505–1508.

104. Eskicioglu, C., Prorot, A., Marin, J., Droste, R. L., Kennedy, K. J. (2008). Synergetic pretreatment of sewage sludge by microwave irradiation in presence of $H_2O_2$ for enhanced anaerobic digestion. *Water Research, 42*(18), 4674–4682.

105. Prasannakumar, B. R., Regupathi, I., Murugesan, T. (2009). An optimization study on microwave irradiated decomposition of phenol in the presence of $H_2O_2$. *Journal of Chemical Technology and Biotechnology, 84*(1), 83–91

106. Ravera, M., Buico, A., Gosetti, F., Cassino, C., Musso, D., Osella, D. (2009). Oxidative degradation of 1, 5-naphthalenedisulfonic acid in aqueous solutions by microwave irradiation in the presence of $H_2O_2$. *Chemosphere, 74*(10), 1309–1314.

107. Remya, N., Lin, J. G. (2011). Carbofuran degradation by the application of MW-assisted $H_2O_2$ process. *Journal of Environmental Science and Health Part B, 46*(4), 350–359.

108. Asgari, G., Seidmohammadi, A., Chavoshani, A., Rahmani, A. R. (2013). Microwave/$H_2O_2$ efficiency in pentachlorophenol removal from aqueous solutions. *Journal of Research in Health Sciences, 14*(1), 36–39.

109. Zeng, H., Lu, L., Liang, M., Liu, J., Li, Y. (2012). Degradation of trace nitrobenzene in water by microwave-enhanced $H_2O_2$-based process. *Frontiers of Environmental Science and Engineering, 6*(4), 477–483.

110. Ju, Y., Yang, S., Ding, Y., Sun, C., Gu, C., He, Z., Qin, C., Hea, H., Xu, B. (2009). Microwave-enhanced $H_2O_2$-based process for treating aqueous malachite green solutions: Intermediates and degradation mechanism. *Journal of Hazardous Materials, 171*(1), 123–132.

111. Lee, Y. C., Lo, S. L., Chiueh, P. T., Chang, D. G. (2009). Efficient decomposition of per-fluorocarboxylic acids in aqueous solution using microwave-induced persulfate. *Water Research*, *43*(11), 2811–2816

112. Vilhunen, S., Sillanpää, M. (2010). Recent developments in photochemical and chemical AOPs in water treatment: A mini-review. *Reviews in Environmental Science and Bio/Technology*, *9*(4), 323–330.

113. Chemat, S., Aouabed, A., Bartels, P. V., Esveld, D. C., Chemat, F. (1999). An original microwave-ultraviolet combined reactor suitable for organic synthesis and degradation. *Journal of Microwave Power and Electromagnetic Energy*, *34*(1), 55–60.

114. Chen, C., Ma, W., Zhao, J. (2010). Semiconductor-mediated photodegradation of pollutants under visible-light irradiation. *Chemical Society Reviews*, *39*(11), 4206–4219.

115. Falciglia, P. P., Catalfo, A., Finocchiaro, G., Vagliasindi, F. G., Romano, S., De Guidi, G. (2017). Microwave heating coupled with UV-A irradiation for PAH removal from highly contaminated marine sediments and subsequent photo-degradation of the generated vaporized organic compounds. *Chemical Engineering Journal*, *334*, 172–183.

116. Trovó, A. G., Melo, S. A. S., Nogueira, R. F. P. (2008). Photodegradation of the pharmaceuticals amoxicillin, bezafibrate and paracetamol by the photo-Fenton process—Application to sewage treatment plant effluent. *Journal of Photochemistry and Photobiology A: Chemistry*, *198*(2), 215–220.

117. Zhang, X., Wang, Y., Li, G. (2005). Effect of operating parameters on microwave assisted photocatalytic degradation of azo dye X-3B with grain $TiO_2$ catalyst. *Journal of Molecular Catalysis A: Chemical*, *237*(1), 199–205.

118. Ki, S. J., Jeon, K. J., Park, Y. K., Jeong, S., Lee, H., Jung, S. C. (2017). Improving removal of 4-chlorophenol using a $TiO_2$ photocatalytic system with microwave and ultraviolet radiation. *Catalysis Today*, *293*, 15–22.

119. Horikoshi, S., Hidaka, H., Serpone, N. (2003). Hydroxyl radicals in microwave photocatalysis. Enhanced formation of OH radicals probed by ESR techniques in microwave-assisted photocatalysis in aqueous $TiO_2$ dispersions. *Chemical Physics Letters*, *376*(3), 475–480.

120. Horikoshi, S., Hidaka, H., Serpone, N. (2002). Environmental remediation by an integrated microwave/UV-illumination method. 1. Microwave-assisted degradation of rhodamine-B dye in aqueous $TiO_2$ dispersions. *Environmental Science and Technology*, *36*(6), 1357–1366.

121. Horikoshi, S., Saitou, A., Hidaka, H., Serpone, N. (2003). Environmental remediation by an integrated microwave/UV illumination method. V. Thermal and nonthermal effects of microwave radiation on the photocatalyst and on the photodegradation of rhodamine-B under UV/V is radiation. *Environmental Science and Technology*, *37*(24), 5813–5822.

122. Horikoshi, S., Hidaka, H., Serpone, N. (2002). Environmental remediation by an integrated microwave/UV-illumination method II: Characteristics of a novel UV–VIS–microwave integrated irradiation device in photodegradation processes. *Journal of Photochemistry and Photobiology A: Chemistry*, *153*(1), 185–189.

123. Horikoshi, S., Hidaka, H., Serpone, N. (2002). Environmental remediation by an integrated microwave/UV illumination technique. 3. A microwave-powered plasma light source and photoreactor to degrade pollutants in aqueous dispersions of $TiO_2$ illuminated by the emitted UV/visible radiation. *Environmental Science and Technology*, *36*(23), 5229–5237.

124. Horikoshi, S., Hojo, F., Hidaka, H., Serpone, N. (2004). Environmental remediation by an integrated microwave/UV illumination technique. 8. Fate of carboxylic acids, aldehydes, alkoxycarbonyl and phenolic substrates in a microwave radiation field in the presence of $TiO_2$ particles under UV irradiation. *Environmental Science and Technology*, *38*(7), 2198–2208.

125. Horikoshi, S., Sakai, F., Kajitani, M., Abe, M., Serpone, N. (2009). Microwave frequency effects on the photoactivity of $TiO_2$: Dielectric properties and the degradation of 4-chlorophenol, bisphenol A and methylene blue. *Chemical Physics Letters*, *470*(4), 304–307.

126. Horikoshi, S., Abe, M., Serpone, N. (2009). Influence of alcoholic and carbonyl functions in microwave-assisted and photo-assisted oxidative mineralization. *Applied Catalysis B: Environmental*, *89*(1), 284–287.

127. Horikoshi, S., Hidaka, H., Serpone, N. (2003). Environmental remediation by an integrated microwave/UV-illumination technique: IV. Non-thermal effects in the microwave-assisted degradation of 2,4-dichlorophenoxyacetic acid in UV-irradiated $TiO_2$/$H_2O$ dispersions. *Journal of Photochemistry and Photobiology A: Chemistry*, *159*(3), 289–300.

128. Horikoshi, S., Kajitani, M., Sato, S., Serpone, N. (2007). A novel environmental risk-free microwave discharge electrodeless lamp (MDEL) in advanced oxidation processes: Degradation of the 2,4-D herbicide. *Journal of Photochemistry and Photobiology A: Chemistry*, *189*(2), 355–363.

129. Horikoshi, S., Miura, T., Kajitani, M., Serpone, N. (2008). Microwave discharge electrodeless lamps (MDEL). III. A novel tungsten-triggered MDEL device emitting VUV and UVC radiation for use in wastewater treatment. *Photochemical and Photobiological Sciences*, *7*(3), 303–310.

130. Horikoshi, S., Tsuchida, A., Sakai, H., Abe, M., Serpone, N. (2011). Microwave discharge electrodeless lamps (MDELs). VI. Performance evaluation of a novel microwave discharge granulated electrodeless lamp (MDGEL)—Photoassisted defluorination of perfluoroalkoxy acids in aqueous media. *Journal of Photochemistry and Photobiology A: Chemistry*, *222*(1), 97–104.

131. Zhanqi, G., Shaogui, Y., Na, T., Cheng, S. (2007). Microwave assisted rapid and complete degradation of atrazine using $TiO_2$ nanotube photocatalyst suspensions. *Journal of Hazardous Materials*, *145*(3), 424–430.

132. Karthikeyan, S., Gopalakrishnan, A. N. (2017). Effect of microwaves on photocatalytic degradation of phenol and cresols in a microwave-ultraviolet reactor. *Environmental Engineering and Management Journal*, *16*(2).

133. Yang, S., Fu, H., Sun, C., Gao, Z. (2009). Rapid photocatalytic destruction of pentachlorophenol in F–Si-comodified $TiO_2$ suspensions under microwave irradiation. *Journal of Hazardous Materials*, *161*(2), 1281–1287.

134. Moreira, A. J., Borges, A. C., Gouvea, L. F., MacLeod, T. C., Freschi, G. P. (2017). The process of atrazine degradation, its mechanism, and the formation of metabolites using UV and UV/MW photolysis. *Journal of Photochemistry and Photobiology A: Chemistry*, *347*, 160–167.

135. Moreira, A. J., Pinheiro, B. S., Araújo, A. F., Freschi, G. P. (2017). Evaluation of atrazine degradation applied to different energy systems. *Environmental Science and Pollution Research*, *24*(7), 6398–6398.

136. Parolin, F., Nascimento, U. M., Azevedo, E. B. (2013). Microwave-enhanced UV/$H_2O_2$ degradation of an azo dye (tartrazine): Optimization, colour removal, mineralization and ecotoxicity. *Environmental Technology*, *34*(10), 1247–1253.

137. Chen, H., Bramanti, E., Longo, I., Onor, M., Ferrari, C. (2011). Oxidative decomposition of atrazine in water in the presence of hydrogen peroxide using an innovative microwave photochemical reactor. *Journal of Hazardous Materials*, *186*(2), 1808–1815.

138. Gromboni, C. F., Kamogawa, M. Y., Ferreira, A. G., Nóbrega, J. A., Nogueira, A. R. A. (2007). Microwave-assisted photo-Fenton decomposition of chlorfenvinphos and cypermethrin in residual water. *Journal of Photochemistry and Photobiology A: Chemistry*, *185*(1), 32–37.

139. Wang, A., Zhang, Y., Zhong, H., Chen, Y., Tian, X., Li, D., Li, J. (2018). Efficient mineralization of antibiotic ciprofloxacin in acid aqueous medium by a novel

photoelectro-Fenton process using a microwave discharge electrodeless lamp irradiation. *Journal of Hazardous Materials, 342*, 364–374

140. Shokri, A., Joshagani, A. H. (2016). Using microwave along with $TiO_2$ for degradation of 4-chloro-2-nitrophenol in aqueous environment. *Russian Journal of Applied Chemistry, 89*(12), 1985–1990.

141. Barik, A. J., Kulkarni, S. V., Gogate, P. R. (2016). Degradation of 4-chloro 2-aminophenol using combined approaches based on microwave and photocatalysis. *Separation and Purification Technology, 168*, 152–160.

142. Lee, H., Park, S. H., Park, Y. K., Kim, S. J., Seo, S. G., Ki, S. J., Jung, S. C. (2015). Photocatalytic reactions of 2,4-dichlorophenoxyacetic acid using a microwave-assisted photocatalysis system. *Chemical Engineering Journal, 278*, 259–264.

143. Seo, S. G., Park, Y. K., Kim, S. J., Lee, H., Jung, S. C. (2014). Photodegradation of HCFC-22 using microwave discharge electrodeless mercury lamp with $TiO_2$ photocatalyst balls. *Journal of Chemistry, 2014*, 584693.

144. Ju, Y., Qiao, J., Peng, X., Xu, Z., Fang, J., Yang, S., Sun, C. (2013). Photodegradation of malachite green using UV–vis light from two microwave-powered electrodeless discharge lamps (MPEDL-2): Further investigation on products, dominant routes and mechanism. *Chemical Engineering Journal, 221*, 353–362.

145. Xiong, Z., Xu, A., Li, H., Ruan, X., Xia, D., Zeng, Q. (2012). Highly efficient photodegradation of Alizarin Green in $TiO_2$ suspensions using a microwave powered electrodeless discharged lamp. *Industrial and Engineering Chemistry Research, 52*(1), 362–369.

146. Genuino, H. C., Hamal, D. B., Fu, Y. J., Suib, S. L. (2012). Synergetic effects of ultraviolet and microwave radiation for enhanced activity of $TiO_2$ nanoparticles in degrading organic dyes using a continuous-flow reactor. *The Journal of Physical Chemistry C, 116*(26), 14040–14051.

147. Cai, Y. J., Lin, L. N., Xia, D. S., Zeng, Q. F., Zhu, H. L. (2011). Degradation of reactive brilliant red X-3B dye by microwave electrodeless UV irradiation. *CLEAN – Soil, Air, Water, 39*(1), 68–73.

148. Ju, Y., Fang, J., Liu, X., Xu, Z., Ren, X., Sun, C., Yang, S., Ren, Q., Ding, Y., Yu, L., Wei, Z., Wang, L. (2011). Photodegradation of crystal violet in $TiO_2$ suspensions using UV–vis irradiation from two microwave-powered electrodeless discharge lamps (EDL-2): Products, mechanism and feasibility. *Journal of Hazardous Materials, 185*(2), 1489–1498.

149. Shin, H. C., Park, S. H., Ahn, H. G., Chung, M., Kim, S. J., Seo, S. G., Jung, S. C. (2011). The effect of microwave-assisted for photo-catalytic degradation of rhodamine B in aqueous nano $TiO_2$ particles dispersions. *Journal of Nanoscience and Nanotechnology, 11*(2), 1597–1600.

150. Gao, Z., Yang, S., Sun, C., Hong, J. (2007). Microwave assisted photocatalytic degradation of pentachlorophenol in aqueous $TiO_2$ nanotubes suspension. *Separation and Purification Technology, 58*(1), 24–31.

151. Gu, D., Huang, L., Shao, C., Fang, H., Zhang, R., Hou, H. (2007). Photodegradation of rhodamine B dye using a microwave electrodeless UV lamp (MWUVL). *Frontiers of Chemistry in China, 2*(4), 436–441.

152. Jeong, S., Lee, H., Park, H., Jeon, K. J., Park, Y. K., Jung, S. C. (2018). Rapid photocatalytic degradation of nitrobenzene under the simultaneous illumination of UV and microwave radiation fields with a $TiO_2$ ball catalyst. *Catalysis Today, 307*, 65–72.

153. Zhang, D., Sun, B., Duan, L., Tao, Y., Xu, A., Li, X. (2016). Photooxidation of guaiacol to organic acids with hydrogen peroxide by microwave discharge electrodeless lamps. *Chemical Engineering and Technology, 39*(1), 97–101.

154. Park, S. H., Kim, S. J., Seo, S. G., Jung, S. C. (2010). Assessment of microwave/UV/ $O_3$ in the photo-catalytic degradation of bromothymol blue in aqueous nano TiO 2 particles dispersions. *Nanoscale Research Letters, 5*(10), 1627.

155. Rezaee, M., Assadi, Y., Hosseini, M. R. M., Aghaee, E., Ahmadi, F., Berijani, S. (2006). Determination of organic compounds in water using dispersive liquid–liquid microextraction. *Journal of Chromatography A*, *1116*(1), 1–9.

156. Hashemi, B., Zohrabi, P., Kim, K. H., Shamsipur, M., Deep, A., Hong, J. (2017). Recent advances in liquid-phase microextraction techniques for the analysis of environmental pollutants. *TrAC Trends in Analytical Chemistry*, *97*, 83–95.

157. Trujillo-Rodríguez, M. J., Rocío-Bautista, P., Pino, V., Afonso, A. M. (2013). Ionic liquids in dispersive liquid-liquid microextraction. *TrAC Trends in Analytical Chemistry*, *51*, 87–106.

158. Kang, H., Wang, X., Zhang, Y., Wu, J., Wang, H. (2015). Simultaneous extraction of bisphenol A and tetrabromobisphenol A from milk by microwave-assisted ionic liquid microextraction. *RSC Advances*, *5*(19), 14631–14636.

159. Wang, Z., Hu, J., Du, H., He, S., Li, Q., Zhang, H. (2016). Microwave-assisted ionic liquid homogeneous liquid–liquid microextraction coupled with high performance liquid chromatography for the determination of anthraquinones in *Rheum palmatum* L. *Journal of Pharmaceutical and Biomedical Analysis*, *125*, 178–185.

160. Feng, X. D., Liang, F. H., Su, R., Wu, L. J., Li, X. Y., Wang, X., Zhang, H., Yu, A. M. (2013). Ionic liquid-based microwave-assisted liquid-liquid microextraction and high performance liquid chromatography determination of sulfonamides from animal oils. *Chemical Research in Chinese Universities*, *29*(4), 647–652.

161. Xu, X., Su, R., Zhao, X., Liu, Z., Zhang, Y., Li, D., Li, X., Zhang, H., Wang, Z. (2011). Ionic liquid-based microwave-assisted dispersive liquid–liquid microextraction and derivatization of sulfonamides in river water, honey, milk, and animal plasma. *Analytica Chimica Acta*, *707*(1), 92–99.

162. Xu, X., Liu, Z., Zhao, X., Su, R., Zhang, Y., Shi, J., Zhao, Y., Wu, L., Ma, Q., Zhou, X., Ziming, W., Zhang, H. (2013). Ionic liquid-based microwave-assisted surfactant-improved dispersive liquid–liquid microextraction and derivatization of aminoglycosides in milk samples. *Journal of Separation Science*, *36*(3), 585–592.

163. Xu, X., Su, R., Zhao, X., Liu, Z., Li, D., Li, X., Zhang, H., Wang, Z. (2011). Determination of formaldehyde in beverages using microwave-assisted derivatization and ionic liquid-based dispersive liquid–liquid microextraction followed by high-performance liquid chromatography. *Talanta*, *85*(5), 2632–2638.

164. Wang, R., Su, P., Yang, Y. (2013). Optimization of ionic liquid-based microwave-assisted dispersive liquid–liquid microextraction for the determination of plasticizers in water by response surface methodology. *Analytical Methods*, *5*(4), 1033–1039.

165. Wang, J., Xiong, J., Baker, G. A., JiJi, R. D., Baker, S. N. (2013). Developing microwave-assisted ionic liquid microextraction for the detection and tracking of hydrophobic pesticides in complex environmental matrices. *RSC Advances*, *3*(38), 17113–17119.

166. Su, R., Li, D., Wu, L., Han, J., Lian, W., Wang, K., Yang, H. (2017). Determination of triazine herbicides in juice samples by microwave-assisted ionic liquid/ionic liquid dispersive liquid–liquid microextraction coupled with high-performance liquid chromatography. *Journal of Separation Science*, *40*(14), 2950–2958.

167. Zhong, Q., Su, P., Zhang, Y., Wang, R., Yang, Y. (2012). *In-situ* ionic liquid-based microwave-assisted dispersive liquid–liquid microextraction of triazine herbicides. *Microchimica Acta*, *178*(3–4), 341–347.

168. Gao, S., You, J., Zheng, X., Wang, Y., Ren, R., Zhang, R., Bai, Y., Zhang, H. (2010). Determination of phenylurea and triazine herbicides in milk by microwave assisted ionic liquid microextraction high-performance liquid chromatography. *Talanta*, *82*(4), 1371–1377.

169. Kokosa, J. M. (2015). Recent trends in using single-drop microextraction and related techniques in green analytical methods. *TrAC Trends in Analytical Chemistry*, *71*, 194–204.

170. Gholivand, M. B., Piryaei, M., Abolghasemi, M. M., Papzan, A. (2013). Comparison of microwave-assisted headspace single-drop microextraction (MA-HS-SDME) with hydrodistillation for the determination of volatile compounds from *Prangos uloptera*. *Journal of Essential Oil Research*, 25(1), 49–54.

171. Vidal, L., Domini, C. E., Grané, N., Psillakis, E., Canals, A. (2007). Microwave-assisted headspace single-drop microextration of chlorobenzenes from water samples. *Analytica Chimica Acta*, 592(1), 9–15.

172. Mansour, F. R., Danielson, N. D. (2017). Solidification of floating organic droplet in dispersive liquid-liquid microextraction as a green analytical tool. *Talanta*, 170, 22–35.

173. Song, Y., Wu, L., Lu, C., Li, N., Hu, M., Wang, Z. (2014). Microwave-assisted liquid–liquid microextraction based on solidification of ionic liquid for the determination of sulfonamides in environmental water samples. *Journal of Separation Science*, 37(23), 3533–3538.

174. Song, Y., Wu, L., Li, N., Hu, M., Wang, Z. (2015). Utilization of a novel microwave-assisted homogeneous ionic liquid microextraction method for the determination of Sudan dyes in red wines. *Talanta*, 135, 163–169.

175. Hu, M., Wu, L., Song, Y., Li, Z., Ma, Q., Zhang, H., Wang, Z. (2015). Microwave-assisted liquid–liquid microextraction based on solidification of floating organic droplet for the determination of triazines in honey samples. *Analytical Methods*, 7(21), 9114–9120.

176. Zhang, Y., Guo, W., Yue, Z., Lin, L., Zhao, F., Chen, P., Wu, W., Zhu, H., Yang, B., Kuang, Y., Wang, J. (2017). Rapid determination of 54 pharmaceutical and personal care products in fish samples using microwave-assisted extraction—Hollow fiber—Liquid/solid phase microextraction. *Journal of Chromatography B*, 1051, 41–53.

177. Alsharaa, A., Basheer, C., Sajid, M. (2015). Single-step microwave assisted headspace liquid-phase microextraction of trihalomethanes and haloketones in biological samples. *Journal of Chromatography B*, 1007, 43–48.

178. Shi, Y. A., Chen, M. Z., Muniraj, S., Jen, J. F. (2008). Microwave-assisted headspace controlled temperature liquid-phase microextraction of chlorophenols from aqueous samples for gas chromatography-electron capture detection. *Journal of Chromatography A*, 1207(1), 130–135.

179. Kumar, P. V., Jen, J. F. (2011). Rapid determination of dichlorodiphenyltrichloroethane and its main metabolites in aqueous samples by one-step microwave-assisted headspace controlled-temperature liquid-phase microextraction and gas chromatography with electron capture detection. *Chemosphere*, 83(2), 200–207.

180. Li, H. P., Li, G. C., Jen, J. F. (2003). Determination of organochlorine pesticides in water using microwave assisted headspace solid-phase microextraction and gas chromatography. *Journal of Chromatography A*, 1012(2), 129–137.

181. Li, J., Wang, Y. B., Li, K. Y., Cao, Y. Q., Wu, S., Wu, L. (2015). Advances in different configurations of solid-phase microextraction and their applications in food and environmental analysis. *TrAC Trends in Analytical Chemistry*, 72, 141–152.

182. Gholivand, M. B., Piryaei, M., Abolghasemi, M. M. (2013). Analysis of volatile oil composition of *Citrus aurantium* L. by microwave-assisted extraction coupled to headspace solid-phase microextraction with nanoporous based fibers. *Journal of Separation Science*, 36(5), 872–877.

183. Wu, M. W., Yeh, P. C., Chen, H. C., Liu, L. L., Ding, W. H. (2013). A microwave-assisted headspace solid-phase microextraction for rapid determination of synthetic polycyclic and nitro-aromatic musks in fish samples. *Journal of the Chinese Chemical Society*, 60(9), 1169–1174.

184. Sha, Y., Huang, D., Zheng, S., Liu, B., Deng, C. (2013). Development of microwave-assisted headspace solid-phase microextraction followed by gas chromatography-mass spectrometry for the analysis of phenol in a cigarette pad. *Analytical Methods*, 5(18), 4655–4659.

185. Hsieh, P. C., Jen, J. F., Lee, C. L., Chang, K. C. (2015). Determination of polycyclic aromatic hydrocarbons in environmental water samples by microwave-assisted headspace solid-phase microextraction. *Environmental Engineering Science*, *32*(4), 301–309.

186. Wei, M. C., Jen, J. F. (2007). Determination of polycyclic aromatic hydrocarbons in aqueous samples by microwave assisted headspace solid-phase microextraction and gas chromatography/flame ionization detection. *Talanta*, *72*(4), 1269–1274.

187. Wu, Y. P., Wang, Y. C., Ding, W. H. (2012). Rapid determination of alkylphenols in aqueous samples by *in situ* acetylation and microwave-assisted headspace solid-phase microextraction coupled with gas chromatography–mass spectrometry. *Journal of Separation Science*, *35*(16), 2122–2130.

188. Tsao, Y. C., Wang, Y. C., Wu, S. F., Ding, W. H. (2011). Microwave-assisted headspace solid-phase microextraction for the rapid determination of organophosphate esters in aqueous samples by gas chromatography-mass spectrometry. *Talanta*, *84*(2), 406–410.

189. Li, H. P., Lin, C. H., Jen, J. F. (2009). Analysis of aqueous pyrethroid residuals by one-step microwave-assisted headspace solid-phase microextraction and gas chromatography with electron capture detection. *Talanta*, *79*(2), 466–471.

190. Huang, Y., Yang, Y. C., Shu, Y. Y. (2007). Analysis of semi-volatile organic compounds in aqueous samples by microwave-assisted headspace solid-phase microextraction coupled with gas chromatography–electron capture detection. *Journal of Chromatography A*, *1140*(1), 35–43.

191. Wu, S. F., Liu, L. L., Ding, W. H. (2012). One-step microwave-assisted headspace solid-phase microextraction for the rapid determination of synthetic polycyclic musks in oyster by gas chromatography–mass spectrometry. *Food Chemistry*, *133*(2), 513–517.

192. Wu, S. F., Ding, W. H. (2010). Fast determination of synthetic polycyclic musks in sewage sludge and sediments by microwave-assisted headspace solid-phase microextraction and gas chromatography–mass spectrometry. *Journal of Chromatography A*, *1217*(17), 2776–2781.

193. Wang, Y. C., Ding, W. H. (2009). Determination of synthetic polycyclic musks in water by microwave-assisted headspace solid-phase microextraction and gas chromatography-mass spectrometry. *Journal of Chromatography A*, *1216*(40), 6858–6863.

194. Płotka-Wasylka, J., Szczepańska, N., de la Guardia, M., Namieśnik, J. (2015). Miniaturized solid-phase extraction techniques. *TrAC Trends in Analytical Chemistry*, *73*, 19–38.

195. Sajid, M., Basheer, C., Narasimhan, K., Choolani, M., Lee, H. K. (2015). Application of microwave-assisted micro-solid-phase extraction for determination of parabens in human ovarian cancer tissues. *Journal of Chromatography B*, *1000*, 192–198.

196. Feng, X. F., Jing, N., Li, Z. G., Wei, D., Lee, M. R. (2014). Ultrasound-microwave hybrid-assisted extraction coupled to headspace solid-phase microextraction for fast analysis of essential oil in dry traditional Chinese medicine by GC–MS. *Chromatographia*, *77*(7–8), 619–628.

197. Yang, Y., Chu, G., Zhou, G., Jiang, J., Yuan, K., Pan, Y., Song, Z., Li, Z., Xia, Q., Lu, X., Xiao, W. (2016). Rapid determination of the volatile components in tobacco by ultrasound-microwave synergistic extraction coupled to headspace solid-phase microextraction with gas chromatography-mass spectrometry. *Journal of Separation Science*, *39*(6), 1173–1181.

4 Functional Rare Earth-Based Micro/ Nanomaterials

*Fast Microwave Preparation and Their Properties*

Lei Wang and Shengliang Zhong

## CONTENTS

In this chapter, microwave (MW) preparation of functional rare earth (RE)-based micro/nanomaterials including inorganic compounds, coordination polymers (CPs), and composite materials were reviewed. Their applications were also summed up.

In the first part, the properties of functional RE-based micro/nanomaterials were briefly introduced and were compared with their corresponding bulk materials. Methods employed for the preparation of RE-based micro/nanomaterials were briefly introduced. Also, the advantages of the MW heating method were summarized.

In the second part, the MW preparation of RE-based inorganic compounds such as hydroxides, fluorides, phosphates, oxides, etc., was reviewed. Also, their properties were reviewed and were compared with the materials prepared by conventional methods.

In the third part, CPs were firstly introduced. Secondly, the characteristics of RE-based CPs and classifications were introduced. The MW preparation of RE-based coordination polymer was reviewed and their applications were discussed.

In the fourth part, the MW preparation of RE-based composite materials was reviewed. Their properties were also introduced.

In the end, the MW synthesis of RE-based micro/nanomaterials was summarized and forecasted.

## 4.1   INTRODUCTION

Rare earth (RE) elements contain lanthanide (Ln) elements with atomic number from 57 to 71, and scandium and yttrium, which have similar electronic structures and chemical properties to Ln elements. These elements are typically split into two sub-groups, the cerium sub-group of "Light" RE elements (LREEs) which includes La to Eu and the yttrium sub-group of "Heavy" RE elements (HREEs) which include the remaining Ln elements, Gd to Lu, as well as yttrium. Scandium, when it is classified as a RE element, is not included in either the LREE or HREE classifications (Gupta and Krishnamurthy 1992). RE bulk materials form a large family of functional materials with diverse applications in electric, magnetic, optical, and catalytic fields, originating from their unique 4f electrons, which have inspired great research interest in their nanoscale counterparts. Micro- and nanostructured materials generally exhibit new material properties as compared to their bulk counterparts. Also, these properties change with their size and shape. Surface effects (causing smooth properties scaling due to the fraction of atoms at the surface) and quantum effects (showing discontinuous behavior due to quantum confinement effects in materials with delocalized electrons) are the two primary factors that cause nanomaterials to behave significantly differently than bulk materials. These factors affect not only their mechanical, optical, electric, and magnetic properties, but also their chemical reactivity of materials (Roduner 2006). The typical RE nanomaterials exhibit certain size-insensitive luminescence behaviors, which is different from quantum dots. Their quasi-line emissions mostly rely on the intra-4f electron transitions and have much longer decay lifetimes. Furthermore, the apparent quenching concentration of luminescent dopants in RE nanomaterials is higher compared to that for corresponding bulk phosphors, which is usually due to the large surface effect of nanocrystalline materials (Yan et al. 2011). As a result, the design and the production of RE micro/nanostructures are of significant importance in order to fabricate materials with new properties and functions. Recently, novel properties have been discovered and exploited in many different areas.

Conventionally, RE-based phosphors, complex oxides, and sulfides are prepared via solid state reactions at high temperatures. Nevertheless, solution-based routes featuring low temperatures, simple apparatus, easy control, and versatile post-treatments are always thought to be the most important strategies. Especially, hydro- and solvothermal methods, and precipitation from high-boiling solvents are widely used. The experimental conditions, including temperature, acidity, surfactant, concentration, and reaction time, can be fine-tuned to regulate the RE nanostructures (Yan and Yan 2008).

MWs are high-frequency electromagnetic waves occupying the region of the spectrum ranging from 300–300,000 MHz and with wavelengths of between 1 mm and 1 m, which are between infrared and radio frequency waves in the electromagnetic spectrum. To avoid interference with telecommunication wavelengths, the wavelengths used by industrial and domestic MW apparatuses are regulated at both national and international levels. The commonly used frequency in laboratories and homes for MW heating is 2.45 GHz (with a wavelength of about 12.24 cm) (Zhu and Chen 2014). MWs possess some unique advantages for materials synthesis. The MW-assisted method, which can shorten chemical reaction times from hours to minutes, has attracted considerable attention. This high-speed synthesis effect results from the efficient in-core volumetric heating with the help of MW radiation, leading to heating directly inside the sample. Compared with conventional heating for chemical reactions, the MW dielectric heating offers the advantages of fast and uniform heating without thermal gradients, superheating of the solvents, and selective heating properties. Hence, the energy-efficient MW-assisted method has developed into a useful and rapid approach for the synthesis of nanosized inorganic materials (Gai et al. 2014). Owing to the ability of MWs to interact directly with the sample and to turn on and off quickly, which makes the particle growth of these materials much more controlled and uniform, and results in an improvement in the desired property of these materials (Collins 2010).

It has been demonstrated that MW heating can significantly increase the rates of chemical reactions by several orders of magnitude. Chemical reactions can be completed in very short periods of time (minutes) under MW irradiation. Otherwise, they would take hours or even days using the conventional heating methods. Even chemical reactions that do not occur under conventional heating conditions can be performed under similar conditions by MW heating. These experimental results cannot be explained by the effect of rapid MW heating alone, leading to the interpretation of the existence of the "specific MW effects" or "nonthermal MW effects" (Zhu and Chen 2014).

More and more RE-based functional materials have been prepared by the MW heating method during the past decade. Especially, materials at micro- and nanoscale have drawn intense interests during the past decade. In this chapter, we have summarized the comprehensive research on the preparation of RE-based micro- and nanomaterials via the MW-assisted method.

## 4.2 MW PREPARATION OF RE-BASED INORGANIC COMPOUNDS

As an efficient, energy-saving, and clean synthesis method, MW heating has been used for the synthesis of inorganic materials among which more and more RE inorganic structures (such as hydroxides, fluorides, phosphates, oxides, cobaltites, chromites, etc.) have been prepared by this method (Zhu and Chen 2014; Baghbanzadeh et al. 2011; Prado-Gonjal et al. 2013a; Gutierrez Seijas et al. 2017).

### 4.2.1 RE Oxides

RE oxides ($RE_2O_3$), the most stable RE compounds, have important applications in luminescent devices, catalysts, biochemical probes, and other functional materials due to their outstanding optical, electronic, and chemical properties resulting from their 4f sub-shell electrons (Majeed and Shivashankar 2014a; Shen et al. 2008; Reynolds et al. 2000; Quazi et al. 2016).

As one of the most frequently investigated RE compounds in recent years, ceria ($CeO_2$), a cubic fluorite-type oxide in which each cerium site is surrounded by eight oxygen sites in fcc arrangement and each oxygen site has a tetrahedron cerium site, is drawing intensive attention. It has been considered as a potential RE oxide for applications in several key areas including catalysts (Montini et al. 2016), fuel cells (Molenda et al. 2017), heavy ion removal (Yu et al. 2015), hydrogen storage materials (Sohlberg et al. 2001; Sangsefidi et al. 2017), gas sensors (Charbgoo et al. 2017); (He et al. 2015), ultraviolet absorbers (He et al. 2014; Saadat-Monfared et al. 2012), optical devices (He et al. 2015; Porosnicu et al. 2017), polishing materials (Shchukin and Caruso 2004; Peng et al. 2014), etc. (Yuan ct al. 2009; Tang et al. 2011). A variety of strategies have been applied to prepare $CeO_2$ nanoparticles (NPs). Among these various fabrication methods, MW possesses many benefits such as rapid and homogeneous volumetric heating, high reaction rates, high reaction selectivity, high product yield, excellent control of reaction parameters, and energy saving. For example, Lehnen et al. reported the synthesis of highly crystalline $CeO_2$ quantum dots (QDs) with a narrow size distribution from $[Ce(OtBu)_2L_2]$ (precursor 2) [L= DmoxCHC($CF_3$)O] by a MW-assisted solvothermal (MAS) process at 150°C within 15 min (Figure 4.1) (Lehnen et al. 2014). It was found that an expedient choice of precursor enabled the replacement of additives like bases or another mineralizer paves the way for preparing $CeO_2$ in a single-source MW synthesis, which is previously undisclosed.

Deus et al. prepared crystalline $CeO_2$ nanospheres by the MW-assisted hydrothermal (MAH) method at a lower temperature (100°C) costing less time (8 min) (Deus et al. 2013). The results showed that $CeO_2$ synthesized under MAH conditions using KOH and NaOH as mineralizer agents were well dispersed and homogeneously distributed while $CeO_2$ synthesized by MAH under $NH_4OH$ revealed

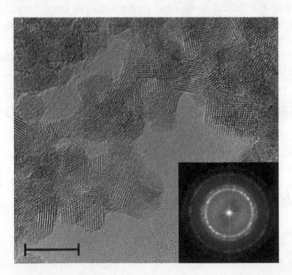

**FIGURE 4.1**    HR-TEM image (scale bar = 10 nm) and electron diffraction pattern of the sample prepared within 15 min using precursor 2. (Reprinted from Lehnen, T., J. Schlafer, and S. Mathur, 2014. Rapid Microwave Synthesis of $CeO_2$ Quantum Dots. *Zeitschrift fur Anorganische und Allgemeine Chemie* 640 (5): 819–825. With permission. Copyright 2014 Wiley-VCH.)

agglomerate particles (Figure 4.2). Muñoz and coworkers synthesized nanostructured $Gd_{0.1}Ce_{0.9}O_{1.95}$ (GDC10), $Gd_{0.2}Ce_{0.8}O_{1.90}$(GDC20), $Pr_{0.1}Ce_{0.9}O_{2-\delta}$ (PrDC10), and $Pr_{0.2}Ce_{0.8}O_{2-\delta}$ (PrDC20) spheres by MAH homogeneous co-precipitation (Figure 4.3) (Muñoz et al. 2015). Liu et al. obtained Core–shell $CeO_2$ micro/nanospheres through a MAS process followed by calcination in air (Liu et al. 2017).

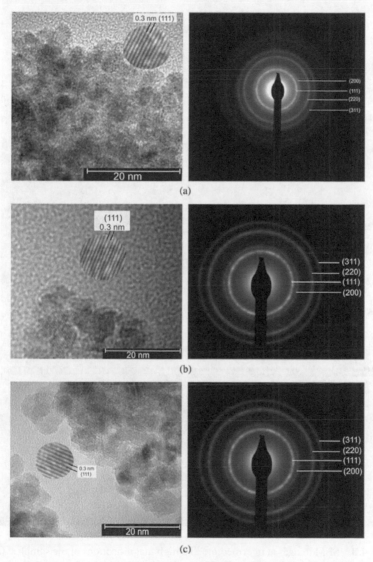

(a)

(b)

(c)

FIGURE 4.2   TEM images of $CeO_2$ nanoparticles synthesized at 100°C for 8 min in the MAH method under different mineralizer agents: (a) KOH; (b) NaOH, and (c) $NH_4OH$. Reprinted from Deus, R. C., M. Cilense, C. R. Foschini, M. A. Ramirez, E. Longo, and A. Z. Simoes, 2013. Influence of mineralizer agents on the growth of crystalline $CeO_2$ nanospheres by the microwave-hydrothermal method. *Journal of Alloys and Compounds* 550: 245–251. With permission. Copyright 2012 Elsevier.)

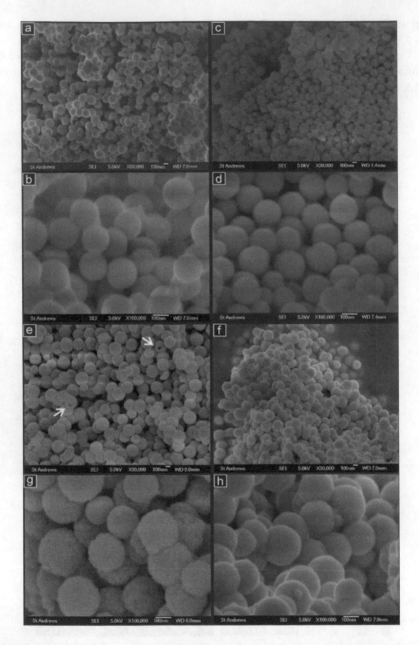

**FIGURE 4.3** SEM images at intermediate and high magnifications of the samples: (a and b) GDC10; (c and d) PrDC10; (e and g) GDC20; and (f and h) PrDC20. Arrows indicate spheres with visible pores. (Reprinted from Muñoz, F. F., L. M. Acuna, C. A. Albornoz, A. G. Leyva, R. T. Baker, and R. O. Fuentes, 2015. Redox properties of nanostructured lanthanide-doped ceria spheres prepared by microwave assisted hydrothermal homogeneous co-precipitation. *Nanoscale* 7 (1): 271–281. With permission. Copyright 2015 Royal Chemical Society.)

Through the MAH method, Araújo et al. obtained $CeO_2$ nanoparticles with controllable morphologies from nanosphere to nanorod by increasing synthesis temperature (Figure 4.4) (Araújo et al. 2012). Godinho and coworkers demonstrated that the use of MW heating during hydrothermal treatment drastically decreases the treatment time required to obtain gadolinium-doped ceria nanorods (Godinho et al. 2008).

Nanocrystalline $CeO_2$ particles with an average size of 7 nm were synthesized by Goharshadi and coworkers using the MW-assisted heating method in the presence of a set of ionic liquids based on the bis(trifluoromethylsulfonyl) imide anion and different cations of 1-alkyl-3-methyl-imidazolium (ILs) (Goharshadi et al. 2011). This method is found to be convenient, rapid, and efficient for the preparation of nanocrystalline $CeO_2$ particles. It is reported that changing the precursor (Ce(IV) or Ce(III)) led to great effects on the size, band-gap energy, reaction time, and the optical properties of the prepared $CeO_2$ nanoparticles synthesized by the MW-assisted method using NaOH as mineralizer agent (Samiee and Goharshadi 2012). As it is demonstrated that the cerium ammonium nitrate and cerium sulfate precursors (Ce(IV) salts) gave similar results while cerium nitrate led to different results. Sm- and Gd-doped $CeO_2$ ($Ce_{0.85}RE_{0.15}O_{1.925}$ (RE=Gd, Sm), $Ce_{0.8}(Gd_{0.1}Sm_{0.1})O_{1.9}$)

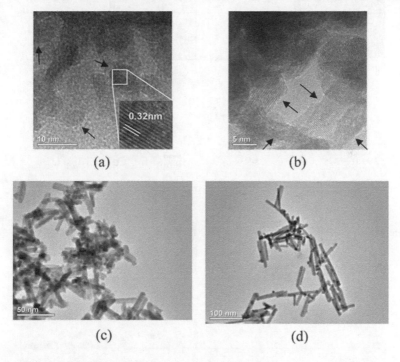

(a)                              (b)

(c)                              (d)

**FIGURE 4.4** HRTEM images of the as-obtained samples synthesized at 80°C reveal that the samples are composed of (a) nanospheres with an average size of 5 nm and (b) nanostructures with a poor morphology. Black arrows indicate these structures. TEM images of samples synthesized at (c) 120 and (d) 160°C. (Reprinted from Araújo, V. D., W. Avansi, H. B. de Carvalho, et al, 2012. CeO2 nanoparticles synthesized by a microwave-assisted hydrothermal method: evolution from nanospheres to nanorods. *Crystengcomm* 14 (3): 1150-1154. With permission. Copyright 2012 Royal Chemical Society.)

and undoped $CeO_{2-\delta}$ nano-powders with a large surface area were also synthesized by the MW-assisted hydrothermal method in a time and energy efficient way (Figure 4.5) (Prado-Gonjal et al. 2012b). Deconvolution of grain boundary (GB) and intrinsic bulk ionic conductivity revealed that GBs constitute barriers for ionic charge transport, with the Sm-doped Ceria $Ce_{0.85}Sm_{0.15}O_{1.925}$ exhibiting the highest GB ionic conductivity. Similar results were also obtained from MW-assisted synthesized $Sm^{3+}$- and $Ca^{2+}$-doped ceria compositions reported by Srivastava and coworkers (Srivastava et al. 2014). Bondioli et al. prepared nanocrystalline Pr-doped $CeO_2$ powders for the first time by a MAH route at about 14 atm (Bondioli et al. 2005). Natile et al. synthesized three nanostructured $CeO_2$ powders by two different synthetic routes: two samples were obtained by precipitation from a basic solution of cerium nitrate and treated at 523 and 923 K, respectively, and the third one was prepared by the MW-assisted heating hydrolysis method and treated at 523 K (Natile et al. 2005). Their results showed that $CeO_2$ obtained by precipitation and treated

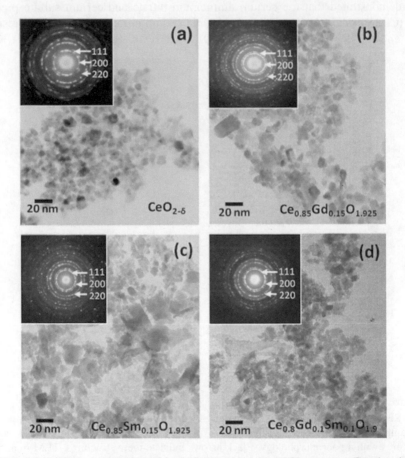

**FIGURE 4.5** TEM micrographs and in the top corners the SAED patterns. (Reprinted from Prado-Gonjal, J., R. Schmidt, D. vila, U. Amador, and E. Morn. 2012a. Structural and physical properties of microwave synthesized orthorhombic perovskite erbium chromite $ErCrO_3$. *Journal of the European Ceramic Society* 32 (3):611–618. With permission. Copyright 2012 Elsevier.)

at 523 K does not oxidize methanol, even at a higher temperature, while traces of oxidation products are noted on the sample treated at 923 K. Methanol oxidation is favored in the $CeO_2$ prepared by MW irradiation: the main oxidation products are formate species and inorganic carboxylate. Using the MW sol-gel synthesis method, Polychronopoulou and coworkers obtained more active $CeO_2$ and $Ce_{0.5}Sm_{0.5}O_2$ catalysts than those obtained by the conventional sol-gel method for CO oxidation, which is demonstrated by $T_{50}$ (temperature where 50% CO conversion is achieved), being reduced by 131°C and 59°C, respectively (Polychronopoulou et al. 2017).

According to Mendiuk and Kepinski's report (Mendiuk and Kepinski 2014), the use of MW radiation enabled the synthesis of the $Ce_{1-x}Er_xO_{2-y}$ nanocubes in a significantly shorter time (3 h), while the hydrothermal method took 24 h. But rod-shape particles of the mixed $(Er, Ce)(OH)_3$ always occur as the second phase.

$Y_2O_3$ is popularly known as a matrix for ion doping of other RE elements and shows promise for applications in high-quality phosphors, catalysts, up-conversion materials, and fine ceramics for its chemical and physical properties, such as optical, catalytic properties, and high chemical and thermal stability. $Y_2O_3$ and rare-earth-doped $Y_2O_3$ can also be synthesized by the MW method (Serantoni et al. 2010; Kaszewski et al. 2016). $Yb:Y_2O_3$ submicrometric particles were obtained through co-precipitation of Yb and Y nitrate in water by Costa et al. (2010). MW heating allowed a fast heating rate and the consequent controlled release of ammonia through urea decomposition at 90°C leading to the formation of disaggregated, monosized spherical particles of carbohydroxy-nitrate precursors with a diameter of 200 nm. Pure crystalline $Yb:Y_2O_3$ powder that preserved the described morphology was obtained after calcination in air at 800°C for 30 min. The authors also carried out control experiments to investigate the effect of the addition of Yb and heating methods on the morphology of the products. The results showed that the particle size of the Yb-doped $Y_2O_3$ powder was smaller than that of no doped $Y_2O_3$ powder, which was thought to be caused by a slower kinetic of growth in the presence of Yb dopant. Notably, the $Y_2O_3$ powder synthesized by traditional heating was composed of submicrometric spheres and large, hard aggregates that typically resulted from a multistep nucleation mechanism (Figure 4.6).

According to Chang et al., spherical and red nanophosphors of $Y_2O_3:Eu^{3+}$ were synthesized using urea hydrolysis that needed a long coprecipitation time (more than 4 h) and post-annealing at 800°C for 2 h. However, when MW was used as the heating

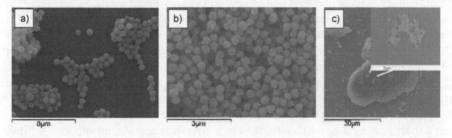

**FIGURE 4.6** SEM micrographs referring to calcined powders: (a) $Yb-Y_2O_3$ by MW heating; (b) $Y_2O_3$ by MW heating; and (c) $Y_2O_3$ by traditional heating. (Reprinted from Costa et al, 2010. Microwave assisted synthesis of $Yb:Y_2O_3$ based materials for laser source application. *Advanced Engineering Materials* 12(3): 205–209. With permission. Copyright 2009 Wiley-VCH.)

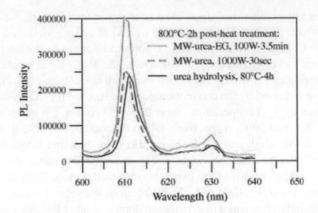

**FIGURE 4.7** Photoluminescent (PL) emission spectra under 466 nm excitation for urea hydrolysis at 80°C for 4 h (black curve), MW-urea hydrolysis with 1000 W for 30 s (dashed curve), and MW-urea-EG solvolysis with 100 W for 3.5 min (light curve). (Reprinted from Chang, H. Y., C. Y. Wu, K. Huang, Y. Lo, I. Shen, and H. Wang, 2014. Preparation and characterization of blue-light-excited nanophosphors using an economically low-energy process. *Journal of Alloys and Compounds* 598 (0): 217–223. With permission. Copyright 2014 Elsevier.)

method, the reaction time was shortened to several minutes or even tens of seconds as the operating power changed (Chang et al. 2014). Besides, the three MW processing routes of MW-urea hydrolysis, MW-urea-EG solvolysis, and MW-urea-EG + KOH solvolysis they used to prepare $Y_2O_3$:$Eu^{3+}$ were found efficient and energy-saving and led to the high photoluminescence (PL) efficiency of the products (Figure 4.7).

Chemical durability, thermal stability, and a low phonon energy cut-off ($\approx 600$ $cm^{-1}$) make gadolinium oxide ($Gd_2O_3$) a good host matrix for the doping of luminescent RE ions, such as $Eu^{3+}$, $Dy^{3+}$, and $Er^{3+}$ (Guo et al. 2004; Majeed and Shivashankar 2014). Meanwhile, the thermodynamic stability, large band-gap (5.9 eV) and high dielectric constant ($\cdot=18$) of $Gd_2O_3$ make it a promising candidate dielectric for complementary metal oxide semiconductor (CMOS) technology of the future and for various nanoelectronics applications (Badylevich et al. 2007). $Gd_2O_3$ can also be used for magnetic resonance imaging (MRI) contrast imaging due to its good proton relaxation properties (Mekuria et al. 2017). Majeed and coworkers prepared ultra-small nanocrystals of undoped and Eu-doped $Gd_2O_3$ with an average diameter of 5.2 nm in 10 min using a simple rapid MW-assisted method, applying benzyl alcohol as the reaction medium without any surface-directing agents (Majeed and Shivashankar 2014). The obtained $Gd_2O_3$ nanocrystals were paramagnetic with a higher relaxation rate than the standard Magnevist (Gd-DTPA) used in MRI imaging (Figure 4.8).

Zhou and coworkers successfully synthesized $Gd_2O_3$:$Eu^{3+}$ rods by MW heating the as-prepared solid substance templated by surfactant assemblies of gadolinium hydroxide and europium hydroxide (Zhou et al. 2010). This novel method required a very short heating time of 10 min, thus reducing the energy consumption. A bright white light emission nanophosphor $Gd_2O_3$:$Dy^{3+}$(1.90%):$Tb^{3+}$(0.1%) was synthesized by Dutta and Tyagi using MW irradiation method in 30 min (Figure 4.9) (Dutta and Tyagi 2012). Majeed and coworkers reported a novel, rapid, and low-temperature method for the synthesis of undoped and Eu-doped GdOOH spherical hierarchical

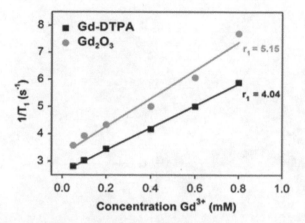

**FIGURE 4.8** Variation of experimentally measured longitudinal relaxation rate $R_1$ ($1/T_1$) with the concentration of $Gd^{3+}$ ions in $Gd_2O_3$ nanocrystals and Gd–DTPA. The solid lines represent the linear fit (linear regression) of experimentally obtained relaxation rates. The regression shows a goodness of fit of $r^2 > 0.98$ for $Gd_2O_3$ nanocrystals and $r^2 > 0.99$ for the Gd–DTPA complex. A lower goodness of fit for $Gd_2O_3$ nanocrystals may be due to nanoparticle agglomeration. (Reprinted from Majeed, S., and S. A. Shivashankar, 2014a. Rapid, microwave-assisted synthesis of $Gd_2O_3$ and $Eu:Gd_2O_3$ nanocrystals: Characterization, magnetic, optical and biological studies. *Journal of Materials Chemistry B* 2 (34): 5585–5593. With permission. Copyright 2014 Royal Chemical Society.)

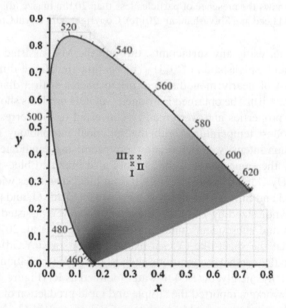

**FIGURE 4.9** Chromaticity diagram of $Gd_2O_3:Dy^{3+}(1.90\%):Tb^{3+}(0.1\%)$ (I) solid, (II) DCM solution, and (III) spin coated on quartz slide. (Reprinted from Dutta, D. P., and A. K. Tyagi, 2012. White light emission from microwave synthesized spin coated $Gd_2O_3:Dy:Tb$ nano phosphors. *Proceedings of the National Academy of Sciences India Section A-Physical Sciences* 82 (1): 53–57. With permission. Copyright 2012 Springer.)

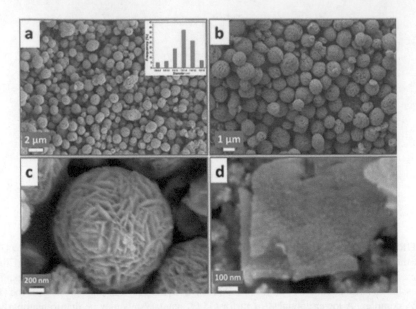

**FIGURE 4.10** SEM images of the as-prepared product (a and b) showing the presence of spherical 3D-spheres and the inset in (a) shows their size distribution. A high-magnification SEM image of a single sphere measuring 1.5 µm in diameter shows the presence of 2D-nanoflakes giving it a flowery appearance (c). A high-magnification SEM image of a single nanoflake shows the presence of particles less than 20 nm in size (d). (Reprinted with permission from Majeed and Shivashankar (2014c). Copyright 2014 Royal Chemical Society.)

structures, without using any surfactants, through the MW-assisted solution-based route (Majeed and Shivashankar 2014). The as-prepared three-dimensional (3D) structures consist of nearly monodisperse microspheres with a diameter of about 1–1.6 µm (Figure 4.10). The obtained Eu-doped GdOOH products showed good photoluminescence properties and were readily converted to the corresponding oxide structures at modest temperatures with the spherical morphology remaining and enhanced emission intensity, which made them promising candidates for phosphor applications. By the same method, the authors also obtained blue–green-emitting microspherical Dy:GdOOH and Dy:Gd$_2$O$_3$ hierarchical structures with similar morphology (Majeed and Shivashankar 2014). Sheet-like (Y,Gd)$_2$O$_3$ and (Y,Gd)$_2$O$_3$:Eu$^{3+}$ particles were synthesized by Dai and coworkers via a MW-assisted mixed-solvent thermal process and subsequent annealing at 550°C (Dai et al. 2010). The results demonstrated that the sheet-like (Y,Gd)-oxalate precursors ((Y,Gd)$_2$(C$_2$O$_4$)$_3$·nH$_2$O) were sensitive to the synthetic conditions such as the MW irradiation power and the volume ratio of ethylene glycol (EG) to water (defined as R) (Figure 4.11).

Panda and coworkers reported the simple and rapid production of uniform, high-quality nanorods, nanowires, and nanoplates of RE oxides (M$_2$O$_3$, M=Pr, Nd, Sm, Eu, Gd, Tb, Dy) using the MW irradiation (MWI) method (Panda et al. 2007). The results demonstrated that the size and shape of the RE oxide nanostructures can be controlled by varying the MWI reaction time and the relative concentrations of the organic surfactants. This MWI method provides a unique opportunity for

**FIGURE 4.11** SEM images of (Y,Gd)-oxalate precursors prepared under different MW powers, (a) 2 W, (b) 5 W, (c) 30 W, and (d) 100 W for 10 min at $R=1$. (Reprinted from Dai, S. H., Y. F. Liu, Y. N. Lu, and H. H. Min, 2010. Microwave solvothermal synthesis of $Eu_{3+}$-doped $(Y,Gd)_2O_3$ microsheets. *Powder Technology* 202 (1–3): 178–184. With permission. Copyright 2010 Elsevier.)

the large-scale synthesis of RE nanostructures without suffering thermal gradient effects and a scalable and flexible approach for device applications.

## 4.2.2 RE Hydroxides

RE hydroxides ($RE(OH)_3$) themselves provide scarce applications, due to their instability in the presence of $CO_2$. Besides, the presence of $OH^-$ results in deterioration effects for photoluminescence emissions. However, $RE(OH)_3$ can be readily converted into plenty of other RE compounds via dry and solution chemical routes without destroying the morphology. A selection of typical works on the conversion of $RE(OH)_3$ nanostructures is listed in Table 4.1. $RE(OH)_3$ proved to be applicable as intermediates for the synthesis of a number of nanocrystals of RE oxides, sulfides, fluorides, oxysulfide, oxyfluoride, and other RE compounds (Yan et al. 2011).

Reprinted with permission from Yan et al. (2011). Copyright 2011 Elsevier.

Via fast MW-assisted method, our group successfully prepared $Y(OH)_3$ hexagonal microprisms on a large scale (Figure 4.12) (Zhong et al. 2010). Our group also successfully prepared self-assisted 3D urchin-like and flower-like $La(OH)_3$ nanostructures for the first time via a facile and fast MW-assisted solution-phase chemical method in 15 min (Wang et al. 2010a). Urchin-like and flower-like $La_2O_3$ nanostructures were obtained after calcining the $La(OH)_3$ precursor at 800°C for 4 h. Urchin-like and flower-like $La_2O_3:Eu^{3+}$ precursors were also prepared by the same method and their photoluminescence (PL) properties were investigated. It was demonstrated that amount of tetraethyl ammonium bromide (TEAB) played a vital

**TABLE 4.1** The Conversion from RE Hydroxide Nanocrystals

| Morphology | Target compound | Conversion method | Conversion condition | Crystal relations |
|---|---|---|---|---|
| Hydroxide, nanowires, nanotubes | $R_2O_3$ | Heating dehydration | 500 °C | Not reported |
| Hydroxide, nanowires, nanotubes | $R_2O_2S$ | Sulfur | 700 °C, 2 h | Not reported |
| Hydroxide, nanowires, nanotubes | $R(OH)_xF_{3-x}$ | F, hydrothermal | 120 °C | Not reported |
| Tb(OH)₃, Y(OH)₃, Dy(OH)₃ nanotubes | Oxide, $Tb_4O_7$, $Y_2O_3$, $Dy_2O_3$ | Heating in air | 450 °C | [001] to [011] for Dy |
| Ce(OH)₃ nanotubes | Oxide, $CeO_2$ | Heating in air | 450 °C | [100] to [110] |
| La(OH)₃, Nd(OH)₃ nanowires | Sulfide, $LnS_2$ | Boron–sulfur method | 400–450 °C, 24 h, more S | Not observed |
| La(OH)₃, Nd(OH)₃ nanowires | Oxysulfide, $Ln_2O_2S$, $Ln_2O_2S_2$ | Boron–sulfur method | 500 °C, 10 min, with Ln(OH)₃/ S = 1 or 0.5 | Not observed |
| Gd(OH)₃ nanorods | Vanadate, $GdVO_4$ | Hydrothermal reaction with $Na_3VO_4$ | 180 °C 18 h, pH 13 | [001] to [001] |
| Eu(OH)₃ nanorods | EuO | Heating in air to $Eu_2O_3$ and | | Polycrystalline |

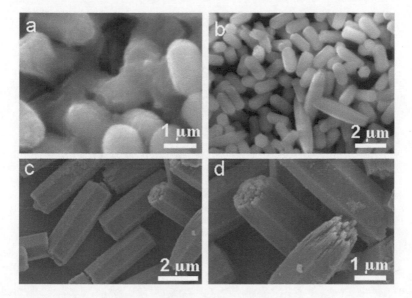

**FIGURE 4.12** SEM images of the as-synthesized Y(OH)$_3$ prepared after different micro-wave heating time in the presence of sodium citrate: (a) 5 min; (b) 15 min; (c) and (d) 30 min. (Reprinted with permission from Zhong et al. (2010). Copyright 2009 Elsevier.)

role in the morphology of the product (Figure 4.13). Ma et al. reported the synthesis of Pr(OH)$_3$ nanorods by the MW-assisted method and their thermal conversion to oxide nanorods (Figure 4.14) (Ma et al. 2007). The results showed that, compared with the conventional heating method, MW irradiation heating could reduce reaction time significantly due to the fast and volumetric heating effect.

**FIGURE 4.13** SEM images of the as-prepared products obtained after MW irradiation at 150 W for 15 min: (a–c) without the addition of TEAB; (d–f) with the addition of TEAB. (Reprinted from Wang, H. Q., R. D. Tilley, and T. Nann, 2010a. Size and shape evolution of upconverting nanoparticles using microwave assisted synthesis. *Crystengcomm* 12 (7):1993–1996. With permission. Copyright 2010 Elsevier.)

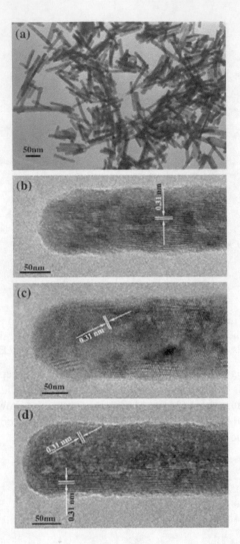

**FIGURE 4.14** (a) TEM image of the as-prepared $Pr(OH)_3$ nanorods and (b–d) HRTEM images of the individual $Pr(OH)_3$ nanorod. (Reprinted from Ma, L., W. X. Chen, J. Zhao, Y. F. Zheng, X. Li, and Z. D. Xu, 2007. Microwave-assisted synthesis of praseodymium hydroxide nanorods and thermal conversion to oxide nanorod. *Materials Letters* 61 (8–9): 1711–1714. With permission. Copyright 2007 Elsevier.)

### 4.2.3 RE FLUORIDES

RE fluorides are the most excellent hosts for optical applications, such as industrial lighting, display, and biological imaging by both downconversion (DC) and upconversion (UC) process (Fouassier 2000; Zhao et al. 2014). Due to a high refractive index and low phonon energy of RE fluorides, low probability of non-radiative decay is realized, which usually results in a higher luminescence quantum yield than in oxide hosts and in most inorganic matrices (Li et al. 2011). In

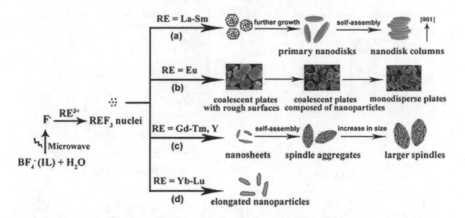

**FIGURE 4.15** Possible formation mechanisms of REF$_3$ nano/microcrystals with multiform morphologies. (Reprinted from Li, C., et al., 2011. Fine structural and morphological control of rare earth fluorides REF$_3$ (RE = La-Lu, Y) nano/microcrystals: Microwave-assisted ionic liquid synthesis, magnetic and luminescent properties. *Crystengcomm* 13 (3):1003–1013. With permission. Copyright 2011 Royal Chemical Society.)

recent years, the synthesis and optical properties of Ln-doped RE fluorides have attracted great interest.

Our group prepared water-soluble YF$_3$ and YF$_3$:Ln$^{3+}$ (Ln=Ce, Tb, Eu) nanocrystals in ethanol solution by employing MW heating for 15 min (Zhong et al. 2010). Lin group reported a fast, facile, and environmentally friendly MW-assisted synthesis of RE fluoride REF$_3$ (RE=La–Lu, Y) nano/microcrystals with multiform crystal structures (hexagonal and orthorhombic) and morphologies (nanodisks, secondary aggregates constructed from nanoparticles, and elongated nanoparticles) (Li et al. 2011). The possible formation mechanisms for REF$_3$ nano/microcrystals with diverse well-defined morphologies were proposed (Figure 4.15). Li et al. successfully synthesized ultra-small monodisperse water-soluble LaF$_3$:Ln$^{3+}$ nanocrystals (NCs) with a size range of 9–12 nm through a MW-assisted modified polyol process applying polyvinylpyrrolidone (PVP) as an amphiphilic surfactant (Li et al. 2013a). By changing the doped RE ions, a series of LaF$_3$:Ln$^{3+}$ NCs (Ln=Eu, Nd, Ce, Tb, Yb, Er, Yb, Ho, Yb, and Tm) with the unique feature of up–down conversion from visible to NIR emission were obtained (Figures 4.16 and 4.17). Wang et al. also synthesized a variety of monodisperse colloidal GdF$_3$:Yb, Er upconversion NCs with different shapes, sizes, and dopants by MW-assisted synthesis (Wang and Nann 2011).

RE ions doped AREF$_4$ (A, alkali metal) compounds with unique luminescent, insulating/magnetic, and piezoelectric properties attract great attention for their applications in the fields of solid-state lasers (Huang et al. 2001), bioanalysis (Chen et al. 2011; Mi et al. 2012), and display technologies (Downing et al. 1996). Highly crystalline, strongly luminescent NaYF$_4$:Yb, Er upconversion nanoparticles were first established with RE acetate as the precursor by Mi et al. through a MW-assisted solvothermal process within a significantly shortened reaction time (only 1 h) compared with the traditional methods (Mi et al. 2011). The results proved that the as-obtained nanoparticles meet the needs for bioapplications, and the MW-assisted solvothermal approach is an effective method for the synthesis of UCNPs which

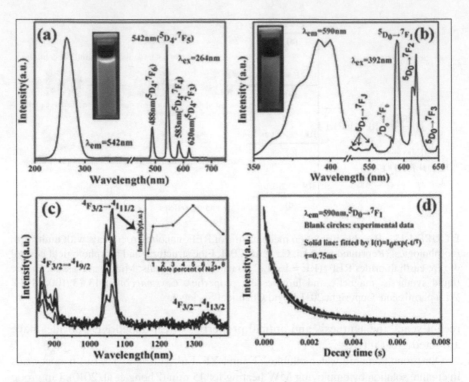

**FIGURE 4.16**  Fluorescence spectra of the DC LaF$_3$:Ln$^{3+}$ NCs. (a) Excitation (left) and emission (right) spectra of the LaF$_3$:Ce$^{3+}$,Tb$^{3+}$ NCs. (b) Excitation (left) and emission (right) spectra of the LaF$_3$:Eu$^{3+}$ NCs. (c) Emission spectra of the LaF$_3$:Nd$^{3+}$ NCs. (d) Decay curve of the Eu$^{3+}$ luminescence in the LaF$_3$:Eu$^{3+}$ NCs. The insets show: (a and b) the corresponding luminescence photographs of the samples under a UV lamp. (c) The effect of Nd$^{3+}$ concentration (x) on the emission intensity of the LaF$_3$:Nd$^{3+}$ NCs. (Reprinted from Li, F. F., C. G. Li, J. H. Liu, et al., 2013a. Aqueous phase synthesis of upconversion nanocrystals through layer-by-layer epitaxial growth for in vivo X-ray computed tomography. *Nanoscale* 5 (15): 6950–6959. With permission. Copyright 2013 Royal Chemical Society.)

can be expanded for the synthesis of other nanomaterials. Wang et al. reported the synthesis of highly luminescent and monodisperse upconverting RE ions doped NaYF$_4$ nanocrystals by a MW-assisted synthesis approach for 5 min in a sealed vessel (Wang and Nann 2009). This fast, homogeneous, inexpensive, and highly reproductive synthesis method resulted in upconverting nanocrystals with different sizes and shapes and provided superior reaction control (Figure 4.18). The authors also studied the size and shape evolution of upconverting AYF$_4$:Yb, Er (with A = Na, Li) nanocrystals using MW-assisted synthesis (Wang et al. 2010b). It was found that low reactant concentrations always resulted in very small and monodisperse spherical nanocrystals, whereas the nanocrystal size could be controlled by changing the reaction time (Figure 4.19). Niu et al. described a MW reflux method for the synthesis of pure hexagonal NaYF$_4$:Yb$^{3+}$, Ln$^{3+}$ (Ln = Er$^{3+}$, Tm$^{3+}$, Ho$^{3+}$) crystals for the first time at a relatively low temperature (160°C) and atmospheric pressure with a much shorter reaction time (50 min) (Niu et al. 2012). The upconversion emission intensities of

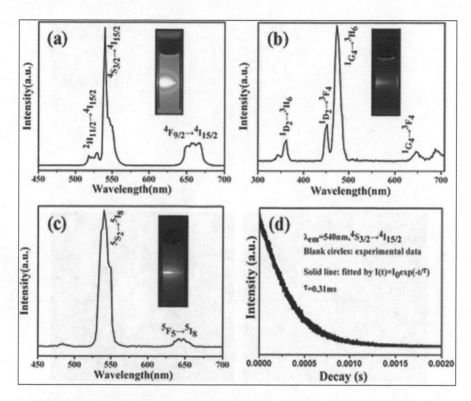

**FIGURE 4.17** Fluorescence spectra of the UC NCs (a) $LaF_3:Yb^{3+}$, $Er^{3+}$; (b) $LaF_3:Yb^{3+}$, $Tm^{3+}$; (c) $LaF_3:Yb^{3+}$, $Ho^{3+}$ NCs; The insets show: (a–c) the corresponding UC luminescence photographs of the annealed samples under a 980 nm laser; (d) decay curve of the $Er^{3+}$ luminescence ($\lambda em = 540$ nm) in the annealed $LaF_3:Yb^{3+}$, $Er^{3+}$ NCs under 980 nm NIR excitation. (Reprinted from Li, F. F., C. G. Li, J. H. Liu, et al., 2013a. Aqueous phase synthesis of upconversion nanocrystals through layer-by-layer epitaxial growth for in vivo X-ray computed tomography. *Nanoscale* 5 (15): 6950–6959. With permission. Copyright 2013 Royal Chemical Society.)

hexagonal $aYF_4:Yb^{3+}$,$Ln^{3+}$ (Ln = $Er^{3+}$, $Tm^{3+}$, $Ho^{3+}$) are increased by 10–12 times compared with those of the cubic phase. Li et al., for the first time, prepared highly water-soluble $NaYF_4:Yb$, $Er@NaGdF_4$ multilayer core-shell upconversion nanocrystals (UCNCs) by a MW-assisted method (Figure 4.20) (Li et al. 2013b). It was revealed that the obtained UCNCs showed low cytotoxicity and long circulation time in vivo. More notably, multilayer core-shell UCNCs provided a much higher efficacy compared to the monolayer core-shell agent. Ding et al. successfully prepared pure hexagonal $\beta$-$NaYF_4:Yb^{3+}$, $Ln^{3+}$ (Ln = Er, Tm, Ho) microrods via a rapid MW-assisted flux cooling method. Multicolor visible emissions including yellow, blue, and green could be obtained by tuning the dopant's species (Ding et al. 2014). The $KYF_4$ nanopowders, non-doped and doped with $Ce^{3+}$ or $Tb^{3+}$, having well-crystallized, unaggregated, monodisperse ($\pm15\%$) nanoparticles with a cubic (the size in the range from 15 to 30 nm) or hexagonal (from 30 to 50 nm) crystal structure have been successfully synthesized through a MW hydrothermal treatment of as-precipitated gels

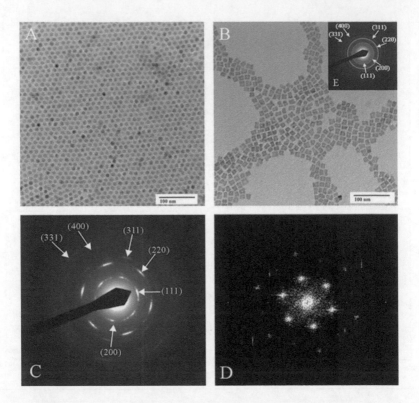

**FIGURE 4.18** (a and b) TEM images of $NaY_{0.78}F_4:Yb^{3+}_{0.2}$, $Er^{3+}_{0.02}$ and $NaY_{0.78}F_4:Yb^{3+}_{0.2}$, $Tm^{3+}_{0.02}$ nanocrystals. (c) Selected area electron diffraction pattern (SAED) of $NaY_{0.78}F_4:Yb^{3+}_{0.2}$, $Er^{3+}_{0.02}$ nanocrystals. (d) Fast-Fourier transform (FFT) image of picture A. (e) SAED of $NaY_{0.78}F_4:Yb^{3+}_{0.2}$, $Tm^{3+}_{0.02}$ nanocrystals. (Reprinted from Wang, H. Q., and T. Nann. 2009. Monodisperse upconverting nanocrystals by microwaveassisted synthesis. *Acs Nano* 3 (11): 3804–3808. With permission. Copyright 2009 American Chemical Society.)

by Makhov and coworkers (Makhov et al. 2013). Olesiak-Banska et al. reported the synthesis of water-soluble $NaYF_4$ nanophosphors co-doped with $Eu^{3+}$ and $Tb^{3+}$ via a hydrothermal MW-assisted technique (Olesiak-Banska et al. 2011).

Barium yttrium fluoride ($BaYF_5$) is expected to be a promising host material in the energy transfer of RE. RE-doped $BaYF_5$ materials synthesized by a MW-assisted method have been reported. For example, Zhang's group has synthesized high hydrophilic Ce, Tb-doped $BaYF_5$ fluorescent nanocrystals with a uniform size by a MW-assisted route (Lei et al. 2011). This group also successfully synthesized hydrophilic $BaYF_5$:Ln (Ln=Yb/Er, Ce/Tb) nanocrystals with down- and up-conversion luminescent properties via a facile and fast MW-assisted method (Pan et al. 2013).

### 4.2.4 RE OXYSALTS

#### 4.2.4.1 RE Orthophosphates

Mostly depending on the cationic radius of RE elements, there are four different phases of RE orthophosphate ($REPO_4$): monazite (monoclinic, dehydrate, for light Lns),

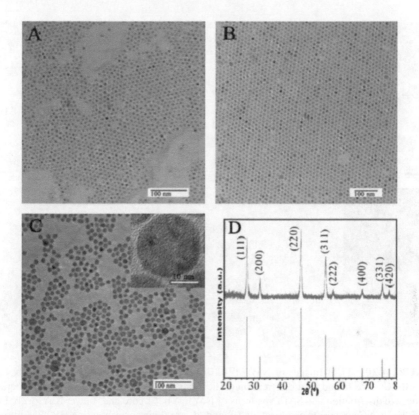

**FIGURE 4.19** Characterization data for time-dependent $NaYF_4$:20%$Yb^{3+}$, 2%$Er^{3+}$ nanocrystals. TEM images of upconverting nanocrystals synthesized using (4.8 mM $Na^+$, 3.5 mM $Y^{3+}$, 0.7 mM $Yb^{3+}$, 0.07 mM $Er^{3+}$) precursors at 290°C, reacted for (a) 10 min, (b) 40 min, and (c) 70 min respectively (inset: HRTEM of one single nanocrystal). (d) Experimental X-ray diffraction pattern (top) of the synthesized nanocrystals (reacted 40 min) and the calculated line pattern for α-$NaYF4$ (bottom). (Reprinted from Wang, H. Q., R. D. Tilley, and T. Nann, 2010a. Size and shape evolution of upconverting nanoparticles using microwave assisted synthesis. *Crystengcomm* 12 (7): 1993–1996. With permission. Copyright 2010 Royal Chemical Society.)

xenotime (also typed as zircon, tetragonal, dehydrate or hydrate, for heavy Lns and $Y^{3+}$), rhabdophane (hexagonal, mostly hydrate, across the series), and churchite (also named weinschenkite, monoclinic, hydrate, for heavy Lns and $Y^{3+}$). Notably, the so-called orthorhombic phase is probably composed of xenotime and rhabdophane (Yan et al. 2011). $REPO_4$ nanomaterials have a variety of potentially beneficial properties, including very low solubility in water ($K_{sp} = 10^{-25}$ to $10^{-27}$) (Firsching and Brune 1991), very high thermal stability with melting points around 2300°C (Hikichi and Nomura 1987), low thermal conductivity, high refractive indexes ($n \approx 1.5$) (Jellison et al. 2000), high luminescence quantum efficiency, chemical capability of containing Lns and actinides, and unusual magnetic behaviors. Because of these unique properties, $REPO_4$ materials find applications in various fields, such as luminescent or laser materials (Fang et al. 2003; Riwotzki et al. 2001), proton conductors

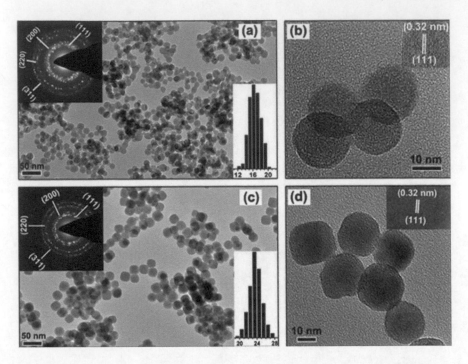

**FIGURE 4.20** TEM images of NaYF$_4$:Yb, Er core (a and b) and 3-layer NaYF$_4$:Yb, Er@ NaGdF$_4$ UCNCs (c and d) at different magnifications. Insets in (a) and (c) display histograms of the size distribution and SAED patterns of NaYF$_4$:Yb, Er core and 3-layer NaYF$_4$:Yb, Er@ NaGdF$_4$ UCNCs, respectively. Insets in (b) and (d) are their corresponding HRTEM images. (Reprinted from Li, F. F., et al., 2013a. Aqueous phase synthesis of upconversion nanocrystals through layer-by-layer epitaxial growth for in vivo X-ray computed tomography. *Nanoscale* 5 (15): 6950–6959. With permission. Copyright 2013 Royal Chemical Society.)

(Norby and Christiansen 1995), catalysts (Sarala Devi et al. 2002), magnets (Carini et al. 1997), ceramics (Hikichi and Nomura 1987), hosts for radioactive nuclear waste (Ordoñez-Regil et al. 2002), and heat-resistant materials (Riwotzki et al. 2000). The most significant applications for REPO$_4$ are the optical ones, such as the scintillators for X-ray and γ-ray detection for medical imaging (Neal et al. 2006), thermo luminescence phosphors (Iacconi et al. 2001), and highly efficient commercial green phosphor ($^5D_4$-$^7F_5$ transition of Tb$^{3+}$ at 543 nm) used in fluorescent lamps (Riwotzki et al. 2001). Ocaña and coworkers prepared mesoporous tetragonal RE:YPO$_4$ nanophosphors (RE=Eu, Ce, Tb, and Ce+Tb) with a lenticular morphology, variable mean size, and high surface area by an homogeneous precipitation procedure at low temperature (80–120°C) in a MW oven (Rodriguez-Liviano et al. 2012). This synthesis approach possesses important advantages including simplicity, rapidness (reaction time=7 min), and high reaction yields. By the MW-assisted method, this group also synthesized multifunctional tetragonal Eu:GdPO$_4$ nanocubes with high transverse relaxivity (r$_2$) values. These nanoparticles are nontoxic for cells and suitable as biolabels for in vitro optical imaging and as negative contrast agents for magnetic resonance imaging (Figure 4.21) (Rodriguez-Liviano et al. 2013b). Biocompatible

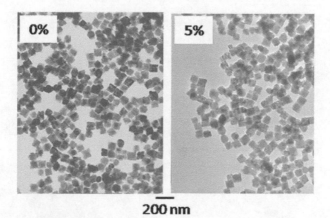

**200 nm**

**FIGURE 4.21** TEM images of the Eu:GdPO$_4$ nanoparticles prepared by heating at 120°C for 1 h in a MW oven and BG solutions containing H$_3$PO$_4$ (0.15 mol dm$^{-3}$) and Eu(acac)$_3$ and Gd(acac)$_3$ (Gd+Eu=0.02 mol dm$^{-3}$) with different Eu content: 0% (GdPO$_4$) and 5% (Eu$_{0.05}$Gd$_{0.95}$PO$_4$). (Reprinted from Rodriguez-Liviano, S., F. J. Aparicio, A. I. Becerro, et al. 2013a. Synthesis and functionalization of biocompatible Tb:CePO$_4$ nanophosphors with spindle-like shape. *Journal of Nanoparticle Research* 15 (2): 14. With permission. Copyright 2012 American Chemical Society.)

monoclinic Tb:CePO$_4$ nanophosphors with a spindle-like morphology and tailored size (in the nanometer and micrometer range) were also prepared by this group through a very simple MW-assisted synthesis procedure (Rodriguez-Liviano et al. 2013a). Patra et al. successfully synthesized rhabdophane-type hexagonal Ln orthophosphate, LnPO$_4$·nH$_2$O (Ln=La, Ce, Nd, Sm, Eu, Gd, and Tb, n=0 to 0.6), and body-centered tetragonal ErPO$_4$·nH$_2$O nanowires/nanorods in high yield (>95%) by simple MW heating of an aqueous solution of Ln(III) nitrate and NH$_4$H$_2$PO$_4$ in the pH range 1.8–2.2 (Figure 4.22) (Patra et al. 2005). The synthesis of LaPO$_4$:RE (RE=Ce$^{3+}$,Eu$^{3+}$,Tb$^{3+}$) nanorods was reported by Ma and coworkers (Ma et al. 2009). Colomer et al. obtained single-crystal nanorods of hexagonal rhabdophane-type La$_{1-x}$Sr$_x$PO$_{4-x/2}$·nH$_2$O (x=0 or 0.02) by a one-pot MW-assisted hydrothermal synthesis that is simple, soft, fast, and energy efficient (Colomer et al. 2014). Li et al. prepared photoluminescent LaPO$^4$:Eu$^{3+}$,Li$^+$ nanophosphors via a MW-assisted sol-gel synthesis process (Li and Lee 2008). Che et al. synthesized high-bright LaPO$_4$:Eu nanocrystals by a MW-assisted method via controlling the nucleation and growth process (Che et al. 2014). Huong et al. prepared (Eu,Tb)PO$_4$·H$_2$O nanorods/nanowires by MW technique and studied their luminescent properties (Huong et al. 2011). Bühler et al. synthesized luminescent LaPO$_4$:Ce, Tb nanocrystals in ionic liquid. The concept of a MW-assisted synthesis of luminescent nanocrystals in ionic liquid media was thought to be presented for the first time by the authors (Bühler and Feldmann 2006). In 2007, the authors fabricated transparent layers containing the highly luminescent nanophosphors LaPO$_4$:Ce,Tb and LaPO$_4$:Eu, both of which were prepared via a MW-assisted synthesis in ionic liquids (Bühler and Feldmann 2007). This group also reported the synthesis of non-agglomerated and very uniform LaPO$_4$:Ce,

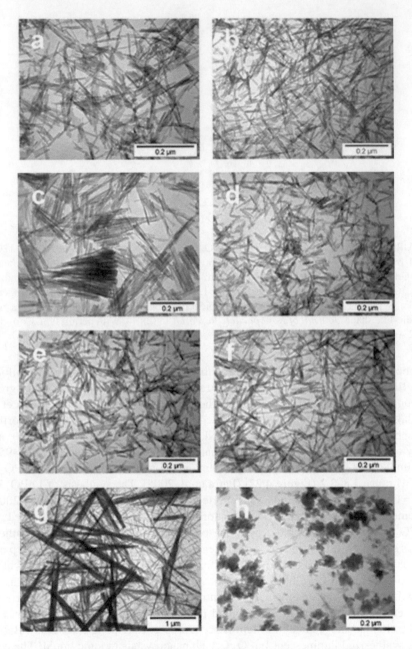

**FIGURE 4.22** LRTEM micrographs of as-synthesized $LnPO_4 \cdot nH_2O$ nanostructures obtained after MW heating: (a) La, (b) Ce, (c) Nd, (d) Sm, (e) Eu, (f) Gd, (g) Tb, and (h) Er. (Reprinted from Patra, C. R., et al., 2005. Microwave approach for the synthesis of rhabdophane-type lanthanide orthophosphate (Ln = La, Ce, Nd, Sm, Eu, Gd and Tb) nanorods under solvothermal conditions. *New Journal of Chemistry* 29 (5): 733–739. With permission. Copyright 2005 Royal Chemical Society.)

Tb nanoscale phosphor with very high quantum yields via MW-accelerated synthesis in ionic liquid (Zharkouskaya et al. 2008).

### 4.2.4.2 Other RE Oxysalts

Besides RE orthophosphates, there are other RE oxysalts synthesized by the MW-assisted method, such as vanadates, borates, chromites, cobaltates, molybdates, orthoborates, carbonate, ferrates, and so on.

Ln orthovanadates ($LnVO_4$) exhibit unique optical, magnetic, catalytic, and electrical properties for various applications and have attracted extensive interests and been widely investigated. Ekthammathat et al. synthesizedcerium orthovanadate ($CeVO_4$) nanostructures using a MW-assisted method without the use of any catalysts or templates (Ekthammathat et al. 2013). Ln orthovanadates, $CeVO_4$, $PrVO_4$, and $NdVO_4$, of nanodimensions, were prepared by Mahapatra and coworkers under MW radiation, and the photocatalytic activities of these compounds were investigated (Mahapatra et al. 2008). Parhi et al. reported the synthesis of $LaVO_4$ and $YVO_4$ through a solid-state metathesis approach driven by MW energy (Parhi and Manivannan 2008). Hu et al. synthesized a series of RE orthovanadates $REVO_4$ (RE=Sm, Eu, Dy, Gd, Tb, Tm, Yb, Lu) with the tetragonal zircon structures via a facile sonochemical and the MW co-assisted method (SMC) at a low temperature (70°C) (Hu and Wang 2014).

MW-assisted synthesis of $Eu^{3+}$-doped $LaBO_3$ was reported by Badan and coworkers (Badan et al. 2012). Prado-Gonjal et al. synthesized single-phase orthorhombic perovskite $ErCrO_3$ by a MW-assisted process (Prado-Gonjal et al. 2012a). They also prepared full RE chromites series $(RE)CrO_3$ with an orthorhombic distorted (Pnma) perovskite structure by the same method (Figure 4.23) and studied their microstructure and physical properties (Prado-Gonjal et al. 2013b). Farhadi et al. reported the preparation of pure and single-phase nanoparticles of perovskite-type $LaCoO_3$ via MW-assisted solid-state decomposition of $La[Co(CN)_6]\cdot 5H_2O$ precursor in the presence of CuO powder as a strong MW absorber within a very short reaction time of 10 min (Farhadi and Sepahvand 2010). Synthesis of $LaCoO_3$ nanoparticles by a MW-assisted process was also reported by Jung and their photocatalytic activity under visible light irradiation was investigated (Jung and Hong 2013). Nanocrystalline $La_2Mo_2O_9$ oxide-ion conductor synthesized by MW-assisted combustion method within a very short time duration using aspartic acid as the newer fuel in a domestic MW oven was reported by Saradha and coworkers (Saradha et al. 2008). By the MW-assisted decomposition method, Farhadi et al. also prepared nanosized and pure single-phase perovskite-type $LaFeO_3$ nanoparticles from bimetallic $La[Fe(CN)_6]\cdot 5H_2O$ compound (Farhadi et al. 2009). Ding and coworkers synthesized perovskite-type samples $ReFeO_3$ (Re: La, Sm, Eu, Gd) by a MW-assisted method (Ding et al. 2010). Laurustinus shaped $NaY(WO_4)_2$ and $NaY(WO_4)_2$:$Er^{3+}$/ $Yb^{3+}$ micro-particles assembled by nanosheets were synthesized via a MW-assisted hydrothermal (MH) route by Zheng and coworkers (Zheng et al. 2014). Rocha et al. prepared polycrystalline ceramic $BiREWO_6$, where RE=Y, Gd, and Nd by MW-assisted solid-state synthesis in air at 900–1100°C for 10 min (Rocha et al. 2013). Trujillano et al. synthesized $A_2Sn_2O_7$(A=Eu or Y) with the pyrochlore structure through a MW-assisted hydrothermal synthesis process (Trujillano et al. 2015).

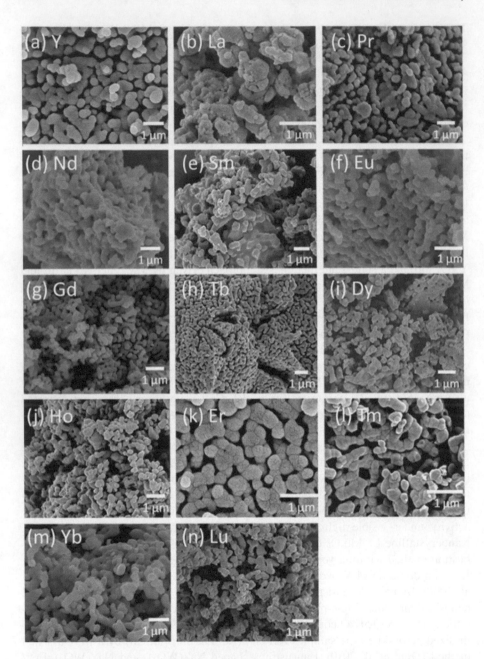

**FIGURE 4.23**  SEM micrographs of (RE)CrO₃ powder. (Reprinted from Prado-Gonjal, J., R. Schmidt, J. J. Romero, D. Vila, U. Amador, and E. Morn, 2013. Microwave-assisted synthesis, microstructure, and physical properties of rare-earth chromites. *Inorganic Chemistry* 52 (1): 313–320. With permission. Copyright 2012 American Chemical Society.)

## 4.3 MW PREPARATION OF RE-BASED CPS

Ln-based CPs have captivated the attention of many researchers due to their characteristic coordination preferences and unique optical and magnetic properties arising from 4f electrons. Compared to the first-row transition metal ions, Ln ions, as functional metal centers, display higher and variable coordination numbers and more flexible coordination geometries, which results in more facile routes to densely packed solids and greater opportunity in creating potentially useful structures with desired optical, magnetic, and structural properties. In recent years, MW-assisted method has been reported for the preparation of RE-based CPs. For example, Silva et al. synthesized a microporous cationic Ln-organic framework, $[Ce_2(pydc)_2(Hpydc)(H_2O)_2]Cl\cdot(9+y)H_2O$ (where $pydc^{2-}$ is the diprotonated residue of 2,5-pyridinedicarboxylic acid) in 30 min (total reaction time) by applying MW heating (Silva et al. 2010). Two Ln complexes, $[Ln(NO_3)_2(H_2O)_3(L)_2](NO_3)(H_2O)$ (Ln = Eu (1), Tb (2); $L = (4-pyH)^+CH_2CH_2-SO_3^-$), were prepared in 40 s under MW-heating (700 W) conditions by Zheng and coworkers (Zheng and Lee 2011). The series of Ln complexes with 2,6-naphthalenedicarboxylic acid and DMF molecules synthesized by MW-assisted solvothermal method was reported by Łyszczek and coworkers (Lyszczek and Lipke 2013). Lin et al. reported the synthesis of a series of isostructural microporous Ln metal–organic frameworks (MOFs) formulated as $[Ln_2(TPO)_2(HCOO)]\cdot(Me_2NH_2)\cdot(DMF)_4\cdot(H_2O)_6$ (Ln = Y (1), Sm (2), Eu (3), Gd (4), Tb (5), Dy (6), Ho (7), Er (8),Tm (9), Yb (10), and Lu (11); $H_3TPO$ = tris-(4-c arboxyphenyl)phosphineoxide; DMF = N,N-dimethylformamide) via a MW-assisted method in 30 min (Lin et al. 2012). A fast, mild and high-yield MW synthesis of novel 1D Ln-polyphosphonate CPs was presented by Vilela and coworkers (Vilela et al. 2013). By MW-assisted method, our group has synthesized the series of RE-based coordination polymer submicrospheres which were readily converted to RE oxide $(RE_2O_3)$ by calcining the precursors in air at 550°C for 4 h (Figure 4.24) (Zhong et al. 2014; Zhao et al. 2012).

## 4.4 MW PREPARATION OF RE-BASED COMPOSITES MATERIALS

Composite materials are designed to serve specific purposes and exhibit desirable properties and functions. Incorporation of a variety of functional materials with RE materials has been proposed recently to combine the merits and mitigate the shortcomings of both the components, and usually results in a multifunctional material (Su et al. 2013; Feng and Zhang 2013; Yan 2012; Li et al. 2014). The preparation of RE-based coordination polymer composites have also been reported (Zheng et al. 2012; Liu et al. 2014; Kim et al. 2014; Huang et al. 2016). However, there are only a few reports about MW-assisted synthesis of RE-based composite materials. For example, $CNTs/La(OH)_3$ nanocomposite was first synthesized via a MW-assisted preparation process by Fang and coworkers, and the nanocomposite modified glassy carbon electrode which not only improved the current but also decreased the overvoltage potential exhibited excellent electrocatalytic activity for oxidation of adenine and guanine (Figure 4.25) (Fang et al. 2008). Yu et al. successfully synthesized $Pd-La(OH)_3/C$ composite nanomaterial via an intermittent MW heating-glycol reduction method

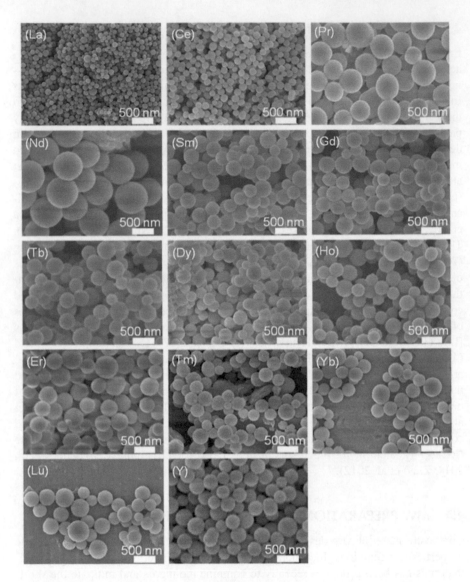

**FIGURE 4.24** SEM images of RE-CPs obtained after MW irradiation at 150 W for 5 min. (Reprinted with permission from Zhong et al. (2014). Copyright 2014 American Chemical Society.)

(Figure 4.26) (Yu et al. 2014). It is revealed that Pd-La(OH)$_3$/C showed significantly higher activity and stability than that of Pd/C with the same Pd loading of 0.1 mg cm$^{-2}$. Miyazaki et al. successfully obtained nanosized Ce$_{0.5}$Hf$_{0.5}$O$_2$/carbon clusters composite material by MW-irradiated calcination of a Ce(acac)$_3$/Hf(acac)$_4$/epoxy resin complex (Miyazaki et al. 2009). Meng and coworkers prepared Mg$_{17}$Ni$_{1.5}$Ce$_{0.5}$ hydrogen storage composites with different contents of graphite were prepared by a mechanical milling and subsequent microwave sintering process (Meng et al. 2013).

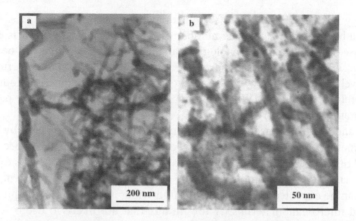

**FIGURE 4.25** TEM image of pure CNTs (a) and CNTs/La(OH)$_3$ composite synthesized under microwave condition (b). (Reprinted from Fang, B., W. Zhang, G. F. Wang, H. Y. Liu, and S. P. Wei., 2008. Microwave-assisted preparation of a carbon nanotube/La(OH)$_3$ nano-composite, and its application to electrochemical determination of adenine and guanine. *Microchimica Acta* 162 (1–2): 175–180. With permission. Copyright 2007 Springer.)

**FIGURE 4.26** TEM images of PdLa(OH)$_3$/C$_{11}$. (Reprinted from Yu, H. Y., D. B. Zhou, and H. M. Zhu, 2014. Synthesis and characterization of Pd-La(OH)$_3$/C electrocatalyst for direct ethanol fuel cell. *Journal of Solid State Electrochemistry* 18 (1): 125–131. With permission. Copyright 2013 Springer.)

Their results demonstrated that the obtained composite with 5 wt pct graphite exhibited a hydrogen storage capacity of 5.34 wt pct and could absorb 3 wt pct within 80 s at 623 K (350°C), which was 6 times faster than the Mg$_{17}$Ni$_{1.5}$Ce$_{0.5}$ alloy prepared with the same method.

## 4.5 CONCLUSION AND PROSPECTS

With MW heating technology, an alternative heat source, chemical reactions, and materials preparation can be completed in minutes, instead of the hours or even days that is usually required by the conventional heating methods. Therefore, MW heating technology has been receiving an exponential increase in adoption rate, which is illustrated by a rapidly increasing number of publications each year. It is expected that the use of MW heating technology as an alternative heat source will be adopted in many research laboratories and even in industry. It is expected that the future role of the novel MW technology will be able to scale up for the industrial production in the near future, although MW-assisted applications are still dominantly performed on a laboratory (milliliter) scale for now.

More and more RE-based micro/nanomaterials have been prepared by the MW heating method and their properties have been investigated. It has been shown that MW often led to small size, better dispersity, and better properties over other methods when they were employed in the preparation RE-based micro/nanomaterials. It can be expected that MW would find wider application in the preparation RE-based micro/nanomaterials. A few things that are likely to occur with MW technology are expected as follows: (1) better understanding of the detailed formation process of RE-based micro/nanomaterials under MW irradiation; (2) studies on some controversial issues between MW heating and conventional heating through rigorously design experiments to broaden the application of MW heating technology; (3) research on MW effects and MW heating mechanisms via simulations to guide the design of experiments efficiently; (4) design and synthesis of RE-based micro/nanomaterials with certain composition, structure, size, and morphology with the assistance of MW; (5) attempts to synthesize RE-based micro/nanomaterials in large scale; and (6) investigation on properties and promising applications of RE-based micro/nanomaterials prepared with the assistance of MW. In short, there is still much room for the development of MW heating technology.

## REFERENCES

Araújo, V. D., W. Avansi, H. B. de Carvalho, et al. 2012. $CeO_2$ nanoparticles synthesized by a microwave-assisted hydrothermal method: Evolution from nanospheres to nanorods. *Crystengcomm* 14 (3):1150–1154.

Badan, C., O. Esenturk, and A. Yilmaz. 2012. Microwave-assisted synthesis of $Eu^{3+}$ doped lanthanum orthoborates, their characterizations and luminescent properties. *Solid State Sciences* 14 (11–12):1710–1716.

Badylevich, M., S. Shamuilia, V. V. Afanas'ev, et al. 2007. Investigation of the electronic structure at interfaces of crystalline and amorphous $Gd_2O_3$ thin layers with silicon substrates of different orientations. *Applied Physics Letters* 90 (25):252101.

Baghbanzadeh, M., L. Carbone, P. D. Cozzoli, and C. O. Kappe. 2011. Microwave-assisted synthesis of colloidal inorganic nanocrystals. *Angewandte Chemie-International Edition* 50 (48):11312–11359.

Bondioli, F., A. M. Ferrari, L. Lusvarghi, et al. 2005. Synthesis and characterization of praseodymium-doped ceria powders by a microwave-assisted hydrothermal (MH) route. *Journal of Materials Chemistry* 15 (10):1061–1066.

Bühler, G., and C. Feldmann. 2006. Microwave-assisted synthesis of luminescent $LaPO_4$: Ce, Tb nanocrystals in ionic liquids. *Angewandte Chemie International Edition* 45 (29):4864–4867.

Bühler, G., and C. Feldmann. 2007. Transparent luminescent layers via ionic liquid-based approach to $LaPO_4$: RE (RE=Ce, Tb, Eu) dispersions. *Applied Physics a-Materials Science and Processing* 87 (4):631–636.

Carini, G., G. D'Angelo, G. Tripodo, A. Fontana, F. Rossi, and G. A. Saunders. 1997. Low-energy magnetic excitations in the Pr metaphosphate glass. *EPL (Europhysics Letters)* 40 (4):435.

Chang, H. Y., C. Y. Wu, K. Huang, Y. Lo, I. Shen, and H. Wang. 2014. Preparation and characterization of blue-light-excited nanophosphors using an economically low-energy process. *Journal of Alloys and Compounds* 598 (0):217–223.

Charbgoo, F., M. Ramezani, and M. Darroudi. 2017. Bio-sensing applications of cerium oxide nanoparticles: Advantages and disadvantages. *Biosensors and Bioelectronics* 96:33–43.

Che, D., X. Zhu, P. Liu, et al. 2014. A facile aqueous strategy for the synthesis of high-brightness $LaPO_4$:Eu nanocrystals via controlling the nucleation and growth process. *Journal of Luminescence* 153:369–374.

Chen, F., W. Bu, S. Zhang, et al. 2011. Positive and negative lattice shielding effects co-existing in Gd (III) ion doped bifunctional upconversion nanoprobes. *Advanced Functional Materials* 21 (22):4285–4294.

Collins, M. J., Jr. 2010. Future trends in microwave synthesis. *Future Medicinal Chemistry* 2 (2):151–155.

Colomer, M. T., I. Delgado, A. L. Ortiz, and J. C. Farinas. 2014. Microwave-assisted hydrothermal synthesis of single- crystal nanorods of rhabdophane-type Sr-doped $LaPO_4$ center dot $nH_2O$. *Journal of the American Ceramic Society* 97 (3):750–758.

Costa, A. L., M. Serantoni, M. Blosi, et al. 2010. Microwave assisted synthesis of $Yb:Y_2O_3$ based materials for laser source application. *Advanced Engineering Materials* 12 (3):205–209.

Dai, S. H., Y. F. Liu, Y. N. Lu, and H. H. Min. 2010. Microwave solvothermal synthesis of $Eu^{3+}$-doped $(Y,Gd)_2O_3$ microsheets. *Powder Technology* 202 (1–3):178–184.

Deus, R. C., M. Cilense, C. R. Foschini, M. A. Ramirez, E. Longo, and A. Z. Simoes. 2013. Influence of mineralizer agents on the growth of crystalline $CeO_2$ nanospheres by the microwave-hydrothermal method. *Journal of Alloys and Compounds* 550:245–251.

Ding, J., X. Lue, H. Shu, J. Xie, and H. Zhang. 2010. Microwave-assisted synthesis of perovskite $ReFeO_3$ (Re: La, Sm, Eu, Gd) photocatalyst. *Materials Science and Engineering B-Advanced Functional Solid-State Materials* 171 (1–3):31–34.

Ding, M., C. Lu, Y. Ni, and Z. Xu. 2014. Rapid microwave-assisted flux growth of pure beta-$NaYF_4$:$Yb^{3+}$, $Ln^{3+}$ (Ln = Er, Tm, Ho) microrods with multicolor upconversion luminescence. *Chemical Engineering Journal* 241:477–484.

Downing, E., L. Hesselink, J. Ralston, and R. Macfarlane. 1996. A three-color, solid-state, three-dimensional display. *Science* 273 (5279):1185–1189.

Dutta, D. P., and A. K. Tyagi. 2012. White light emission from microwave synthesized spin coated $Gd_2O_3$:Dy:Tb nano phosphors. *Proceedings of the National Academy of Sciences India Section A-Physical Sciences* 82 (1):53–57.

Ekthammathat, N., T. Thongtem, A. Phuruangrat, and S. Thongtem. 2013. Synthesis and characterization of $CeVO_4$ by microwave radiation method and its photocatalytic activity. *Journal of Nanomaterials* 2013:1–7.

Fang, B., W. Zhang, G. F. Wang, H. Y. Liu, and S. P. Wei. 2008. Microwave-assisted preparation of a carbon nanotube/$La(OH)_3$ nanocomposite, and its application to electrochemical determination of adenine and guanine. *Microchimica Acta* 162 (1–2):175–180.

Fang, Y. P., A. W. Xu, R. Q. Song, et al. 2003. Systematic synthesis and characterization of single-crystal lanthanide orthophosphate nanowires. *Journal of the American Chemical Society* 125 (51):16025–16034.

Farhadi, S., Z. Momeni, and M. Taherimehr. 2009. Rapid synthesis of perovskite-type $LaFeO_3$ nanoparticles by microwave-assisted decomposition of bimetallic La $Fe(CN)_6$·$5H_2O$ compound. *Journal of Alloys and Compounds* 471 (1–2):L5–L8.

Farhadi, S., and S. Sepahvand. 2010. Microwave-assisted solid-state decomposition of La[Co(CN)$_6$]·5H$_2$O precursor: A simple and fast route for the synthesis LaCoO$_3$ nanoparticles. *Journal of Alloys and Compounds* 489 (2):586–591.

Feng, J., and H. Zhang. 2013. Hybrid materials based on lanthanide organic complexes: A review. *Chemical Society Reviews* 42 (1):387–410.

Firsching, F. H., and S. N. Brune. 1991. Solubility products of the trivalent rare-earth phosphates. *Journal of Chemical and Engineering Data* 36 (1):93–95.

Fouassier, C. 2000. Chapter 10—Luminescent properties of fluorides. In *Advanced Inorganic Fluorides*, edited by T. Nakajima, B. Žemva, and A. Tressaud. Switzerland: Elsevier.

Gai, S., C. Li, P. Yang, and J. Lin. 2014. Recent progress in rare earth micro/nanocrystals: Soft chemical synthesis, luminescent properties, and biomedical applications. *Chemical Reviews* 114 (4):2343–2389.

Godinho, M., C. Ribeiro, E. Longo, and E. R. Leite. 2008. Influence of microwave heating on the growth of gadolinium-doped cerium oxide nanorods. *Crystal Growth and Design* 8 (2):384–386.

Goharshadi, E. K., S. Samiee, and P. Nancarrow. 2011. Fabrication of cerium oxide nanoparticles: Characterization and optical properties. *Journal of Colloid and Interface Science* 356 (2):473–480.

Guo, H., N. Dong, M. Yin, W. Zhang, L. Lou, and S. Xia. 2004. Visible upconversion in rare earth ion-doped Gd$_2$O$_3$ nanocrystals. *The Journal of Physical Chemistry B* 108 (50):19205–19209.

Gutierrez Seijas, J., J. Prado-Gonjal, D. A. Brande, I. Terry, E. Moran, and R. Schmidt. 2017. Microwave-assisted synthesis, microstructure, and magnetic properties of rare-earth cobaltites. *Inorganic Chemistry* 56 (1):627–633.

He, G. P., H. Q. Fan, and Z. W. Wang. 2014. Enhanced optical properties of heterostructured ZnO/CeO$_2$ nanocomposite fabricated by one-pot hydrothermal method: Fluorescence and ultraviolet absorption and visible light transparency. *Optical Materials* 38: 145–153.

He, L. Y., Y. M. Su, L. H. Jiang, and S. K. Shi. 2015. Recent advances of cerium oxide nanoparticles in synthesis, luminescence and biomedical studies: A review. *Journal of Rare Earths* 33 (8):791–799.

Hikichi, Y., and T. Nomura. 1987. Melting temperatures of monazite and xenotime. *Journal of the American Ceramic Society* 70 (10):C-252–C-253.

Hu, J., and Q. Wang. 2014. New synthesis for a group of tetragonal LnVO$_4$ and their Luminescent properties. *Materials Letters* 120 (0):20–22.

Huang, M. H., S. Mao, H. Feick, et al. 2001. Room-temperature ultraviolet nanowire nanolasers. *Science* 292 (5523):1897–1899.

Huang, S., H. L. Xu, M. Y. Wang, S. L. Zhong, and C. H. Zeng. 2016. Rapid microwave synthesis and photoluminescence properties of rare earth-based coordination polymer core-shell particles. *Optical Materials* 62:538–542.

Huong, N. T., N. D. Van, D. M. Tien, et al. 2011. Structural and luminescent properties of (Eu,Tb) PO$_4$·H$_2$O nanorods/nanowires prepared by microwave technique. *Journal of Rare Earths* 29 (12):1170–1173.

Iacconi, P., M. Junker, B. Guilhot, and D. Huguenin. 2001. Thermoluminescence of a mixed rare earth phosphate powder La$_{1-x-y}$Ce$_x$Tb$_y$PO$_4$. *Optical Materials* 17 (3):409–414.

Jellison, G. E. Jr, L. A. Boatner, and Chi Chen. 2000. Spectroscopic refractive indices of metalorthophosphates with the zircon-type structure. *Optical Materials* 15 (2):103–109.

Jung, W. Y., and S. S. Hong. 2013. Synthesis of LaCoO$_3$ nanoparticles by microwave process and their photocatalytic activity under visible light irradiation. *Journal of Industrial and Engineering Chemistry* 19 (1):157–160.

Kaszewski, J., M. M. Godlewski, B. S. Witkowski, et al. 2016. Y$_2$O$_3$:Eu nanocrystals as biomarkers prepared by a microwave hydrothermal method. *Optical Materials* 59:157–164.

Kim, C. R., T. Uemura, and S. Kitagawa. 2014. Sol-gel synthesis of nanosized titanium oxide in a porous coordination polymer. *Microporous and Mesoporous Materials* 195:31–35.

Lehnen, T., J. Schlafer, and S. Mathur. 2014. Rapid microwave synthesis of $CeO_2$ quantum dots. *Zeitschrift fur Anorganische und Allgemeine Chemie* 640 (5):819–825.

Lei, Y. Q., M. Pang, W. Q. Fan, et al. 2011. Microwave-assisted synthesis of hydrophilic $BaYF_5$:Tb/Ce, Tb green fluorescent colloid nanocrystals. *Dalton Transactions* 40 (1): 142–145.

Li, C., P. Ma, P. Yang, et al. 2011. Fine structural and morphological control of rare earth fluorides $REF_3$ (RE=La-Lu, Y) nano/microcrystals: Microwave-assisted ionic liquid synthesis, magnetic and luminescent properties. *Crystengcomm* 13 (3):1003–1013.

Li, F. F., C. G. Li, J. H. Liu, et al. 2013a. Aqueous phase synthesis of upconversion nanocrystals through layer-by-layer epitaxial growth for in vivo X-ray computed tomography. *Nanoscale* 5 (15):6950–6959.

Li, F., C. Li, X. Liu, et al. 2013b. Microwave-assisted synthesis and up-down conversion luminescent properties of multicolor hydrophilic $LaF_3$:$Ln^{3+}$ nanocrystals. *Dalton Transactions* 42 (6):2015–2022.

Li, R., L. Li, Y. Han, S. Gai, F. He, and P. Yang. 2014. Core-shell structured $Gd_2O_3$:Ln@ $mSiO_2$ hollow nanospheres: Synthesis, photoluminescence and drug release properties. *Journal of Materials Chemistry B* 2 (15):2127–2135.

Li, W., and J. Lee. 2008. Microwave-assisted sol-gel synthesis and photoluminescence characterization of $LaPO_4$:$Eu^{3+}$,$Li^+$ nanophosphors. *Journal of Physical Chemistry C* 112 (31):11679–11684.

Lin, Z. J., Z. Yang, T. F. Liu, Y. B. Huang, and R. Cao. 2012. Microwave-assisted synthesis of a series of lanthanide metal-organic frameworks and gas sorption properties. *Inorganic Chemistry* 51 (3):1813–1820.

Liu, H. W., H. F. Liu, and X. Y. Han. 2017. Core-shell $CeO_2$ micro/nanospheres prepared by microwave-assisted solvothermal process as high-stability anodes for Li-ion batteries. *Journal of Solid State Electrochemistry* 21 (1):291–295.

Liu, J. W., Y. Zhang, X. W. Chen, and J. H. Wang. 2014. Graphene oxide-rare earth metal-organic framework composites for the selective isolation of hemoglobin. *Acs Applied Materials and Interfaces* 6 (13):10196–10204.

Lyszczek, R., and A. Lipke. 2013. Microwave-assisted synthesis of lanthanide 2,6-naphthalenedicarboxylates: Thermal, luminescent and sorption characterization. *Microporous and Mesoporous Materials* 168:81–91.

Ma, L., W. X. Chen, J. Zhao, Y. F. Zheng, X. Li, and Z. D. Xu. 2007. Microwave-assisted synthesis of praseodymium hydroxide nanorods and thermal conversion to oxide nanorod. *Materials Letters* 61 (8–9):1711–1714.

Ma, L., L. M. Xu, W. X. Chen, and Z. D. Xu. 2009. Microwave-assisted synthesis and characterization of $LaPO_4$: Re (Re=$Ce^{3+}$, $Eu^{3+}$, $Tb^{3+}$) nanorods. *Materials Letters* 63 (18–19):1635–1637.

Mahapatra, S., S. K. Nayak, G. Madras, and T. N. G. Row. 2008. Microwave synthesis and photocatalytic activity of nano lanthanide (Ce, pr, and Nd) orthovanadates. *Industrial and Engineering Chemistry Research* 47 (17):6509–6516.

Majeed, S., and S. A. Shivashankar. 2014a. Rapid, microwave-assisted synthesis of Gd2O3 and Eu:Gd2O3 nanocrystals: Characterization, magnetic, optical and biological studies. *Journal of Materials Chemistry B* 2 (34):5585–5593.

Majeed, S., and S. A. Shivashankar. 2014b. Microspherical, hierarchical structures of blue–green-emitting Dy:GdOOH and Dy:$Gd_2O_3$. *Materials Letters* 125 (0):136–139.

Majeed, S., and S. A. Shivashankar. 2014c. Novel spherical hierarchical structures of GdOOH and Eu:GdOOH: Rapid microwave-assisted synthesis through self-assembly, thermal conversion to oxides, and optical studies. *Journal of Materials Chemistry C* 2 (16):2965–2974.

Makhov, V. N., A. S. Vanetsev, N. M. Khaidukov, et al. 2013. Intrinsic and impurity luminescence of rare earth ions doped $KYE_4$ nanophosphors. *Radiation Measurements* 56:393–396.

Mekuria, S. L., T. A. Debele, and H. C. Tsai. 2017. Encapsulation of gadolinium oxide nanoparticle ($Gd_2O_3$) contrasting agents in PAMAM dendrimer templates for enhanced magnetic resonance imaging in vivo. *Acs Applied Materials and Interfaces* 9 (8): 6782–6795.

Mendiuk, O., and L. Kepinski. 2014. Synthesis of $Ce_{1-x}Er_xO_{2-y}$ nanoparticles by the hydrothermal method: Effect of microwave radiation on morphology and phase composition. *Ceramics International* 40 (9):14833–14843.

Meng, J., X. L. Wang, K. C. Chou, and Q. Li. 2013. Hydrogen storage properties of graphite-modified Mg-Ni-Ce composites prepared by mechanical milling followed by microwave sintering. *Metallurgical and Materials Transactions a-Physical Metallurgy and Materials Science* 44A (1):58–67.

Mi, C. C., Z. Tian, B. Han, C. Mao, and S. Xu. 2012. Microwave-assisted one-pot synthesis of water-soluble rare-earth doped fluoride luminescent nanoparticles with tunable colors. *Journal of Alloys and Compounds* 525:154–158.

Mi, C., Z. Tian, C. Cao, Z. Wang, C. Mao, and S. Xu. 2011. Novel microwave-assisted solvothermal synthesis of $NaYF_4$:Yb,Er upconversion nanoparticles and their application in cancer cell imaging. *Langmuir* 27 (23):14632–14637.

Miyazaki, H., H. Matsui, H. Kitakaze, S. Karuppuchamy, S. Ito, and M. Yoshihara. 2009. Synthesis and electronic behaviors of $Ce_{0.5}Hf_{0.5}O_2$/carbon clusters composite material. *Materials Chemistry and Physics* 113 (1):21–25.

Molenda, J., J. Kupecki, R. Baron, et al. 2017. Status report on high temperature fuel cells in Poland – Recent advances and achievements. *International Journal of Hydrogen Energy* 42 (7):4366–4403.

Montini, T., M. Melchionna, M. Monai, and P. Fornasiero. 2016. Fundamentals and catalytic applications of $CeO_2$-based materials. *Chemical Reviews* 116 (10):5987–6041.

Muñoz, F. F., L. M. Acuna, C. A. Albornoz, A. G. Leyva, R. T. Baker, and R. O. Fuentes. 2015. Redox properties of nanostructured lanthanide-doped ceria spheres prepared by microwave assisted hydrothermal homogeneous co-precipitation. *Nanoscale* 7 (1):271–281.

Natile, M. M., G. Boccaletti, and A. Glisenti. 2005. Properties and reactivity of nanostructured $CeO_2$ powders: Comparison among two synthesis procedures. *Chemistry of Materials* 17 (25):6272–6286.

Neal, J. S., L. A. Boatner, M. Spurrier, P. Szupryczynski, and C. L. Melcher. 2006. Cerium-doped mixed-alkali rare-earth double-phosphate scintillators for X- and gamma-ray detection – art. no. 631907. In *Hard X-Ray and Gamma-Ray Detector Physics and Penetrating Radiation Systems VIII*, edited by L. A. Franks, A. Burger, R. B. James, H. B. Barber, F. P. Doty, and H. Roehrig. San Diego, CA: SPIE.

Niu, N., F. He, S. Gai, et al. 2012. Rapid microwave reflux process for the synthesis of pure hexagonal $NaYF_4$:$Yb^{3+}$, $Ln^{3+}$, $Bi^{3+}$ ($Ln^{3+}$=$Er^{3+}$, $Tm^{3+}$, $Ho^{3+}$) and its enhanced UC luminescence. *Journal of Materials Chemistry* 22 (40):21613–21623.

Norby, T., and N. Christiansen. 1995. Proton conduction in Ca- and Sr-substituted LaPO4. *Solid State Ionics* 77 (0):240–243.

Olesiak-Banska, J., M. Nyk, D. Kaczmarek, K. Matczyszyn, K. Pawlik, and M. Samoc. 2011. Synthesis and optical properties of water-soluble fluoride nanophosphors co-doped with $Eu^{3+}$ and $Tb^{3+}$. *Optical Materials* 33 (9):1419–1423.

Ordoñez-Regil, E., R. Drot, E. Simoni, and J. J. Ehrhardt. 2002. Sorption of uranium(VI) onto lanthanum phosphate surfaces. *Langmuir* 18 (21):7977–7984.

Pan, S. H., R. P. Deng, J. Feng, et al. 2013. Microwave-assisted synthesis and down- and up-conversion luminescent properties of $BaYF_5$:Ln (Ln=Yb/Er, Ce/Tb) nanocrystals. *Crystengcomm* 15 (38):7640–7643.

Panda, A. B., G. Glaspell, and M. S. El-Shall. 2007. Microwave synthesis and optical properties of uniform nanorods and nanoplates of rare earth oxides. *Journal of Physical Chemistry C* 111 (5):1861–1864.

Parhi, P., and V. Manivannan. 2008. Novel microwave initiated solid-state metathesis synthesis and characterization of lanthanide phosphates and vanadates, $LMO_4$ (L=Y, La and M=V, P). *Solid State Sciences* 10 (8):1012–1019.

Patra, C. R., G. Alexandra, S. Patra, et al. 2005. Microwave approach for the synthesis of rhabdophane-type lanthanide orthophosphate (Ln=La, Ce, Nd, Sm, Eu, Gd and Tb) nanorods under solvothermal conditions. *New Journal of Chemistry* 29 (5): 733–739.

Peng, W. Q., C. L. Guan, and S. Y. Li. 2014. Material removal mechanism of ceria particles with different sizes in glass polishing. *Optical Engineering* 53 (3).

Polychronopoulou, K., A. F. Zedan, M. S. Katsiotis, et al. 2017. Rapid microwave assisted sol-gel synthesis of $CeO_2$ and $Ce_xSm_{1-x}O_2$ nanoparticle catalysts for CO oxidation. *Molecular Catalysis* 428:41–55.

Porosnicu, I., D. Avram, B. Cojocaru, M. Florea, and C. Tiseanu. 2017. Up-conversion luminescence of Er(Yb)-$CeO_2$: Status and new results. *Journal of Alloys and Compounds* 711:627–636.

Prado-Gonjal, J., R. Schmidt, D. Ávila, U. Amador, and E. Morán. 2012a. Structural and physical properties of microwave synthesized orthorhombic perovskite erbium chromite $ErCrO_3$. *Journal of the European Ceramic Society* 32 (3):611–618.

Prado-Gonjal, J., R. Schmidt, J. Espíndola-Canuto, P. Ramos-Alvarez, and E. Morán. 2012b. Increased ionic conductivity in microwave hydrothermally synthesized rare-earth doped ceria $Ce_{1-x}RE_xO_{2-(x/2)}$. *Journal of Power Sources* 209 (0):163–171.

Prado-Gonjal, J., R. Schmidt, J. J. Romero, D. Ávila, U. Amador, and E. Morán. 2013 a. Microwave-assisted synthesis, microstructure, and physical properties of rare-earth chromites. *Inorganic Chemistry* 52 (1):313–320.

Quazi, M. M., M. A. Fazal, Asma Haseeb, F. Yusof, H. H. Masjuki, and A. Arslan. 2016. Effect of rare earth elements and their oxides on tribo-mechanical performance of laser claddings: A review. *Journal of Rare Earths* 34 (6):549–564.

Reynolds, C. H., N. Annan, K. Beshah, et al. 2000. Gadolinium-loaded nanoparticles: New contrast agents for magnetic resonance imaging. *Journal of the American Chemical Society* 122 (37):8940–8945.

Riwotzki, K., H. Meyssamy, A. Kornowski, M. Haase. 2000. Liquid-phase synthesis of doped nanoparticles: Colloids of luminescing $LaPO_4$:Eu and $CePO_4$:Tb particles with a narrow particle size distribution. *The Journal of Physical Chemistry B* 104 (13):2824–2828.

Riwotzki, K., H. Meyssamy, H. Schnablegger, A. Kornowski, and M. Haase. 2001. Liquid-phase synthesis of colloids and redispersible powders of strongly luminescing $LaPO_4$:Ce, Tb nanocrystals. *Angewandte Chemie International Edition* 40 (3):573–576.

Rocha, G. N., L. F. L. Melo, M. C. Castro, Jr., A. P. Ayala, A. S. de Menezes, and P. B. A. Fechine. 2013. Structural characterization of bismuth rare earth tungstates obtained by fast microwave-assisted solid-state synthesis. *Materials Chemistry and Physics* 139 (2–3):494–499.

Rodriguez-Liviano, S., F. J. Aparicio, A. I. Becerro, et al. 2013a. Synthesis and functionalization of biocompatible Tb:$CePO_4$ nanophosphors with spindle-like shape. *Journal of Nanoparticle Research* 15 (2):14.

Rodriguez-Liviano, S., F. J. Aparicio, T. C. Rojas, A. B. Hungria, L. E. Chinchilla, and M. Ocana. 2012. Microwave-assisted synthesis and luminescence of mesoporous REDoped $YPO_4$ (RE=Eu, Ce, Tb, and Ce plus Tb) nanophosphors with lenticular shape. *Crystal Growth and Design* 12 (2):635–645.

Rodriguez-Liviano, S., A. I. Becerro, D. Alcantara, V. Grazu, J. M. de la Fuente, and M. Ocana. 2013b. Synthesis and properties of multifunctional tetragonal Eu:GdPO$_4$ nanocubes for optical and magnetic resonance imaging applications. *Inorganic Chemistry* 52 (2):647–654.

Roduner, E. 2006. Size matters: Why nanomaterials are different. *Chemical Society Reviews* 35 (7):583–592.

Saadat-Monfared, A., M. Mohseni, and M. H. Tabatabaei. 2012. Polyurethane nanocomposite films containing nano-cerium oxide as UV absorber. Part 1. Static and dynamic light scattering, small angle neutron scattering and optical studies. *Colloids and Surfaces a-Physicochemical and Engineering Aspects* 408:64–70.

Samiee, S., and E. K. Goharshadi. 2012. Effects of different precursors on size and optical properties of ceria nanoparticles prepared by microwave-assisted method. *Materials Research Bulletin* 47 (4):1089–1095.

Sangsefidi, F. S., M. Salavati-Niasari, H. Khojasteh, and M. Shabani-Nooshabadi. 2017. Synthesis, characterization and investigation of the electrochemical hydrogen storage properties of CuO-CeO$_2$ nanocomposites synthesized by green method. *International Journal of Hydrogen Energy* 42 (21):14608–14620.

Saradha, T., S. Muzhumathi, and A. Subramania. 2008. Microwave-assisted combustion synthesis of nanocrystalline La$_2$Mo$_2$O$_9$ oxide-ion conductor and its characterization. *Journal of Solid State Electrochemistry* 12 (2):143–148.

Sarala Devi, G., D. Giridhar, and B. M. Reddy. 2002. Vapour phase O-alkylation of phenol over alkali promoted rare earth metal phosphates. *Journal of Molecular Catalysis A: Chemical* 181 (1–2):173–178.

Serantoni, M., E. Mercadelli, A. L. Costa, M. Blosi, L. Esposito, and A. Sanson. 2010. Microwave-assisted polyol synthesis of sub-micrometer Y$_2$O$_3$ and Yb-Y$_2$O$_3$ particles for laser source application. *Ceramics International* 36 (1):103–106.

Shchukin, D. G., and R. A. Caruso. 2004. Template synthesis and photocatalytic properties of porous metal oxide spheres formed by nanoparticle infiltration. *Chemistry of Materials* 16 (11):2287–2292.

Shen, J., L. D. Sun, and C. H. Yan. 2008. Luminescent rare earth nanomaterials for bioprobe applications. *Dalton Transactions* (42):5687–5697.

Silva, P., A. A. Valente, J. Rocha, and F. A. A. Paz. 2010. Fast microwave synthesis of a microporous lanthanide organic framework. *Crystal Growth and Design* 10 (5):2025–2028.

Sohlberg, K., S. T. Pantelides, and S. J. Pennycook. 2001. Interactions of hydrogen with CeO$_2$. *Journal of the American Chemical Society* 123 (27):6609–6611.

Srivastava, M., K. Kumar, N. Jaiswal, N. K. Singh, D. Kumar, and O. Parkash. 2014. Enhanced ionic conductivity of co-doped ceria solid solutions and applications in IT-SOFCs. *Ceramics International* 40 (7):10901–10906.

Su, W., M. He, J. Xing, Y. Zhong, and Z. Li. 2013. Facile synthesis of porous bifunctional Fe$_3$O$_4$@Y$_2$O$_3$:Ln nanocomposites using carbonized ferrocene as templates. *RSC Advances* 3 (48):25970–25975.

Tang, Z. R., Y. Zhang, and Y. J. Xu. 2011. A facile and high-yield approach to synthesize one-dimensional CeO$_2$ nanotubes with well-shaped hollow interior as a photocatalyst for degradation of toxic pollutants. *RSC Advances* 1 (9):1772–1777.

Trujillano, R., V. Rives, M. Douma, and E. H. Chtoun. 2015. Microwave hydrothermal synthesis of A$_2$Sn$_2$O$_7$ (A=Eu or Y). *Ceramics International* 41 (2):2266–2270.

Vilela, S. M. F., A. D. G. Firmino, R. F. Mendes, et al. 2013. Lanthanide-polyphosphonate coordination polymers combining catalytic and photoluminescence properties. *Chemical Communications* 49 (57):6400–6402.

Wang, H. Q., and T. Nann. 2009. Monodisperse upconverting nanocrystals by microwave-assisted synthesis. *Acs Nano* 3 (11):3804–3808.

Wang, H. Q., and T. Nann. 2011. Monodisperse upconversion GdF$_3$:Yb, Er rhombi by microwave-assisted synthesis. *Nanoscale Research Letters* 6:5.

Wang, H. Q., R. D. Tilley, and T. Nann. 2010a. Size and shape evolution of upconverting nanoparticles using microwave assisted synthesis. *Crystengcomm* 12 (7):1993–1996.

Wang, S. P., Y. L. Zhao, J. J. Chen, R. Xu, L. F. Luo, and S. L. Zhong. 2010b. Self-assembled 3D La(OH)$_3$ and La$_2$O$_3$ nanostructures: Fast microwave synthesis and characterization. *Superlattices and Microstructures* 47 (5):597–605.

Yan, B. 2012. Recent progress in photofunctional lanthanide hybrid materials. *RSC Advances* 2 (25):9304–9324.

Yan, C. H., Z. G. Yan, Y. P. Du, J. Shen, C. Zhang, and W. Feng. 2011. Chapter 251—Controlled synthesis and properties of rare earth nanomaterials. In *Handbook on the Physics and Chemistry of Rare Earths*, edited by J.-C. G. B. Karl A. Gschneidner, and K. P. Vitalij. Switzerland: Elsevier.

Yan, Z. G., and C. H. Yan. 2008. Controlled synthesis of rare earth nanostructures. *Journal of Materials Chemistry* 18 (42):5046–5059.

Yu, H. Y., D. B. Zhou, and H. M. Zhu. 2014. Synthesis and characterization of Pd-La(OH)$_3$/C electrocatalyst for direct ethanol fuel cell. *Journal of Solid State Electrochemistry* 18 (1):125–131.

Yu, X. F., J. W. Liu, H. P. Cong, et al. 2015. Template- and surfactant-free synthesis of ultrathin CeO$_2$ nanowires in a mixed solvent and their superior adsorption capability for water treatment. *Chemical Science* 6 (4):2511–2515.

Yuan, Q., H. H. Duan, L. L. Li, L. D. Sun, Y. W. Zhang, and C. H. Yan. 2009. Controlled synthesis and assembly of ceria-based nanomaterials. *Journal of Colloid and Interface Science* 335 (2):151–167.

Zhao, D., L. Wang, Y. Li, L. Zhang, Y. Lv, and S. Zhong. 2012. Uniform europium-based infinite coordination polymer submicrospheres: Fast microwave synthesis and characterization. *Inorganic Chemistry Communications* 20 (0):97–100.

Zhao, Q., Z. H. Xu, and Y. G. Sun. 2014. Rare earth fluoride nano-/microstructures: Hydrothermal synthesis, luminescent properties and applications. *Journal of Nanoscience and Nanotechnology* 14 (2):1675–1692.

Zharkouskaya, A., C. Feldmann, K. Trampert, W. Heering, and U. Lemmer. 2008. Ionic liquid based approach to luminescent LaPO$_4$:Ce,Tb nanocrystals: Synthesis, characterization and application. *European Journal of Inorganic Chemistry* (6):873–877.

Zheng, H., B. J. Chen, H. Q. Yu, et al. 2014. Preparation and characterization of Gd$_2$O$_3$:Eu$^{3+}$ rods by surfactant assemblies-microwave heating. *Journal of Colloid and Interface Science* 420:27–34.

Zheng, S. T., T. Wu, C. Chou, A. Fuhr, P. Feng, and X. Bu. 2012. Development of composite inorganic building blocks for MOFs. *Journal of the American Chemical Society* 134 (10):4517–4520.

Zheng, Z. N., and S. W. Lee. 2011. Microwave-assisted preparation, structures, and photoluminescent properties of [Ln(NO$_3$)$_2$(H2O)$_3$(L)$_2$](NO$_3$)(H$_2$O) {Ln=Tb, Eu; L=2-(4-pyridylium)ethanesulfonate, (4-pyH)$^+$-CH$_2$CH$_2$-SO$_3^-$}. *Bulletin of the Korean Chemical Society* 32 (6):1859–1864.

Zhong, S., J. Chen, S. Wang, Q. Liu, Y. Wang, and S. Wang. 2010. Y$_2$O$_3$:Eu$^{3+}$hexagonal microprisms: Fast microwave synthesis and photoluminescence properties. *Journal of Alloys and Compounds* 493 (1–2):322–325.

Zhong, S., H. Jing, Y. Li, S. Yin, C. Zeng, and L. Wang. 2014. Coordination polymer submicrospheres: Fast microwave synthesis and their conversion under different atmospheres. *Inorganic Chemistry* 53 (16):8278–8286.

Zhong, S., Y. Lu, Z. Huang, S. Wang, and J. Chen. 2010. Microwave-assisted synthesis of water-soluble YF$_3$ and YF$_3$:Ln$^{3+}$ nanocrystals. *Optical Materials* 32 (9):966–970.

Zhou, L. Y., W. Wang, J. L. Huang, et al. 2010. Preparation and characterization of Gd$_2$O$_3$:Eu$^{3+}$ rods by surfactant assemblies-microwave heating. *Optik* 121 (16):1516–1519.

Zhu, Y. J., and F. Chen. 2014. Microwave-assisted preparation of inorganic nanostructures in liquid phase. *Chemical Reviews* 114 (12):6462–6555.

# 5 Microwave-Assisted Synthesis and Functionalization of Six-Membered Oxygen Heterocycles

*Neha Batra, Rahul Panwar, Ramendra Pratap, and Mahendra Nath*

## CONTENTS

## 5.1 INTRODUCTION

Heterocycles belong to one of the most significant and fascinating classes of organic compounds because of their ability to act as both biomimetics and reactive pharmacophores [1]. These molecules have attracted researchers around the globe because of their exclusive pharmaceutical profiles. Among this class of compounds, oxygen-containing six-membered heterocycles are more remarkable due to their diverse pharmacological activities such as antioxidants [2], diuretics [3], anti-coagulant [4], anticancer [5], anti-HIV [6], antitumor [7], anti-inflammatory [8], anti-Alzheimer's [9], antileukemic [10], antibacterial [11], antimalarial [12], emetic [13], and anti-tubercular activity [14]. In addition, some of these oxygen heterocycles have also been associated with cosmetics and pigments [15] and used as dyes [16], sensors, and probes due to their remarkable fluorescent properties [17]. By considering the biological and material significance of these compounds, a number of synthetic methodologies have been developed in the past several decades by using a wide range of substrates, reagents, catalysts, and solvents.

Among reported protocols, microwave-assisted organic synthesis (MAOs) has drawn considerable attention in recent years. The use of microwave irradiations in organic synthesis was first reported by Gedye and coworkers in the year 1986 [18], which was found to be more beneficial than the conventional methods. It offers various advantages such as increased yields, shorter reaction times, reproducibility, cleaner reactions, and easy workup [19, 20]. This technique can also be considered as an eco-friendly strategy for producing various compounds since the generation of toxic waste is reduced by many folds during the reaction [21]. Moreover, due to the selective and effective absorption of microwave energy by the polar molecules, this approach has been successful in synthesizing the products that are difficult to construct by using traditional heating protocols [22].

The following sections of this chapter describe microwave-assisted syntheses and functionalization of diverse six-membered oxygen heterocycles.

## 5.2 MICROWAVE-ASSISTED SYNTHESES OF SIX-MEMBERED OXYGEN HETEROCYCLES

### 5.2.1 SYNTHESIS OF 1,3-DIOXANES

Dioxane rings are present in various biologically active natural products including (+)-dactylolide (a cytotoxic agent) [23] and (+)-SCH 351448 (activates the low density lipoprotein receptors) [24]. Additionally, the 2-substituted-1,3-dioxanes act as an antimuscarinic agent [25] and efficient modulators for multidrug resistance [26]. However, limited approaches are available for the synthesis of 1,3-dioxanes. Gaina et al. synthesized phenothiazine tethered 1,3-dioxanes in moderate yields by microwave-assisted acetalization reaction in the presence of PEG or water under superheated conditions [27] (Scheme 5.1).

On the other hand, Flink et al. synthesized monoprotected acetals of glutaraldehyde and their symmetrical dimethyl derivatives under microwave irradiation. The selectivity for mono-protection in this protocol has been attributed to the microwave irradiation rather than the conventional heating [28] (Scheme 5.2).

In another methodology, Polshettiwar et al. introduced a 1,3-dioxane ring to the third position of 2-phenylchroman-4-one via bis-aldol reaction with paraformaldehyde in the presence of a catalytic amount of polystyrenesulfonic acid (PSSA) in aqueous media. The resulting product was obtained in excellent yield [29] (Scheme 5.3).

A microwave-assisted approach is not only used for the synthesis of 1,3-dioxanes, but it has also been found useful in accelerating the rate of various other reactions

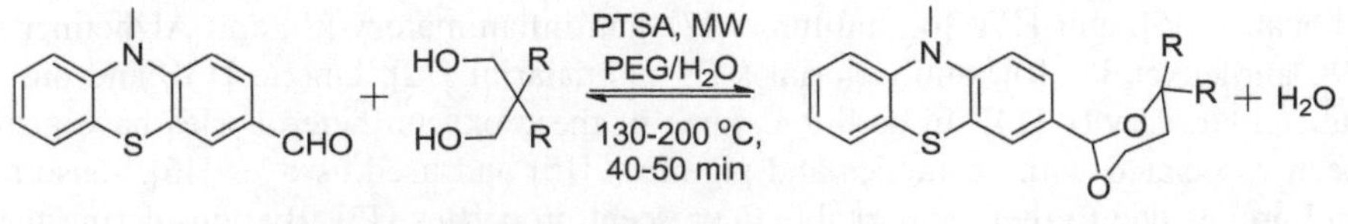

**SCHEME 5.1**  Synthesis of phenothiazine tethered 1,3-dioxanes.

where 1,3-dioxane derivatives are one of the major reactants. To this end, Meldrum's acid (2,2-dimethyl-1,3-dioxane-4,6-dione) reacted with isatin and malononitrile or ethylcyanoacetate in the presence of ZnO nanoparticles in absolute ethanol to provide novel spiro[indoline-pyranodioxine] derivatives in 81–88% yields within 8–11 minutes [30] (Scheme 5.4).

Similarly, the experiments were carried out by reacting Meldrum's acid with aromatic aldehydes and 1*H*-indazol-5-amine in ethylene glycol to produce pyrazolo[4,3-f]quinolin-7-one derivatives in appreciable yields [31] (Scheme 5.5).

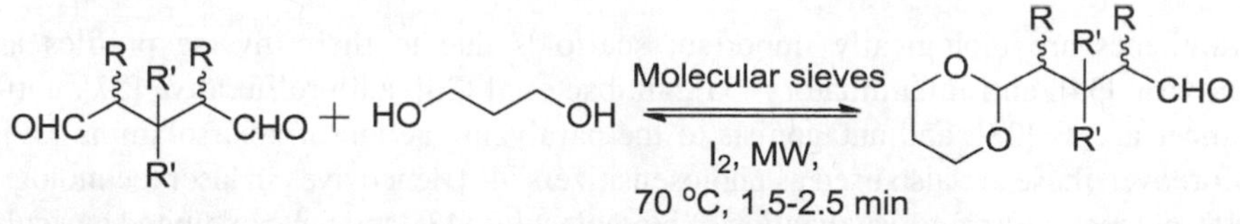

**SCHEME 5.2** Microwave-assisted synthesis of monoprotected acetals.

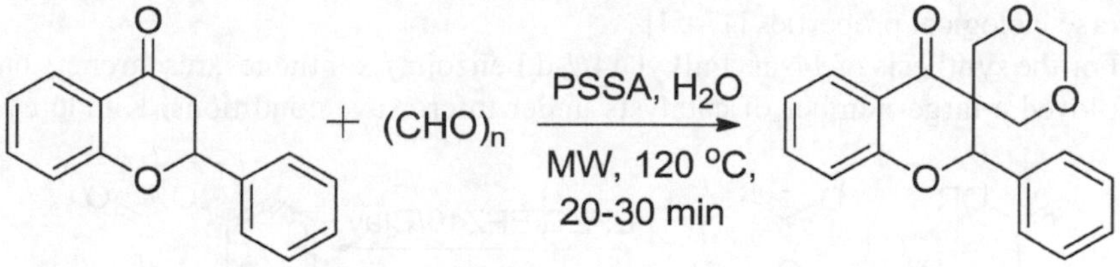

**SCHEME 5.3** PSSA-catalyzed synthesis of 1,3-dioxanes.

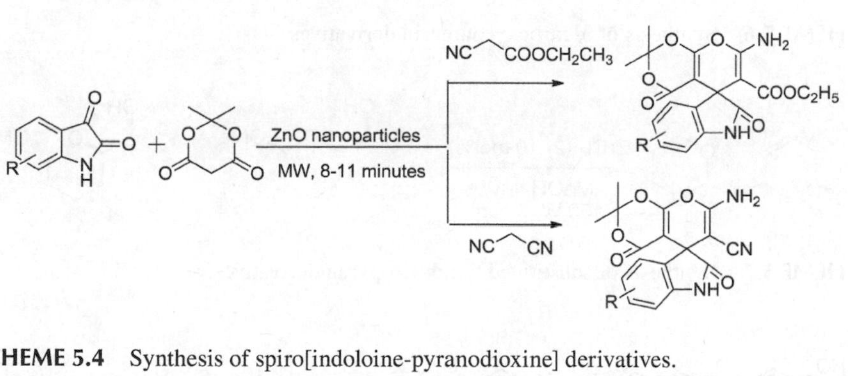

**SCHEME 5.4** Synthesis of spiro[indoloine-pyranodioxine] derivatives.

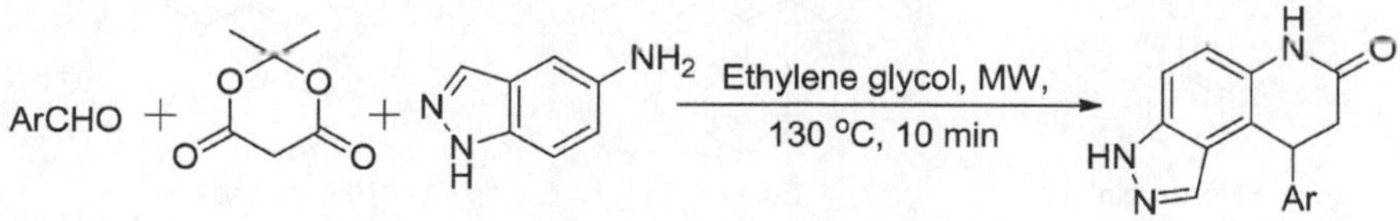

**SCHEME 5.5** Synthesis of pyrazolo[4,3-f]quinolin-7-one derivatives.

In addition, the one-pot reaction of 2-hydroxy- or 2-methoxybenzaldehydes or acetophenones and Meldrum's acid in the presence of Envirocats (EPZ10 and EPZG) or natural kaolinite clay as a catalyst under the microwave irradiation afforded 3-carboxycoumarin analogs [32] (Scheme 5.6).

In addition, certain 1,3-dioxane derivatives were recyclized on deprotection in the presence of $ZrCl_4$ as a catalyst. This protocol was successfully applied for the asymmetric synthesis of (+)-exo and (+)-endo-brevicomin [33] (Scheme 5.7).

## 5.2.2 Synthesis of Xanthenes and Xanthones

Xanthenes are biologically important scaffolds due to their diverse profiles as antiviral [34], anti-inflammatory [35], antibacterial [36], antiproliferative [37], anticancer agents [38], and antagonists to the paralyzing action of zoxazolamine [39]. Moreover, these are also used as photosensitizers [40], leuco dyes in laser technology [41], pH regulators for visualization of biomolecules [42], and have exhibited agricultural bactericidal activity [43]. The xanthene framework has also been a part of various other dyes such as fluorescein [44], eosins [45], and rhodamines [46] (Figure 5.1).

Similarly, xanthones are also considered as "privileged structures" owing to their diverse biological properties [47–51].

For the synthesis of 14-aryl/alkyl-14H-dibenzo[a,j]xanthenes, researchers have employed a large number of catalysts under microwave conditions. Kundu et al.

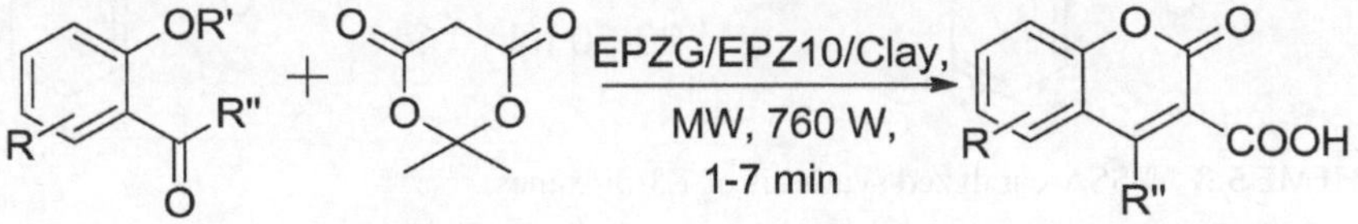

**SCHEME 5.6**  Synthesis of 3-carboxycoumarin derivatives.

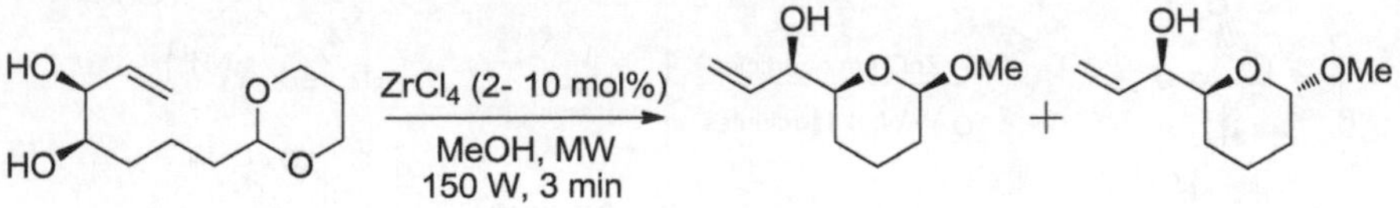

**SCHEME 5.7**  Synthesis of substituted tetrahydropyran derivatives.

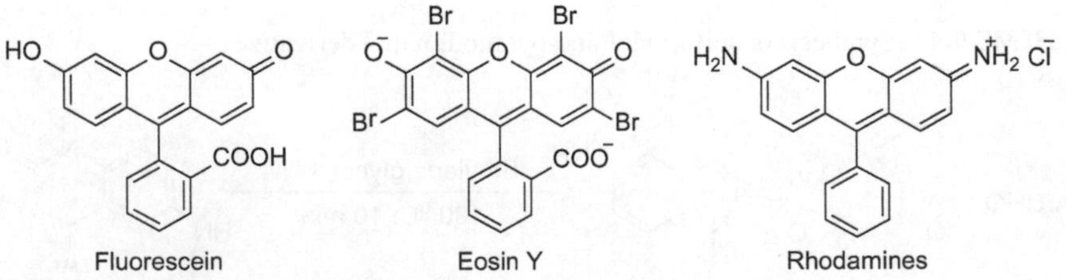

Fluorescein                    Eosin Y                    Rhodamines

**FIGURE 5.1**  The structures of dyes containing xanthenes scaffold.

condensed 2-naphthol with both aliphatic and aromatic aldehydes in presence of (±)-camphor-10-sulfonic acid (CSA) at 25°C to afford a mixture of 1,1-bis-(2-hydroxynaphthyl)phenylmethane and 14-phenyldibenzoxanthenes [52] whereas in the case of aliphatic aldehydes, solely 14-alkyldibenzoxanthenes were obtained (Scheme 5.8). However, Prasad et al. used p-dodecylbenzenesulfonic acid (DBSA) as a Bronsted acid catalyst to obtain single dibenzoxanthene products in excellent yields from both aromatic as well as aliphatic aldehydes (Table 5.1, entry 1) [53]. In the presence of iodine, Ding et al. completed the reaction within 15–20 minutes to produce the desired products (Table 5.1, entry 2) [54], whereas Bhattacharya etal. were able to synthesize dibenzoxanthenes within 1–4 minutes in the presence of methanesulfonic acid (Table 5.1, entry 3) as an acidic catalyst [55]. Nagarapu et al. performed this reaction in the presence of potassium dodecatungstocobaltate

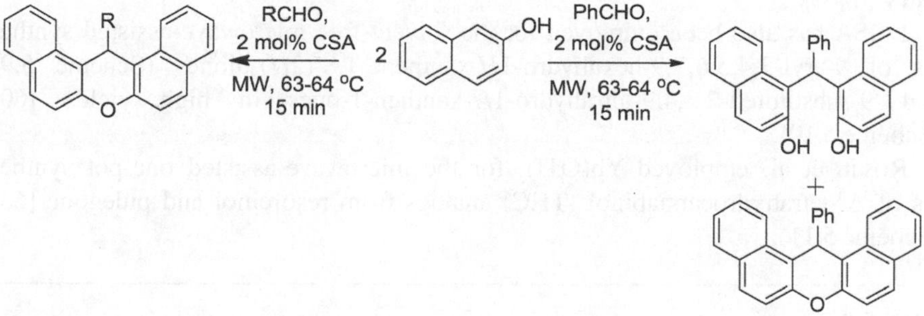

**SCHEME 5.8** One-pot synthesis of 14-alkyl/aryl-14H-dibenzo[a,j]xanthenes.

## TABLE 5.1
## Reaction Conditions Employed by Different Research Groups for the Generation of 14-aryl/alkyl-14H-Dibenzo[a,j]xanthenes

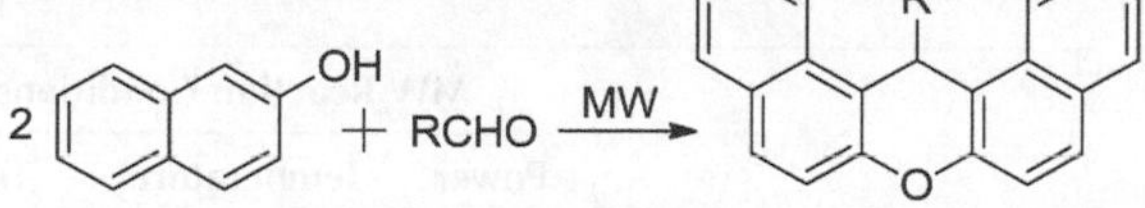

| | | MW Reaction Conditions | | | |
|---|---|---|---|---|---|
| S. No. | Catalysts | Power (W) | Temperature (°C) | Time (min) | Yields (%) |
| 1. | DBSA (2 mol%) | 300 | 125 | 4–5 | 78–99 |
| 2. | Iodine (10 mol%) | 90–180 | a | 15–20 | 75–93 |
| 3. | MeSO$_3$H (10 mol%) | 2350 | a | 1–4 | 65–95 |
| 4. | K$_5$CoW$_{12}$O$_{40}$·3H$_2$O | 300 | a | 1.5–3 | 72–94 |
| 5. | NaHSO$_4$·H$_2$O (30 mol%) | 850 | a | 1.8–4.5 | 83–94 |
| 6. | NanoSPA (2 mol%) | b | a | 6 | 70–92 |

a Temperature and b power have not been mentioned.

trihydrate ($K_5CoW_{12}O_{40} \cdot 3H_2O$) (Table 5.1, entry 4) [56] whereas Shaterian et al. used $NaHSO_4 \cdot H_2O$ to generate the title compounds (Table 5.1, entry 5) [57]. Nanosilica phosphoric acid has also been used for the synthesis of xanthenes (Table 5.1, entry 6) [58].

Similarly, the synthesis of tetrahydrobenzo[a]xanthenes-11-ones has also been reported under microwave conditions. Roa et al. achieved such xanthenes in high yields by condensing β-naphthol with aromatic aldehydes and 1,3-dicarbonyl compounds in the presence of $Sc(OTf)_3$ (Table 5.2, entry 1) [59]. Preetam et al. preferred DBSA as a Bronsted acid catalyst for this transformation (Table 5.2, entry 2) [60], while Bamoniri et al. used nanosilica phosphoric acid (NanoSPA) (Table 5.2, entry 3) [58]. Chereddy et al. displayed the role of ytterbium perfluorooctanoate [$Yb(PFO)_3$] (Table 5.2, entry 4) [61] and Kundu et al. carried out this reaction in presence of sulfonic acid functionalized ionic liquid under solvent-free conditions (Table 5.2, entry 5) [62].

DBSA has also been employed for the solvent-free microwave-assisted synthesis of 9-aryl-3,4,5,6,7,9-hexahydro-1H-xanthene-1,8-(2H)-diones (Scheme 5.9) and 9-substituted-2,3,4,9-tetrahydro-1H-xanthen-1-ones in high yields [60] (Scheme 5.10).

Rosati et al. employed $Yb(OTf)_3$ for the microwave-assisted one-pot synthesis of $\Delta^3$-tetrahydrocannabinol (THC) analogs from resorcinol and pulegone [63] (Scheme 5.11).

---

**TABLE 5.2**
**Synthesis of Tetrahydrobenzo[a]xanthene-11-ones**

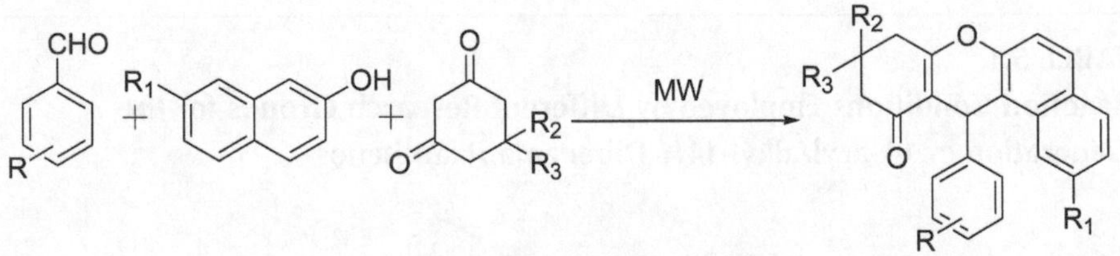

| S. No. | Catalysts | MW Reaction Conditions | | | Yield (%) |
|---|---|---|---|---|---|
| | | Power (W) | Temperature (°C) | Time (min) | |
| 1. | $Sc(OTf)_3$ (10 mol%) | 300 | 120 | 5 | 71–97 |
| 2. | DBSA (2 mol%) | 150 | 50 | 15 | 74–90 |
| 3. | Nano SPA | a | b | 5–18 | 62–96 |
| 4. | $Yb(PFO)_3$ (5 mol%) | a | 80 | 5 | 84–93 |
| 5. | 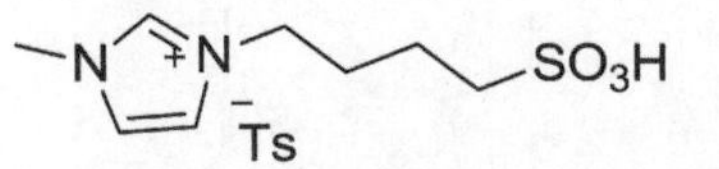 (5 mol%) | 240 | b | 8–15 | 72–89 |

a Power and b temperature have not been mentioned.

In addition, lignin sulfonic acid (LSA) was used to catalyze the synthesis of 1,8-dioxo-octahydroxanthenes under the effect of microwave irradiation in the presence of tetrabutylammonium bromide (TBAB) [64] (Scheme 5.12).

Villemin et al. reported the first synthesis of overcrowded bistricyclic aromatic ethylenes under microwave irradiation [65]. The reaction was carried out

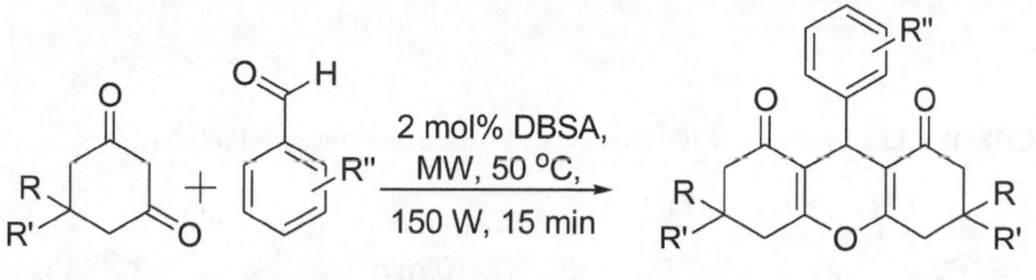

**SCHEME 5.9**  DBSA-catalyzed one-pot synthesis of 9-aryl-3,4,5,6,7,9-hexahydro-1*H*-xanthene-1,8-(2*H*)-diones under microwave irradiation.

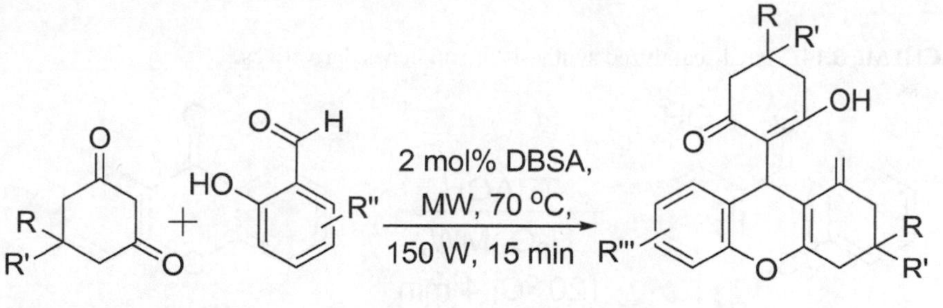

**SCHEME 5.10**  Microwave-assisted one-pot synthesis of 9-substituted-2,3,4,9-tetrahydro-1*H*-xanthen-1-ones using DBSA as a Bronsted acid catalyst.

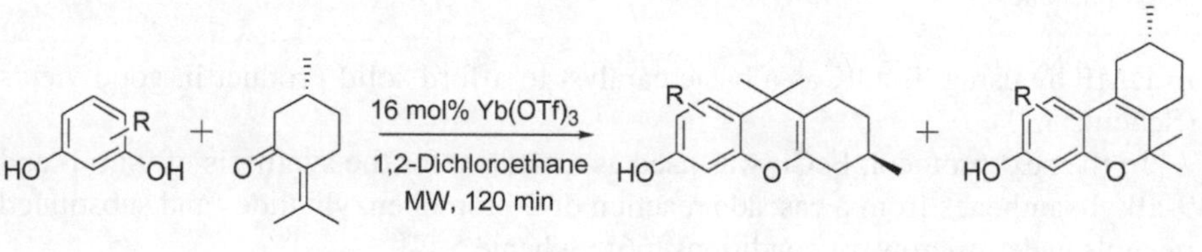

**SCHEME 5.11**  Yb(OTf)$_3$-catalyzed one-pot synthesis of $\Delta^3$-tetrahydrocannabinol derivatives.

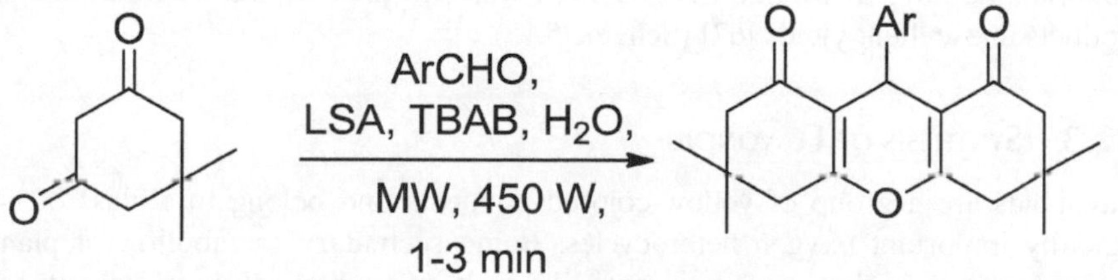

**SCHEME 5.12**  LSA-catalyzed synthesis of 1,8-dioxo-octahydroxanthenes under microwave conditions.

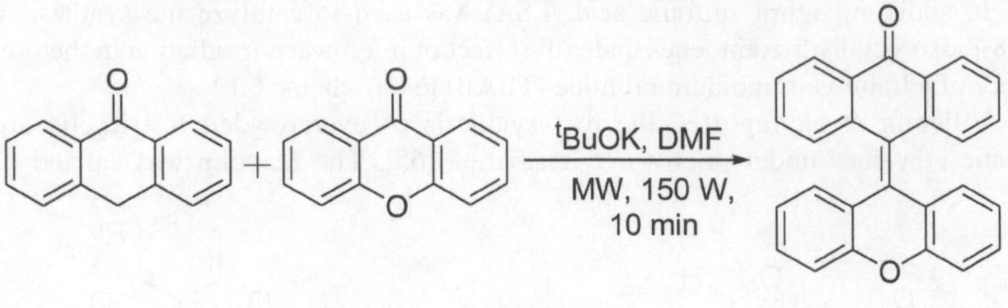

**SCHEME 5.13** Synthesis of 10-(9H-xanthen 9 ylidene)anthracen-9(10H)-one.

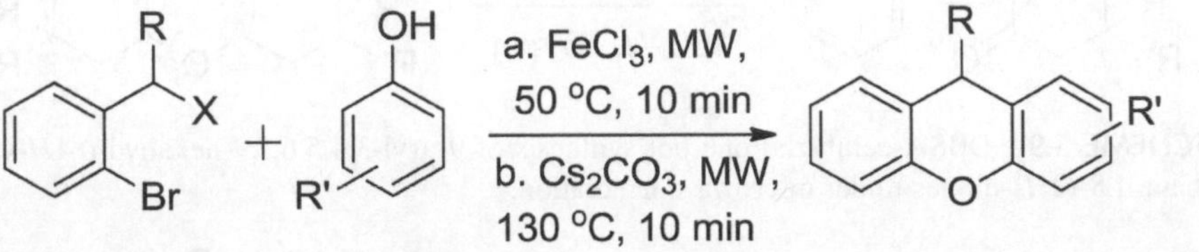

**SCHEME 5.14** FeCl₃-catalyzed synthesis of xanthenes derivatives.

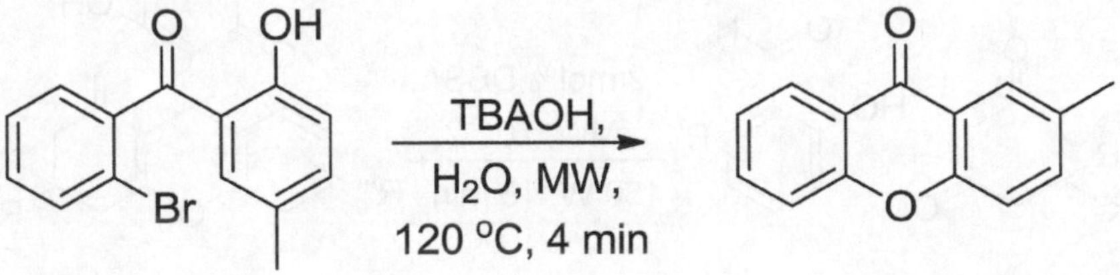

**SCHEME 5.15** Microwave-accelerated synthesis of xanthones from 2-bromo-2′-hydroxy-benzophenone.

in DMF by using ᵗBuOK as a basic catalyst to afford solid product in good yields (Scheme 5.13).

In the next protocol, FeCl₃ was used as a catalyst for the synthesis of 9-aryl- and 9-alkyl-xanthenes from a cascade reaction of 2-bromobenzylhalides and substituted phenols under microwave conditions [66] (Scheme 5.14).

Besides these, the microwave-assisted Ullmann-type intramolecular O-arylation of 2-halo-2′-hydroxybenzophenones has also been reported in the presence of tet-rabutylammonium hydroxide (TBAOH) in water to provide the desired xanthone products in excellent yields [67] (Scheme 5.15).

### 5.2.3 Synthesis of Flavonoids

Flavonoids are a group of yellow-colored pigments and belong to a class of bio-logically important oxygen heterocycles. Being secondary metabolites of plants, they are present in dietary components like fruit, vegetables, olive oil, tea, and red wine [68, 69]. Flavonoids are further divided into six subclasses including flavones, flavanones, flavanols, and isoflavones etc. [70]. These molecules possess diverse

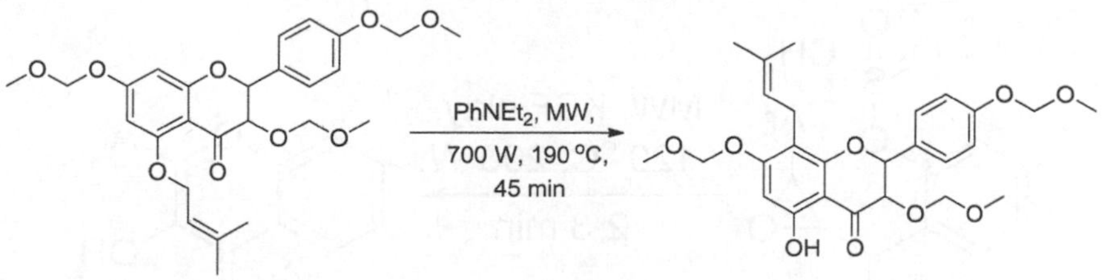

**SCHEME 5.16**   Microwave-assisted Claisen rearrangement of 5-$O$-prenylflavonoid.

biological activities such as antioxidant [71], anxiolytic [72], anti-inflammatory [73], antiviral [74], antiprotozoal [75], anticarcinogenic [76], antibiotic [77], antidiarrheic [78], antiallergic [79], vasodilator [80], antitumor [81], antiulcer [82] agents. The biological efficacies associated with these compounds are due to the presence of free hydroxyl groups which bind with the certain targets and act as a hydrogen bond acceptor (HBA) or a hydrogen bond donor (HBD) [83, 84]. Owing to the diverse pharmacological potential of this class of oxygen heterocycles, a large number of synthetic protocols have been devised in the past years.

Nguyen et al. reported the first multistep synthesis of sophoflasecenol, flaveno-chromane C, and citrusinol [85]. Finally, regioselective microwave-assisted Claisen rearrangement of 5-$O$-prenylflavonoid led to the formation of 8-prenylflavonoid (Scheme 5.16).

Belsare et al. synthesized a series of 3-hydroxyflavone analogs *via* intramolecular cyclization of 2'-hydroxychalcones in ethanol containing alkaline hydrogen peroxide [86] (Scheme 5.17).

Babu et al. demonstrated the effect of montmorillonite KSF clay under microwave irradiation for the formation of flavonols [87] (Scheme 5.18).

Further, thioether substituted flavonoids were synthesized by the reaction of a 2-chloromethylflavonoid with 2-mercaptotriazolopyrimidine analog in an aqueous basic medium under microwave conditions [88] (Scheme 5.19).

In the next protocol, regioselective acylation of quercetin-3-O-glucoside with different long-chain fatty acids was performed in a microwave by using Novozyme 435 to generate the corresponding esters in appreciable yield [89] (Scheme 5.20).

Mokle et al. displayed the oxidative cyclization of chalcones by using iodine under the effect of microwave irradiation to yield the novel flavones [90] (Scheme 5.21).

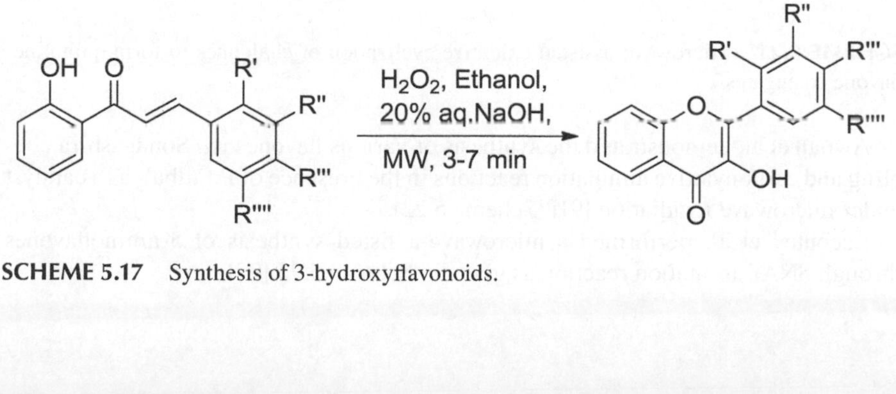

**SCHEME 5.17**   Synthesis of 3-hydroxyflavonoids.

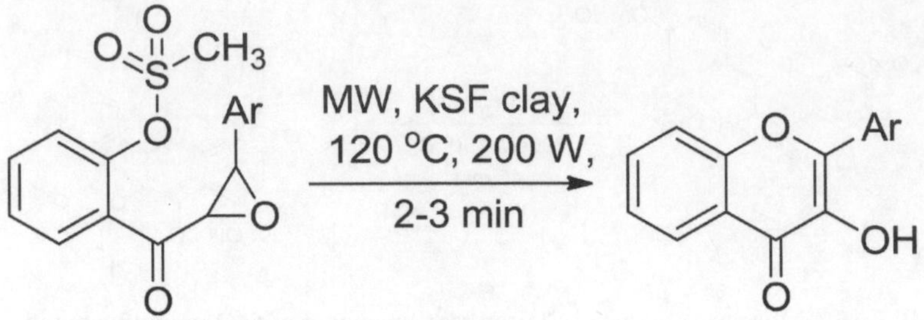

**SCHEME 5.18**  Synthesis of flavonol derivatives.

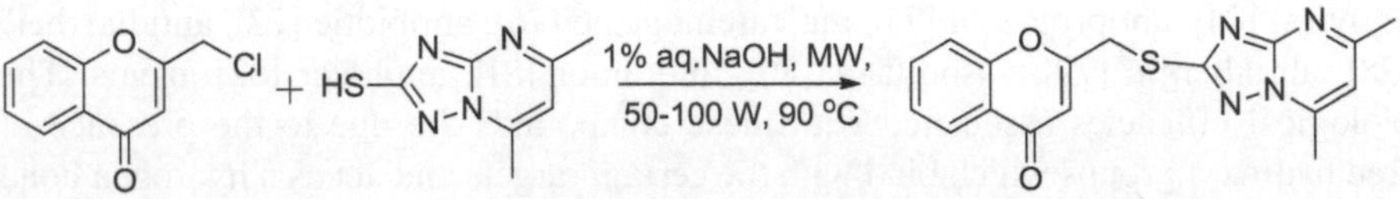

**SCHEME 5.19**  Microwave-assisted synthesis of thioether substituted flavonoids.

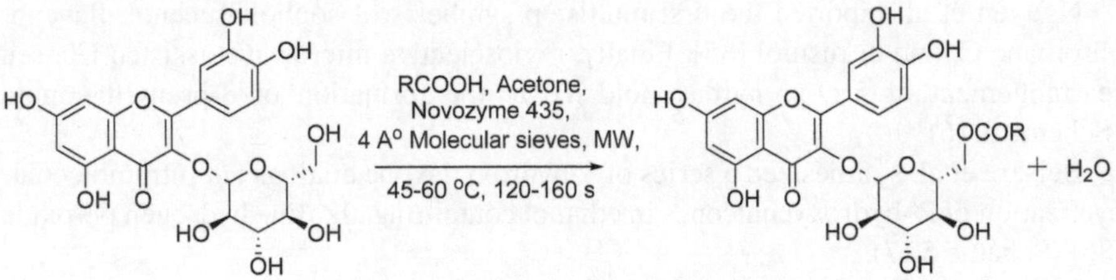

**SCHEME 5.20**  Regioselective synthesis of acylated derivatives of flavonoid glycosides.

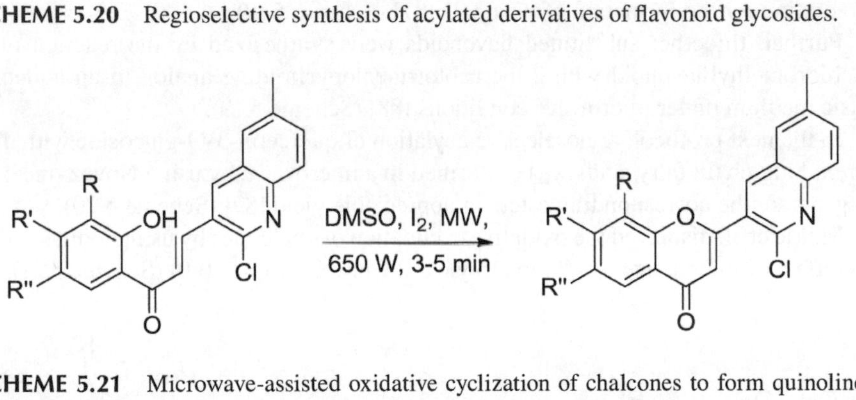

**SCHEME 5.21**  Microwave-assisted oxidative cyclization of chalcones to form quinoline-flavone conjugates.

Awuah et al. demonstrated the synthesis of various flavones *via* Sonagashira coupling and carbonylative annulation reactions in the presence of $Pd_2(dba)_3$ as a catalyst under microwave irradiation [91] (Scheme 5.22).

Lecoutey et al. performed a microwave-assisted synthesis of 5-aminoflavones through SNAr amination reaction as given in Scheme 5.23 [92].

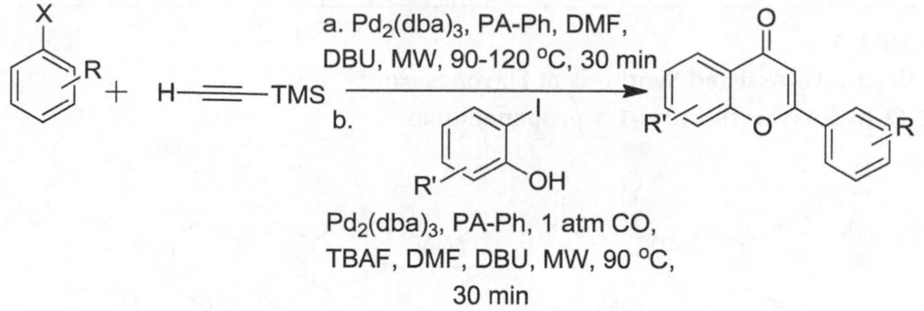

**SCHEME 5.22**   Microwave-assisted synthesis of flavones.

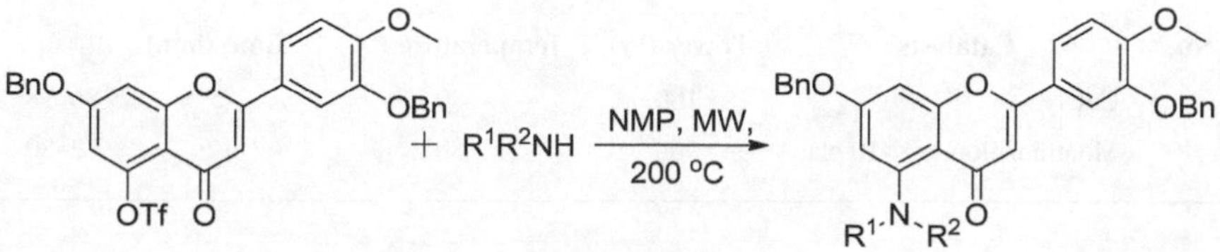

**SCHEME 5.23**   Synthesis of novel 5-aminoflavones.

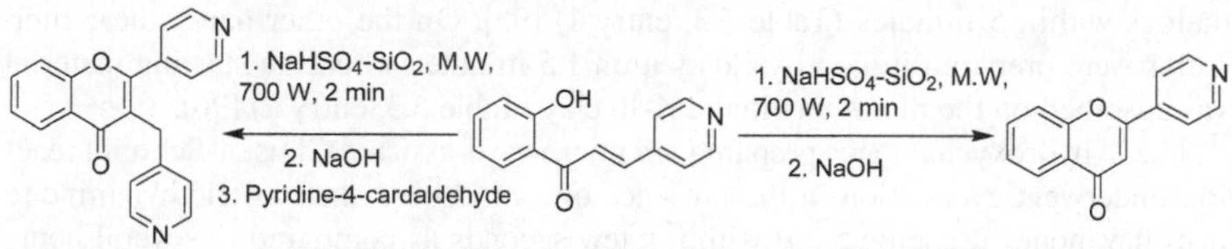

**SCHEME 5.24**   Synthesis of 4'-azaflavone derivatives.

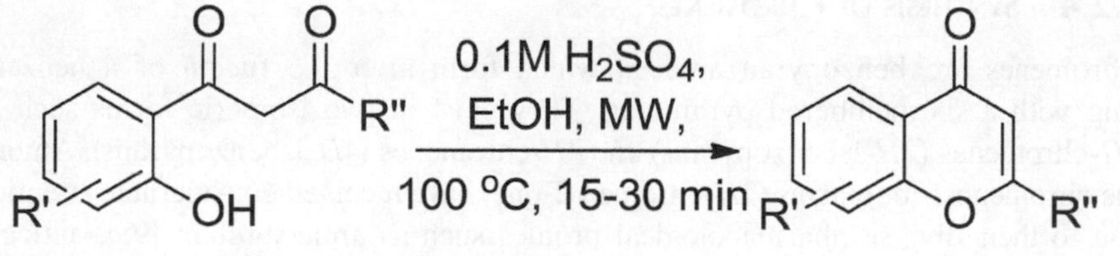

**SCHEME 5.25**   Synthesis of novel flavonoids *via* microwave-assisted Baker-Venkataraman rearrangement.

Ahmet et al. cyclized 3-hydroxy-1-(2-hydroxyphenyl)-3-(pyridine-4-yl)propan-1-one, (*E*)-2'-hydroxy-4-azachalcone, and 2'-hydroxy-2-[(hydroxy)(pyridin-4-yl)methyl]-4'-azachalcone in a neat condition using silica-supported NaHSO₄ under microwave irradiation to give 4'-azaflavone and 3-[(pyridin-4-yl)methyl]-4'-azaflavone respectively [93] (Scheme 5.24).

Ghani et al. performed a microwave-assisted modified Baker-Venkataraman rearrangement to synthesize various antifungal flavonoids [94] (Scheme 5.25).

**TABLE 5.3**
**Microwave-Assisted Synthesis of Flavones from 1-(2-hydroxyaryl)-3-aryl-1,3-propanediones**

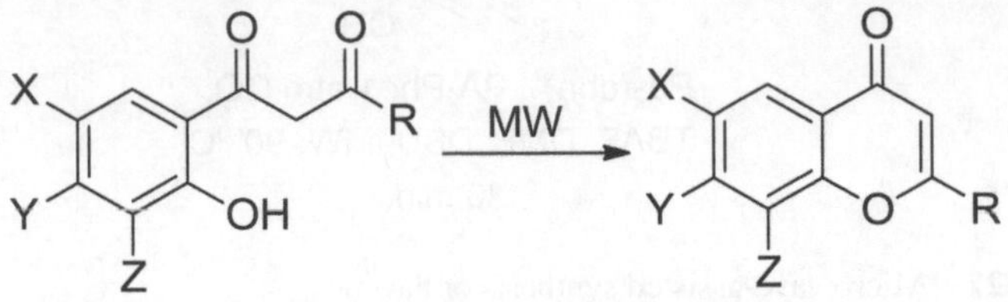

| S. No. | Catalysts | MW Reaction Conditions | | | Yield (%) |
| | | Power (W) | Temperature (°C) | Time (min) | |
|---|---|---|---|---|---|
| 1. | CuCl$_2$ | 100 | 80 | 5 | 86–98 |
| 2. | Montmorillonite K-10 clay | 900 | 80 | 1.5 | 72–80 |

Similarly, microwave-assisted cyclization of 1-(2-hydroxyaryl)-3-aryl-1,3-propanediones in the presence of a catalytic amount of CuCl$_2$ yielded a series of flavones analogs within 5 minutes (Table 5.3, entry 1) [95]. On the other hand, these molecules were prepared in good yields within 1.5 minutes when the starting material was adsorbed on the montmorillonite K-10 clay (Table 5.3, entry 2) [96].

The 2′-hydroxychalcones prepared *via* microwave-assisted Claisen-Schmidt reaction underwent cyclization in the presence of a mild base such as triethylamine to form flavanones (Scheme 5.26) within a few seconds as compared to several hours by using conventional heating [97].

### 5.2.4 SYNTHESIS OF CHROMENES

Chromenes are benzopyran analogs which form after the fusion of a benzene ring with a six-membered pyran ring. They exist in two isomeric forms such as 2H-chromenes (2H-1-benzopyrans) and 4H-chromenes (4H-1-benzopyrans). Among the chromenes, coumarins (2H-chromen-2-one) have received considerable attention due to their diverse pharmacological profiles such as anticoagulant [98], anticancer [99], anti-HIV [100], antitumor [101], anti-inflammatory [102], anti-Alzheimer's [103], antileukemic [104], antibacterial [105], antimalarial [106], antitubercular [107],

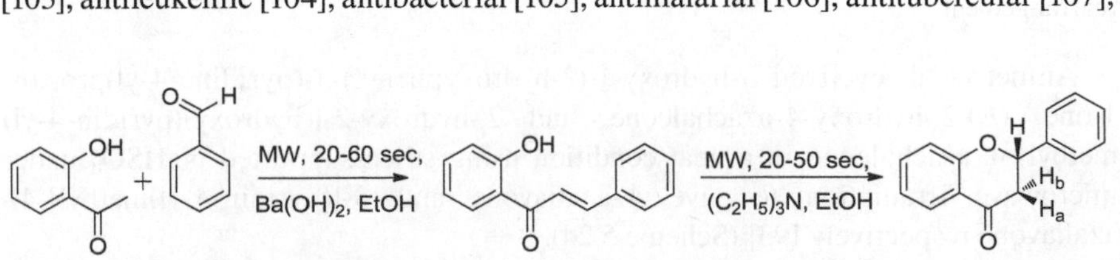

**SCHEME 5.26**  Base-promoted cyclization of 2′-hydroxychalcones to form flavanones.

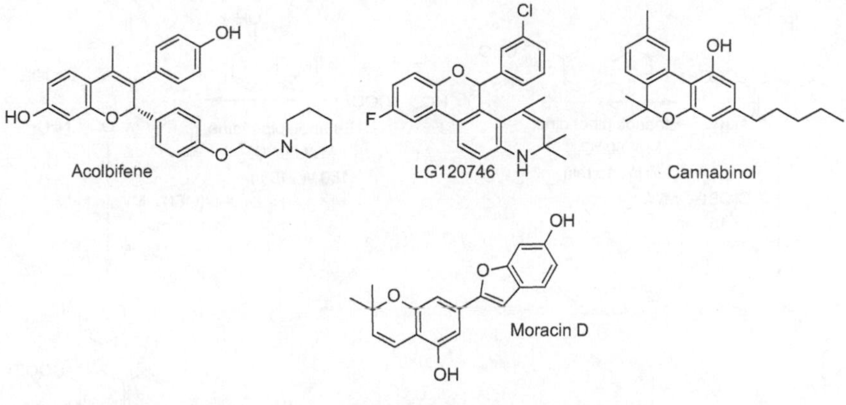

**FIGURE 5.2**    The structures of biologically relevant coumarin analogs.

platelet antiaggregator [108], antioxidant [109], anti-HCV [110], hypoglycaemic [111], antifungal [112], and antifilarial agents [113]. The antitumor effect of coumarins is associated with the metabolite 7-hydroxycoumarin [114]. Coumarins are also associated with strong coronary vasodilating effects [115]. The structures of some of the biologically relevant coumarins are presented in Figure 9.2. LG120746 acts as progesterone receptor modulator [116], acolbifene used as estrogen receptor modulator [117], cannabinol exhibited affinity towards the $CB_1$ and $CB_2$ [118, 119] whereas moracin D has demonstrated fungicidal effect [120]. Besides these, coumarins are also used in cosmetics [121] and agrochemicals [122]. They are useful as dyes [123], sensors, and probes due to significant fluorescent properties [124]. Therefore, a large number of synthetic methods have been developed from time to time to construct these molecules for a variety of applications (Figure 5.2).

Apart from conventional methods, microwave-assisted protocols have attracted the researchers for the synthesis of 4*H*-chromenes. Surpur et al. used a solid base catalyst Mg/Al hydrotalcite for the synthesis of 2-amino-3-cyano-4-arylchromenes [125]. TBAB has also been reported in the literature for catalyzing the synthesis of various 2-aminochromene derivatives [126] under microwave irradiation at 320W. Chaker et al. employed a three-step methodology for the synthesis of chromeno[2,3-*d*] pyrimidinone derivatives [127]. Similarly, El-Agrody and coworkers reported a multicomponent one-pot microwave synthesis of 2-amino-4-arylchromenes [128] (Schemes 5.27 to 5.29).

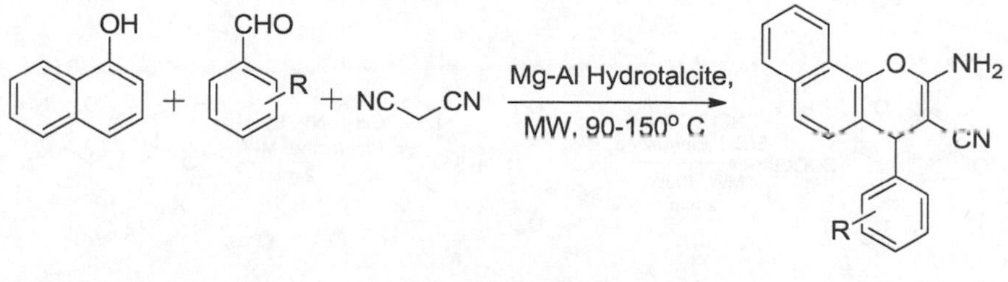

**SCHEME 5.27**    Microwave-assisted synthesis of 2-amino-3-cyano-4-phenylchromenes.

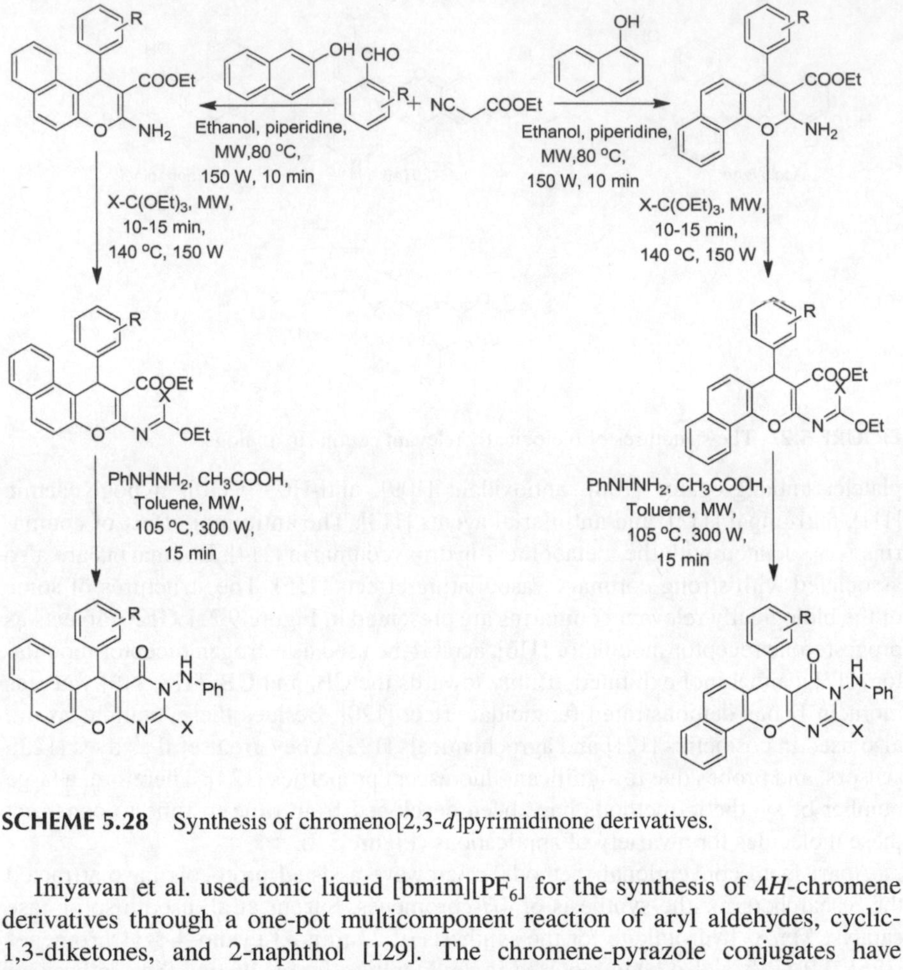

**SCHEME 5.28**  Synthesis of chromeno[2,3-*d*]pyrimidinone derivatives.

Iniyavan et al. used ionic liquid [bmim][PF$_6$] for the synthesis of 4*H*-chromene derivatives through a one-pot multicomponent reaction of aryl aldehydes, cyclic-1,3-diketones, and 2-naphthol [129]. The chromene-pyrazole conjugates have also been synthesized by reacting 5-phenoxypyrazole-4-carbaldehydes, malono-nitrile, and dimedone [130], whereas 4,4′-(1,4-phenylene)bis(4,5-dihydropyran o[3,2-c]chromene was synthesized by condensing terephthaldehyde, malononi-trile, and 4-hydroxycoumarin under microwave irradiations [131]. This synthetic strategy was further extended for the preparation of 2-amino-4-(2-chloroquinol in-3-yl)-5-oxo-4,5-dihydropyrano[3,2-c]chromene-3-carbonitrile and 2-amino-4-(2-chloroquinolin-3-yl)-7-hydroxy-4*H*-chromene-3-carbonitrile by the reaction of

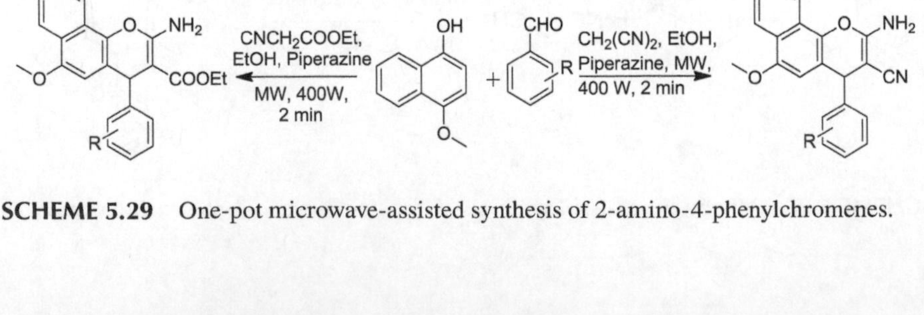

**SCHEME 5.29**  One-pot microwave-assisted synthesis of 2-amino-4-phenylchromenes.

2-chloroquinolin-3-carbaldehyde with malononitrile and resorcinol or 4-hydroxy-coumarin in the presence of $K_2CO_3$ [132, 133] (Scheme 5.30 to 5.33).

Yang et al. synthesized trifluoromethyl-substituted coumarin hydrazones from ethyl-2-hydroxy-2-(trifluoromethyl)-2H-chromene-3-carboxylates in the presence of 30 mol% of silica supported TsOH under microwave irradiation [134] (Scheme 5.34).

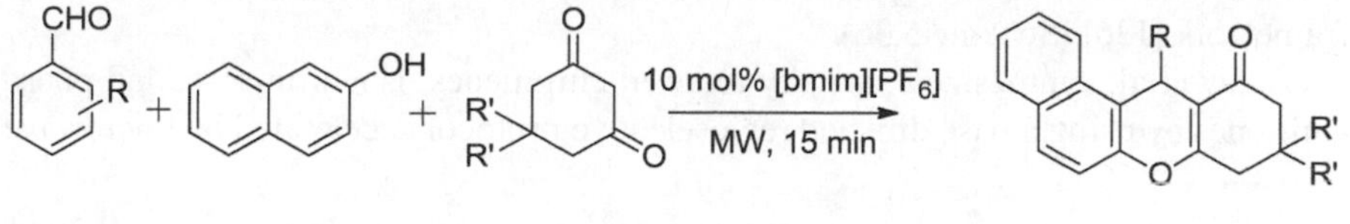

**SCHEME 5.30**    Ionic liquid-catalyzed synthesis of 4H-chromenes.

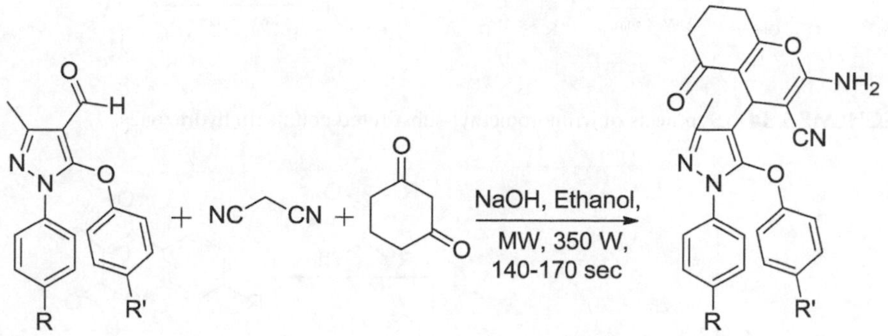

**SCHEME 5.31**    Microwave-assisted synthesis of 4H-chromene derivatives.

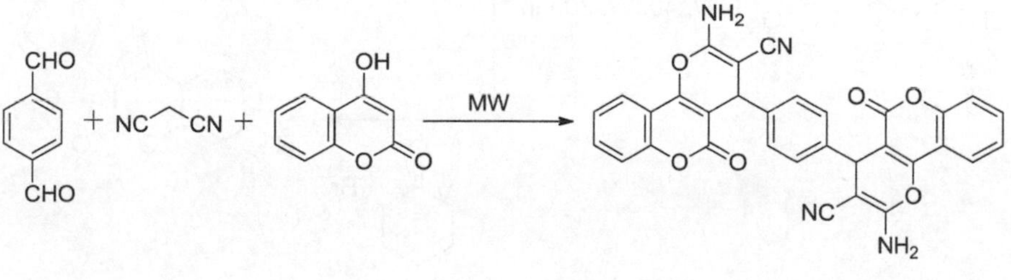

**SCHEME 5.32**    Microwave-assisted synthesis of 4,4′-(1,4-phenylene)bis(4,5-dihydropyrano [3,2-c]chromene) derivative.

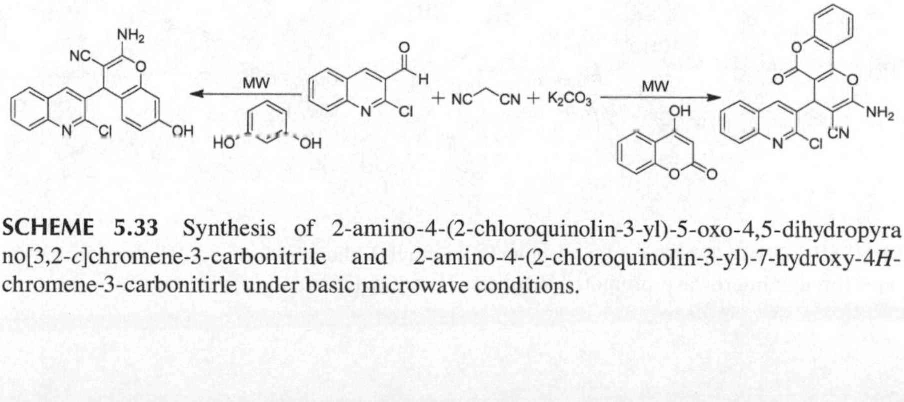

**SCHEME 5.33**    Synthesis of 2-amino-4-(2-chloroquinolin-3-yl)-5-oxo-4,5-dihydropyra no[3,2-c]chromene-3-carbonitrile and 2-amino-4-(2-chloroquinolin-3-yl)-7-hydroxy-4H-chromene-3-carbonitirle under basic microwave conditions.

Zhang et al. explored the scope of microwave technology for the condensation of 4-hydroxy coumarins with α-chloroketones and α-bromocyclohexanone to generate various furo[3,2-*c*]chromen-4-ones and 7,8,9,10-tetrahydro-6*H*-benzofuro[3,2-*c*] chromen-4-ones [135] (Scheme 5.35).

Ashok et al. performed a Vilsmeier-Haack reaction in a microwave to synthesize hybrid molecules containing chromene and coumarin moieties in a single molecular framework [136] (Scheme 5.36).

Yadav et al. synthesized partially reduced chromenes, isochromenes, and phenanthrenes by using a base directed regioselective protocol accelerated by microwave

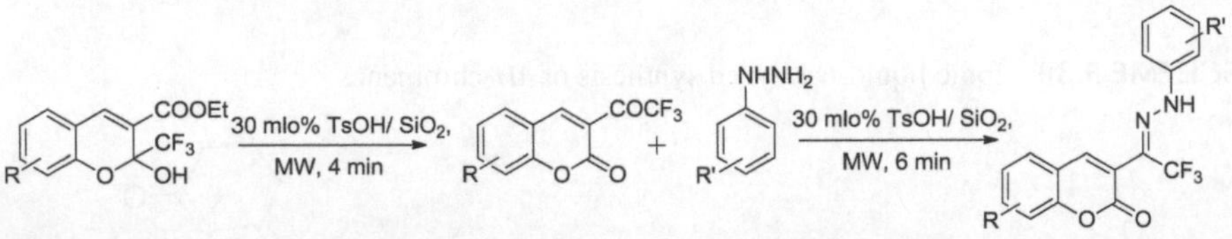

**SCHEME 5.34** Synthesis of trifluoromethyl-substituted coumarin hydrazones.

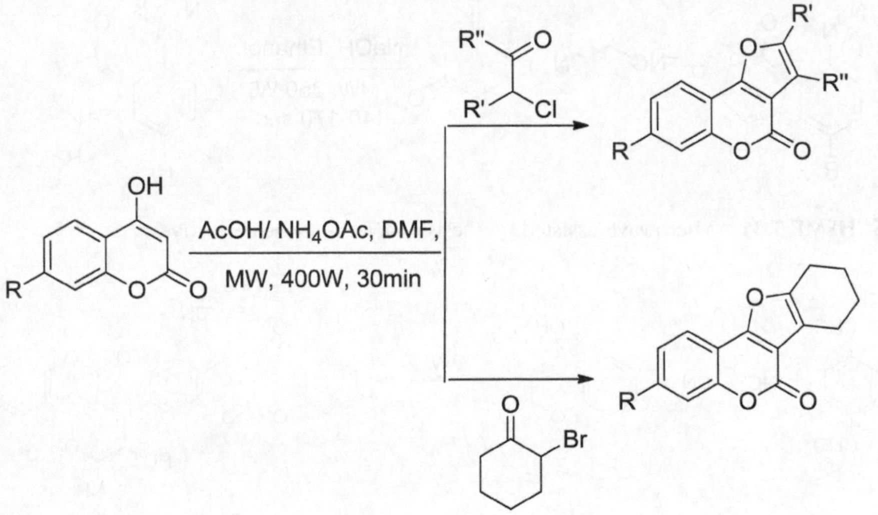

**SCHEME 5.35** Microwave-assisted condensation of 4-hydroxycoumarins with α-haloketones.

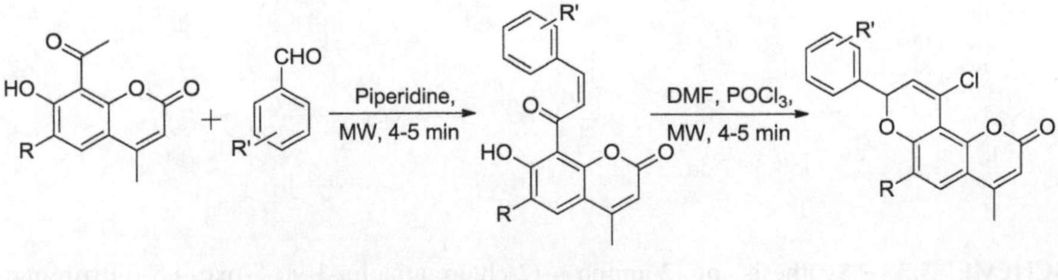

**SCHEME 5.36** Synthesis of 4-chloro-8-methyl-2-phenyl-1,5-dioxa-2*H*-phenanthren-6-ones through microwave-promoted Vilsmeier-Haack reaction.

irradiation [137]. 4-(Piperidin-1-yl)-5,6-dihydro-2*H*-benzo[*h*]-chromen-2-one-3-carbonitriles react with a series of acetophenones in the presence of KOH to give (*Z*)-2-(2-aryl-5,6-dihydro-4*H*-benzo[*f*]isochromen-4-ylidene)acetonitriles, whereas NaH in DMF results in the formation of 3-aryl-1-(piperidin-1-yl)-9,10-dihydroph enanthrene-2-carbonitriles (Scheme 5.37).

Moreover, (*Z*)-2-(3,4,7,8-tetrahydro-1*H*-naphtho[2,1-*c*]chromen-6(2*H*)-ylidene) acetonitrile was obtained when cyclohexanone was used as a nucleophile (Scheme 5.38).

By using microwave heating, Kang et al. were able to synthesize an alkyl ester of an anti-HIV compound daurichromenic acid in good yield [138]. This approach involved a microwave-assisted condensation followed by intramolecular $S_N2$ cyclization to create a 2*H*-benzopyran core within 20 minutes. However, only 15% of the desired product was obtained after 4 days under reflux conditions (Scheme 5.39).

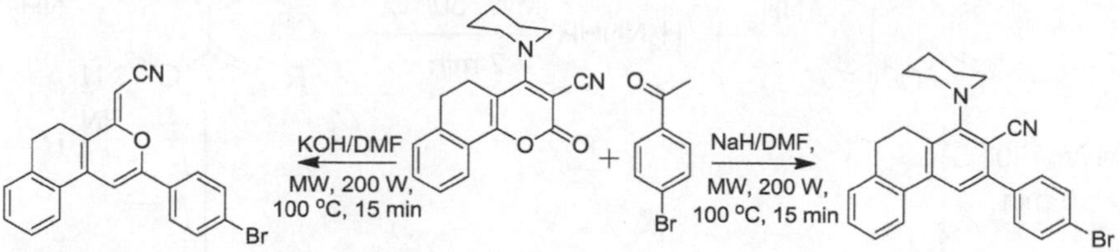

**SCHEME 5.37** Synthesis of 3-(4-bromophenyl)-1-(piperidin-1-yl)-9,10-dihydrophenan threne-2-carbonitrile and (*Z*)-2-(2-(4-bromophenyl)-5,6-dihydro-4*H*-benzo[*f*]isochromen-4-ylidene) acetonitrile.

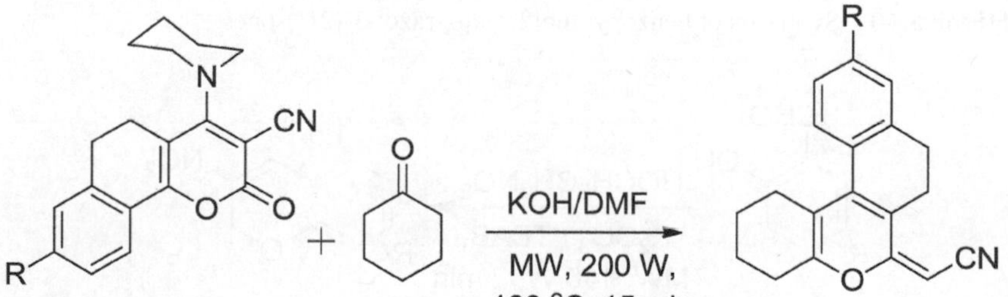

**SCHEME 5.38** Synthesis of (*Z*)-2-(3,4,7,8-tetrahydro-1*H*-naphtho[2,1-*c*]chromen-6(2*H*)-ylidene)acetonitrile.

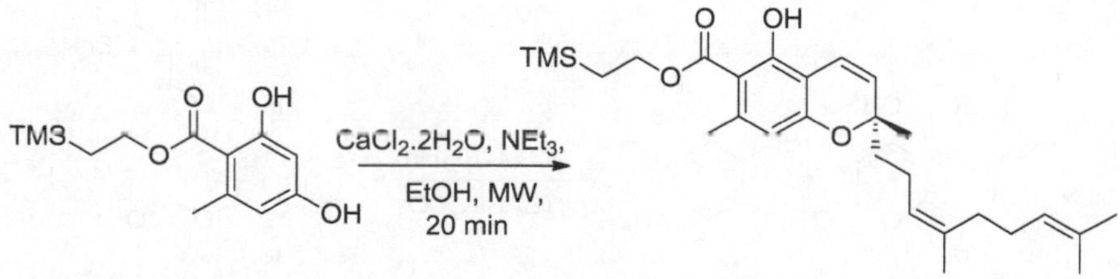

**SCHEME 5.39** Synthesis of daurichromenic acid derivative.

Borisov et al. utilized microwave technology for the efficient synthesis of benzopyrano[2,3-c]pyrazol-3-(2H)-one derivatives by following the protocol presented [139] (Scheme 5.40).

3-Nitro-2H-chromene derivatives were synthesized through a facile phase transfer microwave-assisted protocol (Scheme 5.41).

Tetrabutylammonium bromide (TBAB) has been used as a phase transfer catalyst with anhydrous potassium carbonate for reacting substituted 2-hydroxybenzaldehydes and 2-nitroethanol [140]. This methodology has also been used for the synthesis of 4-(2-bromobenzyloxy)benzopyran-2-ones, by irradiating 4-hydroxy benzopyran-2-ones and 2-bromobenzyl bromides in the presence of TBAB and potassium carbonate for 4–10 minutes in a microwave oven [141] (Scheme 5.42).

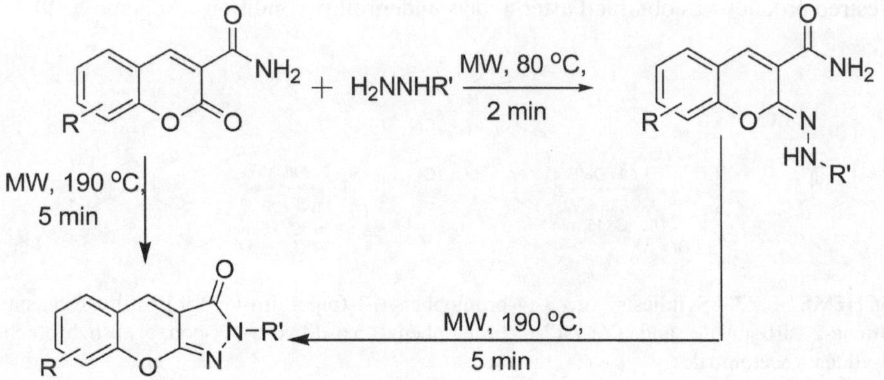

**SCHEME 5.40**  Synthesis of benzopyrano[2,3-c]pyrazol-3-(2H)-ones.

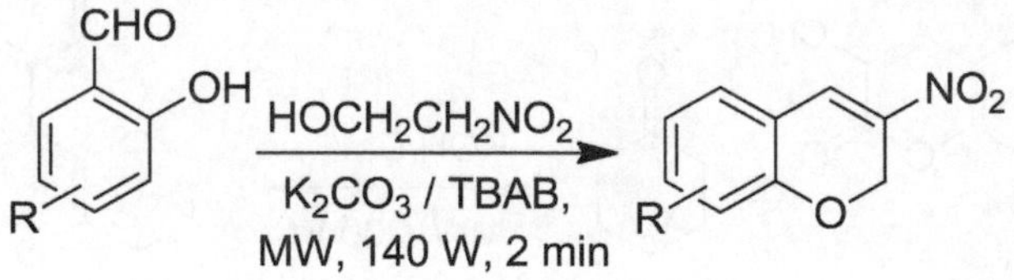

**SCHEME 5.41**  Synthesis of 3-nitro-2H-chromene derivatives.

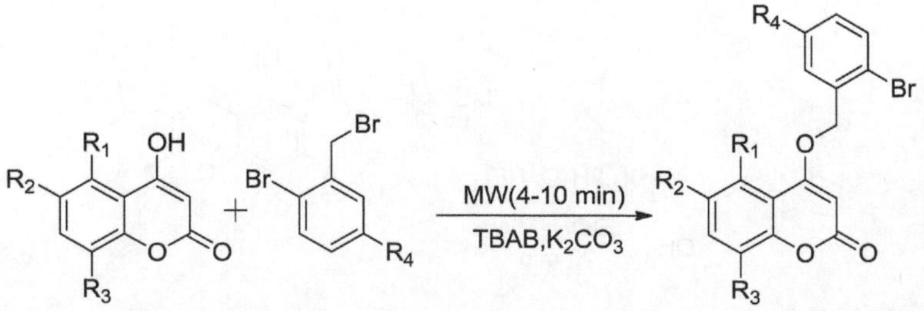

**SCHEME 5.42**  Microwave-assisted synthesis of 4-(2-bromobenzyloxy)benzopyran-2-ones.

Thasana et al. performed microwave-assisted coupling of a series of 2-halobi-arylcarboxylic acids in the presence of Cu (I) salts such as CuI, CuBr, and CuCl to form a phenyl-fused coumarin analog [142]. Based on this protocol, isolamellarin, a new class of pyrroloisoquinoline alkaloid was prepared from a dihydroisoquinoline derivative under a microwave condition (Schemes 5.43 and 5.44).

Zhang and coworkers reacted phenols with cinnamoyl chloride in presence of montmorillonite K-10 to produce dihydrocoumarin derivatives at 160°C in chlo-robenzene (Scheme 5.45) [143], whereas 6-oxo-6$H$-pyran-3-carboxylic acid was formed when a mixture of malic acid and sulfuric acid was heated in a domestic oven [144] (Scheme 5.46).

Microwave-assisted Pechmann reaction was carried out by reacting phenols with 3-oxo-butyric acid methyl ester in the presence of a catalytic amount of graphite/K-10

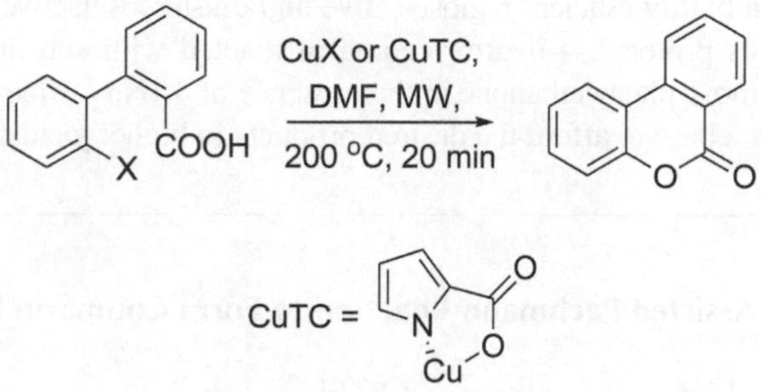

**SCHEME 5.43** Microwave-assisted intramolecular cyclization of 2-halobiarylcarboxylic acids.

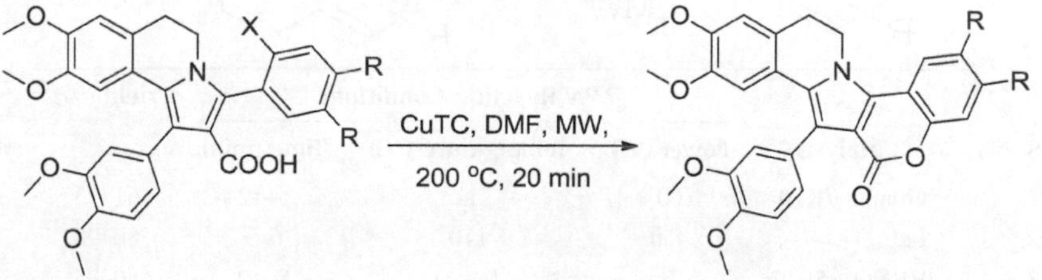

**SCHEME 5.44** Synthesis of isolamellarin (pyrroloisoquinoine alkaloid) derivatives.

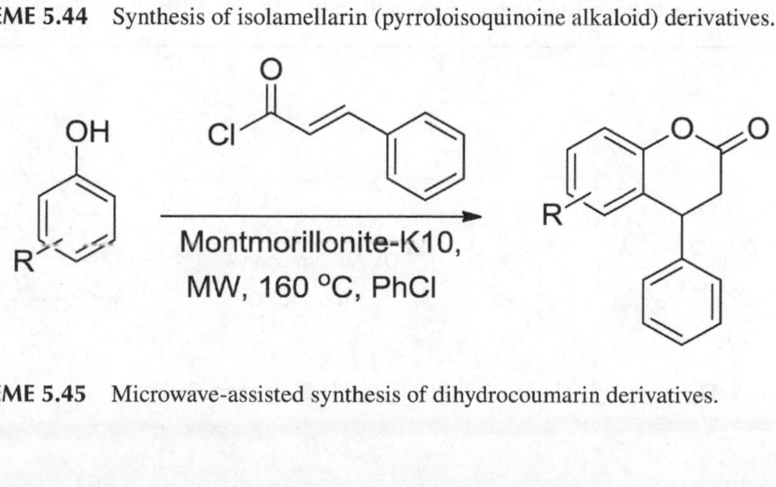

**SCHEME 5.45** Microwave-assisted synthesis of dihydrocoumarin derivatives.

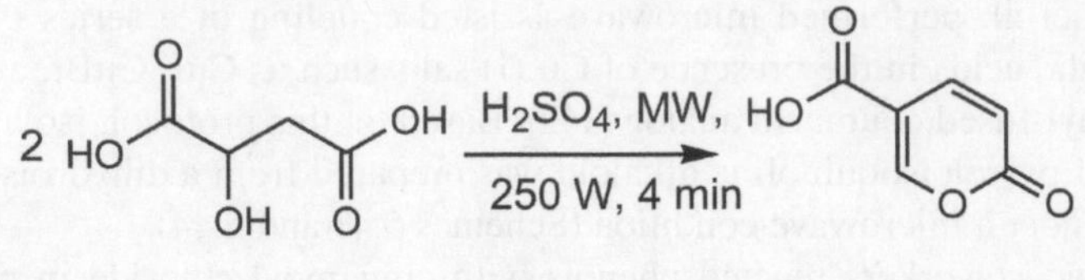

**SCHEME 5.46** Microwave-assisted synthesis of 6-oxo-6*H*-pyran-3-carboxylic acid.

montmorillonite mixture to afford a series of coumarin derivatives (Table 5.4, entry 1) [145]. Besides, the Pechmann condensation was also performed by using FeF$_3$ and bismuth nitrate as a catalyst under solvent-free microwave conditions (Table 5.4, entries 2 and 3) [146, 147].

Karanjule and coworkers generated trans 2,3-dihydro-furo[3,2-*c*]coumarins in high yields by using a highly efficient regioselective and diastereoselective protocol [148]. According to this protocol, 4-hydroxycoumarin reacted with substituted benzaldehydes and 2-bromo-1-phenylethanone in the presence of 4-(N,N-dimethylamino)pyridine (DMAP) as a base to afford the desired products in higher yields (Scheme 5.47).

---

## TABLE 5.4
## Microwave-Assisted Pechmann Reaction to Form Coumarin Derivatives

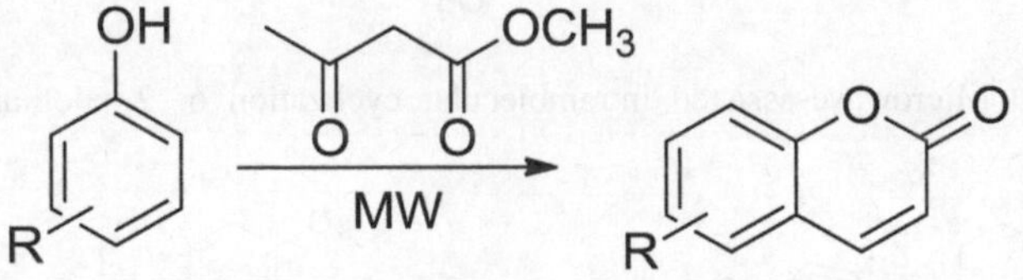

| | | MW Reaction Conditions | | | Yield (%) |
|---|---|---|---|---|---|
| S. No. | Catalysts | Power (W) | Temperature (°C) | Time (min) | |
| 1. | Graphite/K10 | 60 | a | 5–12 | 61–75 |
| 2. | FeF$_3$ | 450 | 110 | 6–9 | 86–98 |
| 3. | Bi(NO)$_3$·5H$_2$O | b | a | 8–10 | 80–85 |

ᵃ Temperature and ᵇ Power have not been mentioned.

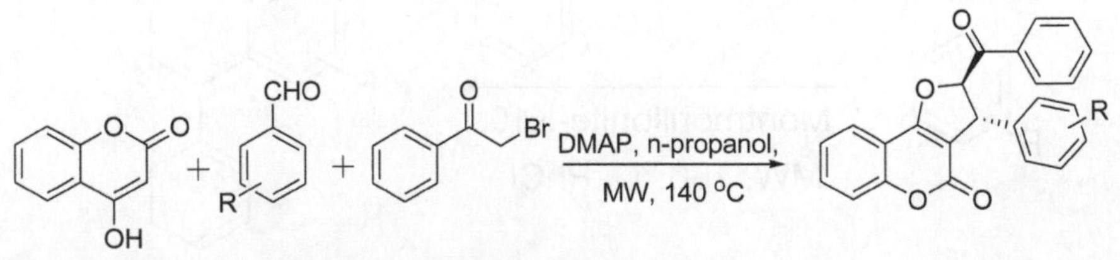

---

**SCHEME 5.47** Synthesis of trans 2,3-dihydrofuro[3,2-*c*]coumarins.

In contrast, 3-aryl-4-hydroxycoumarins prepared from 2-(2-arylacetoxy)benzoates undergo degradation in alkaline medium to form 2-hydroxydeoxybenzoins [149] (Scheme 5.48).

Thakrar et al. performed a microwave-assisted one-pot multicomponent synthesis of 2-amino-3-cyanopyridine-coumarin hybrids in the presence of $Fe^{3+}K$-10 montmorillonite clay or HY- zeolite as an acidic catalyst. It is interesting to note that both the catalysts are found to be recyclable and moreover present a greener approach to making the target molecules [150] (Scheme 5.49).

Microwave-assisted Cadogen cyclization of 3-arylcoumarins produced the corresponding indolo[3,2-c]coumarins in higher yields [151] (Scheme 5.50).

Kumar et al. used ZnO nanoparticles for the synthesis of coumarins by Knovenagel reaction under microwave conditions [152] whereas another report employed $ZrOCl_2 \cdot 8H_2O$ as a catalyst for the preparation of coumarins from salicylaldehydes [153] (Table 5.5).

Kaneria et al. synthesized 3-aryl-furo[3,2-c]coumarins by using two different microwave-assisted methodologies [154]. In the first protocol, they have reacted various 4-hydroxycoumarins with 2-aryl-1-nitroethenes under Nef reaction conditions whereas, in another protocol, 4-hydroxycoumarins were reacted with aroylmethylbromides under Feist-Benary reaction conditions to yield various 3-aryl-furo[3,2-c] coumarins (Scheme 5.51).

A microwave-assisted one-pot multicomponent reaction of N-allylquinolones, cyclic β-diketones, and 4-hydroxy-6-methyl-2H-pyran-2-one or 4-hydroxy coumarin in the presence of ceric ammonium nitrate under solvent-free conditions resulted in the synthesis of pyrano[4,3-b]chromene and benzopyrano[3,2-c]chromene derivatives [155] (Scheme 5.52).

Olomola et al. synthesized 3-methylcoumarins and coumarin-3-carbaldehydes by a series of microwave-assisted steps [156]. The reaction proceeded via Baylis-Hilman reaction of substituted salicylaldehydes in the presence of tert-butylacrylate and DABCO under microwave irradiation followed by one-pot hydrogen iodide-mediated cyclization and reduction to give corresponding 3-methylcoumarins which

---

**TABLE 5.5**

**Microwave-Assisted Knovenagel Reaction**

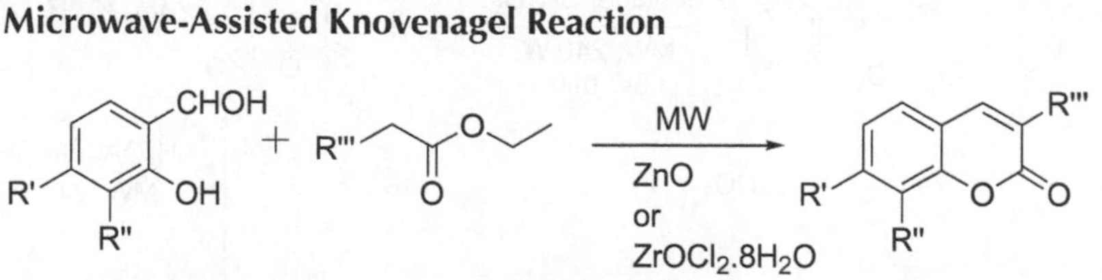

| | | MW Reaction Conditions | | | Yield (%) |
|---|---|---|---|---|---|
| S. No. | Catalyst | Power (W) | Temperature (°C) | Time (min) | |
| 1. | ZnO | 300 | 130 | 3–7 | 86–98 |
| 2. | $ZrOCl_2 \cdot 8H_2O$ | – | – | 6 | 72–82 |

[a] Power and [b] temperature have not been mentioned.

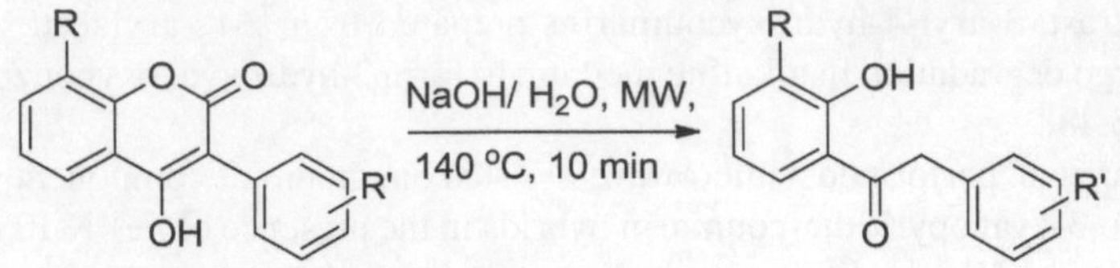

**SCHEME 5.48**  Synthesis of 2-hydroxybenzoin derivatives.

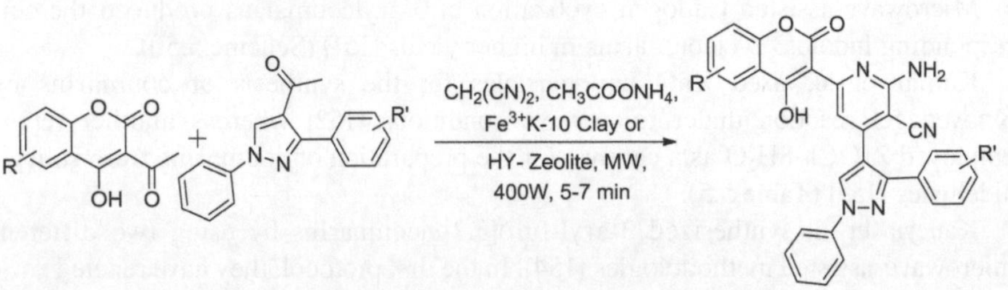

**SCHEME 5.49**  One-pot microwave-promoted synthesis of 2-amino-3-cyanopyridine-coumarin hybrids.

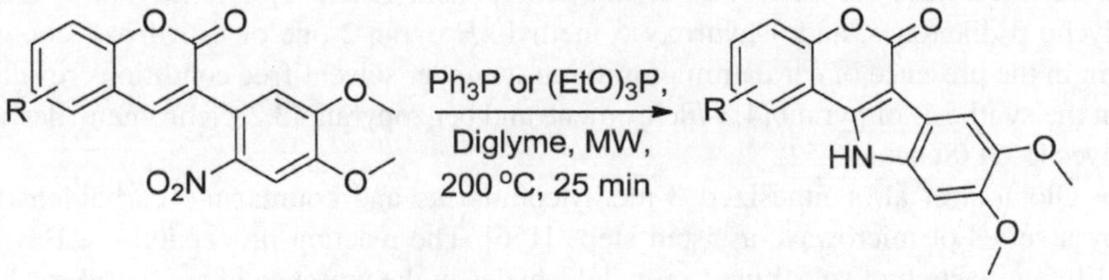

**SCHEME 5.50**  Microwave-assisted Cadogen cyclization to form indolo[3,2-c]coumarins.

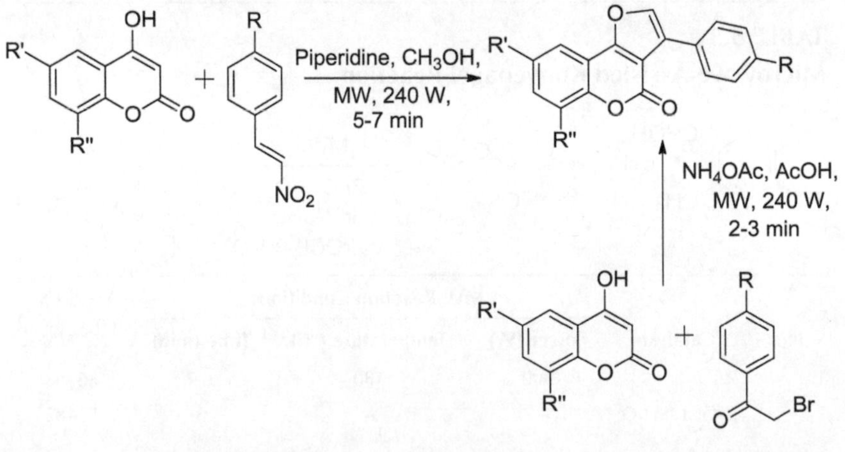

**SCHEME 5.51**  Synthesis of 3-aryl-furo[3,2-c]coumarins.

on oxidation in the presence of selenium dioxide under microwave conditions at 170°C afforded desired coumarin-3-carbaldehyde analogs (Scheme 5.53).

In addition, various pyranocoumarins were also generated *via* the pseudo-multicomponent condensation reaction of 4-chloro-3-formylcoumarin and 4-methylquinoline in the presence of acetic anhydride under microwave irradiation [157] (Scheme 5.54).

Furthermore,   3-[6-(2-amino-ethoxy)-3-methyl-5-(3-morpholin-4-yl-propionyl)-benzofuran-2-carbonyl]-chromen-2-one   was   produced   by   the   reaction   of   (5-acetyl-6-hydroxy-3-methylbenzofuran-2-carbonyl)chromen-2-one,   paraformaldehyde, and morpholine in the presence of TBAHS [158]. The reaction was performed in a microwave at 450 W for 8–10 minutes. The Mannich base generated

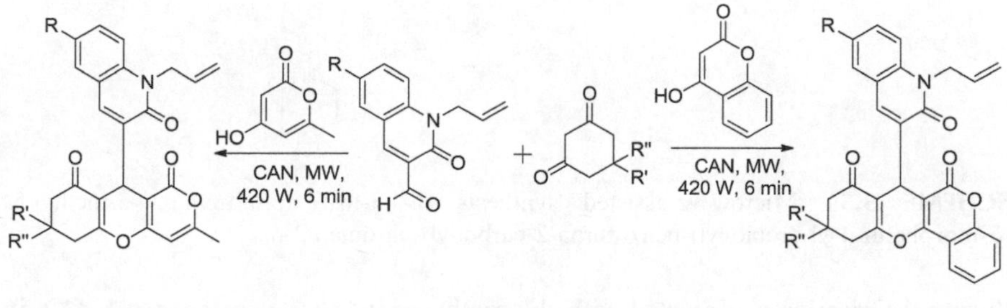

**SCHEME 5.52**   Synthesis of pyrano[4,3-*b*]chromenes and benzopyrano[3,2-*c*]chromenes.

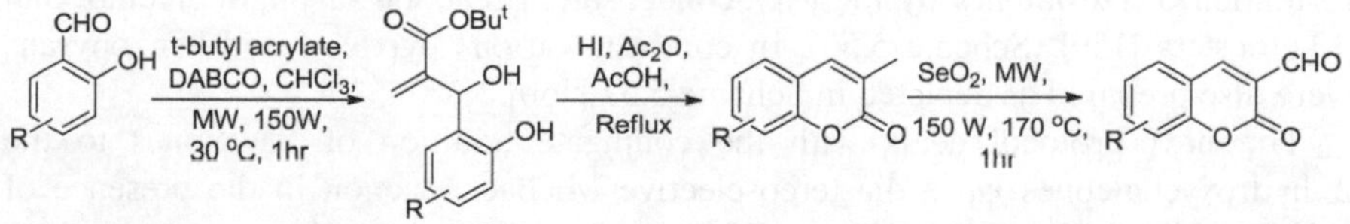

**SCHEME  5.53**   Microwave-assisted  synthesis  of  3-methylcoumarins  and  coumarin-3-carbaldehydes.

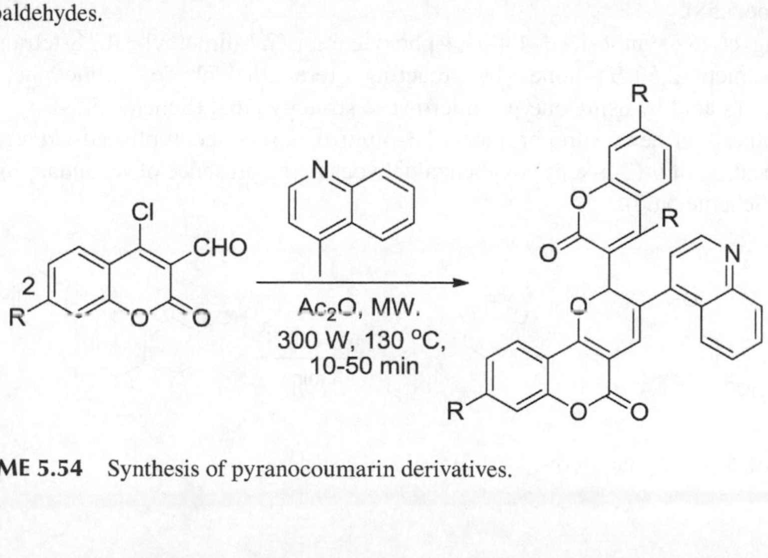

**SCHEME 5.54**   Synthesis of pyranocoumarin derivatives.

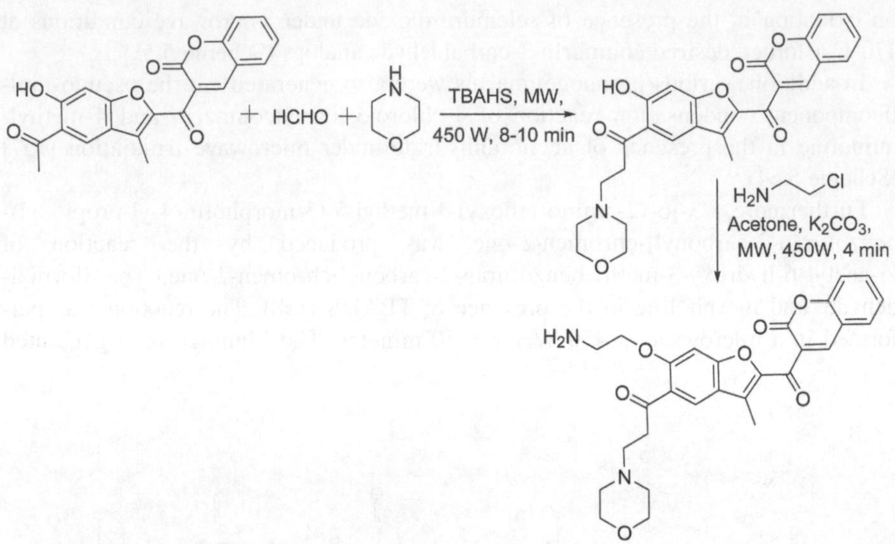

**SCHEME 5.55** Microwave-assisted synthesis of 3-[6-(2-amino-ethoxy)-3-methyl-5-(3-morpholin-4-yl-propionyl)-benzofuran-2-carbonyl]-chromen-2-one.

during the reaction was reacted with chloroethylamine in the presence of $K_2CO_3$ in acetone to afford a collection of products (Scheme 5.55).

The microwave-assisted approach has also been extended to accelerate the formation of chromones by the cyclocondensation reaction of phloroglucinol and β-ketoesters [159] (Scheme 5.56). In contrast, various pyrrole-fused benzopyrans were also prepared as depicted in Scheme 5.57 [160].

The next protocol deals with the conjugate addition of flavanones to the 2′-hydroxychalcones *via* a diastereoselective Michael reaction in the presence of 1,5-diazabicyclo[5.4.0]undec-7-ene (DBU) under solvent-free microwave conditions to afford the formation of various diastereomeric benzopyran-4-one derivatives [161] (Scheme 5.58).

Feng et al. synthesized 4,4′-(1,4-phenylene)bis(7,7-dimethyl-3,4,7,8-tetrahydro-2*H*-chromene-2,5(6*H*)-dione) by reacting terephthaldehyde, dimedone, and Meldrum's acid by using one-pot microwave strategy [162] (Scheme 5.59).

Synthesis of hexahydrochromeno[4,3-*b*]pyrroles was accomplished *via* a microwave heating of o-(3-alkenyl)oxybenzaldehydes in the presence of secondary amines [163] (Scheme 5.60).

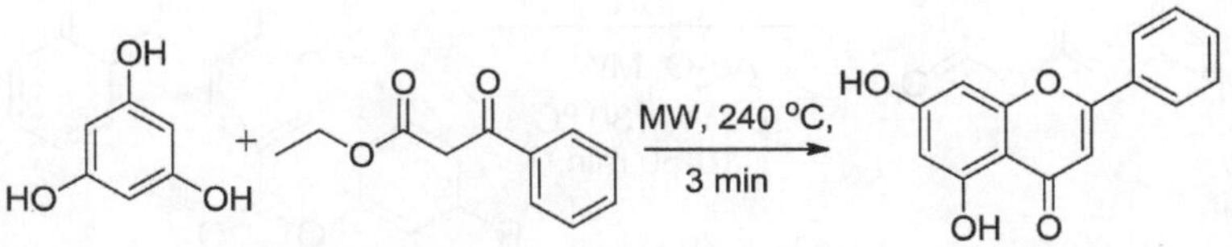

**SCHEME 5.56** Synthesis of 5,7-dihydroxy-2-phenylchromones.

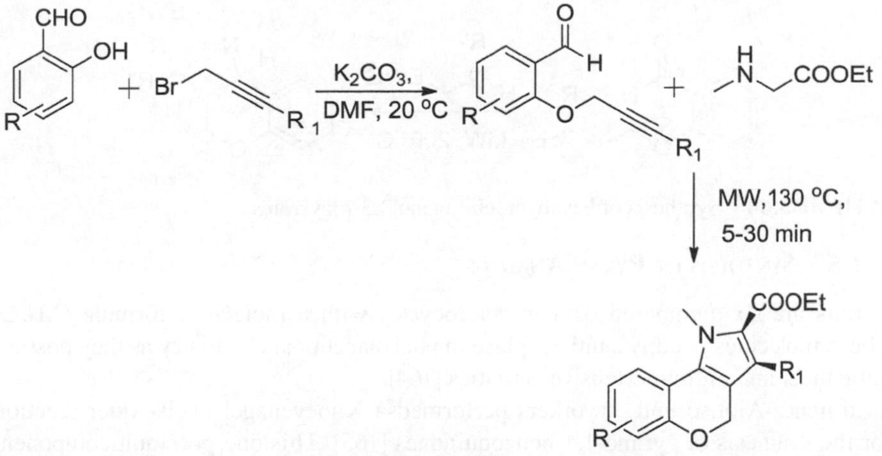

**SCHEME 5.57**  Microwave-assisted cycloaddition for the synthesis of pyrrole fused benzopyrans.

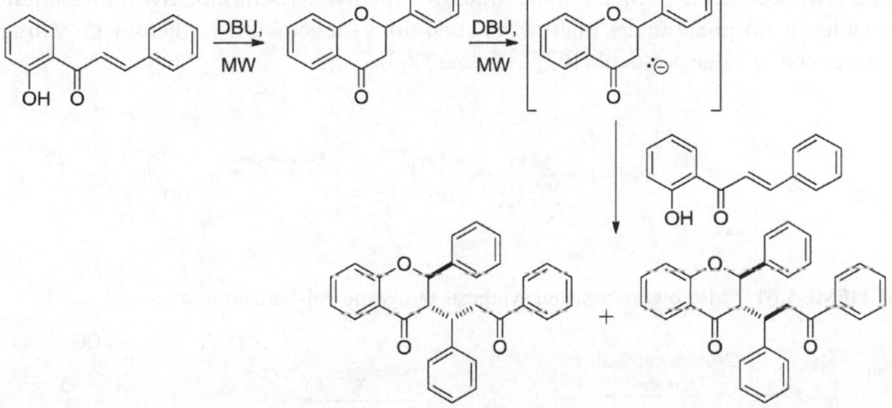

**SCHEME 5.58**  Synthesis of diastereomeric benzopyran-4-one derivatives.

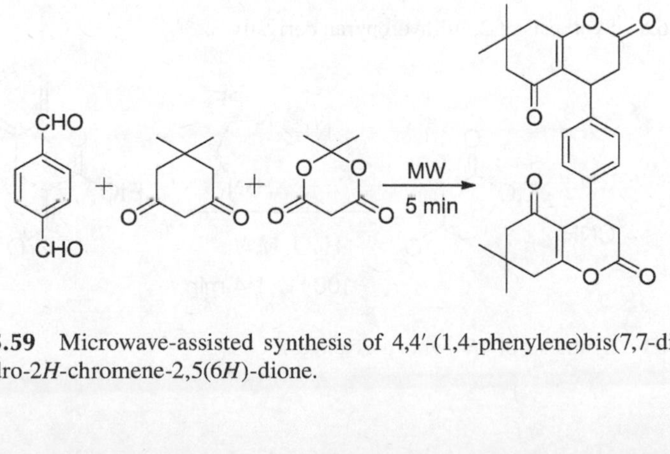

**SCHEME 5.59**  Microwave-assisted synthesis of 4,4'-(1,4-phenylene)bis(7,7-dimethyl-3,4, 7,8-tetrahydro-2H-chromene-2,5(6H)-dione.

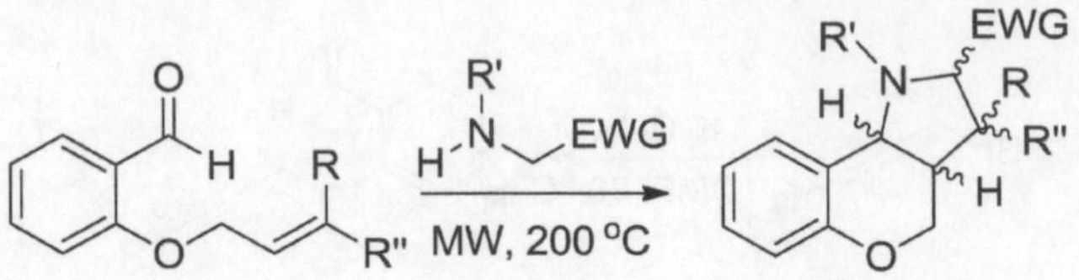

**SCHEME 5.60**   Synthesis of hexahydrochromeno[4,3-*b*]pyrroles.

### 5.2.5  SYNTHESIS OF PYRAN ANALOGS

Pyrans are six-membered oxygen heterocycles with a molecular formula $C_5H_6O$. These molecules occupy a unique place in pharmaceutical chemistry as they possess anticancer and antihypertensive activities [164].

Jimenez-Alonso and coworkers performed a Knoevenagel Diels-Alder reaction for the synthesis of pyrano-1,4-benzoquinones [165]. This one-pot, multicomponent reaction was performed in ethanol at 120°C for 20 minutes (Scheme 5.61).

Similarly, various 2,3-dihydropyrans were produced *via* an enyne-cross hetero Diels-Alder reaction using second generation Grubb's catalyst [166] (Scheme 5.62).

Peng and Song synthesized 4*H*-pyrans by reacting benzaldehyde, malononitrile, and ethylacetoacetate in an ionic liquid, 1-methyl-3-(2-aminoethyl)imidazolium hexafluorophosphate under microwave conditions (Scheme 9.64), thereby providing a greener and safer protocol [167] (Scheme 5.63).

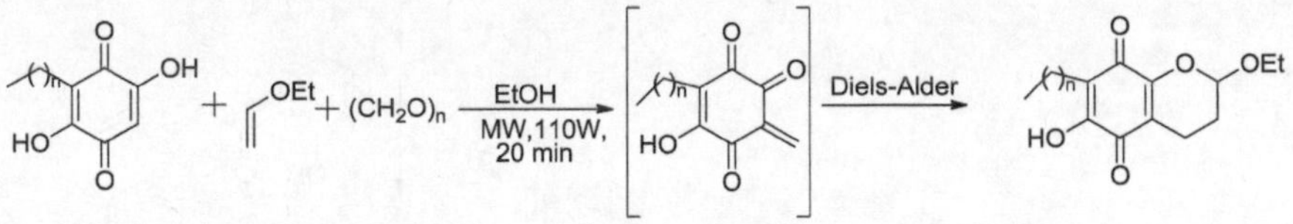

**SCHEME 5.61**   Microwave-assisted synthesis of pyrano-1,4-benzoquinones.

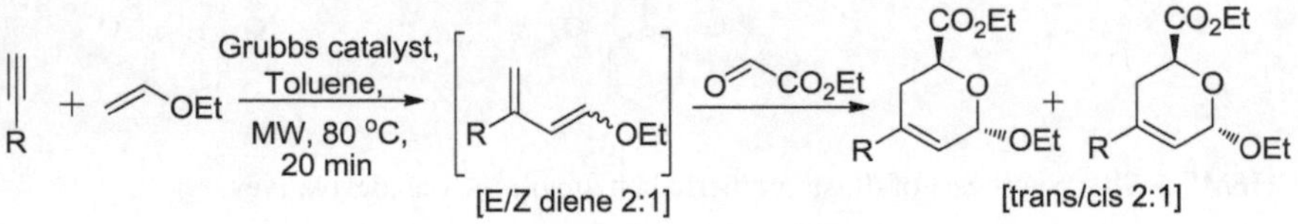

**SCHEME 5.62**   Synthesis of 2,3-dihydropyran derivatives.

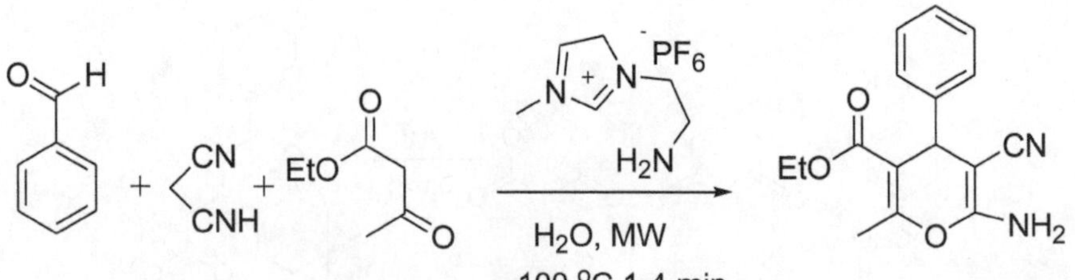

**SCHEME 5.63**   Synthesis of 4*H*-pyran derivatives.

Clubbing dimedone with various other reactants in the presence of a number of catalysts is also one way of producing pyran rings [168]. In the following protocol, dimedone was reacted under microwave irradiation with a number of different reagents to yield the corresponding pyran derivatives [169–171] (Scheme 5.64).

The Prins type reaction has also been performed under the influence of microwave irradiation [172]. Homoallylic alcohols reacted with aldehydes in presence of $BiCl_3$ produced tetrahydropyrans (Scheme 5.65).

Wu et al. reported the microwave-assisted synthesis of lactones [173–175] by irradiating triene-esters at 160–230°C for 2–4 hours in MeCN or PEG under microwave conditions to give 75–99% of the desired product (Scheme 5.66).

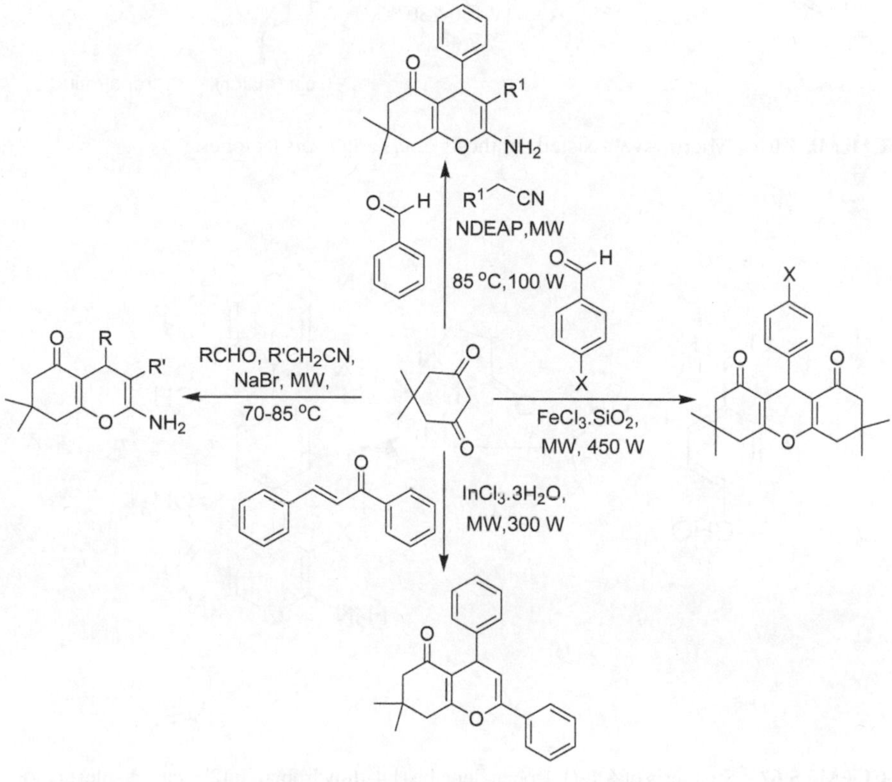

**SCHEME 5.64** Microwave-assisted synthesis of pyran derivatives.

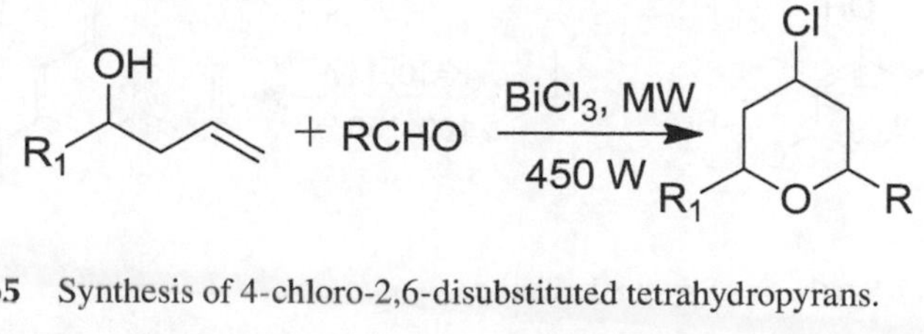

**SCHEME 5.65** Synthesis of 4-chloro-2,6-disubstituted tetrahydropyrans.

Terephthaldehyde, 3-methyl-1-phenyl-2-pyrazolin-5-one and malononitrile or methyl cyanoacetate were irradiated in microwave to produce 4,4′-(1,4-phenylene) bis(1,4-dihydropyrano[2,3-c]pyrazole) derivatives [176] (Scheme 5.67).

K-10 montmorillonite has also been used in a catalytic amount for the generation of pyran derivatives by reacting 1,3-dihydroxy-2-methyl-xanthen-9-one and 1-bromo-3-methyl-but-2-ene at 100–150°C [177] (Scheme 5.68).

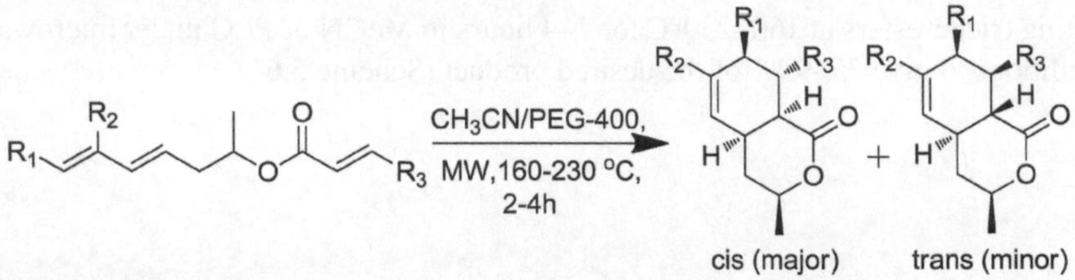

SCHEME 5.66   Microwave-assisted synthesis of *cis* and *trans* lactones.

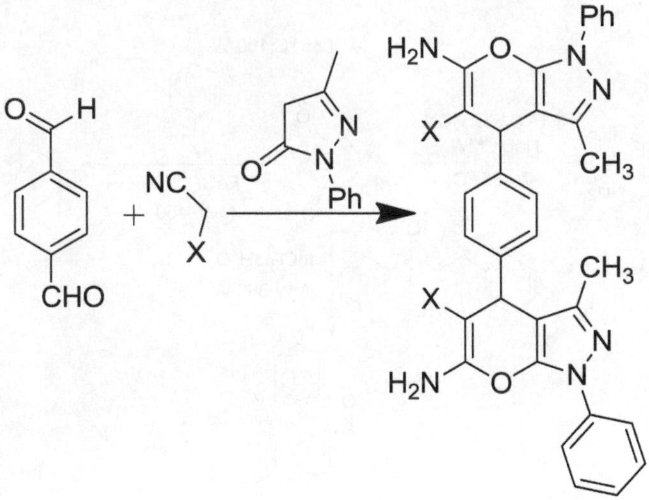

SCHEME 5.67   Synthesis of 4,4′-(1,4-phenylene)bis(1,4-dihydropyrano[2,3-c]pyrazole) analogs.

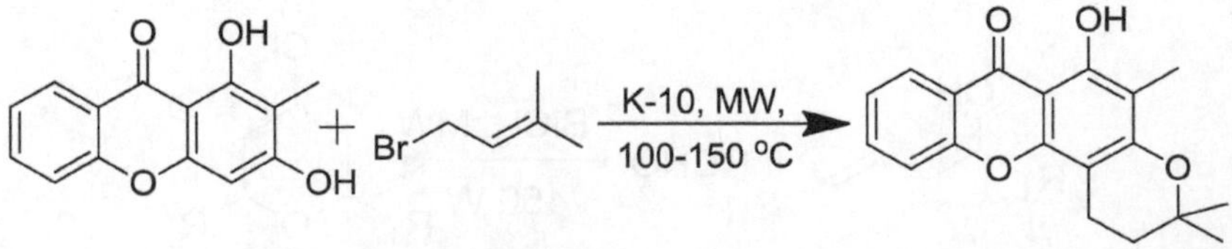

SCHEME 5.68   Synthesis of pyran derivatives.

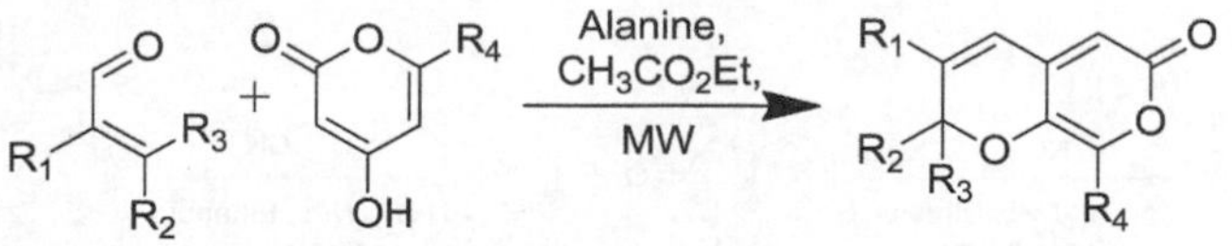

**SCHEME 5.69** Synthesis of 2*H*,5*H*-pyrano[4,3-*b*]pyran-5-one derivatives.

Leutbecher et al. generated a series of substituted 2*H*,5*H*-pyrano[4,3-*b*]pyran-5-one derivatives by microwave-accelerated domino reaction of an α,β-unsaturated aldehyde with 6-substituted-4-hydroxy-2*H*-pyran-2-ones [178] (Scheme 5.69).

## 5.2.6 SYNTHESIS OF SPIROPYRANS

Spiropyrans are widely accessed by reacting ninhydrin with malononitrile and other variants in the presence or absence of a catalyst under microwave dielectric heating conditions as shown in [179] (Scheme 5.70).

Sachdeva et al. performed Knoevenagel condensation for the synthesis of novel spiro[indoline-3,4'-pyrano[2,3-*c*]thiazolecarbonitriles and thiazolo[5″,4″:5′,6′]pyrano-[4′,3′:3,4]furo[2,3-*b*]indole derivatives by using NiO nanoparticles [180]. This is a one-pot protocol in which 1*H*-indole-2,3-diones, an activated methylene reagent, and

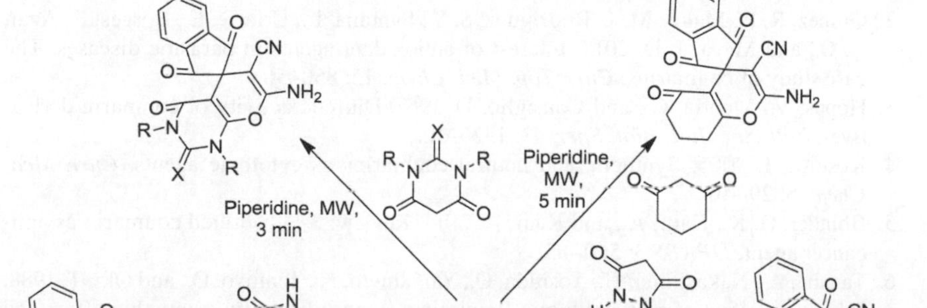

**SCHEME 5.70** Microwave-assisted synthesis of spiropyran analogs.

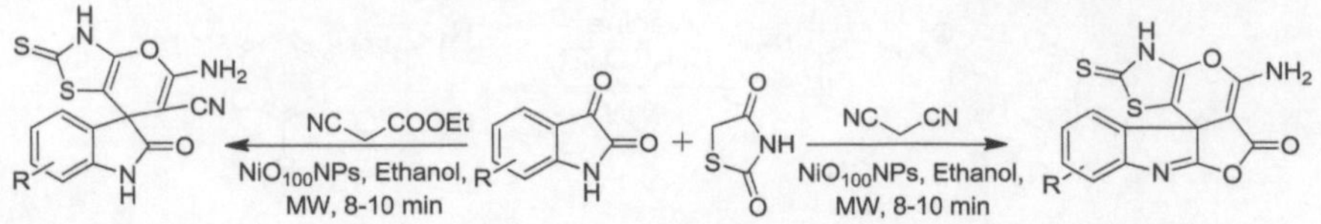

**SCHEME 5.71** Microwave synthesis of novel spiro[indoline-3,4'-pyrano[2,3-*c*]thiazolecarbonitriles and thiazolo[5″,4″:5′,6′]pyrano[4′,3′:3,4]furo[2,3-*b*]indole derivatives.

2-thioxo-4-thiazolidinone were irradiated in presence of NiO nanoparticles for the synthesis of respective spiro derivatives (Scheme 5.71).

## 5.4 CONCLUSIONS

In summary, this chapter gives an overall view of the microwave-assisted synthesis and functionalization of a variety of six-membered oxygen heterocycles. The discussion presented herein may be useful for the development of a novel series of biologically important oxygen-containing heterocyclic scaffolds.

## REFERENCES

1. Sperry, J. B., and Wright, D. L. 2005. Furans, thiophenes and related heterocycles in drug discovery. *Curr. Opin. Drug. Discov. Dev.* 8: 723–40.
2. Guinez, R. F., Matos, M. J., Rodriguez, S. V., Santana, L., Uriarte, E., Borges, F., Azar, C. O., and Maya, J. D. 2015. Interest of antioxidant agents in parasitic diseases. The case study of coumarins. *Curr. Top. Med. Chem.* 15: 850–56.
3. Hopps, V., Mantia, G., and Consiglio, D. 1967. Diuretic activity of coumarin derivatives. *Boll. Soc. Ital. Biol. Sper.* 43: 1526–30.
4. Kostova, I. 2005. Synthetic and natural coumarins as cytotoxic agents. *Curr. Med. Chem.* 5: 29–46.
5. Bhinder, C. K., Kaur, A., and Kaur, A. 2014. Review: 3-Substituted coumarin as anticancer agent. *IJPRBS* 3: 560–85.
6. Tanabe, A., Nakashima, H., Yoshida, O., Yamamoto, N., Tenmyo, O., and Oki, T. 1988. Inhibitory effect of new antibiotic Pradimicin A on infectivity, cytopathic effect and replication of human immunodeficiency virus *in vitro*. *J. Antibiot.* 41: 1708–10.
7. Zhu, T., Chen, R., Yu, H., Feng, Y., Chen, J., Lu, Q., Xie, J., Ding, W., and Ma, T. 2014. Antitumor effect of a copper (II) complex of a coumarin derivative and phenanthroline on lung adenocarcinoma cells and the mechanism of action. *Mol. Med. Rep.* 10: 2477–82.
8. Balaji, P. N., Lakshmi, L. K., Mohan, K., Revathi, K., Chamundeswari, A., and Indrani, P. M. 2012. *In-vitro* anti-inflammatory and antimicrobial activity of synthesized some novel pyrazole derivatives from coumarin chalcones. *Der. Pharm. Sin.* 3: 685–89.
9. Gasque, M. J. O., Gonzalez, M. P., Pena, J. P., Font, N. G., Romero, A., Pino, J. D., Ramos, E., Litina, D. H., Soriano, E., Chioua, M., Samadi, A., Raghuvanshi, D. S., Singh, K. N., and Contelles, J. M. 2014. Toxicological and pharmacological evaluation, antioxidant, ADMET and molecular modeling of selected racemic chromenotacrines {11-amino-12-aryl-8,9,10,12-tetrahydro-7*H*-chromeno[2,3-*b*]quinolin-3-ols} for the potential prevention and treatment of Alzheimer's disease. *Eur. J. Med. Chem.* 74: 491–501.
10. Chou, T. C., Tzeng, C. C., Su, T. L., and Wu T. S. 1992. Antileukemic activities of coumarin derivatives *in vitro*. *Zhonghua Yaoxue Zazhi* 44: 147–52.

11. Azab, I. H. E., Youssef, M. M., and Amin, M. A. 2014. Microwave-assisted synthesis of novel 2H-chromene derivatives bearing phenylthiazolidinones and their biological activity assessment. *Molecules* 19: 19648–64.
12. Harel, D., Schepmann, D., Brun, R., Schmidt, T. J., and Wunsch, B. 2013. Enantioselective synthesis of encecaline-derived potent antimalarial agents. *Org. Biomol. Chem.* 11: 7342–49.
13. Cannon, J. G., and Khonji, P. R. 1975. Centrally acting emetics. 9. Hofmann and Emde degradation products of nuciferine. *.J Med. Chem.* 18: 110–12.
14. Xu, Z. Q., Pupek, K., Suling, W. J., Enache, L., and Flavin, M. T. 2006 Pyranocoumarin, a novel anti-TB pharmacophore: Synthesis and biological evaluation against *Mycobacterium tuberculosis. Bioorg. Med. Chem.* 14: 4610–26.
15. Carola, C., Huber, S., Rosskopf, R., and Buchholz, H. 2005. Synthesis and use of chromene-4-one derivatives for the care of skin and hair. *Eur. Pat. Appl.: EP* 1508327 A1 20050223.
16. Nadaf, Y. F., and Renuka, C. G. 2015. Analysis of rotational diffusion of coumarin laser dyes. *Can. J. Phys.* 93: 3–6. doi: 10.1139/cjp-2014-20.
17. Zhang, Y. R., Chen, X. P., Shao, J., Zhang, J. Y., Yuan, Q., Miao, J. Y., and Zhao, B. X. 2014. A ratiometric fluorescent probe for sensing HOCl based on a coumarin–rhodamine dyad. *Chem. Commun.* 50: 14241–44.
18. Gedye, R., Smith, F., Westaway, K., Ali, H., Baldisera, L., Laberge, L., and Rousell, J. 1986. The use of microwave ovens for rapid organic synthesis. *Tetrahedron Lett.* 27: 279–82.
19. Sarko, C. R. 2005. Timesavings associated with microwave-assisted synthesis: A quantitative approach. In *Microwave Assisted Organic Synthesis*, Tierney J. P. and Lidström P. (Eds.), Oxford, Blackwell: 222–36.
20. Mavandadi, F., and Pilotti, A. 2006. The impact of microwave-assisted organic synthesis in drug discovery. *Drug Discov. Today* 11: 165–74.
21. Grewal, A. S., Kumar, K., Redhu, S., and Bhardwaj, S. 2013. Microwave assisted synthesis: A Green Chemistry approach. *Int. Res. J. Pharm. Appl. Sci.* 3: 278–85.
22. Liu, M., Wang, X., Sun, X., and He, W. 2014. Highly selective N-allylation of anilines under microwave irradiation. *Tetrahedron Lett.* 55: 2711–14.
23. Aubele, D. L., Wan, S., and Floreancig, P. E. 2005. Total synthesis of (+)-Dactylolide through an efficient sequential Peterson olefination and Prins cyclization reaction. *Angew. Chem. Int. Ed.* 44: 3485–88.
24. Chan, K. P., Ling, Y. H., and Loh, T. P., 2007 Formal synthesis of (+)-SCH 351448: The Prins cyclization approach. *Chem. Commun.* 9: 939–41.
25. Marucci, G., Piero, A., Brasili, L., Buccioni, M., Giardina, D., Gulini, U., Piergentili, A., and Sagratini, G. 2005. Synthesis and antimuscarinic activity of derivatives of 2-substituted-1,3-dioxolanes. *Med. Chem. Res.* 14: 274–96.
26. Schmidt, M., Ungvari, J., Glode, J., Dobner, B., and Langner, A. 2007. New 1,3-dioxolane and 1,3-dioxane derivatives as effective modulators to overcome multidrug resistance. *Bioorg. Med. Chem.* 15: 2283–97.
27. Gaina, L., Gal, E., Popa, L. M., Porumb, D., Nicolescu, A., Cristea, C., and Dumitrescu, L. S. 2012. Synthesis, structural investigations and DFT calculations on novel 3-(1,3-dioxan-2-yl)-10-methyl-10H-phenothiazine derivatives with fluorescence properties. *Tetrahedron* 68: 2465–70.
28. Flink, H., Putkonen, T., Sipos, A., and Jokela, R. 2010. Microwave-assisted selective protection of glutaraldehyde and its symmetrical derivatives as monoacetals and –thioacctals. *Tetrahedron* 66: 887–90.
29. Polshettiwar, V., and Varma, R. S. 2007. Tandem bis-aldol reaction of ketones: A facile one-pot synthesis of 1,3-dioxanes in aqueous medium. *J. Org. Chem.* 72: 7420–22.
30. Sachdeva, H., Saroj, R., and Dwivedi, D. 2014. Nano-ZnO catalyzed multicomponent one-pot synthesis of novel spiro(indoline-pyranodioxine) derivatives. *Sci. World J.* 2014: 1–10.

31. Peng, J., Hao, W., Wang, X., Tu, S., Ma, N., and Zhang, G. 2009. Microwave-assisted synthesis of pyrazolo[4,3-f]quinolin-7-one derivatives *via* multi-component reactions. *Chin. J. Chem.* 27: 1707–10.

32. Bandgar, B. P., Uppalla, L. S., and Kurule, D. S. 1999. Solvent-free one-pot rapid synthesis of 3-carboxycoumarins using focused microwaves. *Green Chem.* 1: 243–45.

33. Singh, S., and Guiry, P. J. 2009. Microwave-assisted synthesis of substituted tetrahydropyrans catalyzed by $ZrCl_4$ and its application in the asymmetric synthesis of *exo*- and *endo*-brevicomin. *J. Org. Chem.* 74: 5758–61.

34. Schinazi, R. F., and Floyd, R. A. 1990. Antiviral therapy using thiazine and xanthene dyes. *PCT Int. Appl.*: WO 9013296 A1 19901115.

35. Hafez, H. N., Hegab, M. I., Ahmed-Farag, I. S., and El-Gazzar, A. B. A. 2008. A facile regioselective synthesis of novel *spiro*-thioxanthene and *spiro*-xanthene-9′,2-[1,3,4] thiadiazole derivatives as potential analgesic and anti-inflammatory agents. *Bioorg. Med. Chem. Lett.* 18: 4538–43.

36. Kaya, M., Demir, E., and Bekci, H. 2013. Synthesis, characterization and antimicrobial activity of novel xanthene sulfonamide and carboxamide derivatives. *J. Enzyme Inhibit. Med. Chem.* 28: 885–93.

37. Kumar, A., Sharma, S., Maurya, R. A., Sarkar, J. 2010. Diversity oriented synthesis of benzoxanthene and benzochromene libraries via one-pot, three-component reactions and their anti-proliferative activity. *J. Comb. Chem.* 12: 20–24.

38. Bhattacharya, A. K., Rana, K. C., Mujahid, M., Sehar, I., and Saxena, A. K. 2009. Synthesis and *in vitro* study of 14-aryl-14H-dibenzo[a.j]xanthenes as cytotoxic agents. *Bioorg. Med. Chem. Lett.* 19: 5590–93.

39. Ruf, S. G., Hieu, H. T., and Poupelin, J. P. 1975. The effect of dibenzoxanthenes on the paralyzing action of zoxazolamine. *Naturwissenschaften* 62: 584–85.

40. Macrae, P. E., and Wright, T. R. 1974. Xanthene-dye photo-sensitized decomposition of a diazoniurn salt. *J. Chem. Soc. Chem. Commun.* 21: 898–99.

41. Tisseh, Z. N., Azimi, S. C., Mirzaei, P., and Bazgir, A. 2008. The efficient synthesis of aryl-5H-dibenzo[b,i]xanthene-5,7,12,14 (13H)-tetraone leuco-dye derivatives. *Dyes Pigm.* 79: 273–75.

42. Knight, C. G., and Stephens, T. 1989. Xanthene-dye-labelled phosphatidylethanolamines as probes of interfacial pH. *Biochem. J.* 258: 683–89.

43. Li, C., Ding, W., Lin, Y., and She, Z. 2011. Xanthene derivative from fermentation broth of endophytic fungi of marine plants and its preparation and application. *Faming Zhuanli Shenqing* CN 102180857 A 20110914.

44. Sandin, R. B., Gillies, A., and Lynn, S. C. 1939. The structure of fluorescein, sulfonefluorescein and some of their halogenated derivatives. *J. Am. Chem. Soc.* 61: 2919–21.

45. Hirano, K. 1983. Electronic structure and spectra of organic dye anions of Uranine and Eosin Y. *Bull. Chem Soc. Jpn.* 56: 850–54.

46. Viktorova, E. N., and Gofman, I. A. 1965. Fluorescence of a series of rhodamine dyes. *Zh. Fiz. Khim.* 39: 2643–49.

47. Baguley, B. C. 2003. Antivascular therapy of cancer DMXAA. *Lancet Oncol.* 4: 141–48.

48. Gutierrez-Orozco, F., Chitchumroonchokchai, C., Lesinski, G. B., Suksamrarn, S., and Failla, M. L. J. 2013. α-Mangostin anti-inflammatory activity and metabolism by human cell. *Agric. Food Chem.* 61: 3891–900.

49. Bell, J. 2005. Amlexanox for the treatment of recurrent aphthous ulcers. *Clin. Drug. Investig.* 25: 555–66.

50. Nguyen, H. T., Lallemand, M. C., Boutefnouchet, S., Michel, S., and Tillequin, F. 2009. Antitumor *Psoropermum* Xanthones and *Sarcomelicope* Acridones: Privileged structures implied in DNA alkylation. *J. Nat. Prod.* 72: 527–39.

51. Rankov, G., Sasaki, K., and Fukuda, M. 1990. Pharmacodynamics of Amlexanox (AA-673) in normal and anaphylactic rat conjunctiva and its effect on histamine concentration. *Ophthalmic Res.* 22: 359–64.

52. Kundu, K., and Nayak, S. K. 2014. Camphor-10-sulfonic acid catalyzed condensation of 2-naphthol with aromatic/aliphatic aldehydes to 14-aryl/alkyl-14H-dibenzo[a,j]xanthenes. *J. Serb. Chem. Soc.* 79: 1051–58.

53. Prasad, D., Preetam, A., and Nath, M. 2012. Microwave-assisted green synthesis of dibenzo[a,j]xanthenes using p-dodecylbenzenesulfonic acid as an efficient Bronsted acid catalyst under solvent-free conditions. *C. R. Chim.* 15: 675–78.

54. Ding, F. Q., An, L. T., and Zou, J. P. 2007. Iodine catalyzed microwave-assisted synthesis of 14-aryl(alkyl)-14H-dibenzo[a,j]xanthenes. *Chin. J. Chem.* 25: 645–48.

55. Bhattacharya, A. K., and Rana, K. C. 2007. Microwave-assisted synthesis of 14-aryl-14H-dibenzo[a,j]xanthenes catalysed by methanesulfonic acid under solvent-free conditions. *Mendeleev Commun.* 17: 247–48.

56. Nagarapu, L., Kantevari, S., Mahankhali, V. C., and Apuri, S. 2007. Potassium dodeca-tungstocobaltate trihydrate (K$_5$CoW$_{12}$O$_{40}$.3H$_2$O): A mild and efficient reusable catalyst for the synthesis of aryl-14H-dibenzo[a,j]xanthenes under conventional heating and microwave irradiation. *Catal. Commun.* 8: 1173–77.

57. Shaterian, R. H., Doostmohammadi, R., and Ghashang, M. 2008. Sodium hydrogen sulfate as effective and reusable heterogeneous catalyst for the one-pot preparation of 14H-[(Un)substituted phenyl]-dibenzo[a,j]xanthene leuco-dye derivatives. *Chin. J. Chem.* 26: 338–42.

58. Bamoniri, A., Mirjalili, B. B. F., and Nazemian, S. 2013. Microwave-assisted solvent-free synthesis of 14-aryl/alkyl-14H-dibenzo[a,j]xanthenes and tetrahydrobenzo[a]xanthen-11-ones catalyzed by nano silica phosphoric acid. *Curr. Chem. Lett.* 2: 27–34.

59. Rao, M. S., Chhikara, B. S., Tiwari, R., Shirazi, A. N., Parang, K., and Kumar, A. 2012. Microwave-assisted and scandium triflate catalyzed synthesis of tetrahydrobenzo[a]xanthen-11-ones. *Monatsh. Chem.* 143: 263–68.

60. Preetam, A., Prasad, D., Sharma, J. K., and Nath, M. 2015. Facile one-pot synthesis of oxo-xanthenes under microwave irradiation. *Curr. Microwave Chem.* 2:15–23.

61. Chereddy, S. S., Kunda, U. M R, Nemallapudi, B. R., Mudumala, V. N. R., Sthanikam, S. P., and Cirandur, S. R. 2012. Ytterbium perfluorooctanoate [Yb(PFO)$_3$]: A novel and efficient catalyst for the synthesis of tetrahydrobenzo[a]xanthene-11-ones under microwave irradiation. *Catal. Sci. Technol.* 2: 1382–85.

62. Kundu, D., Majee, A., and Hajra, A. 2011. Task-specific ionic liquid catalyzed efficient microwave-assisted synthesis of 12-alkyl or aryl-8,9,10,12-tetrahydrobenzo[a]xanthen-11-ones under solvent-free conditions. *Green Chem. Lett. Rev.* 4: 205–09.

63. Rosati, O., Messina, F., Pelosi, A., Curini, M., Petrucci, V., Gertsch, J., and Chicca, A. 2014. One-pot heterogeneous synthesis of Δ$^3$-tetrahydrocannabinol analogues and xanthenes showing differential binding to CB$_1$ and CB$_2$ receptors. *Eur. J. Med. Chem.* 85: 77–86.

64. Sagar, A. D., Chamle, S. N., and Yadav, M. V. 2013. Microwave assisted rapid synthesis of 1, 8-dioxo-octahydroxanthenes using lignin sulphonic acid. *J. Chem. Pharm. Res.* 5: 156–60.

65. Villemin, D., Hachemi, M., and Hammadi, M. 2003. An easy synthesis of thermochromic ethylenes under microwave irradiation. *J. Chem. Res. (S):* 260–61.

66. Xu, X., Xu, X., Li, H., Xie, X., and Li, Y. 2010. Iron-catalyzed, microwave-promoted, one-pot synthesis of 9-substituted xanthenes by a cascade benzylation-cyclization process. *Org. Lett.* 12: 100–03.

67. Zhang, X., Yang, L., Wu, Y., Du, J., Mao, Y., Wang, X., Luan, S., Lei, Y., Li, X., Sun, H., and You, Q. 2014. Microwave-assisted transition-metal-free intramolecular Ullmann-type O-arylation in water for the synthesis of xanthones and azaxanthones. *Tetrahedron Lett.* 55: 4883–87.

68. Wu, X., Beecher, G. R., Holden, J. M., Haytowitz, D. B., Gebhardt, S. E., and Prior, R. L. 2006. Concentration of anthocyanins in common foods in the United States and estimation of normal consumption. *J. Agric. Food Chem.* 54: 4069–75.

69. Haytowitz, D. B., Pehrsson, P. R., and Holden, J. M. 2002. The identification of key foods for food composition research. *J. Food Comp. Anal.* 15: 183–94.

70. Lunte, S. M. 1987. Structural classification of flavonoids in beverages by liquid chromatography with ultraviolet-visible and electrochemical detection. *J. Chromatogr.* 384: 371–82.

71. Habbu, P. V., Mahadevan, K. M., Kulkarni, P. V., Singh, C. D., Veerapur, V. P., and Shastry, R. A. 2010. Adaptogenic and *in vitro* antioxidant activity of flavonoids and other fractions of *Argyreia speciosa* (Burm.f) Boj in acute and chronic stress paradigms in rodents. *Indian J. Exp. Biol.* 48: 53–60.

72. Udut, V. V., Vengerovskii, A. I., Suslov, N. I., Shilova, I. V., Kaigorodtsev, A. V., Polomeeva, N. Y., and Dygai, A. M. 2012. Anxiolytic activity of biologically active compounds from *Filipendula Vulgaris*. *Pharm. Chem. J.* 46: 27–29.

73. Ye, H., Xie, C., Wu, W., Xiang, M., Liu, Z., Li, Y., Tang, M., Li, S., Yang, J., Tang, H., Chen, K., Long, C., Peng, A., and Chen, L. 2014. *Millettia pachycarpa* exhibits anti-inflammatory activity through the suppression of LPS-induced NO/iNOS expression. *Am. J. Chin. Med.* 42: 949–65.

74. Amagon, K. I., Wannang, N. N., Iliya, H. A., Ior, L. D., and Otubor, G. O. C. 2012. Flavonoids extracted from fruit pulp of *Cucumis metuliferus* have antiviral properties. *Br. J. Pharm. Res.* 2: 249–58.

75. Bautista, E., Calzada, F., Ortega, A., and Mulia, L. Y. 2011. Antiprotozoal activity of flavonoids isolated from *Mimosa tenuiflora* (Fabaceae-Mimosoideae). *J. Mex. Chem. Soc.* 55: 251–53.

76. Mariangela, M., Francesco, M., and Filomena, C. 2015. A comparative study of *Zingiber officinale* Roscoe pulp and peel: phytochemical composition and evaluation of antitumour activity. *Nat. Prod. Res.* 29: 2045–49. doi:10.1080/14786419.2015.1020491.

77. Xie, Y., Yang, W., Tang, F., Chen, X., and Ren, L. 2015. Antibacterial activities of flavonoids: Structure-activity relationship and mechanism. *Curr. Med. Chem.* 22: 132–49.

78. Saini, N. K., Singhal, M., Awasthi, A., and Mishra, G. 2013. Total tannin content and antidiarrheal activity of *Tecomaria capensis* leaves. *Nat. Prod. J.* 3: 218–23.

79. Sato, A., and Tamura, H. 2015. High antiallergic activity of 5,6,4′-trihydroxy-7,8,3′- trimethoxyflavone and 5,6-dihydroxy-7,8,3′,4′- tetramethoxyflavone from eau de cologne mint (*Mentha × piperita citrata*). *Fitoterapia* 102: 74–83.

80. Perez, A., Manzano, S. G., Jimenez, R., Abud, R. P., Haro, J. M., Osuna, A., Buelga, C. S., Duarte, J., and Vizcaino, F. P. 2014. The flavonoid quercetin induces acute vasodilator effects in healthy volunteers: Correlation with beta-glucuronidase activity. *Pharmacol. Res.* 89: 11–18.

81. Gopal, T. K., Chamundeeswari, D., Sathiya, S., and Babu, C. S. 2015. *In vitro* anti-cancer activity of quercetin and Kaempferol against human epithelial malignant melanoma cells (A375). *Int. J. Pharm. Res. Scholars* 4: 157–62.

82. Nagulsamy, P., Ponnusamy, R., and Thangaraj, P. 2015. Evaluation of antioxidant, anti-inflammatory, and antiulcer properties of *Vaccinium leschenaultii Wight*: A therapeutic supplement. *J. Food Drug. Anal.* 23: 376–83. doi: 10.1016/j.jfda.2014.11.003.

83. Gao, H., and Kawabata, J. 2005. α-Glucosidase inhibition of 6-hydroxyflavones. Part 3: Synthesis and evaluation of 2,3,4-trihydroxybenzoyl-containing flavonoid analogs and 6-aminoflavones as a-glucosidase inhibitors. *Bioorg. Med. Chem.* 13: 1661–671.

84. Cushman, M., Zhu, H., Geahlen, R. L., and Kraker, A. J. 1994. Synthesis and biochemical evaluation of a series of aminoflavones as potential inhibitors of protein-tyrosine kinases p56[lck], EGFr, and p60[v-src]. *J. Med. Chem.* 37: 3353–62.

85. Van-Son, N., Lin-Pei, D., Sheng-Chun, W., and Qiuan, W. 2015. The first total synthesis of Sophoflavescenol, Flavenochromane C, and Citrusinol. *Eur. J. Org. Chem.* 2297–302. doi: 10.1002/ejoc.201403689.

86. Belsare, D. P., and Kazi A 2013. Microwave-assisted synthesis of flavones and their comparative study with conventional method. *IOSR Journal of Pharmacy* 3: 23–27.

87. Mariappan B, Kasi P, and Penugonda R 2013. An expeditious synthesis of flavonols promoted by Montmorillonite KSF clay and assisted by microwave irradiation under solvent-free conditions. *Helv Chim Acta* 96: 1269–72.

88. Wei, H., Qiong, C., Wen-Chao, Y., and Guang-Fu, Y. 2013. Efficient synthesis and antiproliferative activity of novel thioether-substituted flavonoids. *Eur J Med Chem* 66: 161–70.

89. Ziaullah, and Rupasinghe, H. P. V. 2013. An efficient microwave-assisted enzyme-catalyzed regioselective synthesis of long chain acylated derivatives of flavonoid glycosides. *Tetrahedron Lett* 54: 1933–37.

90. Mokle, S. S., and Vibhute, Y. B. 2009. Synthesis of some new biologically active chalcones and flavones. *Der Pharma Chemica* 1: 145–52.

91. Awuah, E., and Capretta, A. 2009. Access to flavones via a microwave-assisted, one-pot sonogashira-carbonylation-annulation reaction. *Org. Lett.* 11: 3210–13.

92. Lecoutey, C., Fossey, C., Demuynck, L., Lefoulon, F., Fabis, F., and Rault, S. 2008. A convenient microwave-assisted 5-amination of flavones. *Tetrahedron* 64: 11243–48.

93. Yasar, A., Akpmar, K., Burnaz, N. A., Kucuk, M., Karaoglu, S. A., Dogan, N., and Yayli, N. 2008. Microwave-assisted synthesis of 4'-azaflavones and their N-alkyl derivatives with biological activities. *Chem. Biodivers.* 5: 830–38.

94. Ghani, S. B. A., Weaver, L., Zidan, Z. H., Ali, H. M., Keevil, C. W., and Brown, R. C. D. 2008. Microwave-assisted synthesis and antimicrobial activities of flavonoid derivatives. *Bioorg. Med. Chem. Lett.* 18: 518–22.

95. Kabalka, G. W., and Mereddy, A. R. 2005. Microwave-assisted synthesis of functionalized flavones and chromones. *Tetrahedron Lett.* 46: 6315–17.

96. Varma, R. S., Saini, R. K., and Kumar, D. 1998. An expeditious synthesis of flavones on Montmorillonite K 10 clay with microwaves. *J. Chem. Res.* (S): 348–49.

97. Kamboj, R. C., Sharma, G., Kumar, D., Arora, R., Sharma, C., and Aneja, K. R. 2011. An environmentally sound approach for the synthesis of some flavanones and their antimicrobial activity. *Int. J. Chem. Tech. Res.* 3: 901–10.

98. Sagar, J., Kumar, V., Shah, D. K., and Bhatnagar, A. 2006. Spontaneous intra-peritoneal bleeding secondary to warfarin, presenting as an acute appendicitis: A case report and review of literature. *BMC Blood Disorders* 6: 7.

99. Rohini, K., and Srikumar, P. S. 2013. *In Silico* approach of anticancer activity of phytochemical coumarins against cancer target JNKS. *Int. J. Pharm. Pharm. Sci.* 5: 741–42.

100. Govindappa, M., Umashankar, T., Ramachandra, Y. L., and Charles Chee-Jen, C. 2013. *In vitro* anti-HIV properties of coumarins from *Crotalaria pallida* Aiton. *Trade Sci. Inc Biochem.* 7: 114–21.

101. Zhu, T., Chen, R., Yu, H., Feng, Y., Chen, J., Lu, Q., Xie, J., Ding, W., and Ma, T. 2014. Antitumor effect of a copper (II) complex of a coumarin derivative and phenanthroline on lung adenocarcinoma cells and the mechanism of action. *Mol. Med. Rep.* 10: 2477–82.

102. Kayal, G., Jain, K., Malviya, S., and Kharia, A. 2014. Comparative SAR of synthetic coumarin derivatives for their anti-inflammatory activity. *IJPSResearch* 5: 3577–83.

103. Patil, P. O., Bari, S. B., Firke, S. D., Deshmukh, P. K., Donda, S. T., and Patil, D. A. 2013. A comprehensive review on synthesis and designing aspects of coumarin derivatives as monoamine oxidase inhibitors for depression and Alzheimer's disease. *Bioorg. Med. Chem.* 21: 2434–50.

104. Chou, T. C., Tzeng, C. C., Su, T. L., and Wu, T. S. 1992. Antileukemic activities of coumarin derivatives *in vitro. Zhonghua Yaoxue Zazhi* 44: 147–52.

105. Patel, K. S., Patel, R. B., and Patel, R. N. 2013. Biological activity of newly synthesized M(II) heterochelates of coumarin derivative and enrofloxacin. *Heterocycl. Lett.* 3: 493–504.

106. Adesanwo, J. K., Ekundayo, O., Shode, F. O., Njar, V. C. O., Van den Berge, and A. J. J., Oludahunsi, A. T. 2004. Eniotorin, An antimalarial coumarin from the root bark of *Quassia undulate. Nig. J. Nat. Prod. Med.* 8: 69–73.

107. Giri, R. R., Lad, H. B., Bhila, V. G., Patel, C. V., and Brahmbhatt, D. I. 2015. Modified pyridine-substituted coumarins: A new class of antimicrobial and antitubercular agents. *Synth. Commun.* 45: 363–75.

108. Vilar, S., Quezada, E., Santana, L., Uriarte, E., Yanez, M., Fraiz, N., Alcaide, C., Cano, E., and Orallo, F. 2006. Design, synthesis, and vasorelaxant and platelet antiaggregatory activities of coumarin–resveratrol hybrids. *Bioorg. Med. Chem. Lett.* 16: 257–61.

109. Patil, R. B., Sawant, S. D., Reddy, K. V., and Shirsat, M. 2015. Synthesis, docking studies and evaluation of antioxidant activity of some chromenone derivatives. *RJPBCS* 6: 381–91.

110. Wu, J., Yang, Z., Fathi, R., and Zhu, Q. 2004. Preparation of 4-thiosubstituted coumarin derivatives as anti-HCV agent. US Pat Appl Publ, US 20040180950 A1 20040916.

111. Ojewole, J. A. O. 2002. Hypoglycaemic effect of *Clausena anisata* (Willd) Hook methanolic root extract in rats. *J. Ethnopharmacol.* 81: 231–37.

112. Marcondes, H. C., De Oliveira, T. T., Taylor, J. G., Hamoy, M., Neto, A. L., De Mello, V. J., and Nagem, T. J. 2015. Antifungal activity of coumarin mammeisin isolated from species of the *Kielmeyera* Genre (Family: Clusiaceae or Guttiferae). *J. Chem.* 2015: 1–5.

113. Tripathi, R. P., Tripathi, R., Bhaduri, A. P., Singh, S. N., Chatterjee, R. K., and Murthy, P. K. 2000. Antifilarial activity of some 2*H*–1-benzopyran-2-ones (coumarins). *Acta Tropica* 76: 101–06.

114. Lacy, A., and Kennedy, R. O. 2004. Studies on coumarins and coumarin-related compounds to determine their therapeutic role in the treatment of cancer. *Curr. Pharm. Des.* 10: 3797–811.

115. Bariana, D. S. 1970. Coumarin derivatives as coronary vasodilators. *J. Med. Chem.* 13: 544–46.

116. Edwards, J. P., Zhi, L., Pooley, C. L. F., Tegley, C. M., West, S. J., Wang, M. W., Gottardis, M. M., Pathirana, C., Schrader, W. T., and Jones, T. K. 1998. Preparation, resolution, and biological evaluation of 5-aryl-1,2-dihydro-5*H*-chromeno[3,4-*f*]quinolines: Potent, orally active, nonsteroidal progesterone receptor agonists. *J. Med. Chem.* 41: 2779–85.

117. Jain, N., Kanojia, R. M., Xu, J., Zhong, G. J., Pacia, E., Lai, M. T., Du, F., Musto, A., Allan, G., Hahn, D. W., Lundeen, S., and Sui, Z. 2006. Novel chromene-derived selective estrogen receptor modulators useful for alleviating hot flushes and vaginal dryness. *J. Med. Chem.* 49: 3056–59.

118. Herring, A. C., Koh, W. S., and Kaminski, N. E. 1998. Inhibition of the cyclic AMP signaling cascade and nuclear factor binding to CRE and κB elements by cannabinol, a minimally CNS-active cannabinoid. *Biochem. Pharmacol.* 55: 1013–23.

119. Mahadevan, A., Siegel, C., Martin, B. R., Abood, M. E., Beletskaya, I., and Razdan, R. K. 2000. Novel cannabinol probes for CB1 and CB2 cannabinoid receptors. *J. Med. Chem.* 43: 3778–85.

120. Takasugi, M., Nagao, S., Ueno, S., Masamune, T., Shirata, A., and Takahashi, K. 1978. Studies on phytoalexins of the Moraceae. 2. Moracin C and D, new phytoalexins from diseased mulberry. *Chem. Lett.* 11: 1239–40.

121. Zhao, X., Fu, X., Wang, P., Li, J., and Hu, X. 2011. Determination of coumarins in cosmetics with high performance liquid chromatography. *Fenxi Kexue Xuebao* 27: 49–52.

122. Lei, C., Xu, W., Wu, J., Wang, S., Sun, J. Q., Chen, Z., and Yang, G. 2015. Coumarins from the roots and stems of *Nicotiana tabacum* and their anti-tobacco mosaic virus activity. *Chem. Nat. Cmpd.* 51: 43–46.

123. Muñoz-Garcia, A. B., and Pavone, M. 2015. Structure and energy level alignment at the dye–electrode interface in p-type DSSCs: new hints on the role of anchoring modes from *ab initio* calculations. *Phys. Chem. Chem. Phys.* 17: 12238–46. doi: 10.1039/c5cp01020a.

124. Fan, J., Xu, Q., Zhu, H., and Peng, X. 2014. Synthesis and properties of an aza-coumarin based sensor for detection of $Zn^{2+}$, $Cd^{2+}$ and $Cu^{2+}$. *Youji Huaxue* 34: 1623–29.

125. Surpur, M. P., Kshirsagar, S., and Samant, S. D. 2009. Exploitation of the catalytic efficacy of Mg/Al hydrotalcite for the rapid synthesis of 2-aminochromene derivatives via a multicomponent strategy in the presence of microwaves. *Tetrahedron Lett.* 50: 719–22.

126. Pasha, M. A., and Jayashankara, V. P. 2007. An efficient synthesis of 2-aminobenzochromene derivatives catalysed by tetrabutylammoniumbromide (TBAB) under microwave irradiation in aqueous medium. *Indian J. Chem. Sec. B: Org. Chem. Incl. Med. Chem.* 46B: 1328–31.

127. Chaker, A., Najahi, E., Nepveu, F., and Chabchoub, F. 2013. Microwave-assisted synthesis of chromeno[2,3-*d*]pyrimidinone derivatives. *Arabian J. Chem.* 10: S3040–47 doi: 10.1016/j.arabjc.2013.11.045.

128. El-Agrody, A. M., Al-Dies, A. A. M., and Fouda, A. M. 2014. Microwave assisted synthesis of 2-amino-6-methoxy-4*H*-benzo[*h*]chromene derivatives. *Eur. J. Chem.* 5: 133–37.

129. Iniyavan, P., Sarveswari, S., and Vijayakumar, V. 2014. Microwave-assisted clean synthesis of xanthenes and chromenes in [bmim][PF$_6$] and their antioxidant studies. *Res. Chem. Intermed.* 41: 7413–26. doi: 10.1007/s11164-014-1821-4.

130. Sangani, C. B., Shah, N. M., Patel, M. P., and Patel, R. G. 2012. Microwave-assisted synthesis of novel 4*H*-chromene derivatives bearing phenoxypyrazole and their antimicrobial activity assessment. *J. Serb. Chem. Soc.* 77: 1165–74.

131. Shaker, R. M. 1996. Synthesis and reactions of some new 4*H*-pyrano[3,2-*c*]benzopyran-5-one derivatives and their potential biological activities. *Pharmazie* 51: 148–51.

132. Kidwai, M., and Saxena, S. 2006. Convenient preparation of pyranobenzopyranes in aqueous media. *Synth. Commun.* 36: 2743.

133. Kidwai, M., Saxena, S., Khan, R. M. K., and Thukral, S. S. 2005. Aqua mediated synthesis of substituted 2-amino-4*H*-chromenes and *in vitro* study as antibacterial agents. *Bioorg. Med. Chem. Lett.* 15: 4295–98.

134. Yang, G. Y., Yang, J. T., Wang, C. X., Fan, S. F., Xie, P. H., and Xu, C. L. 2014. Microwave-assisted TsOH/SiO$_2$ catalyzed one-pot synthesis of novel fluoro-substituted coumarin hydrazones under solvent-free conditions. *J. Fluorine Chem.* 168: 1–8.

135. Zhang, R., Xu, Z., Yin, W., Liu, P., and Zhang, W. 2014. Microwave-assisted synthesis and antifungal activities of polysubstituted furo[3,2-*c*]chromen-4-ones and 7,8,9,10-Tetrahydro-6*H*-benzofuro[3,2-*c*]chromen-6-ones. *Synth. Commun.* 44: 3257–63.

136. Ashok, D., Vijaya Lakshmi, B., Ravi, S., and Ganesh, A. 2014. Microwave-assisted synthesis of substituted 4-chloro-8-methyl-2-phenyl-1,5-dioxa-2*H*-phenanthren-6-ones and their antimicrobial activity. *Med. Chem. Res.* 24: 1487–95. doi:10.1007/s00044-014-1204-9.

137. Yadav, P., Singh, S., Sahu, S. N., Hussain, F., and Pratap, R. 2014. Microwave assisted base dependent regioselective synthesis of partially reduced chromenes, isochromenes and phenanthrenes. *Org. Biomol. Chem.* 12: 2228–34.

138. Kang, Y., Mei, Y., Du, Y., and Jin, Z. 2003. Total synthesis of the highly potent anti-HIV natural product daurichromenic acid along with its two chromane derivatives, Rhododaurichromanic Acids A and B. *Org. Lett.* 5: 4481–84.

139. Borisov, A. V., Gorobets, N. Y., Yermolayev, S. A., Zhuravel, I. O., Kovalenko, S. M., and Desenko, S. M. 2007. One-pot microwave-assisted synthesis of a Benzopyrano[2,3-c] pyrazol-3(2H)-one library. *J. Comb. Chem.* 9: 909–11.

140. Texier-Boullet, F., and Foucaud, A. 1982. Synthesis of 1-Hydroxyalkanephosphonic esters on alumina. *Synthesis* 11: 916.

141. Basu, P. K., and Ghosh, A. 2011. Microwave-assisted improved regioselective synthesis of 12H-benzopyrano[3,2-c][1]benzopyran-5-ones by radical cyclisation. *Org. Chem. Int.* 15:1–6. doi: 10.1155/2011/394619.

142. Thasana, N., Worayuthakarn, R., Kradanrat, P., Hohn, E., Young, L., and Ruchirawat, S. 2007. Copper(I)-mediated and microwave-assisted $C_{aryl}$-$O_{carboxylic}$ coupling: Synthesis of benzopyranones and isolamellarin alkaloids. *J. Org. Chem.* 72: 9379–82.

143. Zhang, Z., Ma, Y., and Zhao, Y. 2008. Microwave-assisted one-pot synthesis of dihydrocoumarins from phenols and cinnamoyl chloride. *Synlett* 7: 1091–95.

144. Rajasekhar, K. K., Ananth, V. S., Nithiyananthan, T. S., Hareesh, G., Kumar, P. N., and Reddy, R. S. P. 2010. Comparative study of conventional and microwave induced synthesis of selected heterocyclic molecules. *Int. J. ChemTech Res.* 2: 592–97.

145. Frere, S., Thiery, V., and Besson, T. 2001. Microwave acceleration of the Pechmann reaction on graphite/montmorillonite K10: application to the preparation of 4-substituted 7-aminocoumarins. *Tetrahedron Lett.* 42: 2791–94.

146. Vahabi, V., and Hatamjafari, F. 2014. Microwave assisted convenient one-pot synthesis of coumarin derivatives via Pechmann Condensation catalyzed by $FeF_3$ under solvent-free conditions and antimicrobial activities of the products. *Molecules* 19: 13093–103.

147. Hector, A., Anupama, R., and Bimal, B. K. 2011. Microwave-induced bismuth nitrate-catalyzed Pechmann reaction under solventless condition. *Heterolett. Org.* 1: 95–96.

148. Karanjule, N. B., and Samant, S. D. 2014. Microwave assisted, 4-dimethylaminopyridine (DMAP) mediated, one-pot, three-component, regio- and diastereoselective synthesis of trans-2,3-dihydrofuro[3,2-c]coumarins. *Curr. Microwave Chem.* 1: 135–41.

149. Zhou, Z. Z., Yan, G. H., Chen, W. H., and Yang, X. M. 2013. Microwave-assisted efficient synthesis of 2-Hydroxydeoxybenzoins from the alkali degradation of readily prepared 3-aryl-4-hydroxycoumarins in water. *Chem. Pharm. Bull.* 61: 1166–72.

150. Thakrar, S., Bavishi, A., Radadiya, A., Vala, H., Parekh, S., Bhavsar, D., Chaniyara, R., and Shah, A. 2014. An efficient microwave-assisted synthesis and antimicrobial activity of novel 2-amino 3-cyano pyridine derivatives using two reusable solid acids as catalysts. *J. Heterocycl. Chem.* 51: 555–61.

151. Irgashev, R. A., Karmatsky, A. A., Slepukhin, P. A., Rusinov, G. L., and Charushin, V. N. 2013. A convenient approach to the design and synthesis of indolo[3,2-c]coumarins via the microwave-assisted Cadogan reaction. *Tetrahedron Lett.* 54: 5734–38.

152. Kumar, P. N. P., Naik, H. S. B., Harish, K. N., and Viswanath, R. 2013. Effect of surfactant-assisted and pH dependent ZnO nanoparticle-catalyzed for the rapid synthesis of coumarin by Knoevenagel condensation under microwave irradiation. *Arch. Appl. Sci. Res.* 5: 132–37.

153. Mirjafary, Z., Saeidian, H., and Moghaddam, F. M. 2009. Microwave-assisted synthesis of 3-substituted coumarins using $ZrOCl_2 \cdot 8H_2O$ as an effective catalyst. *Trans. C Chem. Chem. Eng.* 16: 1–10.

154. Kaneria, A. R., Giri, R. R., Bhila, V. G., Prajapati, H. J., and Brahmbhatt, D. I. 2013. Microwave assisted synthesis and biological activity of 3-aryl-furo[3,2-c]coumarins. *Arabian J. Chem.* 10: S1100–04. doi:10.1016/j.arabjc.2013.01.017.

155. Jardosh, H. H., and Patel, M. P. 2013. Microwave-assisted CAN-catalyzed solvent-free synthesis of N-allyl quinolone-based pyrano[4,3-b]chromene and benzopyrano[3,2-c] chromene derivatives and their antimicrobial activity. *Med Chem Res* 22: 905–15.

156. Olomola, T. O., Klein, R., and Kaye, P. T. 2012. Convenient synthesis of 3-methylcoumarins and coumari-3-carbaldehydes. *Synth. Commun.* 42: 251–57.

157. Li, K. T., Lin, Y. B., and Yang, D. Y. 2012. One-pot synthesis of pyranocoumarins via microwave-assisted pseudo multicomponent reactions and their molecular switching properties. *Org. Lett.* 14: 1190–93.
158. Kuarm, B. S., Kumar, V. N., Madhav, J. V., and Rajitha, B. 2011. Synthesis of 3-[6-(2-aminoethoxy)-3-methyl-5-(3-morpholin-4-yl-propionyl)-benzofuran-2carbonyl]-chromen-2-one under microwave irradiation conditions. *Green Chem. Lett. Rev.* 4: 97–101. doi:10.1080/17518253.2010.506450.
159. Seijas, J. A., Vazquez-Tato, M. P., and Carballido-Reboredo, R. 2005. Solvent-free synthesis of functionalized flavones under microwave irradiation. *J. Org. Chem.* 70: 2852–55.
160. Bashiardes, G., Safir, I., Mohamed, A. S., Barbot, F., and Laduranty, J. 2003. Microwave-assisted [3 + 2] cycloadditions of azomethine ylides. *Org. Lett.* 5: 4915–18.
161. Patonay, T., Varma, R. S., Vass, A., Levaia, A., and Dudas, J. 2001. Highly diastereoselective Michael reaction under solvent-free conditions using microwaves: Conjugate addition of flavanone to its chalcone precursor. *Tetrahedron Lett.* 42: 1403–06.
162. Feng, Y. J., Miao, C. B., Gao, Y., Tu, S. J., Fang, F., and Shi, D. Q. 2004. Dialdehyde in heterocyclic synthesis: Synthesis of compounds containing two 4*H*-benzopyran building blocks under microwave irradiation. *Chin. J. Chem.* 22: 622–26.
163. Neuschl, M., Bogdal, D., and Potacek, M. 2007. Microwave-assisted synthesis of substituted hexahydropyrrolo[3,2-*c*]quinolines. *Molecules* 12: 49–59.
164. Madda, J., Venkatesham, A., Bejjanki, N. K., Kommu, N., Pombala, S., Kumar,C. G., Rao, T. P., and Nanubolu, J. B. 2014. Synthesis of novel chromeno-annulated cis-fused pyrano[3,4-*c*]benzopyran and naphtho pyran derivatives via domino aldol-type/hetero Diels–Alder reaction and their cytotoxicity evaluation. *Bioorg. Med. Chem. Lett.* 24: 4428–34.
165. Jimenez-Alonso, S., Chavez, H., Estevez-Braun, A., Ravelo, A. G., Feresin, G., and Tapia, A. 2008. An efficient synthesis of embelin derivatives through domino Knoevenagel hetero Diels-Alder reaction under microwave irradiation. *Tetrahedron* 64: 8938–42.
166. Castagnolo, D., Botta, L., and Botta, M. 2009. One-pot multicomponent synthesis of 2,3-dihydropyrans: new access to furanose-pyranose 1,3-C–C-linked-disaccharides. *Tetrahedron Lett.* 50: 1526–28.
167. Peng, Y., and Song, G. 2007. Amino-functionalized ionic liquid as catalytically active solvent for microwave-assisted synthesis of 4*H*-pyrans. *Catal. Commun.* 8: 111–14.
168. Shaterian, H. R., Hosseinian, A., and Ghashang, M. 2008. Reaction in dry media: Silica gel supported ferric chloride catalyzed synthesis of 1,8-dioxo-octahydroxanthene derivatives. *Phosphorus Sulfur Silicon Relat. Elem.* 183: 3136–44.
169. Hu, X. Y., Fan, X. S., Zhang, X. Y., Qu, G. R., and Li, Y. Z. 2006. Solvent-free synthesis of 5-oxo-5,6,7,8-tetrahydro-4*H*-benzo-[*b*]-pyran derivatives under microwave irradiation. *Can. J. Chem.* 84: 1054–57.
170. Devi, I., and Bhuyan, P. J. 2004. Sodium bromide catalysed one-pot synthesis of tetrahydrobenzo[*b*]pyrans via a three-component cyclocondensation under microwave irradiation and solvent free conditions. *Tetrahedron Lett.* 45: 8625–27.
171. Hagiwara, H., Numamae, A., Isobe, K., Hoshi, T., and Suzuki, T. 2006. Microwave-promoted sequential three-component synthesis of tetrahydrobenzo[*b*]pyran in water catalyzed by heterogeneous amine grafted on silica. *Heterocycles* 68: 889–91.
172. Yadav, J. S., Reddy, B. V. S., Venugopal, C., Srinivas, R., and Ramalingam, T. 2002. Microwave-accelerated synthesis of 4-chlorotetrahydropyrans by Bismuth(III)chloride. *Synth. Commun.* 32: 1803–08.
173. Wu, J., Sun, L., and Dai, W. M. 2006. Microwave-assisted tandem Wittig–intramolecular Diels–Alder cycloaddition. Product distribution and stereochemical assignment. *Tetrahedron* 62: 8360–72.

174. Wu, J., Yu, H., Wang, Y., Xing, X., and Dai, W. M. 2007. Unexpected epimerization and stereochemistry revision of IMDA adducts from sorbate-related 1,3,8-nonatrienes. *Tetrahedron Lett.* 48: 6543–47.

175. Wu, J., Jiang, X., Xu, J., and Dai, W. M. 2011. Tandem Wittig intramolecular Diels Alder cycloaddition of ester-tethered 1,3,9-decatrienes under microwave heating. *Tetrahedron* 67: 179–92.

176. Raafat, M. S. 2012. Synthesis of 1,4-phenylene bridged bis-heterocyclic compounds. *Arkivoc* 1: 1–44.

177. Castanheiro, R. A. P., Pinto, M. M. M., Cravo, S. M. M., Pinto, D. C. G. A., Silva, A. M. S., and Kijjoa, A. 2009. Improved methodologies for synthesis of prenylated xanthones by microwave irradiation and combination of heterogeneous catalysis (K10 clay) with microwave irradiation. *Tetrahedron* 65: 3848–57.

178. Leutbecher, H., Conrad, J., Klaiber, I., and Beifuss, U. 2004. Microwave assisted domino Knoevenagel condensation/6π-electron electrocyclization reactions for the rapid and efficient synthesis of substituted 2*H*,5*H*-pyrano[4,3-*b*]pyran-5-ones and related heterocycles. *QSAR Comb. Sci.* 23: 895–98.

179. Shaker, R. M., Mahmoud, A. F., and Abdel-Latif, F. F. 2005. Facile one pot microwave assisted solvent-free synthesis of novel spiro-fused pyran derivatives via the three-component condensation of ninhydrin with malononitrile and active methylene compounds. *J. Chin. Chem. Soc.* 52: 563–67.

180. Sachdeva, H., Dwivedi, D., Bhattacharjee, R. R., Khaturia, S., and Saroj, R. 2013. NiO nanoparticles: An efficient catalyst for the multicomponent one-pot synthesis of novel spiro and condensed indole derivatives. *J. Chem.* 2013: 1–10. doi: 10.1155/2013/606259.

# 6 Selectivities in the Microwave-Assisted Organic Reactions

*Jiaxi Xu*

## CONTENTS

## 6.1 SELECTIVITY UNDER MICROWAVE IRRADIATION CONDITIONS

During the last three decades, microwave-assisted organic reactions have been widely developed and applied in organic and medicinal syntheses. Almost all types of organic reactions have been investigated under microwave irradiation conditions [1–5].

The results indicate that microwave irradiation can accelerate most organic reactions and even realize some organic reactions that do not occur under conventional heating conditions. Recently, studies in microwave chemistry have focused on the origin of the microwave acceleration of organic reactions and the application of microwave irradiation in highly efficient preparation of chemicals, rather than on different organic reactions as was previously the case [4–6]. On one hand, chemists and microelectronics experts have focused on the role of the microwave effect in assisting organic reactions and discussed special microwave effect, called the athermal effect. On the other hand, organic chemists turned their interests to the effects of microwave irradiation on the different selectivities, including chemoselectivity, regioselectivity, syn/trans selectivity, diastereoselectivity, and enantioselectivity, in organic reactions [7–9]. Compared with numerous reports on microwave-accelerated organic reactions, studies on the selectivities in microwave-assisted organic reactions only appeared during recent decades [7–9]. Chemists explored the application of microwave in chemoselective and stereoselective organic reactions and the influence of the microwave effect on these selectivities. This chapter focuses on the influence of microwave irradiation on the selectivities in organic reactions. Only examples with different and compared results on the selectivities under conventional heating conditions and microwave irradiation conditions are included. This chapter excludes the microwave-accelerated organic reactions without any changes in selectivities under conventional heating conditions and microwave irradiation conditions.

## 6.2 CHEMOSELECTIVITY UNDER MICROWAVE IRRADIATION CONDITIONS

In traditional organic synthetic chemistry, organic chemists generally regulate or change selectivity through changing reaction temperature, solvent, catalyst, and/or with photo irradiation instead of heating. Since the microwave has been widely applied in organic synthesis, and especially shows high efficiency in accelerating organic reactions, which are slow under conventional heating conditions, organic chemists have paid much attention to the influence of microwave irradiation on the chemoselectivity in organic reactions and found some interesting examples.

Patonay and co-workers found that 2-hydroxychalcone produced flavanone in 3.0% yield, the Michael adduct of flavanone to chalcone in 28% yield and its stereoisomer in 13% yield, and double Michael addition and subsequent aldol condensation product in 11% under conventional heating, while the reaction gave rise to the Michael adduct of flavanone to chalcone in 56% yield and its stereoisomer in 17% yield only under microwave irradiation when they conducted the reaction under heating and microwave irradiation conditions (Scheme 6.1) [10].

The reaction process was proposed as a tandem mechanism of the Michael addition of flavanone with two molecules of 2-hydroxychalcone and subsequent aldol condensation as following (Scheme 6.2).

Raghunathan and co-workers documented that the tandem Knoeveagel and heteroatom-Diels–Alder reaction of 4-hydroxycoumarin and 2-(3-methyl-2-butenoxy) benzaldehyde generated (±)-coumarin derivative and (±)-chromone derivative in 57% yield with a ratio of 68:32 under refluxing in ethanol for 4 hours under conventional

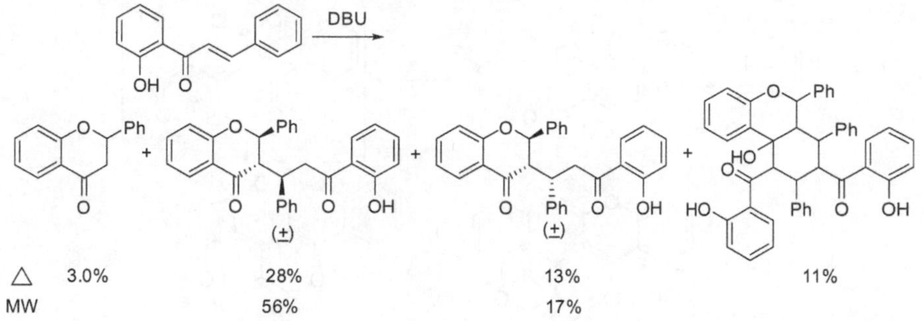

**SCHEME 6.1** Reactions of 2-hydroxychalcone in the presence of DBU under heating and microwave irradiation.

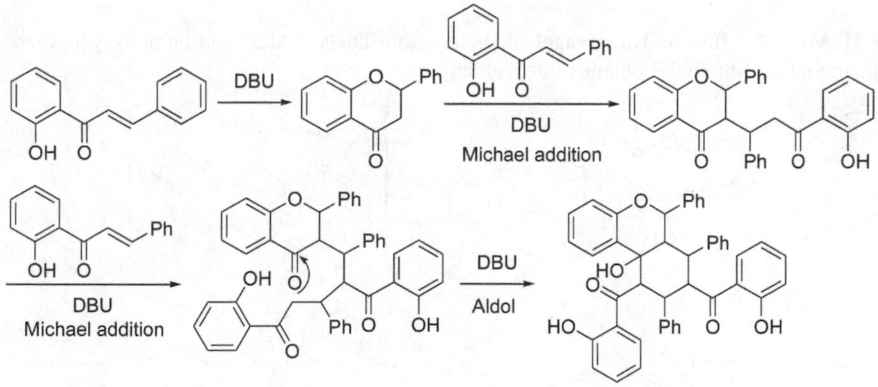

**SCHEME 6.2** Proposed tandem reaction process on the reaction of 2-hydroxychalcone in the presence of DBU.

heating, while the efficiency, selectivity, and yield were obviously improved under microwave irradiation, 82% yield with a ratio of 93:7 is achieved under microwave irradiation (800 W) for 15 seconds. The similar results were also observed in the reactions with benzocoumarin and/or naphthalenecarbaldehyde instead of coumarin and benzaldehyde (Scheme 6.3) [11].

When they investigated the tandem Knoeveagel and heteroatom-Diels–Alder reaction of 4-hydroxycoumarin and citral, they observed that the reaction produced a coumarin derivative and substituted coumarin in a ratio of 58:42 in 55% yield through the tandem Knoeveagel and heteroatom-Diels–Alder reaction and the tandem Knoeveagel and ene reaction, respectively, heating under reflux for 4 hours. However, the two products were obtained in 81% yield with a ratio of 88:12 under microwave irradiation (800 W) for 10 seconds. In the tandem Knoeveagel-ene reaction, a substituted dihydrobenzopyranedione derivative was generated first and tautomerized into substituted coumarin as the main product in the reaction mixture. Similar results were observed as well when coumarin was replaced with benzocoumarin (Scheme 6.4) [11].

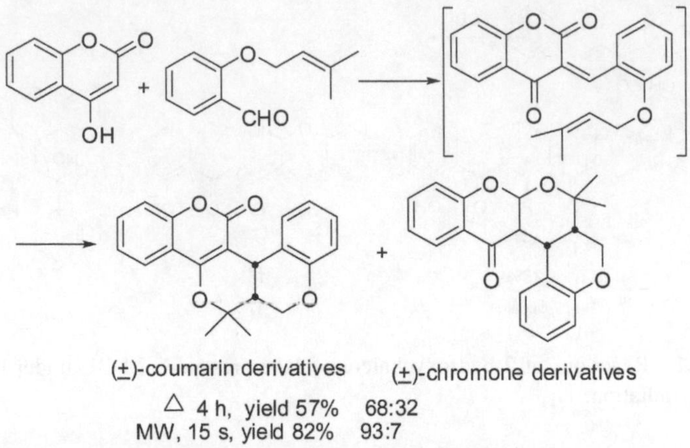

(±)-coumarin derivatives    (±)-chromone derivatives

△ 4 h, yield 57%    68:32
MW, 15 s, yield 82%    93:7

**SCHEME 6.3**   Tandem Knoeveagel and heteroatom-Diels–Alder reaction of 4-hydroxycoumarin and 2-(3-methyl-2-butenoxy)benzaldehyde.

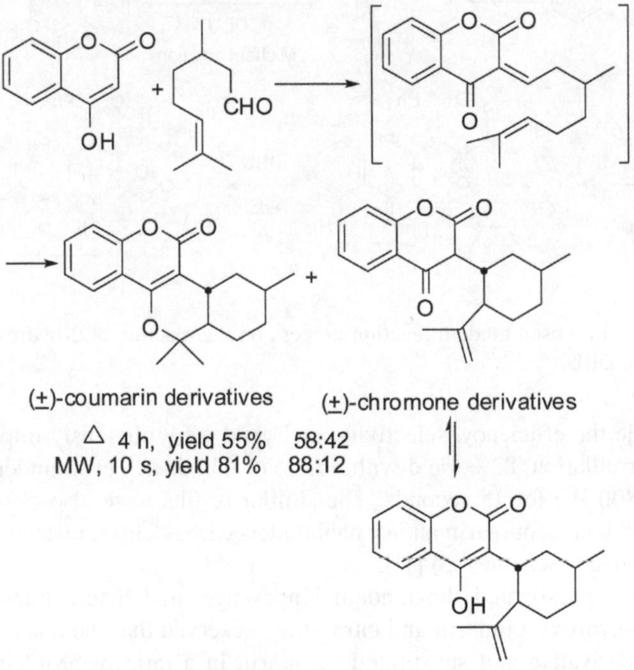

(±)-coumarin derivatives    (±)-chromone derivatives

△ 4 h, yield 55%    58:42
MW, 10 s, yield 81%    88:12

**SCHEME 6.4**   Tandem Knoeveagel and heteroatom-Diels–Alder reaction of 4-hydroxycoumarin and citral.

Four years later, they studied the tandem Knoeveagel and heteroatom-Diels–Alder reactions of 4-hydroxycoumarin and substituted 5-(3-methyl-2-butenylthio) pyrazole-4-carbaldehydes. The reactions produced coumarin derivatives and chromone derivatives in 54 ~ 68% yields with ratios of 51:49 ~ 59:41 under refluxing in ethanol for 4.5 ~ 10 hours under conventional heating, while the reactions gave rise to the corresponding products in 72 ~ 85% yields with ratios of 70:30 ~ 76:24 under microwave irradiation (800 W) for 1 ~ 4.5 minutes. When they absorbed the starting materials on the Montmorillonite K-10 and irradiated with microwave (800 W) for 23 ~ 65 seconds under solvent-free conditions, the corresponding products were obtained in 82 ~ 91% yields with ratios of 89:11 ~ 95:5. The selectivities, yields, and efficiencies were improved obviously (Scheme 6.5) [12].

Dumas and co-workers conducted the asymmetric Michael addition of α-benzyloxycyclohexanone imine with methyl acrylate followed by acidic hydrolysis. They found that the reaction produced the desired Michael adduct in 61% and 85% yields with 97%ee, respectively, when the reaction was conducted at 30°C and 40°C for 7 and 5 days. When the reaction was conducted under microwave irradiation at 100°C for 15 minutes, the desired product was obtained in 85% yield with 96% ee. The reaction rate was improved significantly under microwave irradiation with slight loss of the enantioselectivity. However, the yield decreased to 48% and the benzyl rearrangement occurred to produce α-benzyl-α-hydroxycyclohexanone in 24% yield when the reaction was conducted at 200°C under microwave irradiation for 15 minutes. The results reveal that microwave irradiation can change the chemoselectivity in the Michael addition, especially at a higher temperature (Scheme 6.6) [13].

In 2004, a French group mentioned that the reaction of chalcone and nitromethane yielded different products under conventional heating and microwave irradiation

(±)-coumarin derivatives    (±)-chromone derivatives

R = Me, Ph; R' = H, MeO

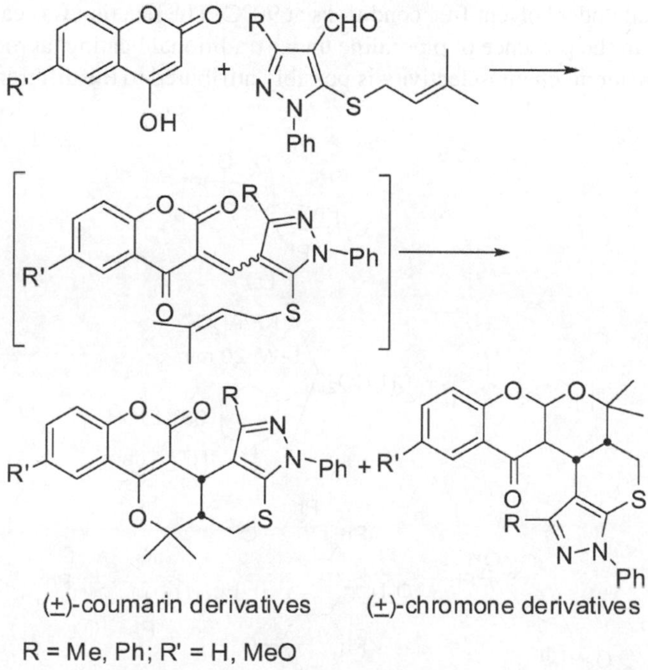

**SCHEME 6.5** Tandem Knoeveagel and heteroatom-Diels–Alder reactions of 4-hydroxycoumarin and substituted 5-(3-methyl-2-butenylthio)pyrazole-4-carbaldehydes.

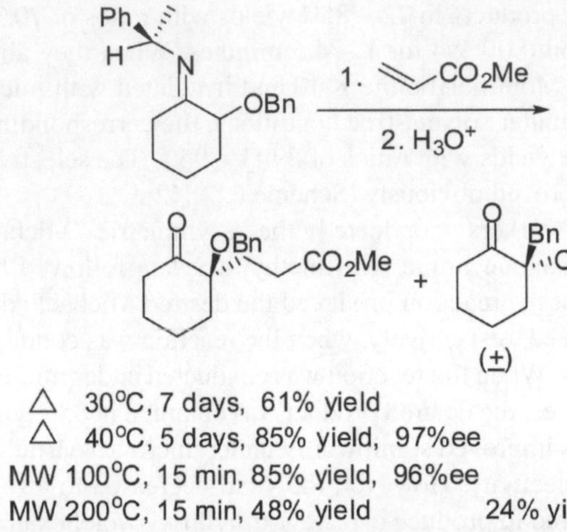

**SCHEME 6.6** Asymmetric Michael addition of α-benzyloxycyclohexanone imine with methyl acrylate.

(Scheme 6.7) [14]. However, besides conventional heating and microwave irradiation, the reaction was actually conducted under different conditions. Under microwave irradiation conditions, the reactants were absorbed on KF/Al₂O₃ and the reaction was conducted under solvent-free conditions at 90°C. The reaction was carried out in 95% ethanol in the presence of piperidine under traditional heating (at room temperature). The different chemoselectivity is possibly attributed to the different catalysts,

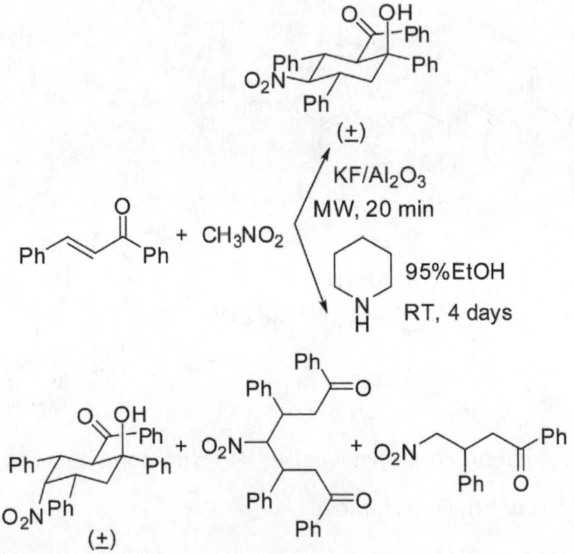

**SCHEME 6.7** Reaction of chalcone and nitromethane.

KF/Al$_2$O$_3$ and piperidine, and reaction temperatures. At room temperature, Michael and double Michael addition products were obtained. Double Michael adduct did not undergo further cyclization possibly due to low reaction temperature. Additional, different stereoisomeric cyclic products generated under conventional heating and microwave irradiation.

One year later, the same group reported another interesting example. The reaction of alkyl cyanoacetates and *N*-arylmethylideneisopropylamines gave rise to a mixture of cyclohexene derivatives and alkyl 2-cyanocinnamates under conventional heating, such as room temperature for 8 days or 80°C for 5 days. However, the reaction generated alkyl 2-cyanocinnamates only under microwave irradiation (Scheme 6.8) [15]. Generally, microwave irradiation accelerates organic reactions. This is a rare example in which the cyclohexene derivatives did not generate under microwave irradiation.

## 6.3 REGIOSELECTIVITY UNDER MICROWAVE IRRADIATION CONDITIONS

### 6.3.1 REGIOSELECTIVITY IN SUBSTITUTIONS UNDER MICROWAVE IRRADIATION CONDITIONS

Kappe and co-workers found that the regioselectivity depended upon the reaction temperature when they conducted the NBS bromination of dihydroquinolinone. 3-Bromo dihydroquinolinone was obtained as the main product at lower temperatures. However, the 6-bromo derivative increased when the reaction temperature was increased. It was obtained as the main product under microwave irradiation, actually higher temperature (100°C) (Scheme 6.9) [16] (Table 6.1).

In the molybdenum-catalyzed asymmetric allylic alkylation, compared with the reactions conducted under the classical heating, microwave irradiation can not

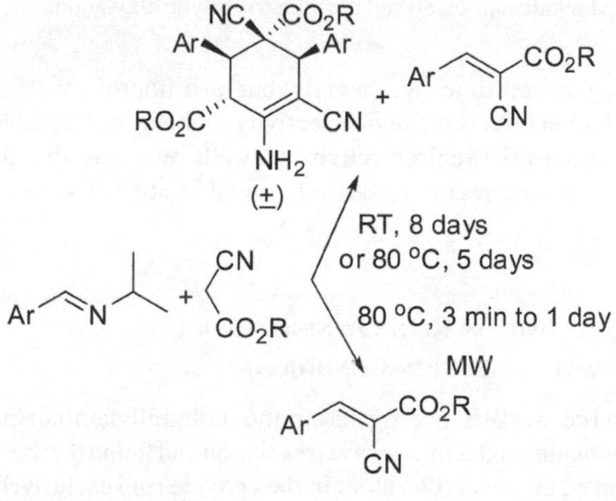

**SCHEME 6.8**  Reaction of alkyl cyanoacetates and *N*-arylmethylideneisopropylamines.

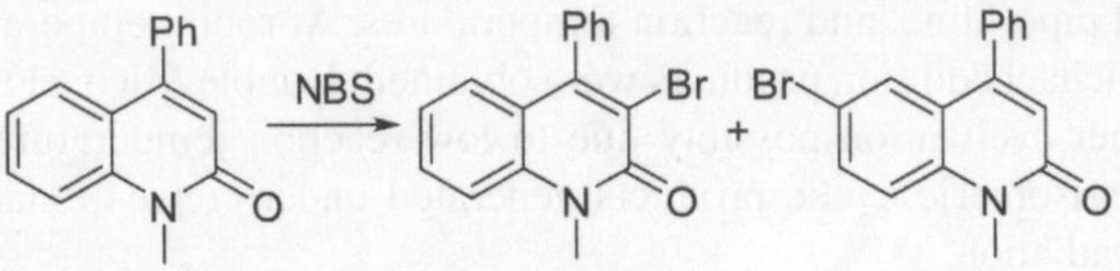

**SCHEME 6.9**   NBS bromination of dihydroquinolinone.

**TABLE 6.1**
**Effect of Classical Heating and Microwave Irradiation on the Regioselectivity of the Bromination of Quinolin-2(1H)-Ones**

| Entry | Time (h) | Solvent | Temp (°C)/Conditions | Ratio (3-Br:6-Br) |
|-------|----------|---------|----------------------|-------------------|
| 1 | 17 | DMF | 0 (Δ) | 92:8 |
| 2 | 4.5 | DMF | 25 (Δ) | 83:17 |
| 3 | 3 | DMF | 50 (Δ) | 55:45 |
| 4 | 0.3 | MeCN | 100 (MWI) | 5:95 |

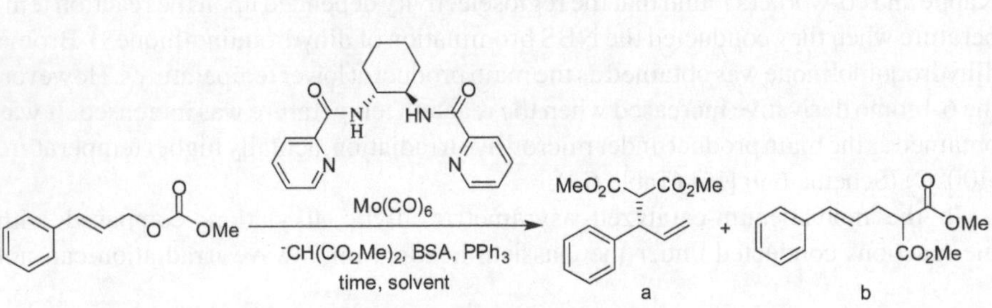

**SCHEME 6.10**   Molybdenum-catalyzed asymmetric allylic alkylation.

only shorten the reaction time significantly, but also improve yields. The power of microwave irradiation affects the regioselectivity (Scheme 6.10) (Table 6.2) [17]. The temperature influences the regioselectivity as well, revealing that the influence of microwave power on the regioselectivity is possibly attributed to the temperature difference.

### 6.3.2   REGIOSELECTIVITY IN RING OPENINGS UNDER MICROWAVE IRRADIATION CONDITIONS

In 2002, Pyne and co-workers investigated ammonia and allylamine ring opening reactions of 2-vinylepoxides under microwave irradiation and found that both ammonia and allyl amine attacked allylic carbon atom in the epoxide ring exclusively to afford ring opening vicinal amino alcohols (Scheme 6.11) [18].

**TABLE 6.2**
**Molybdenum-Catalyzed Asymmetric Allylic Alkylation Under the Classical Heating and Microwave Irradiation Conditions**

| Entry | Temp/Power | Time | Solvent | Yield (%) | Ee (%) | a:b |
|-------|-----------|------|---------|-----------|--------|------|
| 1 | 60 W | 9 min | THF | 86 | 98 | 18:1 |
| 2 | 90 W | 6 min | THF | 86 | 98 | 17:1 |
| 3 | 120 W | 5 min | THF | 86 | 98 | 19:1 |
| 4 | 250 W | 5 min | THF | 87 | 98 | 19:1 |
| 5 | 500 W | 4 min | THF | 85 | 95 | 19:1 |
| 6 | 120 W | 6 min | DME | 92 | 98 | 22:1 |
| 7 | 130 W | 7 min | dioxane | 94 | 98 | 28:1 |
| 8 | 22°C | 20 day | THF | <1 | ND | ND |
| 9 | 80°C | 2 day | THF | 11 | ND | 17:1 |
| 10 | 165°C | 6 min | THF | 59 | 98 | 11:1 |
| 11 | 180°C | 6 min | THF | 70 | 98 | 16:1 |
| 12 | 160°C | 7 min | dioxane | 67 | 98 | 28:1 |

The Fries rearrangement is a general method to prepare *ortho-* or *para*-hydroxy-phenyl ketones. The Fries rearrangement of phenyl acetate shows different regioselectivity under the catalysis of either aluminum trichloride or zinc powder under conventional heating or microwave irradiation, even reverse the regioselectivity under the catalysis of aluminum trichloride (Scheme 6.12) [19, 20].

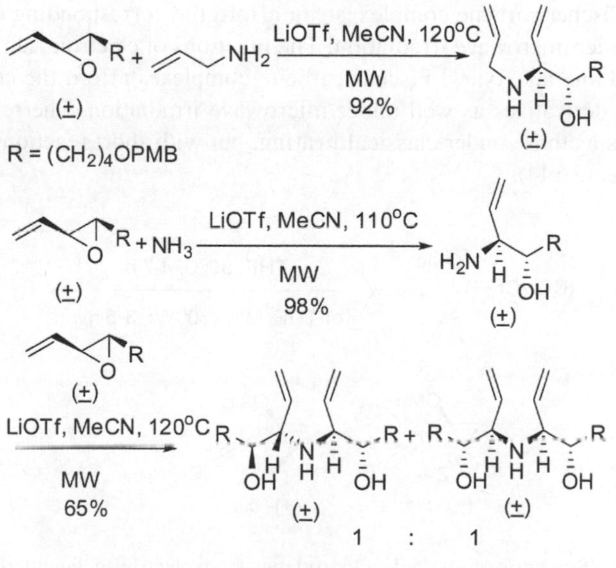

**SCHEME 6.11** Ammonia and allylamine ring opening reactions of 2-vinylepoxides.

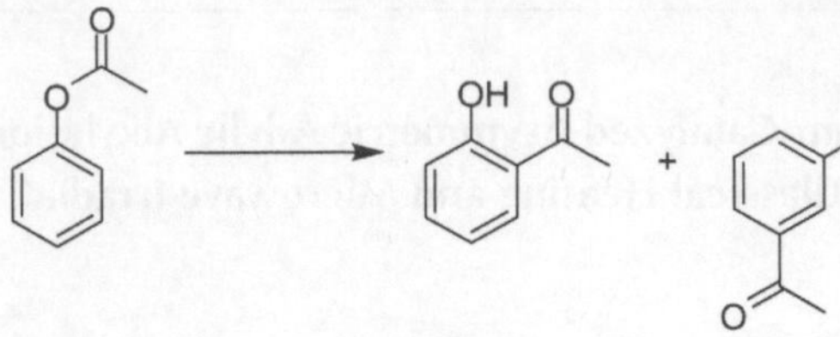

|            |         |       | Yield (%) |     |   |     |
|------------|---------|-------|-----------|-----|---|-----|
| AlCl$_3$   | heating | 5 min | 69        | 0   | : | 100 |
| AlCl$_3$   | MW      | 3 min | 75        | 100 | : | 0   |
| Zn         | heating | 6 h   | 43        | 58  | : | 42  |
| Zn         | MW      | 3 h   | 70        | 73  | : | 27  |

**SCHEME 6.12**   Fries rearrangement of phenyl acetate under the catalysis of aluminum trichloride or zinc powder.

## 6.4   DIASTEREOSELECTIVITY IN CYCLOADDITIONS AND CYCLIZATIONS UNDER MICROWAVE IRRADIATION CONDITIONS

### 6.4.1   DIASTEREOSELECTIVITY IN THE SYNTHESIS OF ALKYNYLCYCLOPROPANES FROM ELECTRON-DEFICIENT ALKENES AND ALKOXYALKYNYL FISCHER CARBENE COMPLEXES

In the synthesis of alkynylcyclopropanes from electron-deficient alkenes and alkoxyalkynyl Fischer carbene complexes, the reactions show slightly lower yields and diastereoselectivities with shorter reaction times under microwave irradiation than under classical heating (Scheme 6.13) (Table 6.3) [21].

The reactions of electron-deficient alkenes with alkoxyalkenyl, alkoxyaryl, and alkoxyalkyl Fischer carbene complexes can afford the corresponding cyclopropane derivatives under microwave irradiation. The reactions of electron-rich alkenes with alkoxyalkenyl and alkoxyaryl Fischer carbene complexes afford the corresponding cyclopropane derivatives as well under microwave irradiation. The reactions show similar results as those under classical heating, but with short reaction time in each of cases (Scheme 6.14).

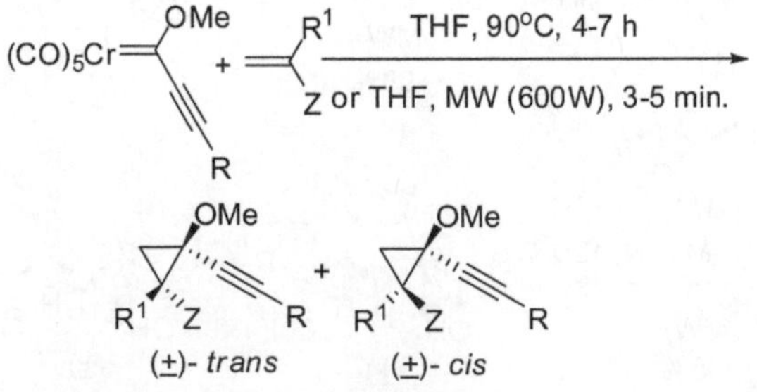

**SCHEME 6.13**   Synthesis of alkynylcyclopropanes from electron-deficient alkenes and alkoxyalkynyl Fischer carbene complexes.

**TABLE 6.3**

**Synthesis of alkynylcyclopropanes from electron-deficient alkenes and alkoxyalkynyl Fischer carbene complexes under classical heating and microwave irradiation conditions**

| Entry | Condition | R | R¹ | Z | Trans:Cis | Yield (%) |
|-------|-----------|-----|-----|--------|-----------|-----------|
| 1 | Δ | Ph | H | CO$_2$Me | 94:6 | 48 |
| 2 | MW | Ph | H | CO$_2$Me | 90:10 | 51 |
| 3 | Δ | Ph | Me | CO$_2$Me | 67:33 | 29 |
| 4 | MW | Ph | Me | CO$_2$Me | 80:20 | 25 |
| 5 | Δ | (Z)-Ph CH = CH | H | CN | >95:<5 | 42 |
| 6 | MW | (Z)-Ph CH = CH | H | CN | 80:20 | 42 |
| 7 | MW | Bu$^t$ | H | CO$_2$Me | 92:8 | 49 |

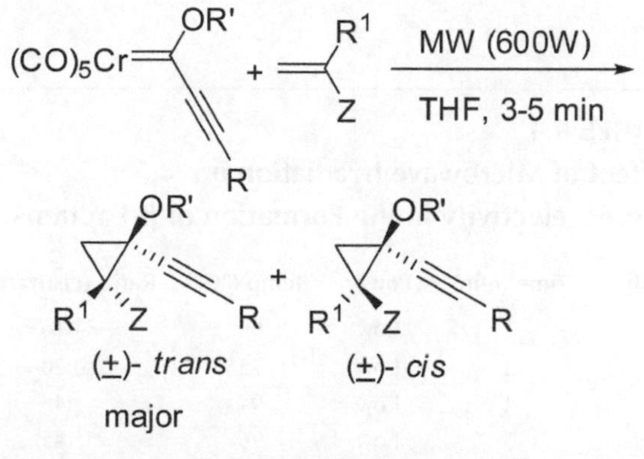

**SCHEME 6.14**  Synthesis of cyclopropane derivatives from alkenes and Fischer carbene complexes.

## 6.4.2  DIASTEREOSELECTIVITY IN THE SYNTHESIS OF β-LACTAMS FROM KETENES AND IMINES

Reactions of acyl chlorides and imines in the presence of an organic base are one of the most important methods for the synthesis of β-lactams [22]. Numerous investigation results indicate that the reaction conditions impact the diastereoselectivity in the reactions. In 1995, Manhas and co-workers studied the influence of microwave irradiation on the diastereoselectivity in the synthesis of β-lactams from benzyloxyacetyl chloride and N-benzylidene methylamine in the presence of N-methylmorpholine (NMM) in chlorobenzene [23]. They found that the microwave power not only speeded up reaction time, but also affected the stereoselectivity in the reaction (Scheme 6.15) (Table 6.4).

This is an early example of studies on the influence the of microwave effect on the stereoselectivity in organic reactions. However, the reaction was conducted in a domestic microwave oven without control and determination of reaction temperature. Thus, it is not clear that the different stereoselectivity was caused by the microwave effect or by different reaction temperature.

In 2000, the same research group reported the influence of the microwave effect on the diastereoselectivity in the synthesis of β-lactams from 3-methyl-2-butenoyl chloride and *N*-benzylidene methylamine in the presence of triethylamine in a domestic microwave oven again (Scheme 6.16) [24]. In the studies, the influence of the reaction temperature was considered. After microwave irradiation, the reaction

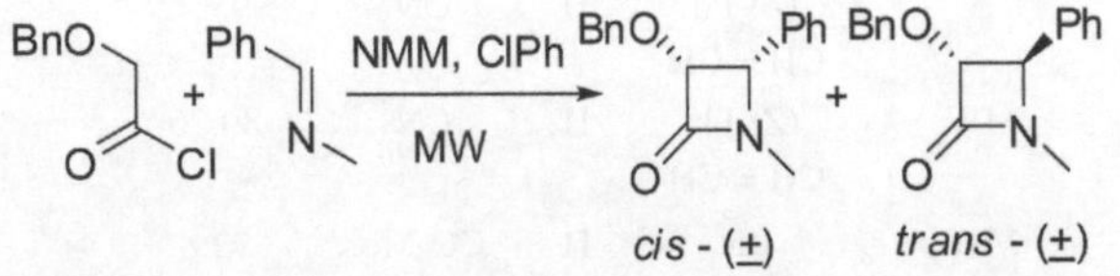

**SCHEME 6.15** Synthesis of β-lactams from benzyloxyacetyl chloride and *N*-benzylidene methylamine.

**TABLE 6.4**
**Effect of Microwave Irradiation on Stereoselectivity in the Formation of β-Lactams**

| Entry | Time (min) | Power | Temp (°C) | Ratio (cis:trans) |
|-------|-----------|-------|-----------|-------------------|
| 1 | 1 | Low | 69 | 84:16 |
| 2 | 2 | Low | 75 | 80:20 |
| 3 | 4 | Low | 94 | 55:45 |
| 4 | 5 | Low | 96 | 55:45 |
| 5 | 4 | High | 112 | 45:55 |

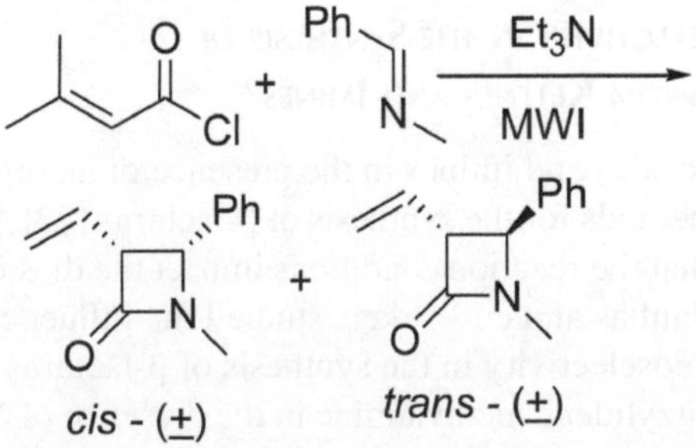

**SCHEME 6.16** Synthesis of β-lactams from 3-methyl-2-butenoyl chloride and *N*-benzylidene methylamine.

mixture was taken out of the oven and determined intermediately. Although the results were compared with those obtained under classical heating, the reactions were not conducted at the exact same temperature. No rational explanation was provided for the influence on the stereoselectivity.

Reactions of $\alpha$-diazomethyl ketones and imines are an alternative method for the synthesis of $\beta$-lactams [25, 26]. $\alpha$-Diazomethyl ketones undergo the Wolff rearrangement to generate ketenes under photo irradiation. The ketenes react with imines to afford $\beta$-lactams. The reaction is named the Staudinger reaction, Staudinger cycloaddition, or Staudinger ketene-imine cycloaddition. Podlech and co-workers realized the stereospecific synthesis of *trans*-$\beta$-lactams from reactions of imines and $\alpha$-diazomethyl ketones, derived from N-Cbz protected amino acids and diazomethane, under microwave irradiation (Scheme 6.17) [27]. Thus, they thought that the microwave-assisted reaction of imines and $\alpha$-diazomethyl ketones afforded *trans*-$\beta$-lactams exclusively.

Our working group conducted the reactions of cyclic imines and $\alpha$-diazomethyl ketones to afford bicyclic *trans*-$\beta$-lactams exclusively. The results indicate that the isomerization of imines is a reasonable explanation of the formation of *trans*-$\beta$-lactams from linear imines (Scheme 6.18) [28]. Our results provide experimental evidence to propose the reaction mechanism and diastereoselectivity.

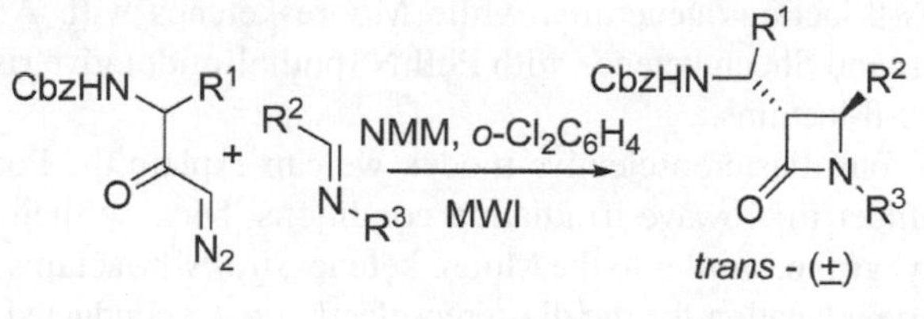

**SCHEME 6.17** Synthesis of *trans*-$\beta$-lactams from imines and $\alpha$-diazomethyl ketones.

R = Ph, *p*-MeOPh, *p*-MePh, *p*-ClPh      27–40%

X = H, Me      55–71%

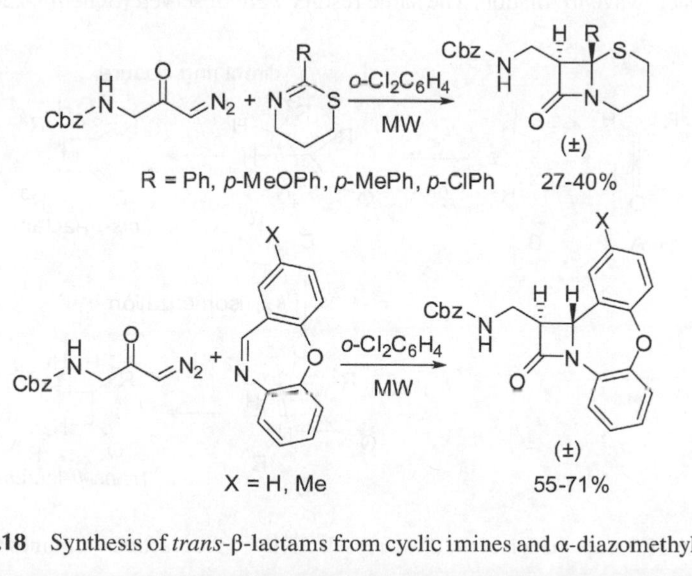

**SCHEME 6.18** Synthesis of *trans*-$\beta$-lactams from cyclic imines and $\alpha$-diazomethyl ketones.

To make clear whether microwave irradiation affects the diastereoselectivity in the synthesis of β-lactams, our group first investigated the origin of controlling diastereoselectivity in the Staudinger cycloaddition and provided a model to explain the diastereoselectivity in the formation of β-lactams from ketenes and imines (Scheme 6.19) [29]. In the Staudinger cycloaddition, imines **B** attack ketenes **A** from their *exo*-side to generate zwitterionic intermediates **C**, which directly undergo conrotatory ring closure to produce *cis*-β-lactams, or isomerize their iminium moiety to yield less steric and more thermodynamic stable intermediates **D**. The intermediates **D** undergo the ring closure to give rise to *trans*-β-lactams. The ratio of *cis*- and *trans*-β-lactams (diastereoselectivity) is controlled by the competition between the rate of the direct ring closure and the rate of the isomerization of the iminium moiety. When the rate of the direct ring closure is larger than the rate of the isomerization, the reactions give *cis*-β-lactams as the main products, while when the rate of the direct ring closure is smaller than the rate of the isomerization, *trans*-β-lactams are obtained as the predominant products. When the rate of the direct ring closure is close to the rate of the isomerization, a mixture of *cis*- and *trans*-β-lactams are generated.

The influence of substituents on the diastereoselectivity and relative rates of the direct ring closure of different ketenes were investigated. The results indicate that the relative rate order of the direct ring closure of different ketenes is as follows: ArO, RO > PhthN > ArS, and RS > Ar, R. Thus, Bose–Evans ketenes with ArO and RO produce *cis*-β-lactams generally, while Moore ketenes with Ar and R prefer *trans*-β-lactams, and Sheen ketenes with PhthN (phthalimido) give rise to a mixture of *cis*- and *trans*-β-lactams.

According to our diastereoselective model, we can explain the Podlech stereoselective results under microwave irradiation conditions, because their ketenes are of the 1-aminoalkyl group, similar to the Moore ketenes, *trans*-β-lactams were obtained.

To verify our explanation for the diastereoselectivity, we conducted the reaction of *S*-phenyl 2-diazoethanethioate and *N*-benzylidene *tert*-butylamine under classical heating and microwave irradiation. The same results were observed (Scheme 6.20) [29]

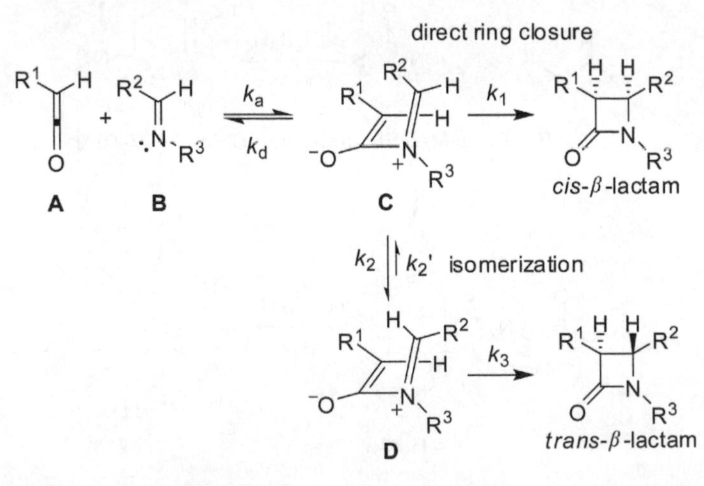

**SCHEME 6.19** Explanation of the diastereoselectivity in the reactions of imines and ketenes.

To further investigate the influence of the microwave effect on the diastereoselectivity in the reactions of imines and $S$-phenyl 2-diazoethanethioate, we conducted the reactions at 80°C under conventional heating and microwave irradiation. The results indicate that only slight differences were observed under carefully controlled temperature conditions (Scheme 6.21) (Table 6.5) [30].

After careful investigation of the influences of reaction conditions, especially temperature, on the diastereoselectivity [31, 32], how the microwave effect alters the diastereoselectivity in the different Staudinger cycloadditions was determined by

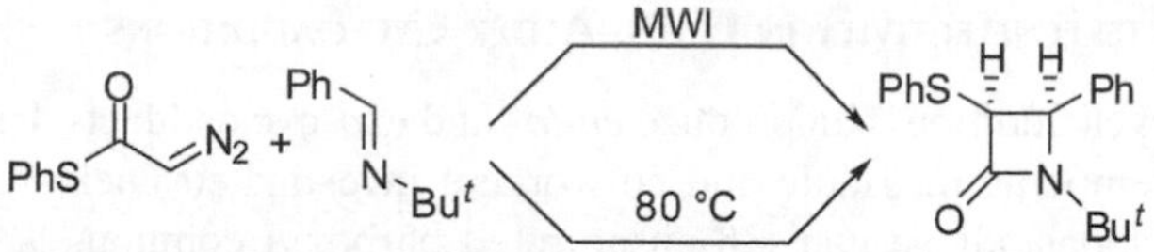

**SCHEME 6.20** Synthesis of *cis*-β-lactam from bulky linear imine and $S$-phenyl 2-diazoethanethioate.

**a: R=MeO, b: R=Me, c: R=H, d: R=Cl, e: R=CF₃, f: R=NO₂**

a: R=MeO, b: R=Me, c: R=H, d: R=Cl, e: $R=CF_3$, f: $R=NO_2$

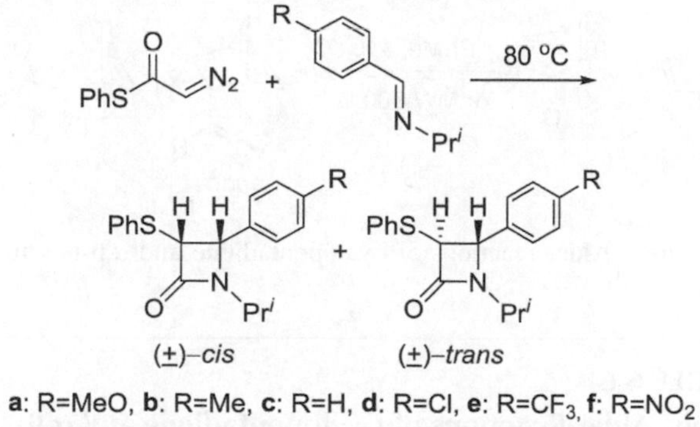

**SCHEME 6.21** Reaction of imines and $S$-phenyl 2-diazoethanethioate under the classical heating and microwave irradiation conditions.

---

## TABLE 6.5
### Reactions of the Formation of β-Lactams from and Imines $S$-Phenyl 2-Diazoethanethioate Under the Classical Heating and Microwave Irradiation Conditions

| Entry | Solvent | Reaction Condition | a | b | c | d | e | f |
|-------|---------|--------------------|------|------|-------|-------|-------|-------|
| | | | | | | *cis:trans* | | |
| 1 | Tol | Δ | 4:96 | 7:93 | 12:88 | 17:83 | 42:58 | 77:23 |
| 2 | Tol | MW | 3:97 | 5:95 | 10:90 | 19:81 | 44:56 | 77:23 |
| 3 | MeCN | Δ | 3:97 | 7:93 | 11:89 | 15:85 | 40:60 | 71:29 |
| 4 | MeCN | MW | 4:98 | 6:94 | 6:94 | 16:84 | 37:63 | 70:30 |

comparison with reactions under conventional heating [33]. The results indicate that different diastereoselectivity under conventional heating and microwave irradiation is attributed to the different reaction temperature, rather than the specific microwave athermal effect [33]. In the Staudinger cycloaddition, polar starting materials imines and ketenes or their precursors (acyl chlorides and α-diazomethyl ketones) are good microwave absorbed species and selectively heated. They generally locate at higher temperatures than determined non-polar or low polar bulk solvent. That is, they locate in so-called "hotspots".

### 6.4.3  DIASTEREOSELECTIVITY IN DIELS–ALDER CYCLOADDITIONS

Diels–Alder cycloadditions can produce *endo-* and *exo*-cycloadducts depending upon the reaction temperature. Haufe and co-workers investigated Diels–Alder cycloadditions of cyclopentadiene and α,β-unsaturated carbonyl compounds (ketones and esters). They found that the *endo-* and *exo* selectivity depended upon the presence or absence of the fluoro substituent in reactants. Only slight differences were observed under conventional heating and microwave irradiation (Scheme 6.22) (Table 6.6) [34].

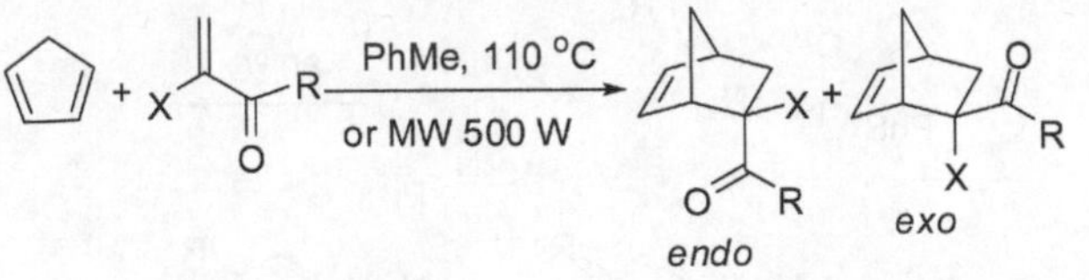

**SCHEME 6.22**   Diels–Alder reactions of cyclopentadiene and α,β-unsaturated ketones or carboxylates.

### TABLE 6.6
### Diels–Alder Reactions of Cyclopentadiene and α,β-Unsaturated Ketones or Carboxylates Under Classical Heating and Microwave Irradiation Conditions

| Entry | X | R | Condition | Time | Yield (%) | Endo/exo |
|-------|---|---|-----------|------|-----------|----------|
| 1 | F | OBn | Δ | 16 h | 73 | 31:69 |
| 2 | F | OBn | MW | 66 min | 54 | 30:70 |
| 3 | F | OBn | MW | 72 min | 44 | 29:71 |
| 4 | H | OBn | Δ | 1.5 h | 91 | 78:22 |
| 5 | H | OBn | MW | 36 min | 88 | 76:24 |
| 6 | H | OBn | MW | 12 min | 91 | 74:26 |
| 7 | F | $C_5H_{11}$ | Δ | 0.9 h | 59 | 27:73 |
| 8 | F | $C_5H_{11}$ | MW | 46 min | 44 | 22:78 |
| 9 | F | $C_5H_{11}$ | MW | 12 min | 48 | 22:78 |
| 10 | H | $C_5H_{11}$ | Δ | 0.5 h | 54 | 75:25 |
| 11 | H | $C_5H_{11}$ | MW | 12 min | 61 | 78:22 |
| 12 | H | $C_5H_{11}$ | MWI | 10 min | 64 | 79:21 |

Jones and his co-workers found that 2-cyclopenten-1-one reacted with (S)-9-(1-methoxyethyl)anthracene to produce different diastereomeric product distribution under traditional heating and microwave irradiation when they investigated Diels–Alder cycloaddition and transformations of 2-cyclopenten-1-one with a chiral anthracene template (Scheme 6.23) [35] (Table 6.7).

### 6.4.4 DIASTEREOSELECTIVITY IN OTHER CYCLIZATIONS

One-pot tandem reactions of N-acylglycines, aromatic aldehydes, and ammonium N-aryldithiocarbamates produced 2-thioxo-1,3-thiazinan-4-ones in the presence of acetic anhydride and sodium acetate with cis:trans ratios of >59:<41 under conventional heating, with cis:trans ratios of >97:<3 under microwave irradiation (Scheme 6.24) [36].

The same group investigated the one-pot cyclization of glycine and acetic anhydride and subsequent addition to imines generated from aromatic aldehydes with 2-amino-1,3,4-oxadiazoles and 2-amino-1,3,4-thiadiazoles, affording 6,7-dihydro-5H-[1,3,4]oxadiazolo[3,2-a]pyrimidin-5-ones and 6,7-dihydro-5H-[1,3,4]thiadiazolo[3,2-a]pyrimidin-5-ones with cis:trans ratios of >70:<30 under conventional heating, with cis:trans ratios of >95:<5 under microwave irradiation (Scheme 6.25) [37].

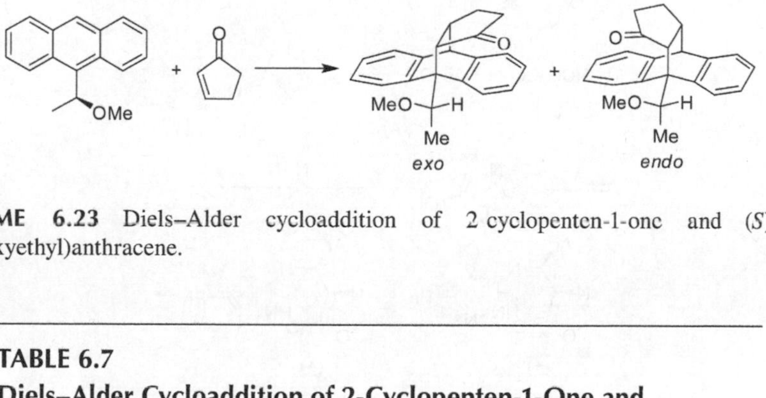

**SCHEME 6.23** Diels–Alder cycloaddition of 2-cyclopenten-1-one and (S)-9-(1-methoxyethyl)anthracene.

### TABLE 6.7
### Diels–Alder Cycloaddition of 2-Cyclopenten-1-One and (S)-9-(1-Methoxyethyl)Anthracene

| | | | Product Distribution (%) | |
|---|---|---|---|---|
| Entry | Reaction Conditions | Conversion (%) | Exo | endo |
| 1 | Xylenes, 160°C, 20 bar, 7 d | 65 | 80 | 20 |
| 2 | DMF, MW, 160°C, 12 h | 15 | 85 | 15 |
| 3 | o-DCB, 140°C, 3 d | 10 | 95 | 5 |
| 4 | o-DCB, 140°C, 4 d | 24 | 85 | 15 |
| 5 | DMF, MW, 140°C, 4 h | 5 | 90 | 10 |
| 6 | DMF, MW, 140°C, 8 h | 23 | 85 | 15 |

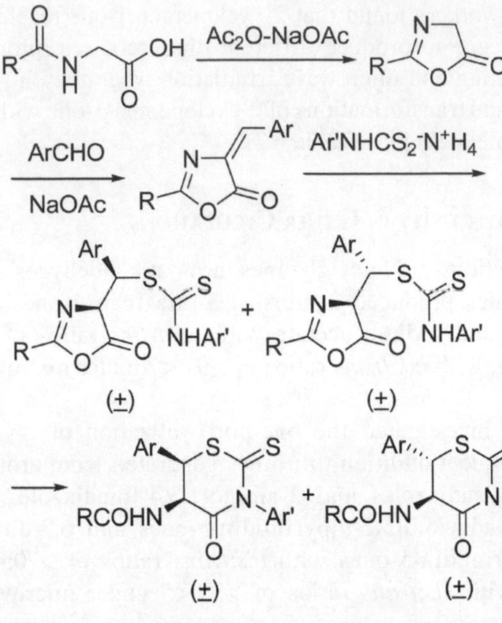

**SCHEME 6.24** One-pot tandem reactions of *N*-acylglycines, aromatic aldehydes, and ammonium *N*-aryldithiocarbamates.

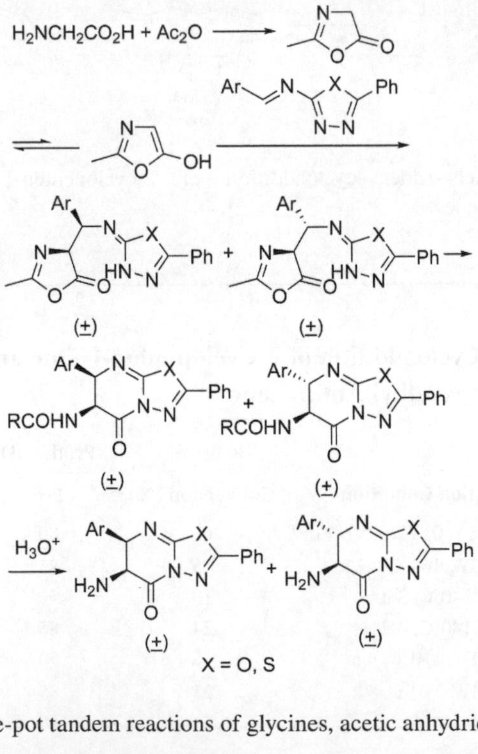

**SCHEME 6.25** One-pot tandem reactions of glycines, acetic anhydride, and imines.

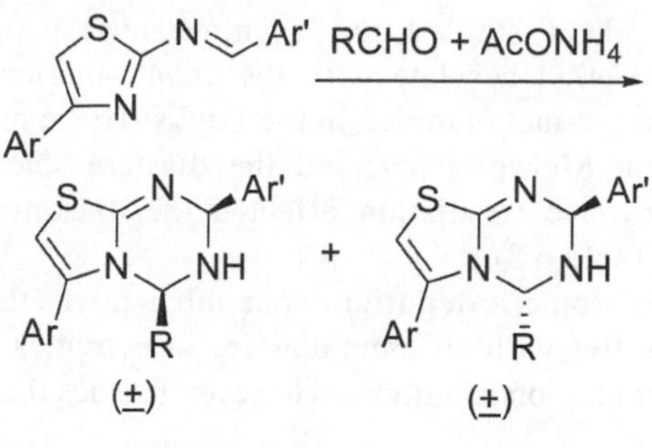

**SCHEME 6.26** One-pot tandem reactions of imines, aldoses, and ammonium acetate.

The one-pot cyclizations of imines, aldoses, and ammonium acetate gave rise to 3,4-dihydro-2*H*-thiazolo[3,2-*a*][1,3,5]triazine derivatives with *cis:trans* ratios of 57 ~ 60:43 ~ 40 under conventional heating, with *cis:trans* ratios of >97:<3 under microwave irradiation (Scheme 6.26) [38].

## 6.4.5 DIASTEREOSELECTIVITY IN ASYMMETRIC REACTIONS

Numerous examples of microwave-assisted asymmetric organic reactions have been explored and documented. The results indicate that microwave irradiation generally accelerated the reaction rates without improvement of diastereoselectivity. Herein, we selected several examples with different diastereoselectivities under classical heating and microwave irradiation conditions.

Michael additions of amines to ethyl acrylate with enantiomerical 1,3-dioxolan-2-yl group afforded β-amino esters in low to moderate yields with good diastereoselectivities (Scheme 6.27) [39].

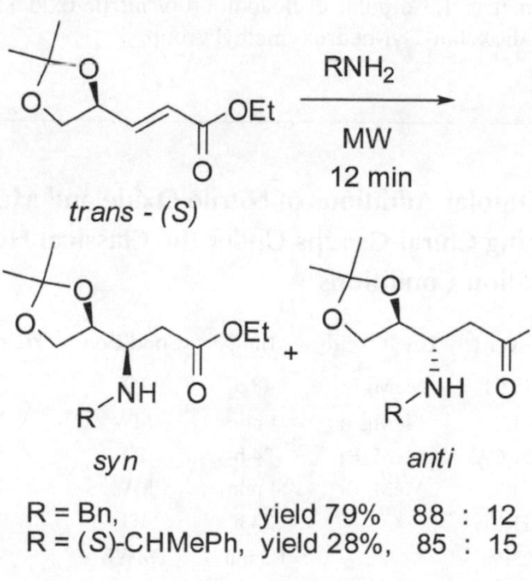

R = Bn,             yield 79%   88 : 12
R = (S)-CHMePh,  yield 28%,   85 : 15

**SCHEME 6.27** Michael additions of amines to ethyl acrylate with enantiomerical 1,3-dioxolan-2-yl group.

Fisera and co-workers studied the asymmetric 1,3-dipolar cycloaddition of nitrile oxides and ethyl acrylate with the enantiomerical 1,3-dioxolan-2-yl-hydroxymethyl group product obtained in the Baylis–Hillman reaction, and found that Grignard reagent MeMgBr inversed the diastereoselectivity in the reaction. However, microwave irradiation affected the diastereoselectivity slightly (Scheme 6.28) [40] (Table 6.8).

Cyclization of enantiomeric tellurium-containing allyl ether afforded bicyclic tetrahydrofuran derivative with different diastereoselectivities under classical heating and microwave irradiation conditions. However, besides the heating mode, other

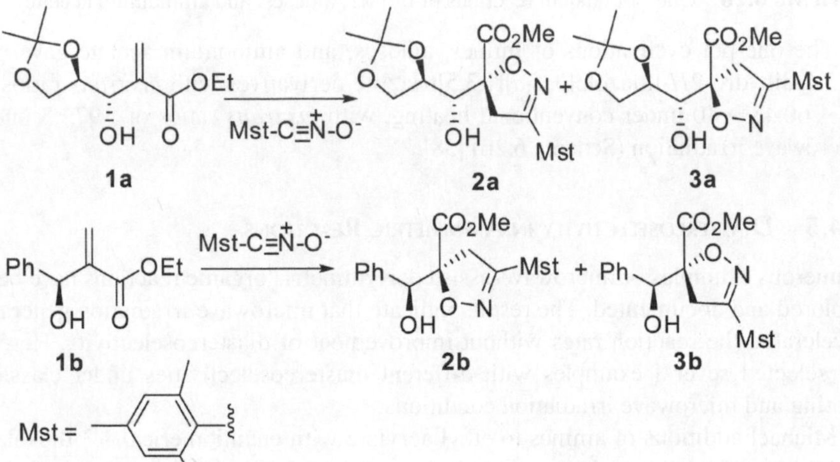

**SCHEME 6.28** Asymmetric 1,3-dipolar cycloaddition of nitrile oxides and ethyl acrylate with enantiomerical 1,3-dioxolan-2-yl-hydroxymethyl group.

**TABLE 6.8**
**Asymmetric 1,3-Dipolar Additions of Nitrile Oxide and Methyl Acrylates Containing Chiral Groups Under the Classical Heating and Microwave Irradiation Conditions**

| Entry | Compd | Solvent | Lewis Acid | Time | Condition | Yield (%) | dr |
|-------|-------|---------|-----------|------|-----------|-----------|-----|
| 1 | 1a | $CH_2Cl_2$ | MeMgBr | 48 h | RT | 50 | >95:<5 |
| 2 | 1a | ClPh | MeMgBr | 4 min | MW | 34 | 78:22 |
| 3 | 1b | $CH_2Cl_2$ | MeMgBr | 24 h | RT | 57 | 61:39 |
| 4 | 1b | ClPh | MeMgBr | 4 min | MW | 40 | 70:30 |
| 5 | 1b | $CH_2Cl_2$ | | 24 h | RT | 96 | 48:52 |
| 6 | 1b | ClPh | | 1.5 min | MWI | 99 | 43:57 |

reaction conditions are different. It is not easy to distinguish between the microwave effect and other reaction conditions in the important role they play in controlling the diastereoselectivity (Scheme 6.29) [41].

Alvarez-Builla and co-workers investigated the diastereoselectivity in the reaction of 2-aminothiophenol and methyl 2,3-epoxy-3-(4-methoxyphenyl)propanoate to synthesize 1,5-benzothiazepine derivatives. They found that the microwave accelerated the reaction efficiently, and microwave power affected the diastereoselectivity obviously. Solvent acetic acid can reverse the diastereoselectivity, possibly due to hydrogen bonding (Scheme 6.30) [42].

The microwave-assisted ring opening-cyclization of enantiopure 4,5-epoxycyclooctanones with ammonia and water afforded bicyclic products stereospecifically (Scheme 6.31) [43]. The nucleophiles ammonia and water first attack the carbonyl

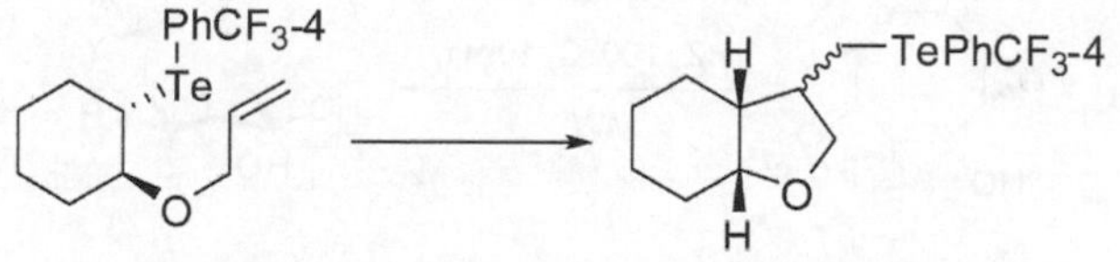

AIBN, AllSnBu$_3$, PhH, refluxing, yield 69%, cis/trans 2:1
MW, 250 °C, HOCH$_2$CH$_2$OH, yield 67%, cis/trans 1:1.1

**SCHEME 6.29**  Cyclization of enantiomeric tellurium-containing allyl ether.

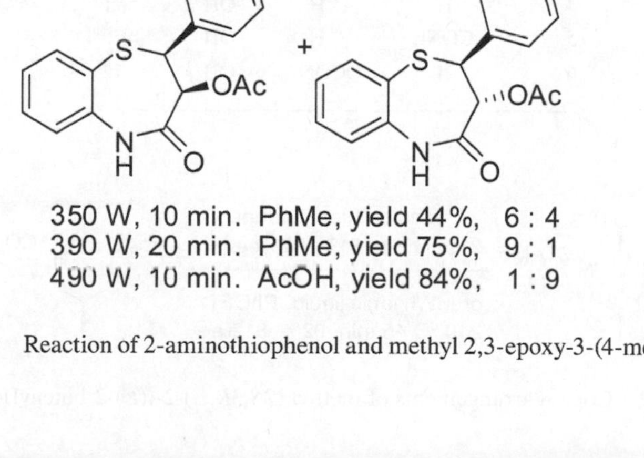

350 W, 10 min.  PhMe, yield 44%,  6 : 4
390 W, 20 min.  PhMe, yield 75%,  9 : 1
490 W, 10 min.  AcOH, yield 84%,  1 : 9

**SCHEME 6.30**  Reaction of 2-aminothiophenol and methyl 2,3-epoxy-3-(4-methoxyphenyl)
propanoate.

group and the generated adducts underwent an intramolecular nucleophilic ring-opening of epoxide to give the bicyclic products (Table 6.9).

The microwave-assisted Cope rearrangements of methyl $(2S,3R,E)$-2-(($E$)-2-butenyl)-3-tributylsilyl-4-hexenoate generated methyl $(4S,5R,E)$-4,5-dimethyl-7-oxo-2-heptanoate with good enantioselectivity. Microwave irradiation improved the yield obviously (Scheme 6.32) [44].

Methyl $(2S,3R,E)$- and $(2S,3R,Z)$-2-(($E$)-2-butenyl)-3-tributylsilyl-4-hexenoates yielded methyl $(4S,5R,E)$- and $(4S,5S,E)$-4,5-dimethyl-7-oxo-2-heptenoates, respectively, with excellent diastereoselectivities and satisfactory enantioselectivities under microwave irradiation (Scheme 6.33) [44]. The olefinic configuration controlled the stereoselectivity because of the Cope rearrangement in the chair conformational transition state.

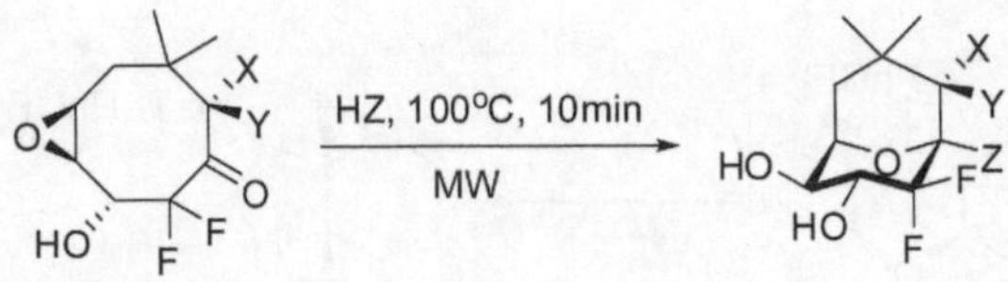

**SCHEME 6.31** Ring opening-cyclization of enantiopure 4,5-epoxycyclooctanones with ammonia and water.

**TABLE 6.9**
**Microwave-Assisted Epoxide Ring-Opening and Intramolecular Cyclization**

| Entry | X | Y | Z | Yield (%) |
|---|---|---|---|---|
| 1 | H | H | NH$_2$ | 60 |
| 2 | OCONEt$_2$ | H | NH$_2$ | 70 |
| 3 | H | OCONEt$_2$ | NH$_2$ | 78 |
| 4 | H | H | OH | 71 |
| 5 | OCONEt$_2$ | H | OH | 75 |
| 6 | H | OCONEt$_2$ | OH | 72 |

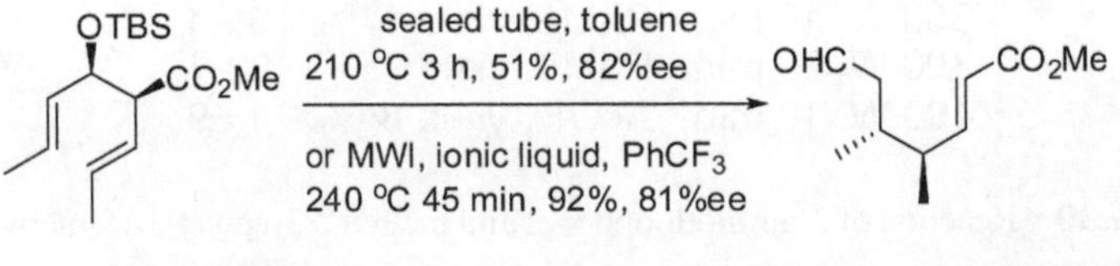

**SCHEME 6.32** Cope rearrangements of methyl $(2S,3R,E)$-2-(($E$)-2-butenyl)-3-tributylsilyl-4-hexenoates.

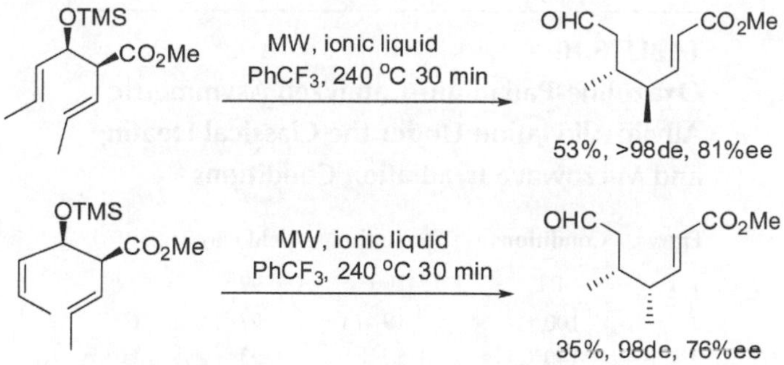

**SCHEME 6.33** Cope rearrangements of methyl 1-((E)-2-butenyl)-3-trimethylsilyl-4-hexenoates.

## 6.5 ENANTIOSELECTIVITY UNDER MICROWAVE IRRADIATION CONDITIONS

Because it is generally considered that microwave irradiation is an alternative heating mode and high enantioselectivity is usually obtained at low temperatures in organic reactions, there are limited examples of the microwave-assisted asymmetric catalysis. Several abnormal influences of temperature on the enantioselectivity were observed in asymmetric catalyses, such as asymmetric borane reduction of ketones [45–47] and asymmetric aziridination of chalcones [48, 49]. Recently, microwave irradiation has been applied in asymmetric catalytic organic reactions.

### 6.5.1 ENANTIOSELECTIVITY IN THE METAL-CATALYZED ORGANIC REACTIONS

Moberg and co-workers first explored the application of microwave irradiation in asymmetric catalytic organic reactions. The 2-(oxazolin-2-yl)quinoline-palladium complex-catalyzed asymmetric allylic alkylation of 1,3-diphenylallyl acetate and dimethyl malonate afford the desired products in short reaction time and excellent yields with a loss of certain enantioselectivity at higher temperatures or under microwave irradiation. However, the microwave-assisted reactions gave higher yields than classical heating reactions. Different microwave powers did not impact the enantioselectivity (Scheme 6.34) [50] (Table 6.10).

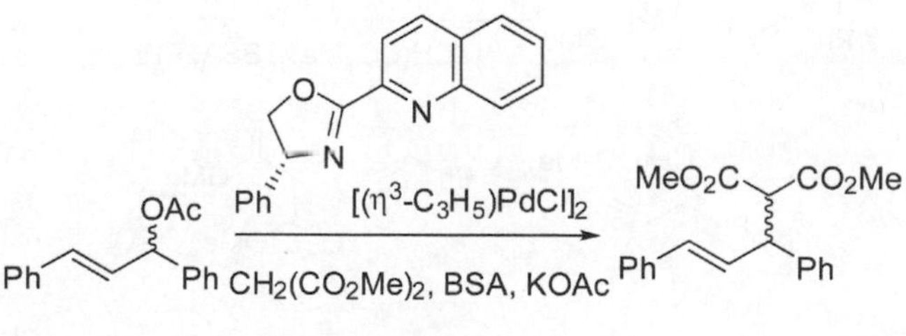

**SCHEME 6.34** Oxazoline-palladium-catalyzed asymmetric allylic alkylation.

**TABLE 6.10**

**Oxazoline-Palladium-Catalyzed Asymmetric Allylic Alkylation Under the Classical Heating and Microwave Irradiation Conditions**

| Entry | Conditions | Time (min) | Yield (%) | Ee (%) |
|-------|-----------|-----------|-----------|--------|
| 1 | RT | 4300 | 99 | 77 |
| 2 | 100°C | 19 | 97 | 62 |
| 3 | 140°C | 6.3 | 93 | 60 |
| 4 | 180°C | 4.5 | 93 | 56 |
| 5 | 35 W | 15 | 99 | 65 |
| 6 | 70 W | 7.5 | 99 | 64 |
| 7 | 120 W | 3.5 | 99 | 63 |
| 8 | 250 W | 3.0 | 99 | 65 |
| 9 | 500 W | 2.0 | 99 | 65 |

One year later, the same group investigated asymmetric allylic alkylation under the catalysis of other chiral ligands and discovered that different microwave powers impacted the enantioselectivity slightly under microwave irradiation [51]. They also explored the influence of the microwave effect on the enantioselectivity in the asymmetric allylic etherification of 4-methoxyphenol and amination of phthalimide. The results also indicate that microwave powers had a slight effect on the enantioselectivity.

In the same year, the same group investigated bisamide-molybdenum-catalyzed intramolecularly asymmetric allylic alkylation of cinnamyl methyl carbonate and discovered that different microwave powers impacted the enantioselectivity slightly, but increased the reaction rate and improved the yield under microwave irradiation (Scheme 6.35) [51]. The microwave power affected the regioselectivity similarly to the influence of temperature on the regioselectivity [17].

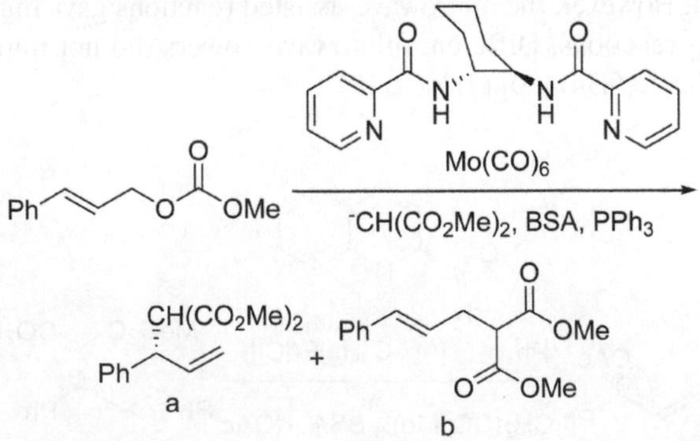

**SCHEME 6.35** Bisamide-molybdenum-catalyzed asymmetric allylic alkylation.

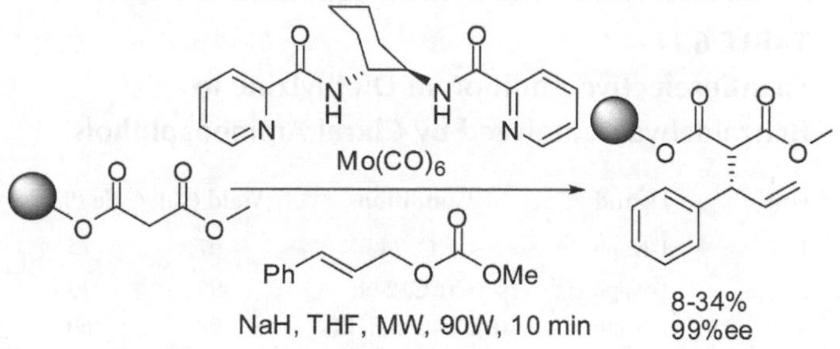

**SCHEME 6.36** Bisamide-molybdenum-catalyzed asymmetric allylic alkylation with a solid immobilized malonate as reagent.

They also attempted the microwave-assisted bisamide-molybdenum-catalyzed asymmetric allylic alkylation with a solid immobilized malonate as a reagent. Compared with homogeneous reaction, it took a long time with low yield without a decrease of the stereoselectivity, possibly due to steric hindrance of solid immobilized malonate and heterogeneous reaction (Scheme 6.36) [17].

Fülöp and his co-workers reported the microwave-assisted highly enantioselective addition of diethylzinc to benzaldehyde catalyzed by chiral aminonaphthols in 2008. They found that relatively lower enantioselectivities were observed under microwave irradiation compared with those under heating possibly due to higher reaction temperature under microwave irradiation (Scheme 6.37). Different catalysts showed similar behavior [52] (Table 6.11).

Braga et al reported that enantioselective addition of phenylboronic acid to *p*-tolualdehyde gave rise to (*S*)-(4-methylphenyl)phenylmethanol in excellent yield and enantioselectivity under the catalysis of aziridine-2-methanol. Microwave irradiation accelerated the reaction efficiently without the loss of enantioselectivity under

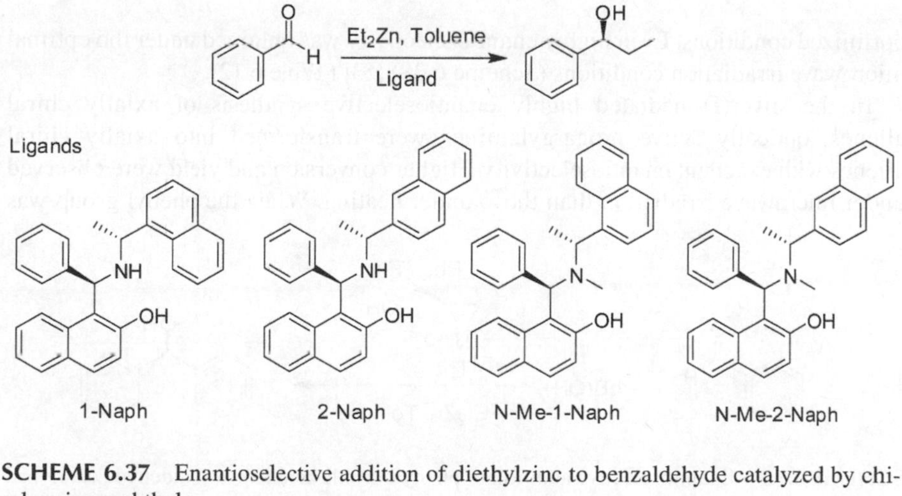

**SCHEME 6.37** Enantioselective addition of diethylzinc to benzaldehyde catalyzed by chiral aminonaphthols.

## TABLE 6.11
## Enantioselective Addition of Diethylzinc to
## Benzaldehyde Catalyzed by Chiral Aminonaphthols

| Entry | Ligand | Conditions | Yield (%) | Ee (%) |
|---|---|---|---|---|
| 1 | 1-Naph | 4°C, 24 h | 97 | 75 |
| 2 | 1-Naph | 20°C, 24 h | 95 | 73 |
| 3 | 1-Naph | 40°C, 40 min, MW | 98 | 60 |
| 4 | 1-Naph | 60°C, 40 min, MW | 95 | 65 |
| 5 | 1-Naph | 60°C, 60 min, MW | 95 | 67 |
| 6 | 1-Naph | 80°C, 40 min, MW | 97 | 10 |
| 7 | 2-Naph | 4°C, 24 h | 85 | 75 |
| 8 | 2-Naph | 20°C, 24 h | 88 | 79 |
| 9 | 2-Naph | 40°C, 40 min, MW | 92 | 75 |
| 10 | 2-Naph | 60°C, 40 min, MW | 94 | 71 |
| 11 | 2-Naph | 60°C, 60 min, MW | 95 | 68 |
| 12 | 2-Naph | 80°C, 40 min, MW | 970 | 70 |
| 13 | N-Me-1-Naph | 4°C, 24 h | 92 | 70 |
| 14 | N-Me-1-Naph | 20°C, 24 h | 97 | 84 |
| 15 | N-Me-1-Naph | 40°C, 40 min, MW | 87 | 72 |
| 16 | N-Me-1-Naph | 60°C, 40 min, MW | 90 | 74 |
| 17 | N-Me-1-Naph | 50°C, 60 min, MW | 97 | 67 |
| 18 | N-Me-2-Naph | 4°C, 24 h | 68 | 42 |
| 19 | N-Me-2-Naph | 20°C, 24 h | 75 | 82 |
| 20 | N-Me-2-Naph | 40°C, 40 min, MW | 85 | 58 |
| 21 | N-Me-2-Naph | 60°C, 40 min, MW | 88 | 73 |
| 22 | N-Me-2-Naph | 50°C, 60 min, MW | 95 | 72 |
| 23 | N-Me-2-Naph | 80°C, 40 min, MW | 90 | 70 |

optimized conditions. Even higher enantioselectivity was obtained under the optimal microwave irradiation conditions (Scheme 6.38) [53] (Table 6.12).

In the silver(I)-mediated highly enantioselective synthesis of axially chiral allenes, optically active progargylamines were transformed into axially chiral allenes with excellent enantioselectivity. Higher conversion and yield were observed under microwave irradiation than those under heating. When the phenyl group was

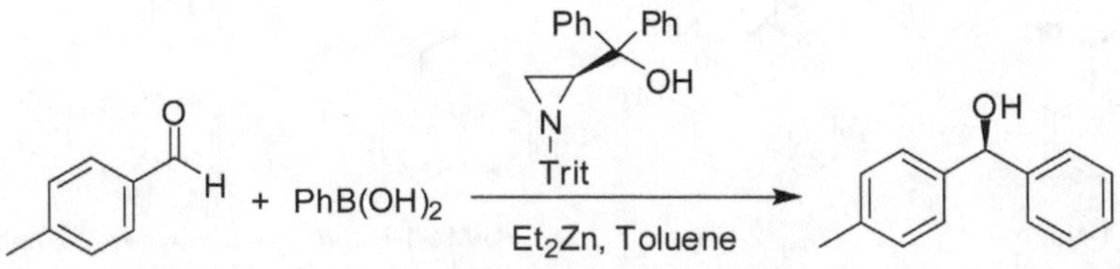

**SCHEME 6.38**  Enantioselective addition of phenylboronic acid to *p*-tolualdehyde catalyzed by chiral aziridine-2-methanol.

displaced by alkyl groups, the reaction gave rise to alkyl axially chiral allenes with excellent enantioselectivities as well (Scheme 6.39) [54].

Yamada and his co-workers investigated how the microwave effect influences catalytic enantioselective Claisen rearrangement under the catalysis of BOX ligand and $Cu(OTf)_2$. They performed their experiments under strictly controlled temperature under both classical heating and microwave irradiation conditions (Scheme 6.40) [55].

## TABLE 6.12
## Enantioselective Addition of Phenylboronic Acid to p-Tolualdehyde Catalyzed by Chiral Aziridine-2-Methanol

| Entry | Conditions | Yield (%) | Ee (%) |
|---|---|---|---|
| 1 | 20°C, 12 h + 12 h | 97 | 96 |
| 2 | 0°C, 12 h + 12 h | 94 | 96 |
| 3 | 60°C, 12 h + 12 h | 96 | 80 |
| 4 | 60°C, 12 h + 10 min, MW | 94 | 96 |
| 5 | 60°C, 12 h + 5 min, MW | 92 | 96 |
| 6 | 80°C, 12 h + 2.5 min, MW | 86 | 95 |
| 7 | 60°C, 20 min, MW + 5 min, MW | 98 | 81 |
| 8 | 60°C, 10 min, MW + 5 min, MW | 97 | 98 |
| 9 | 60°C, 5 min, MW + 5 min, MW | 95 | 97 |
| 10 | 60°C, 2.5 min, MW + 5 min, MW | 90 | 93 |

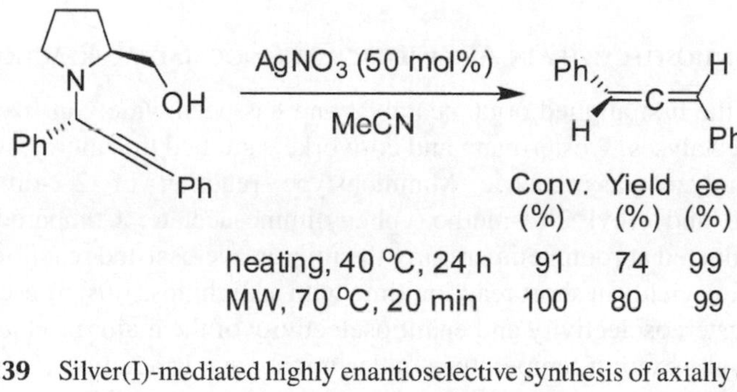

| | Conv. (%) | Yield (%) | ee (%) |
|---|---|---|---|
| heating, 40 °C, 24 h | 91 | 74 | 99 |
| MW, 70 °C, 20 min | 100 | 80 | 99 |

**SCHEME 6.39** Silver(I)-mediated highly enantioselective synthesis of axially chiral allenes.

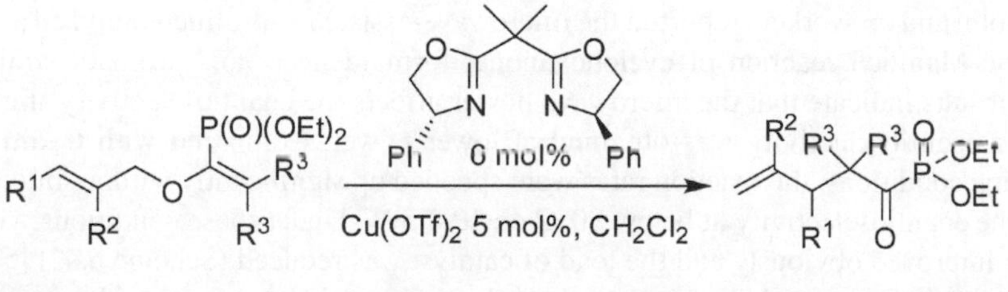

**SCHEME 6.40** Enantioselective Claisen rearrangement under the catalysis of BOX ligand and $Cu(OTf)_2$.

**TABLE 6.13**

**Enantioselective Claisen Rearrangement Under the Catalysis of Box Ligand and Cu(otf)$_2$**

| Entry | R$^1$ | R$^2$ | R$^3$ | Conditions | Time (h) | Yield (%) | Ee (%) |
|-------|-------|-------|-------|------------|----------|-----------|--------|
| 1 | Ph | H | Me | HB 15.0±0.1 | 40 | 86 | 91 |
| 2 | Ph | H | Me | MW 15.4±0.8 | 4 | 84 | 92 |
| 3 | Ph | H | (CH$_2$)$_4$ | HB 15.8±0.4 | 36 | 80 | 86 |
| 4 | Ph | H | (CH$_2$)$_4$ | MW 16.4±0.2 | 4 | 81 | 85 |
| 5 | Ph | H | (CH$_2$)$_5$ | HB 15.2±0.2 | 40 | 81 | 90 |
| 6 | Ph | H | (CH$_2$)$_5$ | MW 15.6±0.5 | 4 | 81 | 89 |
| 7 | Ph | Me | Me | HB 16.5±0.1 | 145 | 84 | 93 |
| 8 | Ph | Me | Me | MW 16.4±0.5 | 4 | 90 | 91 |
| 9 | Ph | Br | Me | HB 27.0±0.2 | 170 | 42 | 91 |
| 10 | Ph | Br | Me | MW 16.4±0.5 | 20 | 87 | 90 |
| 11 | Hex | H | Me | HB 15.5±0.1 | 112 | 84 | 90 |
| 12 | Hex | H | Me | MW 16.1±0.7 | 8 | 87 | 89 |

Although they mentioned that no obviously different enantioselectivity was observed, the enantioselectivities under microwave conditions are lower than those under classical heating in most compared experiments, possibly because hotspots exist under microwave conditions (Table 6.13).

### 6.5.2 ENANTIOSELECTIVITY IN ASYMMETRIC ORGANOCATALYTIC REACTIONS

L-Proline is the first applied organocatalyst and has been widely utilized in many asymmetric catalyses. Westermann and co-workers studied the microwave-assisted L-proline-catalyzed asymmetric Mannich-type reaction of 2,2-dimethyl-1,3-dioxan-5-one and ethyl 2-(4-methoxyphenylimino)acetate. Compared with the reaction conducted at room temperature, the microwave-assisted reactions gave the products in low yields in short reaction time with a slight loss of stereoselectivities, including diastereoselectivity and enantioselectivity of the major product, possibly due to the high reaction temperature under microwave irradiation (Scheme 6.41) (Table 6.14) [56].

Bolm and co-workers reported the microwave-assisted L-proline-catalyzed asymmetric Mannich reaction of cyclohexanone, formaldehyde, and aromatic amines. The results indicate that the microwave power affects the enantioselectivity slightly. Good enantioselectivity was obtained at lower power. Compared with traditional heating conditions, the reaction rates were speeded up significantly without influencing the enantioselectivity at below 80°C on 10–25 W. Under these conditions, yields were improved obviously and the load of catalyst was reduced (Scheme 6.42) [57].

Pozo and Fustero's group conducted a bifunctional squaramide-catalyzed intramolecular aza-Michael addition reaction of α,β-unsaturated N-acyl pyrazole

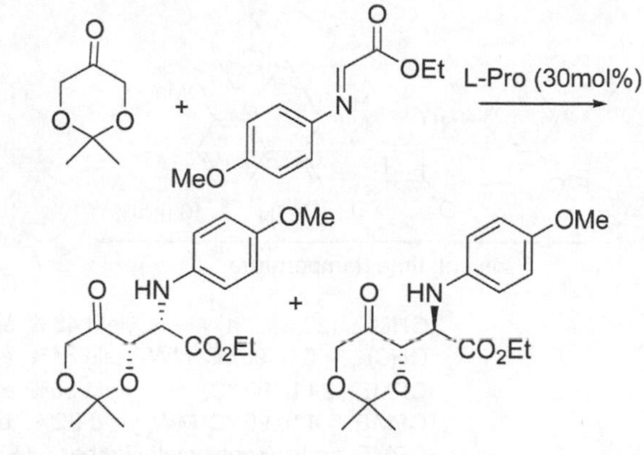

**SCHEME 6.41** L-Proline-catalyzed asymmetric Mannich-type reaction of 2,2-dimethyl-1,3-dioxan-5-one and ethyl 2-(4-methoxyphenylimino)acetate.

## TABLE 6.14
## L-Proline-Catalyzed Asymmetric Mannich-Type Reaction Under the Classical Heating and Microwave Irradiation Conditions

| Entry | Conditions | Time | Solvent | Yield (%) | Ee (%) | Dr |
|-------|-----------|------|---------|-----------|--------|------|
| 1 | RT | 20 h | TFE | 72 | 99 | 97:3 |
| 2 | 300 W | 5 min | TFE | 63 | 94 | 89:11 |
| 3 | 300 W | 10 min | TFE | 72 | 94 | 90:10 |
| 4 | 100 W | 10 min | TFE | 64 | 95 | 90:10 |
| 5 | RT | 20 h | DMSO | 37 | 82 | 91:9 |
| 6 | 300 W | 10 min | DMSO | 18 | ND | 92:8 |
| 7 | RT | 20 h | HCONH₂ | 54 | 96 | 95:5 |
| 8 | 300 W | 5 min | HCONH₂ | 40 | ND | 83:17 |
| 9 | 300 W | 10 min | HCONH₂ | 38 | ND | 80:20 |

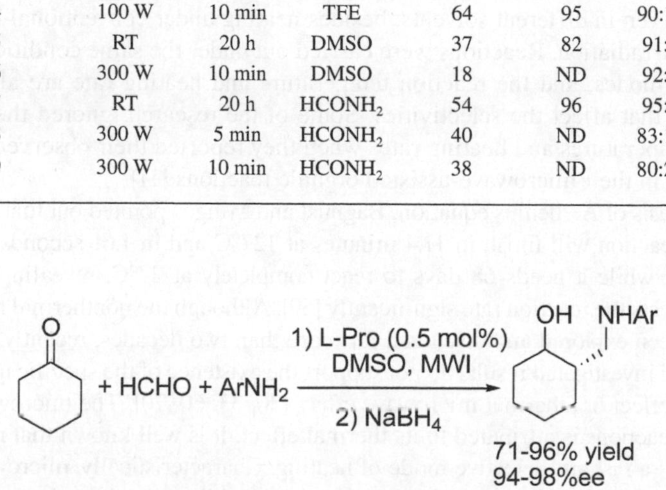

**SCHEME 6.42** L-proline-catalyzed asymmetric Mannich reaction of cyclohexanone, formaldehyde, and aromatic amines.

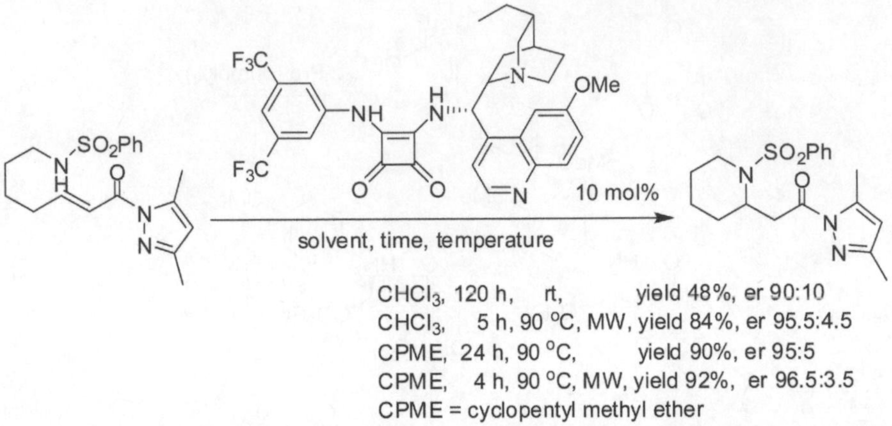

**SCHEME 6.43** Bifunctional squaramide-catalyzed intramolecular aza-Michael addition reaction of α,β-unsaturated N-acyl pyrazole.

(Scheme 6.43) [58]. They found that microwave irradiation not only accelerated the reaction, but improved the enantioselectivity in both chloroform and cyclopentyl methyl ether solvents. The results are indeed interesting.

## 6.6 RATIONALE ON THE INFLUENCE OF MICROWAVE IRRADIATION ON THE SELECTIVITY IN ORGANIC REACTIONS

From some of the previous reported results, one can see that some observed different selectivities were obtained under different reaction conditions, such as solvent-free versus in solution, absence versus presence of catalyst [reactant(s) absorbed on the solid substance, such as silica gel, $Al_2O_3$, etc. The solid substance may play a role as a catalyst], even in different solvents, besides heating under conventional heating or microwave irradiation. Reactions were carried out under the same conditions except for heating modes, and the reaction temperature and heating rate are also important factors that affect the selectivities. Some of the research ignored the different reaction temperatures and heating rates when they reported their observed different selectivities in their microwave-assisted organic reactions [11].

On the basis of Arrhenius equation, Baghust and Mingos pointed out that a classical first-order reaction will finish in 11.4 minutes at 127°C and in 1.61 seconds at 227°C, respectively, while it needs 68 days to react completely at 27°C, revealing that temperature affects the reaction rate significantly [59]. Although the nonthermal microwave effect has been explored and discussed for more than two decades, recently, more and more careful investigated results do not support the existence of the specific nonthermal microwave effect or athermal microwave effect [30, 33, 60–70]. The microwave effect in organic reactions is attributed to its thermal effect. It is well known that microwave irradiation is a fast and selective mode of heating. Characteristically, microwaves generate rapid and intense heating of polar substances while nonpolar substances do not absorb the radiation and are not heated at all or only heated slowly [71]. Conventional heating heats the whole reaction system without obvious temperature gradient under stirring because it generally heats slowly via conduction. The reactants and solvent(s) are heated together. However, only polar substances are heated rapidly under microwave irradiation. Thus, for the reactions involving polar reactant(s) or intermediate(s) in nonpolar solvent(s), the reactive species would obtain more energy from the radiation and locate in a higher energetic region (so-called hotspots). The reactive species heat the solvents. Obvious temperature gradients may exist in the reaction mixture, even under stirring, under microwave irradiation because microwave heating is very fast, resulting in difficulty in determining reaction temperature accurately. The reactions actually occur at a relatively higher temperature than the determined ones, resulting in different rates, selectivities, etc. However, for the reactions involving nonpolar reactant(s) in polar solvent(s), mass solvents are heated and transfer the energy to the reactant(s). In these cases, the reaction conditions are generally similar to those under conventional heating. Thus, only slight differences in the reaction rates and selectivities are observed in most cases. Different microwave powers caused different rates and selectivities because different microwave powers provide different energetic intensity and further result in different temperature gradients. Thus, we are prone to consider that the influences of microwave irradiation on the reaction rates and selectivities are attributed to the microwave thermal effect, that is, temperature difference. From the viewpoint of application, microwave irradiation indeed provides an alternative mode to regulate the selectivities in organic reactions and will be utilized widely in organic and medicinal synthesis.

## 6.7 CONCLUSIONS

Although the microwave has been widely applied in assisted organic reactions during the last four decades and the influence of microwave irradiation on the selectivities, including chemo-, regio-, diastereo-, and enantioselectivities, in organic reactions has also been investigated [7, 8], only limited examples were reported with strictly compared results obtained under conventional heating and microwave irradiation at accurately equal temperatures. Most of the results were not conducted strictly under the same conditions [12, 14]. After analysis of the strictly compared results at the same temperature, one can conclude that reaction conditions actually control the selectivities in most cases. Microwave irradiation generally provides an alternative heating mode for organic reactions with some specific properties, such as selective, fast, and volume heating. If reactions were conducted at exactly the same temperature with the same catalyst and/or additive in the same solvent under conventional heating and microwave irradiation, the same selectivities would be obtained, except the reaction rate would be increased under microwave irradiation due to its selective and fast heating. On the other hand, even if one controls the reactions at the same setting temperature, polar reactants absorb the microwave energy selectively and predominantly and still locate at higher temperatures (so-called hotspots) in the reactions with polar reactants in nonpolar solvents [72]. Thus, microwave-assisted reactions are actually conducted at a higher temperature compared to the same reactions under conventional heating. In some cases, obviously different results, such as yield, reaction rate, selectivity, etc [73]. are observed, especially for the reactions which are sensitive to temperature. We call it the microwave effect because the microwave can selectively heat polar substances, but, the microwave effect is a thermal effect, rather than nonthermal or athermal effect. We hope this chapter provides some useful information for organic and medicinal chemists who are interested in microwave-assisted organic reactions, especially controlling selectivities, including chemo-, regio-, diastereo-, and enantioselectivities, in organic reactions with microwave irradiation.

# REFERENCES

1. Lidstroem, P., Tierney, J., Wathey, B., and Westman, J. 2001. Microwave assisted organic synthesis—A review. *Tetrahedron* 57: 9225–9283.
2. Loupy, A. Ed. 2002. *Microwaves in Organic Synthesis.* Weinheim: Wiley-VCH.
3. Hayes, B.L. 2004. Recent advances in microwave-assisted synthesis. *Aldrichim. Acta* 37: 66–77.
4. Kappe, C.O. 2004. Controlled microwave heating in modern organic synthesis. *Angew. Chem. Int. Ed.* 43: 6250–6284 and cited therein.
5. Kappe, C.O., and Stadler, A. 2005. *Microwaves in Organic and Medicinal Chemistry,* Vol. 25, From the series, Methods and Principles in Medicinal Chemistry, R. Mannhold, H. Kubinyi, and G. Folker, eds. Weinheim: Wiley-VCH.
6. Yin, W., Ma, Y., Xu, J.X., and Zhao, Y.F. 2006. Microwave-assisted one-pot synthesis of 1-indanones from arenes and α,β-unsaturated acyl chlorides. *J. Org. Chem.* 71: 4312–4315.
7. Perreux, L., and Loupy, A. 2001. A tentative rationalization of microwave effects in organic synthesis according to the reaction medium, and mechanistic considerations. *Tetrahedron* 57: 9199–9223.
8. De La Hoz, A., Díaz-Ortiz, A., and Moreno, A. 2004. Selectivity in organic synthesis under microwave irradiation. *Curr. Org. Chem.* 8: 903–918.
9. Xu, J.X. 2007. Microwave irradiation and selectivities in organic reactions. *Prog. Chem.* 19: 700–712.
10. Patonay, T., Varma, R.S., Vass, A., Levai, A., and Dudas, J. 2001. Highly diastereoselective Michael reaction under solvent-free conditions using microwaves: conjugate addition of flavanone to its chalcone precursor. *Tetrahedron Lett.* 42: 1403–1406.
11. Shanmugasundaram, M., Manikandan, S., and Raghunathan, R. 2002. High chemoselectivity in microwave accelerated intramolecular domino Knoevenagel hetero Diels–Alder reactions—an efficient synthesis of pyrano[3-2c]coumarin frameworks. *Tetrahedron* 58: 997–1003.
12. Jayashankaran, J., Manian, R.D.R.S., and Raghunathan, R. 2006. An efficient synthesis of thiopyrano[5,6-c]coumarin/[6,5-c]chromones through intramolecular domino Knoevenagel hetero Diels–Alder reactions. *Tetrahedron Lett.* 47: 2265–2270.
13. Camara, C., Keller, L., and Dumas, F. 2003. Microwave activation of an asymmetric Michael reaction: unexpected behaviour of chiral α-alkoxy imines. *Tetrahedron: Asymmetry* 14: 3263–3266.
14. Correc, O., Guillu, K., Hamelin, J., Paquin, L., Texier-Boullet, F., and Toupet, L. 2004. Microwave solvent-free synthesis of nitrocyclohexanols. *Tetrahedron Lett.* 45: 391–395.
15. Paquin, L., Toupet, L., Hamelin, J., and Texier-Boullet, F. 2005. Microwave and classical activation of solvent-free reactions of *N*-isopropyl arylidene amines with alkyl cyanoacetates. *Lett. Org. Chem.* 2: 334–339.
16. Glasnov, T.N., Stadlbauer, W., and Kappe, C.O. 2005. Microwave-assisted multistep synthesis of functionalized 4-arylquinolin-2(1*H*)-ones using palladium-catalyzed cross-coupling chemistry. *J. Org. Chem.* 70: 3864–3870.
17. Kaiser, N.F.K., Bremberg, U., Larhed, M., Moberg, C., and Hallberg, A. 2000. Fast, convenient, and efficient molybdenum-catalyzed asymmetric allylic alkylation under noninert conditions: An example of microwave-promoted fast chemistry. *Angew. Chem. Int. Ed.* 39: 3596–3598.

18. Lindsay, K.B., Tang, M.Y., and Pyne, S.G. 2002. Diastereoselective synthesis of polyfunctional-pyrrolidines via vinyl epoxide aminolysis/ring-closing metathesis: Synthesis of chiral 2,5-dihydropyrroles and (1*R*,2*S*,7*R*,7a*R*)-1,2,7-trihydroxypyrrolizidine. *Synlett* 731–734.
19. Paul, S., and Gupta, M 2004. Selective Fries rearrangement catalyzed by zinc powder *Synthesis* 1789–1792.
20. Khadilkar, B.M. and Madyar, V.R. 1999. Fries rearrangement at atmospheric pressure using microwave irradiation. *Synth. Commun.* 29: 1195–1200.
21. Barluenga, J., Fernandez-Rodriguez, M.A., Garcia-Garcia, P., Aguilar, E., and Merino, I. 2006. Synthesis of donor–acceptor alkynylcyclopropanes by diastereoselective cyclopropanation of electron-deficient alkenes with alkoxyalkynyl Fischer carbene complexes. *Chem. Eur. J.* 12: 303–313.
22. Georg, G.I., 1993. *The Organic Chemistry of β-Lactams*. New York: Verlag Chemie.
23. Bose, A.K., Banik, B.K., and Manhas, M.S. 1995. Stereocontrol of β-lactam formation using microwave irradiation. *Tetrahedron Lett.* 36: 213–216.
24. Manhas, M.S., Banik, B.K., Mathur, A., Vincent, J.E., and Bose, A.K. 2000. Vinyl-β-lactams as efficient synthons. Eco-friendly approaches via microwave assisted reactions. *Tetrahedron* 56: 5587–5601.
25. Podlech, J., and Linder, M.L. 1997. Cycloadditions of ketenes generated in the Wolff rearrangement. Stereoselective synthesis of aminoalkyl-substituted β-lactams from α-amino acids. *J. Org. Chem.* 62: 5873–5883.
26. Jiao, L., Liang, Y., Zhang, Q.F., Zhang, S.W., and Xu, J.X. 2006. Catalyst-free, high-yield, and stereospecific synthesis of 3-phenylthio β-lactam derivatives. *Synthesis* 37: 659–665.
27. Lindar, M.L., and Podlech, J. 2001. Synthesis of β-lactams from diazoketones and imines: The use of microwave irradiation. *Org. Lett.* 3: 1849–1851.
28. Liang, Y., Jiao, L., Zhang, S.W., and Xu, J.X. 2005. Microwave- and photoirradiation-induced Staudinger reactions of cyclic imines and ketenes generated from α-diazoketones. A further investigation into the stereochemical process. *J. Org. Chem.* 70: 334–337.
29. Jiao, L., Liang, Y., and Xu, J.X. 2006. Origin of the relative stereoselectivity of the β-lactam formation in the Staudinger reaction. *J. Am. Chem. Soc.* 128: 6060–6069.
30. Wang, Y.K., Du, D.M., and Xu, J.X. 2006. Effect of microwave irradiation on the *syn/trans* selectivity in the formation of β-lactams from ketenes and imines. *J. Liaoning Univ. Petroleum Chem. Tech.* 26: 94–95.
31. Wang, Y.K., Liang, Y., Jiao, L., Du, D.M., and Xu, J.X. 2006. Do reaction conditions affect the stereoselectivity in the Staudinger reaction? *J. Org. Chem.* 71: 6983–6990.
32. Li, B.N., Wang, Y.K., Du, D.M., and Xu, J.X. 2007. Notable and obvious ketene substituent-dependent effect of temperature on the stereoselectivity in the Staudinger reaction. *J. Org. Chem.* 72: 990–997.
33. Hu, L.B., Wang, Y.K., Li, B.N., Du, D.M., and Xu, J.X. 2007. Diastereoselectivity in the Staudinger reaction: A useful probe for investigation on nonthermal microwave effects. *Tetrahedron* 63: 9387–9392.
34. Essers, M., Muck-Lichtenfeld, C., and Haufe, G. 2002. Diastereoselective Diels–Alder reactions of α-fluorinated α,β-unsaturated carbonyl compounds: Chemical consequences of fluorine substitution. 2. *J. Org. Chem.* 67: 4715–4721.
35. Adams, H., Jones, S., Meijer, A.J.H.M., Najah, Z., Ojea-Jimenez, I., and Reeder, A.T. 2011. Diels–Alder reactions and transformations of 2-cyclopenten-1-one with a chiral anthracene template. *Tetrahedron: Asymmetry* 22: 1620–1625.
36. Yadav, L.D.S., and Singh, A. 2003. Microwave activated solvent-free cascade reactions yielding highly functionalised 1,3-thiazines. *Tetrahedron Lett.* 44: 5637–5640.
37. Yadav, L.D.S., and Singh, A. 2003. Novel one-pot microwave assisted Gewald synthesis of 2-acyl amino thiophenes on solid support. *Synlett* 2003: 63–66.

38. Yadav, L.D.S., and Kapoor, R. 2003. Solvent-free microwave activated three-component synthesis of thiazolo-s-triazine C-nucleosides. *Tetrahedron Lett.* 44: 8951–8954.

39. Romanova, N.N., Gravis, A.G., Leshcheva, I.F., and Bundel, Y.G. 2001. Solvent-free stereoselective synthesis of β-aryl-β-amino acid esters by the Rodionov reaction using microwave irradiation. *Mendeleev Commun.* 11: 26–27.

40. Micuch, P., Fisera, L., Cyranski, M., and Krygowski, T.M. 1999. Reversal of stereoselectivity of Mg(II) catalysed 1,3-dipolar cycloaddition. Acceleration of cycloaddition by microwave irradiation. *Tetrahedron Lett.* 40: 167–170.

41. Ericsson, C., and Engman, L. 2004. Microwave-assisted group-transfer cyclization of organotellurium compounds. *J. Org. Chem.* 69: 5143–5146.

42. Vega, J.A., Cueto, S., Ramos, A., Vaquero, J.J., Garcia-Navio, J.L., and Alvarez-Builla, J. 1996. A microwave synthesis of the *cis* and *trans* isomers of 3-hydroxy-2-(4-methoxyphenyl)-2,3-dihydro-1,5-benzothiazepin-4(5*H*)-one: The influence of solvent and power output on the diastereoselectivity. *Tetrahedron Lett.* 37: 6413–6416.

43. Fawcett, J., Griffith, G.A., Percy, J.M., and Uneyama, E. 2004. Synthesis of β-lactams from diazoketones and imines: The use of microwave irradiation. *Org. Lett.* 6: 1277–1280.

44. Davies, H.M.L., and Beckwith, R.E.J. 2004. Catalytic asymmetric reactions for organic synthesis: The combined C–H activation/siloxy-Cope rearrangement. *J. Org. Chem.* 69: 9241–9247.

45. Xu, J.X., Wei, T.Z., and Zhang, Q.H. 2003. Effect of temperature on the enantioselectivity in the oxazaborolidine-catalyzed asymmetric reduction of ketones. Noncatalytic borane reduction, a nonneglectable factor in the reduction system. *J. Org. Chem.* 68: 10146–10151.

46. Xu, J.X., Wei, T.Z., Lin, S.S., and Zhang, Q.H. 2005. Rationale on abnormal effect of temperature on the enantioselectivity in the asymmetric borane reduction of ketones catalyzed by prolinol. *Helv. Chim. Acta* 88: 180–186.

47. Liu, H., and Xu, J.X. 2006. (*S*)-2-Aryl-4,4-diphenyl-3,1,2-oxazaboro[3.3.0]octanes: Efficient catalysts for the asymmetric borane reduction of electron-deficient ketones. *J. Mol. Cat. A: Chem.* 244: 68–72.

48. Xu, J.X., Ma, L.G., and Jiao, P. 2004. Asymmetric aziridination of chalcones catalyzed by a novel backbone 1,8-bisoxazolinylanthracene (AnBOX)-copper complex. *Chem. Commun.* 14: 1616–1617.

49. Ma, L.G., Jiao, P., Zhang, Q.H., and Xu, J.X. 2005. Rigid backbone 1,8-anthracene-linked bis-oxazolines (AnBOXes): design, synthesis, application and characteristics in catalytic asymmetric aziridination. *Tetrahedron: Asymmetry* 16: 3718–3734.

50. Moberg, C., Bremberg, U., Hallman, K., Svensson, M., Norrby, P.O., Hallberg, A., Larhed, M., and Csoregh, I. 1999. Selectivity and reactivity in asymmetric allylic alkylation. *Pure Appl. Chem.* 71: 1477–1483.

51. Bremberg, U., Lutsenko, S., Kaiser, N.F., Larhed, M., Hallberg, A., and Moberg, C. 2000. Rapid and stereoselective C-C, C-O, C-N and C-S couplings via microwave accelerated palladium-catalyzed allylic substitutions. *Synthesis* 7: 1004–1008.

52. Szatmári, I., Sillanpää, R., and Fülöp, F. 2008. Microwave-assisted, highly enantioselective addition of diethylzinc to aromatic aldehydes catalyzed by chiral aminonaphthols. *Tetrahedron: Asymmetry* 19: 612–617.

53. Braga, A.L., Paixão, M.W., Westermann, B., Schneider, P.H., and Wessjohann, L.A. 2008. Acceleration of arylzinc formation and its enantioselective addition to aldehydes by microwave irradiation and Aziridine-2-methanol catalysts. *J. Org. Chem.* 73: 2879–2882.

54. Lo, V.K.-Y., Zhou, C.Y., Wong, M.-K., and Che, C.-M. 2010. Silver (I)-mediated highly enantioselective synthesis of axially chiral allenes under thermal and microwave-assisted conditions. *Chem. Commun.* 46: 213–215.

55. Nushiro, K., Kikuchi, S., and Yamada, T. 2013. Microwave effect on catalytic enanti-oselective Claisen rearrangement. *Chem. Commun.* 49: 8371–8373.
56. Westermann, B., and Neuhaus, C. 2005. Dihydroxyacetone in amino acid catalyzed Mannich-type reactions. *Angew. Chem. Int. Ed.* 44: 4077–4079.
57. Rodriguez, B., and Bolm, C. 2006. Thermal effects in the organocatalytic asymmetric Mannich reaction. *J. Org. Chem.* 71: 2888–2891.
58. Sánchez-Roselló, M., Mulet, C., Guerola, M., del Pozo, C., and Fustero, S. 2014. Microwave-assisted tandem organocatalytic peptide-coupling intramolecular aza-Michael reaction: α,β-Unsaturated *N*-acyl pyrazoles as Michael acceptors. *Chem. Eur. J.* 20: 15697–16701.
59. Baghurst, D.R., and Mingos, D.M.P. 1991. Applications of microwave dielectric heating effects to synthetic problems in chemistry. *Chem. Soc. Rev.* 20: 1–47.
60. Pollington, S.D., Bond, G., Moyes, R.B., Whan, D.A., Candlin, J.P., and Jennings, J.R. 1991. The influence of microwaves on the rate of reaction of propan-1-ol with ethanoic acid. *J. Org. Chem.* 56: 1313–1314.
61. Raner, K.D., and Strauss, C.R. 1992. Influence of microwaves on the rate of esterifica-tion of 2,4,6-trimethylbenzoic acid with 2-propanol. *J. Org. Chem.* 57: 6231–6234.
62. Laurent, R., Laporterie, A., Dubac, J., Berlan, J., Lefeuvre, S., and Audhuy, M. 1992. Specific activation by microwaves: Myth or reality? *J. Org. Chem.* 57: 7099–7102.
63. Stuerga, D., Gonon, K., and Lallemant, M. 1993. Microwave heating as a new way to induce selectivity between competitive reactions. application to isomeric ratio control in sulfonation of naphthalene. *Tetrahedron* 49: 6229–6234.
64. Raner, K.D., Strauss, C.R., Vyskoc, F., and Mokbel, L. 1993. A comparison of reaction kinetics observed under microwave irradiation and conventional heating. *J. Org. Chem.* 58: 950–953.
65. Raner, K.D., Strauss, C.R., Trainor, R.W., and Thorn, J.S. 1995. A new microwave reac-tor for batchwise organic synthesis. *J. Org. Chem.* 60: 2456–2460.
66. Bagnell, L., Cablewski, T., Strauss, C.R., and Trainor, R.W. 1996. Reactions of allyl phenyl ether in high-temperature water with conventional and microwave heating. *J. Org. Chem.* 61: 7355–7359.
67. Bagnell, L., Bliese, M., Cablewski, T., Strauss, C.R., and Tsanaktsidis, J. 1997. Environmentally benign procedures for the preparation and isolation of 3-methylcyclo-pent-2-en-1-one. *Aust. J. Chem.* 50: 921–926.
68. Kabza, K.G., Chapados, B.R., Gestwicki, J.E., and McGrath, J.L. 2000. Microwave-induced esterification using heterogeneous acid catalyst in a low dielectric constant medium. *J. Org. Chem.* 65: 1210–1214.
69. Kuhnert, N. 2002. Microwave-assisted reactions in organic synthesis—Are there any nonthermal microwave effects? *Angew. Chem. Int. Ed.* 41: 1863–1866.
70. Strauss, C.R. 2002. Microwave-assisted reactions in organic synthesis—Are there any nonthermal microwave effects? Response to the highlight by N. Kuhnert. *Angew. Chem. Int. Ed.* 41: 3589–3590.
71. Gabriel, C., Gabriel, S., Grant, E.H., Halstead, B.S.J., and Mingos, D. M. P. 1998. Dielectric parameters relevant to microwave dielectric heating. *Chem. Soc. Rev.* 27: 213–223.
72. Li, X. H., and Xu, J. X. 2016. Determination on temperature gradient of different polar reactants in reaction mixture under microwave irradiation with molecular probe. *Tetrahedron* 72: 5515–5520.
73. Li, X. H., and Xu, J. X. 2017. Effects of the microwave power on the microwave assisted esterification. *Curr. Microw. Chem.* 4: 158–162.

# 7 Microwave-Assisted Hirao and Kabachnik–Fields Phosphorus–Carbon Bond Forming Reactions
## A Recent Update*

*Goutam Brahmachari*

## CONTENTS

## 7.1 INTRODUCTION

At present, organophosphorus chemistry is an exciting and enormously explored field of potential research, and, due to multifaceted applications, P-functionalized compounds have become important synthetic targets in modern organic synthesis (Engel and Cohen, 2004; Hartly, 1996; Jablonkai and Keglevich, 2014a; Savignac and Iorga, 2003). Among these, aryl- and vinylphosphonates are used as practical applications (Bock et al., 2007; Ebdon et al., 2000) and also as valuable synthons in organic synthesis (Beletskaya et al., 2006; Jiao and Bentrude, 2003; Maffei, 2004; Minami and Motoyoshiya, 1992; Zhou and Chen, 2001). Arylphosphonate scaffolds are used in

---

* Dedicated to Professor György Keglevich in honor of his paramount contribution to the field of organophosphorus chemistry

designing fuel cell membranes (Bock et al., 2007) and materials with special optical properties (Belfield et al., 1997; Ogawa et al., 1998; Zakeeruddin et al., 1997). Aryl- and vinylphosphonates are widely used for the synthesis of mixed organic-inorganic materials (Lelievre et al., 1996; Schull et al., 1995), polymer technology (Jin and Gonsalves, 1998), and building blocks for promising heterocycles (Dembitsky et al., 2005; Minami et al., 2001). It has been demonstrated that a P-C bond may offer different kinds of biological activities and, accordingly, there has been a growing interest in these classes of organophosphorus compounds in medicinal (Holstein et al., 1998; Jiang et al., 2006; Quntar et al., 2004; Ullrich et al., 1997) and nucleic acid chemistry (Abbas et al., 2001; Harnden et al., 1993; Lazrek et al., 1998; Zmudzka et al., 2003). Similarly, phosphine oxides are also important intermediates for chemical synthesis because they are precursors to free phosphines, which are essential for transition metal catalysis (Engels and Parsch, 2005) and for organoca-talysis (Benaglia and Rossi, 2010; Denmark et al., 2007; Lu et al., 2001; Morimoto et al., 2008; Van Leeuwen, 2004; Zhu et al., 2006). Water-soluble phosphines and phosphine oxides have special advantages in industrial processes based on homo-geneous catalysts since they allow easier separation of the metal catalyst from the reaction mixture by using two-phase water/organic solvent systems (Herrmann and Kohlpaintner, 1993; Pinault and Bruce, 2003; Shaughnessy, 2009).

α-Aminophosphonates and related derivatives are desirable targets in biochem-istry due to their versatile bioactivity of which synthesis and use of such organo-phosphorus compounds have received great attention during the last two decades. As mimics of natural amino acids and of low mammalian toxicity, the functional-ized α-aminophosphonate derivatives are currently attracting a great deal of interest in industrial as well as medicinal chemistry (Bird et al., 1994; Fields, 1999; Giannousi and Bartlett, 1987; Grembecka et al., 2003; Kukhar and Hudson, 2000; Liu et al., 2002; Razaei, 2009; Wuggenig et al., 2011). These compounds are reported to exhibit a vari-ety of biological activities, such as antifungal activity (Ouimette and Coffey, 1989), antibacterial activity (Sonar et al., 2010), antitumor effects (Bloemink et al., 1999; Jin, 2006; Kiran et al., 2008; Koteswara Rao et al., 2011), antiviral activity (Xu et al., 2006), enzyme inhibitors, such as rennin (Allen et al., 1989), HIV, and serine prote-ase (Kafarski and Lejczak, 2001; Siénczyk and Oleksyszyn, 2009), PTP1B inhibitors (Ghotas et al., 2009), and pharmacological agents (Atherton et al., 1968; Kaboudin and Moradi, 2005). Dialkylglycine decarboxylase (Mucha et al., 2011) and leucine ami-nopeptidase (Grembecka et al., 2003) are also inhibited by α-aminophosphonates. A good number of such compounds have been found to possess effective *in vitro* growth inhibitory activity against the malaria parasite *Plasmodium falciparum* (Rawls, 1998), and also against single-celled parasites such as *Toxoplasma* that causes opportunis-tic infections in AIDS patients (Roberts et al., 1998). Certain α-aminophosphonates were proved to be suitable for the design of continuous drug release devices due to their ability to increase the membrane permeability of a hydrophilic probe molecule (Danila et al., 2008). Additionally, α-functionalized phosphonic acid esters can serve as synthetic intermediates in phosphonate chemistry for many organic compounds and dyes, and also be used in laser technology and as fluorescent materials for the visualization of biomolecules (Janardhan Rao et al., 2010; Naydenova et al., 2010; Orsini et al., 2010; Veeranarayana Reddy et al., 2007, 2010).

Owing to such potential and useful applications of these organophosphorus scaffolds, synthetic endeavors to access them are manifold. Access to aryl-/vinylphosphonates and related compounds was improved considerably when, in addition to the traditional synthetic methods (Freedman and Doak, 1957; Holstein et al., 1998; Minami et al., 2001), Hirao and co-workers (Hirao et al., 1980, 1981, 1982) introduced a Pd-catalyzed cross-coupling reaction of H-phosphonates with aryl and vinyl halides. Since its first report, Hirao reaction has attained a good deal of growth, as documented in a number of reviews (Jablonkai and Keglevich, 2014b,c; Prim et al., 2002; Schwan, 2004). Again, for accessing α-aminophosphonates, there are several methodologies, such as Pudovik reaction, Mannich-type reactions, and Kabachnik–Fields reaction. Among them, Kabachnik–Fields (also known as phospha-Mannich condensation) has been well studied so far and appears to be a powerful one-pot three-component reaction for synthesizing α-aminophosphonates of potential interest (Cherkasov and Galkin, 1998; Fields, 1952; Kabachnik and Medved, 1952; Keglevich and Bálint, 2012; Zefirov and Matveeva, 2008).

Currently, with the increasing trend of practicing green chemistry approaches, applications of energy-efficient green chemistry tools, such as microwave, ultrasound, visible and ultraviolet light, and ball-milling technique, are of fascinating choice (Brahmachari, 2014, 2015a)! In recent years, application of microwaves to synthetic organic chemistry has become increasingly popular, and attracted considerable practical (Giguere et al., 1986; Hayes, 2004; Kranjc and Kočevar, 2010) and theoretical attention (Bandyopadhyay and Banik, 2014). It was found on numerous occasions that reaction times are comparatively reduced from hours to minutes and the percentage yields of products are found to be higher when compared to conventional heating methods. Another important aspect of microwave-assisted synthesis is that it uses 50% less power than electric furnaces of equivalent capacity (Keglevich, 2014). As a part of the continuing development of various synthetically important reactions using microwave technique, both Hirao and Kabachnik–Fields reactions also have attained quite significant advancements so far. This chapter is aimed at offering an up-to-date overview of such developments in the formation of a phosphorus-carbon bond, thereby affording functionalized diverse kinds of phosphonates, phosphine oxides and α-aminophosphonates of potential interest.

## 7.2 ACCESS OF DIVERSE ARYL/VINYL-PHOSPHONATES AND PHOSPHINE OXIDES THROUGH MICROWAVE-ASSISTED HIRAO REACTION

As mentioned in the earlier section, the Hirao reaction has become an important P-C bond forming strategy since its discovery about 35 years ago in the early 1980s (Hirao et al., 1980). In their report, Hirao and co-workers discussed the P-C cross-coupling reaction of aryl and vinyl halides with dialkyl phosphites in the presence of tetrakis(triphenylphosphine)palladium, leading to the formation of aryl and vinyl phosphonates (Hirao et al., 1980, 1981, 1982). This reaction has been gradually extended to more aryl and vinyl derivatives and to a variety of >P(O)H species, thus making phosphonic, phosphinic and phosphine oxide derivatives available

(Jablonkai and Keglevich, 2014a). Recently, the present author has published a lecture note on this reaction, highlighting on the mechanistic aspects and applications of the Hirao reaction (Brahmachari, 2015b). The present section offers vivid details on the progress of microwave-assisted Hirao reaction.

Hirao et al. (1982) reported that the reaction between an aryl halide and dialkyl phosphite undergoes smoothly under the catalysis of palladium(0) complexes but requires a lot of time (on average 10 hours) at conventional heating. This reaction was modified after one and half decades by Villemin et al. (1997), particularly in regards to expeditious rate enhancement as well as easy work-up procedure by the application of microwave irradiation — aryl halides (1) and diethyl phosphite (2) on microwave irradiation in the presence of bis(triphenylphosphine)palladium dichloride as a catalyst in triethyl amine (Et₃N) in an inert atmosphere underwent rapid cross-coupling to furnish the desired diethyl arylphosphonates (3) in moderate to good yields within minutes (Scheme 7.1). Aryl iodides, other than *ortho*-substituted sterically-hindered derivatives, gave good yields (up to 97%); however, a very poor yield of only 13% was obtained from 4-iodobenzoic acid, possibly due to the strong electron-withdrawing effect of the carboxylic acid function. Although both iodides and bromides took part in this process, C-Cl bond activation under the reaction conditions was not observed at all with chlorobenzene. However, chlorobenzenetricarbonylchromium complex was found to smoothly undergo the C-Cl bond activation under the reaction conditions, resulting in the desired product with 80% yield after just eight minutes; a phosphonation reaction of this complex was also demonstrated based on an Arbusov-type reaction under classical heating with the same yield (81%) after seven hours at reflux (Chauvin, 1990). Like aryl chloride, aryl triflate did not undergo the reaction under microwave irradiation; only a poor yield (31% after 10 minutes) was isolated (Villemin et al., 1997).

However, in 2006, Jiang et al. successfully cross-coupled aryl triflates with disubstituted phosphates/phosphine oxides as a key step during the multi-step synthesis

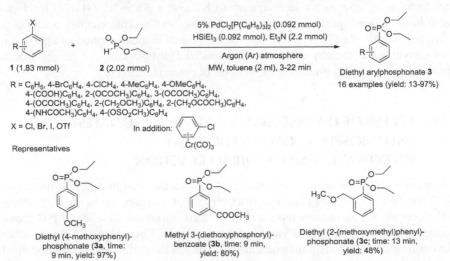

**SCHEME 7.1**   Synthesis of diethyl arylphosphonates by palladium-catalyzed phosphonation of aryl halides.

of a new series of phosphorus-functionalized 11β-aryl-substituted steroids (**6**) with promising progesterone receptor antagonist activity. For this purpose, the investigators developed a modified Pd-catalyzed microwave-assisted Hirao reaction under Pd(OAc)$_2$/dppp/DIEA/dioxane conditions so as to carry out the P-C cross-coupling between synthetically-derived 11β-4-(((trifluoromethyl)sulfonyl)oxy)phenyl-substituted steroid intermediates (**4**) and disubstituted phosphates/phosphine oxides (**5**) in a facile mode (Scheme 7.2). It is worth mentioning that the stereochemical features of the substrate molecule remained intact within the product.

All the synthesized compounds **6** were evaluated for both their progesterone receptor (PR) antagonist (T47D cell-based assay) and glucocorticoid receptor (GR) antagonist activity (A549 cell-based assay), and their corresponding activities were compared with those of mifepristone, the first competitive progesterone antagonist drug. A selective progesterone receptor (PR) antagonist has potential utility as a contraceptive and also in treating reproductive disorders, such as uterine leiomyomas and endometriosis, and hormone-dependent tumors (Jiang et al., 2006). Hence, the search for some novel class of selective progesterone receptor modulators (PRMs) having only the anti-progestational activity remains highly desirable, both in terms of clinical applications and basic endocrine research. Most of the compounds were potent PR antagonists (nanomolar range), with some showing better selectivity than mifepristone (IC$_{50}$'s: 0.2 nM (T47D), 2.6 nM (A549); selectivity ratio: 13). Among the series, the most potent molecules were found to be **6a** ((IC$_{50}$'s: 9.9 nM (T47D), 237.7 nM (A549); selectivity ratio: 24), **6b** (IC$_{50}$'s: 3.58 nM

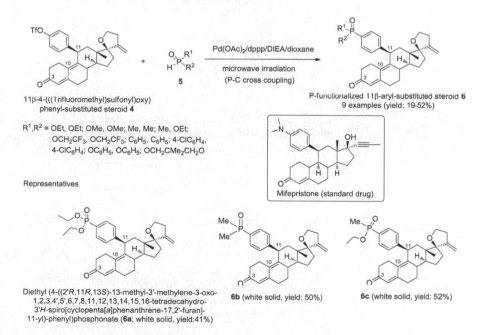

SCHEME 7.2 Palladium-catalyzed synthesis of phosphorus-functionalized 11β-aryl-substituted steroids (**6**) as progesterone receptor antagonists.

(T47D), 660.15 nM (A549); selectivity ratio: 184), and **6c** (IC$_{50}$'s: 1.6 nM (T47D), 270.86 nM (A549); selectivity ratio: 169). In addition, some selected compounds showed modest oral progestin antagonist activity in rat uterus. From the view-point of structure-activity relationship, it was evident that when R$^1$ and R$^2$ are both alkoxy groups, the change in their size did not affect potencies, while a change in the electronics reduced the potency with a more electron-withdrawing group. It was also observed that substitution at the phenyl ring did not significantly change the potency (Jiang et al., 2006).

In 2007, Julienne et al. synthesized a series of vinylphosphine-borane complexes (**9**) via palladium-catalyzed C-P cross-coupling of diverse vinyl triflates (**7**) with secondary diaryl-, dialkyl-, and alkylarylphosphine-boranes (**8**) under conventional heating as well as microwave irradiation. The investigators observed that the microwave-irradiated reactions undergo smoothly with comparable yields to those obtained under conventional heating but with shortened reaction times. The overall reactions and effect of microwave irradiation in shortening the reaction-time are exemplified in Scheme 7.3.

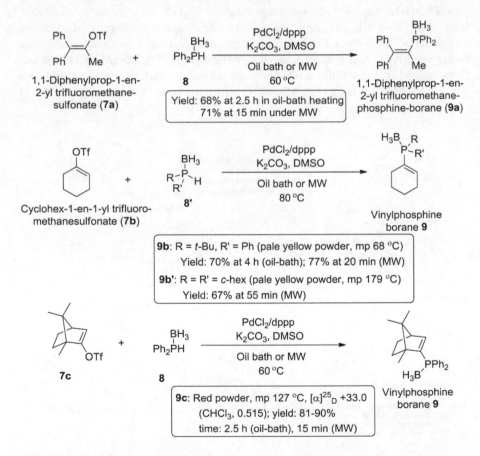

**SCHEME 7.3**    Palladium-catalyzed synthesis of vinylphosphine-borane complexes.

Vinylphosphine boranes find applications as useful building blocks for organic and organometallic chemistry.

In the following year, Stawinski and co-workers (Kalek et al., 2008) developed a general and efficient method for the microwave-assisted preparation of a broad spectrum of aryl- and vinylphosphonates (**15** and **17**, respectively) based on a Pd-catalyzed cross-coupling using Pd(PPh$_3$)$_4$ as a prevalent palladium catalyst and either cesium carbonate or triethylamine as a base in THF; the reaction was completed within ten minutes, giving good yields (Scheme 7.4). It is also noteworthy that the reaction is completely stereospecific with retention of configuration at the phosphorus center and, at the same time, the configuration in the vinyl moiety remains preserved.

The investigators used their protocol to synthesize a few more complex P-compounds bearing nucleoside and cholesteryl moieties with high yields, viz. **16** and **18** (Scheme 7.5)

Andaloussi et al. (2009) used palladium acetate for the Pd(II)-catalyzed P-C coupling reaction between aryl/vinyl boronic acids or trifluoroborates (**19**) and dialkyl phosphites (**2**) in the presence of a rigid bidentate ligand dmphen (dmphen = 2,9-dimethyl-1,10-phenanthroline) and p-benzoquinone as an oxidant to have arylphosphonate diester **20**/vinylphosphophate diester **22** without the addition of base or acid under microwave irradiation in DMF solvent (Scheme 7.6). The investigators showed that the reaction takes place with excellent chemoselectivity, and they extended this feature in synthesizing a *Mycobacterium tuberculosis* glutamine

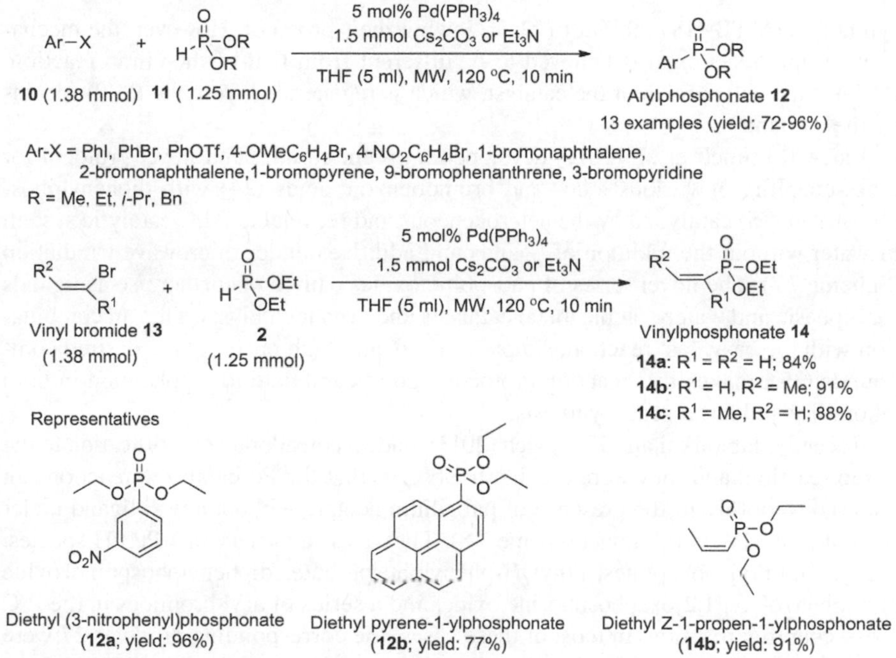

**SCHEME 7.4** Palladium-catalyzed synthesis of aryl- and vinylphosphonate diesters.

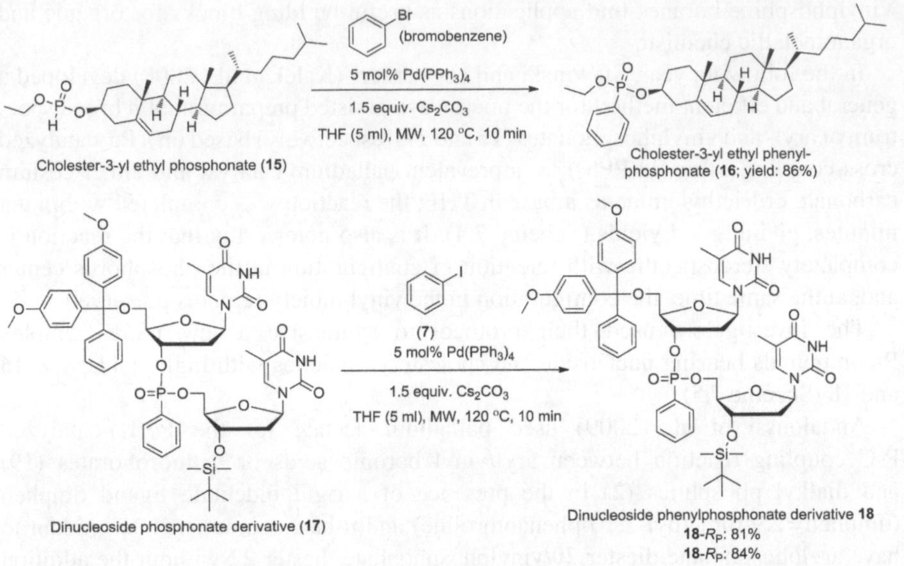

**SCHEME 7.5**  Palladium-catalyzed synthesis of phosphonate diesters linked with nucleoside and cholesteryl moieties.

synthetase (MTB-GS) inhibitor (**23**) following their protocol. However, the mechanism of this reaction was believed to be different from that of the Hirao reaction; Pd(II) is the active form of the catalyst, which is regenerated from the Pd(0) species by the oxidant.

Later, Rummelt et al. (2012) developed a green, simple, and novel protocol for cross-coupling of various iodo- and bromobenzoic acids (**24**) with diphenylphosphine oxide (**5**) catalyzed by the heterogeneous and recyclable Pd/C catalytic system in water without the addition of ligands and additives under microwave irradiation (Scheme 7.7). The novel series of phosphine oxides can be of further use as ligands for biphasic and water-soluble metal catalysis and organocatalysis. This, in combination with the easy, fast reaction, simple workup, and high purity of the desired compounds (**25**), makes this reaction protocol a good candidate for application in both laboratory and large-scale synthesis.

Recently, Jablonkai and Keglevich (2013) made a considerable modification to the Hirao reaction, and they were the first to observe that the Pd-catalyzed reaction can undergo smoothly in the presence of palladium acetate without any *P*-ligand under solvent-free MW conditions (Scheme 7.8). They used a variety of >P(O)H species, such as dialkyl phosphites, ethyl *H*-phenylphosphinate, diphenylphosphine oxide and dibenzo[*c*,*e*][1,2]oxaphosphorine oxide, and a series of aryl bromides in the P-C cross-coupling reaction. In most of these cases, the corresponding products **27** were obtained in high yields (73–95%). The investigators also compared the MW-assisted reactions with those carried out under thermal conditions, and the results revealed

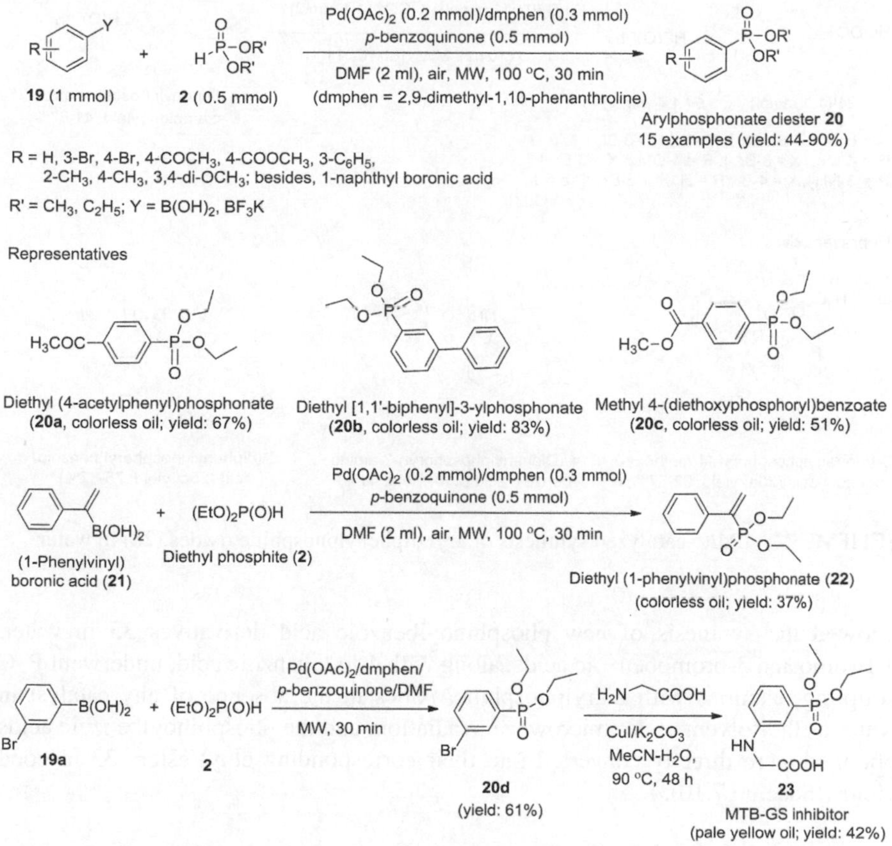
Pd(OAc)2 (0.2 mmol)/dmphen (0.3 mmol)
p-benzoquinone (0.5 mmol)
DMF (2 ml), air, MW, 100 °C, 30 min
(dmphen = 2,9-dimethyl-1,10-phenanthroline)
19 (1 mmol)
2 (0.5 mmol)
Arylphosphonate diester 20
15 examples (yield: 44-90%)
R = H, 3-Br, 4-Br, 4-COCH3, 4-COOCH3, 3-C6H5,
2-CH3, 4-CH3, 3,4-di-OCH3; besides, 1-naphthyl boronic acid
R' = CH3, C2H5; Y = B(OH)2, BF3K
Representatives
Diethyl (4-acetylphenyl)phosphonate
(20a, colorless oil; yield: 67%)
Diethyl [1,1'-biphenyl]-3-ylphosphonate
(20b, colorless oil; yield: 83%)
Methyl 4-(diethoxyphosphoryl)benzoate
(20c, colorless oil; yield: 51%)
Pd(OAc)2 (0.2 mmol)/dmphen (0.3 mmol)
p-benzoquinone (0.5 mmol)
DMF (2 ml), air, MW, 100 °C, 30 min
(1-Phenylvinyl)
boronic acid (21)
Diethyl phosphite (2)
Diethyl (1-phenylvinyl)phosphonate (22)
(colorless oil; yield: 37%)
Pd(OAc)2/dmphen/
p-benzoquinone/DMF
MW, 30 min
19a
2
20d
(yield: 61%)
CuI/K2CO3
MeCN-H2O,
90 °C, 48 h
23
MTB-GS inhibitor
(pale yellow oil; yield: 42%)

**SCHEME 7.6** Pd(II)-catalyzed synthesis of aryl/vinylphosphonate diesters (**20** and **22**).

the specific influence of MW irradiation. Under thermal conditions, lower conversions and yields were observed. The investigators claimed their report as the first occasion of *P*-ligand-free Hirao reaction.

In 2014, the Keglevich group extended the same reaction with more variety of both the substrates, and the reaction seems to be of general value (Keglevich et al., 2014); thus, such a *P*-ligand-free variation demands to be a generally useful and environmentally friendly method. In the following year, the same group of investigators (Jablonkai et al., 2015) demonstrated a new P-ligand-free nickel-catalyzed and MW-assisted variation of the Hirao reaction where the P-C coupling reaction of bromoarenes and >P(O)H species, such as dialkyl phosphites, alkyl phenyl-*H*-phosphinates, and diarylphosphine oxides afforded the dialkyl arylphosphonates, alkyl diphenylphosphinates, and triaryl phosphine oxides, respectively, with a few exceptions, in yields of 75–92% (Scheme 7.9).

In the same year, Jablonkai and Keglevich (2015) demonstrated an environmentally friendly catalyst-free and MW-assisted variation of the Hirao reaction, which

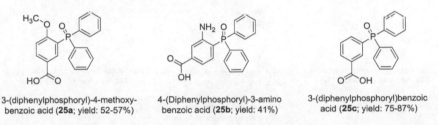

**24** (1.1 mmol)      **5** ( 1.4 mmol)

Aryl diphenylphosphine oxide **25**
6 examples (yield: 41-87%)

R = H; X = 4-Br, Cl, I; R = H; X = 3-Br, I
R = 4-Me;  X = 3-Br, I; R = 4-OMe;  X = 3-Br, I
R = 3-NH₂; X = 4-Br; R = H; X' = 3-Br, X" = 5-I

Representatives

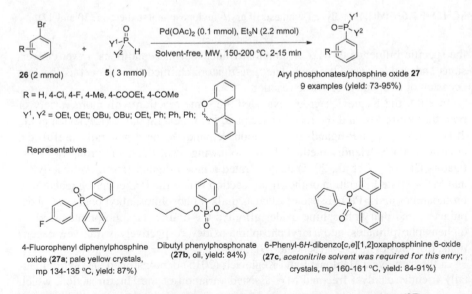

3-(diphenylphosphoryl)-4-methoxy-
benzoic acid (**25a**; yield: 52-57%)

4-(Diphenylphosphoryl)-3-amino
benzoic acid (**25b**; yield: 41%)

3-(diphenylphosphoryl)benzoic
acid (**25c**; yield: 75-87%)

**SCHEME 7.7**   Pd/C-catalyzed synthesis of aryl diphenylphosphine oxides (**28**) in water.

allowed the synthesis of new phosphinoylbenzoic acid derivatives **33** in water. 4-Bromo and 3-bromobenzoic acids, along with 4-iodobenzoic acid, underwent P–C coupling reactions with diarylphosphine oxides in the absence of any catalyst in water as the solvent under microwave irradiation, and the phosphinoylbenzoic acids obtained were directly converted into their corresponding ethyl esters **33** in good yields (Scheme 7.10).

**26** (2 mmol)      **5** ( 3 mmol)

Aryl phosphonates/phosphine oxide **27**
9 examples (yield: 73-95%)

R = H, 4-Cl, 4-F, 4-Me, 4-COOEt, 4-COMe

Y¹, Y² = OEt, OEt; OBu, OBu; OEt, Ph; Ph, Ph;

Representatives

4-Fluorophenyl diphenylphosphine
oxide (**27a**; pale yellow crystals,
mp 134-135 °C, yield: 87%)

Dibutyl phenylphosphonate
(**27b**, oil, yield: 84%)

6-Phenyl-6*H*-dibenzo[*c,e*][1,2]oxaphosphinine 6-oxide
(**27c**, *acetonitrile solvent was required for this entry*;
crystals, mp 160-161 °C, yield: 84-91%)

**SCHEME 7.8**   P-Ligand-free synthesis of arylphosphonates/phosphine oxides (**27**).

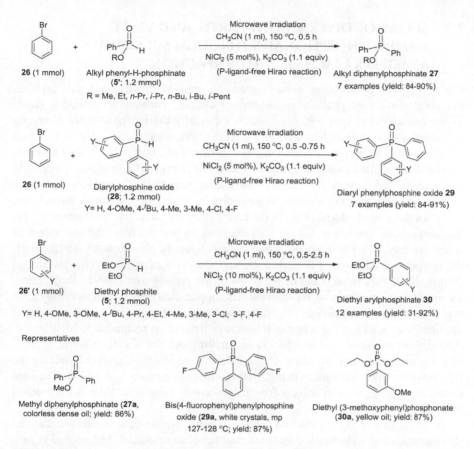

**SCHEME 7.9** Microwave-assisted nickel chloride-catalyzed synthesis of diverse phosphinates/phosphine oxides via P-ligand-free Hirao reaction.

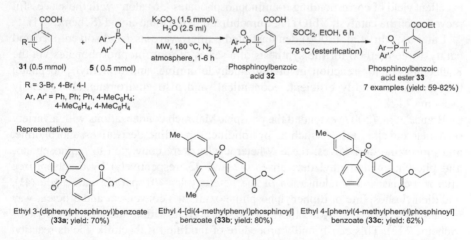

**SCHEME 7.10** Catalyst-free synthesis of phosphinoylbenzoate esters (**33**) in water via ligand-free Hirao reaction under microwave irradiation.

## 7.3   ACCESS OF DIVERSE ALKYL-, ARYL-, AND VINYL α-AMINOPHOSPHONATES THROUGH MICROWAVE-ASSISTED KABACHNIK–FIELDS REACTION

Kabachnik–Fields reaction (also called phospha-Mannich reaction) involves the condensation of primary or secondary amines, carbonyl compounds (aldehydes and ketones), and >P(O)H species, especially dialkyl phosphites affording α-aminophosphonates (Cherkasov and Galkin, 1998; Fields, 1952; Kabachnik and Medved, 1952; Keglevich and Bálint, 2012; Zefirov and Matveeva, 2008). This reaction has been studied enormously and extended to different >P(O)H species, including cyclic phosphites, acyclic and cyclic H-phosphinates, as well as secondary phosphine oxides, and also varying amines, to have diverse α-aminophosphonic, phosphinic, and phosphine oxide derivatives, in the recent past. As far as the mechanism of this reaction is concerned, it is still not fully clarified. In general, these three-component reactions may take place via an imine or an α-hydroxy-phosphonate intermediate, depending on structures of the substrates and also on the conditions (Keglevich and Bálint, 2012); this issue is out of the scope of the present overview. In this section, an up-to-date development on the synthetic applications of the Kabachnik–Fields reaction in generating diverse α-aminophosphonates of promising multidirectional applications under the influence of microwave-irradiation technique is highlighted. Application of microwave irradiation in carrying out the Kabachnik–Fields reaction has appeared to be the most straightforward and effective protocol yielding the desired α-aminophosphonates in high yields under moderate conditions, and, in most cases, the synthesis under solvent-free microwave-assisted conditions is the method of choice.

In 2002, Lee et al. reported on the use of ionic liquids in a microwave assisted three-component Kabachnik–Fields reaction between aromatic aldehydes (**34**), aniline (**35**), and diethyl phosphite (**2**) in [bmim]BF$_4$ in the presence of ytterbium triflate as a catalyst; the reaction proceeded very fast (within 2 minutes), resulting in an excellent yield of corresponding α-aminophosphonates (**36**) along with the successful recycling of the catalyst, Yb(OTf)$_3$, immobilized in the ionic-liquid (Scheme 7.11).

Later on, Mu et al. (2006) demonstrated a microwave-assisted solvent-free and catalyst-free method for the synthesis of a series of α-aminophosphonates via the Kabachnik–Fields reaction involving aldehyde, amine, and dimethyl phosphate; the process is highly efficient, economical, and also environmentally friendly (Scheme 7.12).

Prauda et al. (2007) extended the phospha-Mannich condensations with a variety of N-heterocyclic amines, such as pyrrolidine, piperidine derivatives, morpholine, and piperazine derivatives; these N-heterocycles were converted to N-phosphono- and phosphinoxidomethyl derivatives (**42** and **43**, respectively) by a solvent-free microwave-assisted condensation of the heterocycle (**40**), paraformaldehyde (**41**), and diethyl phosphite or diphenylphosphine oxide in a convenient and efficient way in the presence of 37% hydrochloric acid as a catalyst, with moderate to good yields (Scheme 7.13). This eco-friendly procedure of modified Kabachnik–Fields reaction under solvent-free microwave conditions can obviously be extended to other types of phospha-Mannich reactions as well.

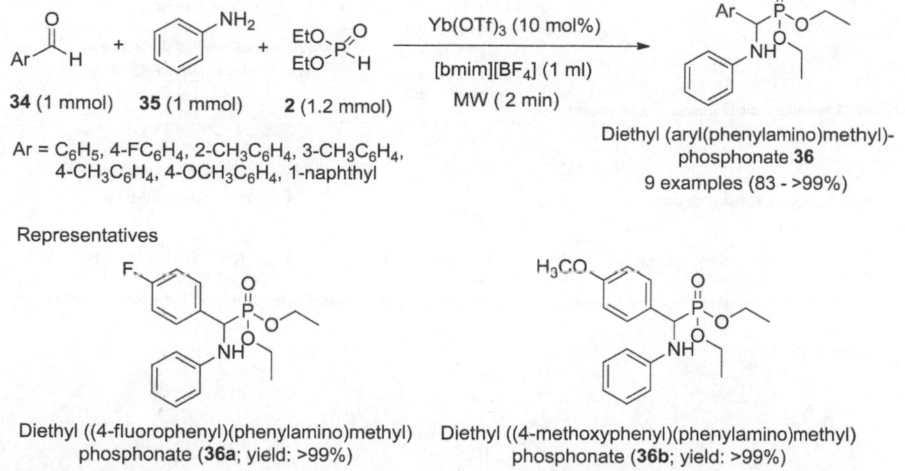
Yb(OTf)3 (10 mol%)
[bmim][BF4] (1 ml)
MW ( 2 min)
34 (1 mmol) 35 (1 mmol) 2 (1.2 mmol)
Ar = C6H5, 4-FC6H4, 2-CH3C6H4, 3-CH3C6H4, 4-CH3C6H4, 4-OCH3C6H4, 1-naphthyl
Diethyl (aryl(phenylamino)methyl)-phosphonate 36
9 examples (83 - >99%)
Representatives
Diethyl ((4-fluorophenyl)(phenylamino)methyl) phosphonate (36a; yield: >99%)
Diethyl ((4-methoxyphenyl)(phenylamino)methyl) phosphonate (36b; yield: >99%)

**SCHEME 7.11** Synthesis of α-aminophosphonates in [bmim][$BF_4$] ionic-liquid in the presence of $Yb(OTf)_3$ as catalyst.

In the same year, Lv et al. (2007) also reported a fast, efficient, and solvent-free one-pot method for the synthesis of diverse fluorinated pharmaceutically promising α-aminoalkylphosphonates (**46**) in good yields by a three-component reaction of fluorinated aromatic aldehydes (**44**), anilines (**45**), and diethyl phosphite in the presence of silica-gel support under microwave irradiation (Scheme 7.14).

Earlier in 2001, Kaboudin and Nazari also developed a simple, efficient, and general method for the synthesis of 1-aminoalkylphosphonates on a solid surface under microwave irradiation; acidic alumina under solvent-free conditions was capable of producing high yields of 1-aminoalkylphosphonates (**47**) from the one-pot reaction

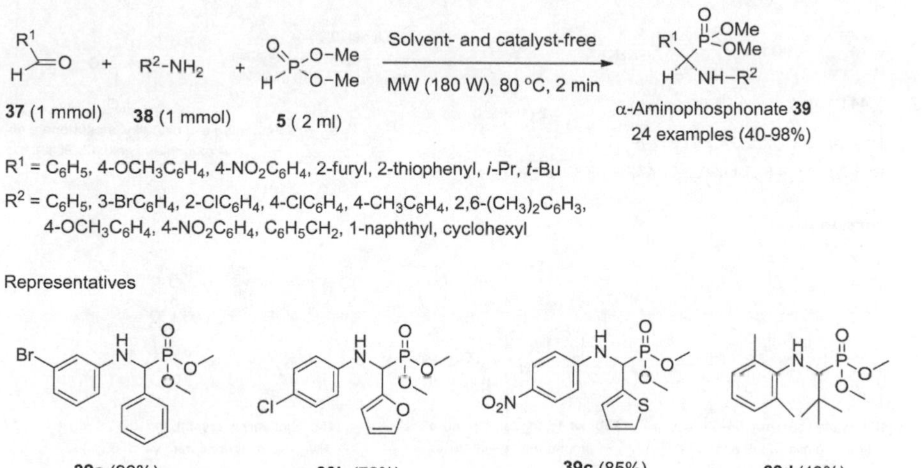
Solvent- and catalyst-free
MW (180 W), 80 °C, 2 min
37 (1 mmol) 38 (1 mmol) 5 ( 2 ml)
α-Aminophosphonate 39
24 examples (40-98%)
R1 = C6H5, 4-OCH3C6H4, 4-NO2C6H4, 2-furyl, 2-thiophenyl, i-Pr, t-Bu
R2 = C6H5, 3-BrC6H4, 2-ClC6H4, 4-ClC6H4, 4-CH3C6H4, 2,6-(CH3)2C6H3, 4-OCH3C6H4, 4-NO2C6H4, C6H5CH2, 1-naphthyl, cyclohexyl
Representatives
39a (96%)
39b (76%)
39c (85%)
39d (40%)

**SCHEME 7.12** Solvent-free and catalyst-free synthesis of α-aminophosphonates (**39**).

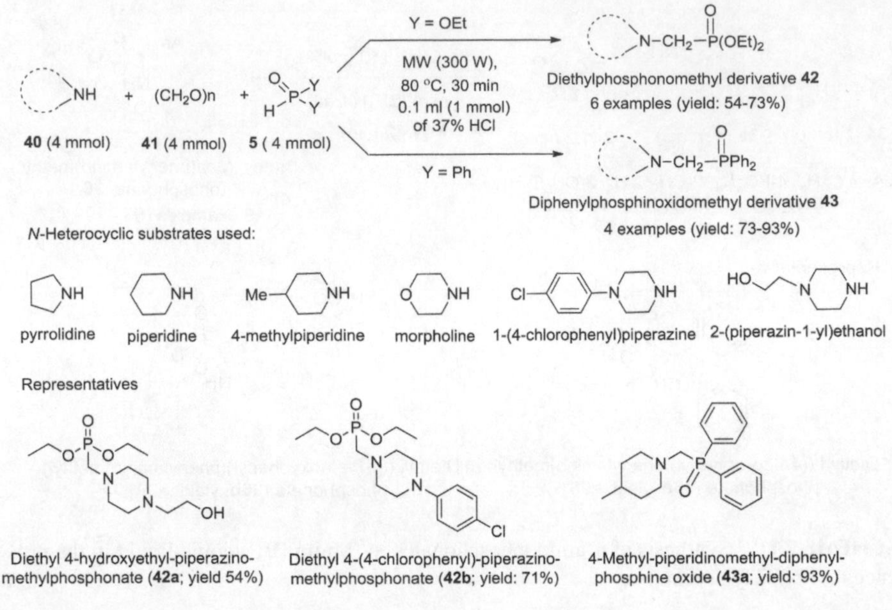

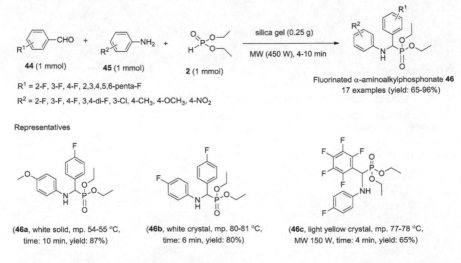

**SCHEME 7.13** Solvent-free synthesis of phosphonomethyl- and phosphinoxidomethyl *N*-heterocycles (**42/43**).

of aldehydes, amines, and diethyl phosphonate under mild reaction conditions (Scheme 7.15).

Amberlite-IR 120 was also found to be an efficient catalyst for the one-pot reaction of aldehydes, amines, and diethyl phosphite to afford α-aminophosphonates in good to excellent yields. Recently, Bhattacharya and Rana (2008) applied this resin

**SCHEME 7.14** Solvent-free solid-supported synthesis of fluorine-containing α-aminophosphonates (**46**).

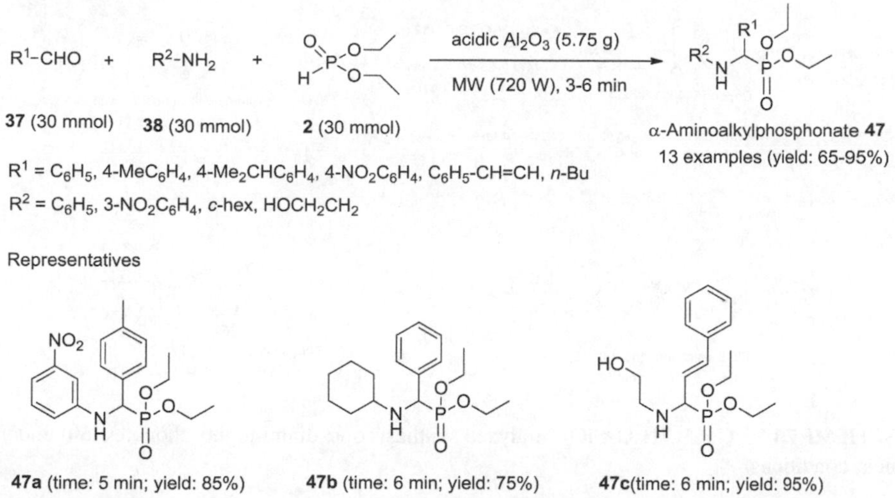

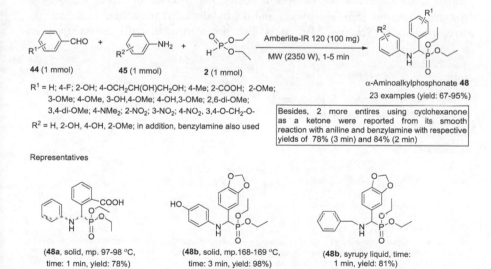

$R^1$-CHO + $R^2$-NH$_2$ + [phosphite] $\xrightarrow{\text{acidic Al}_2\text{O}_3 \text{ (5.75 g)}}$ [product]

$\qquad\qquad\qquad\qquad\qquad\qquad$ MW (720 W), 3-6 min

**37** (30 mmol) $\quad$ **38** (30 mmol) $\quad$ **2** (30 mmol)

α-Aminoalkylphosphonate **47**
13 examples (yield: 65-95%)

$R^1 = C_6H_5$, 4-MeC$_6$H$_4$, 4-Me$_2$CHC$_6$H$_4$, 4-NO$_2$C$_6$H$_4$, C$_6$H$_5$-CH=CH, *n*-Bu

$R^2 = C_6H_5$, 3-NO$_2$C$_6$H$_4$, *c*-hex, HOCH$_2$CH$_2$

Representatives

**47a** (time: 5 min; yield: 85%) $\qquad$ **47b** (time: 6 min; yield: 75%) $\qquad$ **47c** (time: 6 min; yield: 95%)

**SCHEME 7.15** Solvent-free solid-supported synthesis of α-aminophosphonates (**47**).

substance as a useful catalyst in carrying out the one-pot three-component transformations, mainly with a variety of aromatic aldehydes and anilines, under the influence of microwave irradiation in the absence of any solvent (Scheme 7.16). The main advantages of the present synthetic protocol are mild and solvent-free conditions, eco-friendly catalyst, and easy reaction work-up procedures.

In 2013, Devineni et al. reported on the synthesis of α-diaminophosphonates (**50**) having *in vitro* antifungal and antioxidant activities using hydrated ceric chloride supported on silica (CeCl$_3$·7H$_2$O-SiO$_2$) as a heterogeneous, efficient, and recyclable

$R^1$-⟨⟩-CHO + $R^2$-⟨⟩-NH$_2$ + [phosphite] $\xrightarrow{\text{Amberlite-IR 120 (100 mg)}}$ [product]

$\qquad\qquad\qquad\qquad\qquad\qquad$ MW (2350 W), 1-5 min

**44** (1 mmol) $\quad$ **45** (1 mmol) $\quad$ **2** (1 mmol)

α-Aminoalkylphosphonate **48**
23 examples (yield: 67-95%)

$R^1$ = H; 4-F; 2-OH; 4-OCH$_2$CH(OH)CH$_2$OH; 4-Me; 2-COOH; 2-OMe;
$\quad$ 3-OMe; 4-OMe, 3-OH,4-OMe; 4-OH,3-OMe; 2,6-di-OMe;
$\quad$ 3,4-di-OMe; 4-NMe$_2$; 2-NO$_2$; 3-NO$_2$; 4-NO$_2$, 3,4-O-CH$_2$-O-

$R^2$ = H, 2-OH, 4-OH, 2-OMe; in addition, benzylamine also used

Besides, 2 more entires using cyclohexanone as a ketone were reported from its smooth reaction with aniline and benzylamine with respective yields of 78% (3 min) and 84% (2 min)

Representatives

(**48a**, solid, mp. 97-98 °C, time: 1 min, yield: 78%) $\qquad$ (**48b**, solid, mp.168-169 °C, time: 3 min, yield: 98%) $\qquad$ (**48b**, syrupy liquid, time: 1 min, yield: 81%)

**SCHEME 7.16** Solvent-free amberlite-IR 120 catalyzed synthesis of α-aminophosphonates (**48**).

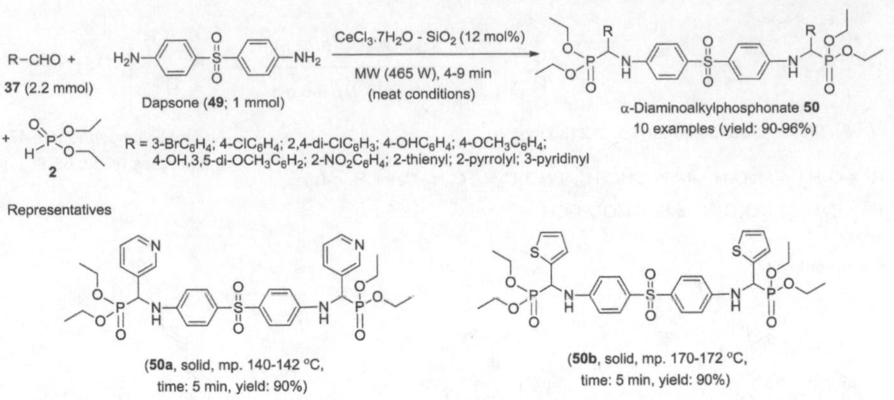

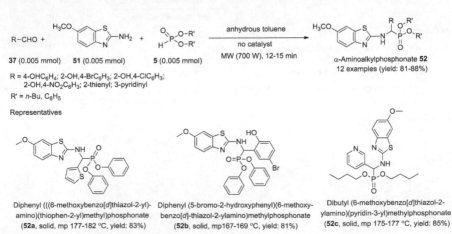

**SCHEME 7.17** CeCl$_3$.7H$_2$O-SiO$_2$ catalyzed synthesis of α-diaminophosphonates (**50**) under neat conditions.

catalyst from the one-pot reaction between aldehydes (**37**), dapsone (**49**), and diethyl phosphite under microwave irradiation (Scheme 7.17). The method is fast, efficient, eco-friendly, and high yielding (90–96%).

Another series of novel α-aminophosphonates (**52**) from a one-pot condensation reaction between substituted aromatic/heterocyclic aldehydes (**37**), 2-amino-6-methoxy-benzothiazole (**51**), and dibutyl/diphenyl phosphites in toluene was achieved by Janardhan Rao et al. (2010) using microwave irradiation in the absence of any catalyst (Scheme 7.18). The synthesized compounds showed promising antimicrobial, anti-oxidant activities depending on the nature of bioactive groups at the α-carbon.

Reddy et al. (2014) prepared a series of substituted (3,5-dichloro-4-hydroxyphenyl)amino methylphosphonate derivatives (**54**) using amberlyst-15 as a heterogeneous catalyst from a solvent-free one-pot condensation reaction of 3,5-dichloro-4-hydroxyphenyl amine (**53**), aldehydes, and diethyl/dimethyl phosphate under microwave irradiation (Scheme 7.19). It is also noteworthy that all the synthesized

**SCHEME 7.18** Catalyst-free synthesis of α-aminophosphonates (**52**) in toluene.

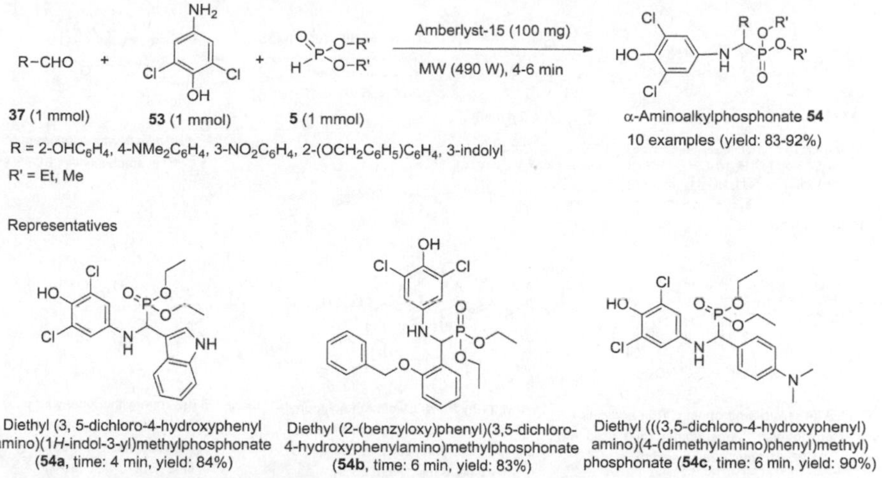

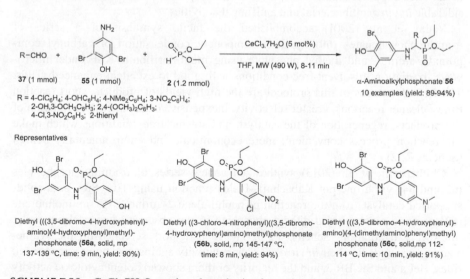

**SCHEME 7.19** One pot solvent-free synthesis of α-aminophosphonates (**54**) using amberlyst-15 as a heterogeneous catalyst.

compounds exhibited considerable antioxidant activity, as assessed from their DPPH radical scavenging, nitric oxide (NO) scavenging, and reducing power assays.

Recently, Varalashmi et al. (2014) prepared a new class of diethyl(3,5-dibromo-4-hydroxyphenylamino) (substituted phenyl/heterocyclic) methylphosphonates (**56**) by a one-pot three component Kabachnik–Fields reaction of 4-amino-2,6-dibromophenol, substituted heterocyclic/phenyl aldehydes, and diethyl phosphite using hydrated ceric chloride (5 mol%) under microwave irradiation in THF (Scheme 7.20). The reaction is rapid and highly efficient. Moreover, the investigators evaluated *in*

**SCHEME 7.20** CeCl$_3$.7H$_2$O catalyzed synthesis of α-aminophosphonates (**56**) in THF.

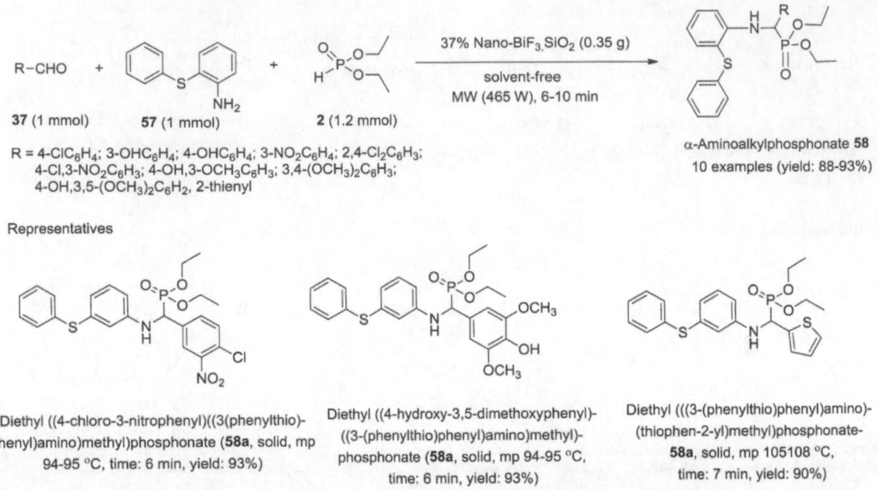

SCHEME 7.21  Nano-BiF$_3$-SiO$_2$ catalyzed solvent-free synthesis of α-aminophosphonates (**58**).

*vitro* antimicrobial activity for the synthesized compounds and found that most of them exhibited potent to moderate activity against the tested pathogens.

The same group of investigators (Rajasekhar et al., 2013) also synthesized a new class of diethyl α-aryl/2-thienyl-α-[2-(phenylthio)phenylamino]methylphosphonates (**58**) via a solvent-free three-component Kabachnik–Fields reaction of 2-aminodiphenylsulfide (**57**), substituted phenyl/heterocyclic aldehydes, and diethyl phosphate in the presence of heterogeneous nano-silica supported nano-BF$_3$.SiO$_2$ under microwave irradiation (Scheme 7.21). The procedure has several advantages, such as short reaction time, low loading of catalyst, good yields, and reusability of the heterogamous silica-supported nano-catalyst. The title compounds were found to exhibit considerable *in vitro* antibacterial and antifungal activities.

Yadav et al. (2001) accomplished the facile synthesis of a series of α-aminophosphonates (**60**) by three-component condensation of carbonyl compounds, amines, and diethyl phosphite using montmorillonite-KSF under microwave irradiation in solvent-free conditions with good to excellent enhanced yields. The key advantages of this protocol are the high yields of products, short reaction times, cleaner reactions, greater selectivity, inexpensive catalyst, ease of isolation of the products, regeneration of the catalyst, and solvent-free conditions which make the reaction process convenient, more economical, and environmentally benign (Scheme 7.22).

Chithaparthi et al. (2013) synthesized a new series of α-aminophosphonates (**62** and **64**) via a one-pot Kabachnik–Fields reaction using TiO$_2$-SiO$_2$ as a solid-supported catalyst from the reaction of naphthalene-2-amine (**61**)/2-aminofluorene (**63**), varying aldehyes and ditheyl phosphate in the absence of any solvent under microwave irradiation (Scheme 7.23). The investigators screened all the synthesized compounds for their *in vitro* anticancer activity against two human cancer cell lines, HeLa and SK-BR-3, and the majority of them showed potent cytotoxic activity against the cancer cell lines tested.

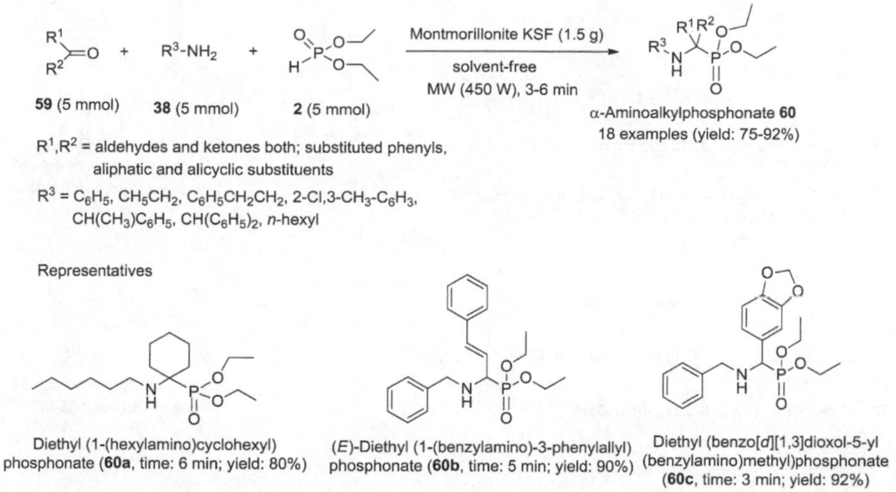

59 (5 mmol)   38 (5 mmol)   2 (5 mmol)

$R^1, R^2$ = aldehydes and ketones both; substituted phenyls, aliphatic and alicyclic substituents

$R^3$ = $C_6H_5$, $CH_5CH_2$, $C_6H_5CH_2CH_2$, 2-Cl,3-$CH_3$-$C_6H_3$, $CH(CH_3)C_6H_5$, $CH(C_6H_5)_2$, n-hexyl

α-Aminoalkylphosphonate 60
18 examples (yield: 75-92%)

Representatives

Diethyl (1-(hexylamino)cyclohexyl) phosphonate (60a, time: 6 min; yield: 80%)

(E)-Diethyl (1-(benzylamino)-3-phenylallyl) phosphonate (60b, time: 5 min; yield: 90%)

Diethyl (benzo[d][1,3]dioxol-5-yl (benzylamino)methyl)phosphonate (60c, time: 3 min; yield: 92%)

**SCHEME 7.22** Clay-catalyzed one-pot solvent-free synthesis of α-aminophosphonates (60).

Recently, Gangireddy et al. (2014) elaborated the synthetic scope for the 2-aminofluorene-substituted phosphate derivatives (64) using a variety of aliphatic/alicyclic/aromatic aldehydes via a three-component one-pot Kabachnik–Fields reaction under solvent-free microwave irradiation technique and by replacing the catalyst with polystyrene-supported p-toluenesulfonic acid (PS/PTSA). This procedure is also a simple, high yielding, straightforward, and environmentally friendly method for the synthesis of biologically relevant α-aminophosphonate scaffolds (Scheme 7.24).

In continuation of such developments, a catalyst-free and solvent-free method for the synthesis of a wide range of both α-aminophosphonates and α-aminophosphine oxides was developed by Keglevich and Szekrényi (2008) out of the three-component one-pot Kabachnik–Fields reaction of carbonyl compounds, amines, and dialkyl phophites or diphenylphosphine oxide under microwave irradiation (Scheme 7.25). This method is quite general and straightforward in the absence of any catalyst and solvent. The investigators (Keglevich and Szekrényi, 2008) were successful in extending such phospha-Mannich reaction conveniently via the one-pot solvent-microwave-assisted condensation of simple N-heterocycles with paraformaldehyde and diethyl phosphite or diphenylphosphine oxide, thereby furnishing diverse phosphonomethyl- and phosphinoxidomethyl N-heterocycle derivatives (Scheme 7.26).

Keglevich et al. (2008) investigated extensively on the phospha-Mannich reaction to modify the reaction conditions as well as to extend the scope. In another report, they demonstrated a catalyst-free synthesis of cyclic aminomethylphosphonates from the microwave-assisted condensation of 1,3,2-dioxaphosphinane 2-oxide (68), paraformaldehyde, and both cyclic and acyclic secondary amines; the reaction was extended to afford also the aminomethyl-arylphosphinic acids from an analogous reaction involving dibenzo[c,e][1,2]oxaphosphinane 2- oxide (68a), instead of the

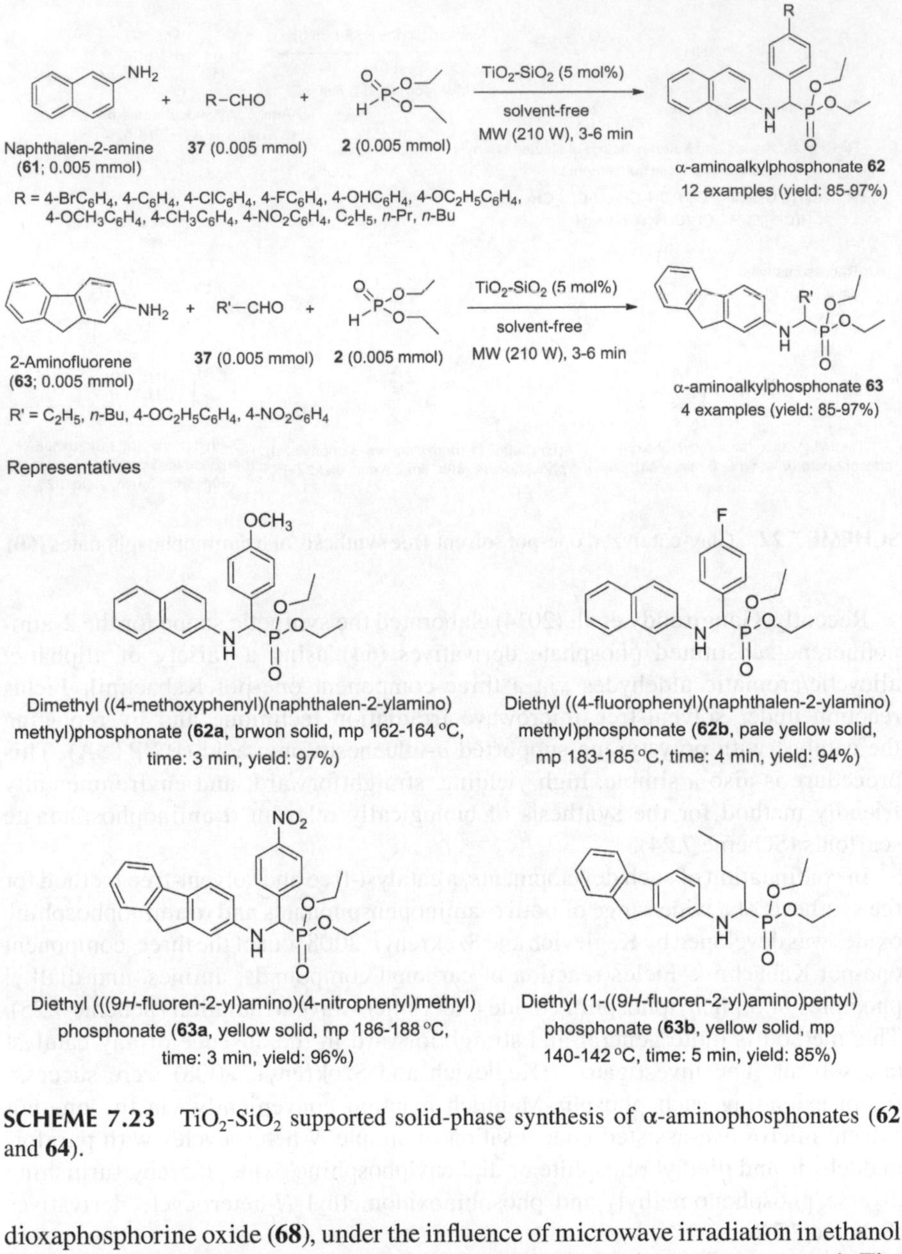

Naphthalen-2-amine  **37** (0.005 mmol)   **2** (0.005 mmol)
(**61**; 0.005 mmol)

R = 4-BrC$_6$H$_4$, 4-C$_6$H$_4$, 4-ClC$_6$H$_4$, 4-FC$_6$H$_4$, 4-OHC$_6$H$_4$, 4-OC$_2$H$_5$C$_6$H$_4$,
4-OCH$_3$C$_6$H$_4$, 4-CH$_3$C$_6$H$_4$, 4-NO$_2$C$_6$H$_4$, C$_2$H$_5$, n-Pr, n-Bu

α-aminoalkylphosphonate **62**
12 examples (yield: 85-97%)

2-Aminofluorene  **37** (0.005 mmol)   **2** (0.005 mmol)
(**63**; 0.005 mmol)

R' = C$_2$H$_5$, n-Bu, 4-OC$_2$H$_5$C$_6$H$_4$, 4-NO$_2$C$_6$H$_4$

α-aminoalkylphosphonate **63**
4 examples (yield: 85-97%)

**Representatives**

Dimethyl ((4-methoxyphenyl)(naphthalen-2-ylamino)
methyl)phosphonate (**62a**, brwon solid, mp 162-164 °C,
time: 3 min, yield: 97%)

Diethyl ((4-fluorophenyl)(naphthalen-2-ylamino)
methyl)phosphonate (**62b**, pale yellow solid,
mp 183-185 °C, time: 4 min, yield: 94%)

Diethyl (((9H-fluoren-2-yl)amino)(4-nitrophenyl)methyl)
phosphonate (**63a**, yellow solid, mp 186-188 °C,
time: 3 min, yield: 96%)

Diethyl (1-((9H-fluoren-2-yl)amino)pentyl)
phosphonate (**63b**, yellow solid, mp
140-142 °C, time: 5 min, yield: 85%)

**SCHEME 7.23**   TiO$_2$-SiO$_2$ supported solid-phase synthesis of α-aminophosphonates (**62**
and **64**).

dioxaphosphorine oxide (**68**), under the influence of microwave irradiation in ethanol
(Scheme 7.27). In this way, "double" heterocyclic derivatives were prepared. The
methodology was simple and offered a straightforward route to have diverse kinds of
target compounds of pharmaceutical interests.

Recently, Bálint et al. (2013) synthesized a series of new N-(2H-pyranonyl)-α-
aminophosphonates or -α-aminophosphine oxides (**72**) through a microwave-assisted

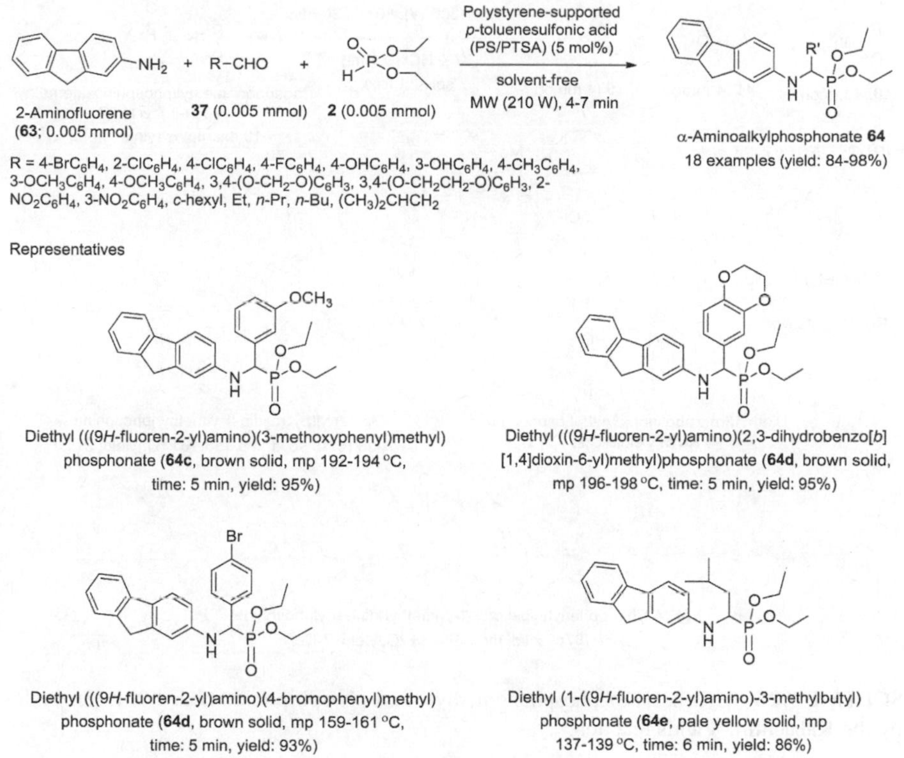

R = 4-BrC$_6$H$_4$, 2-ClC$_6$H$_4$, 4-ClC$_6$H$_4$, 4-FC$_6$H$_4$, 4-OHC$_6$H$_4$, 3-OHC$_6$H$_4$, 4-CH$_3$C$_6$H$_4$, 3-OCH$_3$C$_6$H$_4$, 4-OCH$_3$C$_6$H$_4$, 3,4-(O-CH$_2$-O)C$_6$H$_3$, 3,4-(O-CH$_2$CH$_2$-O)C$_6$H$_3$, 2-NO$_2$C$_6$H$_4$, 3-NO$_2$C$_6$H$_4$, c-hexyl, Et, n-Pr, n-Bu, (CH$_3$)$_2$CHCH$_2$

Representatives

Diethyl (((9H-fluoren-2-yl)amino)(3-methoxyphenyl)methyl) phosphonate (64c, brown solid, mp 192-194 °C, time: 5 min, yield: 95%)

Diethyl (((9H-fluoren-2-yl)amino)(2,3-dihydrobenzo[b] [1,4]dioxin-6-yl)methyl)phosphonate (64d, brown solid, mp 196-198 °C, time: 5 min, yield: 95%)

Diethyl (((9H-fluoren-2-yl)amino)(4-bromophenyl)methyl) phosphonate (64d, brown solid, mp 159-161 °C, time: 5 min, yield: 93%)

Diethyl (1-((9H-fluoren-2-yl)amino)-3-methylbutyl) phosphonate (64e, pale yellow solid, mp 137-139 °C, time: 6 min, yield: 86%)

**SCHEME 7.24** PS/PTSA catalyzed solid-phase synthesis of α-aminophosphonates (64).

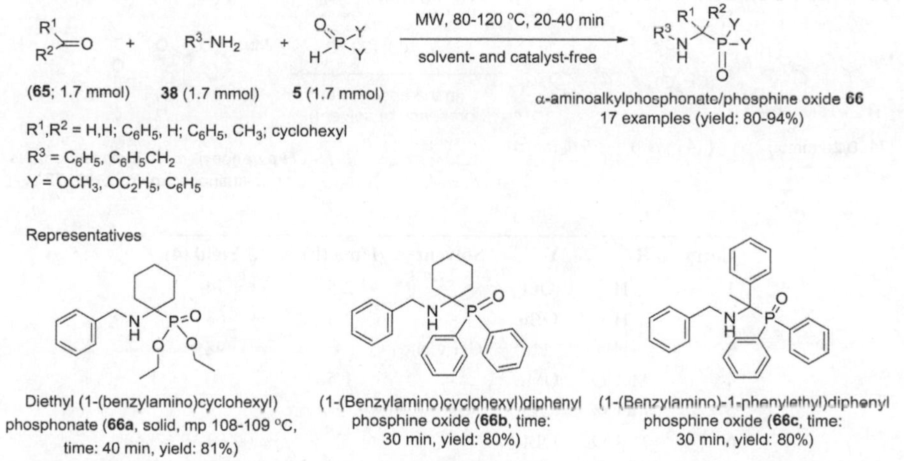

R$^1$,R$^2$ = H,H; C$_6$H$_5$, H; C$_6$H$_5$, CH$_3$; cyclohexyl
R$^3$ = C$_6$H$_5$, C$_6$H$_5$CH$_2$
Y = OCH$_3$, OC$_2$H$_5$, C$_6$H$_5$

Representatives

Diethyl (1-(benzylamino)cyclohexyl) phosphonate (66a, solid, mp 108-109 °C, time: 40 min, yield: 81%)

(1-(Benzylamino)cyclohexyl)diphenyl phosphine oxide (66b, time: 30 min, yield: 80%)

(1-(Benzylamino)-1-phenylethyl)diphenyl phosphine oxide (66c, time: 30 min, yield: 80%)

**SCHEME 7.25** Catalyst- and solvent-free synthesis of α-aminoalkylphosphonates/phosphine oxides (66).

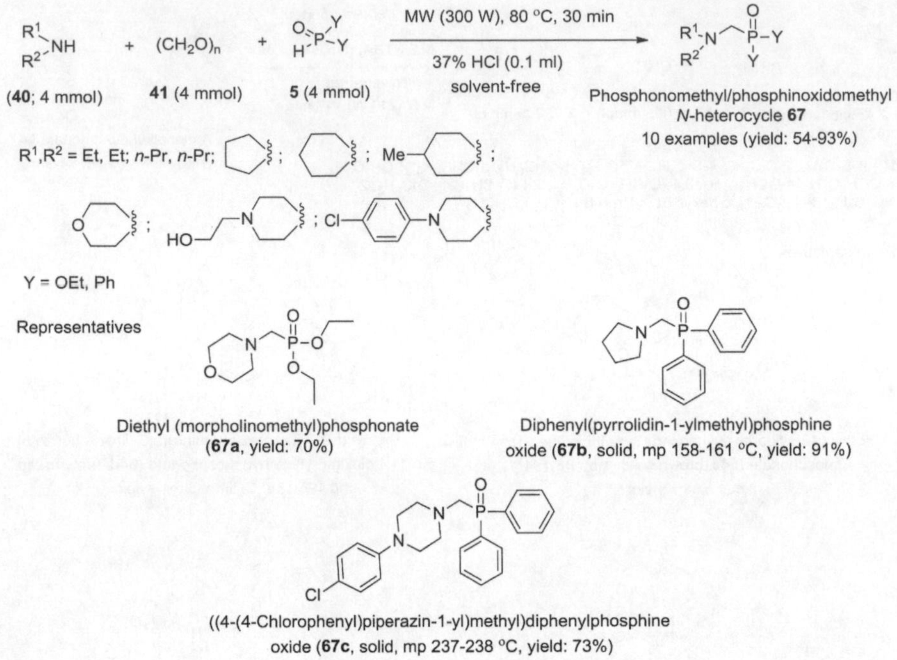

SCHEME 7.26  Synthesis of phosphonomethyl- and phosphinoxidomethyl *N*-heterocycles by the Kabachnik–Fields reaction.

Kabachnik–Fields reaction of 5-unsubstituted and 5-substituted (5-acetyl/benzoyl)-3-amino-6-methyl-2*H*-pyran-2-ones (**71**) with paraformaldehyde and dialkyl phosphites or diphenylphosphine oxide, in most cases, under solvent-free conditions (Scheme 7.28). The protocol is straightforward and high yielding.

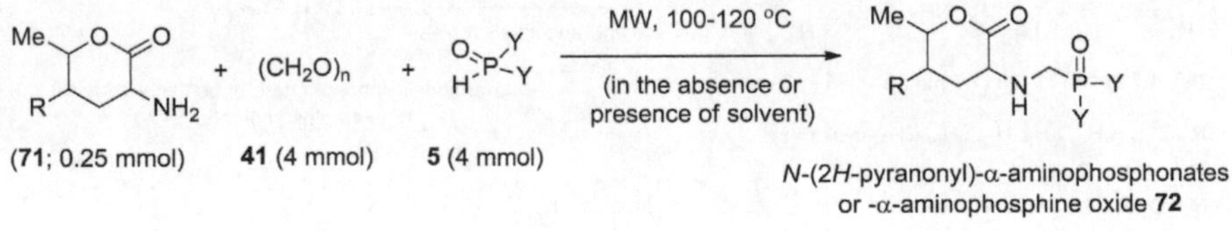

| Entry | R | Y | Solvent | Time (h) | % Yield (4) |
|-------|------|-----|--------------------|----------|-------------|
| 1 | H | OEt | – | 2.5 | 74 |
| 2 | H | OBu | – | 2.5 | 88 |
| 3 | H | Ph | CH$_3$CN | 3 | 98 |
| 4 | MeCO | OMe | – | 1.5 | 90 |
| 5 | MeCO | OEt | – | 2 | 74 |
| 6 | MeCO | OBu | – | 2 | 91 |
| 7 | MeCO | Ph | CH$_3$CN | 2 | 98 |
| 8 | PhCO | OEt | – | 4 | 61 |
| 9 | PhCO | Ph | CH$_3$CN | 2 | 97 |

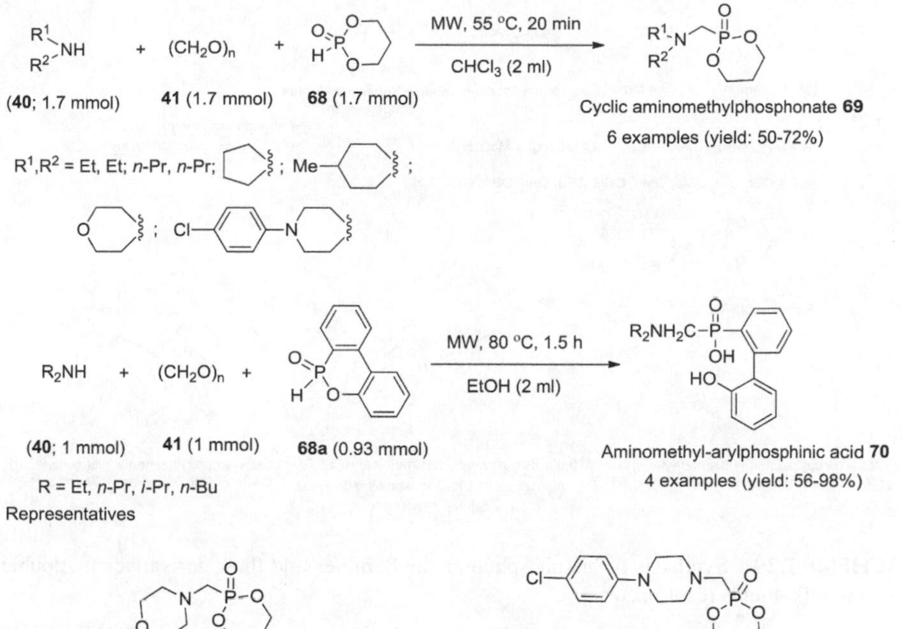

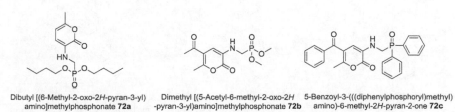

**SCHEME 7.27** Catalyst-free synthesis of cyclic aminomethylphosphonates and amino-methyl-arylphosphinic acids.

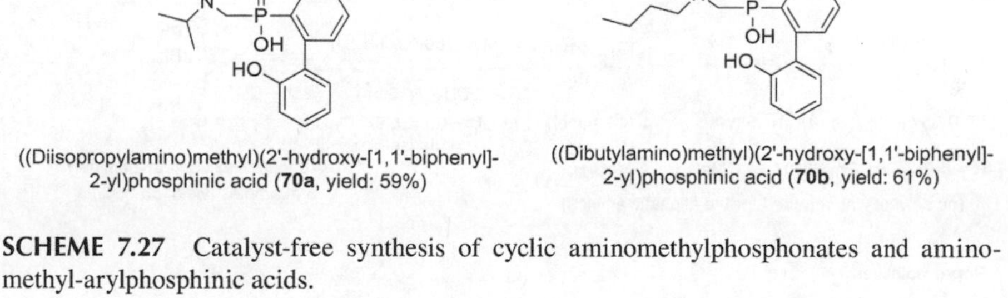

Dibutyl [(6-Methyl-2-oxo-2*H*-pyran-3-yl)amino]methylphosphonate **72a**

Dimethyl [(5-Acetyl-6-methyl-2-oxo-2*H*-pyran-3-yl)amino]methylphosphonate **72b**

5-Benzoyl-3-(((diphenylphosphoryl)methyl)amino)-6-methyl-2*H*-pyran-2-one **72c**

**SCHEME 7.28** Synthesis of *N*-(2*H*-pyranonyl)-substituted α-aminophosphonates and α-aminophosphine oxides.

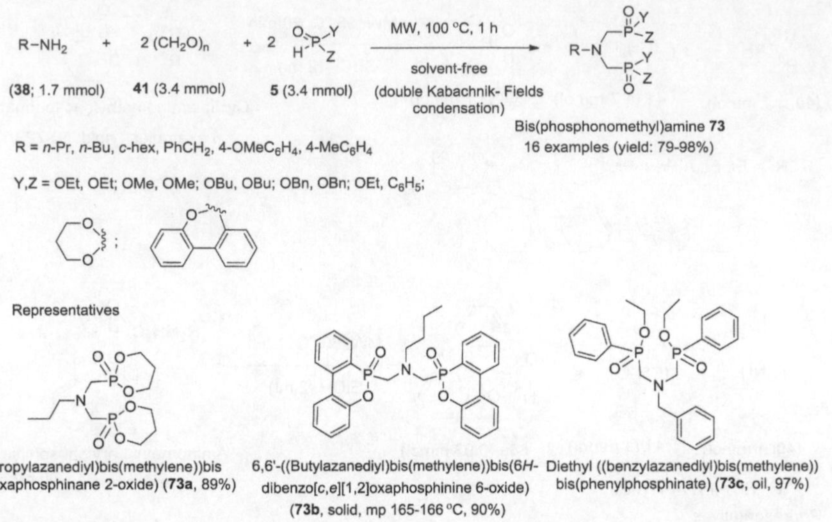
R–NH2 + 2 (CH2O)n + 2
MW, 100 °C, 1 h
solvent-free
(38; 1.7 mmol) 41 (3.4 mmol) 5 (3.4 mmol) (double Kabachnik- Fields condensation)
Bis(phosphonomethyl)amine 73
16 examples (yield: 79-98%)
R = n-Pr, n-Bu, c-hex, PhCH2, 4-OMeC6H4, 4-MeC6H4
Y,Z = OEt, OEt; OMe, OMe; OBu, OBu; OBn, OBn; OEt, C6H5;
Representatives
2,2'-((Propylazanediyl)bis(methylene))bis (1,3,2-dioxaphosphinane 2-oxide) (73a, 89%)
6,6'-((Butylazanediyl)bis(methylene))bis(6H-dibenzo[c,e][1,2]oxaphosphinine 6-oxide) (73b, solid, mp 165-166 °C, 90%)
Diethyl ((benzylazanediyl)bis(methylene)) bis(phenylphosphinate) (73c, oil, 97%)

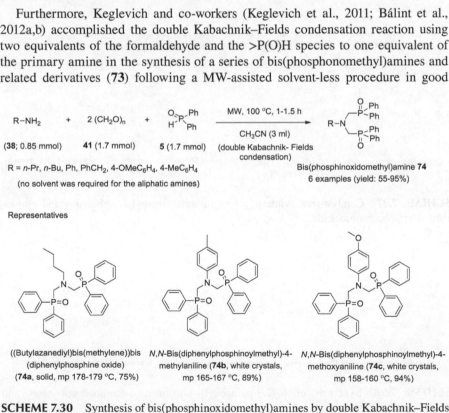

**SCHEME 7.29** Synthesis of bis(phosphonomethyl)amines and their derivatives by double Kabachnik–Fields reaction.

Furthermore, Keglevich and co-workers (Keglevich et al., 2011; Bálint et al., 2012a,b) accomplished the double Kabachnik–Fields condensation reaction using two equivalents of the formaldehyde and the >P(O)H species to one equivalent of the primary amine in the synthesis of a series of bis(phosphonomethyl)amines and related derivatives (**73**) following a MW-assisted solvent-less procedure in good

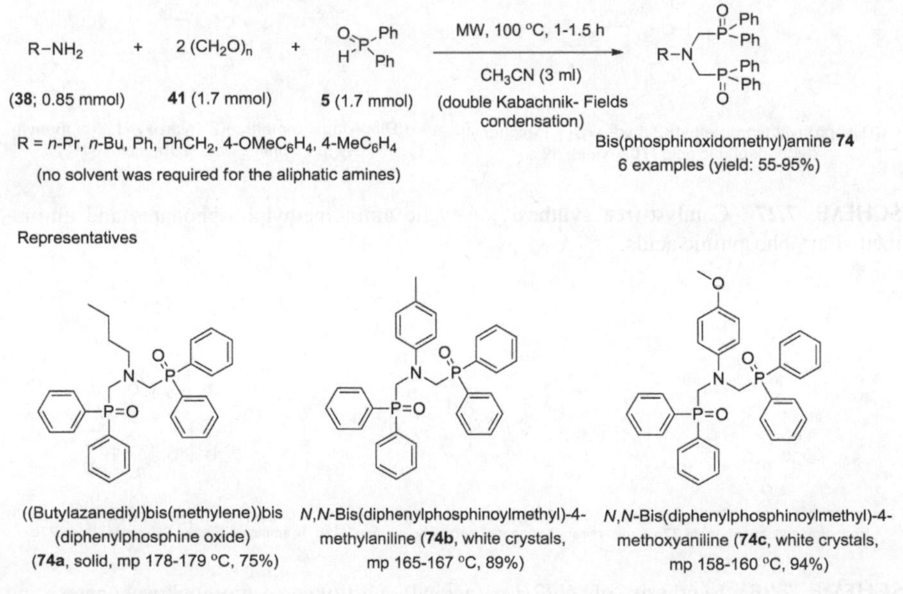
R–NH2 + 2 (CH2O)n +
MW, 100 °C, 1-1.5 h
CH3CN (3 ml)
(38; 0.85 mmol) 41 (1.7 mmol) 5 (1.7 mmol) (double Kabachnik- Fields condensation)
R = n-Pr, n-Bu, Ph, PhCH2, 4-OMeC6H4, 4-MeC6H4
(no solvent was required for the aliphatic amines)
Bis(phosphinoxidomethyl)amine 74
6 examples (yield: 55-95%)
Representatives
((Butylazanediyl)bis(methylene))bis (diphenylphosphine oxide) (74a, solid, mp 178-179 °C, 75%)
N,N-Bis(diphenylphosphinoylmethyl)-4-methylaniline (74b, white crystals, mp 165-167 °C, 89%)
N,N-Bis(diphenylphosphinoylmethyl)-4-methoxyaniline (74c, white crystals, mp 158-160 °C, 94%)

**SCHEME 7.30** Synthesis of bis(phosphinoxidomethyl)amines by double Kabachnik–Fields reaction.

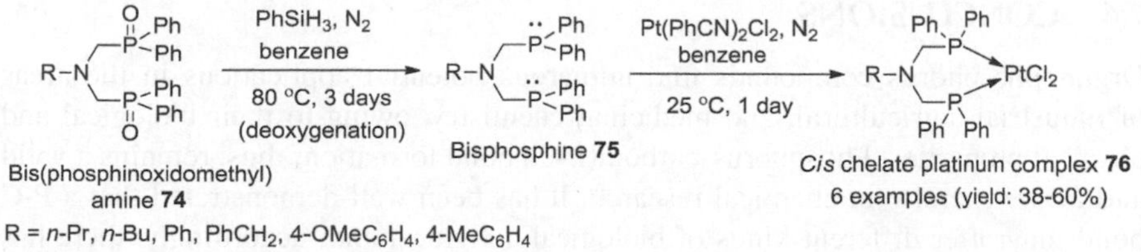

**SCHEME 7.31** Synthesis of ring platinum complexes (**76**) from bis(phosphinoxidomethyl) amines.

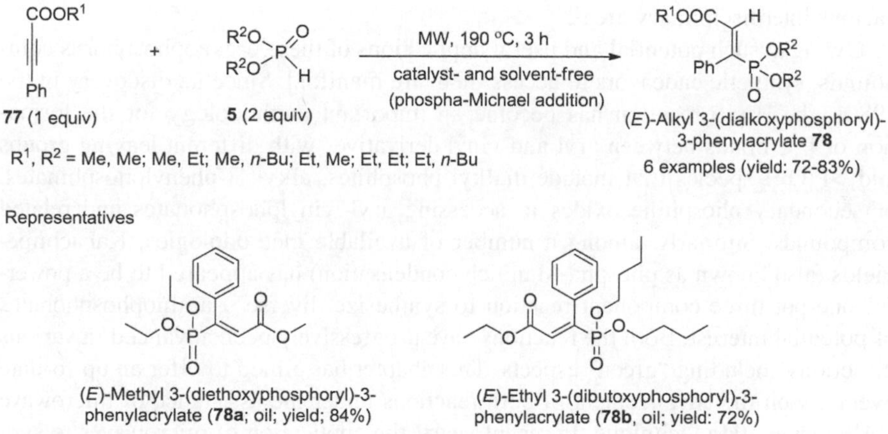

**SCHEME 7.32** Microwave-assisted catalyst- and solvent-free synthesis of (*E*)-alkyl 3-(dialkoxyphosphoryl)-3-phenylacrylates (**78**) via phospha-Michael addition.

yields (Scheme 7.29). The investigators then extended the double Kabachnik–Fields reaction to the synthesis of bis(phosphinoxidomethyl)amines **74**, and, herein, the use of acetonitrile as a solvent was required (Scheme 7.30). The bis(phosphinoylmethyl) amines (**74**) are useful precursors of bidentate P-ligands (**76**) obtained after double deoxygenation by silanes, e.g. trichlorosilane that could be used for the synthesis of ring platinum complexes (**75**) by reacting with half an equivalent of dichlorodiben-zonitriloplatinum (Scheme 7.31). Ring Pt-complexes are special heterocycles that may be regarded potential catalysts in homogeneous catalysis (Keglevich et al., 2011, 2013; Bálint et al., 2012a,b).

Recently, Bálint et al. (2015) demonstrated a catalyst-free diastereoselective synthesis of (*E*)-alkyl 3-(dialkoxyphosphoryl)-3-phenylacrylates (**78**) via phospha-Michael addition of dialkyl phosphites (**5**) to the triple bond of alkyl phenylpropiolates (**77**) upon microwave irradiation in the absence of any solvent (Scheme 7.32). The investigators observed that the mono-additions occur in a diastereoselective manner, and are significantly faster and complete under MW irradiation than in conventional heating.

## 7.4   CONCLUSIONS

Organophosphorus compounds find numerous potential applications in the areas of industrial, agricultural, and medicinal chemistry, owing to their biological and physical properties. Phosphorus-carbon (P-C) bond formation, thus, remains a valid and active exercise in chemical research. It has been well-demonstrated that a P-C bond may offer different kinds of biological activities, and, accordingly, there has been a growing interest in these classes of organophosphorous compounds in medicinal and nucleic acid chemistry. Substituted phosphonates, phosphine oxides, and α-aminophosphonates are a notable class of phosphorus-functionalized compounds which have invoked tremendous interest among researchers, including synthetic chemists, medicinal chemists, pharmacologists, biologists, and others working in various interdisciplinary areas.

Owing to such potential and useful applications of these organophosphorus compounds, synthetic endeavors to access them are manifold. Since its discovery in the 1980s, the Hirao reaction has become an important methodology for the formation of P-C bonds between aryl and vinyl derivatives with different leaving groups and >P(O)H species that include dialkyl phosphites, alkyl H-phenylphosphinates, or secondary phosphine oxides in accessing aryl-/vinylphosphonates and related compounds. Similarly, among a number of available methodologies, Kabachnik–Fields (also known as phospha-Mannich condensation) has appeared to be a powerful one-pot three-component reaction to synthesize diverse α-aminophosphonates of potential interest. Both the reactions have progressively been advanced in various directions, including "green" aspects. This chapter has aimed to offer an up-to-date overview on the advancement of both reactions under the assistance of microwave (MW) irradiation technique. In recent years, the application of microwaves to synthetic organic chemistry has become increasingly popular, and the most common benefits from MW irradiation are the considerable shortening of reaction times and the increase in the selectivities. However, the most valuable benefit is when a reaction can be carried out that is otherwise impossible under traditional thermal conditions.

This chapter has offered an up-to-date and vivid description of the microwave (MW)-assisted P-fictionalization leading to the syntheses of functionalized diverse kinds of phosphonates, phosphine oxides, and α-aminophosphonates of potential biological significance. The author hopes that this overview can boost the on-going research in that direction.

## ACKNOWLEDGMENT

Financial support from the Science and Engineering Research Board (SERB), Department of Science and Technology (DST), New Delhi (Grant No. EMR/2014/001220) is deeply acknowledged.

## ABBREVIATIONS

**[bmim]BF$_4$**   1-butyl-3-methylimidazolium tetrafluoroborate
**DIEA**          *N,N*-diisopropylethylamine

| DMF | *N,N*-dimethylformamide |
| **Dmphen** | 2.9-dimethyl-1,10-phenanthroline |
| **DMSO** | dimethylsulfoxide |
| **DPPH** | 2,2-diphenyl-1-picrylhydrazyl |
| **dppp** | 1,3-bis(diphenylphosphino)propane |
| **Et$_3$N** | triethylamine |
| **MeCN** | acetonitrile |
| **MTB-GS** | *Mycobacterium tuberculosis* glutamine synthetase |
| **MW** | microwave |
| **PS/PTSA** | polystyrene-supported *p*-toluenesulfonic acid |
| **PTC** | phase-transfer catalyst |
| **PTP1B** | protein-tyrosine phosphatase 1B |
| **Yb(OTf)$_3$** | Ytterbium(III) trifluoromethanesulfonate |

# REFERENCES

Abbas, S., R.D. Bertram, and C.J. Hayes. 2001. Commercially available 5′-DMT phosphoramidites as reagents for the synthesis of vinylphosphonate-linked oligonucleic acids. *Org. Lett.* 3:3365–3367.

Allen, M.C., W. Fuhrer, B. Tuck, R. Wade, and J.M. Wood. 1989. Renin inhibitors. Synthesis of transition-state analogue inhibitors containing phosphorus acid derivatives at the scissile bond. *J. Med. Chem.* 32:1652–1661.

Andaloussi, M., J. Lindh, J. Sävmarker, P.J.R. Sjöberg, and M. Larhed. 2009. Microwave-promoted Palladium(II)-catalyzed C-P bond formation by using arylboronic acids or aryltrifluoroborates. *Chem. Eur. J.* 15:13069–13074.

Atherton, F.R., C.H. Hassal, and R.W. Lambert. 1968. Synthesis and structure-activity relationships of antibacterial phosphonopeptides incorporating (1-aminoethyl)phosphonic acid and (aminomethyl) phosphonic acid. *J. Med. Chem.* 29:29–40.

Bálint, E., E. Fazekas, G. Pintér, et al. 2012a. Synthesis and utilization of the bis(>P(O)CH$_2$) amine derivatives obtained by the double Kabachnik–Fields reaction with cyclohexylamine; Quantum chemical and X-ray study of the related bidentate chelate platinum complexes. *Curr. Org. Chem.* 16:547–554.

Bálint, E., E. Fazekas, P. Pongrácz, et al. 2012b. *N*-benzyl and *N*-aryl bis(phospha-Mannich adducts): Synthesis and catalytic activity of the related bidentate chelate platinum complexes in hydroformylation. *J. Organomet. Chem.* 717:75–82.

Bálint, E., J. Takács, L. Drahos, A. Juranovič, M. Kočevar, and G. Keglevich. 2013. α-Aminophosphonates and α-aminophosphine oxides by the microwave-assisted Kabachnik–Fields reactions of 3-amino-6-methyl-2-*H*-pyran-2-ones. *Heteroatom Chem.* 24: 221–225.

Bálint, E., J. Takács, M. Bálint, and G. Keglevich. 2015. The catalyst-free addition of dialkyl phosphites on the triple bond of alkyl phenylpropiolates under microwave conditions. *Curr. Catal.* 2015, 4:57–64.

Bandyopadhyay, D. and B.K. Banik. 2014. Microwave-induced synthesis of heterocycles of medicinal interests. In *Green Synthetic Approaches for Biologically Relevant Heterocycles*, ed. G. Brahmachari, 517–557. Amsterdam: Elsevier.

Beletskaya, I.P., N.B. Karlstedt, E.E. Nifant'ev, et al. 2006. Palladium-catalyzed P-arylation of hydrophosphoryl derivatives of protected monosaccharides. *Russ. J. Org. Chem.* 42:1780–1785.

Belfield, K.D., C. Chinna, and K.J. Schafer. 1997. New NLO stilbene derivatives bearing phosphonate ester electron-withdrawing groups. *Tetrahedron Lett.* 38:6131–6134.

Benaglia, M. and Rossi S. 2010. Chiral phosphine oxides in present-day organocatalysis. *Org. Biomol. Chem.* 8:3824–3830.

Bhattacharya, A.K. and K.C. Rana. 2008. Amberlite-IR 120 catalyzed three-component synthesis of α-aminophosphonates in one-pot. *Tetrahedron Lett.* 49:2598–2601.

Bird, J., R.C. De Mello, G.P. Harper, et al. 1994. Synthesis of novel *N*-phosphonoalkyl dipeptide inhibitors of human collagenase. *J. Med. Chem.* 37:158–169.

Bloemink, M.J., J.J.H. Diederen, J.P. Dorenbos, R.J. Heetebrij, B.K. Keppler, and J. Reedijk. 1999. Calcium ions do accelerate the DNA binding of new antitumor-active platinum aminophosphonate complexes. *Eur. J. Inorg. Chem.* 1655–1657.

Bock, T., H. Moehwald, and R. Muelhaupt. 2007. Arylphosphonic acid-functionalized polyelectrolytes as fuel cell membrane material. *Macromol. Chem. Phys.* 208:1324–1340.

Brahmachari, G. (ed). 2014. *Green Synthetic Approaches for Biologically Relevant Heterocycles.* Amsterdam: Elsevier.

Brahmachari, G. 2015a. Microwave-assisted Hirao reaction: Recent developments. *ChemTexts* 1:15.

Brahmachari, G. 2015b. *Room Temperature Organic Synthesis.* Amsterdam: Elsevier.

Chauvin, R. 1990. Reaction of haloarenetricarbonylchromium with trimethylphosphite: Palladium-catalyzed Arbuzov reaction versus arene displacement by trimethylphosphite. *J. Organomet. Chem.* 387:C1–C4.

Cherkasov, R.A. and V.I. Galkin. 1998. The Kabachnik–Fields reaction: Synthetic potential and the problem of the mechanism. *Russ. Chem. Rev.* 67:857–882.

Chithaparthi R.R., I. Bhatnagar, C.S.R. Gangireddy, S.C. Syama, and S.R. Cirandur. 2013. Green synthesis of α-aminophosphonate derivatives on a solid supported $TiO_2$-$SiO_2$ catalyst and their anticancer activity. *Arch. Pharm.* 356:667–676.

Danila, D.C., X.Y. Wang, H. Hubble, I.S. Antipin, and E. Pinkhassik. 2008. Increasing permeability of phospholipid bilayer membranes to alanine with synthetic alpha-aminophosphonate carriers. *Bioorg. Med. Chem. Lett.* 18:2320–2323.

Dembitsky, V.M., A.A.A. Quntar, A. Haj-Yehia, and M. Srebnik. 2005. Recent synthesis and transformation of vinylphosphonates. *Mini-Rev. Org. Chem.* 2: 91–109.

Denmark, S.E., R.C. Smith, and S.A. Tymonko. 2007. Phosphine oxides as stabilizing ligands for the palladium-catalyzed cross-coupling of potassium aryldimethylsilanolates. *Tetrahedron* 63:5730–5738.

Devineni, S., S. Doddaga, R. Donka, and N.R. Chamarthi. 2013. $CeCl_3.7H_2O$-$SiO_2$: Catalyst promoted microwave assisted neat synthesis, antifungal and antioxidant activities of α-diaminophosphonates. *Chin. Chem. Lett.* 24:759–763.

Ebdon, J.R., D. Price, B.J. Hunt, et al. 2000. Flame retardance in some polystyrenes and poly(methyl methacrylate)s with covalently bound phosphorus-containing groups: Initial screening experiments and some laser pyrolysis mechanistic studies. *Polym. Degrad. Stab.* 69:267–277.

Engel, R. and J.I. Cohen (eds). 2004. *Synthesis of Carbon–Phosphorus Bonds, 2nd edn.* Boca Raton: CRC Press/Taylor & Francis.

Engels, J.W. and J. Parsch. 2005. Nucleic acid drugs. In *Molecular Biology in Medicinal Chemistry*, ed. T. Dingermann, D. Steinhilber, and G. Folkers, 153–178. Weinheim: Wiley-VCH Verlag GmbH & Co.

Fields, E.K. 1952. The synthesis of esters of substituted amino phosphonic acids. *J. Am. Chem. Soc.* 74:1528–1531.

Fields, S.C. 1999. Synthesis of natural products containing a C-P bond. *Tetrahedron* 55:12237–12273.

Freedman L.D. and G.O. Doak. 1957. The preparation and properties of phosphonic acids. *Chem. Rev.* 57:479–523.

Gangireddy, C.S.R., R.R. Chithaparthi, V.R. Mudumala, M. Mamilla, and U.R.S. Arigala. 2014. An efficient green synthesis of a new class of α-aminophosphonates under microwave irradiation conditions in the presence of PS/PTSA. *Heteroat. Chem.* 25:147–156.

Ghotas, E., G.B. Sylvie, J.K. Malcolm, et al. 2009. Synthesis and evaluation of alkoxy-phenylamides and alkoxy-phenylimidazoles as potent sphingosine-1-phosphate receptor subtype-1 agonists. *Bioorg. Med. Chem. Lett.* 19:369–372.

Giannousi, P.P. and P.A. Bartlett. 1987. Phosphorus amino acid analogues as inhibitors of leucine aminopeptidase. *J. Med. Chem.* 30:1603–1609.

Giguere, R.J., T.L. Bray, S.M Duncan, and G. Majetich. 1986. Application of commercial microwave ovens to organic synthesis. *Tetrahedron Lett.* 27:4945–4948.

Grembecka, J., A. Mucha, T. Cierpicki, and P. Kafarski. 2003. The most potent organophosphorus inhibitors of leucine amino peptidase. Structure-based design, chemistry, and activity. *J. Med. Chem.* 46:2641–2655.

Harnden, M.R., A. Parkin, M.J. Parratt, and R.M. Perkins. 1993. Novel acyclonucleotides: Synthesis and antiviral activity of alkenylphosphonic acid derivatives of purines and a pyrimidine. *J. Med. Chem.* 36:1343–1355.

Hartly, R. 1996. *The Chemistry of Organophosphorus Compounds.* New York: John Wiley and Sons.

Hayes, B.L. 2004. Recent advances in microwave-assisted synthesis. *Aldrichimica Acta* 37:66–77.

Herrmann, W.A. and C.W. Kohlpaintner. 1993. Water-soluble ligands, metal complexes, and catalysts: Synergism of homogeneous and heterogeneous catalysis. *Angew. Chem. Int. Ed.* 32:1524–1544.

Hirao, T., T. Masunaga, N. Yamada, and T. Agawa. 1982. Palladium-catalyzed new carbon-phosphorus bond formation. *Bull. Chem. Soc. Jpn.* 55:909–913.

Hirao, T., T. Masunaga, Y. Ohshiro, and T. Agawa. 1980. Stereoselective synthesis of vinylphosphonate. *Tetrahedron Lett.* 21:3595–3598.

Hirao, T., T. Masunaga, Y. Ohshiro, and T. Agawa. 1981. A novel synthesis of dialkyl arenephosphonates. *Synthesis* 56–57.

Holstein, S.A., D.M. Cermak, D.F. Wiemer, K. Lewis, and R.J. Hohl. 1998. Phosphonate and bisphosphonate analogues of farnesyl pyrophosphate as potential inhibitors of farnesyl protein transferase. *Bioorg. Med. Chem.* 6:687–694.

Jablonkai, E. and G. Keglevich. 2013. P-ligand-free, microwave-assisted variation of the Hirao reaction under solvent-free conditions; the P–C coupling reaction of >P(O)H species and bromoarenes. *Tetrahedron Lett.* 54:4185–4188.

Jablonkai, E. and G. Keglevich. 2014a. P-C bond formation by coupling reactions utilizing >P(O)H species as the reagents. *Curr. Org. Synth.* 11:429–453.

Jablonkai, E. and G. Keglevich. 2014b. Advances and new variations of the Hirao reaction. *Org. Prep. Proc. Int.* 46:281–316.

Jablonkai, E. and G. Keglevich. 2014c. P-C bond formation by coupling reactions utilizing >P(O)H species as the reagents. *Curr. Org. Synth.* 11:429–453.

Jablonkai, E. and G. Keglevich. 2015. Catalyst-free P–C coupling reactions of halobenzoic acids and secondary phosphine oxides under microwave irradiation in water. *Tetrahedron Lett.* 56:1638–1640.

Jablonkai, E., L.B. Balázs, and G. Keglevich. 2015. A P-ligand-free nickel-catalyzed variation of the Hirao reaction under microwave conditions. *Curr. Org. Chem.* 19:197–202.

Janardhan Rao, A., P. Visweswara Rao, V. Koteswara Rao, C. M. C. Naga Raju, and C. Suresh Reddy. 2010. Microwave assisted one-pot synthesis of novel α-aminophosphonates and their biological activity. *Bull. Korean Chem. Soc.* 31:1863–1868.

Jiang, W., G. Allan, J.J. Fiordeliso, et al. 2006. New progesterone receptor antagonists: Phosphorus-containing 11β-aryl-substituted steroids. *Bioorg. Med. Chem.* 14: 6726–6732.

Jiao, X.Y. and W.G. Bentrude. 2003. A facile route to vinyl- and arylphosphonates by vinyl and aryl radical trapping with (MeO)₃P. *J. Org. Chem.* 68:3303–3306.

Jin, L., B. Song, and G. Zhang. 2006. Synthesis, X-ray crystallographic analysis, and antitumor activity of *N*-(benzothiazole-2-yl)-1-(fluorophenyl)-*O, O*-dialkyl-α-aminophosphonates. *Bioorg. Med. Chem. Lett.* 16:1537–1543.

Jin, S. and K.E. Gonsalves. 1998. Synthesis and characterization of functionalized poly(Ɛ-caprilactone) copolymers by free-radical polymerization. *Macromolecules* 31:1010–1015.

Julienne, D., J.-F. Lohier, O. Delacroix, and A.-C. Gaumont. 2007. Palladium-catalyzed C-P coupling reactions between vinyl triflates and phosphine-boranes: efficient access to vinylphosphine-boranes. *J. Org. Chem.* 72:2247–2250.

Kabachnik, M.I. and T.Y. Medved. 1952. New synthesis of aminophosphonic acids. *Dokl. Akad. Nauk SSSR* 83:689–692.

Kaboudin, B. and K, Moradi. 2005. A simple and convenient procedure for the synthesis of 1-aminophosphonates from aromatic aldehydes. *Tetrahedron Lett.* 46:2989–2991.

Kaboudin, B. and R. Nazari. 2001. Microwave-assisted synthesis of 1-aminoalkyl phosphonates under solvent-free conditions. *Tetrahedron Lett.* 42:8211–8213.

Kafarski, P. and B. Lejczak. 2001Aminophosphonic acids of potential medical importance. *Curr. Med. Chem. Anticancer Agents* 1:301–312.

Kalek, M., A. Ziadi, and J. Stawinski. 2008. Microwave-assisted palladium-catalyzed cross-coupling of aryl and vinyl halides with H-phosphonate diesters. *Org. Lett.* 10:4637–4640.

Keglevich, G. 2014. Application of microwave irradiation in the synthesis of P-heterocycles. In *Green Synthetic Approaches for Biologically Relevant Heterocycles*, ed. G. Brahmachari, 559–570. Amsterdam: Elsevier.

Keglevich, G. and E. Bálint. 2012. The Kabachnik–Fields reaction: Mechanism and synthetic use. *Molecules* 17:12821–12835.

Keglevich, G. and Szekrényi A. 2008. Eco-friendly accomplishment of the extended Kabachnik–Fields reaction; A solvent- and catalyst-free microwave-assisted synthesis of aminophosphonates and α-aminophosphine oxides. *Lett. Org. Chem.* 5:616–622.

Keglevich, G., A. Szekrényi, Á. Szöllősy, and L. Drahos. 2011. Synthesis of bis(phosphonatomethyl)-,bis(phosphinatomethyl)-, and bis(phosphinoxidomethyl) amines, as well as related ring bis(phosphine) platinum complexes. *Synth. Commun.* 41:2265–2272.

Keglevich, G., A. Szekrényi, M. Sipos, K. Ludányi, and I. Greiner. 2008. Synthesis of cyclic aminomethylphosphonates and aminomethyl-arylphosphinic acids by an efficient microwave mediated phospha-Mannich approach. *Heteroat. Chem.* 19:207–210.

Keglevich, G., E. Jablonkai, and L.B. Balàzs. 2014. A "green" variation of the Hirao reaction: The P–C coupling of diethyl phosphite, alkyl phenyl-*H*–phosphinates and secondary phosphine oxides with bromoarenes using a P-ligand-free Pd(OAc)$_2$ catalyst under microwave and solvent-free conditions. *RSC Adv.* 4:22808–22816.

Keglevich, G., P. Bagi, E. Bálint, and T. Körtvélyesi. 2013. The synthesis of platinum complexes of cyclic phosphines and bisphosphines. In *Platinum: Compounds, Production and Applications*, ed. L. Varennikov and E. Yedemsky, 83–102. New York: Nova Science Publishers Inc.

Kiran, Y.B., C.D. Reddy, D. Gunasekar, C. Suresh Reddy, A. Leon, and L.C.A. Barbosa. 2008. Synthesis and anticancer activity of new class of bisphosphonates/phosphanamidates. *Eur. J. Med. Chem.* 43: 885–892.

Koteswara Rao, V., S. Subba Reddy, B. Sathish Krishna, et al. 2011. Design, synthesis and anti colon cancer activity evaluation of phosphorylated derivatives of lamivudine (3TC). *Lett. Drug Des. Discov.* 8:59–64.

Kranjc, K. and M. Kočevar. 2010. Microwave-assisted organic synthesis: General considerations and transformations of heterocyclic compounds. *Curr. Org. Chem.* 14:1050–1074.

Kukhar, V.P. and H.R. Hudson (eds). 2000. *Aminophosphonic and Aminophosphinic Acids: Chemistry and Biological Activity.* Weinheim: Wiley-VCH Verlag GmbH & Co.

Lazrek, H.B., A. Rochdi, H. Khaider, et al. 1998. Synthesis of (Z) and (E) α-alkenyl phosphonic acid derivatives of purines and pyrimidines. *Tetrahedron* 54:3807–3816.

Lee, S., J.K. Lee, C.E. Song, and D.-C. Kim. 2002. Microwave-assisted Kabachnik–Fields reaction in ionic liquid. *Bull. Korean Chem. Soc.* 23:667–668.

Lelievre, S., F. Mercier, and E. Mathey. 1996. Phosphanorbornadienephosphonates as a new type of water-soluble phosphines for biphasic catalysis. *J. Org. Chem.* 61:3531–3533.

Liu, W., C.J. Royers, A.J. Fisher, and M. Toney. 2002. Aminophosphonate inhibitors of dialkylglycine decarboxylase: Structural basis for slow binding inhibition. *Biochem.* 41:12320–12328.

Lu, X., C. Zhang, and Z. Xu. 2001. Reactions of electron-deficient alkynes and allenes under phosphine catalysis. *Acc. Chem. Res.* 34:535–544.

Lv, X.-Y., J.-M. Zhang, C.-H. Xing, W.-Q. Du, and S.-Z. Zhu. 2007. One-pot preparation of fluorinated α-aminoalkyl phosphonates under microwave irradiation and solvent-free conditions. *Synth. Commun.* 37:343–357.

Maffei, M. 2004. Transition metal-promoted syntheses of vinylphosphonates. *Curr. Org. Synth.* 1:355–375.

Minami, T. and J. Motoyoshiya. 1992. Vinylphosphonates in organic synthesis. *Synthesis* 333–349.

Minami, T., T. Okauchi, and R. Kouno. 2001. α-Phosphonovinyl carbanions in organic synthesis. *Synthesis* 349–357.

Morimoto, H., T. Yoshino, T. Yukawa, G. Lu, S. Matsunaga, and M. Shibasaki. 2008. Lewis base assisted brønsted base catalysis: bidentate phosphine oxides as activators and modulators of Brønsted basic lanthanum–aryloxides. *Angew. Chem. Int. Ed.* 47:9125–9129.

Mu, X.-J., M.-Y. Lei, J.-P. Zou, and W. Zhang. 2006. Microwave-assisted solvent-free and catalyst-free Kabachnik–fields reactions for α-aminophosphonates. *Tetrahedron Lett.* 47:1125–1127.

Mucha, A., P. Kafarski, and L. Berlicki. 2011. Remarkable potential of the α-aminophosphonate/phosphinate structural motif in medicinal chemistry. *J. Med. Chem.* 54:5955–5980.

Naydenova, E.D., P. Todorov, and K. Troev. 2010. Recent synthesis of aminophosphonic acids as potential biological importance. *Amino Acids* 38:23–30.

Ogawa, T., N. Usuki, and N. Ono. 1998. A new synthesis of π-electron conjugated phosphonates and phosphonic bis(diethylamides) and their SHG activities. *J. Chem. Soc. Perkin Trans. 1* 2953–2958.

Orsini, F., G. Sello, and M. Sisti. 2010. Aminophosphonic acids and derivatives. Synthesis and biological applications. *Curr. Med. Chem.* 17:264–289.

Ouimette, D. and M. Coffey. 1989. Comparative antifungal activity of four phosphonate compounds against isolates of nine *Phytophthora* species. *Phytopathology* 79:761–767.

Pinault, N. and D.W. Bruce. 2003. Homogeneous catalysts based on water-soluble phosphines. *Coord. Chem. Rev.* 241:1–25.

Prauda, I., I. Greiner, K. Ludányi, and G. Keglevich. 2007. Efficient synthesis of phosphono- and phosphinoxidomethylated N-heterocycles under solvent-free microwave conditions. *Synth. Commun.* 37:317–322.

Prim, D., J.M. Campagne, D. Joseph, and B. Rioletti. 2002. Palladium-catalyzed reactions of aryl halides with soft, non-organometallic nucleophiles. *Tetrahedron* 58:2041–2075.

Quntar, A., O. Baum, R. Reich, and M. Srebnik. 2004. Recently synthesized class of vinylphosphonates as potent matrix metalloproteinase (MMP-2) inhibitors. *Arch. Pharm.* 337:76–80.

Rajasekhar, D., D. Subba Rao, D. Srinivasulu, C. Naga Raju, and M. Balaji. 2013. Microwave assisted synthesis of biologically active α-aminophosphonates catalyzed by nano-BF$_3$ SiO$_2$ under solvent-free conditions. *Phosphorus Sulfur Silicon Relat. Elem.* 188:1017–1025.

Rawls, R. 1998. Synthetic biology makes its debut. Nucleic acids are one focus of approach based on nonnatural molecules designed to function in biological system. *Chem. Eng. News* 76:12–13.

Razaei, Z., H. Friouzabadi, N. Iranpoor, et al. 2009. Design and one-pot synthesis of α-aminophosphonates and bis(α-aminophosphonates) by iron(III) chloride and cytotoxic activity. *Eur. J. Med. Chem.* 44:4266–4275.

Reddy, G.S., K.U.M. Rao, C.S. Sundar, et al. 2014. Neat synthesis and antioxidant activity of α-aminophosphonates. *Arab. J. Chem.* 7:833–838.

Roberts, F., C.W. Roberts, J.J. Johnson, et al. 1998. Evidence for the shikimate pathway in apicomplexan parasites. *Nature* 393:801–805.

Rummelt, S.M., M. Ranocchiari, and J.A. van Bokhoven. 2012. Synthesis of water-soluble phosphine oxides by Pd/C-catalyzed P-C coupling in water. *Org. Lett.* 14:2188–2190.

Savignac, P. and B. Iorga. 2003. *Modern Phosphonate Chemistry*. Boca Raton: CRC Press/ Taylor & Francis.

Schull, T.L., J.C. Fettinger, and D.A. Knight. 1995. The first examples of an aryl ring substituted by both phosphine and phosphonate moieties: Synthesis and characterization of the new highly water-soluble phosphine ligand $Na_2[Ph_2P(C_6H_4-p-PO_3)].1.5H_2O$ and platinum(II) complexes. *J. Chem. Soc. Chem. Commun.* 1487–1488.

Schwan, A.L. 2004. Palladium catalyzed cross-coupling reactions for phosphorus-carbon bond formation. *Chem. Soc. Rev.* 33:218–224.

Shaughnessy, K.H. 2009. Hydrophilic ligands and their application in aqueous-phase metal-catalyzed reactions. *Chem. Rev.* 109:643–710.

Siénczyk, M. and J. Oleksyszyn. 2009. Irreversible inhibition of serine proteases-design and *in vivo* activity of diaryl alpha-aminophosphonate derivatives. *Curr. Med. Chem.* 16:1673–1687.

Sonar, S.S., S.A. Sadaphal, V.B. Labade, B.B. Shingate, and M.S. Shingare. 2010. An efficient synthesis and antibacterial screening of novel oxazepine α-aminophosphonates by ultrasound approach. *Phosphorus Sulfur Silicon Relat. Elem.* 185:65–73.

Ullrich, K.J., G. Rumrich, T.R. Burke, S.P. Shirazie-Beechey, and G.-L. Lang. 1997. Interaction of alkyl/arylphosphonates, phosphonocarboxylates and diphosphonates with different anion transport systems in the proximal renal tubule. *J. Pharmacol. Exp. Ther.* 283:1223–1229.

Van Leeuwen, P.W.N.M. 2004. *Homogeneous Catalysis*. Dordrecht: Kluwer Academic Publishers.

Varalashmi, M., D. Srinivasulu, D. Rajasekhar, C. Naga Raju, and S. Sreevani. 2014. CeCl₃-7H₂O catalyzed, microwave-assisted high-yield synthesis of α-aminophosphonates and their biological studies. *Phosphorus Sulfur Silicon Relat. Elem.* 189:106–112.

Veeranarayana Reddy, M., B. Siva Kumar, A. Balakrishna, C. Suresh Reddy, S.K. Nayak, and C. Devendranath Reddy. 2007. One-pot synthesis of novel α-aminophosphonates using tetramethylguanidine as a catalyst. *ARKIVOC* 246–254.

Veeranarayana Reddy, M., S. Annar, A. Balakrishna, G. Chandra Sekhar Reddy, and C. Suresh Reddy. 2010. Tetramethyl guanidine (TMG) catalyzed synthesis of novel α-amino phosphonates by one-pot reaction. *Org. Commun.* 3:39–44.

Villemin, D., P.-A. Jaffrés, and F. Siméon. 1997. Rapid and efficient phosphonation of aryl halides catalysed by palladium under microwaves irradiation. *Phosphorus Sulfur Silicon Relat. Elem.* 130:59–63.

Wuggenig, F., A. Schweifer, K. Mereiter, and F. Hammerschmidt. 2011. Chemoenzymatic synthesis of phosphonic acid analogues of l-lysine, l-proline, l-ornithine, and l-pipe-colic acid of 99% ee – Assignment of absolute configuration to (–)-proline. *Eur. J. Org. Chem.* 1870–1879.

Xu, Y., K. Yan, B. Song, et al. 2006. Synthesis and antiviral bioactivities of α-aminophosphonates containing alkoxyethyl moieties. *Molecules* 11:666–676.

Yadav J.S., B.V. Subba Reddy, and C. Madan. 2001. Montmorillonite clay-catalyzed one-pot synthesis of α-aminophosphonates. *Synlett*, 1131–1133.

Zakeeruddin, S.M., M.K. Nazeeruddin, P. Pechy, et al. 1997. Molecular engineering of photosensitizers for nanocrystalline solar cells: Synthesis and characterization of Ru dyes based on phosphonated terpyridines. *Inorg. Chem.* 36:5937–5946.

Zefirov, N.S. and E.D. Matveeva. 2008. Catalytic Kabachnik-Fields reaction: New horizons for old reaction. *ARKIVOC* 1–17.

Zhou, T. and Z.C. Chen. 2001. Hypervalent iodine in synthesis. 52. Palladium-catalyzed arylation of *O, O* -dialkyl phosphites with diaryliodonium salts: A convenient method for synthesis of arylphosphonates. *Synth. Commun.* 31:3289–3294.

Zhu, D., L. Xu, F. Wu, and B. Wan. 2006. A mild and efficient copper-catalyzed coupling of aryl iodides and thiols using an oxime-phosphine oxide ligand. *Tetrahedron Lett.* 47:5781–5784.

Zmudzka, K., T. Johansson, M. Wojcik, et al. 2003. Novel DNA analogues with 2-, 3- and 4-pyridylphosphonate internucleotide bonds: Synthesis and hybridization properties. *New J. Chem.* 27:1698–1705.

# 8 Microwave Synthesis of Materials for Thin-Film Photovoltaic Absorber Layer Application

*Raghunandan Seelaboyina, Manoj Kumar, and Kulvir Singh*

## CONTENTS

## 8.1 INTRODUCTION

Among the renewable energy sources available on earth, the essential form of energy with the highest theoretical potential is the sun. The key benefit of solar energy is that it is non-pollutant and is the cleanest energy source. Electricity from solar energy is generated by various photovoltaic (PV) devices (solar cells) and technologies via an electronic process that occurs naturally in semiconductor materials. As seen in Figure 8.1 first, the PV device (solar cell) absorbing the sunlight creates electron-hole pairs. Second, the electron and hole carriers are separated by an internal electric field of the PV device. Third, the generated charge is distributed to an external load [1].

The various technologies that utilize semiconductors to convert solar energy into electricity can be classified as silicon (crystalline and non-crystalline), compound (CuInGaSe$_2$/S$_2$ [CIGS], copper zinc tin sulfide/selenide [CZTS], GaAs [Gallium Arsenide], and so forth), and organic (DSSC [Dye Sensitized Solar Cells], and so forth) semiconductors [2–6]. Each technology has its advantages and disadvantages. Silicon (Si) solar cells are prepared on Si wafers, which are fragile with very little flexibility. Modules are prepared by connecting the Si solar cells either in series to achieve the required output voltage or in parallel to provide the required current. Si modules are considered the best for grid-based applications like large solar farms and so forth. Despite the remarkable performance of Si technology, it is the

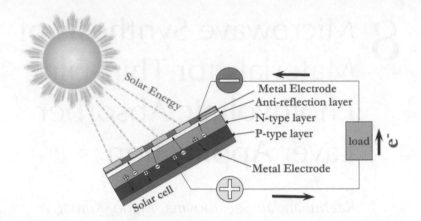

**FIGURE 8.1** Illustration of solar energy conversion to electricity by a solar cell.

physical characteristics of the rigid Si wafers that render them difficult to be used in some applications, especially in the portable power market including the military and civilian sectors (buses, trains, building integrated PV [BIPV], and so forth). For applications where flexibility, portability, and light weight are required, the options are either amorphous silicon (a-Si), $CuInS_2$ (CIS), or CIGS. Compared to a-Si, CIGS and CIS have several advantages including double-digit efficiency of 12.8–20.4% for cells and 10–15% for modules [2–6]. Further, a significant advantage of CIGS and CIS is that they work even in diffused light conditions. For CIGS average relative efficiency reduction of maximum power from an irradiance of 1,000 $W/m^2$ to 200 $W/m^2$ at 25°C is 5% [5].

Rigid or flexible CIGS/CIS solar cells/modules are prepared either by vacuum processes, including sputtering or thermal co-evaporation, or by non-vacuum processes, including printing or electro-deposition [7] (Figure 8.2). For rigid cells/ modules, glass substrates and flexible ones, steel substrates are mainly utilized. The majority of commercially produced CIGS/CIS modules are based on sputtering and co-evaporation techniques [8, 9]. The utilization of the mentioned techniques is related to process reliability, repeatability, and availability of equipment [10–13] for large-scale manufacturing. Vacuum-based methods are energy intensive and expensive. Significant energy input is typically required to evaporate or sputter CIGS from a source, often onto a heated substrate under stringent vacuum conditions which involve expensive systems. However, an advantage of these methods is that they produce energy-efficient (19–20.5%) solar cells and modules [4, 14].

Non-vacuum methods are solution-based approaches that require a low energy input and cost and can deposit a CIGS/CIS/CZTS layer in atmospheric conditions in a clean room/environment. Solution-based approaches based on inks and pastes, such as ink jet printing, spin coating, and doctor blade coating, utilize nanoparticles dispersed in a solvent to form ink. The ink is utilized to coat substrates in the form of films and if these processes are adequately optimized they can lead to high throughput. The advantage of these processes is that they are relatively cost-effective and can

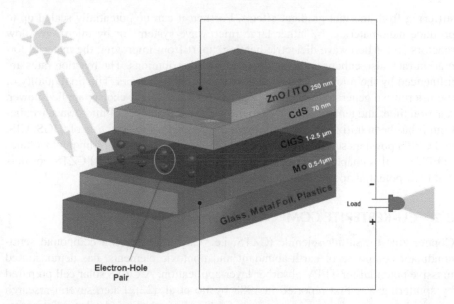

**FIGURE 8.2** Schematic of TFPV solar cell depicting the various layers of the cell and their approximate thickness.

produce efficient (11–20.5%) thin-film solar cells (TFSC) [14]. However, availability of commercial production equipment is a concern.

The primary requirement for both vacuum and non-vacuum methods is the composition of the starting material. In the case of sputtering from a compound target, the composition of target material influences the efficiency of the absorber layer. Compound material targets are prepared by hot isostatic pressing of powders with some binder addition followed by sintering [15]. In the case of inkjet printing, spin coating, and doctor blade methods, powder composition of the ink or paste influences the efficiency of the absorber layer. Apart from composition, other process steps also influence the solar cell efficiency. However, the starting material plays a significant role in solar cell efficiency, and hence it is vital to have suitable quality materials.

The efficiency of the absorber layer is dependent on the band-gap of CIGS/CIS/CZTS, whereas the band-gap of a material is dependent on the particle size and composition. The size-dependent band-gap property is likely to minimize or vanish after nanopowder is deposited and converted into a dense film consisting of micron-sized grains. So, optimization of the composition is essential for band-gap control in solar cell applications [16]. Composition-controlled CIGS nanopowder is prepared by several methods including the hot injection method [16], single source precursor (SSP) [17], Schlenk line [18], and microwave heating [19–21]. Similar methods have also been reported for CIS and CZTS synthesis [22–36]. Of these methods, microwave heating can produce powders of CIGS/CZTS/CIS with relatively uniform phases and sizes.

Microwave energy as a heat source leads to faster chemical reactions and improved product yield and purity compared to conventional heating sources. Microwave synthesis, due to its volumetric heating, is unique in its ability to be scaled up without suffering from thermal gradient effects. Further, it can be potentially scaled up to produce nanomaterials by either large microwave systems or by microwave flow reactors [19]. Microwave dielectric heating, apart from improving the material formation rate, also enhances the quality and size distributions. The reaction rates are influenced by the microwave field and by choice of precursors. The final quality of the microwave generated materials depends on the reactant choice, applied power, reaction time, and temperature. Due to the various advantages of microwave irradiation, it has been utilized by several research groups for the synthesis of CIGS, CIS, and CZTS powders suitable for absorber layer application in thin-film photovoltaics (TFPV). In this chapter, the microwave synthesis of CIGS/CIS and CZTS powders and their potential applications is reviewed.

## 8.2   CU-KESTERITE COMPOUND SYNTHESIS

Copper zinc tin sulfide/selenide (CZTS, i.e., $Cu_2ZnSnS_4/Se_4$), a compound semiconductor composed of earth-abundant and nontoxic elements, has demonstrated massive potential for TFPV absorber layer application. A CZTS solar cell prepared by sputtering was first reported in 1988 by Ito et al. [22]. Later, several research groups demonstrated various techniques for CZTS synthesis and solar cell preparation [23]. Two decades of research and innovation of CZTS has led to the demonstration of ~12.6% [4, 23] efficiency, which is quite remarkable. Many research groups are working to improve the efficiency of CZTS by developing methods to synthesize phase and band-gap controlled CZTS material. One such method includes microwave synthesis, which will be discussed in this section [23–36].

CZTS synthesis by microwave was first reported in 2012 [27, 33, 35]. In most of the reported methods the copper, zinc, and tin (hydrous or anhydrous) sources have been acetates, [24, 27, 33, 36, 37], chlorides [25, 31–32, 35], and nitrates [26, 29, 30]. The sulfur source has been thioacetamide [26, 27, 29, 31, 33] and thiourea [24, 25, 30, 32, 35, 36]. The solvents have been ammonia [27, 33], ethylene glycol [24, 26, 29, 30, 35], ethylenediamine [31, 32], dimethylformamide [25], and oleylamine [37]. Depending on the reaction conditions of temperature, time, and type of surfactant, various shapes of CZTS particles including spherical [24, 30, 35], irregular [26, 27, 29, 31–33, 37], and doughnut [25] were synthesized. In addition, direct film synthesis on substrates [34, 36] has also been reported.

CZTS particle synthesis from acetates as precursors has been reported by Shin et al. [27, 33] and Saraswat et al. [37]. Shin et al. [33] reported the synthesis of irregular shaped 10–70 nm, 1.5 eV CZTS particles which were also sulfurized by annealing in the $H_2S$ environment. Further Shin et al. also studied the effect of Cu concentration on structural, morphological, compositional, chemical, and optical properties of CZTS. By changing the Cu concentration from 0.01 to 0.025 M, the band-gap energy was decreased from 1.65 to 1.28 eV. The irregular shaped CZTS particles may be utilized for preparing inks or pastes for forming CZTS films. The report by Saraswat et al. [37] provides a detailed understanding of CZTS nanocrystal synthesis and the effect of various processing parameters on the CZTS formation. The paper provides extensive information on phase identification, size, and shape, and optical properties analysis. Their results suggest that CZTS formation starts after 8 min of reaction and the uniformity improves for 18 min of reaction time (determined from Raman analysis). However, it should be observed that these parameters will change when the volume of the reaction mixture is changed [37].

CZTS particle synthesis from Cu, Zn, and Sn chlorides has been reported by Lin et al. [32], Kumar et al. [25, 35], and Yan et al. [31]. Shin et al. [33] studied the effect of ethylenediamine (ED) concentration on the particle size of CZTS. The synthesis was carried out at 180 °C for ~1 hour with various ED concentration. It was observed that particle size increased with an increase in ED concentration. However, ED is not an environmentally friendly solvent, so CZTS synthesis by other solvents is better. Kumar et al. [25] reported the synthesis of spherical shaped CZTS particles and studied their structural, morphological, and optical properties. The band-gap was determined to be 1.76 eV. The advantage of the process is that it used a relatively benign ethylene glycol solvent for CZTS synthesis. In another report by Kumar et al. [35], the authors synthesized doughnut and hierarchical shaped CZTS particles (Figure 8.3). The shape variation of CZTS was achieved by utilizing N,N-dimethylformamide and polyvinylpyrrolidone (PVP) as a solvent and a stabilizing agent respectively. Figure 8.3 shows the effect of PVP on the CZTS particle shape. The hierarchical CZTS with a band-gap of 1.54 eV exhibited better light absorption, which was attributed to the increased optical path to capture the light efficiently by scattering between the nano-plates network. The actual effect of the CZTS particle shape for the formation of CZTS absorber film of ~2 μm would have provided more insight into the solar cell characteristics. However, the authors did not prepare any solar cells with CZTS particles. In a report by Yan et al. [31], CZTS synthesis was demonstrated with Cu, Zn, and Sn chlorides, thioacetamide as a sulfur source, and water as a solvent. By varying the reaction time (24 h–1 h) and temperature (200–180°C), kesterite phase irregular shaped CZTS with a band-gap between 1.5 and 1.6 eV was synthesized.

CZTS particle synthesis from Cu, Zn, and Sn nitrates has been reported by Wang et al. [26, 29, 30]. Wang et al. studied the influence of temperature and sulfur sources on the CZTS synthesis from nitrates. By changing the sulfur source from thiourea to l-cysteine, the shape of CZTS particles changed from spherical (or flower-like) to hollow. The formation of hollow particles was attributed to the vesicle-template

**FIGURE 8.3** Effect of PVP on CZTS particle shape (a) with PVP, doughnut shape, and (b) without PVP, hierarchical structure. (Reprinted from Kumar, R. S., Hong, C. H., and Kim, M. D., 2014. Doughnut-shaped hierarchical $Cu_2ZnSnS_4$ microparticles synthesized by cyclic microwave irradiation. *Adv. Powder Technol.*, 25: 1554–1559. With permission. Copyright 2014 Elsevier.)

mechanism, whereas the mechanism for the formation of spherical (or flower-like) particles was not discussed. The band-gaps of CZTS nanoparticles determined by UV absorption measurement were 1.43 eV (flower like) and 1.67 eV (hollow sphere).

In direct film deposition, the advantage is that CZTS films can be directly formed on the substrate without any post-processing required for film deposition, i.e., required for solar cell fabrication. The disadvantage may be that substrate size may be limited to a few centimeters; this may be due to the unavailability of microwave systems which can accommodate large substrates. Recently, Knutson et al. [36] demonstrated CZTS film deposition by microwave irradiation on 1 cm$^2$ Mo-coated glass substrates from acetates of Cu, Zn, and Sn, and thiourea and thioglycolic acid. The preferential heating of the conductive substrates by microwave energy in homogeneous solutions of precursors resulted in relatively uniform CZTS film of 1.4 μm with excellent adhesion to the Mo layer. The average particle size in the film was measured to be 12 nm and increased to 24 nm upon microwave annealing in octadecene at 160°C for 30 min. From XRD (X-ray diffraction) and Raman analysis the formed film on Mo-coated glass substrate was determined to consist of the CZTS phase without any other phases. The facile film deposition technique is a good method to fabricate CZTS absorber films directly onto substrates. Solar cells can be prepared by depositing buffer layers of ZnS either by the microwave [38] or the chemical bath deposition (CBD) method.

Figure 8.4 shows the various analytical techniques that can be utilized to study/ determine the various properties of CIGS/CIS/CZTS absorber layer films. In the case of CZTS, the XRD spectrum of synthesized material should match with standard ICDD card no-26-0575. The typical XRD spectrum of CZTS is shown in Figure 8.5. In addition to XRD, the CZTS phase has to be determined by Raman spectroscopy; this is due to similar diffraction patterns of ZnS and $Cu_2SnS_3$. The peak at 331–338 cm$^{-1}$ in the Raman spectrum can be seen as a sign of complete CZTS phase formation. However, for complete CZTS phase, peaks at 275 cm$^{-1}$ and 352 cm$^{-1}$ attributable to ZnS and at 267 cm$^{-1}$, 303 cm$^{-1}$, and 365 cm$^{-1}$ attributable to $Cu_2SnS_3$ should not be present. Almost all the microwave synthesis [20–28] methods demonstrated the formation of kesterite structure CZTS with a band-gap of 1.4–1.6 eV. The reported

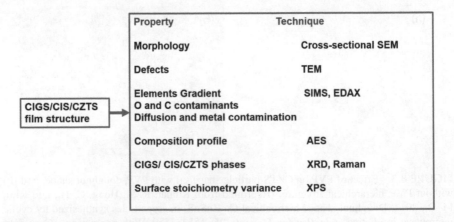

| Property | Technique |
| --- | --- |
| Morphology | Cross-sectional SEM |
| Defects | TEM |
| Elements Gradient O and C contaminants Diffusion and metal contamination | SIMS, EDAX |
| Composition profile | AES |
| CIGS/ CIS/CZTS phases | XRD, Raman |
| Surface stoichiometry variance | XPS |

CIGS/CIS/CZTS film structure

**FIGURE 8.4** Characterization methods for CIGS/CIS/CZTS film analysis.

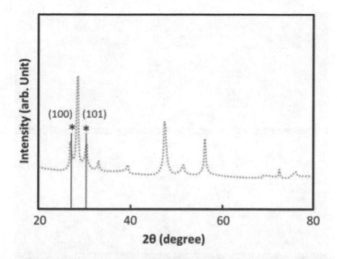

**FIGURE 8.5** XRD pattern of microwave synthesized kesterite structure CZTS (ICDD 26-0575), * wurtzite structure of CZTS (ICDD 36-1450). (Reprinted from Saraswat, P. K., and Free, M. L., 2013. An investigation of rapidly synthesized $Cu_2ZnSnS_4$ nanocrystals. *J. Crystal Growth*, 37: 287–294. With permission. Copyright 2013 Elsevier.)

band-gap is apt for absorbing the incoming solar radiation. The prepared CZTS powder can be utilized in preparing printable inks and sputtering targets.

## 8.3 CU-CHALCOPYRITE COMPOUNDS SYNTHESIS

Among the TFPV, solar cells or modules based on the copper (Cu) chalcopyrite compound semiconductor absorber layer materials, i.e., $CuInSe_2$ (CISe) and $CuInGaSe_2/S_2$ (CIGS), are the most efficient. The highest reported efficiencies for CIGS solar cells and modules prepared by vacuum method are 21.7% and 17.5% [5]; the reported values are edging closer to conventional wafer-based Si PV. The technologically advantageous features of a high absorption coefficient ($10^4$ cm$^{-1}$), excellent radiation hardness, and high efficiency even under diffused light conditions attract considerable scientific interest for the synthesis of these materials by microwave irradiation. The synthesis of $CuInSe_2$ (CISe), $CuInS_2$ (CIS), and $CuInGaS_2/Se_2$ (CIGS) is reviewed in the following sections.

## 8.4 CUINSE$_2$ (CISE) AND CUINS$_2$ (CIS) SYNTHESIS

Hwang et al. [39] reported the synthesis of CISe nanoparticles by the microwave-enhanced solvothermal method using two approaches. The first approach was without pre-treatment of the selenium (Se) powder in ethylenediamine (ED) under microwave heating. To understand the morphology and formation mechanism of CISe nanoparticles without pre-treatment and with pre-treatment of Se powder in ED under microwave heating, Hwang et al. investigated the samples obtained at different reaction times using the scanning electron microscope (SEM) and XRD. SEM images of the as-prepared CISe particles are shown in Figure 8.6, in which plate-like morphologies (Figure 8.6a–c) and rod-like morphologies (Figure 8.6d–f) were observed for two different approaches

**FIGURE 8.6** SEM images of CISe particles synthesized by microwave-assisted solvothermal method (a–c) without pre-treatment and (d–f) with pre-treatment of Se powder at different reaction times. (Reprinted from Wu, C. C., Shiau, C. Y., Ayele, D. W., Su, W. N., Cheng, M. Y., Chiu, C. Y., and Hwang, B. J., 2010. Rapid microwave-enhanced solvothermal process for synthesis of CuInSe₂ particles and its morphologic manipulation. *Chem. Mater.*, 22: 4185–4190. With permission. Copyright 2011 American Chemical Society.)

for 5, 15, and 30 min of reaction, respectively. The formation of rod-like CISe particles was attributed to the one-dimensional (1D) morphology of Se seeds that resulted from the pre-treatment of the Se powder, whereas the plate-like CISe particles were produced without any pre-treatment of Se powder. Figure 8.7 shows the XRD pattern of CISe particles prepared without pre-treatment (Figure 8.7a) and with pre-treatment of Se powder, followed by different microwave reaction times (Figure 8.7b). In both approaches, at shorter reaction times, many diffraction peaks appeared in the XRD patterns, which can be attributed to CuSe and the unreacted selenium, along with the characteristic peaks of CISe particles. However, for longer reaction times (30 min), only the characteristic peaks of CISe are observed, indicating the formation of a pure chalcopyrite phase with an intense peak at $2\theta \approx 26.6°$, which is oriented along the (112) direction. Although the intensity of this peak increases with reaction time, it is higher for CISe particles obtained with a pre-treatment of Se powder than without pre-treatment.

The XRD spectra also show that the starting crystalline phase of the CISe particles prepared with the pre-treatment of Se powders is different from the one without

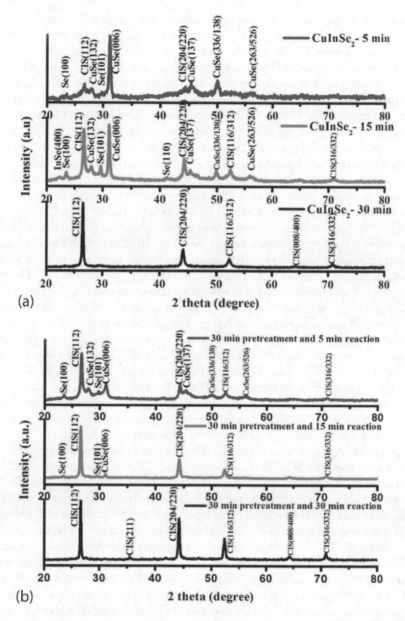

**FIGURE 8.7** XRD spectra of CISe particles synthesized by microwave-assisted solvothermal method (a) without pre-treatment and (b) with pre-treatment of Se powder in ethylenediamine for 30 min with 5, 15, and 30 min of reaction time. (Reprinted from Wu, C. C., Shiau, C. Y., Ayele, D. W., Su, W. N., Cheng, M. Y., Chiu, C. Y., and Hwang, B. J., 2010. Rapid microwave-enhanced solvothermal process for synthesis of CuInSe₂ particles and its morphologic manipulation. *Chem. Mater.*, 22: 4185–4190. With permission. Copyright 2011 American Chemical Society.)

pre-treatment. In the case of without pre-treatment, the initial XRD pattern showed only small signals of CISe particles. However, the CISe peaks became more prominent as the reaction progressed. On the other hand, with a 30-min pre-treatment of the Se and ED mixture, a strong XRD peak for CISe was present after only 5 min of reaction time, suggesting a more dominant presence of the (112) phase of CISe and less impurity. Furthermore, after 15 min of reaction time, pure CISe was almost obtained suggesting that the pre-treatment of Se powders is a favorable condition for the formation of pure CISe particles. Therefore, the critical factor for the manipulation of morphology and formation of chalcopyrite phase of CISe is the pre-treatment of Se powder in ED before reaction with copper and indium precursors. Also, the traditional solvothermal method requires ~48 h to obtain pure chalcopyrite crystalline of CISe particles. Whereas, the microwave-assisted synthesis reduces the reaction time to ~30 min. In this study, a highly microwave-absorbing solvent (i.e., ED) was used to synthesize the CISe nanoparticles. However, weakly microwave-absorbing solvents may allow the selective heating of precursors with higher microwave absorptivity resulting in elevated local temperatures and accelerated reaction rates [40], which may further reduce the reaction time.

Oleksak et al. [41] reported the synthesis of CISe nanoparticles in low absorbing solvents. A low microwave-absorbent solvent system of tri-n-octylphosphine (TOP) and oleic acid (OA) was used to investigate potential benefits of selective heating of the dissolved precursors. From the XRD analysis, Oleksak et al. [41] demonstrated that weakly microwave-absorbing solvents could reduce the reaction time drastically to 5 min. However, in the previous approaches relatively toxic ED and TOP/OA solvents were utilized for CISe synthesis. The toxicity and corrosiveness of the ED may cause environmental problems, as well as a decrease in durability of fabrication equipment. The use of TOP/OA may create a capping agent around the nanoparticles, which may limit the efficiency of the fabricated device. So alternative synthesis methods with environmentally benign solvents are preferred.

In recent years, green chemistry principles applied to nanoparticle synthesis has attracted a lot of interest [42, 43]. Nanoparticle synthesis in aqueous media will reduce utilization of organic solvent, toxic reagents, increasing the yield, reducing purification steps, and mitigating environmental impact. In addition to this, the formation of a capping agent around the nanoparticles could be avoided using aqueous media. Bensebaa et al. [44] reported the synthesis of CISe and CIS nanoparticles using water as a solvent. The synthesis of CIS nanoparticles was carried out utilizing cupric chloride dihydrate, indium chloride, and sodium sulfide ($Na_2S$) as starting precursors. Deionized water (DI) was used as a solvent, and mercapto-acetic acid (MAA) was used as a surfactant. MAA is a water-soluble surfactant and used to avoid agglomeration and obtain size distribution with reasonably uniform diameters. The surfactant MAA can be easily stripped away during the washing steps since it is weakly bonded to the nanoparticles. CISe nanoparticles were also prepared using a similar approach. For the Se source, the $Na_2Se$ precursor was used instead of $Na_2S$. XRD was utilized to characterize the structure of the CIS and CISe nanoparticles. Figure 8.8 shows a typical diffraction pattern of synthesized materials in water solvent. Three relatively broad diffraction peaks are detected around $2\theta = 28°$, $47°$, and $55°$. These three peaks correspond to the (112), (204)/(220), and (116)/(312) planes of the tetragonal structure, respectively (ICDD 27-0159). XRD of CISe nanoparticles also closely matched the reported results

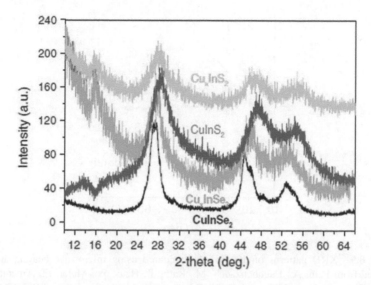

**FIGURE 8.8** XRD spectra of CIS and CISe nanoparticles. (Reprinted from Bensebaa, F., Durand, C., Aouadou, A., Scoles, L., Du, X., Wang, D., and Page, Y. L., 2010. A new green synthesis method of CuInS2 and CuInSe2 nanoparticles and their integration into thin films. *J. Nanopart Res.*, 12: 1897–1903. With permission. Copyright 2010 Springer Nature.)

in the literature. The observed results indicate that CIS and CISe nanoparticles could be synthesized by the microwave route without using any toxic solvents.

In most of the microwave-based approaches, the reaction time for the synthesis of CIS or CISe nanoparticles ranges from 30–60 min. However, Pein et al. [45] demonstrated that CIS nanoparticles could be synthesized in a relatively short reaction time of ~90 s, which is around 1000 times less than conventional heating. In this approach, the CIS nanoparticle was synthesized using CuI, $InCl_3$, and elemental sulfur as precursors and oleylamine as a solvent and capping agent. The reaction temperature used in the oleylamine route was around 220°C for the reaction times of 15 and 60 min, respectively. In addition, an experiment with a very short reaction time of 90 s including 80 s ramping and 10 s heating at the target temperature (high microwave power) was performed; the fast heating ramp is considered to be an especially important asset of microwave chemistry [19]. Figure 8.9 compares the XRD patterns of the particles from these experiments. From the XRD patterns, it is evident that CIS nanoparticles could be synthesized using the very short exposure time of 90 s.

## 8.5 CUINGAS₂/SE₂ (CIGS) SYNTHESIS

It is also well known that the band-gap of CISe can be modified by substituting Ga for In to further improve the absorption and hence the efficiency of the fabricated device. However, incorporation of Ga in CISe lattice is a bit difficult. Juhaiman et al. [46] reported the synthesis of CIGSe (Copper indium gallium (di)selenide) nanoparticles by the microwave route using water as a solvent (green method). The synthesis of $CuIn_{1-x}Ga_xSe_2$ nanoparticles with x = 0, 0.25, and 0.5 were reported by the

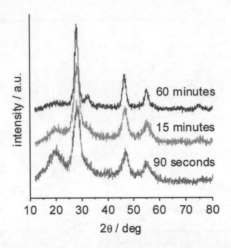

**FIGURE 8.9** XRD patterns of the particles prepared using microwave heating at 220°C. (Reprinted from Pein, A., Baghbanzadeh, M., Rath, T., Haas, W., Maier, E., Amenitsch, H., Hofer, F., Kappe, C. O., and Trimmel, G., 2011. Investigation of the formation of $CuInS_2$ nanoparticles by the oleylamine route: Comparison of microwave-assisted and conventional syntheses. *Inorg. Chem.*, 50:193–200. With permission. Copyright 2011 American Chemical Society.)

authors. In all cases, sub-10 nm nanoparticles with relatively uniform size distribution were obtained. The majority of the $CuInSe_2$ had an average size of about 4 nm, and CIGS with higher Ga are slightly larger. In recent years, various other quaternary $CuIn_xGa_{1-x}S_2$ ($0 \leq x \leq 1$) chalcopyrites have been prepared from molecular single-source precursors (SSPs) via microwave decomposition. Use of SSPs in preparation of nanomaterials presents distinct advantages such as the precise control of reaction conditions and stoichiometry as SSPs contain all necessary elements in a single molecule. Pak et al. [47] fabricated the quaternary $CuIn_xGa_{1-x}S_2$ ($0 \leq x \leq 1$) chalcopyrites with tunable band-gap and solubility on scales of up to 150 g with precise stoichiometric control by selectively decomposing mixtures of two different SSPs. Further, Yousefi et al. [48] and Gardner et al. [49] also prepared the CIS nanoparticles using SSPs via the microwave route, having a particle size in the range of 3–12 nm.

In conclusion, the microwave route using nontoxic solvents in shorter reaction times can be used to fabricate the CISe/CIGSe/CIS/CIGS/CZTS nanoparticles with a low level of impurities and controlled particle size and band-gap, and so forth. The synthesized materials can be further utilized to prepare printable inks/pastes, and they may also be used for preparing sputtering targets.

## REFERENCES

1. Chopra, K. L., and Das, S. R. 1983. *Thin Film Solar Cells*. Plenum Press, New York.
2. Wolden, C. A., Kurtin, J., Baxter, J. B., Repins, I., Shaheen, S. E., Torvik, T. J., Rockett, A. A., Fthenakis, V. M., and Aydil, E. S. 2011. Photovoltaic manufacturing: Present status, prospects, and research needs. *J. Vac. Sci. Technol. A.*, 29(3):030801-1–030801-16.
3. Reinhard, P., Chirila, A., Blosch, P., Pianezzi, F., Nishiwaki, S., Buechler, S., and Tiwari, A. N. 2013. Review of progress toward 20% efficiency flexible CIGS solar cells and manufacturing issues of solar modules. *IEEE J. Photovolt.*, 3(1):72–580.

4. Green, M. A., Emery, K., Hishikawa, Y., Warta, W., and Dunlop, E. D. 2018. Solar cell efficiency tables (version 51). *Prog. Photovolt.: Res. Appl.*, 26(1):3–12.

5. TS CIGS Series: Datasheet page. www.tsmc-solar.com/Assets/downloads/en-US/TS_ CIGS_Series_Datasheet_NA_05-2013.pdf (accessed January 15, 2018).

6. Wuerz, R., Eicke, A., Kessler, F., Patel, S., Efimenko, S., and Schlegel, C. 2012. CIGS thin-film solar cells and modules on enameled steel substrates. *Solar Energy Mater. Solar Cells*, 100:132–137.

7. Thin-Film Photovoltaics page. http://solopower.com/technology/ (accessed January 15, 2018).

8. Next Generation CIS page. www.solar-frontier.com/eng/technology/index.html (accessed January 15, 2018).

9. CIS photovoltaics: efficient solar energy page. www.avancis.de/en/cis-technology/ (accessed January 15, 2018).

10. Midsummer Technology page. http://midsummer.se/technology (accessed January 15, 2018).

11. Singulus Technologies—Developer, Enabler and Supplier for – Thin Film Solar page. www.singulus.com/de/solar/thin-film-solar.html (accessed January 15, 2018).

12. CIGS Solar PV Manufacturing Solutions page. www.xsunx.com/manufacturing-solu-tions13.htm (accessed January 15, 2018).

13. Enabling CIGS mass production—Manz page. www.manz.com/products-services/ manz-cigs-fab (accessed January 15, 2018).

14. Chirila, A., Reinhard, P., Pianezzi, F., Blosch, P., Alexander, R. U., Carolin, F., Lukas, K., Debora, K., Christina, G., Herald, H., Dominik, J., Rolf, E., Nishiwaki, S., Buechler, S., and Tiwari, A. N. 2013. Potassium-induced surface modification of Cu(In,Ga)Se$_2$ thin films for high-efficiency solar cells. *Nat. Mater.*, 12:1107–1111.

15. Tomoya, T., Hideo, T., Masakatsu, I., Masaru, S., and Ryo, S. 2011. Cu-In-Ga-Se quaternary alloy sputtering target. International Patent: WO 2011/058828.

16. Pan, D., Wang, X., Zhou, Z. H., Chen, W., Xu, C. L., and Lu, Y. F. 2009. Synthesis of quaternary semiconductor nanocrystals with tunable band gaps. *Chem. Mater.*, 21:2489–2493.

17. Sun, C., Gardner, J. S., Long, G., Bajracharya, C., Thurber, A., Punnoose, A., Rodriguez, R. G., and Pak, J. J. 2010. Controlled stoichiometry for quaternary CuIn$_x$Ga$_{1-x}$S$_2$ chalcopyrite nanoparticles from single-source precursors via microwave irradiation. *Chem. Mater.*, 22:2699–2701.

18. Tang, J., Hinds, S., Kelley, S. O., and Sargent, E. H. 2008. Synthesis of colloidal CuGaSe$_2$, CuInSe$_2$ and Cu(InGa)Se$_2$. *Chem. Mater.*, 20:6906–6910.

19. Baghbanzadeh, M., Carbone, L., Cozzoli, P. D., and Kappe, C.O. 2011. Microwave-assisted synthesis of colloidal inorganic nanocrystals. *Angew. Chem.*, 50:11312–11359.

20. Seelaboyina, R., Kumar, M., Madiraju, A. V., Taneja, K., Keshri, A. K., Mahajan, S., and Singh, K. 2013. Microwave synthesis of copper indium gallium (di)selenide nanopowders for thin-film solar applications. *J. Renew. Sustain. Energy*, 5: 031608-1–031608-7.

21. Madiraju, A. V., Taneja, K., Seelaboyina, R., Kumar, M., Keshri, A. K., and Mahajan, S. B. 2012. Synthesis of CZTS in aqueous media using microwave irradiation. In: International Conference on Solar Energy Photovoltaics, Bhubaneswar, India, December 2012. Hindawi Publishing Corporation, Conference Papers on Energy, pp. 1–3.

22. Ito, K., and Nakazawa, T. 1988. Electrical and optical properties of stannite type quaternary semiconductor thin films. *Jpn. J. Appl. Phys.*, 27:2094.

23. Wang, W., Winkler, M. T., Gunawan, O., Gokmen, T., Todorov, T. K., Zhu, Y., and Mitzi, D. B. 2014. Device characteristics of CZTSSe thin-film solar cells with 12.6% efficiency. *Adv. Energy Mater.*, 4:1301465-1–1301465-5.

24. Ghediya, P. R., and Chaudhuri, T. K. 2015. Doctor-blade printing of Cu$_2$ZnSnS$_4$ films from microwave-processed ink. *J. Mater. Sci.: Mater. Electron.*, 26:1908–1912.

25. Kumar, R. S., Hong, C. H., and Kim, M. D. 2014. Doughnut-shaped hierarchical $Cu_2ZnSnS_4$ microparticles synthesized by cyclic microwave irradiation. *Adv. Powder Technol.*, 25:1554–1559.

26. Wang W, Shen H, He X, and Li J. 2014. Effects of sulfur sources on properties of $Cu_2ZnSnS_4$ nanoparticles. *J. Nanopart. Res.*, 16:2437-1–2437-8.

27. Shin, S. K., Han, J. H., Park, C. Y., Kim, S. R., Park, Y. C., Agawane, G. L., Moholkar, A. V., Yun, J. H., Jeong, C. H., Lee, J. Y., and Kim, J. H. 2012. A facile and low-cost synthesis of earth-abundant element $Cu_2ZnSnS_4$ (CZTS) nanocrystals: Effect of Cu concentrations. *J. Alloys Compd.*, 541:192–197.

28. Wang, K. C., Chen, P., and Tseng, C. M. 2013. Facile one-pot synthesis of $Cu_2ZnSnS_4$ quaternary nanoparticles using a microwave-assisted method. *Cryst. Eng. Comm.*, 15: 9863–9868.

29. Wang, W., Shen, H., Yao, H., Li, J., and Jiao, J. 2015. Influence of solution temperature on the properties of $Cu_2ZnSnS_4$ nanoparticles by ultrasound-assisted microwave irradiation. *J. Mater. Sci.: Mater. Electron.*, 26:1449–1454.

30. Wang, W., Shen, H., Jiang, F., He, X., and Yue, Z. 2013. Low-cost chemical fabrication of $Cu_2ZnSnS_4$ microparticles and film. *J. Mater. Sci.: Mater. Electron.*, 24:1813–1817.

31. Yan, X., Michael, E., Komarneni, S., Brownson, J. R., and Feng, Y. Z. 2014. Microwave-hydrothermal/solvothermal synthesis of kesterite, an emerging photovoltaic material. *Ceramics Int.*, 40:1985–1992.

32. Lin, Y. H., Das, S., Yang, C. Y., Sung, J. C., and Lu, C. H. 2015. Phase-controlled synthesis of $Cu_2ZnSnS_4$ powders via the microwave assisted solvothermal route. *J. Alloys Compd.*, 632:354–360.

33. Shin, S. W., Han, J. H., Park, C. Y., Moholkar, A. S., Lee, J. Y., and Kim, J. H. 2012. Quaternary $Cu_2ZnSnS_4$ nanocrystals: Facile and low-cost synthesis by microwave-assisted solution method. *J. Alloys Compd.*, 516:96–101.

34. Kagawa, R., Hirata, S., Sasaki, M., and Klenk, R. 2013. Rapid direct preparation of $Cu_2ZnSn(S_{1-x}, Se_x)_4$ films using microwave irradiation. *Phys. Status Solidi C*, 10:1012–1014.

35. Kumar, R. S., Ryu, B. D., Chandramohan, S., Seol, J. K., Lee, S. K., and Hong, C. H. 2012. Rapid synthesis of sphere-like $Cu_2ZnSnS_4$ microparticles by microwave irradiation. *Mater. Lett.*, 86:174–177.

36. Knutson, T. R., Hanson, P. J., Aydil, E. S., and Penn, R. L. 2014. Synthesis of $Cu_2ZnSnS_4$ thin films directly onto conductive substrates via selective thermolysis using microwave energy. *Chem.Commun.*, 50:5902–5904.

37. Saraswat, P. K., and Free, M. L. 2013. An investigation of rapidly synthesized $Cu_2ZnSnS_4$ nanocrystals. *J. Crystal Growth*, 37:287–294.

38. Zhai, R., Wang, S., Xu, H., Wang, H., and Yan, H. 2005. Rapid formation of CdS, ZnS thin films by microwave-assisted chemical bath deposition. *Mater. Lett.*, 59:1497–1501.

39. Wu, C. C., Shiau, C. Y., Ayele, D. W., Su, W. N., Cheng, M. Y., Chiu, C. Y., and Hwang, B. J. 2010. Rapid microwave-enhanced solvothermal process for synthesis of $CuInSe_2$ particles and its morphologic manipulation. *Chem. Mater.*, 22:4185–4190.

40. Bilecka, I., and Niederberger, M. 2010. Microwave chemistry for inorganic nanomaterials synthesis. *Nanoscale*, 2:1358–1374.

41. Oleksak, R. P., Flynn, B. T., Schut, D. M., and Herman, G. S. 2014. Microwave-assisted synthesis of $CuInSe_2$ nanoparticles in low-absorbing solvents. *Phys. Status Solidi A*, 211:219–225.

42. Dahl, J. A., Maddux, B. L. S., and Hutchison, J. E. 2007. Toward greener nano-synthesis. *Chem. Rev.*, 107:2228–2269.

43. Sweeney, S. F., Woehrle, G. H., and Hutchison, J. E. 2006. Rapid purification and size separation of gold nanoparticles via diafiltration. *J. Am. Chem. Soc.*, 128:3190–3197.

44. Bensebaa, F., Durand, C., Aouadou, A., Scoles, L., Du, X., Wang, D., and Page, Y. L. 2010. A new green synthesis method of $CuInS_2$ and $CuInSe_2$ nanoparticles and their integration into thin films. *J. Nanopart Res.*, 12:1897–1903.

45. Pein, A., Baghbanzadeh, M., Rath, T., Haas, W., Maier, E., Amenitsch, H., Hofer, F., Kappe, C. O., and Trimmel, G. 2011. Investigation of the formation of $CuInS_2$ nanoparticles by the oleylamine route: Comparison of microwave-assisted and conventional syntheses. *Inorg. Chem.*, 50:193–200.

46. Juhaiman, L. A., Scoles, L., Kingston, D., Patarachao, B., Wang, D., and Bensebaa, F. 2010. Green synthesis of tunable $Cu(In_{1-x}Ga_x)Se_2$ nanoparticles using non-organic solvents. *Green Chem.*, 12:1248–1252.

47. Sun, C., Gardner, J. G., Long, G., Bajracharya, C., Thurber, A., Punnoose, A., Rodriguez, R. G, and Pak, J. J. 2010. Controlled stoichiometry for quaternary $CuIn_xGa_{1-x}S_2$ chalcopyrite nanoparticles from single-source precursors via microwave irradiation. *Chem. Mater.*, 22:2699–2701.

48. Sabet, M., Niasari, M. S, Ghanbari, D., Amiri, O., and Yousefi, M. 2013. Synthesis of $CuInS_2$ nanoparticles via simple microwave approach and investigation of their behavior in the solar cell. *Mater. Sci. Semiconductor Process.*, 16:696–704.

49. Gardner, J. S., Shurdha, E., Wang, C., Lau,, L. D., Rodriguez, R. D., and Pak, J. J. 2008. Rapid synthesis and size control of $CuInS_2$ semi-conductor nanoparticles using microwave irradiation. *J. Nanopart. Res.*, 10:633–641.

# 9 Microwave-Assisted Transition Metal-Catalyzed Synthesis of Pharmaceutically Important Heterocycles

*Dipti Shukla, Priyank Purohit, and Asit K. Chakraborti*

## CONTENTS

## 9.1 INTRODUCTION

Microwave (MW) is a form of electromagnetic radiation which lies between the infrared and radio waves in the frequency region of 0.3 to 300 GHz. When a molecule is irradiated with microwave, it aligns itself with the applied field. The oscillating electric field affects the particular substance (solvent or substrate), and in consequence the molecule tries to align itself with the oscillating field, and energy is absorbed. The ability of a molecule to convert electromagnetic energy into thermal energy depends upon its dielectric constant, and the larger the dielectric constant the more rapid is the heating [1]. The use of microwaves as an energy source to heat reaction solutions is associated with several advantages, such as the reduction of time of a chemical reaction, instant and uniform heating, feasibility for solvent-free reaction or reaction in an aqueous medium in a green and sustainable manner, dynamic temperature range, and rapid reaction optimization. So, the use of microwave is a boon to organic/medicinal chemists to achieve synthetic goals by drastically reducing reaction times (from hours or days to minutes and seconds) as compared to traditional conductive heating methods [2].

The heterocyclic scaffolds occupy a prominent place in the chemical space of new drug design [3]. The high occurrence of heterocyclic moieties in drugs makes them an important component of the pharmacophoric feature for therapeutic intervention of the pathogenesis of various diseases. Aromatic heterocycles are almost ubiquitous in medicinal chemistry as >70% of medicinally important compounds contain at least one heteroaromatic ring [4]. Thus, construction of heterocyclic scaffolds of pharmaceutical importance remains an active area of chemical research [5]. However, the adverse effect of the chemical manufacturing processes on the environment urges for sustainable synthetic methods. This demands for further improvements to tried and tested (and widely employed) methodologies, with deficiencies in areas such as atom economy and safety to enrich the chemists' toolbox with cost-effective and safe methodologies.

Microwave-Assisted Organic Synthesis (MAOS) has found widespread use in medicinal chemistry due to its ability to offer dramatic rate enhancement and significantly increase the efficiency for timely synthetic support. The methods by which increased efficiency can be achieved in the drug discovery and development process is of enormous importance to the pharmaceutical industry. The benefit of enhanced efficiency includes a reduction in product development lifecycle with a simultaneous increase in the number of new pharmaceuticals introduced to the marketplace. Pharmaceutical companies use MAOS in target discovery, high throughput screening, pharmacokinetics, and production of compound libraries. Heterocyclic scaffolds represent the central framework of many biologically active compounds. The preparation of bioactive heterocyclic compounds in an environmentally safe fashion has become an integral part of the industry [6–8].

Microwave-assisted transition metal-catalyzed reactions represent one of the most interesting transformations for the synthesis of bioactive heterocyclic scaffolds. Many useful reactions which would have been otherwise impossible have been realized through transition metal catalysis [9]. However, the conventional method of the transition metal-catalyzed synthesis of heterocycles is too slow to meet the demand for the timely supply of these compounds [10–13], as these reactions often need hours or days for completion, frequently in an inert atmosphere. With the introduction of microwave heating, the transition metal-catalyzed heterocycles synthesis has been proved to be a convenient procedure in the drug discovery and development process as it provides the product with superior yield in lesser time [14].

The focus of this chapter will be on the development of transition metal-catalyzed microwave-assisted synthesis of bioactive heterocyclic compounds during the past decade with a few specific examples where the transition metal-catalyzed reaction under microwave heating is a key step in the heterocyclic ring formation. Although the synthesis of nitrogen-, oxygen-, and sulfur-containing heterocycles are discussed, the discussion on nitrogen-containing heterocycles covers the largest part of this review as they are abundant in nature.

## 9.2  N-HETEROCYCLES

The nitrogen heterocycles constitute an integral part of life being present in deoxyribonucleic acid (DNA) [15]. They also constitute the essential structural component of many drugs, e.g., quinolone moiety in nalidixic acid used as antibiotic [16], triazole ring system in many antifungal drugs [17], the pyrazolone and indole rings in Non-Steroidal Anti-Inflammatory Drugs (NSAIDs) [18, 19], quinine as potent antitumor agent [20], pyrimidine moiety in ATP-inhibitors for the treatment of tumors [21], and in antidepressant drugs [22]. Thus, synthesis of *N*-heterocycles in a rapid and economical way is a highly demanding task for medicinal chemists, and the use of microwave irradiation appears to have an impact in controlling the economic burden for the synthesis of heterocycles [23].

Van der Eycken has extensively studied [24] the synthesis of medium-sized heterocycles by employing transition metal catalyst under microwave irradiation. The synthesis of 1-substituted 3-benzazepinones **2** has been demonstrated by MW-assisted intramolecular Heck cyclization of propynoic acid amides **1** in the presence of Pd(0) as the catalyst (Scheme 9.1) [25].

Later on, the same research group applied the methodology for the synthesis of 3-benzazepine alkaloid (–)-aphanorphine [26]. Acrylamide **3** was cyclized in the presence of Pd(dppf)Cl$_2$ at 110°C under microwave irradiation for 15 min. The 3-benzazepinone **4** thus obtained was utilized as the intermediate for the total synthesis of (–)-aphanorphine (Scheme 9.2).

More recently, the same authors reported the synthesis of 3-benzazepines **6** through a Heck–Suzuki tandem reaction involving the propargylamide **5** and organoboronic acid or boronic acid pinacol ester in the presence of a Pd catalyst and base under microwave irradiation (Scheme 9.3) [27].

The multicomponent reactions (MCRs) have been recognized as new tools of synthetic organic/medicinal chemists due to their multifold advantages: high atom economy, applicability in the construction of complex structural features from simple and readily available substrates in one-pot, etc. [28]. Thus, the strategy of MCRs is extended for the synthesis of various organic compounds [29–33], which

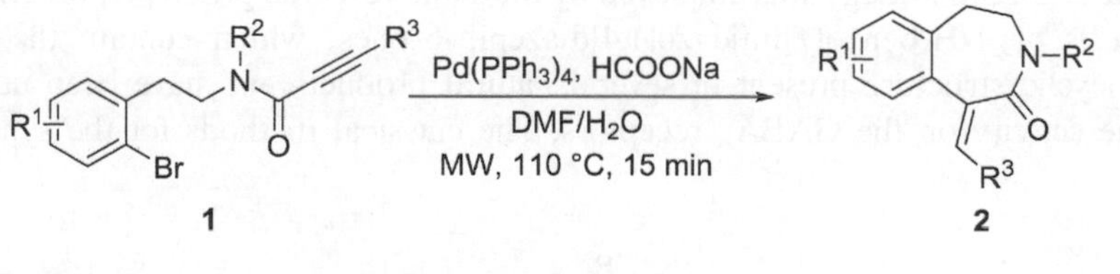

**SCHEME 9.1** MW-Assisted Synthesis of Benzazepinones.

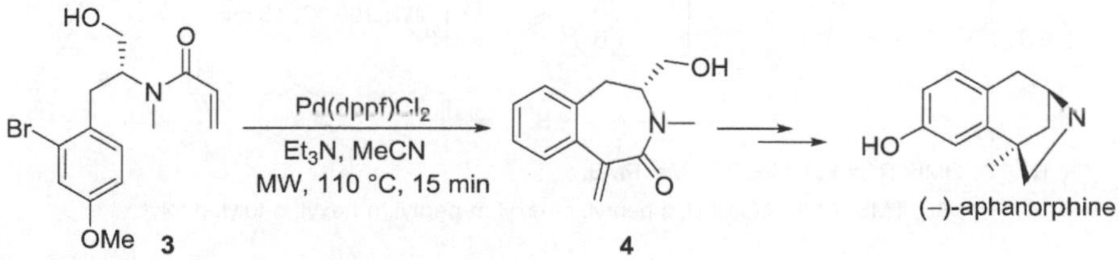

**SCHEME 9.2** Synthesis of 3-benzazepinone.

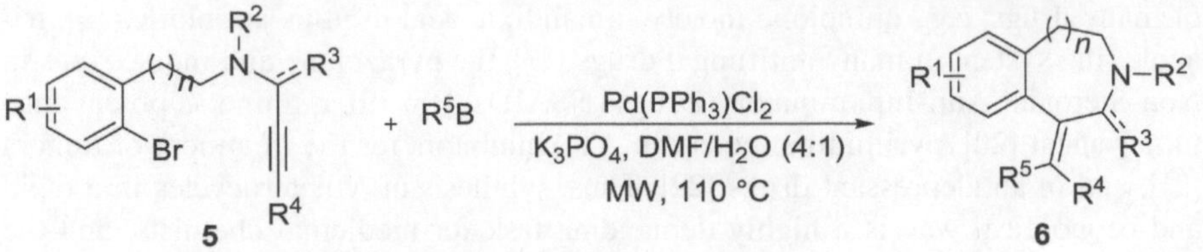
R$^2$
R$^1$
Br
R$^3$
R$^4$
5
+ R$^5$B
Pd(PPh$_3$)Cl$_2$
K$_3$PO$_4$, DMF/H$_2$O (4:1)
MW, 110 °C
N–R$^2$
R$^5$
R$^4$
6

**SCHEME 9.3** Microwave-Assisted Synthesis of 3-benzazepines.

find applications in the generation of new anti-leishmanial agents [34], as important building blocks, and in the construction of bioactive heterocycles [35–38].

Thus, integrating the strategies of MCRs and MAOS would be an elegant approach in organic synthesis. The effort towards this endeavor is observed in the efficient synthesis of dibenzoazepine and dibenzoazocine derivatives **8**, the structural analog of apogalanthamine and buflavin alkaloids. Buflavins are known to have α-adrenolytic and anti-serotonic activities. The first application of intramolecular $A^3$-coupling, a three-component reaction of aldehyde, amine, and alkyne, for the synthesis of seven- and eight-membered ring has been achieved by the use of inexpensive copper catalyst under microwave irradiation. The use of CuBr for the cyclization of the amine, generated in-situ by *N*-Boc deprotection of **7**, in toluene under microwave irradiation for 15 min afforded the corresponding product **8** as single diastereomer in good to excellent yields [39] (Scheme 9.4). However, the cyclization to the eight-membered azocine derivatives resulted in mixture of diastereomers (~1:1, when ring B is phenyl and ~1:4, when ring B is thiophene).

The nine-membered buflavin analogs having the dibenzo[c,e]azonine skeleton **10** was synthesized by ring-closing metathesis of the key biaryl intermediate **9** in the presence of Grubbs' catalyst under microwave irradiation at 150°C for 15 min (Scheme 9.5) [40].

The $A^4$ MCR strategy was employed by the same research group [41] for efficient synthesis of 4-H-benzo[*f*]imidazo[1,4]diazepin-6-ones, which contain the basic heterocyclic structure present in several natural products and have been utilized for the activity on the $GABA_A$ receptors. The classical methods for the synthesis

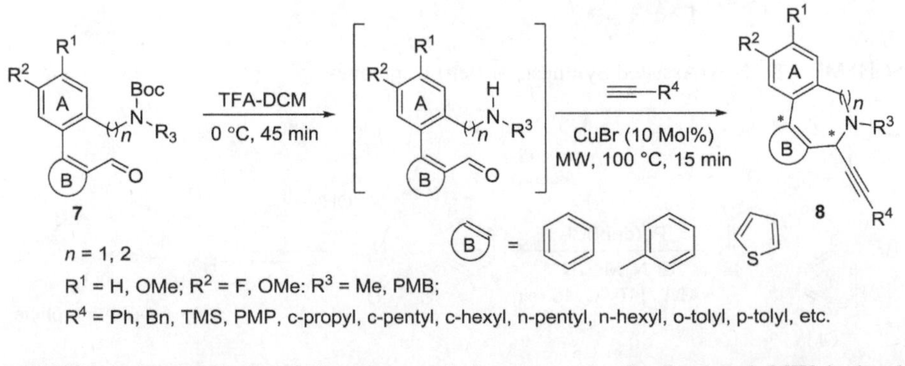
R$^1$
R$^2$
A
Boc
N
R$_3$
B
O
7
TFA-DCM
0 °C, 45 min
H
N
R$^3$
R$^4$
CuBr (10 Mol%)
MW, 100 °C, 15 min
N–R$^3$
8
R$^4$
B =
S
n = 1, 2
R$^1$ = H, OMe; R$^2$ = F, OMe: R$^3$ = Me, PMB;
R$^4$ = Ph, Bn, TMS, PMP, c-propyl, c-pentyl, c-hexyl, n-pentyl, n-hexyl, o-tolyl, p-tolyl, etc.

**SCHEME 9.4** Synthesis of Azepines and Azocines via Cu-Catalyzed MW-Assisted Intramolecular $A^3$-Coupling.

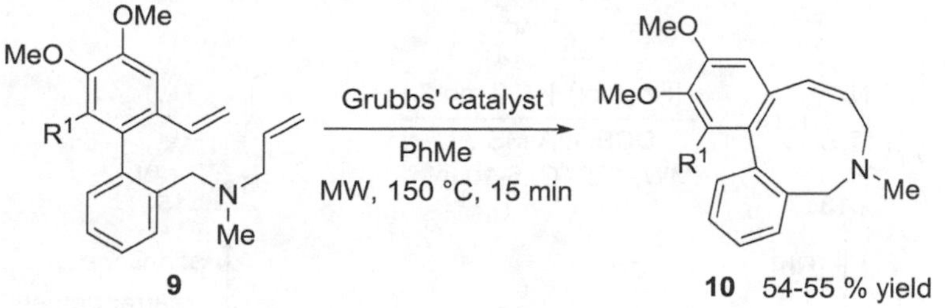

**SCHEME 9.5** Synthesis of Dibenzo[c,e]azonine: Buflavin Analogs.

of this scaffold required harsh reaction conditions with less substrate scope. The synthesis of the target compounds involves copper-catalyzed microwave-assisted intramolecular Ullmann coupling of the Ugi adduct **11**, synthesized through a four-component Ugi-reaction of imidazo-4-carbaldehyde, iodobenzoic acid derivatives, benzylamines and isonitrile derivatives, to form various 4-H-benzo[f]imidazo[4]diazepin-6-ones **12** in moderate to good yields (Scheme 9.6). The reaction conditions tolerate a wide range of substituents on the aromatic rings. The reaction proceeds through coordination of CuI to imine nitrogen **11**, followed by Cu(I) insertion into the aryl iodine bond to provide the cyclic eight-membered Cu intermediate, which upon reductive elimination affords **12**.

A new synthetic route for the synthesis of highly functionalized 2,5-dihydroazepines **15** was reported by Shang et al. through rhodium-catalyzed intermolecular aza-[4+3] cycloaddition of the easily available triazole **13** and (Z)-1-phenyl-1,3-butadiene **14** in 1,2-Dichloroethane (DCE) under microwave irradiation at 120°C [42] through a rhodium(II) iminocarbene intermediate. Although the rhodium(II) iminocarbenes have been recognized as versatile intermediates in the [2+1] and [3+2] cycloadditions, in this report the authors claim for the first time utilization of rhodium(II) iminocarbenes for aza-[4+3] cycloadditions. The reaction was proposed to proceed through the reaction of rhodium(II) iminocarbene with 1,3-diene to generate the cyclopropylaldimine intermediate, which rapidly undergoes aza-Cope-rearrangement to furnish the azepine derivatives **15** (Scheme 9.7). However, conventional heating of reaction with (E)-diene at 140°C provides a formal [3+2] cycloadduct, 2,3-dihydropyrroles, as the sole product.

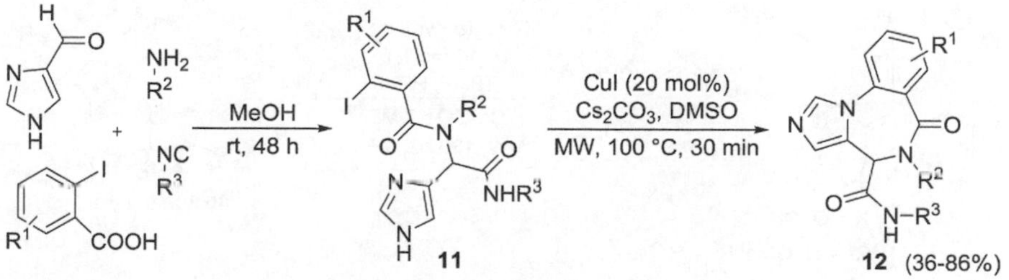

**SCHEME 9.6** Synthesis 4-H-benzo[f]imidazo[1,4]diazepin-6-one Derivatives via A⁴ MCR Strategy.

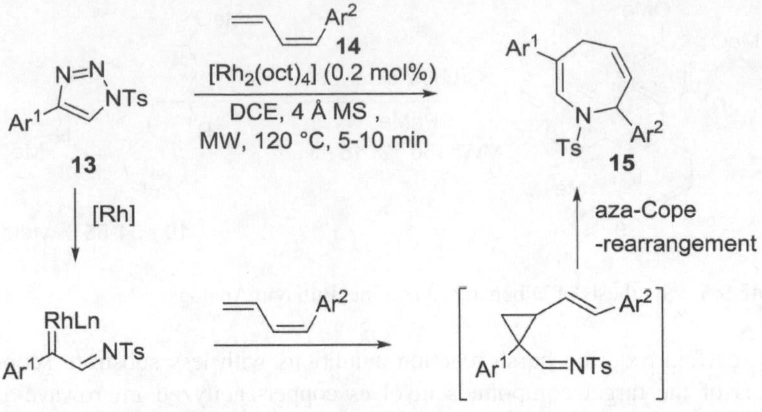

**SCHEME 9.7**   Synthesis of 2,5-dihydroazepines via Rh-Catalyzed MW-Assisted Cyclization.

Following a similar strategy, fused dihydroazepines were synthesized by intramolecular cycloaddition of triazole with tethered diene in the presence of Rh-catalyst [43], where microwave irradiation was utilized in some cases when conventional heating proved to be inefficient to provide the desired product.

Benzimidazole scaffold has drawn the attention of synthetic organic/medicinal chemists for its biological activities such as antimicrobial, antifungal, anticancer, anti-HIV, etc. [44]. Manna et al. utilized Ru catalyst under microwave for the synthesis of fused benzimidazole derivatives **16** [45]. The protocol involves intramolecular domino cyclization between imine and alkyne in the presence of 10 mol% of the Ru catalyst, TBAF (as additive) in DMF, under microwave irradiation for 1.5 h to furnish the fused benzimidazoles **16** in moderate to good yields. The proceeds via the initial formation of the imine, which helps in the coordination of the Ru (II) catalyst to the *ortho*-alkyne for the cyclization to get fused imidazole (Scheme 9.8). Conventional heating took a longer amount of time (15 h) to form the product in low yield.

The pyrido[2,3-*d*]pyrimidine ring system is a key structural feature of several bioactive compounds such as AZD8055 and piritrexim B. Compounds belonging to pyrido[2,3-*d*]pyrimidine class are known to have several other biological activities such as anti-inflammatory, antibacterial, anti-cardiovascular, anti-Parkinson's. etc. There are only a few methods to access this scaffold. Liu et al. exemplified an

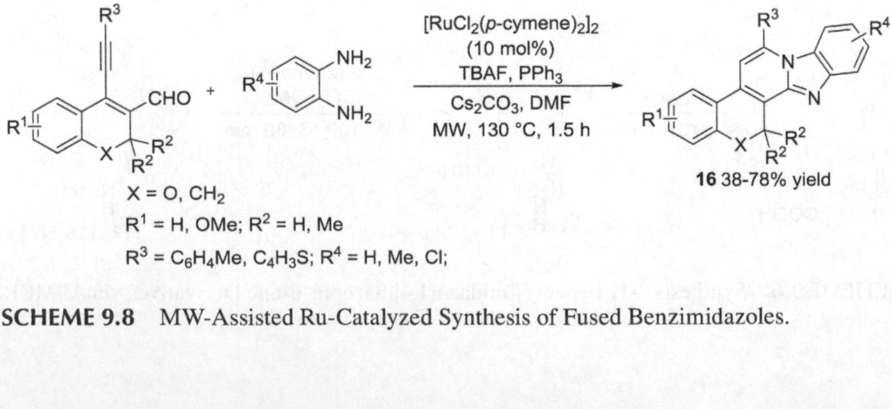

$X = O, CH_2$
$R^1 = H, OMe; R^2 = H, Me$
$R^3 = C_6H_4Me, C_4H_3S; R^4 = H, Me, Cl;$

**16** 38–78% yield

**SCHEME 9.8**   MW-Assisted Ru-Catalyzed Synthesis of Fused Benzimidazoles.

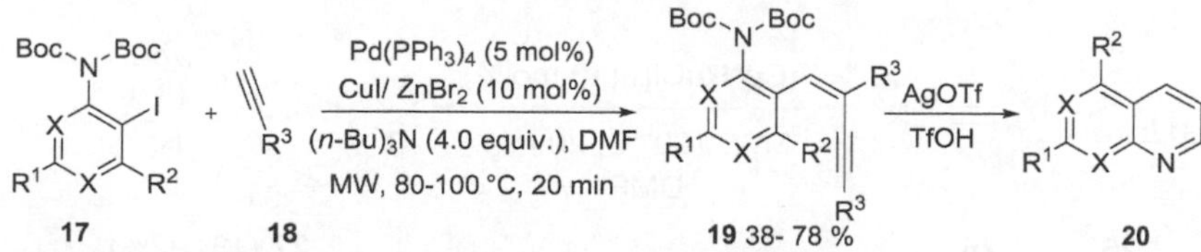

$R^1$ = H, Ph, $CH_3$; $R^2$ = H, OH; $R^3$ = $C_6H_4$-, TMS-; $R^4$ = H, Me, Cl; X = N, CH

**SCHEME 9.9** Synthesis of Pyrido[2,3-d]pyrimidine via a MW-Assisted Pd-Cu Co-catalyzed Sonogashira Coupling.

efficient synthesis of pyrido[2,3-*d*] pyrimidines **20** in moderate to good yields via the enyne **19** intermediate, formed by an unusual Heck–Sonogashira reaction of *N,N-di*-Boc aminoarenes **17** with terminal alkynes **18** (Scheme 9.9) [46]. The Boc protection of the amine functionality in the starting substrate **17** not only improves the solubility but also stabilizes the decisive Pd-C σ-bonded vinyl palladium species.

Liskamp and co-workers synthesized multivalent dendrimeric peptides, which can be useful for synthetic vaccine preparation [47]. These dendrimeric peptides were synthesized by Huisgen 1,3-dipolar cycloaddition using $CuSO_4$ as a catalyst with sodium ascorbate under microwave heating to afford the monovalent peptide in excellent yield compared to the traditional heating. The methyl ester **21** was saponified with Tesser's base to form **22** for coupling with the amino acid azide (AA) **23** in the presence of copper catalyst under microwave heating to furnish the dendrimer peptide **24** (Scheme 9.10). Yoon et al. used 1,3-dipolar cycloaddition for the synthesis of monofunctionalized dendrons and dendrimers. They used benzyl azide and alkyne/azide dendrimer for click reaction with 10 mol% $CuSO_4$. $5H_2O$ and 20 mol% sodium ascorbate in $^t$-BuOH:$H_2O$ (2:1) as the solvent system at 80°C for 20 h. However, the same reaction was achieved in 10 min with 5 mol% $CuSO_4$. $5H_2O$ and 10 mol% Na ascorbate under microwave heating [48]. These dendrimers can be used as career in drug delivery as they possess a peptide like amide backbone and have low toxicity and are non-immunogenic.

Rasmussen et al. devised a method for the synthesis of triazoles, which mimic the amide bond, using [Cp*RuCl]$_4$ as the catalyst under microwave irradiation [49]. While the copper-catalyzed cycloaddition of azide and alkyne (CuAAC) produces

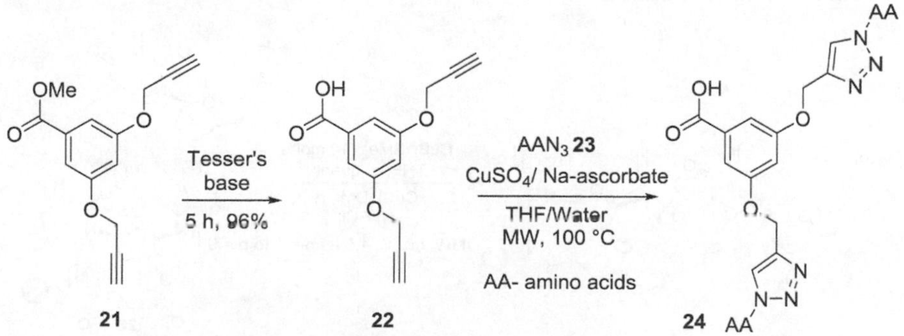

**SCHEME 9.10** Synthesis of Multivalent Dendrimers.

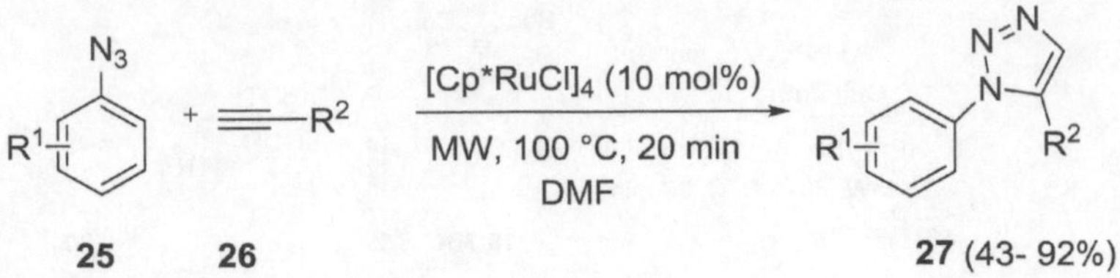

**SCHEME 9.11** Ru-Catalyzed MW-Assisted Synthesis of Bio-Relevant 1,5 Disubstituted Triazoles.

only the 1,4-disubstituted triazole, the Ru-catalyzed azide-alkyne cycloaddition (RuAAC) furnished the 1,5-substituted 1,2,3-triazoles. The reaction was performed under microwave heating at 110°C in DMF, which is known to coordinate with metal, especially Ru, and hence activates the Ru catalyst and stabilizes the complex. Different substituted azides **25** as well alkynes **26** were used for the synthesis of substituted traizoles **27** (Scheme 9.11) in good to excellent yield.

Isobe et al. used copper-catalyzed cycloaddition for designing artificial oligonucleotide **28**, which forms a stable double strand with the complementary strand of natural DNA [50] and represents the first example of artificial oligonucleotide. This shows the wide applicability of azide, alkyne and Cu-catalyzed click chemistry. This reaction was tried in solid phase using resin with the help of microwave irradiation. Moreover, this microwave-assisted Cu-catalyzed reaction was highly efficient for large-scale preparation of longer oligosaccharides (Scheme 9.12).

Indolines, alicyclic *N*-heterocycles, are present in many bioactive compounds and pharmaceuticals. Reported methods for the synthesis of indolines require expensive Pd catalysts and strong oxidants. Miura and co-workers reported the synthesis of indoline derivatives under mild reaction conditions using less toxic, inexpensive Cu catalyst by the intramolecular aromatic C-H amination protocol [51]. The substrate **29** containing picolinamide-type bidentate coordinating group plays a decisive role in the C-H activation by coordinating with the Cu catalyst, which is followed by disproportionation of Cu(II) to Cu(III), and reductive elimination to furnish the cyclized product **30** in good yields. Two methods were developed: in one case Cu(OAc)$_2$ was used in stoichiometric amount without any co-oxidant while in other case Cu(OAc)$_2$ was used in catalytic amount with MnO$_2$ as the oxidant (Scheme 9.13).

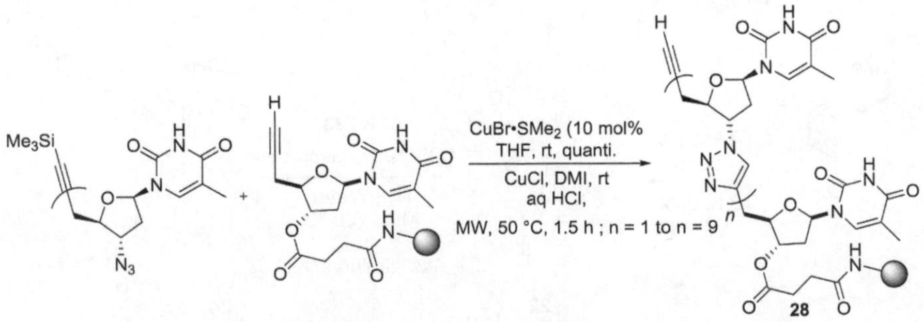

**SCHEME 9.12** Synthesis of Oligosaccharides.

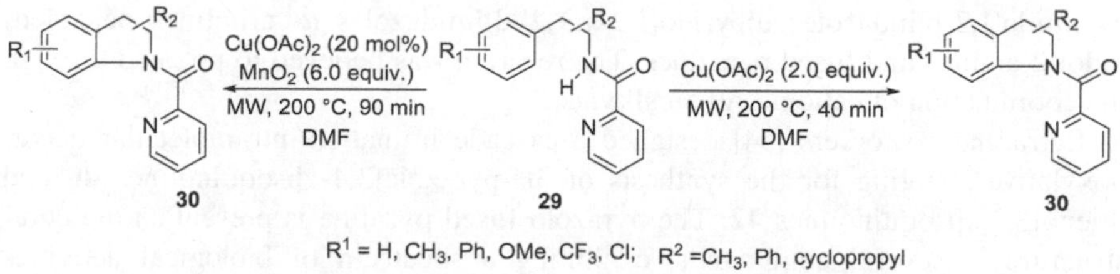

$R^1$ = H, CH₃, Ph, OMe, CF₃, Cl; $R^2$ =CH₃, Ph, cyclopropyl

**SCHEME 9.13** An MW-Assisted Cu-Catalyzed Protocol for Indoline Synthesis.

The fused pyrazole derivatives have been recognized as biologically important scaffold, and among them, the pyrazolo[3,4-b]pyridine, pyrazolo[3,4-b]quinoline, pyrazolo[1,5-a]pyrimidine, and pyrazolo[1,5-a]quinazoline moieties are an integral part of marketed drugs. The pyrazolo[3,4-b]pyridine derivatives **33** were synthesized through Pd-catalyzed microwave-assisted reaction of β-bromovinyl/aryl aldehyde **31** with 5-aminopyrazoles/3-amino pyrazole derivatives **32** and **32a** under solvent-free conditions [52]. The use of 2.5 mol% of Pd(OAc)₂ in neat condition under microwave (700 W) for 15 min was the optimal condition to provide the fused pyrazolo[3,4-b]pyridine **33** in good to excellent yields. The imine side product, which was formed in conventional heating conditions was not present under microwave heating conditions. The optimized reaction condition was used to investigate the scope of 3-aminopyrazole derivatives **32a** to form various substituted pyrazolo[1,5-a]pyrimidines **34** (Scheme 9.14). All synthesized compounds were screened for *in vitro* activities against cervical HeLa cancer cell line and prostate DU 205 cancer line using MTT assay, and four compounds were found to have anticancer activity comparable to that of the marketed drug doxorubicin.

A similar approach for the synthesis of fused azaheterocycles through the gold/acid co-catalyzed and microwave-assisted reaction of propargylic hydroperoxide with amines was reported by Alcaide and Almendros [53]. The reaction of propargylic hydroperoxide **35** with 3-aminopyrazole **36** in the presence of [AuCl(PPh₃)] as the catalyst in DCE under microwave irradiation produced the pyarazolo[1,5-a]pyrimidines **37** in good yield (Scheme 15). The methodology was extended to synthesize

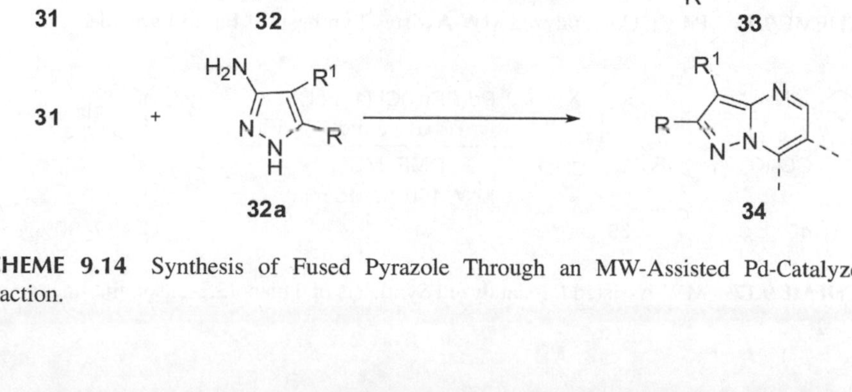

**SCHEME 9.14** Synthesis of Fused Pyrazole Through an MW-Assisted Pd-Catalyzed Reaction.

pyrimido[1,2-b]indazoles, dipyrido[1,2-a:3′,2′-d]imidazoles, α-carbolines, benzo[cd] indol-2-amines and fused pyrazoles. The reaction was believed to proceed through the coordination of cationic Au on alkyne.

Batra and coworkers [54] designed a cascade imination/intramolecular decor-boxylative coupling for the synthesis of 3H-pyrazolo[3,4-c]isoquinolines **40** and thieno[3,2-c]isoquinolines **42**. The pyrazolo fused pyridine is present as the core-structure in several compounds, exhibiting a spectrum of biological activities such as antidepressant, anxiolytic, platelet aggregation inhibition, etc. Synthesis of 3H-pyrazolo[3,4-c]isoquinolines **40** was achieved by the treatment of 5-amino-1-phenyl-1H-pyrazole carboxylic acid derivatives **38** with 2-bromo benzaldehyde derivatives **39** in the presence of Pd-Cu co-catalyst system in DMF under microwave irradiation for 15 min in good yields (Scheme 9.16). However, the formation of the thieno[3,2-c]isoquinolines **42** (Scheme 9.17) requires only the Pd catalyst due to easy delocalization of the lone pair in the heterocyclic substrate **40**, but in the case of pyrazole the presence of two nitrogen causes hindrance in delocalization of electron, requiring dual metallic system.

Pyrrole, a common structural motif found in many natural products such as alka-loids, porphyrins, chlorins, and corrins, exhibits a wide array of biological activities,

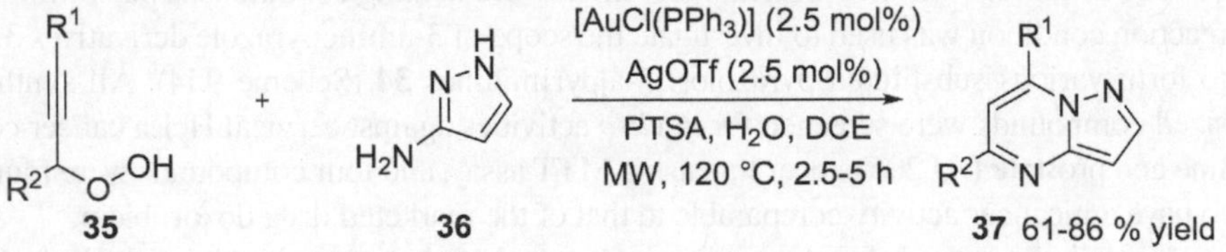

$R^1 = 4\text{-MeOC}_6H_4, R^2 = H; R^1 = Ph, R^2 = H; R^1 = 4\text{-BrC}_6H_4, R^2 = H; R^1 = Ph, R^2 = Me$

**SCHEME 9.15**   Au-Catalyzed MW-Assisted Synthesis of Pyarazolo[1,5-a]pyrimidines.

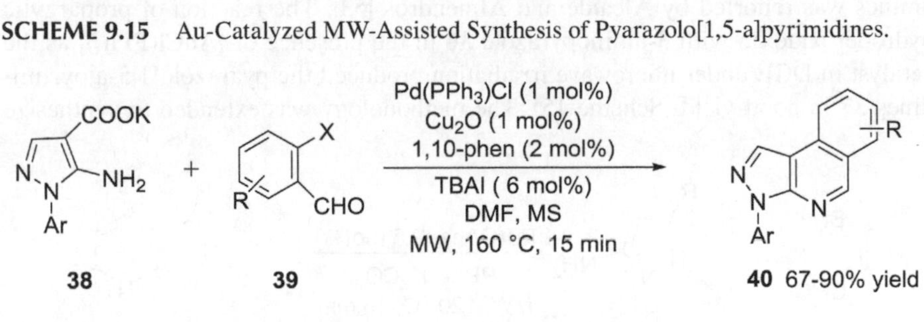

**SCHEME 9.16**   Pd-Cu Co-catalyzed MW-Assisted Synthesis of Fused Pyrazole.

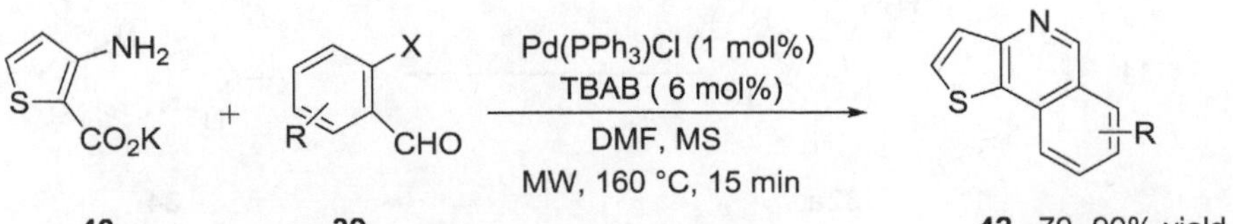

**SCHEME 9.17**   MW-Assisted Pd-Catalyzed Synthesis of Thieno[3,2-c]isoquinolines.

e.g., antifungal, antibacterial, antitumor, anti-inflammatory. Several marketed drugs possess the pyrrole subunit. Pyrrole-2-carboxylate has its own distinct advantage for the biosynthesis of many antibiotics. Imbri et al. synthesized pyrrole-2-carboxylates/carboxamides from one-pot two-step reactions [55]. While the authors were able to synthesize the pyrrole-2-carboxylates under conventional heating conditions, the synthesis of the corresponding carboxamides could be achieved under microwave heating conditions. The scaffold was synthesized by two-step procedure: cyclization followed by oxidation. This reaction involves enones **43**, which could be readily obtained through Claisen–Schmidt condensation [56] catalyzed by alkali-metal hydroxide, and glycine amides **44** as a substrate and this gives enaminone intermediate which underwent 6π-electrocyclization followed by oxidation to provide the pyrrole-2-caboxamide **45** in moderate yields (Scheme 9.18). Different oxidants were tried but DDQ and $Cu^{2+}$ salts gave better results. The Cu salts were used as these are cost-effective and less toxic.

Müller and co-workers synthesized the 5-(3-indolyl)oxazoles framework by consecutive one-pot three-component reaction [57]. This framework was recognized as the key feature of many biologically active alkaloids, which have been used to generate diverse therapeutic agents such as antimicrobial and anticancer drugs etc. The authors utilized the Sonogashira coupling reaction of acetamide **46** and aroyl chloride **47** to generate *in situ* the ynones followed by cycloisomerization with *p*-toluene sulfonic acid (PTSA) and *t*-BuOH to form 1-(hetero)aryl-2-(2-methyl-4-(hetero)aryl-oxazol-5-yl)ethanones under microwave irradiation for 1 h. After complete consumption of the ynone (monitored by TLC), PTSA was added further along with the aryl hydrazine hydrochloride **48** and the reaction was subjected to MW irradiation of 15 min to furnish 5-(3-indolyl)oxazoles products **49** in poor to moderate yields (Scheme 9.19).

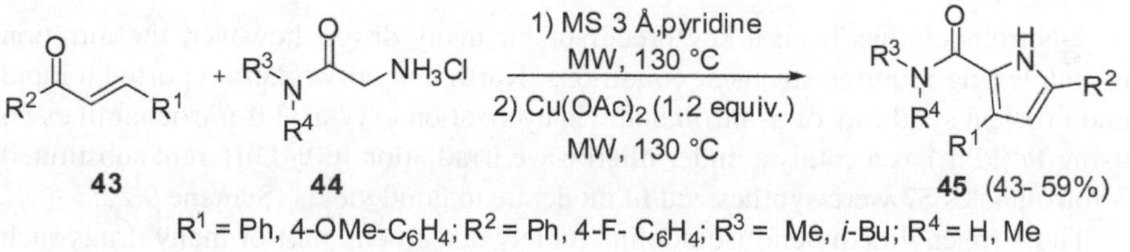

R$^1$ = Ph, 4-OMe-C$_6$H$_4$; R$^2$ = Ph, 4-F- C$_6$H$_4$; R$^3$ = Me, *i*-Bu; R$^4$ = H, Me

**SCHEME 9.18**  Synthesis of Fused Pyrazole-2-carboxamide.

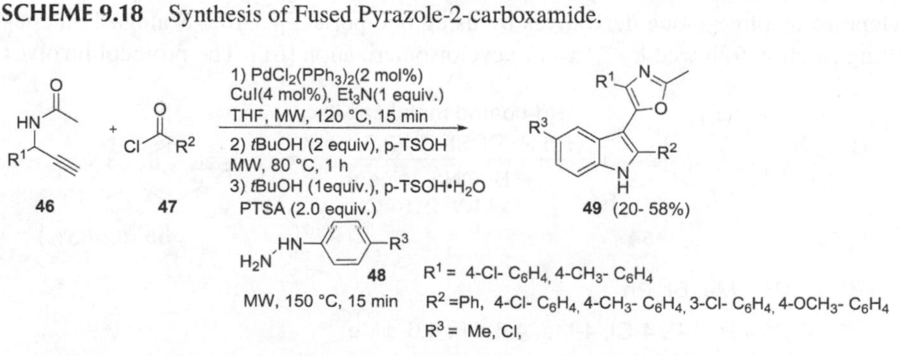

**SCHEME 9.19**  MW-Assisted Three-Component Synthesis of 5-(3-indolyl)oxazoles.

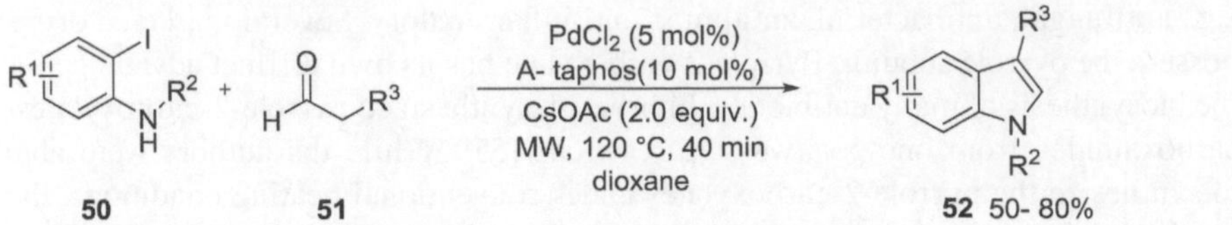

$R^1$ = CH$_3$, Cl, OCF$_3$, F, NO$_2$, OCH$_3$, COOCH$_3$; $R^2$ = H, CH$_3$

$R^3$ = CH$_3$, Et, (CH$_2$)$_2$CH$_3$, (CH$_2$)$_3$CH$_3$, (CH$_2$)$_5$CH$_3$, Ph, CH$_2$Ph, CH$_2$Ph (OCH$_3$)$_3$, (CH$_2$)$_2$CH$_2$-OH

**SCHEME 9.20**   Pd-Catalyzed MW-Assisted Synthesis of Indole Derivatives.

Indoles are important heterocycles in medicinal chemistry because of their abundance in several bio-relevant natural products and marketed drugs. Hence, the development of methodologies for the synthesis of indole derivatives with various substitution is of significant value to medicinal chemists. Muralidharan and co-workers synthesized substituted indoles using palladium catalyst under microwave heating [58]. The protocol involves the annulation reaction of *ortho*-iodo aniline **50** and aldehyde **51** to provide various substituted indole derivatives **52** with good to excellent yield (Scheme 9.20). To investigate the optimum condition, the authors tried different catalytic systems and PdCl$_2$/A-taphos and CsOAc in dioxane under microwave irradiation was found to be the best condition.

A microwave-assisted continuous flow synthesis of indoles **55** through two-step aryl amination/cross-coupling sequence of bromoalkenes **53** and 2-bromoamines **54** was demonstrated by Shore et al. [59]. The methodology requires a metal-lined flow tube and the Pd PEPPSI-IPr catalyst (Scheme 9.21). The presence of both metal capillary and catalyst was essential for the completion of the reaction to form the indoles.

3-Nitroindole has been a key precursor for many drugs; however, the nitration of indole ring requires stringent conditions. Kurth and co-workers reported a rapid and efficient synthesis of 3-nitroindole via cyclization of *N*-aryl β-nitroenamines **56** using Pd(PPh$_3$) as a catalyst under microwave irradiation [60]. Different substituted 3-nitroindoles **57** were synthesized in moderate to good yields (Scheme 9.22).

The 3-(phenylmethylene)isoindoline moiety is a crucial part of many drugs such as local anesthetic and the AChE inhibitors. Hellal et al. synthesized 3-(phenylmeth-ylene)isoindoline-1-one derivatives by using one-pot copper-free Sonogashira coupling reaction followed by 5-*exo-dig* cycloisomerization [61]. The protocol involves

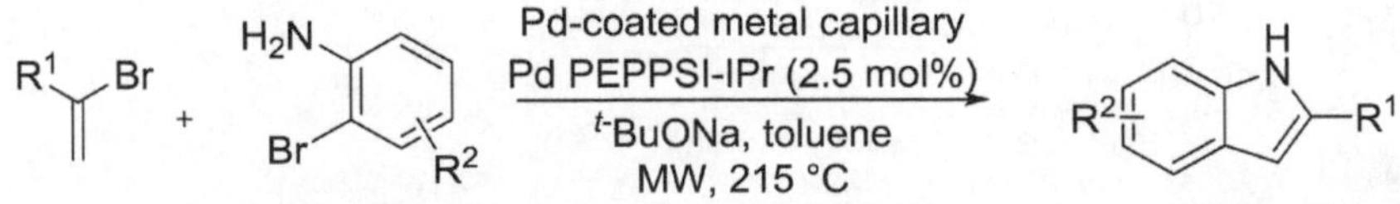

$R^1$ = Me, Et, Ph

$R^2$ = H, 4-F, 4-Cl, 4-Me, 4,6-diF, 4,6-diMe

**SCHEME 9.21**   MW-Assisted Pd-Catalyzed Synthesis of Indoles.

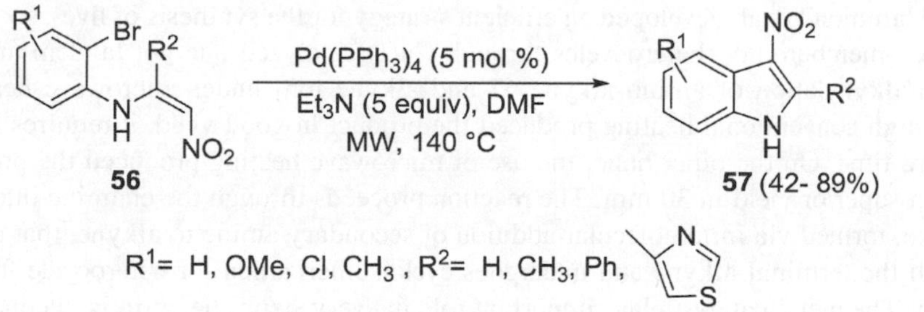

**SCHEME 9.22** MW-Assisted Synthesis of 3-Nitro Indoles.

the reaction of *N*-substituted 2-halobenzamide **58** with phenylacetylene **59** in the presence of $PdCl_2(MeCN)_2$, BINAP as ligand, DBU as base in DMF under the microwave irradiation (120°C for 15 min) to afford different substituted isoindolines **60** with prominent (*Z*)-isomer in moderate to excellent yields (Scheme 9.23). The methodology was further utilized for the synthesis of pyrrolopyridine derivatives. Chauhan et al. also synthesized similar isoindoline scaffold by using $Pd(OAc)_2$ as catalyst and isocyanides in place of terminal alkyne under microwave irradiation [62].

Turner and co-workers developed a microwave-assisted palladium-catalyzed protocol for the synthesis of *N*-substituted oxindoles using two-step reactions [63]. The reaction proceeds through initial amide bond formation between 2-halo-arylacetic acid **61** and the amine **62** under microwave heating, followed by intramolecular amidation in the presence of $Pd(OAc)_2$, and a phosphine ligand **63** and base (NaOH) in toluene/$H_2O$ under microwave irradiation for 30 min to provide substituted oxindole **64** in good to excellent yields (Scheme 9.24). In the case of alkyl amine, the second step was carried out without isolation of the amide intermediate, making it a one-pot process. However, in the case of aryl amines, it was necessary to isolate the amide which was used for next step intramolecular amidation.

Zhu et al. [64] introduced a two-step process involving an Ugi four-component reaction followed by palladium-catalyzed intramolecular Buchwald–Hartwig amidation reaction for the synthesis of the highly functionalized oxindole derivatives. The protocol involves the reaction of amines, 2-iodobenzaldehyde, carboxylic acid, and isonitrile derivatives under classical Ugi conditions to form the amide **65** which, upon palladium-catalyzed Buchwald–Hartwig reaction under microwave irradiation at 100°C, produced the oxindole derivatives **66** (Scheme 9.25).

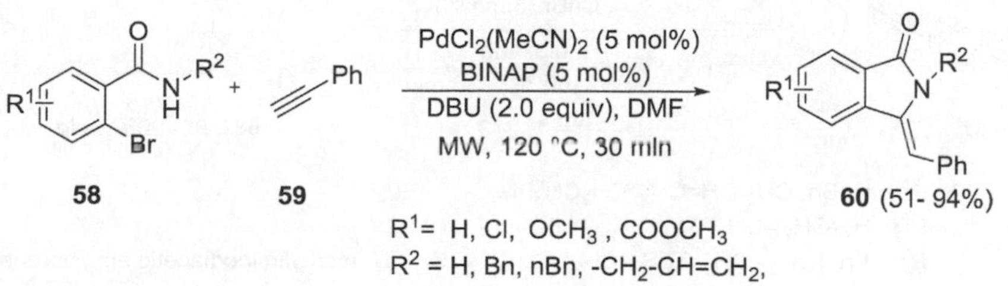

**SCHEME 9.23** Synthesis of 3-(phenylmethylene) Isoindolin-1-ones.

Hammond et al. developed an efficient strategy for the synthesis of five-, six-, and seven-membered *N*- heterocycles through Cu(I)-catalyzed one-pot tandem amination/alkynylation of amino alkyne **67** and alkyne [65] under microwave heating. Though conventional heating produced the product in good yield, it requires much more time. On the other hand, the use of microwave heating produced the product with superior yield in 30 min. The reaction proceeds through the enamine intermediate, formed via intramolecular addition of secondary amine to alkyne, that reacts with the terminal alkyne and undergoes cycloisomerization for heterocycle formation. The metal catalyst plays important role in every step. The optimized condition was used to get different substituted *N*-heterocycles and different ring size *N*- heterocycles **68** (Scheme 9.26).

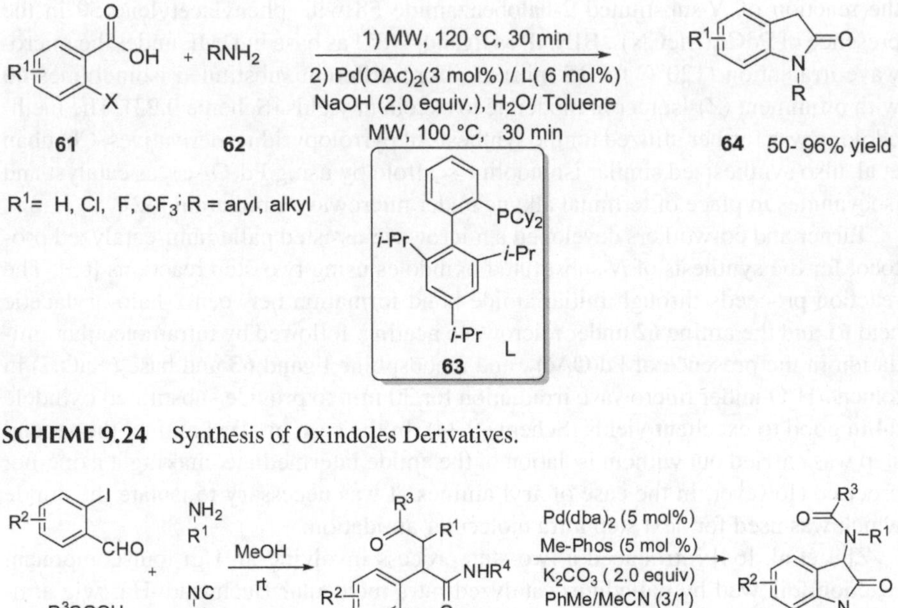

**SCHEME 9.24** Synthesis of Oxindoles Derivatives.

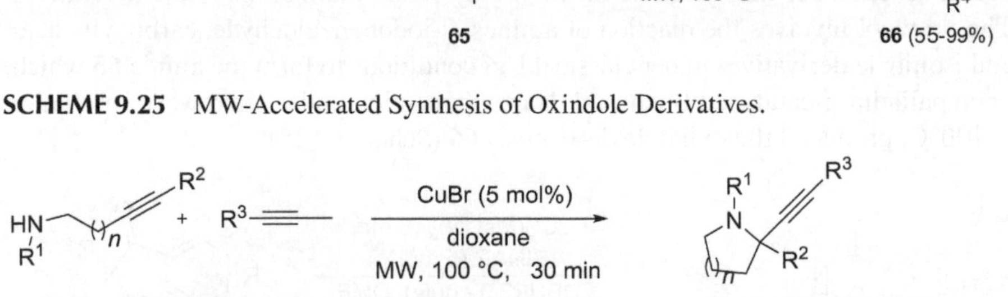

**SCHEME 9.25** MW-Accelerated Synthesis of Oxindole Derivatives.

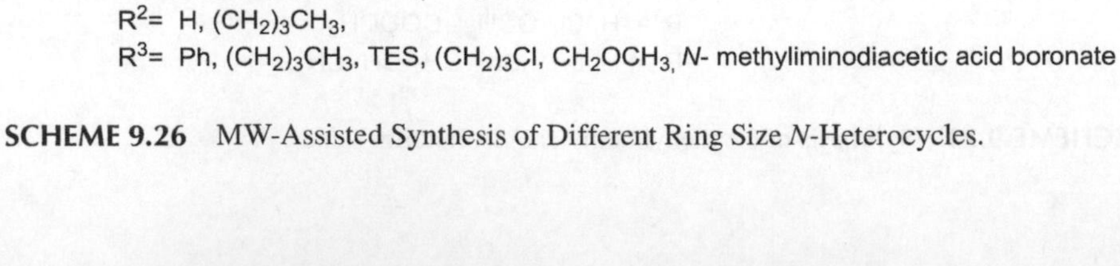

R$^1$= H, Bn, CH$_2$-CH=CH$_2$, CH$_2$CH$_3$Ph,
R$^2$= H, (CH$_2$)$_3$CH$_3$,
R$^3$= Ph, (CH$_2$)$_3$CH$_3$, TES, (CH$_2$)$_3$Cl, CH$_2$OCH$_3$, *N*- methyliminodiacetic acid boronate

**SCHEME 9.26** MW-Assisted Synthesis of Different Ring Size *N*-Heterocycles.

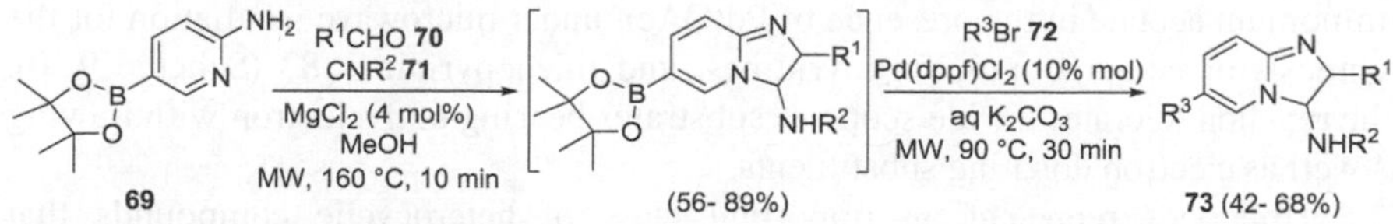

**SCHEME 9.27** Synthesis of 2-, 3-, 6-Substituted Imidazopyridine Derivatives.

Microwave-assisted metal-catalyzed MCR was well demonstrated by Kennedy et al. for the synthesis of imidazo[1,2-a]pyridine [66], the framework that has been well utilized by the pharmaceutical sector. The protocol involves the sequential addition of 2-aminopyridine-5-boronic acid pinacol ester **69**, aldehyde **70**, and the isocyanide **71** in the presence of catalytic amount of MgCl$_2$ under microwave irradiation for 10 min. The reaction proceeds via Ugi-type cyclization followed by Suzuki coupling reaction with the bromide **72** in the presence of Pd(dppf)Cl$_2$ as the catalyst under microwave irradiation to form the variously substituted 2,3,6-imidazo pyridine derivatives **73** (Scheme 9.27).

Furoquinoxalines, featured as important structural components in many biologically relevant compounds, have been prepared by one-pot three-component Cu(II)-catalyzed reaction using *ortho*-phenylenediamine (amine source) in A$^3$ coupling [67]. The protocol involves the coupling of *ortho*-phenyldiamine **74** with ethyl-glyoxalate **75** and terminal alkyne **76** in the presence of Cu(OTf)$_2$ as the catalyst under microwave heating to provide the furoquinoxalines **77** in good yields (Scheme 9.28). The reaction proceeds through propargylamine formation, followed by 5-endo-*dig* cyclization.

The furoquinolines scaffold is an attractive synthetic target for synthetic organic/medicinal chemists as it is present in many bioactive compounds and natural products. Mondal et al. reported [68] the synthesis of furoquinolines **80** by microwave-assisted one-pot tandem copper(I)-catalyzed basic alumina-supported Sonogashira cross-coupling of quinolines **78** with alkynes **79** followed by cyclization (Scheme 9.29).

Chen et al. reported a palladium-catalyzed one-pot microwave-assisted sequential coupling reaction for the synthesis of isoquinolines, furopyridines, and thienopyridines [69]. The reaction proceeds through imination, followed by annulations, where the metal plays an important role in the coupling step. The protocol involves treatment of 2-bromoarylaldehyde derivatives **81** with terminal acetylenes **82** and

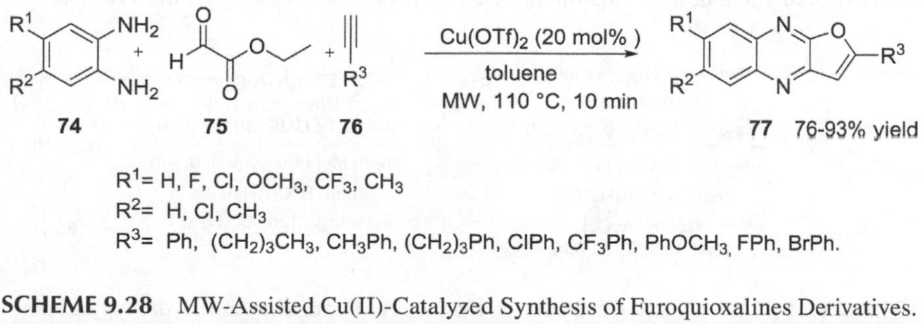

R$^1$= H, F, Cl, OCH$_3$, CF$_3$, CH$_3$
R$^2$= H, Cl, CH$_3$
R$^3$= Ph, (CH$_2$)$_3$CH$_3$, CH$_3$Ph, (CH$_2$)$_3$Ph, ClPh, CF$_3$Ph, PhOCH$_3$, FPh, BrPh.

**SCHEME 9.28** MW-Assisted Cu(II)-Catalyzed Synthesis of Furoquioxalines Derivatives.

ammonium acetate in the presence of Pd(OAc)$_2$ under microwave irradiation for the synthesis of isoquinolines, furopyridines, and thienopyridines **83** (Scheme 9.30). The reaction tolerates a wide scope of substrates bearing both electron withdrawing as well as electron donating substituents.

Pyrimidones represent an important class of heterocyclic compounds that exhibit several biological and pharmacological activities. Müller et al. reported a one-pot three-component synthesis of pyrimid-4(3H)-ones and 1,5-disubstituted 3-hydroxypyrazoles [70]. The acetylenes **84** were reacted with CO$_2$ in DMF in the presence of Cu(I) as the catalyst, phenanthroline as ligand, (Cs$_2$CO$_3$) as base to afford copper(I) propiolate, which was treated with MeI to furnish the methyl propiolate which, upon consecutive Michael addition-cyclocondensation with acetamidine chloride or hydrazine hydrochlorides under microwave heating, provided the corresponding pyrimid-4(3H)-ones **85** or hydroxypyrazoles **86**, respectively, (Scheme 9.31). It is worth mentioning here that traditional heating led to lower yield and led to side product formation.

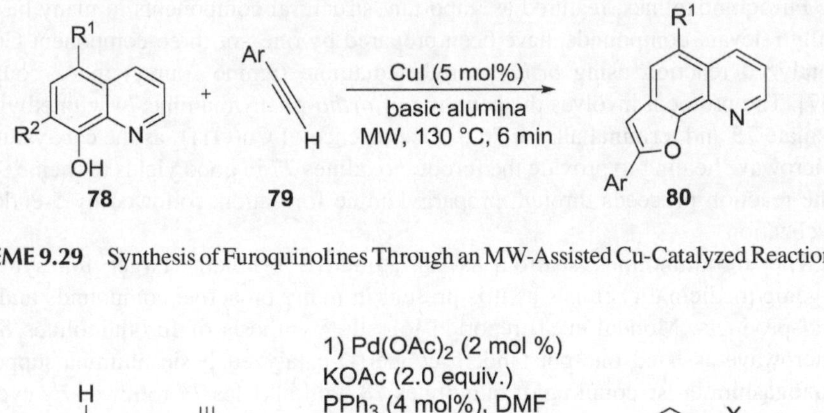

**SCHEME 9.29** Synthesis of Furoquinolines Through an MW-Assisted Cu-Catalyzed Reaction.

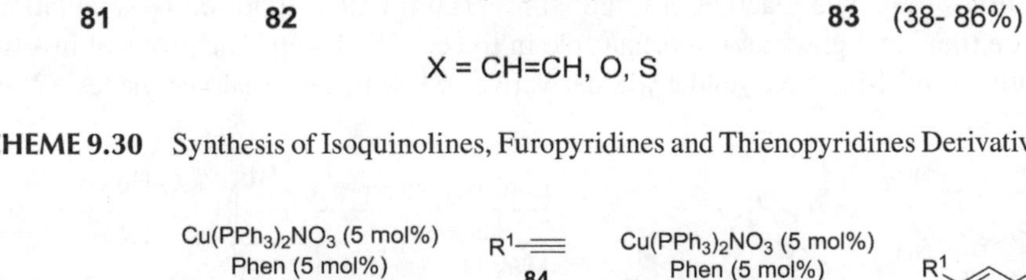

**SCHEME 9.30** Synthesis of Isoquinolines, Furopyridines and Thienopyridines Derivatives.

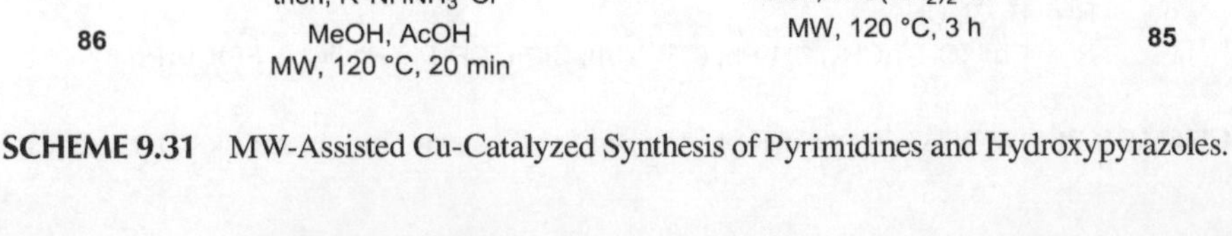

**SCHEME 9.31** MW-Assisted Cu-Catalyzed Synthesis of Pyrimidines and Hydroxypyrazoles.

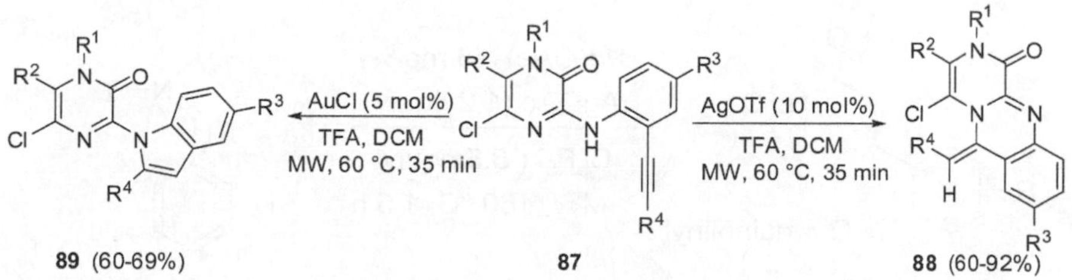

**SCHEME 9.32** Gold or Silver-Catalyzed MW-Assisted Synthesis of Pyrimidones and Hydroxypyrazols.

Van der Eycken demonstrated an efficient synthesis of pyrazino[2,1-b]quinazoline **88** and 3-indolyl-2(1H)-pyrazinone **89** derivatives through cyclization of **87** by microwave-assisted silver- and gold-catalyzed reactions, respectively. The pyrazino quinazoline was produced via a 6-*exo-dig* cyclization and indolyl-pyrazinone was formed via a 5-*endo-dig* cyclization to afford the corresponding product in good to excellent yield [71] (Scheme 9.32).

The quinoxaline ring is part of several bioactive natural products exhibiting a broad spectrum of therapeutic activities, such as anti-tubercular, antibacterial, antiviral, antifungal, anticancer and anti-inflammatory, and this inspired synthetic chemists to develop new methods for the construction of this privileged heterocyclic system [72]. Zhang et al. [73] reported a four-component Ugi reaction to form the Ugi adducts **90** which were cyclized either in the presence of copper or palladium catalyst to afford the indole fused quinoxalines **91** or indole fused quinolones **92**, respectively, in good to excellent yields (Scheme 9.33).

Sun et al. developed an efficient synthesis of β-lactams through easily accessible aminoquinoline carboxamide by using a palladium-catalyzed C(sp³)–H bond activation and intramolecular amination strategy under microwave irradiation in the absence of solvent [74]. The Pd coordinates to both the nitrogen of aminoquinoline

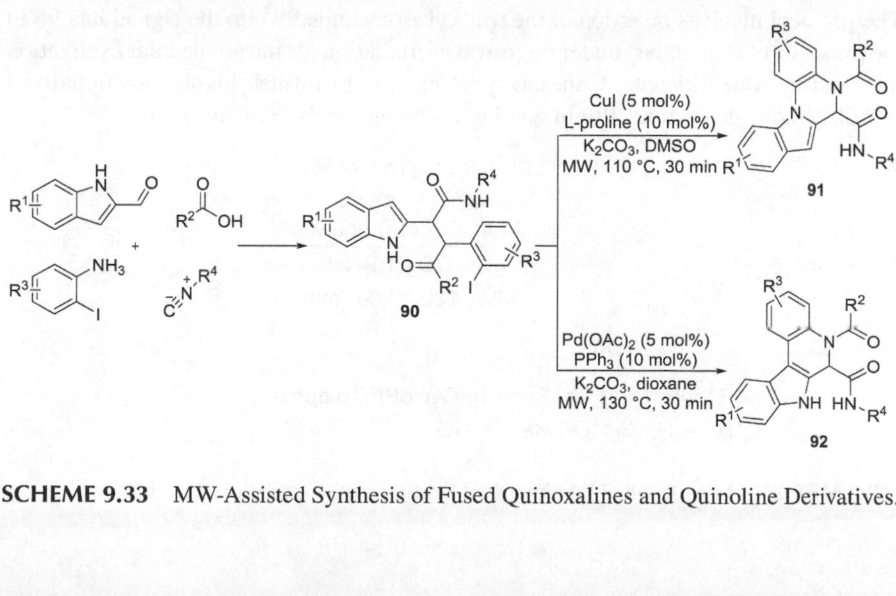

**SCHEME 9.33** MW-Assisted Synthesis of Fused Quinoxalines and Quinoline Derivatives.

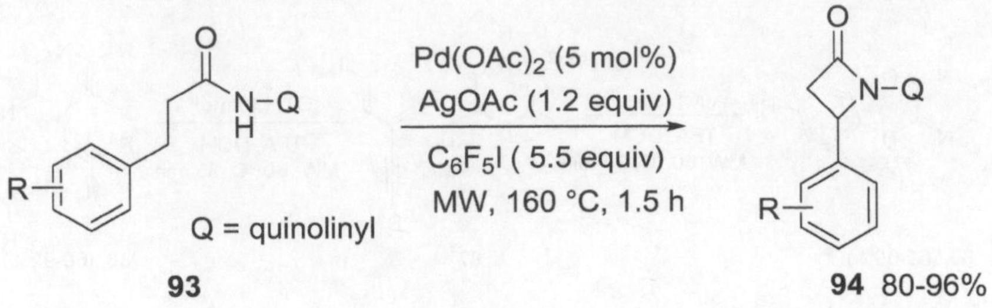

**SCHEME 9.34** MW-Assisted Pd-Catalyzed Synthesis of β-lactam Derivatives.

carboxamide **93**. The $C_6F_5I$ promotes the conversion of Pd(II) to Pd(IV) species, which is followed by reductive elimination to produce a diverse range of β-lactam ring **94** in excellent yields (Scheme 9.34). The methodology was successfully applied for the formal synthesis of a β-lactamase inhibitor MK-8712 (originally investigated by Merck Company).

4-Quinolone derivatives have been recognized as an important scaffold of pharmaceutical application. Larhed and co-workers developed an environmentally friendly method for the synthesis of 4-quinolones [75], where $Mo(CO)_6$ was used as CO source. The reaction involves Sonogashira coupling reaction of terminal alkyne **96** with 2-iodoaniline derivatives **95** in the presence of $Pd_2(dba)_3$ as catalyst, $Cs_2CO_3$ as base, dppf as ligand and diethyl amine as solvent under microwave heating at 120°C for 20 min to form 4-quinolones **97** in moderate to excellent yield (Scheme 9.35). The reaction tolerates a wide range of functional group. The palladium catalyst was essential for the reaction as in the absence of Pd catalyst no conversion takes place.

Isoquinolinone derivatives have attracted attention due to their broad spectrum of biological activities such as antihypertensive, anticancer, and inhibitors of some enzymes such as topoisomerase I, JNK, Rho-kinase, and Lck kinase etc. Chauhan and co-workers demonstrated the synthesis of variously substituted isoquinolinone [76]. The protocol involves insertion of the trialkyl-isocyanide **99** into the Ugi adduct **98** in the presence of Pd-catalyst under microwave irradiation for intramolecular cyclization followed by Mazurkiewitz–Ganesan type reaction to furnish highly functionalized isoquinolinone derivatives **100** in good to excellent yields (Scheme 9.36).

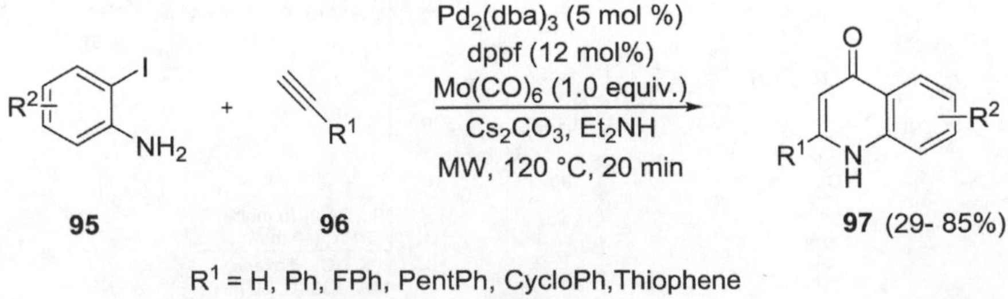

$R^1$ = H, Ph, FPh, PentPh, CycloPh,Thiophene
$R^2$ = H, Cl, COOMe, Cl, $NO_2$

**SCHEME 9.35** MW-Assisted Pd-Catalyzed Synthesis of 4-Quinolones Derivatives.

1,3-dihydrobenzimidazol-2-ones are known to possess important biological activities such as selective vasopressin α receptor antagonists, p38 MAP kinase inhibitor, progesterone receptor antagonist, etc. Liu et al. developed an efficient method for the synthesis of *N*-substituted 1,3-dihydrobenzimidazol-2-ones **102** [77] by Cu(I)-catalyzed cyclization of *N*-substituted *N*-(2-halophenyl)urea **101** in DMSO under microwave heating at 120°C for 20 min in good to excellent yields (Scheme 9.37).

1,2,4-Triazines is a well-versed scaffold used in pharmaceutical research. Moody et al. demonstrated the synthesis of 1,2,4-triazines via metal carbene N-H insertion as the key step [78]. Benzhydrazide **103** was treated with α-diazo-β-ketoester **104** in the presence of catalytic amount of Cu(OAc)$_2$ under microwave heating, followed by reaction of the intermediate **105** (produced by N-H insertion) with NH$_4$OAc in acetic acid to yield the 1,2,4-triazine **106** (Scheme 9.38). The Cu(II) also plays important role in the aerobic oxidation process.

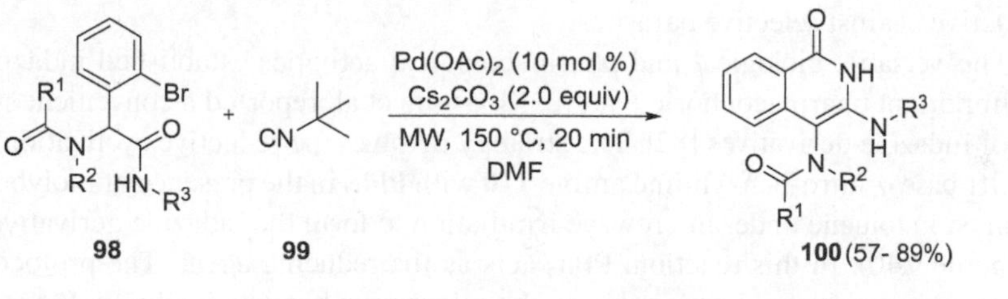

**SCHEME 9.36** MW-Assisted Pd-Catalyzed Ligand-free Synthesis of Isoquinolin-1(2H)-one Derivatives.

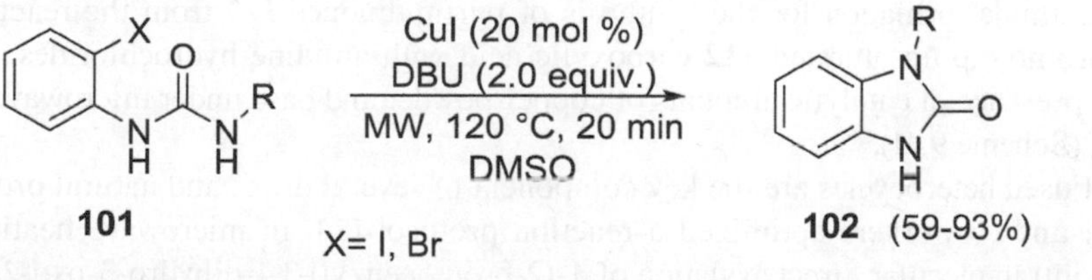

**SCHEME 9.37** Synthesis *N*-Substituted 1,3-Dihydrobenzimidazol-2-ones Derivatives.

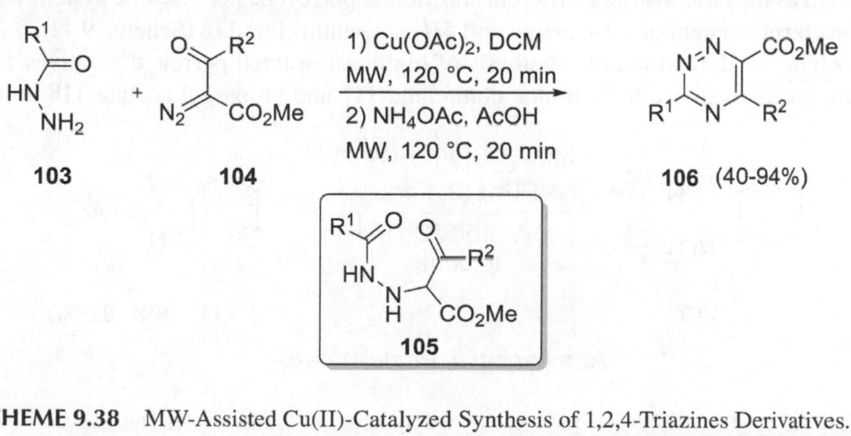

**SCHEME 9.38** MW-Assisted Cu(II)-Catalyzed Synthesis of 1,2,4-Triazines Derivatives.

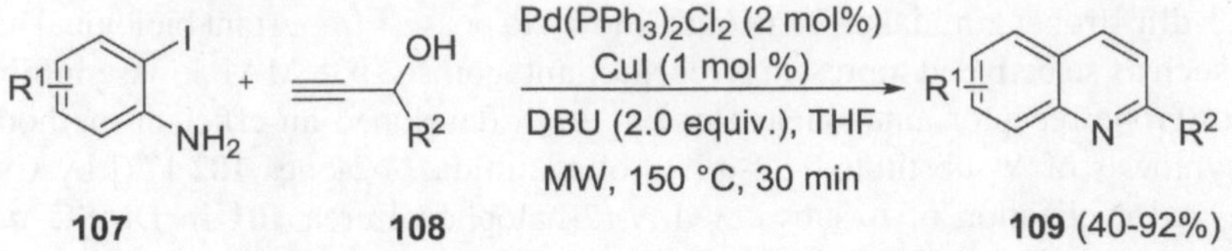

**SCHEME 9.39**  MW-Assisted Synthesis of Quinoline Derivatives.

The Friedländer annulation [79] has been a common strategy to construct the quinolone ring system for therapeutic evaluation of substituted quinolones [80]. Müller and co-workers reported [81] an effective and alternative strategy for the synthesis of quinolines through microwave-assisted Pd-Cu co-catalyzed coupling-isomerization of *o*-amino(hetero) aryl halide **107** and propargyl alcohols **108** to produce the quinolines **109** in good to excellent yields (Scheme 9.39). These quinolines were found to be active against selective parasites.

The versatile biological and pharmacological activities established indazoles as an important pharmacophoric feature. Moustafa et al. reported a convenient synthesis of indazole derivatives [82]. The strategy involves the reductive cyclization of the Schiff base *o*-nitro-benzylidine amine **110** with PPh$_3$ in the presence of molybdenum catalyst in toluene under microwave irradiation to form the indazole derivatives **111** (Scheme 9.40). In this reaction, PPh$_3$ acts as the reducing agent. The protocol also works in conventional heating, but in the microwave heating condition, the reaction rate was 30 times faster than that of the conventional heating condition.

Cho and co-workers [83] reported a tandem aromatic nucleophilic substitution-amide formation for the synthesis of pyrimidinones **114** from the reaction of β-bromo α,β-unsaturated **112** carboxylic acid with amidine hydrochlorides **113** in the presence of catalytic amounts of copper-powder and base under microwave heating (Scheme 9.41).

Fused heterocycles are the key component of several drugs and natural products. Dai and co-workers optimized a reaction protocol [84] in microwave heating for the intramolecular direct arylation of 4-(2-bromonenzyl)-3,4-dihydro-3-oxo-2*H*-1,4-benzoxazines **115**, where Pd(OAc)$_2$ and dppf were used in 1:1 ratio, with K$_2$CO$_3$ (2.0 equiv.) as the base to afford different substituted polycyclic heterocyclic system with a fused ring system of 1,4-oxazine and 5*H*- phenanthridine **116** (Scheme 9.42).

Zhang et al. reported the synthesis of highly substituted pyrrole derivatives **119** from easily available β-enaminoe compound **117** and propargyl acetate **118** under

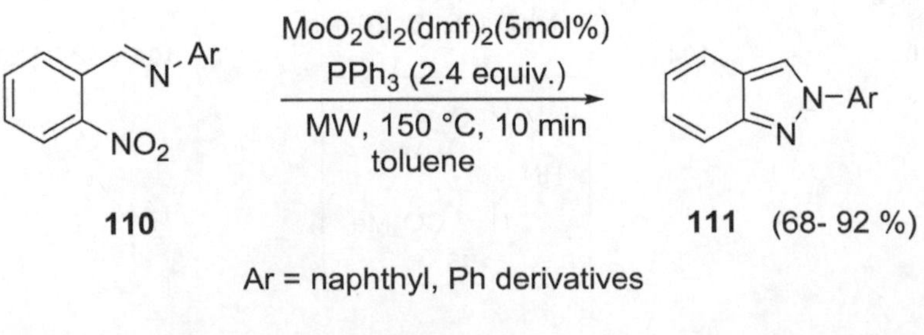

Ar = naphthyl, Ph derivatives

**SCHEME 9.40**  MW-Assisted Mo-Catalyzed Synthesis of 2H-Indazoles Derivatives.

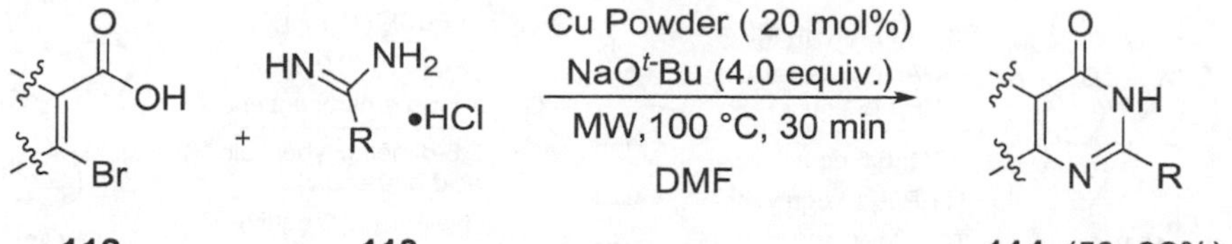


**112** **113** **114** (58- 92%)

R= Ph, Me

**SCHEME 9.41** Cu-Powder-Catalyzed MW-Assisted Synthesis of Pyrimidinones Derivatives.

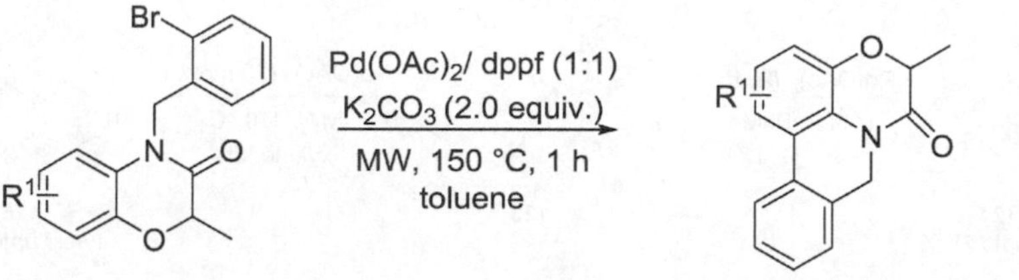


**115** **116** (43- 99%)

**SCHEME 9.42** MW-Assisted Synthesis of Fused Heterocyclic Ring System.

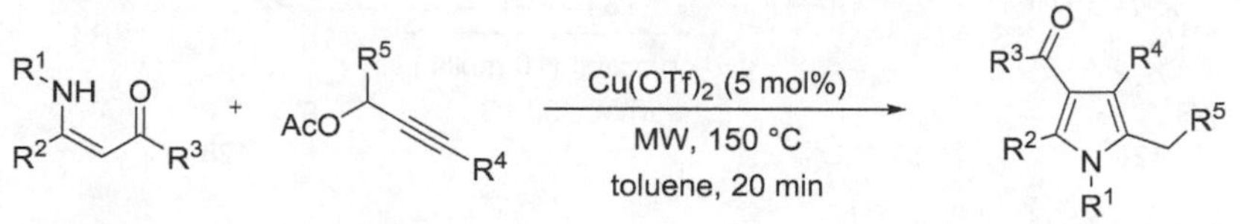


**117** **118** **119** up to 93%

**SCHEME 9.43** MW-Assisted Synthesis of Highly Substituted Pyrrole Serivatives.

microwave irradiation [85]. The reaction is catalyzed by copper(II) and proceeds through the propargylation of **117** followed by alkene azacyclization/isomerization (Scheme 9.43). The authors have successfully scaled the reaction to gram quantity.

Wu and co-workers have reported the synthesis of polycyclic azetidines **121** and pyrrolidines **122** via picolinamide-assisted, palladium-catalyzed, C-H bond activation followed by intramolecular amination of aliphatic amines **120** under microwave irradiation condition [86]. While the formation of the polycyclic azetidines is attributed to γ-C-H bond activation, the formation of the pyrrolidines is realized due to δ-C-H bond activation (Scheme 9.44).

Wang et al. [87] reported an interesting palladium-catalyzed annulation of *N*-allyaldimines **123** where under conventional heating the imidazoles **124** are formed as the main product, but under microwave heating the quinazolines **125** were obtained as the major product (Scheme 9.45).

## 9.3 *O*-HETEROCYCLE

A wide array of medicinally important natural and synthetic molecules contain oxygen heterocycle as the core moiety [88]. The transition metal (TM)-catalyzed

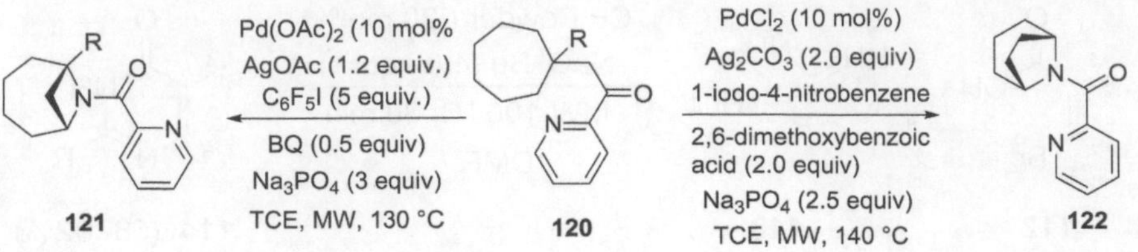

**SCHEME 9.44** Pd-Catalyzed MW-Assisted Synthesis of Azetidine and Pyrrolidine Derivatives.

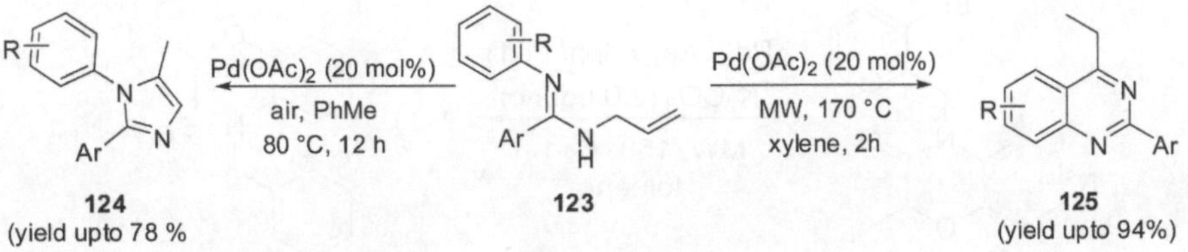

**SCHEME 9.45** Pd-Catalyzed Annulation of *N*-Allylamidines.

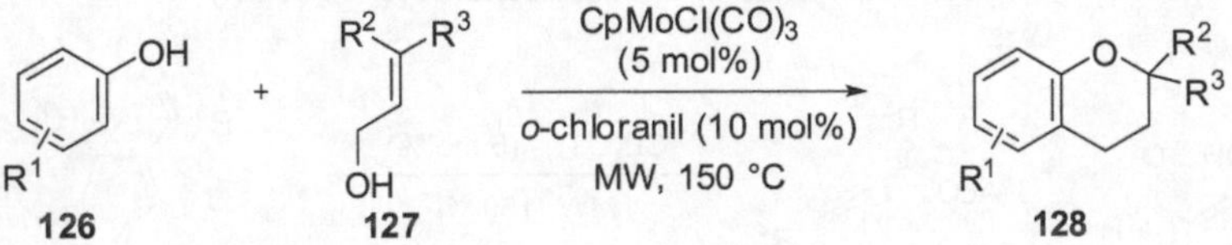

**SCHEME 9.46** MW-Assisted Mo-Catalyzed Synthesis of Chroman Derivatives.

coupling reactions have emerged as new techniques for the preparation of oxygen-containing heterocyclic scaffolds via the formation of C–O bond [89, 90].

The chroman ring is the key motif of several natural and synthetic bioactive compounds such as vitamin E, and its derivatives, and flavonoids. However, the classical methods for the synthesis of this scaffold require a large amount of acid catalysts. Yamamoto and co-workers [91] disclosed an interesting route for the synthesis of the chroman derivatives **128** through the reaction of *o*-cresol **126** with prenyl alcohol **127** in the presence of CpMoCl(CO)$_3$/*o*-chloronil catalytic system under microwave irradiation (Scheme 9.46).

The dibenzopyranone ring system constitutes the core structure in many natural products and biologically active compounds. Dibenzopyranones have also been used as key intermediates for several pharmaceutically important molecules such as progesterone, androgen, glucocorticoid modulators, etc. The synthesis of dibenzo-pyranones has been a challenge for the chemist as, although several protocols have been developed, the overall yields are low, need long reaction time and multiple steps requiring purification in every step, making this method tedious. Vishnumurthy et al. [92] have prepared a library of dibenzopyranones **131** from *ortho*-bromo aryl carboxylates **129** and *o*-hydroxyaryl boronic acids **130** via Suzuki–Miyaura cross-coupling under microwave irradiation (Scheme 9.47).

Thasana et al. [93] devised a method for the microwave-assisted synthesis of dibenzopyranone 133 by a $C_{aryl}$–$O_{carboxylic}$ intramolecular coupling reaction of various 2-halobiarylcarboxylic acids 132 catalyzed by copper(I) salts. Among the copper salts, the use of 2 molar equivalents of CuTC furnished the product under base- and ligand-free conditions. However, the use of CuI led to a drastic reduction in the quantity of the catalyst to 30 mol% to afford the desired pyranones derivatives 133 (Scheme 9.48). When conventional heating was used (as an alternative to microwave heating), the lactonization of the 2-halobiarylcarboxylic acids to form the dibenzopyranone 133 either failed or provided the product in very poor yield. The reaction could be extended for the synthesis of isolamellarin derivatives 134 (Scheme 9.49).

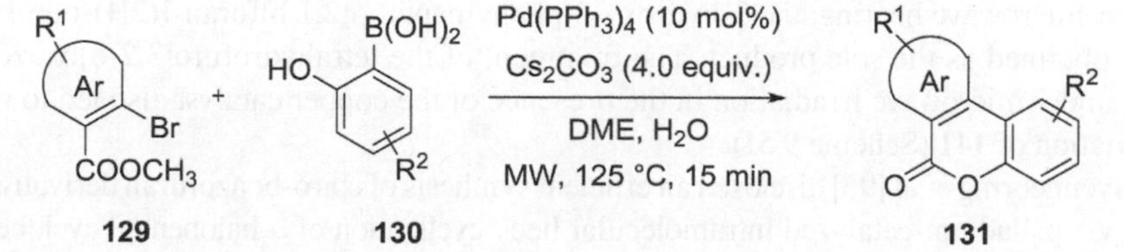

**SCHEME 9.47** Synthesis of Library of Benzopyranones.

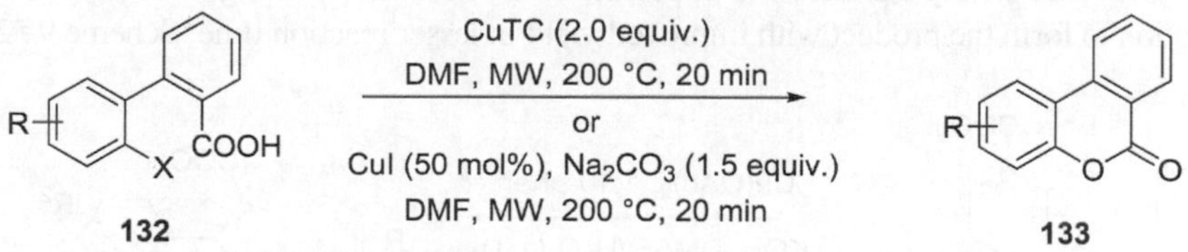

**SCHEME 9.48** MW-Assisted Cu(I)-Catalyzed Synthesis of Dibenzopyranones.

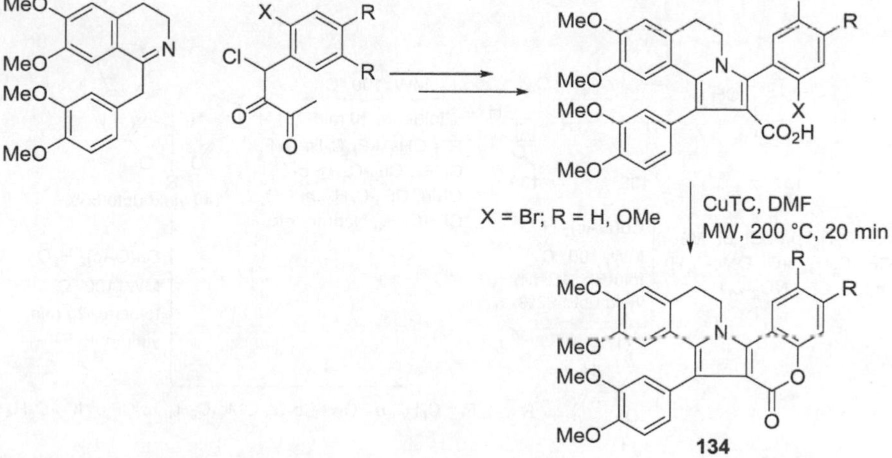

**SCHEME 9.49** Copper-Catalyzed Synthesis of Isolamellarins.

The benzofuran moiety is present in several drugs and natural products. The various methodologies that have been developed for the synthesis of benzofurans involve the use of expensive transition-metal catalysts, require harsh reaction conditions, offer low yields, and have limited substrate scope. Xu and co-workers [94] developed a microwave-assisted Cu-catalyzed intramolecular cyclization of (*E*)-2-(2-bromophenyl)-3-phenylacrylic acids **135** to afford 2-arylbenzofuran-3-carboxylic acids **136** in excellent yields (Scheme 9.50).

Ahmed et al. [95] achieved the synthesis of tetrahydrofuro[3,2-d]oxazole **140** during the reaction of β-naphthol **137**, pyrrolidine **138**, and 2-oxoaldehyde **139** at 100°C under microwave conditions. The reaction proceeds via the Betti base as an intermediate. However, when the same reaction was performed in the presence of Cu(OAc)$_2$ under microwave heating, the 2-hydroxy-2-phenylnaphtho[2,1-b]furan-1(2H)-one **141** was obtained as the sole product. The treatment of the tetrahydrofuro[3,2-d]oxazole **140** under microwave irradiation in the presence of the copper catalyst also led to the formation of **141** (Scheme 9.51).

Svennebring et.al [96] disclosed an efficient synthesis of spiro-benzofuran derivatives **143** via palladium-catalyzed intramolecular heck cyclization of *o*-halobenzyl cyclohexenyl ethers **142**. The reaction was found to be advantageous over the other methods due to its regioselectivity. However, conventional heating required 18 to 24 h and afforded low yield that clearly signifies the importance of microwave heating to accelerate the reaction to form the product with improved yield in lesser reaction time (Scheme 9.52).

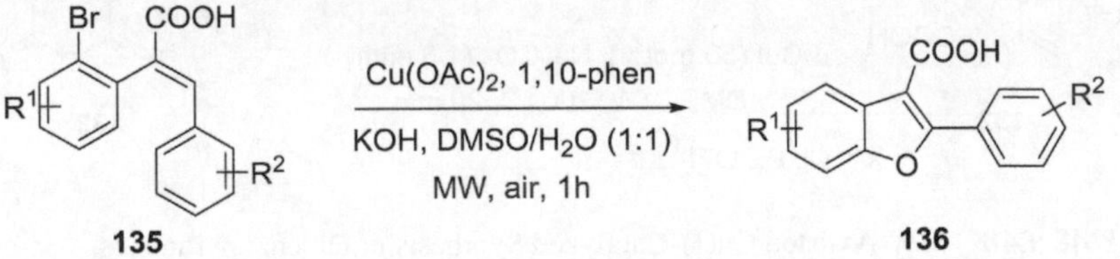

**SCHEME 9.50** MW-Assisted Synthesis Cu(II)-Catalyzed Synthesis of Benzofurans.

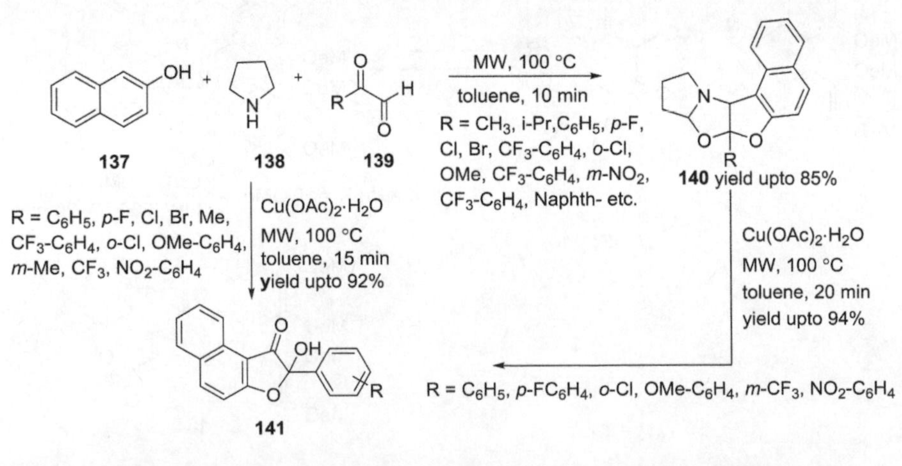

**SCHEME 9.51** Copper-Catalyzed MW-Assisted Synthesis of Naphthofuranones.

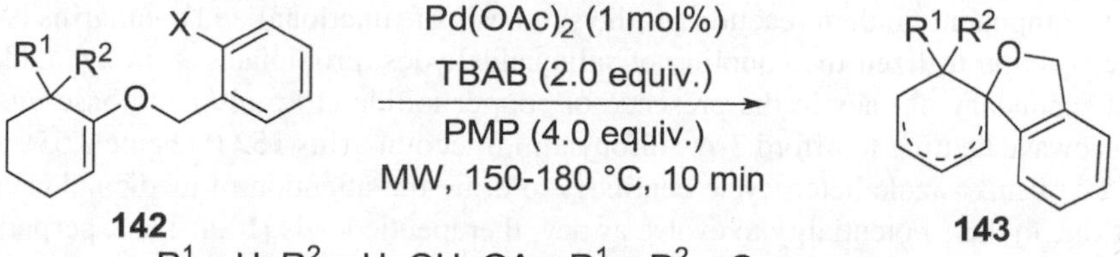


$R^1 = H; R^2 = H, OH, OAc; R^1 = R^2 = O$
X = Br, I

**SCHEME 9.52** MW-Assisted Synthesis of Conformationally Restricted Furans.

Castagnolo et al. developed [97] a microwave-assisted MCR strategy for the synthesis of 2,3-dihydropyrans **147**, bearing the C1-anomeric center, which offers scope for glycosylation reaction. The reaction protocol involves the treatment of the alkyne **144**, ethyl vinyl ether **145**, and ethyl glyoxalate **146** in toluene in the presence of Grubbs' catalyst at 80°C to afford 2,3-dihydropyrans **147**, with a 2:1 *trans/cis* diastereomeric ratio. The reaction proceeds through the formation of a diene intermediate by cross metathesis of alkyne and ethyl vinyl ether, which upon hetero-Diels–Alder reaction with ethyl glyoxalate produces the dihydropyrans **147** in 40–75% yield (Scheme 9.53). The MCR was further utilized to synthesize furanose-pyranose C–C-linked disaccharide.

Flavones, the secondary metabolites of the plants, are known to have a wide spectrum of biological activities. An interesting one-pot synthesis of substituted flavones under milder condition was developed by Capretta [98] through sequential application of microwave-assisted Sonogashira reaction and carbonylative annulation. The initial reaction involves the reaction of the aryl iodide **148** and TMS acetylene **149** in the presence of a Pd/PA-Ph catalyst, and DBU (as base) in DMF under microwave heating for 30 min. The reaction mixture is then transferred to another vessel containing iodophenol **150**, fresh catalyst, DBU, and TBAF (to remove the TMS and generate the terminal alkyne moiety) in DMF, which, upon exposure to CO followed by microwave irradiation for 30 min, produced the corresponding flavones **151** in moderate to good yields (Scheme 9.54).

Coumarins are also a much sought for synthetic targets due to their well-known biological and medicinal values. The classical synthesis of coumarins involves Knoevenagel condensation, which, however, have limited substrate scope. Punniyamurthy et al. reported microwave-assisted copper-catalyzed one-pot

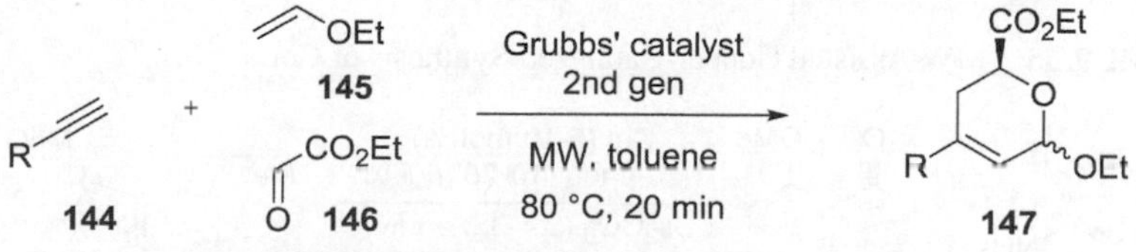


$R = BrCH_2, (EtO)_2CH, Cl(Me)_2C, BocNHCH_2, TMSOCH_2, PMBOCH_2, TMS, Ph$

**SCHEME 9.53** MW-Assisted Synthesis of Dihydropyrans.

four-component tandem reaction for the synthesis of functionalized coumarins [99]. The reaction utilized the coupling of salicylaldehydes, propiolates, sulfonyl azides, and secondary amines in the presence of copper iodide (10 mol%) and base under microwave heating to afford 3-*N*-sulfonylamidinecoumarins **152** (Scheme 9.55).

The benzoxazole heterocycle continues to draw the attention of medicinal chemists due to their potentiality to evolve as new therapeutic leads [100]. Thus, perpetual efforts are directed to functionalize the benzoxazole moiety to increase the diversity and generate more effective leads [101]. The cyclocondensation of *ortho*-aminophenols with carboxylic acid derivatives has long been considered a straightforward approach for the construction of the benzoxazole ring system [102], and the use of microwave irradiation facilitates the condensation of the carboxylic acids as such [103]. Viirre and coworkers [104] demonstrated that the domino reaction of 2-haloanilines **153** with carboxylic acid chlorides **154** in the presence of catalytic amount of CuI, 1,10-phenanthroline as ligand, and $Cs_2CO_3$ in MeCN forms the benzoxazole derivatives **155** under microwave heating (170–210°C) for 5–15 min which, otherwise, takes a prolonged reaction time (24 h) under the conventional heating (95°C) (Scheme 9.56).

An interesting synthesis of dihydroisobenzofurans **158** under microwave irradiation was reported by Najera and co-workers [105]. The reaction involves the use of the oxime palladacycle-catalyst (as palladium nanoparticles) for copper-free

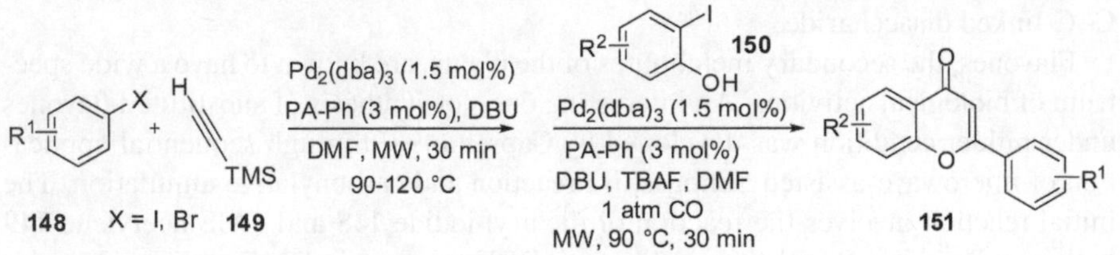

**SCHEME 9.54** Synthesis of Flavones by MW-Assisted Sonagashira Reaction followed by Carbonylative Annulation.

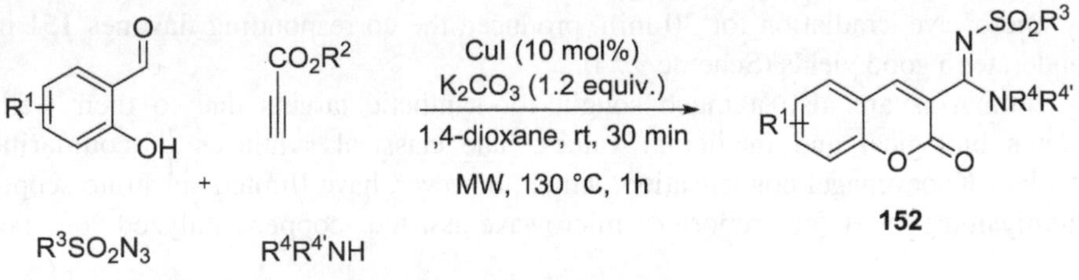

**SCHEME 9.55** MW-Assisted Copper-Catalyzed Synthesis of Coumarins.

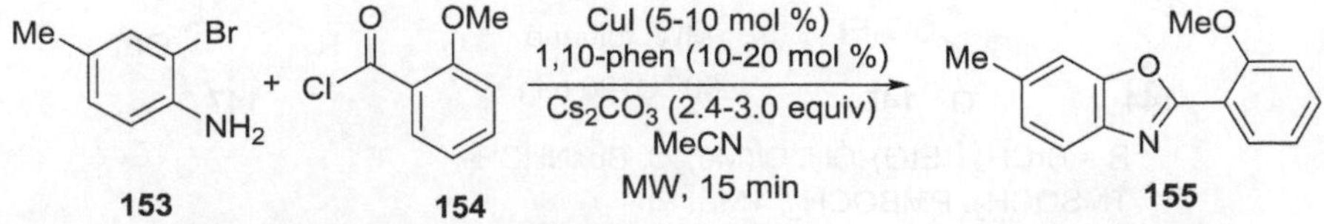

**SCHEME 9.56** Cu-Catalyzed Domino Coupling for the Synthesis of Benzoxazoles.

Sonogashira cross-coupling reaction of 2-(hydroxymethyl)bromo/chlorobenzene **156** with the terminal alkyne **157** followed by cyclization under microwave irradiation. The copper-free reaction generally avoids the homo coupling of alkynes. The reactions were successfully performed at 130°C in 15 min using the Pd catalyst and XPhos as the ligand to furnish the product (Scheme 9.57).

Dihydrofurans exhibit a broad spectrum of biological activities, and the scaffold is used to prepare many pharmaceutically important compounds. The synthesis of dihydrofurans via the cycloaddition of 1,3-dicarbonyl compounds with olefin in the presence of strong oxidants which, however, often promotes olefin polymerization. Xia et al. [106] reported an efficient synthesis of dihydrofurans that involves [3+2] cycloaddition of the 2-diazo-5,5-dimethylcyclohexane-1,3-dione **159** with the ethyl vinyl ether **160** in the presence of environmentally friendly ruthenium(II)-phosphine complex and ionic liquid as additive under microwave irradiation to afford the dihydrofurans **161** in good to excellent yields (Scheme 9.58).

Nicolaus and Schmalz reported [107] the synthesis of seven-membered oxygen heterocycles **164**, analogs of naturally occurring allocolchicine and *N*-acetyl colchinol alkaloids, through a cobalt-catalyzed [2+2 + 2] cycloaddition involving the diynes **162** and the nitriles **163** under microwave assistance (Scheme 9.59).

Benzoxazines represent a class of compound often encountered in many bioactive natural products. The reaction of *ortho*-amino phenols with 1,2-bis-electrophiles

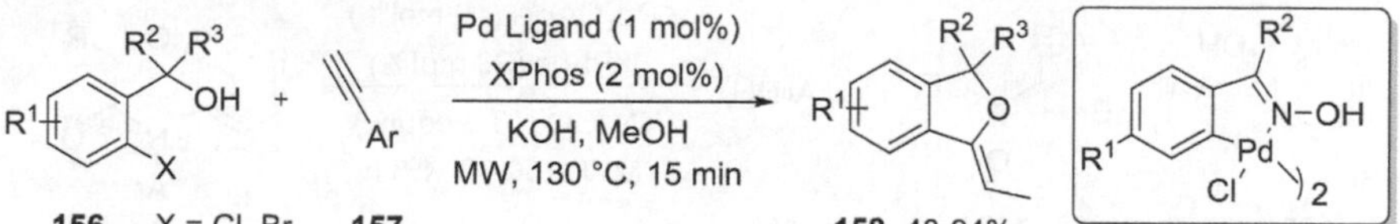


**156** X = Cl, Br **157** **158** 43-94% Pd ligand

**SCHEME 9.57** MW-Assisted Synthesis of Dihydroisobenzofurans.

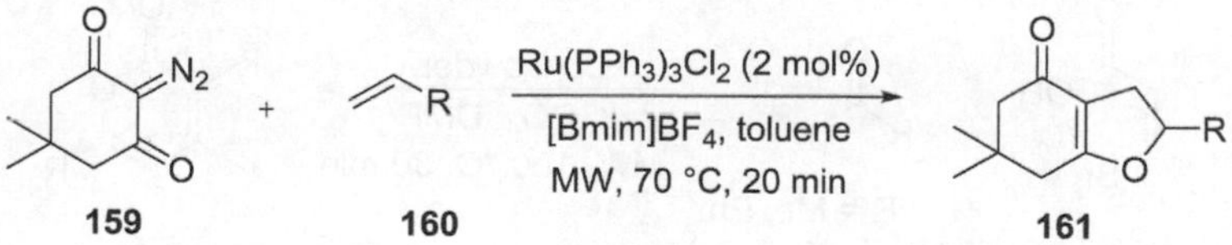


**159** **160** **161**

**SCHEME 9.58** Ru-Catalyzed MW-Assisted Synthesis of Dihydrofurans.

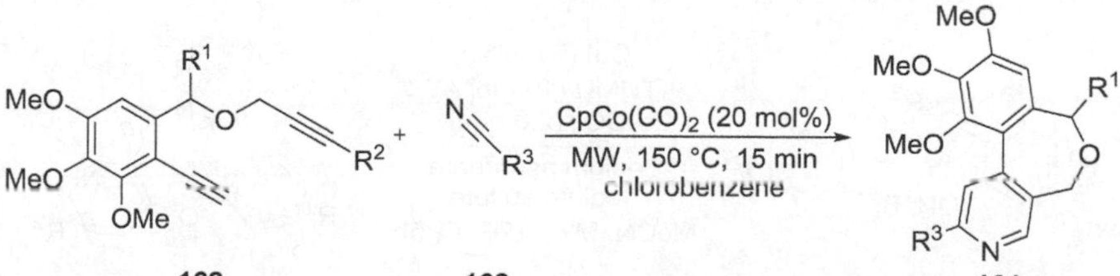


**162** **163** **164**

**SCHEME 9.59** Co-Catalyzed MW-Assisted Synthesis of Seven-Membered Oxygen Heterocycle.

leads to the formation of benzoxazines [108], contrary to the reaction of *ortho*-amino thiophenols with the 1,2-bis-electrophilic agents that form the corresponding benzothiazoles as the five-membered ring closure products [109], and represents a methodology for the direct construction of the benzoxazine ring system. Feng et al. adopted a similar strategy [110] for the synthesis of 3,4-dihydro-3-oxo-2H-1,4-benzoxazines **168** employing Pd-catalyzed three-component one-pot reaction of 2-halophenols **165**, ethyl-2-bromoalkanoates **166,** and aryl amines **167** under microwave irradiation (Scheme 9.60), wherein the *ortho*-amino phenol moiety is generated *in situ* via the Pd-catalyzed Buchwald–Hartwig type coupling of the *ortho*-halo phenol with the amine.

An efficient synthesis of 2H-pyran-2-one **171** was reported by Cho at al [111] through a copper-powder-catalyzed, MW-assisted reaction of β-bromo-α,β-unsaturated carboxylic acid **169** with 1,3-diketone **170** (Scheme 9.61).

Mestichelli et al. reported [112] an efficient synthesis of the tricyclic biaryl ether linked seven-, eight-, and nine-membered aza-heterocycles **173,** which are otherwise difficult to synthesize. Compounds containing this scaffold is well known as Central Nervous System (CNS) active agents and are used for the treatment of pain and inflammation. The authors used cheaper copper(I)-catalyzed, intramolecular cyclization of easily accessible precursor **172** under microwave irradiation. The reaction tolerates a wide substrate scope to provide the cyclized ring in moderate to excellent yields (Scheme 9.62).

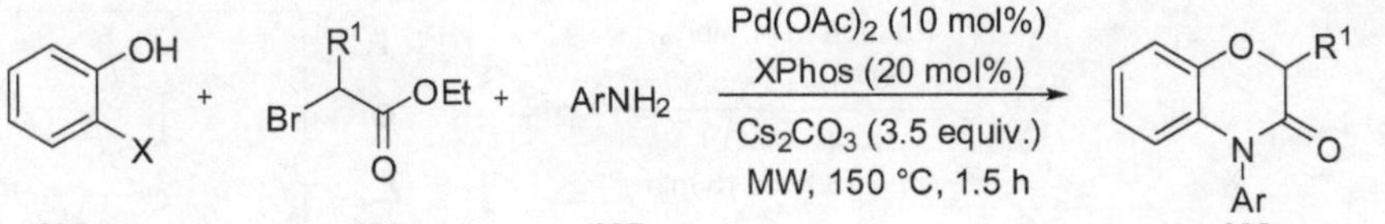

**SCHEME 9.60**   Pd-Catalyzed Synthesis of Benzoxazine.

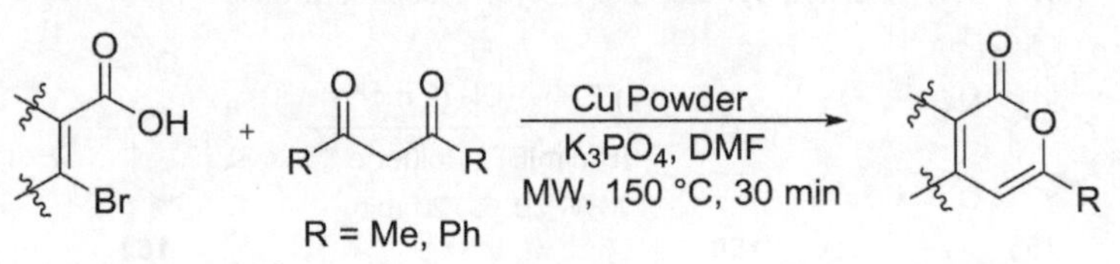

**SCHEME 9.61**   Copper-Powder-Catalyzed Synthesis of 2H-Pyran-2-one.

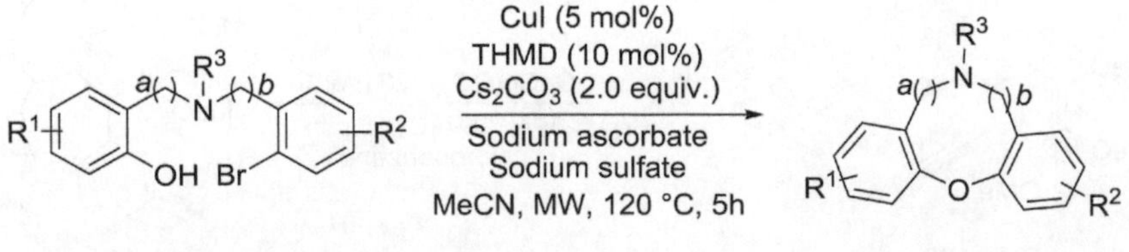

**SCHEME 9.62**   Cu(I)-Catalyzed Synthesis of Fused Medium-Sized Ring Heterocycles.

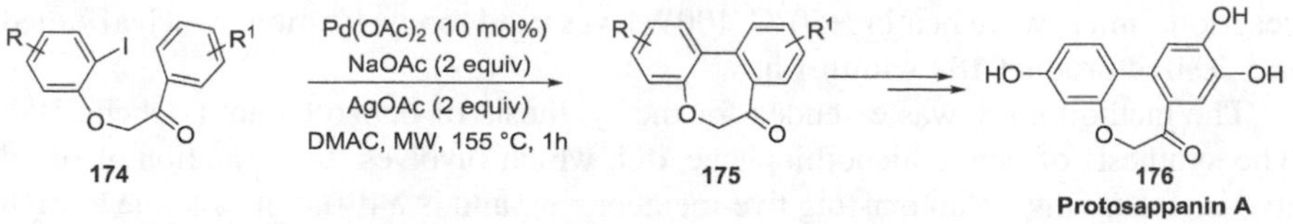

**SCHEME 9.63** Pd-Catalyzed Synthesis of Oxepinones.

Liu et al. [113] reported the total synthesis of a complex natural product protosappanin A **176**. The key intermediate dibenzo[b,d]oxepinones **175** required for the synthesis of the natural product was synthesized by palladium-catalyzed intramolecular biaryl coupling via *ortho* C–H bond activation of the α-(2-iodoaryloxy)-acetophenones **174** under microwave irradiation (Scheme 9.63).

## 9.4 S-HETEROCYCLES

The synthetic and natural sulfur-heterocycles play important role in medicinal and biological chemistry [114] and find applications in the pharmaceutical industry. The benzo[b]thiophene is an important bicyclic heterocycle found in many bioactive molecules such as zileuton, raloxifene, Sertaconazole, and benocyclidine. Recently, Gong et al. reported [115] the synthesis of benzo[b]thiophene **178** and the thiochromone **179** through the thia-Ugi adduct **177** under copper-catalyzed microwave irradiation conditions (Scheme 9.64). The thia-Ugi reaction, a four-component reaction involving an aldehyde or a ketone, isocyanide, amine, and thioacid affords **177**, which undergoes copper-catalyzed cyclization under microwave irradiation.

Raloxifene, a benzo[b]thiophene analog, is a potential drug candidate. Knochel et al. reported [116] copper-catalyzed intramolecular carbomagnesiation of easily available alkynyl(aryl)thioethers for the synthesis of functionalized benzo[b] thiophenes. The reaction of the thioether **180** with *i*PrMgCl·LiCl in tetrahydrofuran (THF) furnished the corresponding arylmagensium chloride **181**, which, upon ring closure with catalytic amount of CuCN·2LiCl at 25°C, provided the magnesiated benzothiophenes **182** in 24 h. Acylation of **182** with carbonyl chloride led to the polyfunctional benzothiophenes **183** (Scheme 9.65). As the ring closing reaction is slow and conventional heating of the reaction mixture led to undesired side

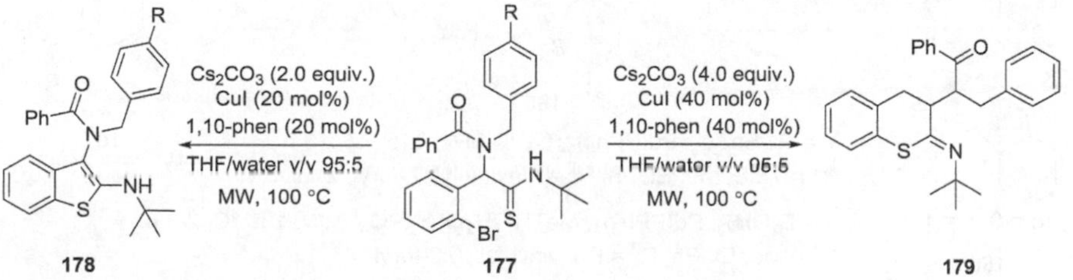

**SCHEME 9.64** Synthesis of Benzo[b]thiophene and Thiochromone Derivatives Through the Thia-Ugi Adduct.

reactions, microwave heating (50°C, 100W) was used for cyclization, which afforded the desired product **182** within 1 h.

The methodology was extended for the synthesis of benzothienothiophene **184**. The synthesis of benzothienothiophene **184**, which involves the formation of fused five-member ring on an existing five-member ring and is a difficult task due to high activation energy needed for the ring closure, was achieved under microwave heating (Scheme 9.66).

The research group of Suffert demonstrated [117] a microwave-assisted synthesis of benzothiolane **186** and isothiochromane **187** via a cyclocarbopalladation/cross-coupling cascade reaction. Both the Stille and Suzuki–Miyaura cross-coupling reactions were utilized for the synthesis of the benzothiolanes **186** and isothiochromanes **187** bearing exocyclic tetrasubstituted double bond via the reaction of **185** with either stannane or boronic acid as the coupling partner (Scheme 9.67). The methodology

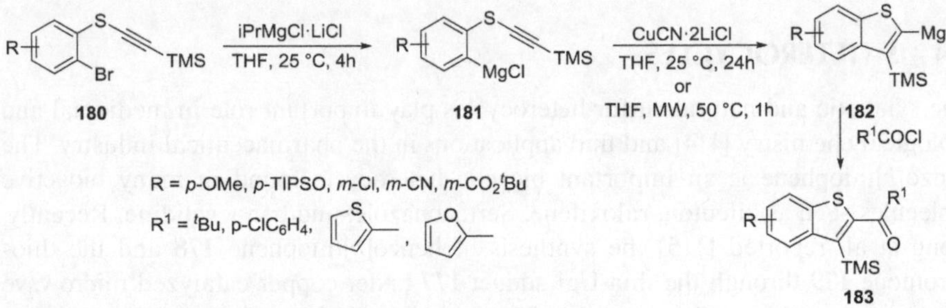

**SCHEME 9.65**   MW-Assisted Synthesis of Benzothiophenes.

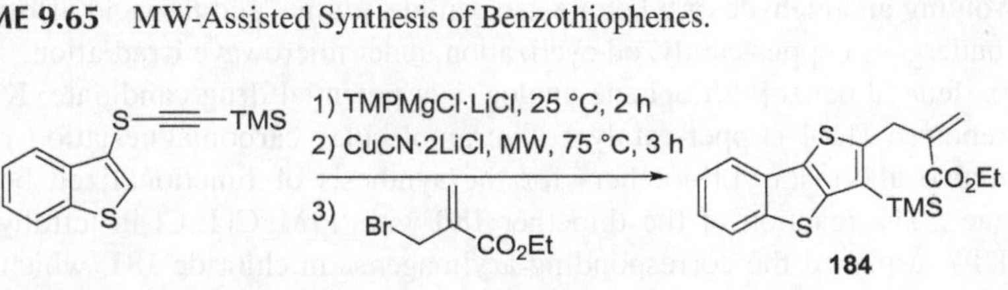

**SCHEME 9.66**   MW-Assisted Synthesis of Benzothienothiophenes.

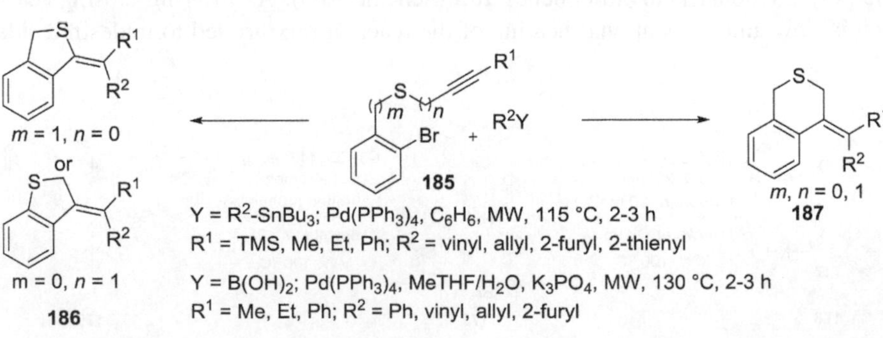

**SCHEME 9.67**   Pd-Catalyzed Stille/Suzuki–Miyaura Coupling for the Synthesis of *S*-Heterocycles.

was successfully utilized for the synthesis of a sulfur analog of Tamoxifen (an anti-cancer drug used to treat breast cancer).

## 9.5 CONCLUSION

The construction of pharmaceutically important heterocyclic scaffolds in a convenient and cost-effective way has been a bottleneck for drug discovery scientists. Various classical methods have been adopted for the synthesis of pharmaceutically important heterocycles, but the requirements of longer steps and harsh condition make such methodologies less attractive. The transition-metal-catalyzed reactions offer efficient alternatives in reducing the number of synthetic steps for the preparation of small and medium ring heterocycles. The combined effects of transition metal catalysis and microwave irradiation provide added advantages in reducing the reaction time from days/hours to minutes/seconds. Thus, the transition-metal-catalyzed reactions under the influence of microwave irradiation have become popular in drug discovery, providing an alternate approach for the synthesis of heterocyclic in a cheaper and convenient way. This chapter highlighted the importance of transition-metal-catalyzed reactions performed under microwave heating for the synthesis of nitrogen-, oxygen-, and sulfur-containing bio-relevant heterocycles. This would bring to the notice of the readers for wider acceptance of the important applications of transition-metal-catalyzed and MW-assisted synthesis in academia and pharmaceutical industries. Moreover, the use of less privileged, cheaper transition metal catalysts under microwave irradiation for the synthesis of bio-relevant heterocyclic compounds have made the methodologies more attractive for the synthesis of lead molecules. It is highly probable that in future there will be more use of these transition metal catalysts in combination with microwave irradiation as a powerful tool in the development of new and efficient methodologies for the synthesis of pharmaceutically important heterocycles in drug discovery. The examples presented in this chapter will inspire more successful research in this area.

## ACKNOWLEDGMENTS

Dipti Shukla would like to thank the Department of Pharmaceuticals (New Delhi, India) for the Research Associateship, and Priyank Purohit would like to thank the National Institute of Pharmaceutical Education and Research (NIPER) for a senior research fellowship.

## ABBREVIATIONS

| | |
|---|---|
| AChE | Acetylcholinesterase |
| A-taphos | [4-(*N,N*-Dimethylamino)phenyl]-*di-tert*-butyl phosphine |
| ATP | Adenosine triphosphate |
| [Bmim]BF$_4$ | 1-Butyl-3-methylimidazolium tetrafluoroborate |
| BINAP | 2,2'-bis(diphenylphosphino)-1,1'-binaphthyl |
| [Cp*RuCl]$_4$ | Pentamethylcyclopentadienyl ruthenium(II) chloride tetramer |
| CuTC | Copper(I)-thiophene-2-carboxylate |
| DBU | 1,8-Diazabicyclo[5.4.0]undec-7-ene. |
| DCE | 1,2-Dichloroethane |

| DCM | Dichloromethane |
|---|---|
| DMF | Dimethylformamide |
| DNA | Deoxyribonucleic acid |
| DOP | Department of Pharmaceuticals |
| Dppf | 1,1′-Bis(diphenylphosphino)ferrocene |
| Et$_3$N | Triethyl amine |
| GABA | Gamma-aminobutyric acid |
| GHz | Gigahertz |
| MACIR | Microwave-Assisted Coupling-Isomerization |
| MAOS | Microwave-Assisted Organic Synthesis |
| MCR | Multicomponent reaction |
| MeCN | Acetonitrile |
| Mephos | 2-Dicyclohexylphosphino-2′-methylbiphenyl, 2-Methyl-2-dicyclo hexylphosphinobiphenyl |
| NMP | N-Methyl-2-pyrrolidone |
| NSAID | Non-Steroidal Anti-Inflammatory Drugs |
| [pmIm]Br | 1-Pentyl-3-methylimidazolium bromide |
| Pd$_2$(dppf)$_3$ | [1,1′-Bis(diphenylphosphino)ferrocene]dichloropalladium(II) |
| PEPPSI™ | [1,3-Bis(2,6-Diisopropylphenyl)imidazol-2-ylidene](3-chlorop yridyl)palladium(II) dichloride |
| Phen | Phenanthroline |
| PMP | (±)-2-(p-Methoxyphenoxy)propionic acid |
| PPA | Polyphosphoric acid |
| PPh$_3$ | Triphenyl phosphine |
| PTSA | p-Toluenesulfonic acid |
| PyBOP | (Benzotriazol-1-yloxy)tripyrrolidinophosphonium hexafluorophosphate |
| RA | Research Associate |
| TBAC | Tetrabutylammonium chloride |
| TBAF | Tetrabutylammonium fluoride |
| TBAI | Tetrabutylammonium iodide |
| TBATB | Tetrabutylammonium tribromide |
| TFA | Trifluoroacetic acid |
| TFA | Trifluoroacetic acid |
| THF | Tetrahydrofuran |
| TM | Transition metal |
| TMS | Tetramethylsilane |
| Xphos | 2-Dicyclohexylphosphino-2′,4′,6′-triisopropylbiphenyl. |

## REFERENCES

1. (a) Adam D (2003) Out of the kitchen. *Nature* 421: 571–572. doi:10.1038/421571a.
(b) Kappe CO, Dallinger D, Murphree SS (eds) (2009) *Practical MW Synthesis for Organic Chemists: Strategies, Instruments and Protocols*, Wiley-VCH: Weinheim, Germany.
2. Kappe CO (2004) Controlled microwave heating in modern organic synthesis. *Angew. Chem. Int. Ed.* 43:6250–6284. doi:10.1002/anie.200400655.

3. (a) Martins MAP, Frizzo CP, Moreira DN, Zanatta N, Bonacorso HG (2008) Ionic liquids in heterocyclic synthesis. *Chem. Rev.* 108: 2015–2050. doi:10.1021/cr078399y. (b) Martins MAP, Frizzo CP, Moreira DN, Buriol L, Machado P (2009) Solvent-free heterocyclic synthesis. *Chem. Rev.* 109: 4140–4182. doi:10.1021/cr9001098. (c) Candeias NR, Branco LC, Gois PMP, Afonso CAM, Trindade AF (2009) More sustainable approaches for the synthesis of N-based heterocycles. *Chem. Rev.* 109: 2703–2802. doi:10.1021/cr800462w.

4. Bacolini G (1996) *Topics Heterocycl Syst Synth React Prop.* 1: 103.

5. Dua R, Shrivastava S, Sonwane SK, Srivastava SK (2011) Pharmacological significance of synthetic heterocycles scaffold: A review. *Adv. Bio. Res.* 5: 120–144. ISSN 1992-0067.

6. (a) Kappe CO, Dallinger D (2006) The impact of microwave synthesis on drug discovery. *Nat. Rev. Drug Discov.* 5: 51–63. doi: 10.1038/nrd1926. (b) Lidstrom P, Tierney J, Wathey B, Westman J (2001) Microwave-assisted organic synthesis: A review. *Tetrahedron* 57: 9225–9283. doi:10.1016/S0040-4020(01)00906-1.

7. (a) Polshettiwar V, Varma RS (2008) Microwave-assisted organic synthesis and transformations using benign reaction media. *Acc. Chem. Res.* 41: 629–639. doi:10.1021/ar700238s. (b) Polshettiwar V, Varma RS (Eds.) (2010) In *Aqueous Microwave Assisted Chemistry: Synthesis and Catalysis RCS Green Chemistry*, doi:10.1002/anie.201006427. (c) Polshettiwar V, Varma RS (2008) Aqueous microwave chemistry: A clean and green synthetic tool for rapid drug discovery. *Chem. Soc. Rev.* 37: 1546–1557. doi:10.1039/B716534J.

8. Shipe WD, Wolkenberg, SE, Lindsley CW (2005) Accelerating lead development by microwave-enhanced medicinal chemistry. *Drug Discov. Today* 2: 155–161. doi:10.1016/j.ddtec.2005.05.002.

9. (a) Tanaka K (Ed.) (2013) Index, in *Transition-Metal-Mediated Aromatic Ring Construction.* Wiley: Hoboken, NJ. (b) de Meijere A, Bräse S, Oestreich M (Eds.) (2014) *Metal-Catalyzed Cross-Coupling Reactions and More.* Wiley-VCH: Weinheim, Germany.

10. (a) Tsuji J (2000) *Transition Metal Reagents and Catalysts: Innovations in Organic Synthesis.* Wiley: Chichester, UK. (b) Tsuji J (2005) *Topics in Organometallic Chemistry: Palladium in Organic Synthesis.* Springer, New York.

11. (a) de Meijere A, Diederich F (2004) *Metal-Catalyzed Cross-Coupling Reactions.* Wiley-VCH: Weinheim, Germany. (b) Beller M, Bolm C (2004) *Transition Metals for Organic Synthesis* (2nd edn). Wiley-VCH: Weinheim.

12. (a) Nilsson P, Olofsson K, Larhed M (2006) Microwave-assisted and metal-catalyzed coupling reactions. *Top. Curr. Chem.* 266: 103–144. doi:10.1007/128_046. (b) Wang K-U, Wang J-X (2013) Microwave-assisted synthesis in C–C and carbon–heteroatom coupling reactions. In *Palladium-Catalyzed Coupling Reactions: Practical Aspects and Future Developments.* Molnár A (Ed.). Wiley-VCH: Weinheim, Germany. doi:10.1002/9783527648283.

13. Magano J, Dunetz JR (2011) Large-scale applications of transition metal catalyzed couplings for the synthesis of pharmaceuticals. *Chem. Rev.* 111: 2177–2250. doi:10.1021/cr100346g.

14. (a) Santagada V, Perissutti E, Caliendo G (2002) The application of microwave irradiation as new convenient synthetic procedure in drug discovery. *Curr. Med. Chem.* 9: 1251–1283. doi:10.2174/0929867023369989. (b) Santagada V, Frecentese F, Perissutti E, Fiorino F, Severino R, Caliendo G (2009) The application of microwave irradiation as new convenient synthetic procedure in drug discovery. *Mini-Rev. Med. Chem.* 9: 340–358. doi: 10.2174/0929867023369989. (c) Appukkuttan P, Van der Eycken EV (2008) Recent developments in microwave-assisted, transition-metal-catalyzed C–C and C–N bond-forming reactions. *Eur. J. Org. Chem.* 7: 1133–1155 doi:10.1002/ejoc.200701056. (d) Bai L, Wang J-X (2005) Environmentally friendly suzuki aryl-aryl cross-coupling reaction. *Curr. Org. Chem.* 9: 535–553. doi:10.2174/1385272053544407.

15. Watson JD, Crick FH (1993) Molecular structure of nucleic acid: A structure for deoxyribose nucleic acid. *JAMA* 269: 1966–1967. doi:10.1001/jama.1993.03500150078030.
16. Lesher GY, Froelich EJ, Gruett, MD, Bailey J H, Brundage RP (1962) 1,8-Naphthyridine derivatives: A new class of chemotherapeutic agents. *J. Med. Pharm. Chem.* 91: 1063–1065. doi:10.1021/jm01240a021.
17. Pasgualooto AC, Thiele KO, Goldani LZ (2010) Novel triazole antifungal drugs: Focus on isavuconazole, ravuconazole and albaconazole. *Curr. Opin. Investig. Drugs* 11: 165–174. PMID: 20112166.
18. Lesyk R, Vladzimirska O, Zimenkovsky B, Horishny V, Nektegayev I, Solyanyk V, Vovk O (1998) New thiazolidones-4 with pyrazolone-5 substituent as the potential NSAIDs. *Boll. Chim. Farm*, 137: 210–217. PMID: 9713155
19. Hart FD, Boardman PL (1963) Indomethacin: A new non-steroidal anti-inflammatory agent. *Br. Med. J.* 2(5363): 965–970. PMID: 14056924.
20. Laura G, Robeti Marinella R, Daniela P (2007) Nitrogen-containing heterocyclic quinine: A class of potential selective antitumor agents. *Mini-Rev. Med. Chem.* 7: 481–489. doi: 10.2174/138955707780619626.
21. Liu Q, Chang JW, Wang J, Kang S A, Thoreen CC, Markhard A, Hur W, Zhang J, Sim T, Sabatani DM, Gray NS (2010) Discovery of 1-(4-(4-Propionylpiperazin-1-yl)-3-(trifluoromethyl)phenyl)-9-(quinolin-3-yl)benzo[h][1,6]naphthyridin-2(1H)-one as a highly potent, selective mammalian target of rapamycin (mTOR) inhibitor for the treatment of cancer. *J. Med. Chem.* 53: 7146–7155. doi:10.1021/jm101144f.
22. Siddiqui N, Andalip, Bawa. S, Ali R, Afzal O, Akhtar MJ, Azad B, Kumar R (2011), Antidepressant potential of nitrogen-containing heterocyclic moieties: An updated review. *J. Pharm. Bioallied Sci.* 3: 194–212. doi:10.4103/0975-7406.80765.
23. Majumder A, Gupta R, Jain A (2013) Microwave-assisted synthesis of nitrogen-containing heterocycles. *Green Chem. Lett. Rev.* 6: 151–182. doi:10.1080/17518253.2012.733032.
24. Sharma A, Appukkuttan P, Eycken EV (2012) Microwave-assisted synthesis of medium-sized heterocycles. *Chem. Commun.* 48: 1623–1637. doi:10.1039/c1cc15238f.
25. Donets PA, Eycken EV (2007) Efficient synthesis of the 3-benzazepine framework via intramolecular heck reductive cyclization. *Org. Lett.* 9: 3017–3020. doi:10.1021/ol071079g.
26. Donets PA, Goeman JL, Van der Eycken J, Robeyns K, Van Meervelt L, Van der Eycken EV (2009), An asymmetric approach towards (–)-Aphanorphine and its analogs. *Eur. J. Org. Chem.* 793–796. doi:10.1002/ejoc.200801175.
27. Peshkov AA, Peshkov VA, Pereshivko OP, Hecke KV Kumar R, Van der Eycken EV (2015) Heck–Suzuki tandem reaction for the synthesis of 3-benzazepines. *J. Org. Chem.* 80: 6598–6608. doi:10.1021/acs.joc.5b00670.
28. Zhu J, Bienaymé H (Eds.) (2005) Multicomponent Reactions. Wiley-VCH: Weinheim, Germany.
29. Bhagat S, Chakraborti AK (2007) An extremely efficient three-component reaction of aldehydes/ketone, amines, and phosphates (Kabachnik–Fields reaction) for the synthesis of α-aminophosphonates catalysed by magnesium perchlorate. *J. Org. Chem.* 72: 1263–1270, doi:10.1021/jo062140i.
30. Bhagat S, Chakraborti AK (2008) Zirconium(IV) compounds as efficient catalysts for synthesis of α-aminophosphonates. *J. Org. Chem.* 73: 6029–6032. doi:10.1021/jo8009006.
31. Bindal S, Kumar D, Kommi DN, Bhatiya S, Chakraborti AK (2011) An efficient organocatalytic dual activation strategy for preparation of the versatile synthons 2(E)-1-aryl/heteroaryl/styryl-3-dimethylamino-2-propen-1-ones and α-(E)-dimethylaminoformylidene cycloalkanones. *Synthesis*. 12: 1930–1935. doi:10.1055/s-0030-1260048.

32. Sarkar A, Raha Roy S, Kumar D, Madaan C, Rudrawar S, Chakraborti AK (2012) Lack of correlation between catalytic efficiency and basicity of amines during the reaction of aryl methyl ketones with DMF-DMA: an unprecedented supramolecular domino catalysis. *Org. Biomol. Chem.* 10: 281–286. doi:10.1039/c1ob06043k.

33. Kumar D, Kommi DN, Patel AR, Chakraborti AK (2012) L-Proline catalysed activation of methyl ketones/active methylene compounds and DMF-DMA for synthesis of (2E)-3-dimethylamino-2-propen-1-ones. *Eur. J. Org. Chem.* 32: 6407–6413. doi:10.1002/ejoc.201200778.

34. Bhagat S, Shah P, Garg SK, Mishra S, Kamal P, Singh S, Chakraborti AK (2014) α-Aminophosphonates as novel antileishmanial chemotypes: synthesis, biological evaluation, and CoMFA studies. *Med. Chem. Commun.* 5: 665–670. doi:10.1039/C3MD00388D.

35. Raha Roy S, Jadhavar PS, Seth K, Sharma KK, Chakraborti AK (2011) Organo-catalytic application of room temperature ionic liquids: [bmim][MeSO4] as a recyclable organocatalyst for one-pot multicomponent reaction for preparation of dihydropyrimidinones and –thiones. *Synthesis.* 14: 2261–2267, doi:10.1055/s-0030-1260067.

36. Kumar D, Kommi, DN, Patel AR, Chakraborti AK (2012) Catalytic procedures for multicomponent synthesis of imidazoles: selectivity control during the competitive formation of tri- and tetra-substituted imidazoles. *Green Chem.* 14: 2038–2049. Doi:10.1039/C2GC35277J.

37. Kumar D, Sonawane M, Pujala B, Jain VK, Bhagat S, Chakraborti AK (2013) Supported protic acid-catalyzed synthesis of 2,3-disubstituted hiazolidine-4-ones: Enhancement of the catalytic potential of protic acid by adsorption on solid support. *Green Chem.* 15: 2872–2884. Doi:10.1039/C3GC41218K.

38. Parikh N, Raha Roy S, Seth K, Kumar A, Chakraborti AK (2016) "On-water" multicomponent reaction for the diastereoselective synthesis of functionalized tetrahydropyridines and mechanistic insight. *Synthesis.* 48: 547–556. doi:10.1055/s-0035-1561296.

39. Bariwal, JB, Ermolat DS, Glasnov TN, Hecke KV, Mehta VP, Meervelt LV, Kappe CO, Van der Eycken EV (2010) Diversity-oriented synthesis of dibenzoazocines and dibenzoazepines via a microwave-assisted intramolecular $A^3$-coupling reaction. *Org. Lett.* 12: 2774–2777. doi. 10.1021/ol1008729.

40. Appukkuttan P, Dehaen W, Van der Eycken EV (2007) Microwave-assisted transition-metal-catalyzed synthesis of N-shifted and ring-expanded buflavine analogs. *Chem. Eur. J.* 13: 6452–6460. doi:10.1002/chem.200700177.

41. Li Z, Legras L, Kumar A, Vachhani DD, Sharma SK, Parmar VS, Van der Eycken EV (2014) Microwave-assisted synthesis of 4H-benzo[f]imidazo[1,4]diazepin-6-ones via a post-Ugi copper-catalyzed intramolecular Ullmann coupling. *Tetrahedron Lett.* 55: 2070–2074. doi:10.1016/j.tetlet.2014.02.023.

42. Shang H, Wang Y, Tian Y, Feng J, Tang Y (2014) The divergent synthesis of nitrogen heterocycles by rhodium(II)-catalyzed cycloadditions of 1-sulfonyl 1,2,3-triazoles with 1,3-dienes. *Angew. Chem. Int. Ed.* 53: 5662–5666. doi:10.1002/anie.201400426.

43. Schultz EE, Lindsay VNG, Sarpong R (2014) Expedient synthesis of fused azepine derivatives using a sequential rhodium(II)-catalyzed cyclopropanation/1-aza-cope rearrangement of dienyltriazoles. *Angew. Chem. Int. Ed.* 53: 9904–9908. doi:10.1002/anie.201405356.

44. Seth, K, Purohit P, Chakraborti AK (2017) Microwave assisted synthesis of biorelevant benzazoles. *Curr. Med. Chem.* 24: 4638–4676. doi:10.2174/0929867323666161025142005.

45. Manna SK, Panda G (2014) Microwave assisted [RuCl$_2$(p-cymene)$_2$]$_2$ catalyzed regioselective endo-tandem cyclization involving imine and alkyne activation: An approach to benzo[4,5]imidazo[2,1-a]pyridine scaffold. *RSC Adv.* 4: 21032–21041. doi:10.1039/C4RA02581D.

46. Liu Y, Jin S, Wang Z, Song L, Hu Y (2014) Microwave assisted tandem Heck–Sonogashira reactions of N,N-di-Boc-protected 6-amino-5-iodo-2-methyl pyrimidin-4-ol in an efficient approach to functionalized pyrido[2,3-d]pyrimidines. *Org. Lett.* 16: 3524–3527. doi: 10.1021/ol501459e.

47. Rijkers DTS, Wilma van Esse G, Merkx R, Brouwer AJ, Jacobs HJF, Pieters RJ, Liskamp RMJ (2005) Efficient microwave-assisted synthesis of multivalent dendrimeric peptides using cycloaddition reaction (click) chemistry. *Chem. Commun.* 4581–4583. doi:10.1039/b507975f.

48. Yoon K, Goyal P, Weck M (2007) Mono functionalization of dendrimers with use of microwave-assisted 1,3-dipolar cycloaddition. *Org. Lett.* 9: 2051–2054 doi:10.1021/ol062949h.

49. Rasmussen LK, Boren BC, Fokin VV (2007) Ruthenium-catalyzed cycloaddition of aryl azides and alkynes. *Org. Lett.* 9: 5337–5339. doi:10.1021/ol701912s.

50. Isobe H, Fujino T, Yamazaki N, Guillot-Nieckowski M, Nakamura E (2008) Triazole-linked analog of deoxyribonucleic acid (TLDNA): Design, synthesis, and double-strand formation with natural DNA. *Org. Lett.* 10: 3729–3732. doi:10.1021/ol801230k.

51. Takamatsu K, Hirano K, Satoh T, Miura M (2015) Synthesis of indolines by copper-mediated intramolecular aromatic C-H amination. *J. Org. Chem.* 80: 3242–3249. doi:10.1021/acs.joc.5b00307.

52. Shekarrao K, Kaishap PP, Saddanapu V, Addlagatta A, Gogoi S, Boruah RC (2014) Microwave-assisted palladium mediated efficient synthesis of pyrazolo[3,4-b]pyridines, pyrazolo[3,4-b]quinolines, pyrazolo[1,5-a]pyrimidines and pyrazolo[1,5-a]quinazolines. *RSC Adv.* 4: 24001–24006. doi:10.1039/C4RA02865A.

53. Alcaide B, Almendros P, Quiros MT (2014) Gold/acid-co-catalyzed direct microwave-assisted synthesis of fused azaheterocycles from propargylic hydroperoxides. *Chem. Eur. J.* 20: 3384–3393, doi:10.1002/chem.201304509.

54. Pandey G, Bhowmik, S, Batra, S (2013) Synthesis of 3H-pyrazolo[3,4-c]-isoquinolines and thieno[3,2-c]-isoquinolines via cascade imination/intramolecular decarboxylative coupling. *Org. Lett.* 15: 5044–5047, doi:10.1021/ol4023722.

55. Imbri, D, Netz N, Kucukdisli M, Kammer LM, Jung P, Kretzschmann A, Opatz T (2014) One-pot synthesis of pyrrole-2-carboxylates and -carboxamides via an electrocyclization/oxidation sequence. *J. Org. Chem.* 79: 11750–11758. doi:10.1021/jo5021823.

56. (a) Bhagat S, Sharma R, Sawant DM, Sharma L, Chakraborti AK (2006) "LiOH·H₂O as a novel dual activation catalyst for highly efficient and easy synthesis of 1,3-diaryl-2-propenones by Claisen–Schmidt condensation under mild conditions. *J. Mol. Cat. A: Chem.* 244: 20–24, doi:10.1016/j.molcata.2005.08.039. (b) Bhagat S, Sharma R, Chakraborti AK (2006) Dual-activation protocol for tandem cross aldol condensation: An easy and highly efficient synthesis of α,α′-bis(arylmethylidene) ketones. *J. Mol. Cat. A.: Chem.* 260: 235–240. doi: 10.1016/j.molcata.2006.07.018.

57. Grotkopp Ol, Ahmad A, Frank W, Müller TJJ (2011) Blue-luminescent 5-(3-indolyl) oxazoles via microwave-assisted three-component coupling–cycloisomerization–Fischer indole synthesis. *Org. Biomol. Chem.* 9: 8130–8140. doi:10.1039/C1OB06153D.

58. Karuvalam RP, Haridas KR, Sajith AM, Muralidharan A (2013) A facile access to substituted indoles utilizing palladium catalyzed annulation under microwave enhanced conditions. *Tetrahedron Lett.* 54: 5126–5129. doi:10.1016/j.tetlet.2013.07.073.

59. Shore G, Morin S, Mallik D, Organ MG (2008) Pd PEPPSI-IPr-mediated reactions in metal-coated capillaries under MACOS: the synthesis of indoles by sequential aryl amination/Heck coupling. *Chem. Eur. J.* 14: 1351–1356. doi:10.1002/chem.200701588.

60. Nguyen HH, Kurth MJ (2013) Microwave-assisted synthesis of 3-nitroindoles from N-aryl enamines via intramolecular arene-alkene coupling. *Org. Lett.* 15: 362–365. doi:10.1021/ol303314x.

61. Hellal M, Cuny GD (2011) Microwave assisted copper-free Sonogashira coupling/5-exo-dig cycloisomerization domino reaction: Access to 3-(phenylmethylene)isoindolin-1-ones and related heterocycles. *Tetrahedron Lett.* 52: 5508–5511. doi:10.1016/j.tetlet.2011.08.070.

62. Tyagi V, Khan S, Chauhan PMS (2013) A simple and efficient microwave-assisted synthesis of substituted isoindolinone derivatives via ligand-free Pd-catalyzed domino C–C/C–N coupling reaction. *Synlett* 24 (5): 645–651. doi:10.1055/s-0032-1318331.

63. Poondra RR, Turner NJ (2005) Microwave-assisted sequential amide bond formation and intramolecular amidation: A rapid entry to functionalized oxindoles. *Org. Lett.* 7: 863–866. doi 10.1021/ol0473804.

64. Bonnaterre F, Bois-Choussy M, Zhu J (2006) Rapid access to oxindoles by the combined use of an Ugi four-component reaction and a microwave-assisted intramolecular Buchwald–Hartwig amidation reaction. *Org. Lett.* 8: 4351–4354. doi:10.1021/ol061755z.

65. Han J, Xu B, Hammond GB (2010) Highly efficient Cu(I)-catalyzed synthesis of N-heterocycles through a cyclization-triggered addition of alkynes. *J. Am. Chem. Soc.* 132: 916–917, doi:10.1021/ja908883n.

66. DiMauro E F, Kennedy JM. (2007) Rapid synthesis of 3-amino-imidazopyridines by a microwave-assisted four-component coupling in one pot. *J. Org. Chem.* 72: 1013–1016. doi:10.1021/jo0622072.

67. Naresh G, Kant R, Narender T (2014) Copper(II) catalyzed expeditious synthesis of furoquinoxalines through a one-pot three-component coupling strategy. *Org. Lett.* 16: 4528–4531. doi:10.1021/ol502072k.

68. Saha P, Naskar S, Paira R, Mondal S, Maity A, Sahu KB, Paira P, Hazra A, Bhattacharya D, Banerjee S, Mondal NB (2010) One-pot tandem synthesis of furo[3,2-h]quinolines by a Sonogashira cross-coupling and cyclization reaction supported by basic alumina under microwave irradiation. *Synthesis* 3: 486–492. doi:10.1055/s-0029-1217143.

69. Yang D, Burugupall S, Daniel D, Chen Y (2012) Microwave-assisted one-pot synthesis of isoquinolines, furopyridines, and thienopyridines by palladium-catalyzed sequential coupling-imination-annulation of 2-bromoarylaldehydes with terminal acetylenes and ammonium acetate. *J. Org. Chem.* 77: 4466–4472. doi:10.1021/jo300494a.

70. Schreiner E, Braun S, Kwasnitschka C, Frank W, Müller TJJ (2014) Consecutive three-component synthesis of 2,6-disubstituted pyrimid-4(3H)-ones and 1,5-disubstituted 3-hydroxypyrazoles initiated by copper(I)-catalyzed carboxylation of terminal alkynes. *Adv. Synth. Catal.* 356: 3135–3142. doi:10.1002/adsc.201400411.

71. Vachhani DD, Mehta VP, Modha SG, Hecke KV, Meervelt LV, Van der Eycken EV (2015) Microwave-assisted synthesis of pyrazino[2,1-b]quinazolines and 3-indolyl-2(1H)-pyrazinones employing a chemoselective silver(I)- and gold(I)-catalyzed reaction. *Adv. Synth. Catal.* 354:1593–1599.doi: 10.1002/adsc.201100881.

72. (a) Kumar D, Seth K, Kommi DN, Bhagat S, Chakraborti AK (2013) Surfactant micelles as microreactors for the synthesis of quinoxalines in water: scope and limitations of surfactant catalysis. *RSC Adv.* 3: 15157–15168, doi:10.1039/C3RA41038B. (b) Jadhavar PS, Kumar D, Purohit P, Pipaliya BV, Kumar A, Bhagat S, Chakraborti AK (2014) Sustainable approaches towards the synthesis of quinoxalines. In *Green Chemistry: Synthesis of bioactive heterocycles*, K. L. Ameta, A. Dandia (Eds.). Springer: New Delhi, India. (c) Tanwar B, Purohit P, Naga Raju B, Kumar D, Kommi DN, Chakraborti AK (2015) An "all-water" strategy for regiocontrolled synthesis of 2-aryl quinoxalines. *RSC Adv.* 5: 11873–11883. doi:10.1039/C4RA16568C.

73. Zhang L, Zhao F, Zheng M, Zhai Y, Wang J, Liu H (2013) Selective synthesis of 5,6-dihydroindolo[1,2-a]quinoxalines and 6,7-dihydroindolo[2,3-c]quinolines by orthogonal copper and palladium catalysis. *Eur. J. Org. Chem.* 25: 5710–5715. doi:10.1002/ejoc.201300667.

74. Sun W-W, Cao P, Mei R-Q, Li Y, Ma Y-L, Wu B (2014) Palladium-catalyzed unacti-
    vated C(sp3)–H bond activation and intramolecular amination of carboxamides: A new
    approach to β-lactams. *Org. Lett.* 16: 480–483. doi:10.1021/ol403364k.
75. Åkerbladh L, Nordeman P, Wejdemar M, Odell LR, Larhed M (2015) Synthesis of
    4-quinolones via a carbonylative Sonogashira cross-coupling using Molybdenum hexa-
    carbonyl as a CO source. *J. Org. Chem.* 80: 1464–1471. doi:10.1021/jo502400h.
76. Tyagi V, Khan S, Giri A Gauniyal HM, Sridhar B, Chauhan PMS (2012) A ligand-free
    Pd-catalyzed cascade reaction: An access to the highly diverse isoquinolin-1(2H)-one
    derivatives via isocyanide and Ugi-MCR synthesized amide precursors. *Org. Lett.* 14:
    3126–3129. doi:10.1021/ol301131s.
77. Li Z, Sun H, Jiang H, Liu H (2008) Copper-catalyzed intramolecular cyclization to
    N-substituted 1,3-dihydrobenzimidazol-2-ones. *Org. Lett.* 10: 3263–3266. doi:10.1021/
    ol8011106.
78. Shi B, Lewis W, Campbell IB, Moody CJ (2009) A concise route to pyridines from
    hydrazides by metal carbene N-H insertion, 1,2,4-triazine formation, and Diels–Alder
    reaction. *Org. Lett.* 11:3686–3688. doi:10.1021/ol901502u.
79. Tanwar B, Kumar D, Kumar A, Ansari MI, Qadri MM, Vaja MD, Singh M, Chakraborti
    AK (2015) Friedländer annulation: Scope and limitations of metal salt Lewis acid cata-
    lysts in selectivity control for the synthesis of functionalised quinolones. *New J. Chem.*
    39: 9824–9833. doi:10.1039/C5NJ02010G.
80. Tanwar B, Kumar A, Yogeeswari P, Sriram D, Chakraborti AK (2016) Design, develop-
    ment of new synthetic methodology, and biological evaluation of substituted quinolines
    as new anti-tubercular leads. *Bioorg. Med. Chem. Lett.* 26: 5960–5966. doi:10.1016/j.
    bmcl.2016.10.082.
81. Schramm OG, Oeser T, Kaiser M, Brun R, Müller TJJ (2008) Rapid one-pot synthesis
    of antiparasitic quinolines based upon the microwave-assisted coupling-isomerization
    reaction (MACIR). *Synlett* 3: 359–362. doi:10.1055/s-2008-1032067.
82. Moustafa AH, Malakar CC, Aljaar N, Merisor E, Conrad J, Beifuss U (2013) Microwave-
    assisted molybdenum-catalyzed reductive cyclization of o-nitrobenzylidene amines to
    2-aryl-2H-indazoles. *Synlett.* 24: 1573–1577. doi:10.1055/s-0033-1339195.
83. Ho SL, Cho CS (2013) Microwave-assisted copper-powder-catalyzed synthesis of
    pyrimidinones from β-bromo α,β-unsaturated carboxylic acids and amidine. *Synlett.*
    24: 2705–2708. doi:10.1055/s-0033-1340283.
84. Wu J, Nie L, Luo J, Dai W-M (2007) Microwave-assisted, palladium-catalyzed intra-
    molecular direct arylation for the synthesis of novel fused heterocycles. *Synlett.* 17:
    2728–27321. doi:10.1055/s-2007-991053.
85. Zhang X-Y, Yang Z-W, Chen Z, Wang J, Yang D-L, Shen Z, Hu L-L, Xie J-W, Zhang J,
    Cui H-L (2016) Tandem copper-catalyzed propargylation/alkyne azacyclization/isom-
    erization reaction under microwave irradiation: synthesis of fully substituted pyrroles.
    *J. Org. Chem.* 81: 1778–1785. doi:10.1021/acs.joc.5b02429.
86. Zhao J, Zhao X-J, Cao P, Liu J-K, and Wu B (2017) Polycyclic azetidines and pyr-
    rolidines via palladium-catalyzed intramolecular amination of unactivated C(sp3)–H
    bonds. *Org. Lett.* 19, 4880–4883. doi:10.1021/acs.orglett.7b02339.
87. Xu L, Li H, Liao Z, Lou K, Xie H, Li H, Wang W (2015) Divergent synthesis of imidaz-
    oles and quinazolines via Pd(OAc)₂-catalyzed annulation of N-allylamidines. *Org. Lett.*
    17: 3434–3437. doi:10.1021/acs.orglett.5b01435.
88. Sperry JB, Wright DL (2005) Furans, thiophenes and related heterocycles in drug dis-
    covery. *Curr. Opin. Drug Discov. Dev.* 8: 723–740. PMID:16312148.
89. Hartwig, JF (1998). Transition metal catalyzed synthesis of aryl amines and aryl ethers
    from aryl halides and triflates: scope and mechanism. *Angew. Chem. Int. Ed.* 37: 2046–
    2067. doi: 10.1002/(SICI)1521-3773(19980817)37:15<2046::AID-ANIE2046>3.0.
    CO;2-L.

90. Prim D, Campagne JM, Joseph D, Andrioletti B (2002) Palladium-catalyzed reactions of aryl halides with soft, non-organometallic nucleophiles. *Tetrahedron* 58: 2041–2075. doi:10.1016/S0040-4020(02)00076-5.

91. Yamamoto Y, Itonaga K (2009) Synthesis of chromans via [3+3] cyclocoupling of phenols with allylic alcohols using a Mo/o-chloranil catalyst system. *Org. Lett.* 11: 717–720. doi:10.1021/ol802800s.

92. Vishnumurthy K, Makriyannis A (2010) Novel and efficient one-step parallel synthesis of dibenzopyranones via Suzuki-Miyaura cross coupling. *J. Comb. Chem.* 12: 664–669. doi:10.1021/cc100068a.

93. Thasana N, Worayuthakarn R, Kradanrat P, Hohn E, Young L, Ruchirawat S (2007) Copper(I)-mediated and microwave-assisted C aryl-O-carboxylic coupling: Synthesis of benzopyranones and isolamellarin alkaloids. *J. Org. Chem.* 72: 9379–9382. doi:10.1021/jo701599g.

94. Xu T, Zhang E, Wang D, Wang Y, Zou Y (2015) Cu-catalyzed consecutive hydroxylation and aerobic oxidative cycloetherification under microwave condition: Entry to 2-arylbenzofuran-3-carboxylic acids. *J. Org. Chem.* 80: 4313–4324. doi:10.1021/jo502802k.

95. Battini N, Battula S, Kumar RR, Ahmed QN (2015) 2-Oxo driven unconventional reactions: Microwave assisted approaches to tetrahydrofuro[3,2-d]oxazoles and furanones. *Org. Lett.* 17: 2992–2995. doi:10.1021/acs.orglett.5b01271.

96. Svennebring A, Nilsson P, Larhed M (2007) Microwave accelerated spiro-cyclizations of o-halobenzyl cyclohexenyl ethers by palladium(0) catalysis. *J. Org. Chem.* 72: 5851–5854. doi:10.1021/jo0708487.

97. Castagnolo D, Botta L, Botta M (2009) One-pot multicomponent synthesis of 2,3-dihydropyrans: new access to furanose-pyranose 1,3-C-C-linked-disaccharides. *Tetrahedron Lett.* 50: 1526–1528. doi:10.1016/j.tetlet.2009.01.047.

98. Awuah E, Capretta A (2009) Access to flavones via a microwave-assisted, one-pot Sonogashira-carbonylation-annulation reaction. *Org. Lett.* 11: 3210–3213. doi:10.1021/ol901043q.

99. Murugavel G, Punniyamurthy T, (2015) Microwave-assisted copper-catalyzed four component tandem synthesis of 3-N-sulfonylamidine coumarins. *J. Org. Chem.* 80: 6291–6299. doi:10.1021/acs.joc.5b00738.

100. Seth K, Garg SK, Kumar R, Purohit P, Meena VS, Goyal R, Banerjee UC, Chakraborti AK (2014) 2-(2-Arylphenyl)benzoxazole as a novel anti-inflammatory scaffold: Synthesis and biological evaluation. *ACS Med. Chem. Lett.* 5: 512–516. doi:10.1021/ml400500e.

101. (a) Seth K, Purohit P, Chakraborti AK (2014) Cooperative catalysis by palladium-nickel binary nanocluster for Suzuki–Miyaura reaction of ortho-heterocycle-tethered sterically hindered aryl bromides. *Org. Lett.* 16: 2334–2337. doi:10.1021/ol500587m. (b) Seth K, Nautiyal M, Purohit P, Parikh N, Chakraborti AK (2015) Palladium catalyzed Csp2-H activation for direct aryl hydroxylation: Unprecedented role of 1,4-dioxane as source of hydroxyl radical. *J. Chem. Soc. Chem. Commun.* 51: 191–194. doi:10.1039/C4CC06864E. (c) Pipaliya BV, Chakraborti AK (2017) Cross dehydrogenative coupling of heterocyclic scaffolds with unfunctionalised aroyl surrogates by palladium(II) catalyzed C(sp2)-H aroylation through organocatalytic dioxygen activation. *J. Org. Chem.* 82: 3767–3780. doi:10.1021/acs.joc.7b00226. (d) Purohit P, Seth K, Kumar A, Chakraborti AK (2017) C–O bond activation by nickel-palladium hetero-bimetallic nano-particles for Suzuki–Miyaura reaction of bioactive heterocycle-tethered sterically hindered aryl carbonates. *ACS Catal.* 5: 2452–2457. doi:10.1021/acscatal.6b02912. (e) Pipaliya BV, Chakraborti AK (2017) Ligand-assisted heteroaryl C(sp2)-H bond activation by cationic ruthenium(II) complex for alkenylation of heteroarenes with alkynes directed by bio-relevant heterocycles. *ChemCatChem.* 9: 4191–4198. doi:10.1002/cctc.201701016.

102. Kumar D, Rudrawar S, Chakraborti, AK (2008) One-pot synthesis of 2-substituted benzoxazoles directly from carboxylic acids. *Aust. J. Chem.* 61: 881–887. doi:10.1071/CH08193.

103. Kumar R, Selvam C, Kaur G, Chakraborti AK (2005) Microwave-assisted direct synthesis of 2-substituted benzoxazoles from carboxylic acids under catalyst and solvent free conditions. *Synlett.* 9: 1401–1404. doi:10.1055/s-2005-868509.

104. Viirre RD, Evindar G, Batey RA (2008) Copper-catalyzed domino annulation approaches to the synthesis of benzoxazoles under microwave-accelerated and conventional thermal conditions. *J. Org. Chem.* 73: 3452–3459. doi:10.1021/jo702145d.

105. Buxaderas E, Alonso DA, Najera C (2014) Synthesis of dihydroisobenzofurans via palladium-catalyzed sequential alkynylation/annulation of 2-bromobenzyl and 2-chlorobenzyl alcohols under microwave irradiation. *Adv. Synth. Catal.* 356: 3415–3421. doi:10.1002/adsc.201400457.

106. Xia L, Lee YR (2013) Efficient one-pot synthesis of multi-substituted dihydrofurans by ruthenium(II)-catalyzed [3+2] cycloaddition of cyclic or acyclic diazodicarbonyl compounds with olefins. *Adv. Synth. Catal.* 355: 2361–2374. doi:10.1002/adsc.201300245.

107. Nicolaus N, Schmalz H-G (2010) Synthesis of novel allocolchicine analogs with a pyridine C-ring through intermolecular Vollhardt diyne–nitrile cyclotrimerization. *Synlett.* 14: 2071–2074. doi:10.1055/s-0030-1258512.

108. Dhameliya TM, Chourasiya SS, Mishra E, Jadhavar PS, Bharatam PV, Chakraborti AK (2017) Rationalisation of benzazole-2-carboxylate vs benzazine-3-one/benzazine-2,3-dione selectivity switch during cyclocondensation of 2-amino thiophenols/phenols/anilines with 1,2-biselectrophiles in aqueous medium. *J. Org. Chem.* 82: 10077–10091. doi:10.1021/acs.joc.7b01548.

109. (a) Pancholia S, Dhameliya TM, Shah P, Jadhavar PS, Sridevi JP, Yogeshwari P, Sriram D, Chakraborti AK (2016) Benzo[*d*]thiazol-2-yl(piperazin-1-yl)methanones as new anti-mycobacterial chemotypes: Design, synthesis, biological evaluation and 3D-QSAR studies. *Eur. J. Med. Chem.* 116: 187–199. doi:10.1016/j.ejmech.2016.03.060. (b) Shah P, Dhameliya TM, Bansal R, Nautiyal M, Kommi DN, Jadhavar PS, Sridevi JP, Yogeeswari P, Sriram D, Chakraborti AK (2014) *N*-Arylalkylbenzo[*d*]thiazole-2-carboxamides as anti-mycobacterial agents: Design, new methods of synthesis and biological evaluation. *Med. Chem. Commun.* 5: 1489–1495, doi:10.1039/C4MD00224E.

110. Feng G, Wang S, Li W, Chen F, Qi (2013) Palladium-catalyzed, microwave-assisted synthesis of 3,4-dihydro-3-oxo-2H–1,4-benzoxazines: An improved catalytic system and multicomponent process. *Synthesis* 45: 2711–2718. Doi:10.1055/s-0033-1338508.

111. Ho SL, Cho CS, Sohn H-S (2015) Microwave-assisted copper-catalyzed coupling and cyclization of β-bromo-α,β-unsaturated carboxylic acids with 1,3-diketone leading to 2H-pyran-2-ones. *Synthesis.* 47: 216–220 doi:10.1055/s-0034-1379103.

112. Mestichelli P, Scott MJ, Galloway WRJD, Selwyn J, Parker JS, Spring DR (2013) Concise copper-catalyzed synthesis of tricyclic biaryl ether-linked aza-heterocyclic ring systems. *Org. Lett.* 15: 5448–5451. doi:10.1021/ol4025259.

113. Liu J, Zhou X, Wang C, Fu W, Chu W, Sun Z (2016) Total synthesis of protosappanin A and its derivatives via palladium catalyzed ortho C–H activation/C–C cyclization under microwave irradiation. *Chem. Commun.* 52: 5152–5155 doi:10.1039/c6cc01149g.

114. Damani LA (1989) In *Sulfur-Containing Drugs and Related Organic Compounds: Chemistry, Biochemistry, and Toxicology*; Vol. 1, Part B. Ellis Horwood Ltd: Chichester, UK.

115. Kim Y-S, Kwak SH, Gong Y-D (2015) Application of thio-Ugi adducts for the preparation of benzo[b]thiophene and S-heterocycle library via copper catalyzed intramolecular C-S bond formation. *ACS Comb. Sci.* 17: 365–373. doi:10.1021/acscombsci.5b00034.

116. Kunz T, Knochel P (2012) Synthesis of functionalized benzo[b]thiophenes by the intramolecular copper-catalyzed carbomagnesiation of alkynyl(aryl)thioethers. *Angew. Chem. Int. Ed.* 51: 1958–1961. doi:10.1002/anie.201106734.
117. Castanheiro T, Donnard M, Gulea M, Suffert J (2014) Cyclocarbopalladation/cross-coupling cascade reactions in sulfide series: Access to sulfur heterocycles. *Org. Lett.* 16: 3060–3063. doi:10.1021/ol501165h.

# 10 Microwaves in Lactam Chemistry

*Debasish Bandyopadhyay and Bimal Krishna Banik*

## CONTENTS

## 10.1 LACTAMS: AN OVERVIEW

Amides are an important class of organic compounds that are basically formed by the reaction between carboxylic acids and amines. Cyclic amides, commonly known as lactams, are considered as a special category of compounds because of their prolonged history of use as medicines to cure several ailments, fascinating chemistry, and the future possibility of becoming effective medicines to combat several other diseases. Depending on ring size, lactams can be classified as follows (Figure 10.1): This ring-size nomenclature is derived from the fact that during hydrolysis the lactam produces the corresponding amino acid e.g. the hydrolysis of α-lactams produces α-amino acids.

1. α-Lactams: Three-membered cyclic amides
2. β-Lactams: Four-membered cyclic amides
3. γ-Lactams: Five-membered cyclic lactams
4. δ-Lactams: Six-membered cyclic amides
5. ε-Lactams: Seven-membered cyclic amides
6. ω-Lactams: Eight-membered to medium-sized cyclic amides
7. Macrolactams: Higher cyclic amides

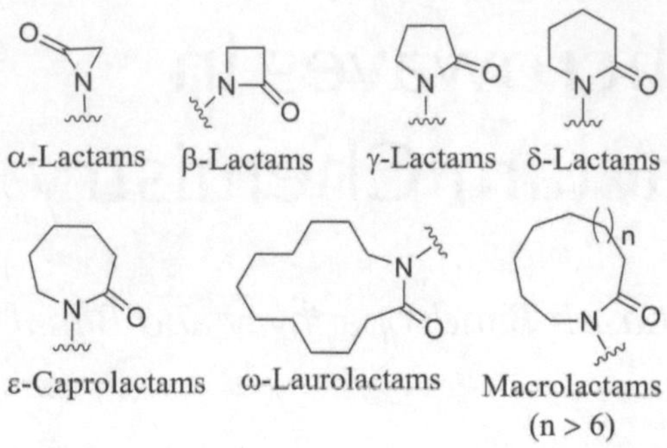

α-Lactams   β-Lactams   γ-Lactams   δ-Lactams

ε-Caprolactams   ω-Laurolactams   Macrolactams
(n > 6)

**FIGURE 10.1**   Classification of lactams.

Caprolactam ($C_6H_{11}NO$), a seven-membered (ε-) lactam, is a precursor of the widely-used synthetic polymer nylon 6. Approximately 6.5 million tons of caprolactam is produced every year globally [1]. Examples of ω-lactam include laurolactam, an industrially important 13-membered (ω-) lactam, is mainly used as a monomer to synthesize nylon 12 and copolyamides.

## 10.2   NATURAL ABUNDANCE OF LACTAMS

Among all the lactam categories, the β-lactams are most prevalent, both natural and synthetic (Figure 10.2). Several γ- and δ-lactams as well as macrolactams are also known. Pukeleimide (A-G) [2, 3], malyngamides [4–6] and isomalyngamides [7], and anatine and isoanatine [8] are biologically significant γ-lactams (Figure 10.3). Examples of natural δ-lactams (Figure 10.3) include gelegamine B [9] and sempervilam [10] whereas a huge number of bioactive macrolactams (Figure 10.4) have been reported [11, 12].

## 10.3   USES OF MICROWAVE IRRADIATION IN LACTAM SYNTHESIS

### 10.3.1   MICROWAVE: A GREEN TECHNIQUE

One of the major goals of green chemistry is to maintain a greener environment by removing/reducing the use of hazardous substances. Unfortunately, most organic synthetic procedures seriously violate the principles of green chemistry [13] because these methods require extensive heating, prolonged reaction time, tedious work-up procedure, use of potentially dangerous catalyst/support/solvent, and so on. Protection of nature is one of the foremost criteria to reduce global warming and to make a better (safer) world for the next generation. Since the last decade of the past century, the protection of our environment has been increasingly emphasized by researchers, both in academia and industry [14]. Under such circumstances, development of green methodologies that involve green techniques (microwave irradiation/sonication/mechanochemical procedures etc.) as well as eco-friendly solvents/catalysts/supports have become an integral part of chemical research in the

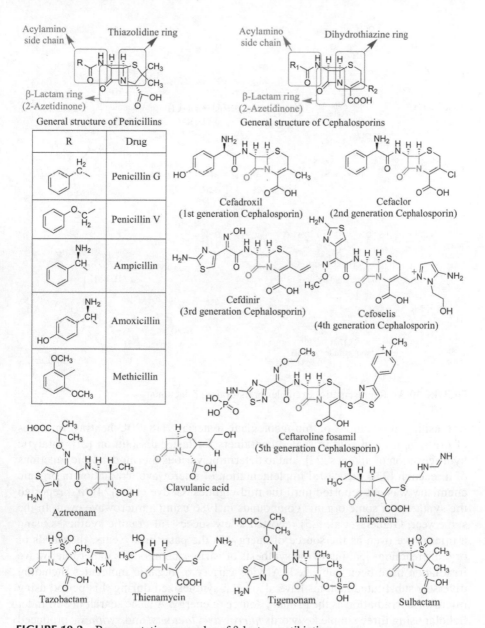

**FIGURE 10.2** Representative examples of β-lactam antibiotics.

modern era. Microwave irradiation has widely been using as one of the major green techniques to accomplish this goal for the past thirty years. Although, the use of microwave (wavelength range: 1 mm to 1 meter; frequency range: 300 GHz to 300 MHz; energy range: 1.24 meV to 1.24 μeV) [15] was introduced in chemistry laboratories in the late sixties to conduct polymerization [16], moisture analysis [17],

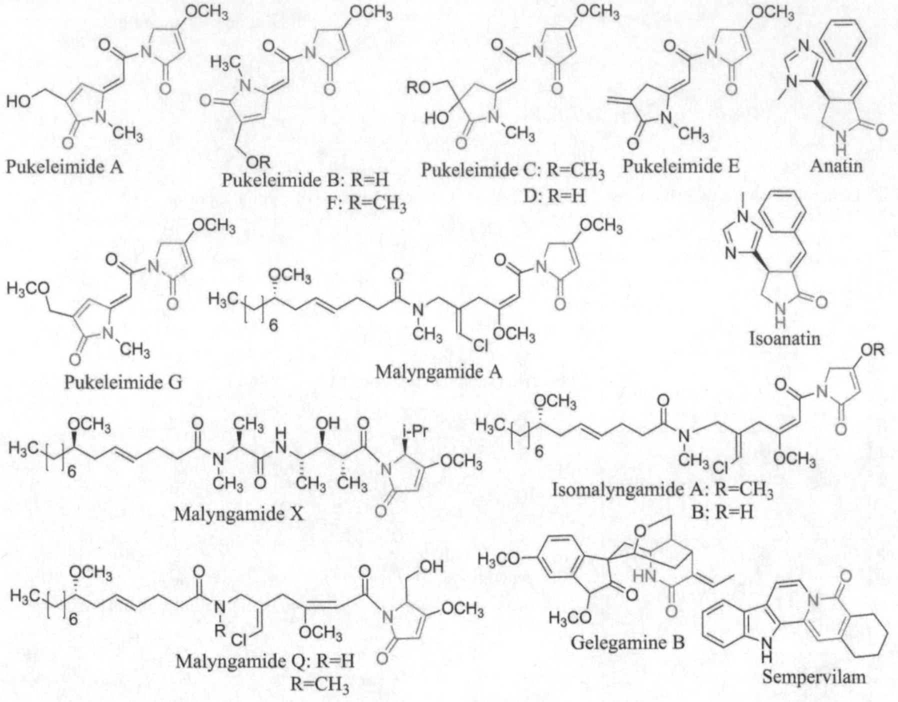

**FIGURE 10.3** Representative examples of natural γ-, δ-lactams.

wet ashing procedures of bio-/geological materials [18–20], heating mixtures of ore samples and acid in sealed containers for faster dissolution [21], catalytic hydrogenation of alkenes [22], and to determine various thermodynamic functions of a reaction [23], successful implementation of microwave irradiation in organic chemistry was not reported until the mid-eighties. Gedye et al. [24] first reported the synthesis of some organic compounds in 1986 using a microwave oven. In the same year, Giguere et al. [25] reported a few successful organic syntheses using a microwave oven as the source of energy. In the past thirty years, thousands of organic reactions leading to the synthesis of various compounds under microwave irradiation have been published. A wide variety of organic molecules including diversely substituted lactams have also been synthesized during this period using microwave irradiation as the primary source of energy. A basic search in SciFinder Scholar using three simple keywords *microwave*, *lactams*, and *synthesis* produced more than 400 microwave-assisted organic syntheses of different lactams between 1987 and 2017 (Figure 10.5).

Microwave-assisted synthesis, also known as MORE (Microwave-assisted Organic Reaction Enhancement) or MEC (Microwave-Enhanced Chemistry) synthesis, of medicinally beneficial lactams is gaining increasing interest because of its several advantages over conventional/classical procedures. Apart from many other advantages, this green technique can reduce the reaction time from days to hours,

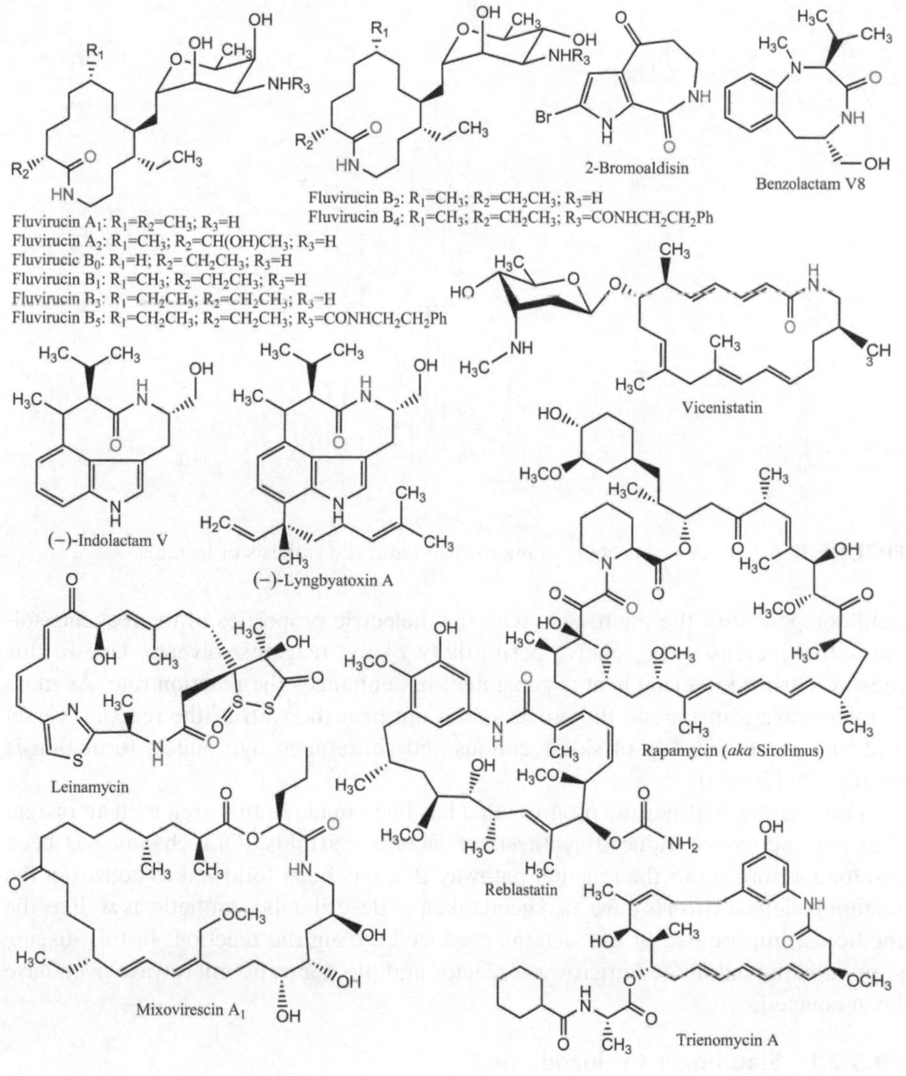

**FIGURE 10.4** Representative examples of bioactive capro-, lauro-, and macrolactams.

hours to minutes, and minutes to seconds and satisfies several principles of green chemistry [13].

## 10.3.2 How Does Microwave Work in Organic Synthesis?

There are thousands of examples demonstrating that reaction rate is enhanced abruptly by dielectric heating. The microwave methodology is highly energy economical as it requires 50% less power than electric furnaces of equivalent capacity and, in general, the rate of organic reactions is enhanced under the influence of microwave irradiation. This is mainly because of the interaction between the electric

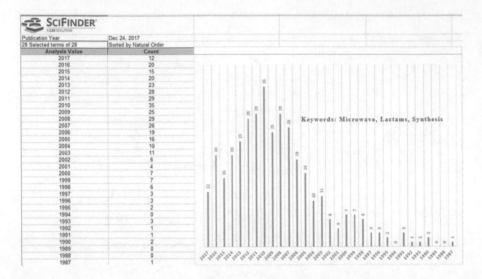

**FIGURE 10.5**    A concise account of microwave-induced synthesis of lactams.

field component of the microwave with the dielectric properties of the reagents/sol-
vents that increases the relative permittivity of the reagents/solvents. Due to this
reason, intrinsic internal heat is generated that enhances the reaction rate. As glass
is microwave transparent, the process does not heat the wall of the reaction vessel
and thus the possibility of side reactions and subsequent byproducts formation is
eliminated [26–35].

This chapter outlines the progress that has been made in this area with an insight
into the microwave-induced synthesis of lactam scaffolds. This chapter has been
grouped according to the reaction pathway that has been followed to construct the
lactam moieties. Efforts have also been taken to describe the synthetic as well as the
medicinal importance of the lactams produced during the reaction. In this discussion both, the automated microwave reactor and the domestic microwave oven have
been counted.

### 10.3.2.1   Staudinger Cycloaddition

Staudinger cycloaddition [36], the formal [2 + 2] ketene-imine cycloaddition reaction, is the most widely used methodology for the synthesis of β-lactam pharmacophores. This strategy has been applied to synthesize, both the structurally unique as
well as medicinally privileged β-lactam derivatives.

### 10.3.2.1.1   Synthesis of Structurally Unique β-Lactams

The first report of microwave-induced synthesis of β-lactams, following Staudinger
cycloaddition strategy, was published by Bose and co-workers in 1991 while they
reported the synthesis of α-vinyl-β-lactams employing microwave irradiation as
the source of energy [37]. The next year, a synthesis of enantiopure α-hydroxy-β-
lactams [38] using a domestic microwave oven was reported from the same laboratory. The Schiff bases (R = Ph, $C_6H_4OMe$-4, $C_6H_4Me$-4, $CH_2Ph$) were prepared from

diisopropylidenemannitol and were treated with PhCH$_2$OCH$_2$COCl in a domestic microwave to give 70–75% azetidinones (**1**, Figure 10.6) which were debenzylated with Pd-HCO$_2$NH$_4$, also under microwave irradiation. Diastereomeric synthesis of β-lactams *via* silyl ketene aldimines in dry media (KF/18 crown-6) under microwave irradiation was realized by Texier-Boullet et al. in 1993 [39]. Banik et al. reported a stereocontrolled synthesis of α-vinyl/alkyl β-lactam (**2**) under microwave irradiation in a kitchen microwave. Initially, *p*-anisidine condensed with diethyl ketomalonate and the product was cyclized with 3-methylbut-2-enoyl chloride following the Staudinger strategy to give two isomeric β-lactams that were decarboxylated to give **2** (Figure 10.6). Both the steps were carried out under microwave irradiation using a domestic microwave oven. These compounds are convenient synthons for various

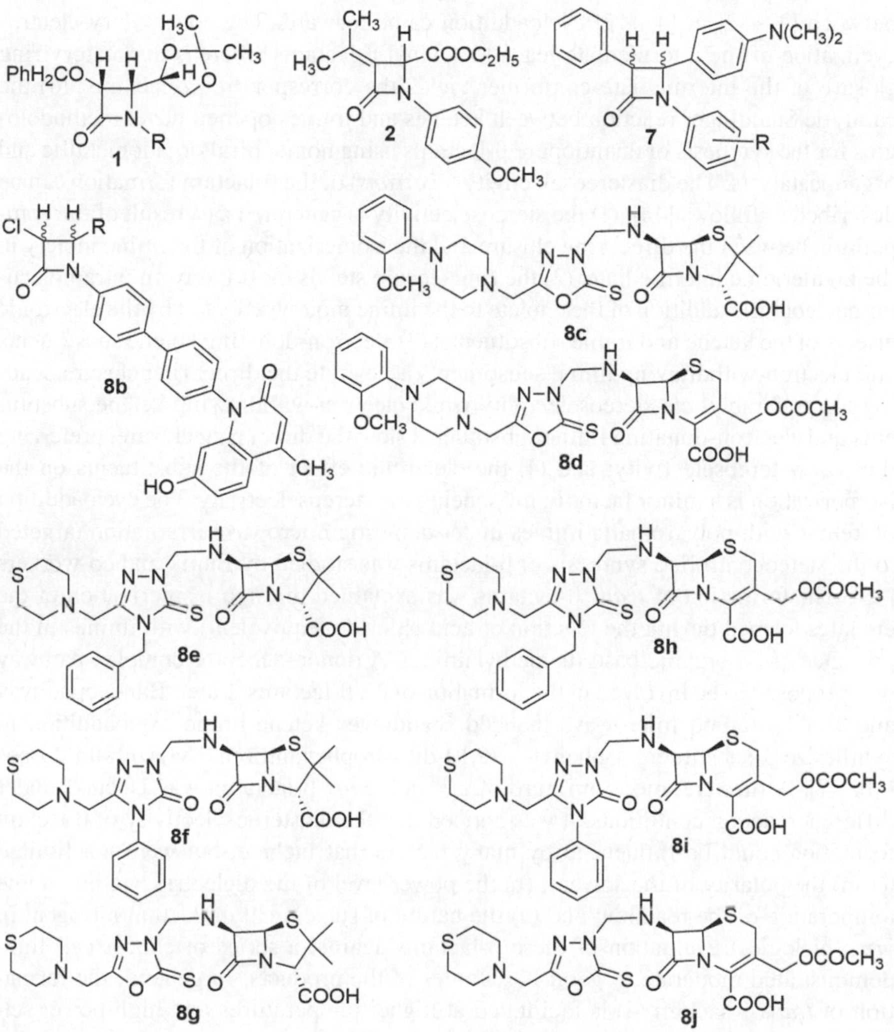

**FIGURE 10.6**   Some bioactive β-lactams synthesized by microwave irradiation.

*cis* and *trans* carbapenems [40]. Hydroxylamine-*O*-sulfonic acid reacts with alicyclic ketones on silica (solid support) under microwave irradiation to give an amino acid salt, which cyclizes to the corresponding lactam in high yield after work up in an alkaline medium, reported by Laurent and co-workers [41]. Synthesis of *N*-tert-butyldimethylsilyl azetidinones was reported following the ring-closure of 1,3-azadienes under microwave irradiation using open-vessel microwave system [42].

The mechanism of the apparently simple Staudinger ketene-imine cycloaddition reaction is controversial [43]. Interesting features of this mechanism can be summarized as (a) the step-wise concerted nature of the cycloaddition; (b) two stereogenic centers may be formed during the reaction and their formation can be effected with complete stereocontrol; and (c) the reaction of ketenes and α,β-unsaturated imines can yield either β- or δ-lactams. Subsequently the question of periselectivity between [2 + 2] *vs* [4 + 2] cycloaddition came forward. The conrotatory electrocyclization of the intermediate leads to 4-vinyl-β-lactams whereas disrotatory ring closure of the intermediate conformer yields the corresponding δ-lactams; (d) the catalytic Staudinger reaction between ketenes and imines opened novel methodologies for the synthesis of enantiopure β-lactams using homochiral organometallic and organocatalysts. The diastereoselectivity (*cis/trans*) of the β-lactam formation can be described as follows [44]: (1) the stereoselectivity is generated as a result of the competition between the direct ring closure and the isomerization of the imine moiety in the zwitterionic intermediate; (2) the ring closure step is most likely an intramolecular nucleophilic addition of the enolate to the imine moiety, affected by the electronic effects of the ketene and imine substituents; (3) electron-donating ketene substituents and electron-withdrawing imine substituents accelerate the direct ring closure, leading to (preferably) *cis*-stereoselectivity while electron-withdrawing ketene substituents and electron-donating imine substituents slow the direct ring closure, preferring the *trans*-stereoselectivity; and (4) the electronic effect of the substituents on the isomerization is a minor factor in influencing the stereoselectivity. The cycloaddition of ketene with polyaromatic imines under domestic microwave irradiation targeted to the stereocontrolled synthesis of β-lactams was studied by Banik and co-workers [45]. The formation of *trans* β-lactams was explained through isomerization of the enolates formed during the reaction of acid chloride (equivalent) with imines in the presence of an organic base (trimethylamine). A donor–acceptor complex pathway was proposed to be involved in the formation of *cis* β-lactams. Later, Bandyopadhyay and Banik studied microwave-induced Staudinger ketene-imine cycloaddition to synthesize 3-substituted 4-phenyl-*N*-(9,10-dihydrophenanthren-3-yl)azetidin-2-ones [46, 47]. A diastereomeric mixture of *cis* and *trans* β-lactams was isolated under different reaction conditions. It was noticed that the diastereoselectivity of β-lactam formation could be influenced by many factors that include, but were not limited to, (a) the polarity of the solvent; (b) the power level of the dielectric heating; (c)the temperature of the reaction; and (d) the nature of substituent on lactam nitrogen. *In vitro* biological evaluation of these β-lactams against a series of cancer cell lines demonstrated moderate to good $IC_{50}$ values of the products. In general, the formation of *trans*-β-lactam (**4**) is facilitated at higher temperatures and high power settings of microwave irradiation. This study [46, 47] reveals that solvent polarity also plays important role in determining the diastereomeric ratio of the product. Less

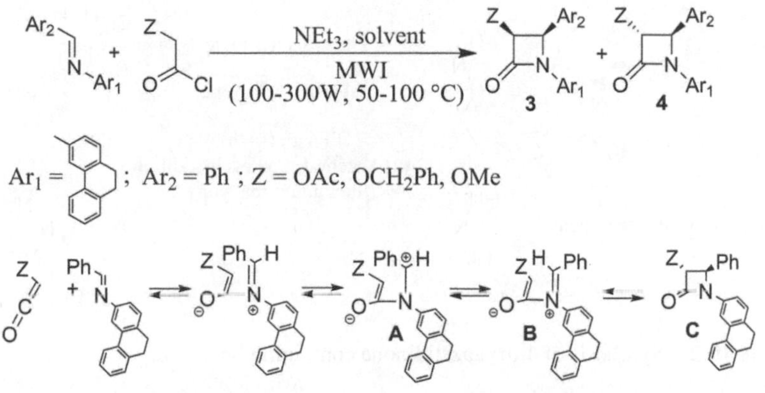

**SCHEME 10.1** Stereoselective synthesis of N-dihydrophenanthrenyl β-lactams.

polar solvent (toluene) accelerates the formation of *trans*-β-lactam (**4**) whereas *cis*-β-lactams (**3**) were found to be the major product when comparatively more polar solvent (dichloromethane) was used. The formation of (**3**) and (**4**) was explained by enolate isomerization (Scheme 10.1, **A** to **B**). The electron-withdrawing dihydrophenanthrene moiety on nitrogen (imine as well as the β-lactam) can stabilize the iminium ion through a rotation of the bond (**A** to **B**) and results in the formation of **C**.

### 10.3.2.1.2 Synthesis of Medicinally Privileged β-Lactams

The β-lactam scaffold (a four-membered cyclic amide, also known as 2-azetidinone) is one of the most widely explored pharmacophores in medicinal chemistry and drug discovery research. Because of extensive (sometimes unnecessary) use of β-lactam antibiotics over the past 75 years a broad range of bacteria have developed resistance mechanisms against many β-lactam antibiotics. The antimicrobial efficacy of many β-lactam antibiotics has been either been lost or significantly reduced due to this multi-drug resistance ability of the bacteria [14]. The primary reason for resistance toward β-lactam antibiotics is the formation of β-lactamase that cleaves the β-lactam ring; consequently, the development of β-lactamase inhibitors has been one of the main strategies in drug development [48]. Nevertheless, efforts are still continuing to develop a new generation of β-lactam antibiotics by (a) various types of derivatization; (b) chemical modification of the substituents on the 2-azetidinone core; and (c) chemical combination of a pharmacologically privileged unit with 2-azetidinones to rule out the possibility of bacterial resistance. Besides anti-infective activities, β-lactams have also been reported as anticancer, antidiabetic, anti-HIV, antiparkinsonian, anti-inflammatory, and so on [49, 50]. Medicinal scientists around the world are being engaged to accomplish this goal. Recently, Dubey et al. reported [51] the synthesis of nine β-lactam derivatives of benzotriazole (Scheme 10.2) by means of microwave irradiation and compared the efficacy of this method with the conventional route. As expected, the microwave-assisted methodology (yield: 84–92%; reaction time: 2.5–5 min) has been proved to be superior to the conventional pathway (yield: 62–72%; time: 4–7 h). A few of these β-lactams (**5**) demonstrated better activity (*in vitro*) than streptomycin against *Mycobacterium tuberculosis* H37Rv genome,

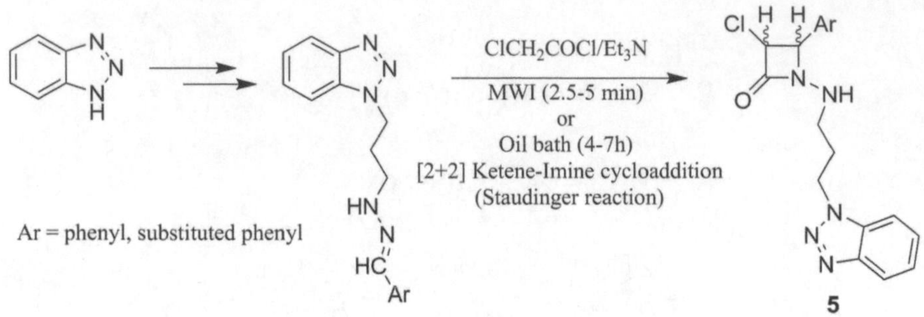

**SCHEME 10.2**    Synthesis of 4-aryl azetidinone containing benzotriazoles.

moderate to good antibacterial activity against *Escherichia coli, Staphylococcus aureus*, and *Streptococcus pneumoeniae* as well as good antifungal activity against *Aspergillus fumigatus, Candida albicans*, and *Asparagus niger* strains. The 3-chloro-2-azetidinone core might have a better potential to bind the target proteins; however, further investigation is required to find out the actual correlation between diastereoselectivity and antimicrobial activity.

Meshram et al. synthesized [52] *N*-thiazole-3-phenyl-4-aryl-azetidin-2-ones (**6**) from 2-aminothiazole following the ketene-imine cycloaddition methodology in dioxane in the presence of an organic base (triethylamine). Both the microwave-assisted (85–95% yield, 1.5–2 min) as well as the classical procedure (oil bath, 62–65% yield, 8–10 h) were studied. The antibacterial activity of eight newly synthesized monocyclic β-lactam derivatives (**6**) was screened against two Gram-positive (*Staphylococcus aureus* and *Proteus vulgaris*) and two Gram-negative (*Pseudomonas aeruginosa* and *Escherichia coli*) bacteria. The compounds showed good activity against all the four strains. It was hypothesized that the presence of more than one pharmacophore site (including β-lactam) was responsible for good antibacterial activity. The difference in charges between two heteroatoms of the same pharmacophore site might facilitate the inhibition of bacteria, more than viruses (Scheme 10.3).

The use of zeolite as a catalyst in Staudinger cycloaddition was reported from the same laboratory [53] although a detailed study regarding the catalytic activity was not carried out. Synthesis of *N*-phenylazetidin-2-ones (**7**, Figure 10.6) (R = H, 3-Cl, 4-MeO, 2-HO$_2$C), were performed by the cycloaddition of imines (derived from *N,N*-dimethylaminobenzaldehyde with diversely substituted aromatic amines) and phenylacetyl chloride in the presence of zeolite under microwave irradiation. All the newly formed β-lactams demonstrated low to moderate antibacterial activity. The authors suggested the activity was due to the presence of C=O and C-N, linkages in 2-azetidinones.

Microwave-assisted cyclo-condensation of imines (**8**) with chloroacetyl chloride to synthesize *N*-amino-3-chloro-4-(3,5-diaryl-2 pyrazoline-2-yl)- azetidine -2-ones (**8a**) was reported by Malhotra et al. [54] (Scheme 10.4), although the stereochemical aspect of β-lactam formation was not considered. The *in vitro* antimicrobial activity of the resulting β-lactams (six) was validated against *Escherichia coli, Staphylococcus aureus, Klebsiella pneumoniae*, and *Proteus vulgaris*. A few of these β-lactams

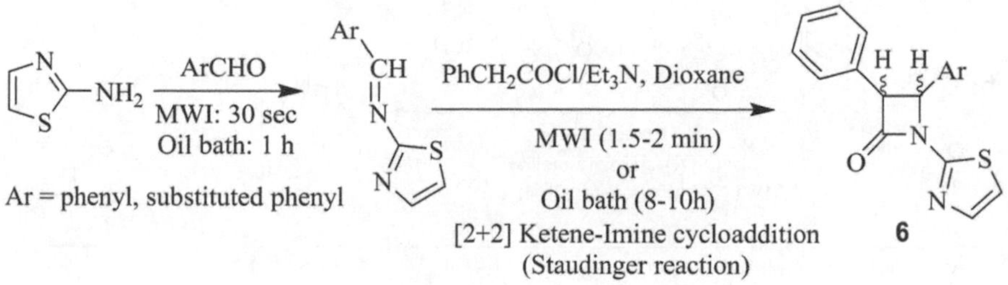

**SCHEME 10.3** Synthesis of 2-aminothiazole bearing β-lactams.

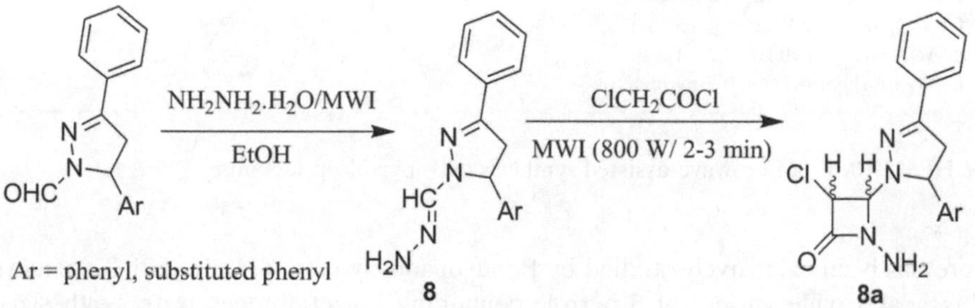

**SCHEME 10.4** Synthesis of N-amino-3-chloro-4-(3,5-diaryl-2-pyrazoline-2-yl-)-azetidine - 2-ones.

were found to be moderately active against the experimental strains. Compounds (**8c**) and (**8d**) were synthesized by one-pot, three component Mannich type condensation between 5-((4-(2-methoxyphenyl)piperazin-1-yl)methyl)-1,3,4-oxadiazo le-2-thiol (1 mmol), formaldehyde (37%, 3 mmol) and 6-aminopenicillanic acid (1 mmol, for **8c**), and 7-aminocephalosporanic acid (1 mmol, for **8d**) under controlled microwave exposure. The β-lactam (**8c**) demonstrated moderate to good activity against *Escherichia coli* and *Mycobacterium smegmatis* [55]. The corresponding thiomorpholinomethyl analogs of 6-aminopenicillanic acid, and 7-amino-cephalosporanic acid (**8e-8j**) (Figure 10.6) were also reported. The cephalosporin derivatives (**8h-8j**) demonstrated about 9-fold higher inhibitory activity (MIC: 0.46 μg/mL) than the positive control streptomycin whereas the corresponding penicillin derivatives (**8e-8g**) exhibited insignificant inhibitory effect (MIC: 31.3 μg/mL) towards *Mycobacterium smegmatis*. This indicates that cephalosporin derivatives are manifold more active than their penicillin analogs. The antimicrobial activity of the β-lactam derivatives (**8e–8j**) was also evaluated against *Escherichia coli*, *ersinia pseudotuberculosis*, *Pseudomonas aeruginosa*, *Staphylococcus aureus*, *Enterococcus faecalis*, *Bacillus cereus*, *Candida albicans*, and *Saccharomyces cerevisiae* but no significant activity was observed [56].

Synthesis of 1-{4′-[3-chloro-2-aryl-4-oxo-azetidin-1-yl]-biphenyl-4-yl}-7-hydr oxy-4-methyl-1*H*-quinolin-2-ones (**8b**) as an anticonvulsant agent has been reported by Pawar et al. [57]. The imines were treated with chloroacetyl chloride in the presence of triethylamine under microwave irradiation to yield the anticonvulsant agents (**8b**, Figure 10.6). Microwave-induced chemical modification on C3 of the β-lactam

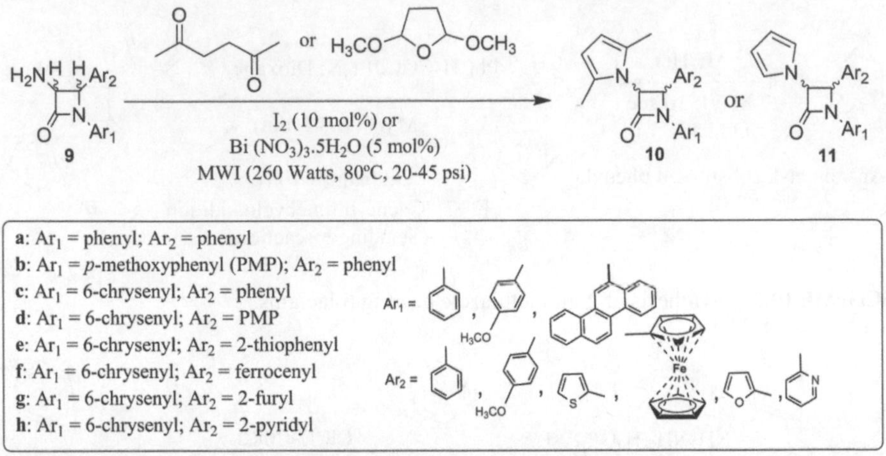

**SCHEME 10.5**  Microwave-assisted synthesis of 3- pyrrolo-β-lactams.

core has been extensively studied by Bandyopadhyay et al. [14, 58–60]. To achieve this goal a wide variety of 3-pyrrole containing 2-azetidinones were synthesized (Scheme 10.5) under microwave irradiation. The reaction of 3-amino β-lactams and 2,5-hexanedione (or 2,5-dimethoxytetrahydrofuran) following the Paal–Knorr (or Clauson–Kaas) procedure under microwave irradiation in the presence of green Lewis activators such as $Bi(NO_3)_3 \cdot 5H_2O$ or iodine (weak Lewis acid) yielded the N-(2-azetidinonyl) 2,5-disubstituted pyrroles (or 3-pyrrolo-β-lactams). *In vitro* anti-tumor evaluation of some of these 2-azetidinones unveiled moderate activity against a small series of human cancer cell lines.

### 10.3.2.2  Diels–Alder Reaction

The Diels–Alder [4 +2 ]-π cycloaddition reaction (1928) is one of the most widely used electrocyclic ring-closure reactions in synthetic organic chemistry that involves the cycloaddition of a conjugated diene (4π) with a dienophile (2π) to form six-membered cyclic framework. The driving force of the reaction is the formation of new σ-bonds, which are energetically more stable than the π-bonds. A chemical modification of the Diels–Alder reaction is the hetero-Diels–Alder, in which either the diene or the dienophile contains a heteroatom(s), most often nitrogen or oxygen. This variation constitutes a powerful method for the synthesis of six-membered ring heterocycles. Under certain circumstances, the Diels–Alder reactions can be reversible; the reverse reaction is known as the retro-Diels–Alder reaction [61].

A convenient strategy for the conversion of [2.2.2]-diazabicyclic alkene structures to 2-pyridone (δ-lactam) aromatic heterocyclic products has been reported [62] that involves the formation of an intermediate 2,5-diketopiperazine derivative (**12**). In the final step, a retro [4 + 2]-Diels–Alder cycloreversion and selective extrusion of cyanate derivatives under microwave (300W, 1h) irradiation produced 2-pyridone (**13**) in moderate yield (Scheme 10.6). This methodology was effectively implemented to synthesize novel anticancer alkaloid Louisianin A (**14**, Scheme 10.6).

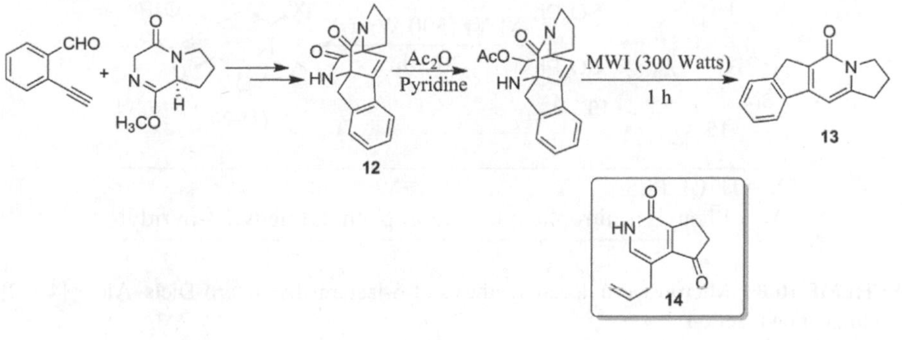

**SCHEME 10.6** Synthesis of 2-pyridones (δ-lactams) by cycloreversion of [2.2.2]-bicycloalkene diketopiperazines.

Louisianin (A–D) are a family of alkaloids reported in 1995, which were isolated from the fermentation broth of *Streptomyces* sp. WK-4028, obtained from a soil sample collected in Louisiana [63]. Louisianin A proved to be a potent inhibitor of testosterone-responsive (androgen dependent) mouse mammary Shionogi carcinoma 115 (SC115) cells.

An interesting observation was made by Long et al. [64] to synthesize 6-aryl substituted δ-lactams (**20**). The authors noticed the formation of a mixture of δ-lactam (**17**) and β-lactam (**18**) from the reaction between azadienes (**15**) and vinylsulfone (**16**) when X=Cl, Br, PhS whereas δ-lactam was isolated as the sole product when X=H (Scheme 10.7).

The X-ray crystallographic analysis of (**17**, X = H, Ar=phenyl) demonstrated a distorted half-chair conformation of the piperidinone ring with the hydrogen atoms bound to C5 and C6 in the *trans* orientation. The δ-lactam, produced through a [4 + 2] HDA reaction, was isolated as the sole product with low to moderate yield. Alternatively, in the reaction between azadienes (**15**, X=Cl, Br, PhS) with vinylsulfone, the formation of β-lactam (**18**) along with δ-lactam was observed in the crude reaction mixture (Scheme 10.7); thus a competitive electrocyclization reaction of the azadiene itself with the formation of the corresponding β-lactam (**18**) was supported. The stereochemical relationship between the substituents in the 3- and 5- and 5- and

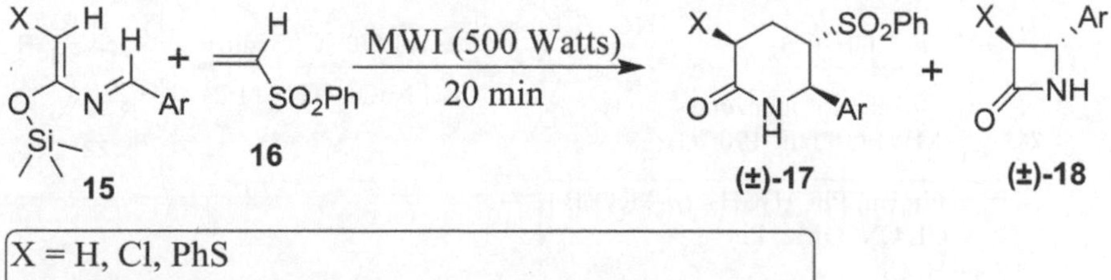

X = H, Cl, PhS
Ar = Phenyl, p-nitrophenyl, p-anisyl, thienyl, 3-pyridyl

**SCHEME 10.7** Concurrent synthesis of δ-lactams (by Hetero-Diels–Alder) and β-lactams (by 2 + 2EC, Staudinger type) reactions.

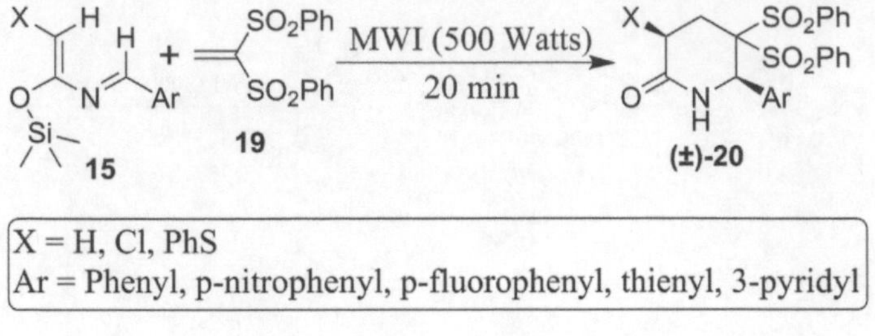

**SCHEME 10.8** Microwave-induced synthesis of δ-lactams by Hetero-Diels–Alder [4 + 2] cycloaddition reaction.

6- positions of the δ-lactam ring (17) was found to be *trans* which is supportive of the competition between a [2 + 2] electrocyclic (EC) reaction (Staudinger type) and a [4 + 2] hetero-Diels–Alder cycloaddition. Only the formation of 3,4-*trans* β-lactams was identified. Results obtained from DFT computations indicate that the competitive formation of δ- and β-lactam depends on the temperature of the reaction and the nature of the substituents of the diene and dienophile. On the other hand, the reaction between (15) and ethene-(1,1-diyldisulfonyldibenzene) (19) yielded δ-lactam as the sole product (Scheme 10.8) irrespective of the substituents at position 3 (20).

DFT studies showed the reactivity of the dienophile (16) was augmented by introducing another sulfone moiety into the vinyl sulfone scaffold (19). As predicted from the computational (DFT) studies, a hetero-Diels–Alder reaction took place and 2-azetidinone was not formed during the process, consequently, δ-lactam (20) was isolated as the sole product (Scheme 10.8).

In another study [65] the effect of pressure in microwave-assisted Diels–Alder reaction was compared with the classical heating method (Scheme 10.9). The DA reaction of substituted 2(1*H*)-pyrazinones (21) with ethylene under dielectric heating was more efficient when conducted in a pre-pressurized condition, than conventional conditions [63]. To determine the influence of pressure, the compound (21) [$R_1$ = Ph, $R_2$ = Cl, $R_3$ = $R_4$ = H] was subjected to the reaction with ethylene in *o*-dichlorobenzene under conventional conditions while the product *bis*-δ-lactam (22) was isolated only in a 12% yield. In contrast, in a pre-pressurized reaction vessel with ethylene to 5 bar at 190°C, an 87% yield of the corresponding hydrolyzed product (as NaOH

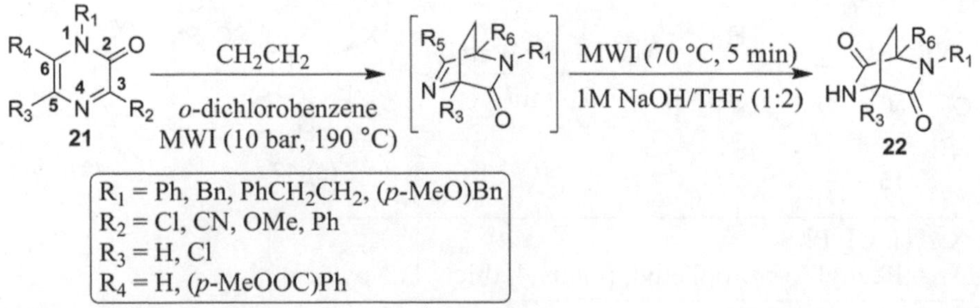

**SCHEME 10.9** Microwave-induced synthesis of δ-lactams through Diels–Alder cycloaddition: the effect of pressure.

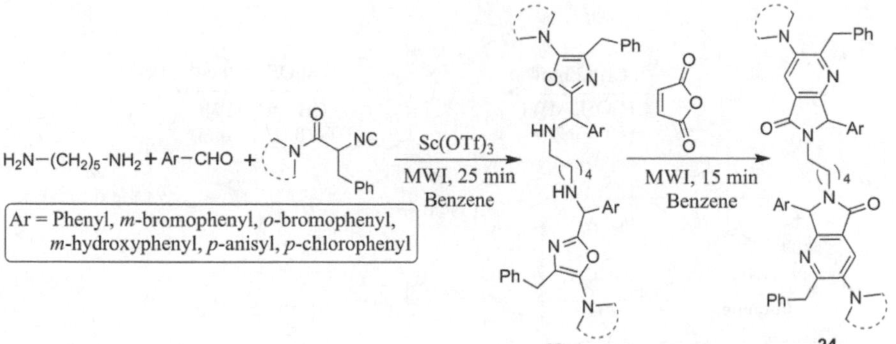

**SCHEME 10.10** Microwave-induced synthesis of fused γ-lactams through the aza-Diels–Alder route.

was present in the reaction) was isolated within 30 min. The cycloaddition reaction in a sealed vessel under dielectric heating dramatically sped up the overall process.

It was found that EWG (CN) accelerated the cycloaddition whereas EDG ($OCH_3$) reduced the reaction rate and the yield of the product. The presence of chlorine atom at C5 was found to be crucial for the success of the reaction.

A microwave-induced oxazole-maleic anhydride aza-Diels–Alder reaction was explored as the key step to synthesize a series of cyclic analogs of fused γ-lactams (**24**) [66]. In the first step, an aliphatic diamine, aromatic aldehyde, and isonitrile derivative (Ugi type reaction) reacted in the presence of the Lewis activator (scandium triflate) under dielectric heating (α-addition) to synthesize the oxazole intermediate (**23**) that eventually underwent aza-Diels–Alder cycloaddition and subsequent ring-opening with maleic anhydride and produced a moderate yield of fused γ-lactam derivatives (**24**). (Scheme 10.10).

### 10.3.2.3 Ugi Reaction

The Ugi four-component condensation/reaction (U-4CC/R) between aldehydes (or ketone), amines, carboxylic acid, and isocyanide allows the rapid preparation of α-aminoacyl amide (*bis*-amide) derivatives. The Ugi reaction products can typify a wide variety of substitution patterns and constitute peptidomimetics that have potential pharmaceutical applications. The mechanism is believed to involve a prior formation of an imine by condensation of the amine with the aldehyde, followed by the addition of the carboxylic acid oxygen and the imino carbon across the isocyanide carbon; the resulting acylated isoamide rearranges by acyl transfer to generate the final product. This exothermic reaction has an inherent high atom economy as only a molecule of water is lost and, in general, the yield is high.

#### 10.3.2.3.1 Synthesis of Medicinally Privileged Lactams

Till date, numerous studies have been conducted to get a clear understanding about the harmful effects of tobacco on human health and the investigation to reveal the relationship between the central nervous system (CNS) and nicotine (and its secondary metabolites) is still ongoing. Interestingly, it has been found that cotinine

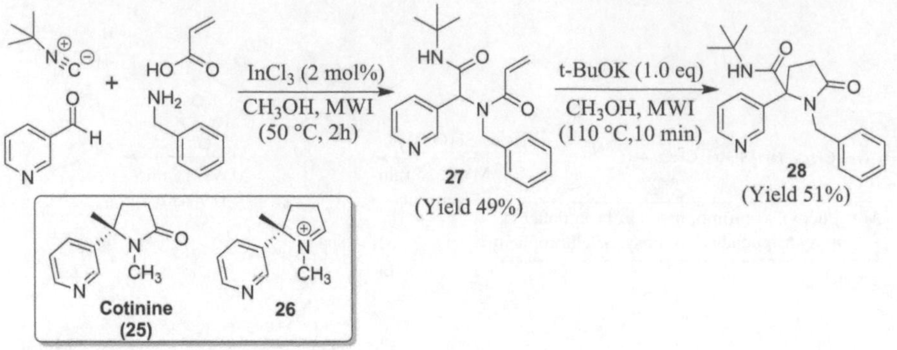

**SCHEME 10.11** Microwave-induced synthesis of cotinine (γ-lactam) analog *via* formal 5-*endo* cycloisomerization.

[**25**, the major secondary metabolite of nicotine (a γ-lactam)] showed a relatively safer pharmacological profile in human. The toxicity of cotinine is about 100 times lesser than nicotine but plasma half-life is about 8–10 times longer than nicotine. Studies showed that, in human, cotinine is almost non-toxic to the cardiovascular system despite the chemical structures of these two molecules (cotinine and nicotine) being similar and cotinine being the major secondary metabolite of nicotine (about 70–80% conversion). Initially, nicotine is converted to nicotine-$\Delta^{1'\ (5')}$-iminium ion (**26**, Scheme 10.11), that exists in equilibrium with 5'-hydroxynicotine. Subsequently, following an enzyme-catalyzed route the compound (**26**) is converted to cotinine [(5S)-1-methyl-5-(3-pyridyl)-pyrrolidin-2-one, **25**] [67, 68]. Pre-clinical studies (in mouse models) showed that cotinine could facilitate the elimination of fear memories and improve attention and working memory for Alzheimer disease (AD) [69, 70], reduce fear and anxiety of post-traumatic stress disorder (PTSD) [71, 72], and might demonstrate antipsychotic drug-like properties [73] and stimulate nicotinic cholinergic receptors [74]. A convenient base-mediated two-step synthesis of cotinine analogs and a one-pot base-free synthesis of *iso*-cotinine derivatives featuring an Ugi-4CR/cyclization protocol have been reported (Scheme 10.11). The mentioned approach demonstrated a facile construction of the γ-lactam core [68].

Generation of the quaternary spiro center is considered as an art in organic synthesis and spiro-lactams are considered as a medicinally privileged pharmacophore. Spiro-cyclohexadienoyl γ- and δ-lactams have been synthesized under microwave exposure following radical spirocyclization of xanthate-containing Ugi-4CR adducts (Scheme 10.12). A sequential application of Ugi-4CR and radical spirocyclization procedures under microwave have been carried out to synthesize the products in moderate to good yield [75].

### 10.3.2.3.2  Synthesis of Structurally Unique β-Lactams

A microwave-assisted solventless synthesis of γ- and δ-lactams *via* four centers-three components Ugi procedure (U-4C-3CR) has been reported by Deprez and co-workers [76]. Either levulinic acid or 5-ketohexanoic acid was used as ketoacid component in this reaction. All the reactions were performed under neat conditions

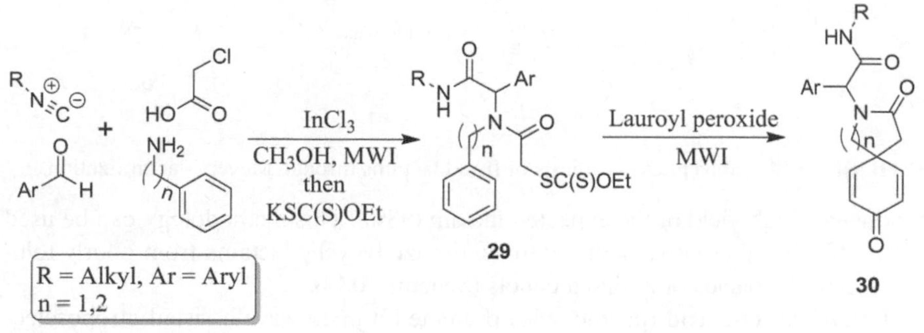

**SCHEME 10.12** Microwave-induced sequential Ugi-4CR and radical spirocyclization to synthesize spiro γ-, and δ-lactams.

at 100°C for 3 min at 75 Watts power level. A set of structurally diverse amines and isocyanides were combined with ketoacid under optimized conditions to give 24 lactam derivatives. A Design of Experiments (DoE) approach has been reported by Tye et al. [77] for lactam synthesis *en route* to a microwave assisted U-4C-3CR procedure using levulinic acid as the acid source. The reported method produced lactam derivatives in moderate to excellent yields (17–90%) within 30 min in comparison to the conventional methodology which required up to 48 h.

### 10.3.2.4 Meyers' Lactamization

Meyers' lactamization [78–84] is a classic bielectrophile-binucleophile reaction (BiE-BiNu) reaction that produces quaternary centers, mostly in a stereoselective manner (Scheme 10.13). This reaction is usually performed in toluene (non-polar/less polar aprotic solvent), with or without using a Dean–Stark water separator. This reaction is frequently carried out using the reagents ketoacid (normally γ-ketoacid or ester) and aminoalcohol (usually chiral). Besides γ-ketoacids, δ- and ω-ketoacids were also reported. It is a well-known tool for the synthesis of natural products with a special attention to the stereoselective synthesis of alkaloids.

Microwave-assisted solventless Meyers' lactamization has been reported by Laconde and co-workers [84] to synthesize Meyers' five-membered fused chiral lactams (**33**) in good yield and diastereoselectivity. A wide variety of β-amino alcohols (**31**) and γ-keto acids (**32**) were used to generate diversely-substituted products (Scheme 10.14). In the case of δ-ketoacid, a comparatively lower yield (60%) was observed. Interestingly, indole containing aminoalcohol (tryptophanol) also

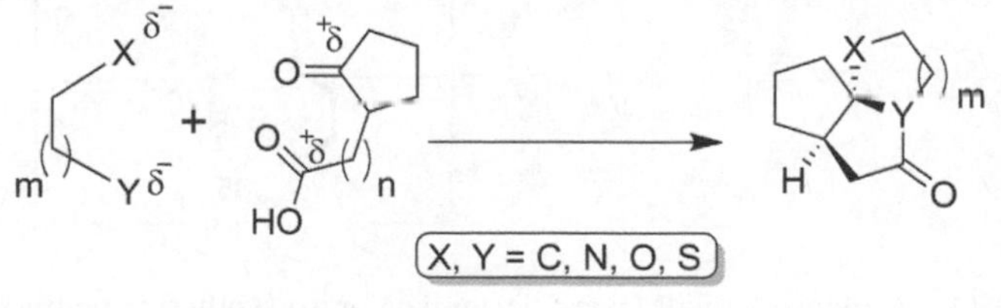

**SCHEME 10.13** Meyers' lactamization (bielectrophile-binucleophile) reaction.

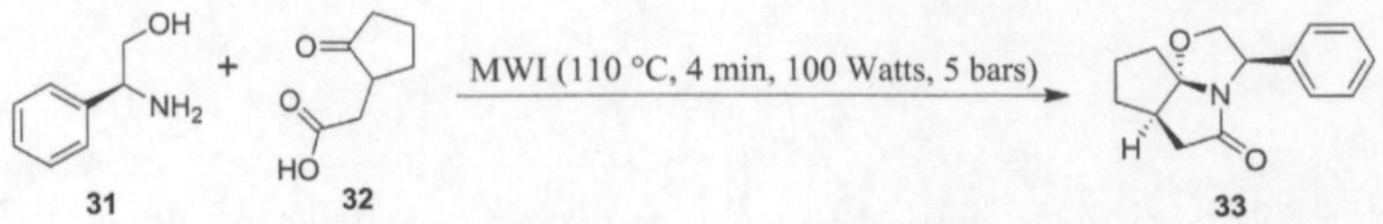

**SCHEME 10.14**  Solventless synthesis of fused lactams through Meyers' lactamization.

produced a high yield of the expected lactam (98%). This methodology can be used as an effective alternative pathway to synthesize bicyclic lactams from poorly toluene-soluble ketoacids or amino alcohols (Scheme 10.14).

Trimethylacetic acid (pivalic acid) promoted a microwave-assisted atroposelective Meyers' lactamization procedure to synthesize dibenzo(di)azepines (**34**) through a center to axial chirality transfer principle, as reported by Levacher and co-workers [85]. In the first step, a palladium-based organometallic catalyst was used for the cross-coupling reaction. The product of this process subsequently underwent pivalic acid-promoted Meyers' lactamization in an atroposelective manner to yield the pharmacophore analogs (**34**). Excellent diastereoselection ($dr > 96{:}4$) was observed in this process (Scheme 10.15). Biheteroaryl frameworks using pyrrole and indole (**35**) have also been synthesized following a similar route.

### 10.3.2.5  The Kinugasa Reaction

The Kinugasa reaction is a type of [3 + 2] dipolar cycloaddition reaction, in which nitrones (dipole) react with terminal alkynes in the presence of catalytic or stoichiometric amounts of copper (I) salts to produce β-lactams. It is well known that 1,3-dipolar cycloaddition of nitrone to olefin is one of the most convenient routes to synthesize isoxazoline core. Besides isoxazoline synthesis, nitrones are also used to synthesize diversely substituted β-lactams through the copper(I)-catalyzed Kinugasa reaction. In fact, it is a reaction between nitrones and copper(I) acetylides and in most cases high atom economy is maintained. Microwave-induced diastereoselective

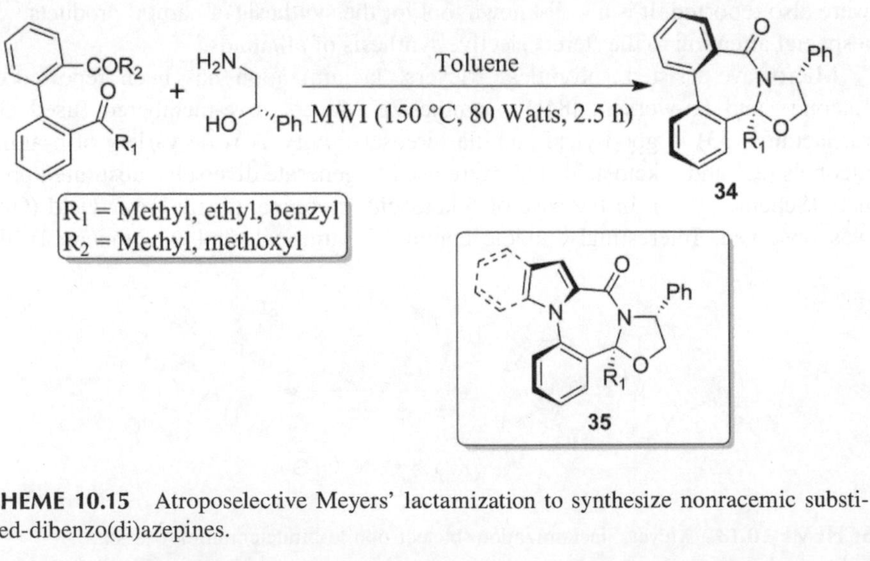

$R_1$ = Methyl, ethyl, benzyl
$R_2$ = Methyl, methoxyl

**SCHEME 10.15**  Atroposelective Meyers' lactamization to synthesize nonracemic substituted-dibenzo(di)azepines.

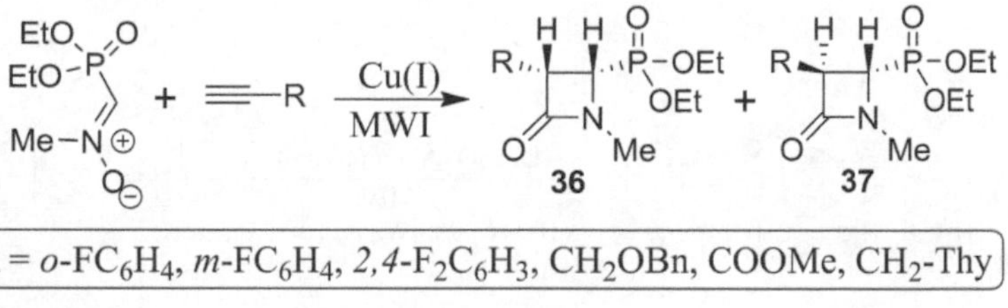

**SCHEME 10.16** Synthesis of 4-phosphonylated β-lactams *via* Kinugasa reaction.

synthesis of 4-phosphonylated β-lactams from *N*-methyl-*C*-(diethoxyphosphonyl) nitrone with a small series of diversely substituted terminal alkynes *via* the Kinugasa route has been reported [86]. The application of microwave irradiation significantly shortened the reaction time. Both the *cis*- (**36**) and *trans*- (**37**) β-lactams were isolated (Scheme 10.16).

### 10.3.2.6 The Smiles Rearrangement

The Smiles rearrangement is an intramolecular $S_NAr$ reaction in which a nucleophilic attack takes place on an aromatic system bearing an electron-withdrawing group at the *ortho*- or *para*-position to the reaction center connected to a heteroatom. This rearrangement is accompanied by the migration of an aromatic ring from the heteroatom binding to the reaction center to a more nucleophilic heteroatom. In general, the rearrangement is facilitated by the presence of moderately strong EWG and the steric hindrance arising from a substituent at a particular position in the aromatic ring.

Medicinally privileged optically active aryl-fused δ-lactams (**38** and **39**) were prepared *via* the Smiles rearrangement in conventional as well as microwave irradiation pathways in a one-pot fashion using (2*S*)-2-chloro-*N*-(phenylmethyl)propanamide with 2,4-dichlorophenol, 2,3-dichlorobenzenethiol, 2,5-dichlorobenzenethiol, and so on (Scheme 10.17). Most of the compounds displayed good inhibition of Gram-positive bacteria (*Staphylococcus aureus* and *Bacillus subtilis*), Gram-negative bacteria (*Escherichia coli* and *Micrococcus luteus*), and fungi (*Candida albicans*) in the antibacterial/antifungal evaluations [87].

### 10.3.2.7 Cascade/Domino/Tandem Reactions

From green chemistry perspective, a multicomponent reaction (MCR) is considered one of the most promising emerging strategies in organic synthesis. In MCR, three or more reactants combine in a consecutive manner through a network of reaction equilibria that eventually leads to a high final product yield. This approach is rapid,

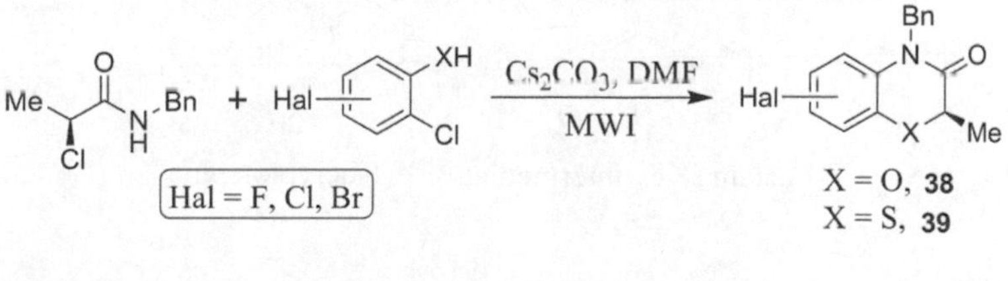

**SCHEME 10.17** Synthesis of δ-lactams *via* Smiles rearrangement.

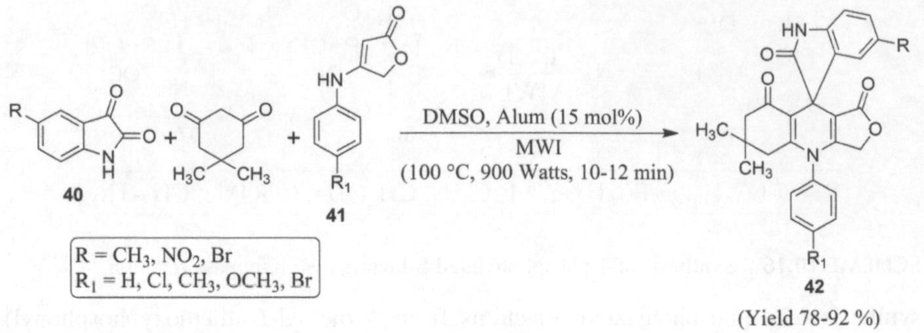

R = CH₃, NO₂, Br
R₁ = H, Cl, CH₃, OCH₃, Br

**SCHEME 10.18** Knoevenagel condensation-Michael addition-intramolecular cyclization cascade to synthesize spirocyclic γ-lactams.

less power-consuming, atom-economical and in most cases does not produce waste or minimum waste is produced. In MCR, the reactants assemble *in situ* succeeding the cascade/domino approach and its effectiveness is evaluated by the number of newly formed bonds, the structural novelty of the product, and its use(s) in the broader area. One-pot multicomponent synthesis of targeted pharmacophores *via* a green route is a challenge for the synthetic as well as medicinal chemists in the present era.

Alum [KAl(SO₄)₂·12H₂O]-catalyzed microwave-enhanced synthesis of spirocyclic γ-lactams (**42**, 78–92%) was carried out in DMSO [88] *en route* to the Knoevenagel condensation-Michael addition-intramolecular cyclization cascade using isatins (**40**), dimedones, and anilinolactones (**41** (Scheme 10.18).

Total synthesis (in seven steps) of isocryptolepine (**44**) (an alkaloid isolated from *Cryptolepis sanguinolenta*) was reported with an overall yield of 57% [89]. The key step of this synthesis consisted of microwave-assisted tandem Curtius rearrangement with DPPA and electrocyclic ring closure of an aza 6π-electron system. The target compound (**44**) was synthesized from the tetracyclic δ-lactam (**43**, key intermediate) in four-steps (Figure 10.7).

A microwave-assisted cascade comprising of Overman rearrangement-ring closing metathesis-Kharasch cyclization has been reported to synthesize bicyclic γ-lactams (**45**). This significantly faster procedure produced the bicyclic γ-lactams in a cleaner way compared to standard thermal conditions. This new approach

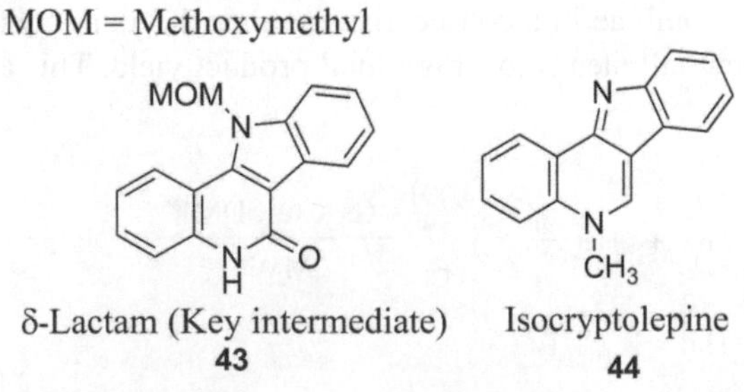

MOM = Methoxymethyl

δ-Lactam (Key intermediate)   Isocryptolepine
**43**                         **44**

**FIGURE 10.7** Isocryptolepine (**44**) and its key intermediate δ-lactam (**43**).

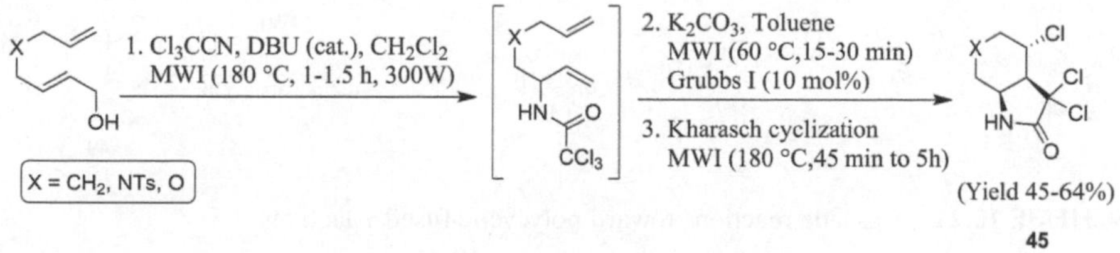

**SCHEME 10.19** Overman rearrangement-ring closing metathesis-Kharasch cyclization sequence for the synthesis of bicyclic γ-lactams.

successfully bypasses the difficulties (possibility of byproduct formation) through thermal Overman rearrangements (Scheme 10.19). The reaction was conducted in a silicon carbide vial using toluene as a solvent under microwave exposure [90].

A microwave-assisted imination/Wolff rearrangement/[2 + 4] cycloaddition cascade for the diastereoselective synthesis of spiro δ-lactams (**46**) has been reported [91]. On a few occasions, a minor product was observed along with the target compound (major) at 140°C whereas only the target compounds (**46**) were obtained at 160–200°C. In the case of aniline derivatives, the subsequent formation of spiro compounds did not take place at all. To overcome this issue a domino aza-Wittig/Wolff rearrangement/[2+4] cycloaddition was carried out that yielded the expected product (**47**) in moderate yields (Scheme 10.20).

Aromatic nucleophilic substitution of 2,4-dinitrochlorobenzene with pyridines yields *N*-arylated pyridinium salts (commonly known as Zincke salts) which on reaction with 2° amine and subsequent base-mediated hydrolysis produces Zincke aldehydes (5-amino-2,4-pentadienals, **48**), a well-studied class of donor-acceptor dienes [92]. A microwave-assisted cascade comprising of *E-Z* alkene isomerization-6π electrocyclic ring closure-[1,5]-sigmatropic hydrogen shift-6π electrocyclic ring-opening-Diels–Alder cycloaddition was reported to synthesize polycyclic fused γ-lactams (**49**) from Zincke aldehydes [93]. Complex polycyclic lactam cores were synthesized with high stereoselectivity following the reported method (Scheme 10.21).

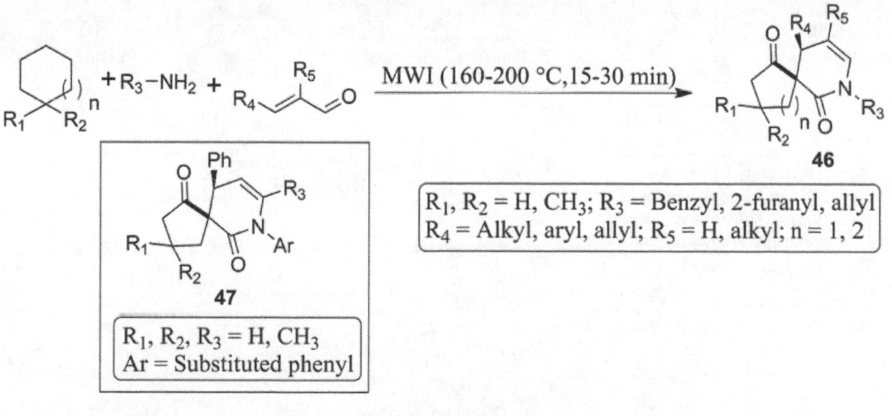

**SCHEME 10.20** Imination/Wolff rearrangement/[2+4] cycloaddition cascade toward *N*-substituted δ-lactams.

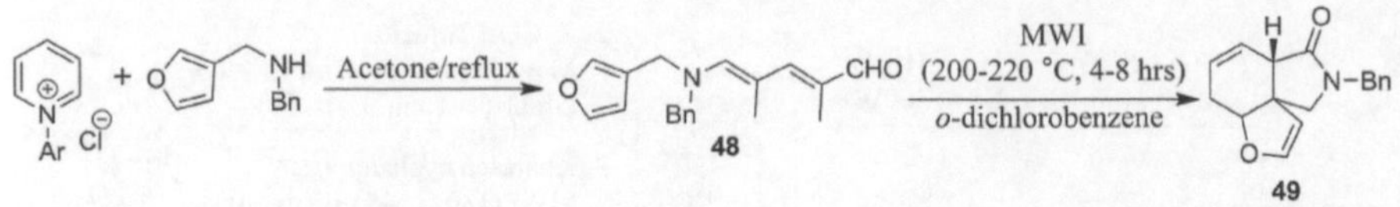

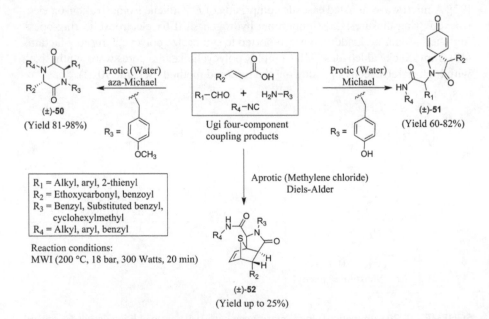

**SCHEME 10.21**    Cascade reactions toward polycyclic fused γ-lactams.

Microwave-promoted domino Ugi-aza-Michael, Ugi-Michael, and Ugi-Diels–Alder reactions have been used to prepare *bis*-δ-lactams (**50**), spiro γ-lactams (**51**), and thiophene-derived fused γ-lactams (**52**) respectively [94]. The steric and solvent effects have also been studied (Scheme 10.22).

Through a microwave-assisted three-component (one-pot) protocol alkylhydroxylamine hydrochlorides (**53**), formaldehyde [or an alkyl glyoxylate (**54**)] and bicyclopropylidene react to yield 3-spirocyclopropanated β-lactams (**55**). Depending on substituents the reaction times vary from 15–120 min to produce moderate to good yields of the product (49–78%) [95]. The plausible mechanism involves a nitrone-alkene 1,3-DC reaction (Scheme 10.23).

A sequential *N*-alkylation (S$_N$2)–aza-Cope (or Claisen) rearrangement–hydrolysis reaction to synthesize fused δ-lactam (**57**) has been reported in which the second step has been carried out under microwave irradiation (Scheme 10.24) [96]. The *N*-alkylated product was irradiated in isopropyl alcohol/water to produce the aldehyde (**56**), which was subjected to ionic aza-Cope rearrangement and further hydrolysis with potassium carbonate produced (**57**).

**SCHEME 10.22**    Steric and solvent effects in microwave-induced synthesis of lactams.

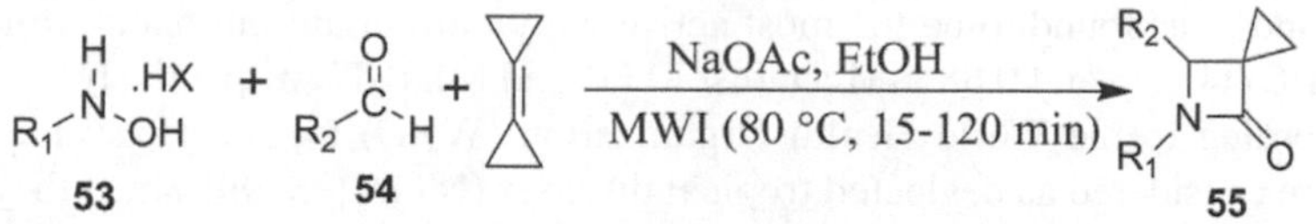


$R_1$ = Benzyl, benzhydryl, *p*-methoxybenzyltet-butyl, methyl
$R_2$ = H, COOEt, COOMe
HX = HCl, $(COOH)_2$

**SCHEME 10.23** Three component cascade reactions toward spiro β-lactams.

### 10.3.2.8 Miscellaneous Reactions

Apart from the previous discussion, several other reactions have been reported wherein microwave irradiation played a crucial role either to complete the key step or to carry out more than one significant step (in multistep syntheses) or to accomplish the entire sequence of reactions (one-pot) to furnish bioactive lactams or structurally unique lactams or lactam-encompassing natural products or its core. A brief account of these novel reactions is summarized hereafter.

#### *10.3.2.8.1 Synthesis of Medicinally Privileged Lactams*

1,3-Dipolar cycloaddition constitutes a proficient approach for the synthesis of five-membered heterocycles. An efficient solvent-free synthesis of anti-tubercular diversely substituted spiro-γ-lactams was reported under microwave irradiation [97]. It is well known that tuberculosis (TB) is contagious and airborne. After the human immunodeficiency virus (HIV), it is the 2nd leading cause of death from a single infectious agent worldwide. In 2014 alone, 9.6 million people fell ill with TB and 1.5 million died. More than 95% of TB deaths occur in low- and middle-income countries. With growing bacterial resistance against even new generation antibiotics, multi-drug resistant *M. tuberculosis* (MDR-TB) is becoming a major health issue [96]. The spiro-γ-lactams that synthesized in the reaction [97] were screened for their *in vitro* activity against *Mycobacterium tuberculosis* H37Rv (MTB), MDR-TB, and *Mycobacterium smegmatis* (MC2) using an agar dilution method. One of these

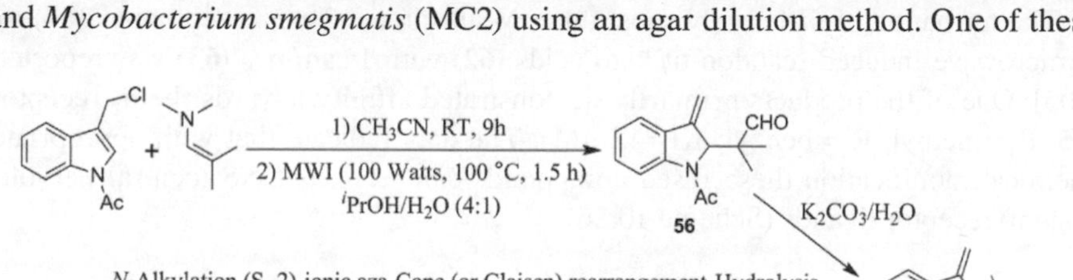


**SCHEME 10.24** *N*- Alkylation and subsequent microwave-irradiated ionic aza-Cope rearrangement-Hydrolysis to synthesize fused δ-lactam.

compounds was found to be the most active with a minimum inhibitory concentration (MIC) of 1.76 and 0.88 mM against MTB and MDR-TB respectively.

According to the World Health Organization (WHO) factsheet, seventeen diseases are considered as neglected tropical diseases (NTD) [98, 99], since they occur mostly in tropical and developing countries in Africa, South America, and Asia. On the other hand, leishmaniasis represents a complex of diseases with an important clinical and epidemiological diversity. Leishmaniasis is endemic in 88 countries, 72 of which are developing countries. There are three main forms of leishmaniasis – visceral (often known as kala-azar and the most serious form of the disease), cutaneous (the most common), and mucocutaneous. Leishmaniasis is caused by the protozoan *Leishmania* parasites (about 20 species) which are transmitted by the bite of infected sand-flies. The disease mainly affects low-income populations suffering from malnutrition and the associated weak immunity and an estimated 1.3 million new cases and 20000 to 30000 deaths occur annually [100]. In the search for new and novel antileishmanial drugs, selected hydrocarbon-stapled peptides were prepared through solid-phase synthesis protocol and optimized through ring-closing metathesis reactions under microwave irradiation [101].

Lactam-bridged cyclic peptides have been utilized as therapeutic agents and biochemical tools. An efficient microwave-assisted multi-step synthetic strategy for the synthesis of lactam-bridged cyclic peptides (inhibitor of melanocortin receptor) has been reported (**58**, Figure 10.8) [102].

Serra et al. reported a new synthetic strategy [103] to obtain a 7/5-fused lactam scaffold, an important intermediate for the synthesis of cRGD-based bioconjugates as possible anticancer compounds. The concurrent formation of the lactam ring and the double bond took place *via* the intramolecular Hosomi–Sakurai reaction. (**59**, Figure 10.8).

A series of (five each) 3-chloro substituted β-lactams (**60**) and sulfur-containing γ-lactams (4-thiazolidinones, **61**) have been synthesized from halo-substituted imines using the classical as well as microwave technique [104]. As expected, the microwave irradiated route produced a much higher yield at a faster rate. The newly synthesized compounds were screened for antibacterial and antifungal activities against *B. subtilis*, *E. coli*, *A. niger*, and *A. flavus*. A few compounds showed potent antimicrobial activity (Scheme 10.25).

The formation of differently substituted N-substituted 3-benzazepinones (**64**) by a microwave-induced reaction of keto acids (**62**) with 1° amines (**63**) was reported [105]. One of the products primarily demonstrated affinity towards the $\sigma_1$ receptor (**65**, $R_1$ = methyl, $R_2$ = benzyl; Ki = 12 nM). The data indicate that with appropriate chemical modification these fused compounds can serve as CNS (central nervous system) receptor ligands (Scheme 10.26).

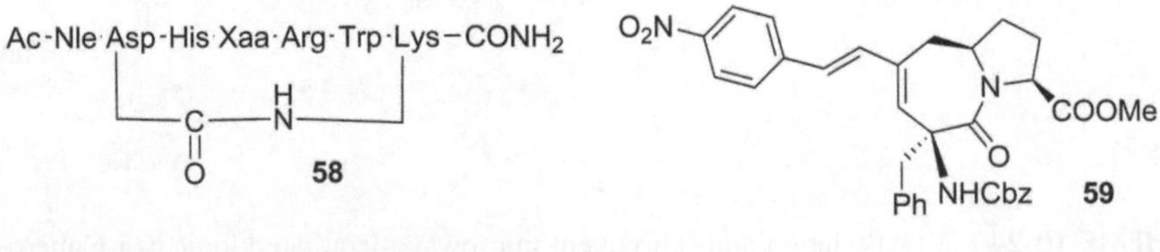

**FIGURE 10.8** Microwave-assisted synthesis of medicinally privileged lactams.

Design and solid-phase synthesis of indole-containing γ-lactams (66 and 67, Figure 10.9) have been reported under dielectric heating [106].

Microwave-assisted synthesis of the compounds (66a–67d) was carried out through 1,3-dipolar cycloaddition of isatin, 1,3-thiazole-4-carboxylic acid, and a Knoevenagel adduct in 2,2,2-trifluoroethanol as a green solvent. All four compounds

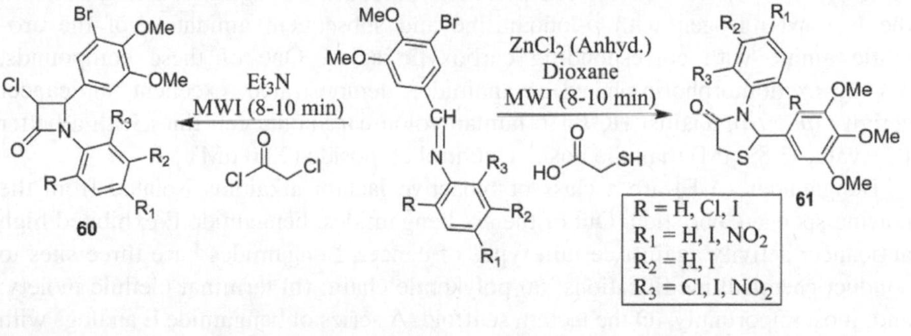

SCHEME 10.25 Microwave-promoted synthesis of antimicrobial lactams.

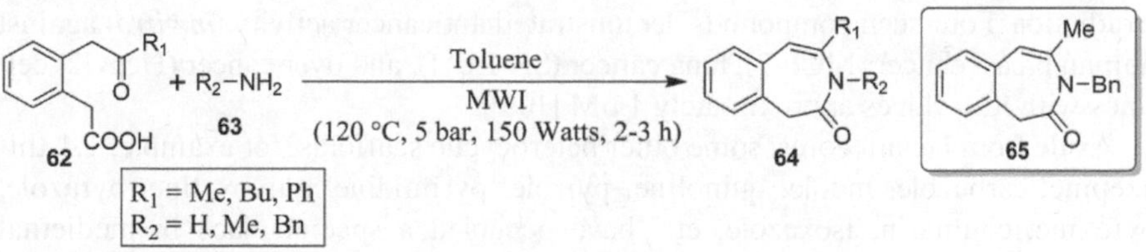

SCHEME 10.26 Microwave-assisted synthesis of diversely substituted 3-benzazepin-2-ones.

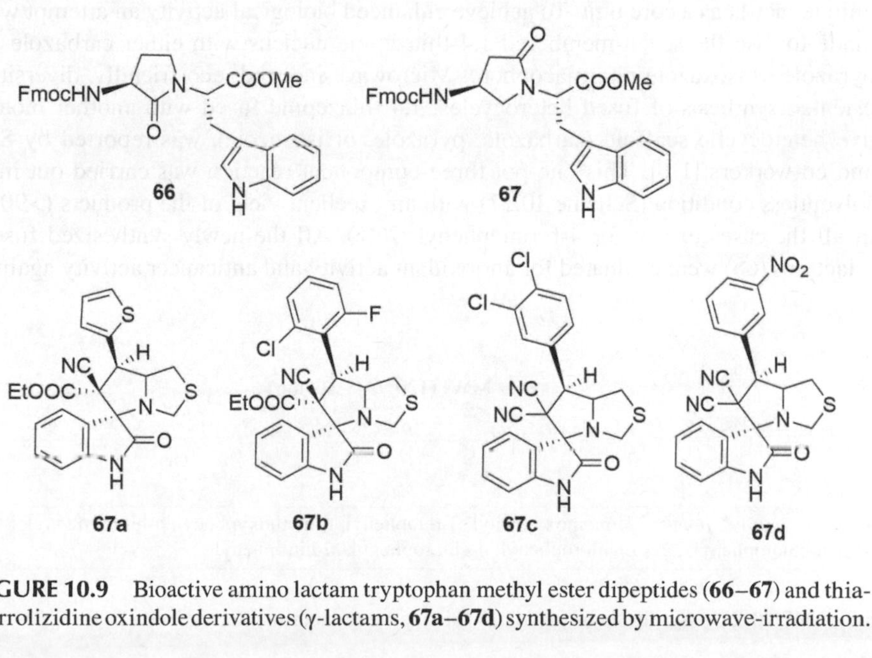

FIGURE 10.9 Bioactive amino lactam tryptophan methyl ester dipeptides (66–67) and thia-pyrrolizidine oxindole derivatives (γ-lactams, 67a–67d) synthesized by microwave-irradiation.

demonstrated high antibacterial activity against *Klebsiella pneumonia* with MIC value ranging from 0.005 to 0.19 μg/mL. In addition, the compound **67c** exhibited *in silico* drugability against the protein New Delhi Metallo-beta-Lactamase-1 (NDM-1; PDB ID: 4HL2) [107].

Microwave-induced synthesis of δ-lactams having C/O/S at position four has recently been reported [108]. The synthesis precedes through a coupling between the lactonyl nitrogen with *p*-iodoaniline and subsequent amidation of the aromatic amine with corresponding carboxylic acids. One of these compounds, *N*-(4-(3-oxotiomorpholin)phenyl)hexanamide, demonstrated excellent anticancer activity (*in vitro*) against HCT-116 human colon carcinoma cell lines with a better IC$_{50}$ value (1.89 μM) than the positive control etoposide (2.80 μM).

Bengamides (A-F) are a class of bioactive lactam alkaloids isolated from the marine sponge Choristida. Out of the six bengamides, bengamide E exhibited high anticancer activity against certain types of cancer. Bengamides have three sites to conduct chemical modifications: (a) polyketide chain; (b) terminal olefinic moiety; and, most importantly, (c) the lactam scaffold. A series of bengamide E analogs with various δ- and capro-lactam cores were synthesized by opening the polyketide chain lactone ring and subsequent condensation with α-aminolactams under microwave irradiation. Four such compounds demonstrated anticancer activity (*in vitro*) against human breast cancer (MCF-7), lung cancer (SK-LU-1), and liver cancer (HepG2) cell lines with IC$_{50}$ values approximately 1 μM [109].

Aside from lactam cores, some other heterocyclic scaffolds, for example, 1,4-thiazepine, carbazole, indole, quinoline, pyrrole, pyrimidine, quinoxaline, pyrazole, pyrazine, coumarin, isoxazole, etc. have occupied a special place in medicinal chemistry and drug discovery research because of their remarkable pharmacological profiles. A huge number of commercial drugs contain one (or more) of these unique motifs as a core unit. To achieve enhanced biological activity an attempt was made to fuse the seven-membered 1,4-thiazepine nucleus with either carbazole or pyrazole or isoxazole pharmacophore. Microwave-induced, eco-friendly, diversity-oriented synthesis of fused heterocycles, 1,4-thiazepine fused with another bioactive heterocyclic scaffold (carbazole, pyrazole, or isoxazole), was reported by Shi and co-workers [110]. This one-pot three-component reaction was carried out in a solventless condition (Scheme 10.27) with an excellent yield of the products (>90% in all the cases except for 4-bromophenyl, 76%). All the newly synthesized fused ε-lactams (**68**) were evaluated for antioxidant activity and anticancer activity against

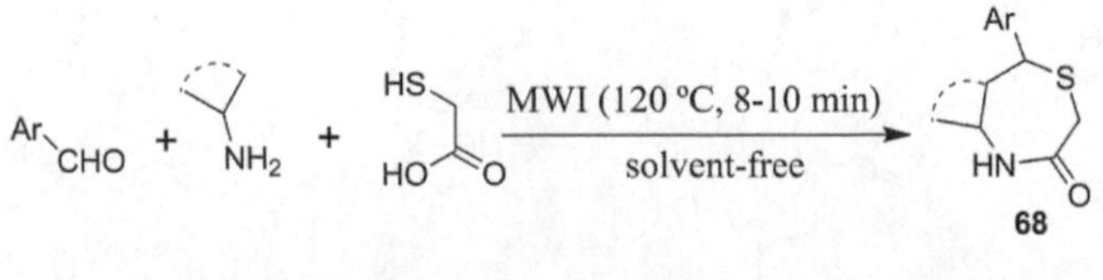

Ar = *p*-tolyl, *p*-anisyl, dimethoxyphenyl, 2-thiophenyl, trimethoxyphenyl, *p*-bromophenyl, *p*-chlorophenyl, , 2,4-dichlorophenyl, *o*-chlorophenyl, *m*-nitrophenyl

**SCHEME 10.27** Synthesis of fused 1,4-thiazepine (ε-lactams) derivatives.

HCT 116 (human colon carcinoma cell lines) and mice lymphocytes. The antioxidant activity was evaluated considering the scavenging capacity of 1,1-diphenyl-2-picryl-hydrazyl free radical (DPPH), superoxide anion, and hydroxyl free radicals. A few tested compounds showed strong capability for scavenging DPPH, superoxide anion, and hydroxyl free radicals in comparison to positive control L-ascorbic acid. Some of these compounds also inhibited the growth of HCT-116 cell lines more selectively than mice lymphocytes.

IRAP (insulin-regulated aminopeptidase) is a protein (a member of the family of zinc-dependent membrane aminopeptidases) which co-localizes and is translocated with GLUT4 (glucose transporter type 4) to the plasma membrane in response to insulin. The extracellular domain of IRAP is cleaved and released into the blood-stream. Therefore, IRAP plasma concentration could be a good marker of insulin sensitivity. It is well known that insulin's dramatic effect on glucose disposal is mediated through its action on insulin-responsive glucose transporter (GLUT4). Impaired IRAP action may thus play a role in the development of complications in type 2 diabetes [111]. Therefore, it is essential to design and synthesis effective IRAP inhibitors.

In medicinal chemistry and drug discovery research, the effect of dielectric heating (green heating) is not only confined to the synthesis of small heterocyclic ring systems, but has also expanded its application in many other areas. Anderson and co-workers [112] designed and synthesized novel 13- and 14-membered macrocyclic tripeptide analogs (macrolactams) of angiotensin IV by an olefin ring-closing metathesis (RCM) reaction through microwave-assisted macrocyclization to afford substituted dioxo-1,4-diazacyclotetradec-7-ene (potent IRAP inhibitors). The cyclization was performed on a preparative scale using HGII (Hoveyda-Grubbs second generation catalyst) (0.15 equiv.) at 150°C for 5 min and repeated once after a second addition of the catalyst (Scheme 10.28). The authors also assessed the impact of the ring size and the type (saturated vs. unsaturated), configuration, and position of the carbon-carbon bridge within the molecule. It was concluded that the ring size generally affects the potency more than the carbon-carbon bond characteristics. Removal of the carboxyl group in the C-terminal slightly reduced the potency. Inhibitors **69** ($K_i = 4.1$ nM) and **70** ($K_i = 1.8$ nM), both encompassing 14-membered ring systems connected to AMPAA, are 10-fold more potent than angiotensin IV and are also more selective over aminopeptidase N (AP-N). Both compounds displayed high stability against proteolysis by metallopeptidases. These both have the potential to become useful drugs in the future.

Microwave-assisted conversion has been used to accomplish a crucial step to synthesize high-affinity hNK1R antagonists, based on a seven-membered ε-lactam core unit [113]. Good selectivity over an affinity for the hERG ion channel (hIKr) was reported. In particular, lactam (**71**) was the most potent compound (*in vivo*) in this series (Figure 10.10).

An expeditious synthesis of a known γ-secretase inhibitor (fused seven-membered ε-lactam, **73**) was performed [114]. The key step of this synthesis, the formation of seven-membered diphenyl ε-lactam (**72**), was carried out under microwave irradiation. The target compound (**73**) was isolated from a mixture of four diastereomers by simple flash silica gel chromatography. Interestingly, there was no chiral chromatography in the entire sequence (Scheme 10.29).

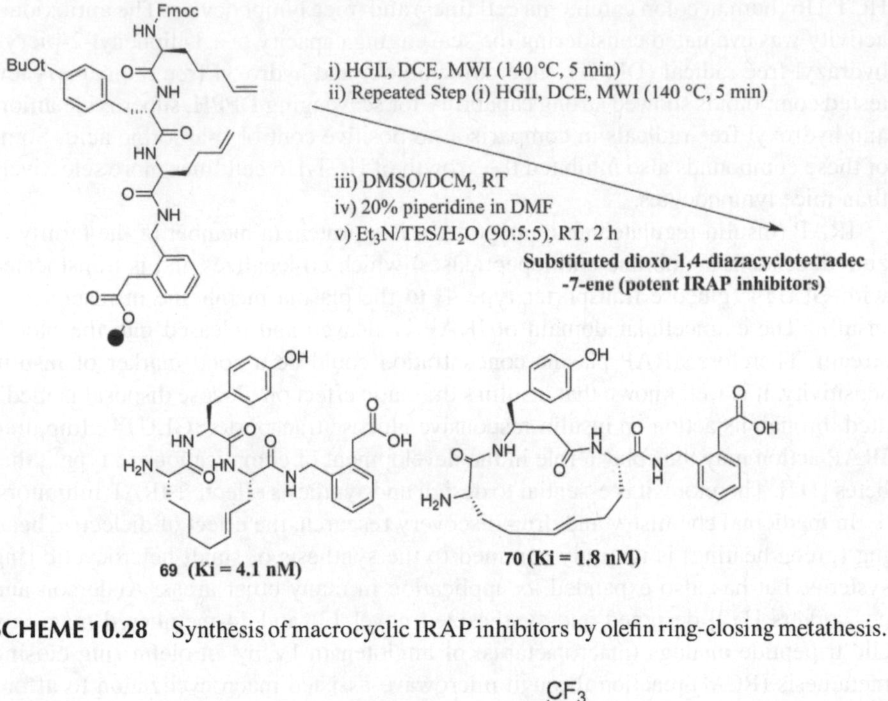
Fmoc
HN
BuOt
NH
HN
NH
i) HGII, DCE, MWI (140 °C, 5 min)
ii) Repeated Step (i) HGII, DCE, MWI (140 °C, 5 min)
iii) DMSO/DCM, RT
iv) 20% piperidine in DMF
v) Et3N/TES/H2O (90:5:5), RT, 2 h
Substituted dioxo-1,4-diazacyclotetradec
-7-ene (potent IRAP inhibitors)
OH
H2N
69 (Ki = 4.1 nM)
70 (Ki = 1.8 nM)

**SCHEME 10.28** Synthesis of macrocyclic IRAP inhibitors by olefin ring-closing metathesis.

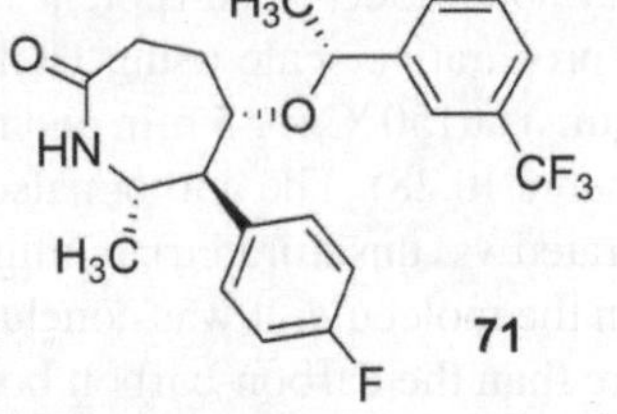
CF3
H3C
HN
CF3
H3C
F
71

**FIGURE 10.10** Seven-membered ε-lactam as hNK1R antagonist.

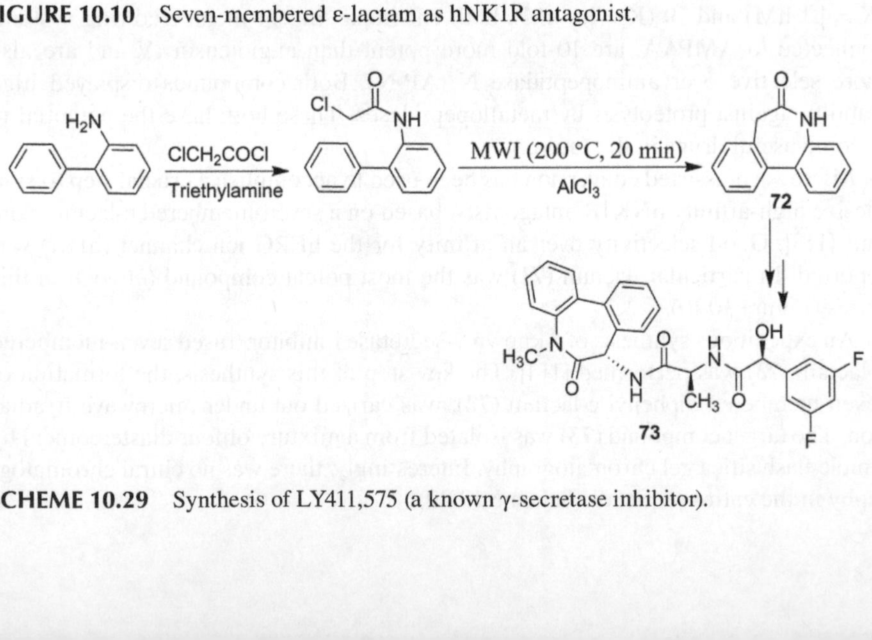
H2N
ClCH2COCl
Triethylamine
Cl
NH
MWI (200 °C, 20 min)
AlCl3
NH
72
H3C
N
OH
F
CH3
73
F

**SCHEME 10.29** Synthesis of LY411,575 (a known γ-secretase inhibitor).

### 10.3.2.8.2 Synthesis of Structurally Unique Lactams

Microwave-promoted, ionic liquid-supported rapid synthesis of δ-lactam-fused aza-pentacycles (77) has been reported [115]. This two-step, three-component proto-col includes Knoevenagel condensation of 2-cyanomethylbenzimidazole (74) with methyl-2-formylbenzoate (75) to produce an adduct (76) that on *cis-trans* isomeriza-tion, followed by [4 + 1] cycloaddition with an isocyanide, subsequent aromatization yielded the final product (Scheme 10.30).

Chen et al. reported the synthesis of δ-glyconolactams from δ-azido sugars through an intramolecular Schmidt–Boyer reaction in moderate to good (61–69%) yields [116] although the mechanism of product formation is not very clear. The reac-tion was conducted in 80% $CF_3COOH$ under microwave irradiation (50°C, 5 min).

Nitrile hydratases (NHase)-catalyzed selective conversion of aziridine-contain-ing nitrile (78) to 1° amide (79) and subsequent microwave-induced intramolecular aziridine-amide rearrangement to δ-lactams (80) *via* 5-*exo-trig* ring closure [117] (Scheme 10.31). It is well-known that enzymes play a major role in the synthesis of amino acids and in synthetic organic chemistry various enzymes are being used to develop numerous chemo-, regio-, and stereoselective processes under mild condi-tions in an aqueous medium and neutral pH [118].

Microwave-assisted $InCl_3$-catalyzed synthesis of N-substituted fused δ-lactams (82) was reported [119]. The uncatalyzed microwave-induced reaction produced 40–50% yield of the target compound whereas the concurrent influence of micro-wave and the Lewis acid ($InCl_3$) produced 81–91% yield. The higher yield of the product was hypothesized by the formation of a stable activated dipolar complex (81). Under the influence of the electromagnetic component of microwave irradia-tion, a dipole-dipole interaction was proposed that provided additional stability to the complex (81), as well as reduced the energy of activation. A plausible mechanism involves (i) condensation of the isatin with hydrazine hydrate to liberate water and

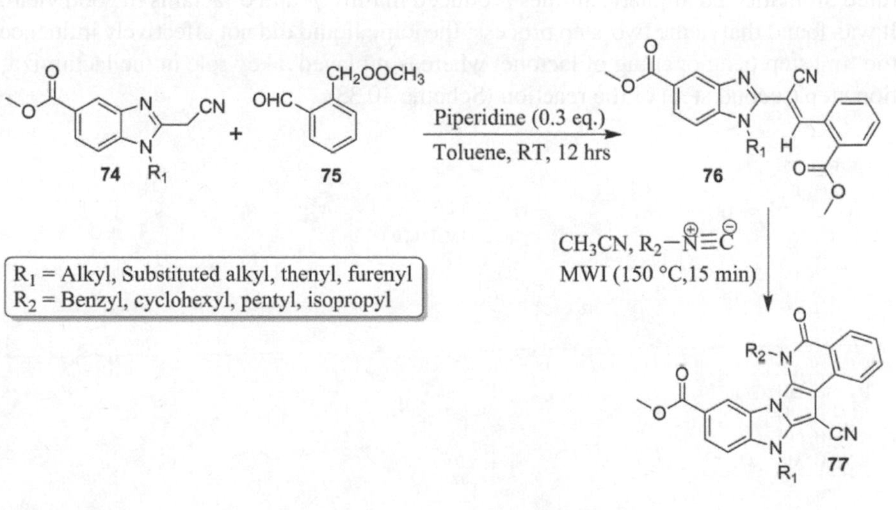

R₁ = Alkyl, Substituted alkyl, thenyl, furenyl
R₂ = Benzyl, cyclohexyl, pentyl, isopropyl

**SCHEME 10.30** Synthesis of δ-lactam-fused azapentacycles.

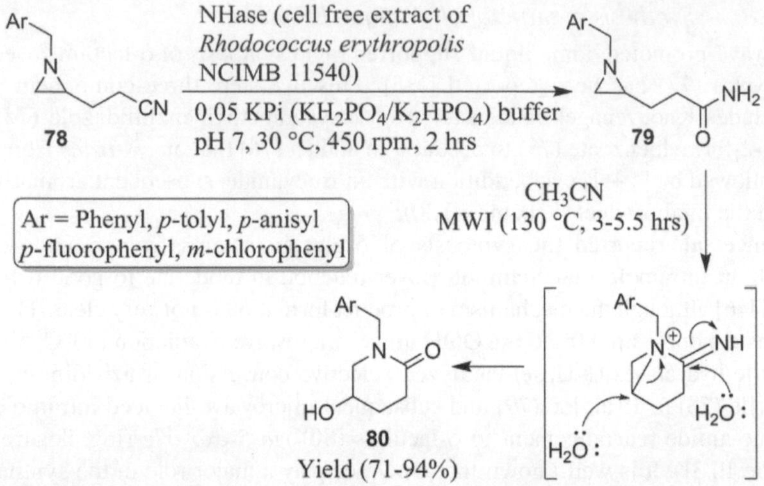

**SCHEME 10.31** Intramolecular aziridine-amide rearrangement to synthesize 5-hydroxy δ-lactams.

(ii) the subsequent dehydrative nucleophilic attack on the lactone carbonyl of coumarin to yield N-substituted lactams (Scheme 10.32).

Microwave-assisted, one-pot, ionic liquid-mediated lactamization of lactones has been reported [120]. A series of reactions were carried out involving primary amines (**83**) with diverse lactones (**84**) in the ionic liquid, bmim [BF₄ (1-butyl-3-methylimidazolium salt)]. No Brønsted acid was required to complete the process. The reported methodology was quite general as acid-sensitive functional groups remained unaffected and a wide range of primary amines (having weak to strong nucleophilicity) produced moderate to excellent yield of the corresponding lactams. Diversely substituted aromatic and aliphatic amines produced mainly γ- and δ-lactams in good yield. It was found that in the two-step process, the ionic liquid did not effectively influence the first step (ring opening of lactone) whereas it played a key role in the lactamization step (second step) of the reaction (Scheme 10.33).

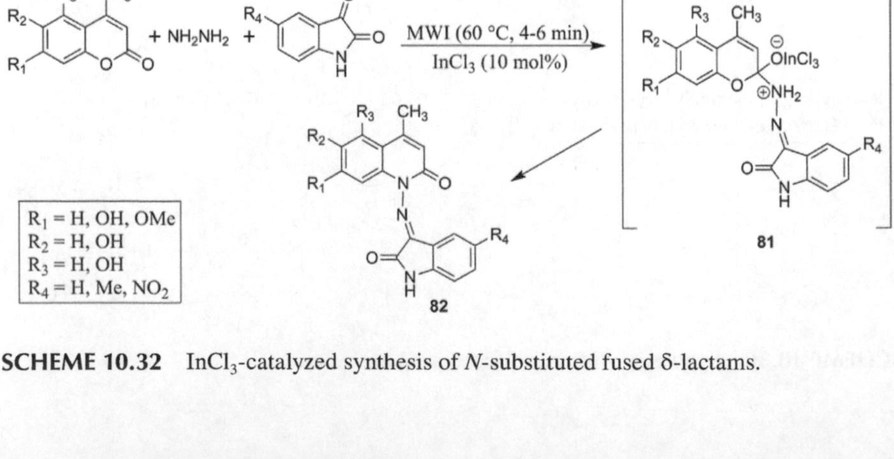

**SCHEME 10.32** InCl₃-catalyzed synthesis of N-substituted fused δ-lactams.

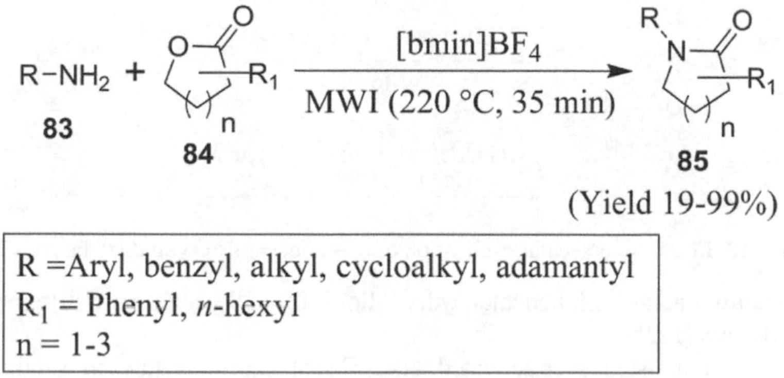

R-NH$_2$ + [lactone **84**, with O-C(=O), R$_1$, n] $\xrightarrow[\text{MWI (220 °C, 35 min)}]{\text{[bmin]BF}_4}$ [product **85**]

**83**      **84**                                              **85**

(Yield 19–99%)

R = Aryl, benzyl, alkyl, cycloalkyl, adamantyl
R$_1$ = Phenyl, n-hexyl
n = 1–3

**SCHEME 10.33** Microwave-assisted lactamization of lactones.

A microwave-assisted, three component domino reaction of aldehydes, enaminones, and malononitrile using ethylene glycol as a solvent was reported to yield fused polycyclic δ-lactams [121]. According to the authors, the volume of the solvent (ethylene glycol) was important in this procedure to achieve a higher yield. The optimum reaction condition includes an equimolar mixture of the three reactants (aldehyde, malononitrile, and enaminone, 1 mmol scale), in 2 mL ethylene glycol at 120°C under microwave exposure. It was observed that aromatic aldehydes with EWG accelerated the reaction, while the presence of EDG decreased the reactivity and took longer to complete the reaction (Scheme 10.34). The plausible mechanism involves condensation of the aldehyde with malononitrile to form 2-arylidenemalononitrile that attacks (electrophilic attack) the enaminone, and subsequent intramolecular cyclization and dehydration yielded the target compound (**86**).

Facile synthesis of 4- (β-lactam) to 10-membered lactam cores by intramolecular cyclization of α-iminoesters in the presence of microwave irradiation has been reported (Scheme 10.35). The substrate α-iminoesters were prepared by condensing

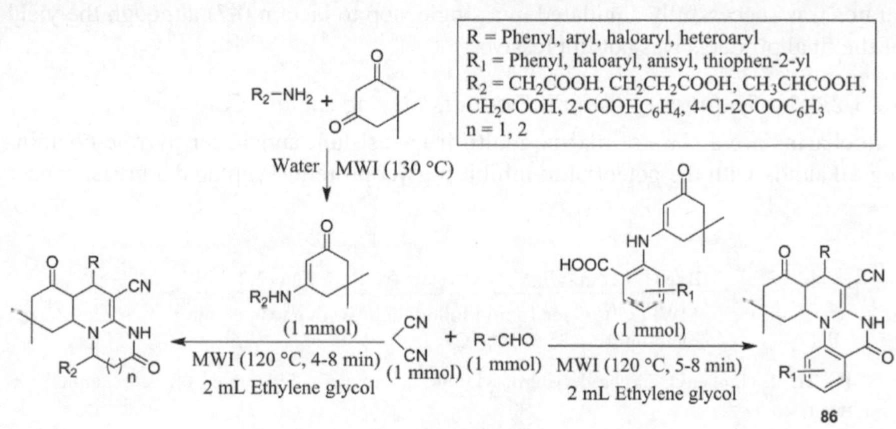

R$_2$–NH$_2$ + [cyclohexanedione]

R = Phenyl, aryl, haloaryl, heteroaryl
R$_1$ = Phenyl, haloaryl, anisyl, thiophen-2-yl
R$_2$ = CH$_2$COOH, CH$_2$CH$_2$COOH, CH$_3$CHCOOH, CH$_2$COOH, 2-COOHC$_6$H$_4$, 4-Cl-2COOC$_6$H$_3$
n = 1, 2

Water | MWI (130 °C)

[R$_2$HN-enaminone intermediate]

[product left, with CN, NH, O, R$_2$, n]  $\xleftarrow[\substack{\text{MWI (120 °C, 4–8 min)} \\ \text{2 mL Ethylene glycol}}]{\text{(1 mmol)}}$  $\underset{\text{(1 mmol)}}{\overset{\text{CN}}{\underset{\text{CN}}{\diagup}}}$ + R–CHO (1 mmol) $\xrightarrow[\substack{\text{MWI (120 °C, 5–8 min)} \\ \text{2 mL Ethylene glycol}}]{\text{(1 mmol)}}$ [product right, with CN, NH, O, R, R$_1$]

**86**

**SCHEME 10.34** Microwave-assisted synthesis of polycyclic δ-lactams.

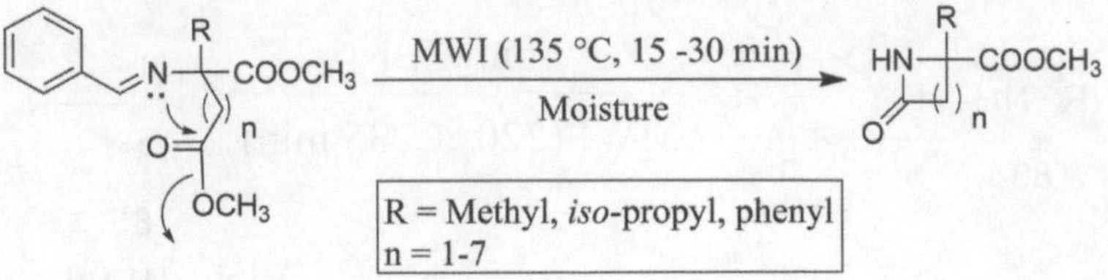

**SCHEME 10.35**   Intramolecular cyclization of α-iminoesters to synthesize lactams.

various amino acids with benzaldehyde, followed by alkylation with different bromoalkylesters [122].

Microwave-assisted over-activated alkali-Norit carbon catalyzed synthesis of N-substituted-γ-lactams under neat condition was studied [123]. A combined effect of the basicity of the carbon catalyst and microwave irradiation reinforced each other to produce the target compound in higher yield.

The use of hydrogen bromide as a catalyst in the synthesis of α,β-unsaturated γ-, δ-, and ε-lactams was studied by Yasui et al. [124]. Through this microwave-assisted procedure, δ-lactams have been synthesized within 5 min at 150°C. Besides δ-lactams, γ-, and ε-lactams were also synthesized in good yields exploring the concurrent catalytic effect of HBr and microwave irradiation.

Microwave-assisted conversion of β-lactams to 4/6-fused δ-lactams was carried out starting from N-vinyl-β-lactams. The plausible mechanism involves initial [3,3] sigmatropic rearrangement to form the eight-membered ring that eventually undergoes intramolecular electrocyclization at higher temperatures (160–200°C) to furnish the desired product. The presence of $Cs_2CO_3$ and a copper source, such as CuI or CuCl, were found to be essential for the reaction [125].

Single-step conversion of 1° amines to various lactams (5- to 7-membered) using a class of amine-derivatizing reagents, termed SPAn (Solid/solution-Phase Annulation) reagents, was reported [126]. Microwave heating was essential to obtain the tandem N-alkylation/intramolecular acylation as the subsequent annulation to yield the lactams (Scheme 10.36). A range of differently substituted 1° amine substrates was successfully annulated in a single step to lactam (**87**) although the yield of the final products was not impressive.

### 10.3.2.8.3   Synthesis of Natural Products Having Lactam Core

Lamellarins are a class of marine multi-drug resistant anticancer pyrrole-containing alkaloids with the potential to inhibit p-glycoprotein (P-gp) and a breast cancer

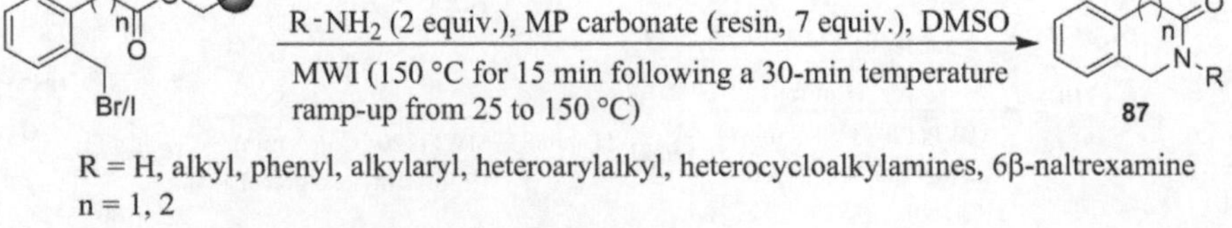

R = H, alkyl, phenyl, alkylaryl, heteroarylalkyl, heterocycloalkylamines, 6β-naltrexamine
n = 1, 2

**SCHEME 10.36**   Synthesis of lactams from SPAn reagents.

resistance protein (BCRP) efflux pump. Up to now, about 70 members (about 50 lamellarins and 20 related alkaloids) of this class have been reported although lamellarin D has been identified as the most effective anticancer agent in this class [127, 128]. Cu$^I$-mediated and microwave-assisted synthesis of azalamellarins (lactam analogs of lamellarins) has been reported [129]. A series of azalamellarins have been synthesized and evaluated against the cancer cell lines HuCCA-1, A-549, HepG2, and MOLT-3. The results showed that certain azalamellarins exhibited good anticancer activities in comparison to the IC$_{50}$ values to their parent lamellarin analog. The most potent compound of this series, azalamellarin D, a polyaromatic fused δ-lactam (**88**) was synthesized following a multi-step procedure among which three steps including the key step (the lactamization step), were accomplished under microwave radiation (Scheme 10.37).

A convenient three steps synthesis of (–)-desethyleburnamonine (**92**), a direct precursor of vindeburnol (antidepressant) was reported [130]. The reaction of 3-(2-bromo-ethyl)indole (**89**) and ethyl pyridin-3-yl acetate (**90**) produced the compound (**91**) that under microwave irradiation in the presence of a tertiary base DBU yielded (–)-desethyleburnamonine. An allylamine–enamine isomerization, followed by a Pictet–Spengler condensation and subsequent lactamization were proposed to obtain the final product (**92**) in 52% overall yield (Scheme 10.38).

Quinocitrinines A and B are natural alkaloids (found in *Penicillium citrinum*) that possess anticancer and antimicrobial activities [131–133]. An attempt to synthesize

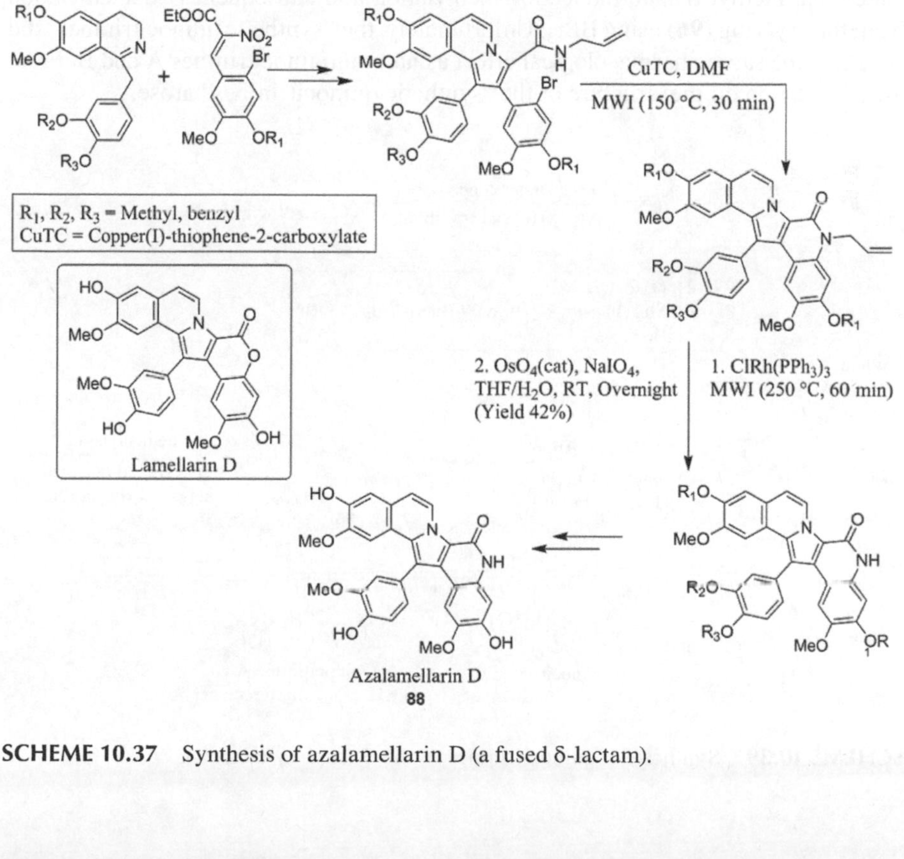

**SCHEME 10.37** Synthesis of azalamellarin D (a fused δ-lactam).

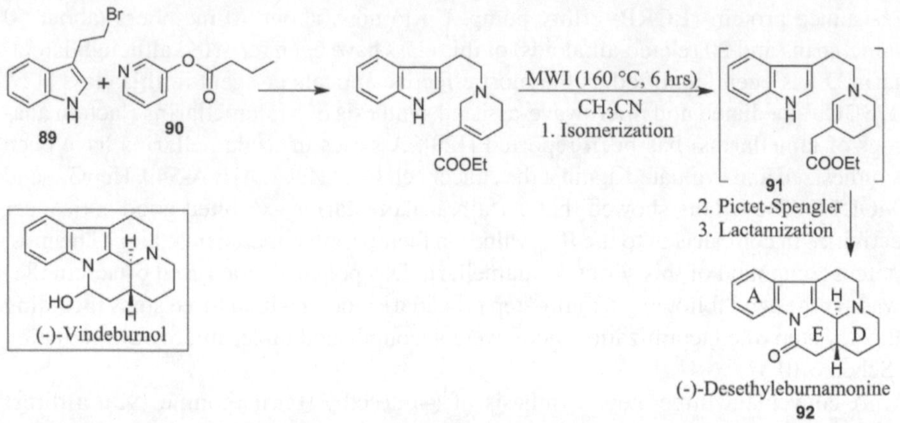

**SCHEME 10.38**   Synthesis of (–)-desethyleburnamonine (a fused δ-lactam).

quinocitrinines A and B has been reported that includes Friedlander condensation, followed by *N*-methylation and *O*-demethylation (Scheme 10.39). Microwave
irradiation was used to accomplish the key step (Friedlander condensation) in this
synthesis [133]. The first step of this three-step procedure comprised of microwave
assisted Friedlander condensation between (**93**) and (**94**) to generate (**95**) which
underwent methyl triflate-induced *N*-methylation and subsequent *O*-demethylation
from the aryl ring (**96**) using $BBr_3$. Unfortunately, the "synthetic quinocitrinines" did
not show the same pharmacological effect as natural quinocitrinines A and B. From
this, a question on the structure of the "synthetic quinocitrinines" arose.

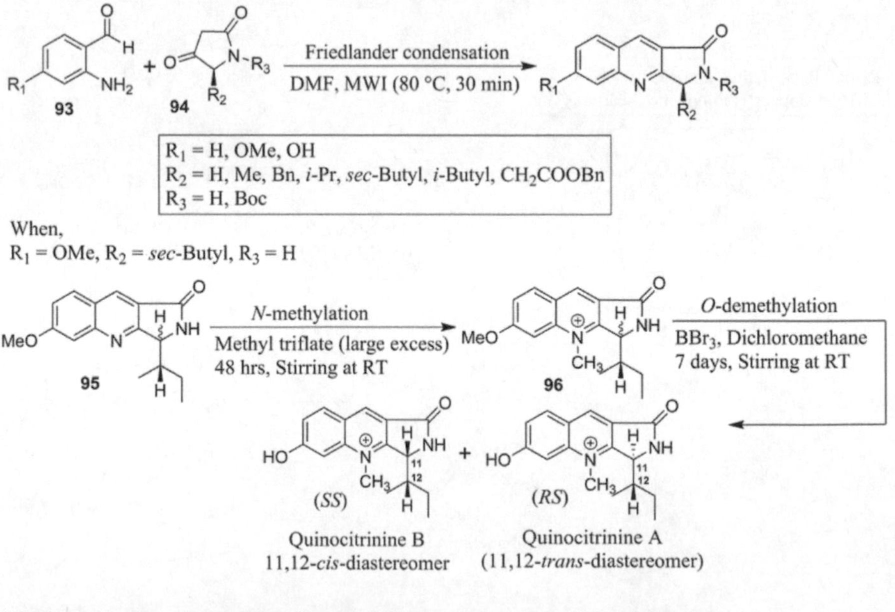

**SCHEME 10.39**   Synthesis of quinocitrinines A and B (fused γ-lactams).

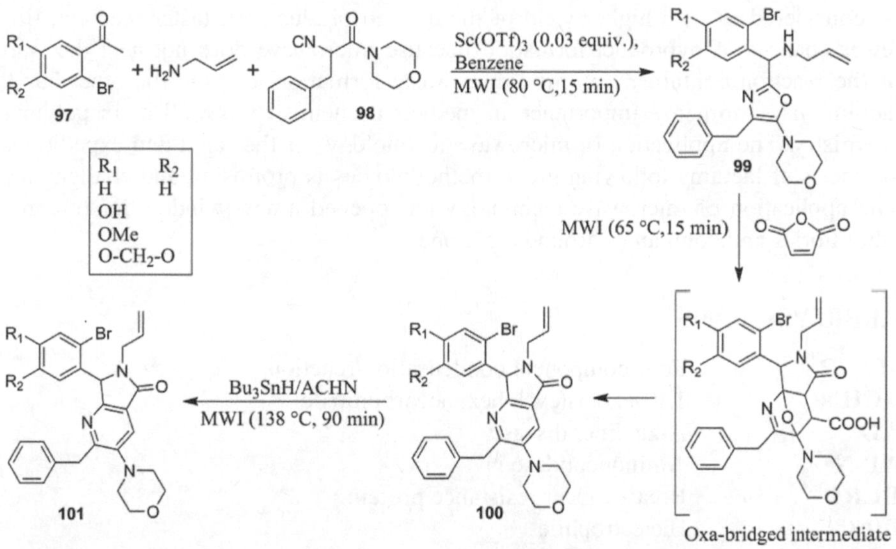

**SCHEME 10.40** Synthesis of aza-nuevamine analogs (fused γ-lactams).

Nuevamine, the first reported isoindoloisoquinoline lactam alkaloid [134], and its derivatives have attracted the interest of medicinal chemists because of their manifold biological activities that include anti-inflammatory, anti-microbial, anti-leukemic, and anti-tumoral [130]. A series of aza-analogs of nuevamine was prepared under microwave irradiation following a four-component sequential reaction that involves aromatic aldehydes, allylamine, isocyanoacetamide (isonitrile), and maleic anhydride in the presence of catalytic amount of Sc(OTf)$_3$ and subsequent intramolecular free radical cyclization using ACHN [1,10-azobis(cyclohexanecarbonitrile)] as the initiator along with tributyltinhydride (Bu$_3$SnH). In this multistep synthesis, the allylamine and o-bromobenzaldehyde (**97**) initially reacted and then the sequential addition of the isonitrile (**98**) yielded (**99**). In the next step, maleic anhydride and (**99**) were irradiated in the presence of Bu$_3$SnH and 1,10-azobis(cyclohexanec arbonitrile) (ACHN) to produce the desired product (**101**) along with a mixture of diastereoisomers.

The Diels–Alder cycloaddition between (**99**) and maleic anhydride under microwave radiation produced pyrrolopyridinone (**100**) *via* the oxa-bridged intermediate (Scheme 10.40) [135].

## 10.4 CONCLUSION

Advancement of green techniques and sustainability of the chemical process are valuable characteristics in green chemistry. Microwave heating, one of the major green techniques being used to synthesize thousands of medicinally-privileged molecules including lactams. A huge number of pharmacologically active lactams have been synthesized during the past 30 years and the application of microwave reactors has drawn growing attention from academia and industry. The primary reasons may be considered as (i) a higher yield of the desired product, (ii) faster reaction, (iii) fewer chances of byproduct formation because microwave does not heat the wall of the reaction container, (iv) minimum waste formation, etc. On the other hand, lactams have immense importance in medicinal chemistry as well as in polymer chemistry. The application of microwave technology for the rapid and expeditious synthesis of lactams following green methodologies is promising and challenging. The application of microwave technology has opened a new window for research laboratories and companies around the globe.

## ABBREVIATIONS

| | |
|---|---|
| **4CC/R** | Four-component condensation/reaction |
| **ACHN** | 1,10-azobis(cyclohexanecarbonitrile) |
| **AD** | Alzheimer disease |
| **AP-N** | Aminopeptidase N |
| **BCRP** | Breast cancer resistance protein |
| **BiE** | Bielectrophile |
| **BiNu** | Binucleophile |
| **CNS** | Central nervous system |
| **CuTC** | Copper(I) thiophene-2-carboxylate |
| **CYP** | Cytochrome P |
| **DA** | Diels–Alder |
| **DBU** | 1,8-diazabicyclo[5.4.0]undec-7-ene |
| **DC** | Dipolar cycloaddition |
| **DFT** | density functional theory |
| **DMF** | Dimethylformamide |
| **DoE** | Design of Experiments |
| **DPPH** | Diphenyl-2-picrylhydrazyl |
| **DMSO** | Dimethyl sulfoxide |
| **EDG** | Electron donating group |
| **eV** | Electron volt |
| **EWG** | Electron withdrawing group |
| **GHRP-6** | Growth hormone-releasing peptide 6 |
| **GHz** | Gigahertz |
| **GLUT4** | Glucose transporter type 4 |
| **HDA** | Hetero Diels–Alder |
| **HGII** | Hoveyda-Grubbs second generation catalyst |
| **HIV** | Human immunodeficiency virus |
| **IC** | Inhibitory concentration |
| **IL** | Ionic liquid |
| **IRAP** | Insulin-regulated aminopeptidase |
| **L-ascorbic acid** | Levo-ascorbic acid |
| **Li-TryR** | Leishmania infantum trypanothione reductase |
| **MCR** | Multicomponent reaction |
| **MDR** | Multidrug-resistant |
| **MEC** | Microwave-enhanced chemistry |

| MHz | Megahertz |
|---|---|
| MIC | Minimum inhibitory concentration |
| MORE | Microwave-assisted organic reaction enhancement |
| MWI | Microwave irradiation |
| NHase | Nitrile hydratases |
| NMR | Nuclear magnetic resonance |
| NTD | Neglected tropical diseases |
| P-gp | P-glycoprotein |
| Ph | Phenyl |
| PTSD | Post-traumatic stress disorder |
| RCM | Ring-closing metathesis |
| SAR | Structure–activity relationship |
| SN | Substitution nucleophilic |
| $S_N Ar$ | Substitution nucleophilic aromatic |
| SPAn | Solid/solution-phase annulation |
| TB | Tuberculosis |
| TFE | 2,2,2-Trifluoroethanol |
| Triflate | Trifluoromethanesulfonate |
| WHO | World Health Organization |

## REFERENCES

1. Zuidhof KT, de Croon, MHJM, Schouten, JC, Tinge JT (2015) Density, viscosity, and surface tension of liquid phase Beckmann rearrangement mixtures. *J. Chem. Eng. Data* 60: 1056–1062.
2. Cardellina JH, Moore RE (1979) The structures of pukeleimides A, B, D, E, F, and G. *Tetrahedron Lett.* 22: 2007–2010.
3. Charles JS, Franz-Josef M, Cardellina JH, Moore RE, Seff K (1979) Pukeleimide C, a novel pyrrolic compound from the marine cyanophyte *Lyngbya majuscula*. *Tetrahedron Lett.* 22: 2003–2006.
4. Cardellina JH, Marner FJ, Moore RE (1979) Malyngamide A, a novel chlorinated metabolite of the marine cyanophyte *Lyngbya majuscule*. *J. Am. Chem. Soc.* 101: 240–242.
5. Milligan KE, Marquez B, Williamson TR, Davies-Coleman M, Gerwick WH (2000) Two new malyngamides from a Madagascan *Lyngbya majuscule*. *J. Nat. Prod.* 63: 965–968.
6. Suntornchashwej S, Suwanborirux K, Koga K, Isobe M (2007) Malyngamide X: The first (7R)-lyngbic acid that connects to a new tripeptide backbone from the Thai sea hare *Bursatella leachii*. *Chem. Asian J.* 2: 114–122.
7. Kan, Y, Sakamoto B, Fujita T, Nagai H (2000) New malyngamides from the Hawaiian cyanobacterium *Lyngbya majuscule*. *J. Nat. Prod.* 63: 1599–1602.
8. Natio T, Honda M, Miyata O, Ninomiya I (1993) Total syntheses of (±)-anantine and (±)-isoanantine via thiyl radical addition-cyclization reaction. *Chem. Pharm. Bull.* 41: 217–219.
9. Zhang Z, Di YT, Wang YH, Zhang Z, Mu SZ, Fang X, Zhang Y, Tan CJ, Zhang Q, Yan XH, Guo J, Li CS, Hao XJ (2009) Gelegamines A-E: five new oxindole alkaloids from *Gelsemium elegans*. *Tetrahedron* 65: 4551–4556.
10. Kogure N, Nishiya C, Kitajima M, Takayama H (2005) Six new indole alkaloids from *Gelsemium sempervirens* Ait. f. *Tetrahedron Lett.* 46: 5857–5861.

11. Xu L, Wu P, Wright S, Du L, Wei X (2015) Bioactive polycyclic tetramate macrolactams from *Lysobacter enzymogenes* and their absolute configurations by theoretical ECD calculations. *J. Nat. Prod.* 78: 1841–1847.

12. Kudo F, Miyanaga A, Eguchi T (2014) Biosynthesis of natural products containing β-amino acids. *Nat. Prod. Rep.* 31: 1056–1073.

13. Anastas PT, Warner JC (1998) *Green Chemistry, Theory and Practice.* Oxford University Press, Oxford, UK.

14. Bandyopadhyay D, Rhodes E, Banik BK (2013) A green, chemoselective, and practical approach toward N-(2-azetidinonyl) 2,5-disubstituted pyrroles. *RSC Adv.* 3: 16756–16764.

15. Pozar DM (1993) *Microwave Engineering.* Addison–Wesley Publishing Company, Boston, USA.

16. Vanderhoff JW (1969). Microwave-induced polymerization. U.S. Pat. 3,432,413 [C.A. 70: 97422v, 1969].

17. Hesek JA, Wilson RC (1974) Practical analysis of high-purity chemicals. X. Use of a microwave oven in in-process control. *Anal. Chem.* 46: 1160–1160.

18. Abu-Samra A, Morris JS, Koirtyohann SR (1975) Wet ashing of some biological samples in a microwave oven. *Tr. Sub. Env.* 9: 297–301.

19. Abu-Samra A, Morris JS, Koirtyohann SR (1975) Wet ashing of some biological samples in a microwave oven. *Anal. Chem.* 47: 1475–1477.

20. Nadkarni RA (1984) Applications of microwave oven sample dissolution in analysis. *Anal. Chem.* 56: 2233–2237.

21. Smith F, Cousins B, Bozic J, Flora W (1985) The acid dissolution of sulfide mineral ores using a microwave oven. IUPAC Symposium of Analytical Chemistry in the Exploration, Mining and Processing Materials, Pretoria.

22. Wan JKS, Wolf K, Heyding RD (1984) Some chemical aspects of the microwave-assisted catalytic hydro-cracking processes. In *Catalysis on the Energy Scene* (S. Kaliaguine and K. Mahay Eds.), pp. 561–568. Elsevier Science Publishers, Amsterdam.

23. Bacci M, Bini M, Checcucci A, Ignesti A, Millanta L, Rubino N, Vanni R (1981) Microwave heating for the rapid determination of thermodynamic functions of chemical reactions. *J. Chem. Soc., Faraday Trans.* 1, 77: 1503–1509.

24. Gedye R, Smith F, Westaway K, Ali H, Baldisera L, Laberge L, Rousell J (1986) The use of microwave ovens for rapid organic synthesis. *Tetrahedron Lett.* 27: 279–282.

25. Giguere RJ, Bray TL, Duncan SM, Majetich G (1986) Application of commercial microwave ovens to organic synthesis. *Tetrahedron Lett.* 27: 4945–4948.

26. Blackwell HE (2003) Out of the oil bath and into the oven—Microwave-assisted combinatorial chemistry heats up. *Org. Biomol. Chem.* 1: 1251–1255.

27. Kidwai, M (2006) Green chemistry trends toward sustainability. *Pure Appl. Chem.* 78: 1983–1992.

28. Bandyopadhyay D, Mukherjee S, Banik BK (2010) An expeditious synthesis of N-substituted pyrroles via microwave-induced iodine-catalyzed reaction under solventless conditions. *Molecules* 15: 2520–2525.

29. Rivera S, Bandyopadhyay D, Banik BK (2009) Facile synthesis of N-substituted pyrroles via microwave-induced bismuth nitrate-catalyzed reaction under solventless conditions. *Tetrahedron Lett.* 50: 5445–5448.

30. Bandyopadhyay D, Mukherjee S, Rodriguez RR, Banik BK (2010) An effective microwave-induced iodine-catalyzed method for the synthesis of quinoxalines *via* condensation of 1,2-dicarbonyl compounds. *Molecules* 15: 4207–4212.

31. Kranjc K, Kočevar M (2010) Microwave-assisted organic synthesis: General considerations and transformations of heterocyclic compounds. *Curr. Org. Chem.* 14: 1050–1074.

32. Bandyopadhyay D, Maldonado S, Banik BK (2012) A microwave-assisted bismuth nitrate-catalyzed unique route toward 1,4-dihydropyridines. *Molecules* 17: 2643–2662.
33. Bandyopadhyay D, Cruz J, Morales LD, Arman, HD, Cuate E, Lee YS, Banik BK, Kim, DJ (2013) A green approach toward quinoxalines and *bis*-quinoxalines and their biological evaluation against A431, human skin cancer cell lines. *Future Med. Chem.* 5: 1377–1390.
34. Martins MAP, Frizzo CP, Moreira DN, Buriol L, Machado P (2009) Solvent-free heterocyclic synthesis. *Chem. Rev.* 109: 4140–4182.
35. Bandyopadhyay D, Mukherjee S, Granados JC, Short JD, Banik BK (2012) Ultrasound-assisted bismuth nitrate-induced green synthesis of novel pyrrole derivatives and their biological evaluation as anticancer agents. *Eur. J. Med. Chem.* 50: 209–215.
36. Staudinger H (1907) Ketenes. 1. *Diphenylketene Liebigs Ann.* 356: 51–123.
37. Bose AK, Manhas MS, Ghosh M, Shah M, Raju VS, Bari SS, Newaz SN, Banik BK, Chaudhary AG, Barakat KJ (1991) Microwave-induced organic reaction enhancement chemistry. 2. Simplified techniques. *J. Org. Chem.* 56: 6968–70.
38. Banik BK, Manhas MS, Kaluza Z, Barakat KJ, Bose AK (1992) Microwave-induced organic reaction enhancement chemistry. 4. Convenient synthesis of enantiopure α-hydroxy-β-lactams. *Tetrahedron Lett.* 33: 3603–3606.
39. Texier-Boullet F, Latouche R, Hamelin J (1993) Synthesis in dry media coupled with microwave irradiation: application to the preparation of β-aminoesters and β-lactams via silyl ketene acetals and aldimines. *Tetrahedron Lett.* 34: 2123–2126.
40. Banik BK, Manhas MS, Newaz SN, Bose AK (1993) Studies on lactams. 92. Facile preparation of carbapenem synthons *via* microwave-induced rapid reaction. *Bioorg. Med. Chem. Lett.* 3: 2363–2368.
41. Laurent A, Jacquault P, Di Martino JL, Hamelin J (1995) Fast synthesis of amino acid salts and lactams without solvent under microwave irradiation. *J. Chem. Soc., Chem. Commun.* 11: 1101–1101.
42. Martelli G, Spunta G, Panunzio M (1998) Microwave-assisted solvent-free organic reactions: synthesis of β-Lactams from 1,3-azadienes. *Tetrahedron Lett.* 39: 6257–6260.
43. Cossio FP, Arrieta A, Sierra MA (2007) The mechanism of the ketene-imine (Staudinger) reaction in its centennial: still an unsolved problem? *Acc. Chem. Res.* 41: 925–936.
44. Jiao L, Liang Y, Xu J (2006) Origin of the relative stereoselectivity of the β-lactam formation in the Staudinger reaction. *J. Am. Chem. Soc.* 128: 6060–6069.
45. Banik BK, Banik I, Becker FF (2005) Stereocontrolled synthesis of anticancer β-lactams *via* the Staudinger reaction. *Bioorg. Med. Chem.* 13: 3611–3622.
46. Bandyopadhyay D, Banik BK (2010) Microwave-induced stereoselectivity of β-lactam formation with dihydrophenanthrenyl imines *via* Staudinger cycloaddition. *Helv. Chim. Acta* 93: 298–301.
47. Bandyopadhyay D, Yañez M, Banik BK (2011) Microwave-induced stereoselectivity of β-lactam formation: effects of solvents. *Heterocycl. Lett.* 1(spl. Issue): 65–67.
48. Mehta PD, Sengar NPS, Pathak AK (2010) 2-Azetidinone—A new profile of various pharmacological activities. *Eur. J. Med. Chem.* 45: 5541–5560.
49. Banik I, Banik BK (2013) Synthesis of β-lactams and their chemical manipulations via microwave-induced reactions. In *Topics in Heterocyclic Chemistry*, Vol. 30 (β-Lactams) (B. K. Banik, Ed.), pp 183–222. Springer, Germany.
50. Banik BK, Banik I, Becker FF (2010) Novel anticancer β-lactams. In *Topics in Heterocyclic Chemistry*, Vol. 22 (Heterocyclic Scaffolds I) (B. K. Banik, Ed.), pp. 349–373. Springer, Germany.
51. Dubey A, Srivastava SK, Srivastava SD (2011) Conventional and microwave assisted synthesis of 2-oxo-4-substituted aryl-azetidine derivatives of benzotriazole: A new class of biological compounds. *Bioorg. Med. Chem. Lett.* 21: 569–573.

52. Ali P, Meshram J, Tiwari V (2010) Microwave mediated cyclocondensation of 2-aminothiazole into β-lactam derivatives: Virtual screening and *in vitro* antimicrobial activity with various microorganisms. *Int. J. ChemTech Res.* 2: 956–964.

53. Pagadala R, Meshram JS, Chopde HN, Jetti V, Udayini V (2011) An expeditious one-pot synthesis of substituted phenylazetidin-2-ones in the presence of zeolite. *J. Heterocycl. Chem.* 48: 1067–1072.

54. Malhotra G, Gothwal P, Srivastava YK (2011) Microwave induced synthesis of some biologically active azetidinones. *Der Chem. Sin.* 2: 47–50.

55. Mermer A, Demirci S, Ozdemir SB, Demirbas A, Ulker S, Ayaz FA, Aksakal F, Demirbas N (2017) Conventional and microwave irradiated synthesis, biological activity evaluation and molecular docking studies of highly substituted piperazine-azole hybrids. *Chin. Chem. Lett.* 28: 995–1005.

56. Demirci S, Mermer A, Ak G, Aksakal F, Colak N, Demirbas A, Ayaz FA, Demirbas, N (2017) Conventional and microwave assisted total synthesis, antioxidant capacity, biological activity, and molecular docking studies of new hybrid compounds. *J. Heterocycl. Chem.* 54(3): 1785–1805.

57. Pawar PY, Gaikwad PM, Balani PH (2011) Microwave-assisted synthesis of N-substituted-7-hydroxy-4-methyl-2-oxo-quinolines as anticonvulsant agents. *Orbital: Electron. J. Chem.* 8: 945–951.

58. Bandyopadhyay D, Cruz J, Banik BK (2012) Novel synthesis of 3-pyrrole substituted β-lactams *via* microwave-induced bismuth nitrate-catalyzed reaction. *Tetrahedron* 68:10686–10695.

59. Bandyopadhyay D, Cruz J, Yadav RM, Banik BK (2012) An expeditious iodine-catalyzed synthesis of 3-pyrrole-substituted 2-azetidinones. *Molecules* 17: 11570–11584.

60. Bandyopadhyay D, Rivera G, Salinas I, Aguilar H, Banik BK (2010) Iodine-catalyzed remarkable synthesis of novel *N*-polyaromatic β-lactams bearing pyrroles. *Molecules* 15: 1082–1088.

61. Diels O, Alder K (1928) Synthesen in der hydroaromatischen Reihe. *Justus Liebigs Ann. Chem.* 460: 98–122.

62. Margrey KA, Hazzard AD, Scheerer JR (2014) Synthesis of 2-pyridones by cycloreversion of [2.2.2]-bicycloalkene diketopiperazines. *Org. Lett.* 16: 904–907.

63. Komiyama K, Takamatsu S, Kim YP, Matsumoto A, Takahashi Y, Hayashi M, Woodruff HB, Omura S (1995) Louisianins A, B, C and D: non-steroidal growth inhibitors of testosterone-responsive SC 115 cells. I. Taxonomy, fermentation, isolation and biological characteristics. *J. Antibiot.* 48: 1086–1089.

64. Long S, Monari M, Panunzio M, Bandini E, D'Aurizio A, Venturini A (2011) Hetero-Diels–Alder (HDA) strategy for the preparation of 6-aryl- and heteroaryl-substituted piperidin-2-one scaffolds: Experimental and theoretical studies. *Eur. J. Org. Chem.* 31: 6218–6225.

65. Kaval N, Dehaen W, Kappe CO, Van der Eycken E (2004) The effect of pressure on microwave-enhanced Diels–Alder reactions. A case study. *Org. Biomol. Chem.* 2: 154–156.

66. Zamudio-Medina A, Garcia-Gonzalez MC, Gutierrez-Carrillo A, Gonzalez-Zamora E (2015) Synthesis of cyclic analogues of hexamethylenebis(3-pyridine(amide (HMBPA) in a one-pot process. *Tetrahedron Lett.* 56: 627–629.

67. Benowitz NL, Hukkanen J, Jacob P (2009) Nicotine chemistry, metabolism, kinetics and biomarkers. *Handb. Exp. Pharmacol.* 192: 29–60.

68. Polindara-Garcia LA, Montesinos-Miguel D, Vazquez A (2015) An efficient microwave-assisted synthesis of cotinine and iso-cotinine analogs from an Ugi-4CR approach. *Org. Biomol. Chem.* 13: 9065–9071.

69. Echeverria V, Zeitlin R, Burgess S, Patel S, Barman A, Thakur G, Mamcarz M, Wang L, Sattelle DL, Kirschner DA, Mori T, Leblanc RM, Prabhakar R, Arendash GW (2011) Cotinine reduces amyloid-β aggregation and improves memory in Alzheimer's disease mice. *J. Alzheimers Dis.* 24: 817–835.

70. Gao J, Adam BL, Terry AV (2014) Evaluation of nicotine and cotinine analogs as potential neuroprotective agents for Alzheimer's disease. *Bioorg. Med. Chem. Lett.* 24: 1472–1478.
71. Echeverria V, Zeitlin R (2012) Cotinine: A potential new therapeutic agent against Alzheimer's disease. *CNS Neurosci. Ther.* 18: 517–523.
72. Grizzell JA, Iarkov A, Holmes R, Mori T, Echeverria V (2014) Cotinine reduces depressive-like behavior, working memory deficits, and synaptic loss associated with chronic stress in mice. *Behav. Brain Res.* 268: 55–65.
73. Terry AV, Hernandez C, Hohnadel EJ, Bouchard KP, Buccafusco JJ (2005) Cotinine, a neuroactive metabolite of nicotine: potential for treating disorders of impaired cognition. *CNS Drug Rev.* 11: 229–252.
74. Terry AV, Callahan PM, Bertrand D (2015) *R*-(+) and *S*-(–) isomers of cotinine augment cholinergic responses *in vitro* and *in vivo*. *J. Pharmacol. Exp. Ther.* 352: 405–418.
75. Gamez-Montano R, Ibarra-Rivera T, El Kaim L, Miranda LD (2010) Efficient synthesis of azaspirodienones by microwave-assisted radical spirocyclization of xanthate-containing Ugi adducts. *Synthesis* 8:1285–1290.
76. Jida M, Malaquin S, Deprez-Poulain R, Laconde G, Deprez B (2010) Synthesis of five- and six-membered lactams via solvent-free microwave Ugi reaction. *Tetrahedron Lett.* 51: 5109–5111.
77. Tye H, Whittaker M (2004) Use of a Design of Experiments approach for the optimization of a microwave assisted Ugi reaction. *Org. Biomol. Chem.* 2: 813–815.
78. Meyers AI, Burgess LE (1991) A simple asymmetric synthesis of 2-substituted pyrrolidines from 3-acylpropionic acids. *J. Org. Chem.* 56: 2294–2296.
79. Burgess LE, Meyers AI (1992) A simple asymmetric synthesis of 2-substituted pyrrolidines and 5-substituted pyrrolidinones. *J. Org. Chem.* 57: 1656–1662.
80. Meyers AI, Snyder L (1993) The synthesis of aracemic 4-substituted pyrrolidinones and 3-substituted pyrrolidines. An asymmetric synthesis of (-)-rolipram. *J. Org. Chem.* 58: 36–42.
81. Meyers AI, Andres CJ, Resek JE, Woodall CC, McLaughlin MA, Lee PH, Price DA (1999) Asymmetric routes to aza sugars from chiral bicyclic lactams. Synthesis of 1,4-dideoxy-1,4-imino-D-lyxitol; L-deoxymannojirimycin; rhamno-1-deoxynojirimycin and 1-deoxy-6-epicastanospermine. *Tetrahedron* 55: 8931–8952.
82. Groaning MD, Meyers AI (1999) An asymmetric synthesis of the key precursor to (-)-indolizomycin. *Tetrahedron Lett.* 40: 4639–4642.
83. Groaning MD, Meyers AI (2000) Chiral non-racemic bicyclic lactams. Auxiliary-based asymmetric reactions. *Tetrahedron* 56: 9843–9873.
84. Jida M, Deprez-Poulain R, Malaquin S, Roussel P, Agbossou-Niedercorn F, Deprez B, Laconde G (2010) Solvent-free microwave-assisted Meyers' lactamization. *Green Chem.* 12: 961–964.
85. Postikova S, Sabbah M, Wightman D, Nguyen IT, Sanselme M, Besson T, Briere JF, Oudeyer S, Levacher V (2013) Developments in Meyers' lactamization methodology: En route to bi(hetero)aryl structures with defined axial chirality. *J. Org. Chem.* 78: 8191–8197.
86. Piotrowska DG, Bujnowicz A, Wroblewski AE, Glowacka IE (2015) A new approach to the synthesis of 4-phosphonylated β-lactams. *Synlett* 26: 375–379.
87. Meng LJ, Zuo H, Vijaykumar BVD, Dupati G, Choi KM, Jang K, Yoon YJ, Shin DS (2013) Microwave synthesis of chiral *N*-benzyl-2-methyl-2*H*-benzo[*b*][1,4]oxazin/thiazin-3(4*H*)-ones *via* Smiles rearrangement and their biological evaluation. *Bull. Korean Chem. Soc.* 34: 585–589.
88. Ghahremanzadeh R, Rashid Z, Zarnani AH, Naeimi H (2014) A rapid and high efficient microwave promoted multicomponent domino reaction for the synthesis of spiro-oxindole derivatives. *J. Ind. Eng. Chem.* 20: 4076–4084.

89. Hayashi K, Choshi T, Chikaraishi K, Oda A, Yoshinaga R, Hatae N, Ishikura M, Hibino S (2012) A novel total synthesis of isocryptolepine based on a microwave-assisted tandem Curtius rearrangement and aza-electrocyclic reaction. *Tetrahedron* 68: 4274–4279.

90. McGonagle FI, Brown L, Cooke A, Sutherland A (2011) Microwave- promoted tandem reactions for the synthesis of bicyclic γ-lactams. *Tetrahedron Lett.* 52: 2330–2332.

91. Presset M, Coquerel Y, Rodriguez J (2010) Periselectivity switch of acylketenes in cycloaddition with 1-azadienes: Microwave-assisted diastereoselective domino three-component synthesis of α-spiro-δ-lactams. *Org. Lett.* 12: 4212–4215.

92. Vanderwal CD (2011) Reactivity and synthesis inspired by the Zincke ring-opening of pyridines. *J. Org. Chem.* 76: 9555–9567.

93. Steinhardt SE, Vanderwal CD (2009) Complex polycyclic lactams from pericyclic cascade reactions of Zincke aldehydes. *J. Am. Chem. Soc.* 131: 7546–7547.

94. Santra S, Andreana PR (2007) A one-pot, microwave-influenced synthesis of diverse small molecules by multicomponent reaction cascades. *Org. Lett.* 9: 5035–5038.

95. Zanobini A, Brandi A, de Meijere A (2006) A new three-component cascade reaction to yield 3-spirocyclopropanated β-lactams. *Eur. J. Org. Chem.* 5:1251–1255.

96. Chen X, Fan H, Zhang S, Yu C, Wang W (2016) Facile installation of 2-reverse prenyl functionality into indoles by a tandem N-alkylation-aza-Cope rearrangement reaction and its application in synthesis. *Chem. Eur. J.* 22: 716–723.

97. Ranjithkumar R, Perumal S, Senthilkumar P, Yogeeswari P, Sriram, D (2009) A facile synthesis and antimycobacterial evaluation of novel spiro-pyrido-pyrrolizines and pyrrolidines. *Eur. J. Med. Chem.* 44: 3821–3829.

98. WHO (2018) Tuberculosis fact sheet. www.who.int/mediacentre/factsheets/fs104/en/ (Accessed on February 20, 2018).

99. WHO (2018) Neglected tropical diseases. www.who.int/neglected_diseases/diseases/en/ (Accessed on February 22, 2018).

100. WHO (2018) Leishmaniasis. www.who.int/mediacentre/factsheets/fs375/en/ (Accessed on February 22, 2018).

101. Sanchez-Murcia PA, Ruiz-Santaquiteria M, Toro MA, de Lucio H, Jimenez MA, Gago F, Jimenez-Ruiz A, Camarasa MJ, Velazquez S (2015) Comparison of hydrocarbon- and lactam-bridged cyclic peptides as dimerization inhibitors of Leishmania infantum trypanothione reductase. *RSC Adv.* 5: 55784–55794.

102. Tala SR, Schnell SM, Haskell-Luevano C (2015) Microwave-assisted solid-phase synthesis of side-chain to side-chain lactam-bridge cyclic peptides. *Bioorg. Med. Chem. Lett.* 25: 5708–5711.

103. Serra M, Tambini SM, Di Giacomo M, Peviani EG, Belvisi L, Colombo L (2015) synthesis of easy-to-functionalize aza-bicycloalkane scaffolds as dipeptide turn mimics en route to cRGD-based bioconjugates. *Eur. J. Org. Chem.* 34: 7557–7570.

104. Chavan S, Zangade S, Vibhute A, Vibhute Y (2013) Synthesis and antimicrobial activity of some novel 2-azetidinones and 4-thiazolidinones derivatives. *Eur. J. Chem.* 4: 98–101.

105. Sarkar S, Husain SM, Schepmann D, Froehlich R, Wuensch B (2012) Microwave-assisted synthesis of 3-benzazepin-2-ones as building blocks for 2,3-disubstituted tetrahydro-3-benzazepines. *Tetrahedron* 68: 2687–2695.

106. Jamieson AG, Boutard N, Beauregard K, Bodas MS, Ong H, Quiniou C, Chemtob S, Lubell WD (2009) Positional scanning for peptide secondary structure by systematic solid-phase synthesis of amino lactam peptides. *J. Am. Chem. Soc.* 131: 7917–7927.

107. Dandia A, Khan S, Soni P, Indora A, Mahawar DK, Pandya P, Chauhan CS (2017) Diversity-oriented sustainable synthesis of antimicrobial spiropyrrolidine/thiapyrrolizidine oxindole derivatives: New ligands for a metallo-β-lactamase from *Klebsiella pneumonia*. *Bioorg. Med. Chem. Lett.* 27(13), 2873–2880.

108. Nunez-Navarro NE, Segovia GF, Burgos RA, Lagos CF, Fuentes-Ibacache N, Faundez MA, Zacconi FC (2017) Microwave assisted synthesis of novel six-membered 4-c, 4-O and 4-S lactams derivatives: characterization and in vitro biological evaluation of cytotoxicity and anticoagulant activity. *J. Braz. Chem. Soc.* 28(2): 203–207.

109. Phi TD, Mai, Huong DT, Tran VH, Vu VL, Truong BN, Tran TA, Chau VM, Pham VC (2017) Synthesis of bengamide E analogues and their cytotoxic activity. *Tetrahedron Lett.* 58(19), 1830–1833.

110. Shi F, Zeng XN, Cao XD, Zhang S, Jiang B, Zheng WF, Tu SJ (2012) Design and diversity-oriented synthesis of novel 1,4-thiazepan-3-ones fused with bioactive heterocyclic skeletons and evaluation of their antioxidant and cytotoxic activities. *Bioorg. Med. Chem. Lett.* 22: 743–746.

111. Keller SR (2004) Role of the insulin-regulated aminopeptidase IRAP in insulin action and diabetes. *Biol. Pharm. Bull.* 27: 761–764.

112. Andersson H, Demaegdt H, Johnsson A, Vauquelin G, Lindeberg G, Hallberg M, Erdelyi M, Karlen A, Hallberg A (2011) Potent macrocyclic inhibitors of insulin-regulated aminopeptidase (IRAP) by olefin ring-closing metathesis. *J. Med. Chem.* 54: 3779–3792.

113. Elliott JM, Carlson EJ, Chicchi GG, Dirat O, Dominguez M, Gerhard U, Jelley R, Jones AB, Kurtz MM, Tsao KL, Wheeldon A (2006) NK1 antagonists based on seven membered lactam scaffolds. *Bioorg. Med. Chem. Lett.* 16: 2929–2932.

114. Fauq AH, Simpson K, Maharvi GM, Golde T, Das P (2007) A multigram chemical synthesis of the γ-secretase inhibitor LY411575 and its diastereoisomers. *Bioorg. Med. Chem. Lett.* 17: 6392–6395.

115. Narhe BD, Tsai MH, Sun CM (2014) Rapid two-step synthesis of benzimidazo[1',2':1,5] pyrrolo[2,3-c]isoquinolines by a three-component coupling reaction. *ACS Comb. Sci.* 16: 421–427.

116. Chen H, Li R, Gao F, Li X (2012) An efficient synthesis of δ-glyconolactams by intramolecular Schmidt-Boyer reaction under microwave radiation. *Tetrahedron Lett.* 53: 7147–7149.

117. Vervisch K, D'hooghe M, Rutjes FPJT, De Kimpe N (2012) Chemical and enzymatic synthesis of 2-(2-carbamoylethyl)- and 2-(2-carboxyethyl)aziridines and their conversion into δ-lactams and γ-lactones. *Org. Lett.* 14: 106–109.

118. Busto E, Gotor-Fernandez V, Gotor V (2011) Hydrolases in the stereoselective synthesis of N-heterocyclic amines and amino acid derivatives. *Chem. Rev.* 111: 3998–4035.

119. Siddiqui IR, Shamim S, Singh A, Srivastava V, Yadav S (2010) Moisture compatible and recyclable indium (III) chloride catalyzed and microwave-assisted efficient route to substituted 1H-quinolin-2-ones. *ARKIVOC* 11: 232–241.

120. Orrling KM, Wu X, Russo F, Larhed M (2008) Fast, acid-free, and selective lactamization of lactones in ionic liquids. *J. Org. Chem.* 73: 8627–8630.

121. Tu S, Li C, Li G, Cao L, Shao Q, Zhou D, Jiang B, Zhou J, Xia M (2007) Microwave-assisted combinatorial synthesis of polysubstituent imidazo[1,2-a]quinoline, pyrimido[1,2-a]quinoline and quinolino[1,2-a]quinazoline derivatives. *J. Comb. Chem.* 9: 1144–1148.

122. Zradni FZ, Hamelin J, Derdour A (2007) Solventless lactam synthesis by intramolecular cyclizations of α-iminoester derivatives under microwave irradiation. *Molecules* 12: 439–454.

123. Calvino-Casilda V, Martin-Aranda RM, Lopez-Peinado AJ (2011) Alkaline carbons as effective catalysts for the microwave-assisted synthesis of N-substituted-gamma-lactams. *Appl. Catal. A* 398: 73–81.

124. Yasui Y, Kakinokihara I, Takeda H, Takemoto Y (2009) Preparation of α,β-unsaturated lactams through intramolecular electrophilic carbamoylation of alkenes. *Synthesis* 23: 3989–3993.

125. Cheung LLW, Yudin AK (2009) Synthesis of aminocyclobutanes through ring expansion of $N$-vinyl-$\beta$-lactams. *Org. Lett.* 11: 1281–1284.
126. Dolle RE, MacLeod C, Martinez-Teipel B, Barker W, Seida PR, Herbertz T (2005) Solid/solution-phase annulation reagents: Single-step synthesis of cyclic amine derivatives. *Angew. Chem. Int. Ed.* 44: 5830–5833.
127. Andersen RJ, Faulkner DJ, Cun-heng H, van Duyne GD, Clardy J (1985) Metabolites of the marine prosobranch mollusk *Lamellaria* sp. *J. Am. Chem. Soc.* 107: 5492–5495.
128. Bailly C (2015) Anticancer properties of lamellarins. *Mar. Drugs* 13: 1105–1123.
129. Boonya-udtayan S, Yotapan N, Woo C, Bruns CJ, Ruchirawat S, Thasana N (2010) Synthesis and biological activities of azalamellarins. *Chem. Asian J.* 5: 2113–2123.
130. Jung-Deyon L, Giethlen B, Mann A (2011) Expeditious route towards (±)-desethyleburnamonine, a precursor of (±)-vindeburnol. *Eur. J. Org. Chem.* 32: 6409–6412.
131. Kozlovsky AG, Zhelifonova VP, Antipova TV, Adanin VM, Ozerskaya SM, Kochkina GA, Schlegel B, Dahse HM, Gollmick FA, Grafe U (2003) Quinocitrinines A and B, new quinoline alkaloids from *Penicillium citrinum* thom 1910, a permafrost fungus. *J. Antibiot.* 56: 488–491.
132. Arinbasarova AY, Medentsev AG, Kozlovskii AG (2007) Effect of quinocitrinines from the fungus Penicillium citrinum on the respiration of yeasts and bacteria. *Appl. Biochem. Microbiol.* 43: 625–628.
133. Machtey V, Gottlieb HE, Byk G (2011) Total synthesis of structures proposed for quinocitrinines A and B and their analogs. Microwave energy as efficient tool for generating heterocycles. *ARKIVOC* 9: 308–324.
134. Valencia E, Freyer AJ, Shamma M, Fajardo V (1984) (±)-Nuevamine, an isoindoloisoquinoline alkaloid, and (±)-lennoxamine, an isoindolobenzazepine. *Tetrahedron Lett.* 25: 599–602.
135. Zamudio-Medina A, Garcia-Gonzalez MC, Padilla J, Gonzalez-Zamora E (2010) Synthesis of a tetracyclic lactam system of nuevamine by four-component reaction and free radical cyclization. *Tetrahedron Lett.* 51: 4837–4839.

# 11 Microwave Synthetic Technology

## An Eco-friendly Approach in Organic Synthesis

*Biswa Mohan Sahoo, Bimal Krishna Banik, and Jnyanaranjan Panda*

## CONTENTS

## 11.1   INTRODUCTION

One of the most attractive concepts in chemistry for sustainability is Green Chemistry, which is the employment of a set of principles that reduces or eliminates the use or production of hazardous substances in the design, manufacture and application of chemical products. It should be noted that the rapid development of Green Chemistry is due to the recognition that environmentally friendly products and processes will be economical in the long term. One of the key areas of Green Chemistry is the elimination of solvents in chemical processes or the replacement of hazardous solvents with environmentally benign solvents. The most important goals of sustainable development are reducing the adverse consequences of the substances that scientists use and generate. The role of chemistry is to ensure that the next generation of chemicals, materials and energy is more sustainable than the current generation. Worldwide demand for environmentally friendly chemical processes and products requires the development of novel and cost-effective approaches to pollution prevention.

At present, microwave technique is considered to be an important approach towards Green Chemistry because this technique is more environmentally friendly. It has the potential to have a large impact on the fields of screening, combinatorial chemistry, organic chemistry, medicinal chemistry and drug development. Conventional methods of organic synthesis usually need longer heating time and tedious apparatus setup, which result in higher cost of process, and the excessive use of solvents or reagents, which leads to environmental pollution. The growth of Green Chemistry holds significant potential for a reduction of the by-product, a reduction in waste production and a lowering of the energy costs. Due to its ability to couple directly with the reactants molecule and to pass thermal conductivity to rapidly rise the temperature, microwave irradiation is used to improve organic synthesis. The application of alternative solvents, such as water, fluorous, ionic liquids and supercritical media, is increasing rapidly. Catalysis remains one of the most important fields of Green Chemistry by providing atom-economical, selective and energy efficient solutions to many industrially important problems. The utilization of inorganic solid-supported reagents has attracted attention because of enhanced selectivity, milder reaction conditions and associated ease of manipulation. Microwave heating has attracted the attention of investigators because it makes it possible to shorten the length of reactions significantly, to increase their selectivity and to increase the product yields, which is particularly important in the case of high-temperature processes that take a long time.

Microwave irradiation is successfully applied in organic chemistry. Spectacular accelerations, higher yields under milder reaction conditions and higher product purities have all been reported. The effect of microwave irradiation in organic synthesis is a combination of thermal effects, arising from the heating rate, superheating or "hot spots" and the selective absorption of radiation by polar substances [1].

### 11.1.1 MILESTONES IN THE DEVELOPMENT OF MICROWAVE CHEMISTRY

Microwave technology originated in 1946, when Dr. Percy Le Baron Spencer, while conducting laboratory tests for a new vacuum tube called a magnetron (a device that generates an electromagnetic field), accidentally discovered that a candy bar in his pocket had melted on exposure to microwave radiation. Dr. Spencer developed the idea further and established that microwaves could be used as a method of heating. Subsequently, he designed the first microwave oven for domestic use in 1947. Since then, the development of microwave radiation as a source of heating has been very gradual, as given in Table 11.1.

### TABLE 11.1
### Development of Microwave Chemistry

| Year Wise Development | Development of Microwave Chemistry |
|---|---|
| 1946 | Microwave radiation is discovered as a method of heating |
| 1947 | The first commercial domestic microwave oven is introduced |
| 1978 | The first microwave laboratory instrument is developed by CEM Corporation to analyze moisture in solids |
| 1980–82 | Microwave radiation is developed to dry organic materials |
| 1983–85 | Microwave radiation is used for chemical analysis |
| 1986 | Robert Gedye, Laurentian University, Canada; George Majetich, University of Georgia, USA; and Raymond Giguere of Mercer University, USA, published papers relating to microwave radiation in chemical synthesis |
| 1990 | Microwave chemistry emerged and developed as a field of study for its applications in chemical reactions |
| 1990 | Milestone S.R.L generated the first high pressure vessel (HPV 80) for performing complete digestion of materials like oxides, oils and pharmaceutical compounds |
| 1992–96 | CEM developed a batch system reactor and a single mode cavity system that is used for chemical synthesis |
| 1997 | Milestone S.R.L and Kingston of Duquesne University culminated a reference book titled *Microwave-Enhanced Chemistry–Fundamentals, Sample Preparation, and Applications*, edited by H. M. Kingston and S. J. Haswell |
| 2000 | The first commercial microwave synthesizer is introduced to conduct chemical synthesis |

## 11.1.2  MICROWAVES

Microwaves are in the form of electromagnetic energy which lie in the electromagnetic spectrum and correspond to wavelengths of 1cm to 1m and frequencies of 30GHz to 300MHz. This places it between infrared radiations, which have shorter wavelengths in the 1–25cm range for radar, whereas the remaining sections are devoted to telecommunication. Microwave energy consists of both an electric as well as magnetic field. Microwaves move at the speed of light and it have less energy than that which is required to break the bond in a chemical molecule, thus, microwaves are a source of energy which will not hamper the structure of a chemical molecule. Microwaves are coherent and polarized in contrast to visible waves (apart from lasers). They obey the laws of optics and can be transmitted, absorbed or reflected depending on the type of material [2].

## 11.1.3  MECHANISM OF MICROWAVE HEATING

Traditionally, organic reactions are heated using an external heat source, such as an oil bath, and, therefore, heat is transferred by conductance. This is a comparatively slow and inefficient method for transferring energy into the system because it depends on the thermal conductivity of the various materials that must be penetrated and on the results in the temperature of the reaction vessel being higher than that of the reaction mixture. By contrast, microwave irradiation produces efficient internal heating by direct coupling of microwave energy with the polar molecules (for example, solvents, reagents and catalysts) that are present in the reaction mixture. Materials respond to microwave radiation differently and some may not be susceptible to microwave heating. Based on their response to microwaves, materials can be broadly classified as follows:

1. Materials that are transparent to microwaves, e.g. sulphur
2. Materials that reflect microwaves, e.g. copper
3. Materials that absorb microwaves, e.g. water

Microwave absorbing materials are of utmost importance for microwave chemistry; three main different mechanisms are involved in their heating, namely: dipolar polarization, conduction mechanism and interfacial polarization [3].

### 11.1.3.1  Dipolar Polarization

For a substance to generate heat when irradiated with microwaves it must possess a dipole moment. It is the electric field component of the microwave radiation, rather than magnetic field component which is responsible for heating, when a dipole tries to reorient itself with respect to an alternating electric field; it loses energy in the form of heat by molecular friction. Dipolar polarization can generate heat by either the interaction between polar solvent molecules, such as water, methanol and ethanol, or the interaction between polar solute molecules, such as ammonia and formic acid. The main requirement for dipolar polarization is that the frequency range of the oscillating field should be appropriate to enable adequate inter-particle interaction. If the frequency range is very high, intermolecular forces will stop the motion of a polar molecule before it tries to follow the field, resulting in inadequate inter-particle interaction. On the other hand, if the frequency range is low, the polar molecule gets sufficient time to align itself in phase with the field. Microwave radiation has the appropriate frequency (0.3–30GHz) to oscillate polar particles and enable enough inter-particle interaction, which makes it an ideal choice for heating polar solutions [4, 5].

### 11.1.3.2 Conduction Mechanism

This mechanism produces heat through resistance to an electric current. The oscillating electromagnetic field generates an oscillation of electrons or ions in a conductor, resulting in an electric current. This current faces internal resistance, which heats the conductor. In a sample of a solution containing ions, or even a single isolated ion with a hydrogen bonded cluster, the ions will move through the solution under the influence of an electric field, resulting in an expenditure of energy due to the assumption that the more polar the solvent, the more readily the microwave irradiation is absorbed and the higher the temperature obtained. Where the irradiated sample is an electrical conductor, the charge carriers (electrons, ions, etc.) are moved through the material under the influence of the electric field, resulting in a polarization. These induced currents will cause heating in the sample due to any electrical resistance. A major limitation of this method is that it is not applicable to materials with high conductivity since such materials reflect most of the energy that falls on them [4].

### 11.1.3.3 Interfacial Polarization

The interfacial polarization method can be considered as a combination of both the conduction and dipolar polarization mechanisms. It is important for heating systems that comprise of a conducting material dispersed in a non-conducting material. For example, consider the dispersion of metal particles in sulphur. Sulphur does not respond to microwaves, and metals reflect most of the microwave energy they are exposed to, but combining the two makes them a good microwave-absorbing material. However, for this to take place, metals have to be used in powder form. This is because, unlike a metal surface, metal powder is a good absorber of microwave radiation. It absorbs radiation and is heated by a mechanism that is similar to dipolar polarization. The environment of the metal powder acts as a solvent for polar molecules and restricts the motion of ions by forces that are equivalent to inter-particle interactions in polar solvents. These restricting forces under the effect of an oscillating field induce a phase lag in the motion of ions, resulting in a random motion of ions and, ultimately, heating of the system [5–7].

### 11.1.4 Benefits of Microwave-assisted Synthesis

Microwaves can accelerate the rate of reaction, provide better yields, higher purity and uniform and selective heating with lower energy usage, achieve greater reproducibility of reactions and help in developing convenient and cleaner synthetic routes [8]. The main advantages of microwave-assisted organic synthesis are:

### 11.1.4.1 Faster Reaction

Based on experimental data it is found that microwave-enhanced chemical reaction rates can be faster than those of conventional heating methods by as much as 1,000-fold. The microwave can use higher temperatures than conventional heating systems, and therefore the reactions are completed in a few minutes instead of hours, for instance, synthesis of fluorescein, which usually takes about 10 hours by conventional heating methods, can be conducted in only 35 minutes by means of microwave heating.

### 11.1.4.2 Better Yield with Higher Purity

Less formation of side products are observed when using microwave irradiation, and the product is recovered in a higher yield. As a result, the purification step is faster and easier. For example, microwave synthesis of aspirin results in an increase in the yield of the reaction from 85% to 97%.

### 11.1.4.3 Energy Saving

Heating by means of microwave radiation is a highly efficient process and results in significant energy saving. This is primarily because microwaves heat up just the sample and not the apparatus, and therefore energy consumption is less.

### 11.1.4.4 Uniform and Selective Heating

In conventional heating, the walls of the oil bath get heated first, and then the solvent. As a result of this distributed heating in an oil bath, there is always a temperature difference between the walls and the solvent. In the case of microwave heating, only the solvent and the solute particles are excited, which results in uniform heating of the solvent. Selective heating is based on the principle that different materials respond differently to microwaves. Some materials are transparent, whereas others absorb microwaves.

### 11.1.4.5 Green Synthesis

Reactions performed using microwaves are cleaner and more eco-friendly than conventional heating methods. Microwaves heat the compounds directly. Therefore, usage of solvents in the chemical reaction can be reduced or eliminated. Synthesis without solvent, in which reagents are absorbed on mineral support, has a great potential as it offers an eco-friendly green protocol in synthesis. The use of microwaves has also reduced the amount of purification required for the end products of chemical reactions involving toxic-reagents.

### 11.1.4.6 Reproducibility

Reactions with microwave heating are more reproducible compared to conventional heating because of uniform heating and better control of process parameters. The temperature of chemical reactions can also be easily monitored [9–12].

## 11.1.5  LIMITATIONS OF MICROWAVE-ASSISTED SYNTHESIS

The yield obtained by using microwave apparatus available in the market is limited to a few grams. Although there have been developments in the recent past, relating to the scalability of microwave equipment, there is still a gap that needs to be spanned to make the technology scalable. The use of microwaves as a source of heating has limited applicability to materials that absorb them. Microwaves cannot heat materials such as sulphur, which are transparent to their radiation. Improper use of microwave heating for the rate enhancement of chemical reactions involving radioisotopes may result in uncontrolled radioactive decay. Certain problems, with dangerous end results, have also been observed while conducting polar acid-based reactions. For example, microwave irradiation of concentrated sulphuric acid may damage the polymer vessel used for heating. Conducting microwave reactions at high pressure conditions may also result in uncontrolled reactions and cause explosions. Health hazards related to microwaves are caused by the penetration of microwaves. While microwaves operating at a low frequency ranges are able to penetrate the human skin, higher frequency-range microwaves can reach body organs. Research has proven that on prolonged exposure microwaves may result in the complete degeneration of body tissues and cells. It has also been established that constant exposure of DNA to high frequency microwaves during a biochemical reaction may result in complete degeneration of the DNA strand [13, 14].

## 11.1.6  MICROWAVE SYNTHESIS APPARATUS

The apparatus for microwave-assisted synthesis include single-mode microwave ovens and multi-mode microwave ovens [15].

### 11.1.6.1  Single-Mode Microwave Apparatus

The important feature of a single-mode apparatus is its ability to create a standing wave pattern. This interface generates an array of nodes where microwave energy intensity is zero and an array of antinodes where the magnitude of microwave energy is at its highest. One of the limitations of single-mode apparatus is that only one vessel can be irradiated at a time. However, the apparatus is user-friendly. An advantage of single-mode apparatus is their high rate of heating. This is because the sample is always placed at the antinodes of the field where the intensity of microwave radiation is the highest. These apparatus can process volumes ranging from 0.2 to about 50 ml under sealed-vessel conditions and volumes around 150 ml under open-vessel conditions. Single-mode microwave ovens are currently used for small-scale drug discovery, automation and combinatorial chemical applications.

### 11.1.6.2  Multi-Mode Microwave Apparatus

An essential feature of a multi-mode apparatus is the deliberate avoidance of generating a standing wave pattern inside it. The goal is to generate as much chaos as possible inside the apparatus. The greater the chaos, the higher the dispersion of radiation, which increases the area that can cause effective heating inside the apparatus. As a result, a multi-mode microwave heating apparatus can accommodate a number of samples simultaneously for heating, unlike single-mode apparatus where only one sample can be irradiated at a time. Owing to this characteristic, a multi-mode heating apparatus is used for bulk heating and carrying out chemical analysis processes, such as ashing, extraction, etc. In large multi-mode apparatus, several liters of the reaction mixture can be processed in both open- and closed vessel conditions. A major limitation of multi-mode apparatus is that heating samples cannot be controlled efficiently because of the lack of temperature uniformity [16–18].

### 11.1.7 APPLICATIONS OF MICROWAVE-ASSISTED SYNTHESIS

Microwave-enhanced synthesis results in faster reactions, higher yields and increased product purity. In addition to this, due to the availability of high-capacity microwave apparatus, the yields of the experiments are now easily scaled up from milligrams to kilograms without the need to alter reaction parameters. Microwave-assisted synthesis can be suitably applied to the drug discovery process [14].

#### 11.1.7.1 Organic Synthesis

Microwave-assisted organic synthesis is the foremost and one of the most researched applications of microwaves in chemical reactions. The literature survey reveals that scientists have successfully conducted a large range of organic reactions. These include Diels-Alder reaction, Ene reaction, Heck reaction, Suzuki reaction, Mannich reaction, hydrolysis, dehydration, esterification, cycloaddition reaction, epoxidation, reductions, condensations, cyclisation reactions, protection and deprotection, etc. Based on reaction conditions, organic synthesis reactions can be conducted in the following techniques [19].

*11.1.7.1.1 Microwave-assisted Organic Synthesis at Atmospheric Pressure*
Microwave-assisted organic synthesis can be most conveniently conducted at atmospheric pressure in reflux conditions, for example, oxidation of toluene to benzoic acid with $KMnO_4$ under normal conditions of refluxing takes 10–12 hours compared to a reaction in microwave conditions, which takes only 5 minutes [20, 21].

Microwave-assisted organic reactions can be most conveniently conducted at atmospheric pressure in reflux conditions. A good example of microwave-assisted organic synthesis at atmospheric pressure is the Diels-Alder reaction of maleic anhydride with anthracene. In the presence of diglyme (boiling point 162°C), this reaction can be completed in a minute with a 90% yield. However, the conventional synthetic route, which uses benzene, requires 90 minutes. It is extremely important to use dipolar solvents for reactions in such conditions. Solvent systems with higher boiling points are preferred in microwave-assisted organic synthetic reactions.

*11.1.7.1.2 Microwave-assisted Organic Synthesis at Elevated Pressure*
Microwaves can be used to directly heat the solvents in sealed microwave-transparent containers. The sealed container helps in increasing the pressure in the reactor, which facilitates the reaction that will take place at much higher temperatures. This results in a substantial increase in the reaction rate of microwave-assisted organic synthesis. However, an increase in the reaction rate of any chemical synthesis

depends on three factors: volume of the vessel, solvent to space ratio and solvent boiling point.

### 11.1.7.1.3 Microwave-assisted Organic Synthesis under Solvent-free Conditions

Microwave-assisted solvent-free organic synthesis has been developed as an environmentally friendly process because it combines the selectivity associated with most reactions carried out under microwaves with solvent and waste-free procedures in which organic solvents are avoided throughout all stages. The solvent-free organic syntheses are of three types: (i) reactions using neat reactants, (ii) reactions using solid-liquid phase transfer catalysis (PTC) and (iii) reactions using solid mineral supports. The microwave-assisted reaction could be completed within two to three minutes, compared to conventional oil-bath heating at 75°C for 40 hours [22].

### 11.1.7.2 Various Types of Microwave-Assisted Organic Reactions

The microwave-assisted organic reactions are broadly classified into two categories: microwave-assisted reactions using solvents and microwave-assisted reactions using solvent-free conditions.

### 11.1.7.2.1 Microwave-assisted Reactions Using Solvents

In the case of the microwave-assisted reactions using (organic) solvents, the reactants are usually dissolved in the solvent, which often couples effectively with microwaves and thus acts as the energy transfer medium. The use of aqueous media for organic reactions is also under active investigation, and temperatures of up to 100°C and above are employed for the syntheses often intended to exploit the hydrophobic effect. Water has a dielectric constant of 78 at 25°C, which decreases to 20 at 300°C; the latter value being comparable with that of the solvents, such as acetone, at ambient temperature. Thus, water at elevated temperature can behave as a pseudo-organic solvent and is a possible environmentally benign replacement for organic solvents. In addition to the environmental advantages of using water instead of the organic solvents, isolation of the products is often facilitated by the decrease of the solubility of the organic material upon post-reaction cooling [23–27].

An alternative method for performing microwave-assisted organic reactions, termed enhanced microwave synthesis (EMS), has also been examined. By externally cooling the reaction vessel with compressed air, while simultaneously administering microwave irradiation, more energy can be directly applied to the reaction mixture. In the conventional microwave synthesis (CMS), the initial microwave power is high, increasing the bulk temperature (BT) to the desired value very quickly. However, upon reaching this temperature, microwave power decreases or shuts off completely in order to maintain the desired bulk temperature without exceeding it.

Recently, the combination of two prominent Green Chemistry principles, namely microwaves and water, has become very popular and received substantial interest. A plethora of very recent synthetic applications describes a variety of new chemistries that can be performed with microwave irradiation but a wide range of microwave-assisted applications is still waiting. Many organic transformations proceed via radical chemistry. As chemists wonder if microwave irradiation can promote radical transformations, microwave-assisted free radical chemistry is increasingly being explored. Microwave irradiation is applicable not only to the solvent phase chemistry, but also to the solid-phase organic synthesis. Following are the examples of microwave-assisted reactions using solvents [28, 29].

1. *Hydrolysis:*
   Hydrolysis of benzyl chloride with water in microwave oven gives 97% yield of benzyl alcohol in 3 minutes, whereas the usual hydrolysis in the conventional way takes about 35 minutes.

$$C_6H_5CH_2Cl \ + \ H_2O \xrightarrow{\text{MW, 3 min}} C_6H_5CH_2OH$$

   Benzyl chloride                                       Benzyl alcohol

   The usual hydrolysis of benzamide takes one hour by conventional synthesis. However, under microwave conditions, the hydrolysis is completed in seven minutes, giving 99% yield of benzoic acid [30].

$$C_6H_5CONH_2 \xrightarrow[\text{MW, 7 min}]{\text{20\% } H_2SO_4} C_6H_5COOH$$

   Benzamide                                             Benzoic acid

2. *Oxidation:*
   Oxidation of toluene with $KMnO_4$ under normal conditions of refluxing takes 10–12 hours as compared to the reaction in microwave conditions, which takes only five minutes [30].

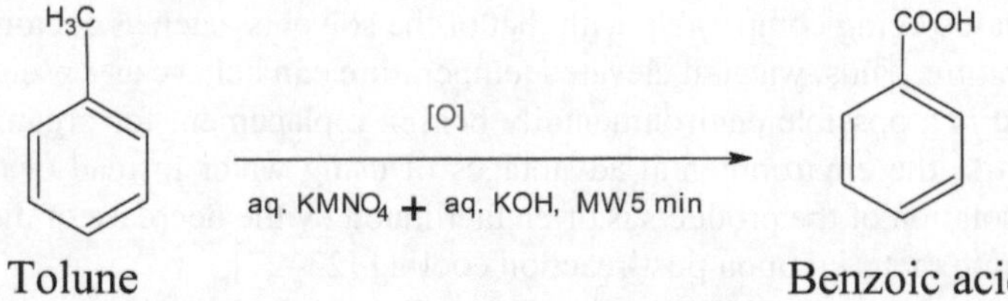

   Tolune                                                Benzoic acid

   A number of primary alcohols can be oxidized to the corresponding carboxylic acid using sodium tungstate as a catalyst in 30% aqueous hydrogen peroxide.

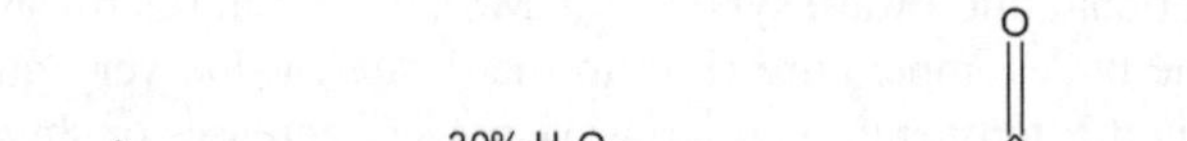

   Primary alcohol                                       Carboxylic acid

3. *Esterification:*
   A mixture of benzoic acid and n-propanol on heating in a microwave oven for six minutes in the presence of a catalytic amount of concentrated sulfuric acid produces propylbenzoate [29].

$$\text{C}_6\text{H}_5\text{COOH} \quad + \quad n\text{C}_3\text{H}_7\text{OH} \xrightarrow[\text{MW, 6 min}]{\text{Conc. H}_2\text{SO}_4} \text{C}_6\text{H}_5\text{COOC}_3\text{H}_7$$

Benzoic acid                 Propylbenzoate

4. *Decarboxylation:*

Conventional decarboxylation of carboxylic acids involves refluxing in quinoline in the presence of copper chromate and the yields are low. However, when the same reaction is carried out by microwaves, decarboxylation takes place in a much shorter time [31].

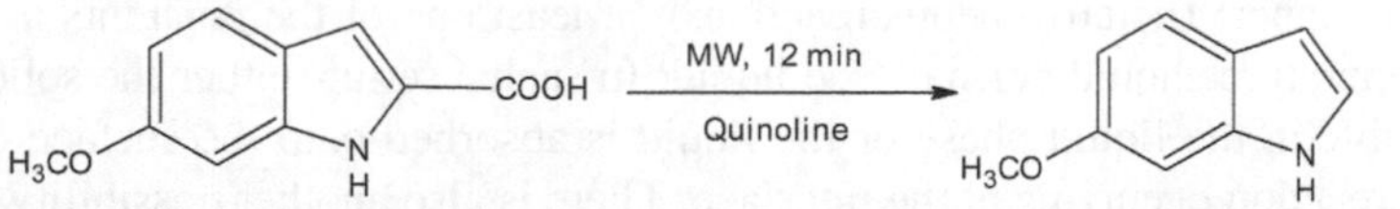

6-Methoxyindol-2-carboxylic acid             6-methoxyindole

5. *Cycloaddition:*

1,3-dipolar cycloadditions are important reactions in organic synthesis. Cycloadducts are prepared by carrying out the reaction between an azide and a substituted amide in toluene. This reaction is carried out under microwave irradiation at 120W at 75°C for one hour. The yield of the product is found to be 70–80% [32].

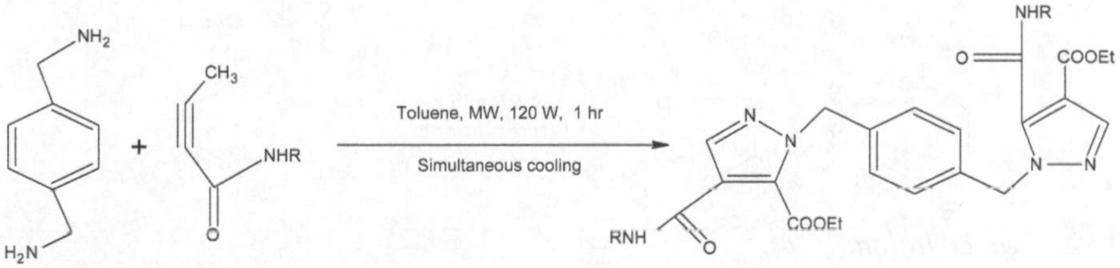

6. *N-Acylations:*

N-acylations are carried out using secondary amines and isocyanate in dichloromethane under microwave irradiation (8–10 minutes), the yield of the product was 94% [33–36].

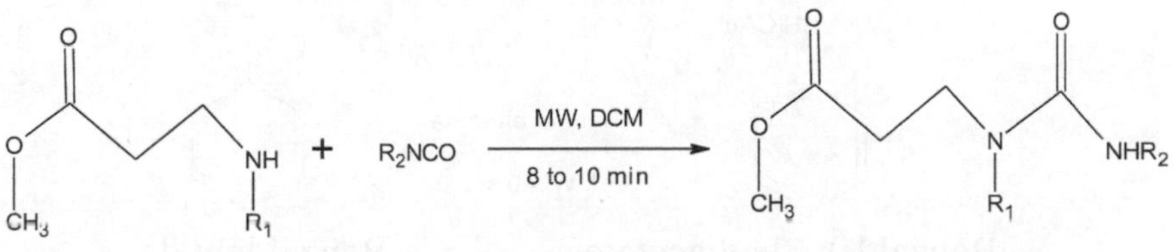

### 11.1.7.2.2 Microwave-assisted Reactions Under Solvent-free Conditions

Due to the environmental concerns, there is an increasing demand for efficient synthetic processes and solvent-free reactions. Some old and new methodologies are being used to diminish and prevent pollution caused by chemical activities. In this context, microwaves have become an important source of energy in many laboratory procedures. Furthermore, microwave-assisted solvent-free organic synthesis (MASFOS) has been developed as an environmentally friendly process because it combines the selectivity associated with most reactions carried out under microwaves with solvent and waste-free procedures in which organic solvents are avoided throughout all stages. In these environmentally conscious times, research and development are being directed towards devising cleaner processes. Environmental hazards and the subsequent degradations are instrumental in the rapid evolution of Green Chemistry concepts involving benign reagents and conditions [37–39].

For carrying out reactions with neat reactants, i.e. without the use of a solvent or a support (heterogeneous reactions), at least one of the reactants at the reaction temperature should normally be liquid. In such a set-up, either the solid is partially soluble in the liquid phase or the liquid is absorbed onto the surface of solid with the reaction occurring at the interface. There is also another possibility, namely that both the reactants are solid. Usually, they melt during the reaction course and then undergo a reaction as described above [40]. Following are the examples of microwave-assisted reactions with neat reactants.

1. *Aromatic Nucleophilic Substitutions:*
   Aromatic nucleophilic substitutions are carried out using sodium phenoxide and 1,3,5-trichlorotriazine under microwave irradiation for 6 minutes. The products, 1,3,5-triarlyoxytriazines, are obtained in 85–90% yields [41, 42].

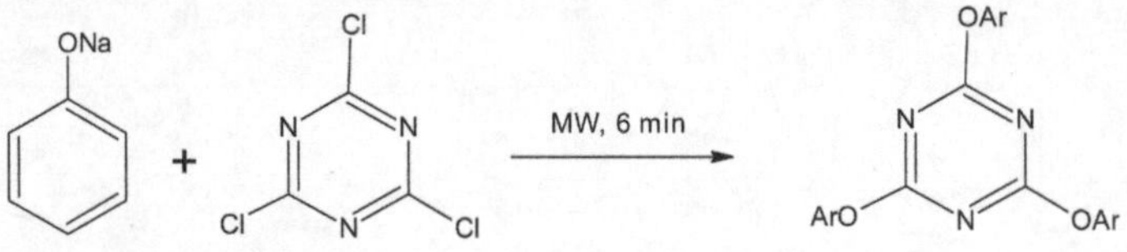

2. *Deacetylation:*
   Aldehydes, phenol and alcohols are protected by acetylation. After the reaction, the deacetylation of the product is carried out, usually under acidic or basic conditions; in conventional synthesis, the process takes a long time, and the yields are low. Use of microwave irradiation reduces the time of deacetylation, and the yields are higher [43].

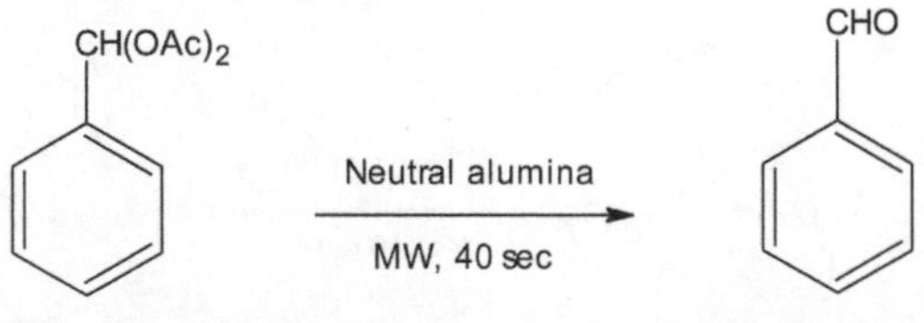

Benzaldehyde diacetate                    Benzaldehyde

### 11.1.7.2.3   Microwave-assisted Reactions Using Solid Liquid Phase

Solid liquid phase transfer catalysis (PTC) is described as an effective method in organic synthesis and is under active investigation. This method is specific for anionic reactions as it involves anionic activation. A catalytic amount of a tetralkyl-ammonium salt or a cation complexing agent is added to the mixture (in equimolar amounts) of both pure reactants.

Reactions occur in the liquid organic phase, which consists here of only the electrophilic R-X. The presence of an additional liquid component is disadvantageous as it induces a dilution of reactants and, consequently, a decrease in reactivity. The electrophile R-X is therefore both the reactant and the organic phase for the reaction. Following are the examples of microwave-assisted reaction using a solid liquid phase.

1. *O-Alkylation:*

   Preparation of ethers is carried out from β-naphthol using benzyl bromide and 1-butyl-3-methyl-imidazolium tetrafluoroborate under microwave irradiation (6–12 minutes); the products are isolated in 75–90% yields.

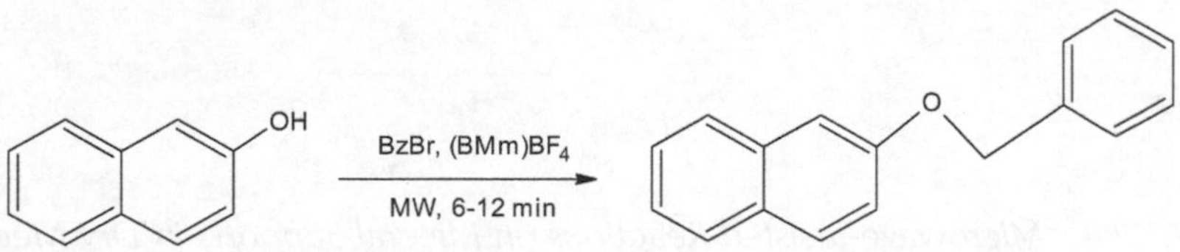

2. *N-Alkylations:*

   N-Alkylations under microwave irradiation using phase transfer catalysts occupy a unique place in organic chemistry. Bogdal and co-workers reported the synthesis of N-alkyl phthalimides using phthalimide, alkyl halides, potassium carbonate and TBAB, giving products in 45–98% yields [44].

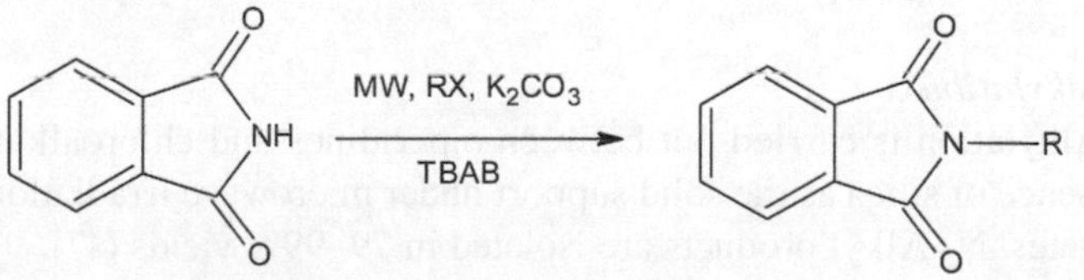

3. *Oxidations:*

   The oxidation of secondary alcohol and benzyl alcohols using phase transfer catalysts is reported by Chakraborty et al.. Oxidation of secondary alcohols to acetone derivatives is carried out using PCC, tetrabutyl ammonium bromide and dichloromethane under microwave irradiation (6–8 minutes); products are isolated in 70–99% yields. Oxidation of benzyl alcohols is conducted using BIFC under microwave irradiation (1–8 minutes), yielding benzaldehyde derivatives in 70–92% yields [45].

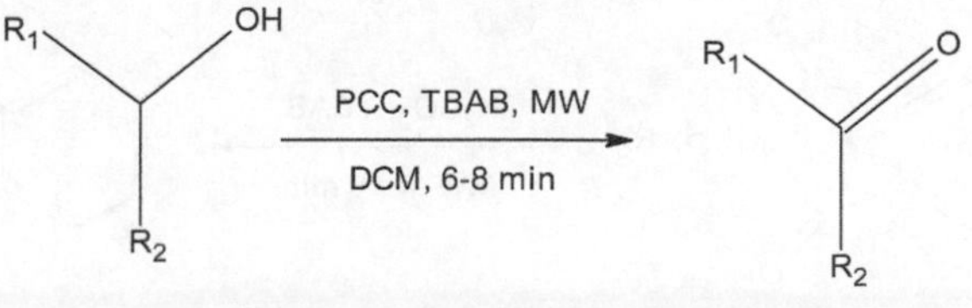

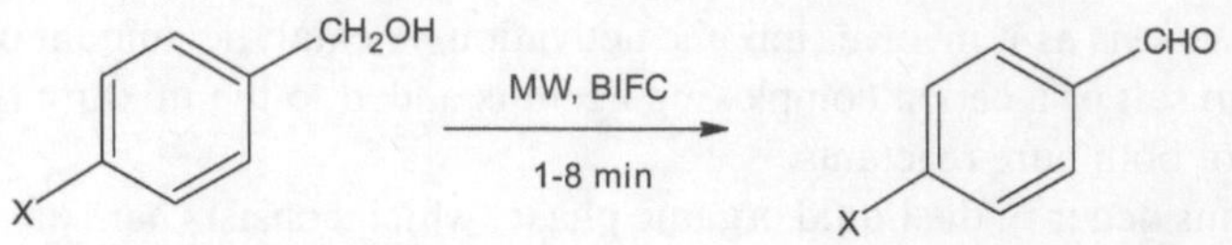

4. *Knoevenagel Condensation:*

Knoevenagel condensation is a well-known organic reaction, also applied in the synthesis of unsaturated acids, which are used as precursors for perfumes and flavonoids and as building blocks for many heterocycles. Gupta and Wakhloo studied Knoevenagel condensation between carbonyl compounds and active methylene compounds, such as malonic acid, tetrabutylammonium bromide and potassium carbonate, in water, forming unsaturated acids with excellent yield and purity under microwave irradiation [46].

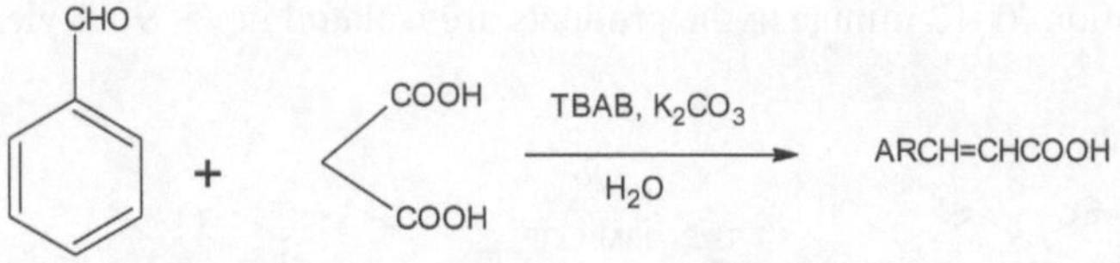

## 11.1.7.2.4   Microwave-assisted Reactions on Mineral Supports in Dry Media

Solid supports are often very poor conductors of heat but behave as very efficient microwave absorbents. This, in turn, results in very rapid and homogeneous heating. Consequently, they display very strong specific microwave effects with significant importance placed on temperature homogeneity and heating rates, enabling faster reactions and less degradation of final products, as compared to classical heating. Following are the examples of the microwave activation with supported reagents.

1. *N-Alkylation:*

N-Alkylation is carried out between piperidines and chloroalkanes in the presence of silica as the solid support under microwave irradiation for 6–10 minutes. N- Alkyl products are isolated in 79–99% yields [47].

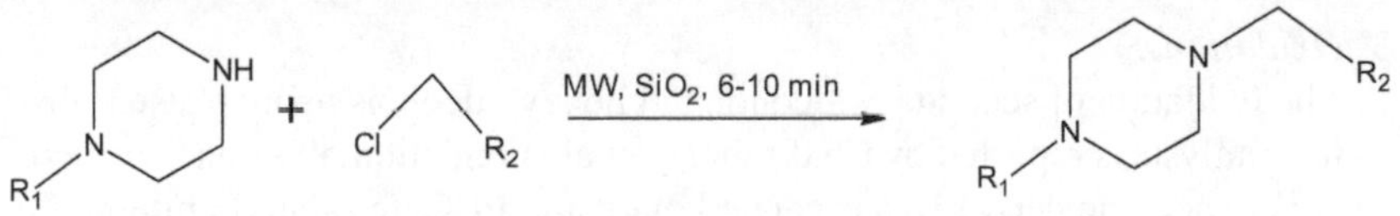

A variety of solvent-free N-alkylation reactions are reported with the use of phase transfer agents, such as tetrabutyl ammonium bromide (TBAB), under microwave irradiation conditions. The important examples are the N-alkylation of phthalimides in the presence of potassium carbonate and TBAB.

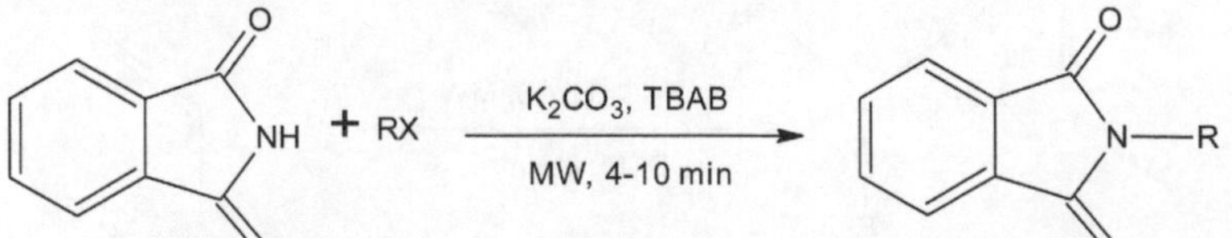

2. *S-Alkylation:*

S-Alkylation is studied and accomplished by carrying out the reaction between mercaptobenzene and alkyl halides using potassium carbonate and alumina under microwave irradiation for 4–10 minutes. Products are isolated in 70–89% yields [48].

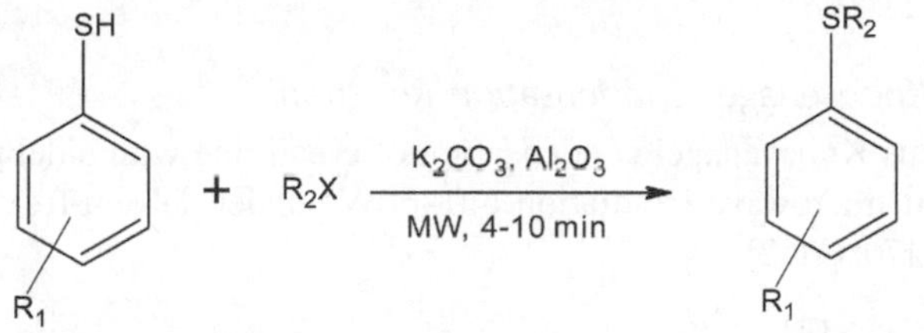

### 11.1.7.3 Rearrangement Reactions

*11.1.7.3.1 Pinacol-Pinacolone Rearrangement*

A solvent-free pinacol-pinacolone rearrangement reaction using microwave irradiation is achieved, involving the irradiation of the gem-diols with Al+3-montmorillonite K-10 clay for 15 minutes to afford the rearrangement product in excellent yields. These results are compared to conventional heating in an oil bath where the reaction times are too long (15 hours) [49].

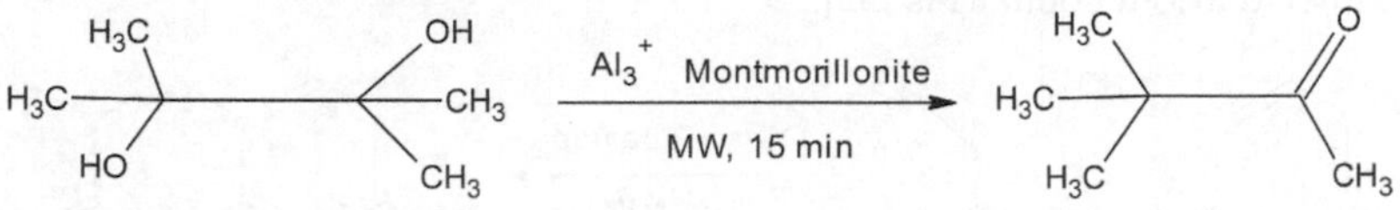

*11.1.7.3.2 Beckmann Rearrangement*

The Beckmann rearrangement of ketoximes with montmorillonite K-10 clay in "dry" media in good yields has been reported [50].

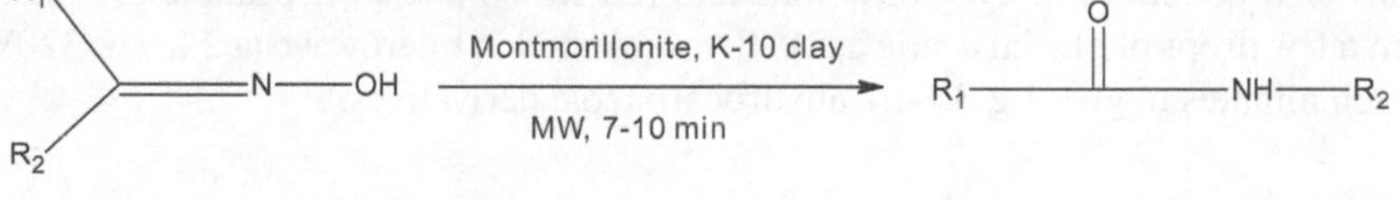

### 11.1.7.4 Condensation Reactions

*11.1.7.4.1 Synthesis of Imines, Enamines and Nitroalkenes*

The driving force in the preparation of imines, enamines and nitroalkenes is the azeotropic removal of water from the intermediate, which is normally catalyzed by p toluenesulfonic acid, titanium (IV) chloride and montmorillonite K-10 clay. Conventionally, a Dean–Stark apparatus is used, which requires a large excess of aromatic hydrocarbons, such as benzene or toluene, for azeotropic water elimination. Microwave-induced acceleration of such dehydration reactions using montmorillonite K-10 clay is demonstrated in a facile preparation of imines and enamines via the reactions of primary and secondary amines with aldehydes and ketones, respectively [51].

### 11.1.7.4.2 Knoevenagel Condensation Reactions

An expeditious Knoevenagel condensation of creatinine with aldehydes is achieved using focused microwave irradiation (40–60W) under solvent-free reaction conditions at 160–170°C [52].

The useful synthesis of coumarins via the microwave-promoted Pechmann reaction is extended to solvent-free systems where salicylaldehydes undergo Knoevenagel condensation with a variety of ethyl acetate derivatives under basic conditions (in piperidine) to afford coumarins [53].

### 11.1.7.4.3 Borsche–Drechsel Cyclization (Synthesis of Tetrahydrocarbazole)

Equimolar quantities of cyclohexanone and redistilled phenyl hydrazine are placed with a few drops of glacial acetic acid, then subjected to microwave at 2 level (320W) for ten minutes to give 1,2,3,4-trtrahydrocarbazole derivatives.

1, 2, 3, 4-Tetrahydrocarbazole

### 11.1.7.4.4 Benzillic Acid Rearrangement (Synthesis of Phenytoin)

A mixture of benzil and urea are taken in ethanol, to this, 30% NaOH solution is added and the reaction is subjected to microwave at 160W for 30 minutes to give diphenyl hydantoin derivatives [54].

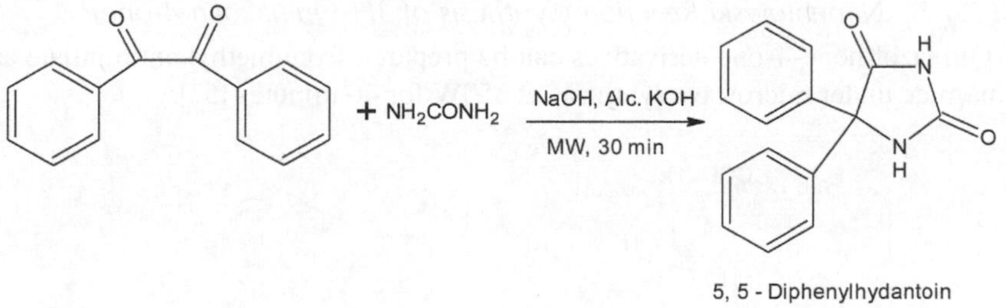

5, 5 - Diphenylhydantoin

## 11.1.7.4.5 Bignelli Condensation (Synthesis of Tetrahydro Pyrimidine)

An equimolar mixture of aromatic aldheyde, ethylacetocaetate and urea was taken. To this, a few drops of concentrated hydrochloric acid is added as a catalyst. The reaction is subjected to microwave at 180W and monitored on TLC (Thin layer chromatography) intermittently for 30 minutes [55].

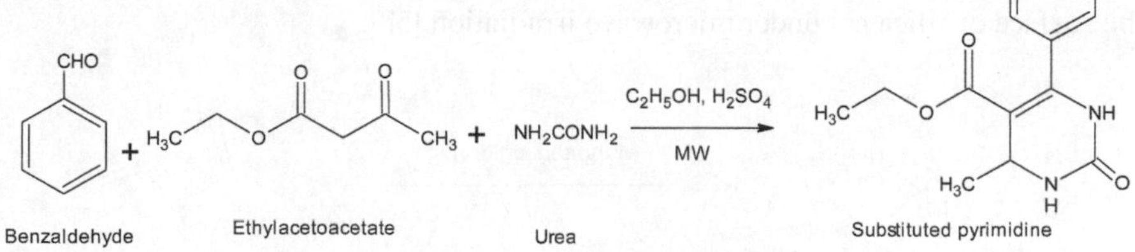

Benzaldehyde    Ethylacetoacetate    Urea    Substituted pyrimidine

## 11.1.7.4.6 Claisen-Smith Reaction (Synthesis of Pyrazoline)

Synthesis of pyrazoline derivatives can be carried out by using phenyl hydrazine and ethanol. The above mixture is subjected to MWI at 320W for 30 minutes [56].

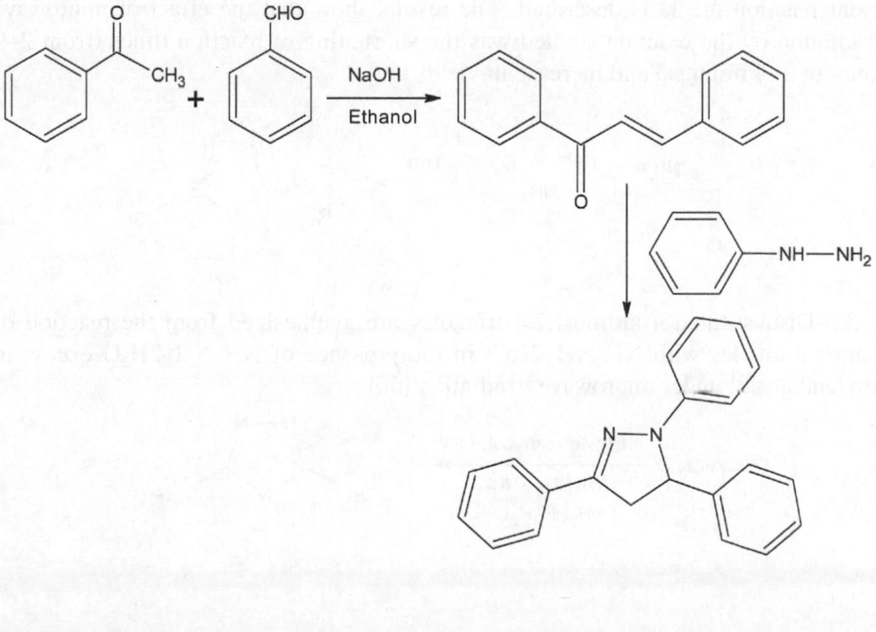

*11.1.7.4.7 Nimentowski Reaction (Synthesis of 3H-Quinazolin-4-one)*

3H-Quinazolinone-4-one derivatives can be prepared from methyl anthranilate and formamide under microwave synthesis at 350W for 40 minutes [57].

Methyl anthranilate      Formamide            3H-quinazoline- 4- one

### 11.1.7.5 Preparation of 1,2,4-Triazoles

4,5-Disubstituted-1,2,4-triazole-3-thiones are prepared in one stage from the reaction of acid hydrazide with alkyl or aryl isothiocyanate in the presence of a KOH (10%) solution on the surface of silica gel as well as on the surface of montmorillonite K10 under microwave irradiation. These triazoles are also prepared from the reaction of 4-substituted-1-aroyl thiosemicarbazides, with a KOH (10%) solution on the surface of silica gel under microwave irradiation [58].

The different types of 4,5-disubstituted 1,2,4-triazole-3-thiones are synthesized by microwave irradiation as well as by a classical method. The beneficial effect of microwave irradiation on the dehydrative cyclization of thiosemicarbazides in different reaction media is described. The results show that the effect of microwave irradiation on the reaction studied was the shortening of reaction times (from 2–9 hours to 2–4 minutes) and increase in yields [59].

3,5-Disubstituted-4-amino-1,2,4-triazoles are synthesized from the reaction of aromatic nitriles with $NH_2NH_2 \cdot 2HCl$ in the presence of $NH_2NH_2 \cdot 2H_2O$ excess in ethylene glycol under microwave irradiation [60].

Condensation of acid hydrazide with S-methylisothioamide hydroiodide and ammonium acetate on the surface of silica gel under microwave irradiation afforded 1,2,4-triazoles [61].

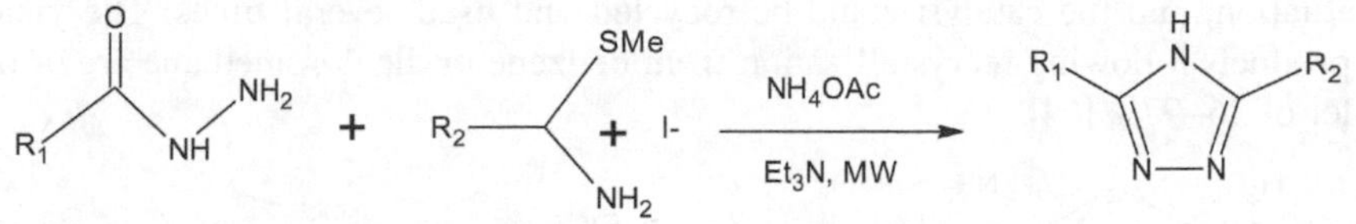

A convenient and efficient one-step, base-catalyzed synthesis of 3,5-disubstituted 1,2,4-triazoles is carried out by the condensation of nitriles and hydrazides under microwave irradiation. Under the reaction conditions, a diverse range of functionality and heterocycles are tolerated. The reactivity of the nitrile partner is relatively insensitive to electronic effects [62].

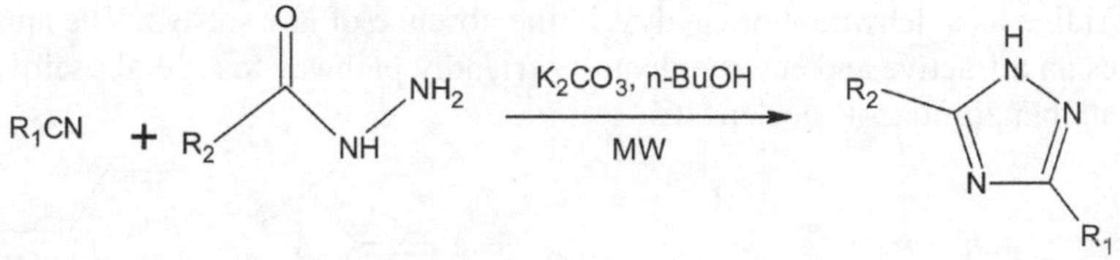

### 11.1.7.6 Preparation of Benzimidazoles

The efficiency of an Ugi/de-Boc/cyclization strategy for the construction of heterocyclic compounds has been improved through the incorporation of microwave and fluorous technologies. In the synthesis of substituted quinoxalinones and benzimidazoles, a fluorous-Boc protected diamine is employed for the Ugi reactions. Both the Ugi and the post-condensation reaction proceed rapidly under microwave irradiation and the reaction mixtures are purified by solid-phase extraction (SPE) over fluoro flash cartridges [63].

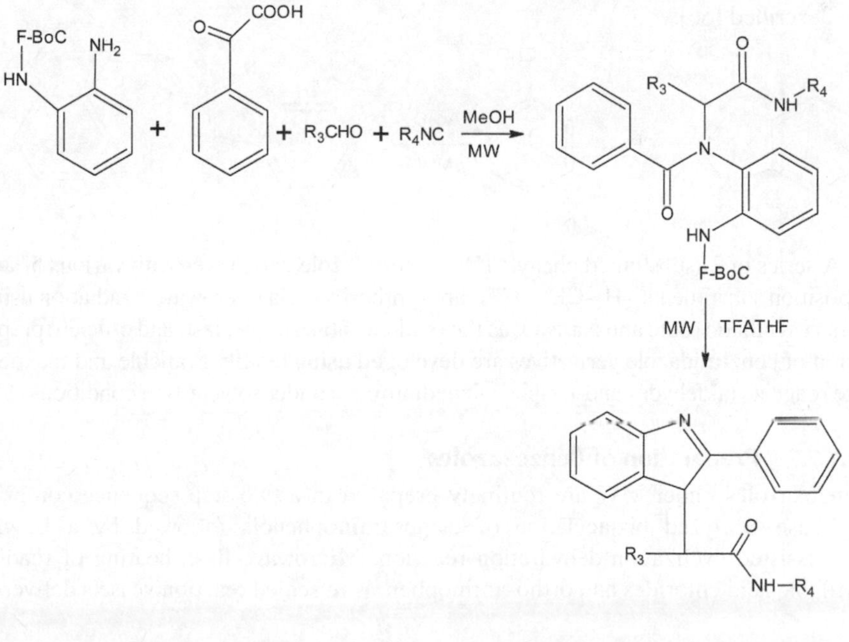

An efficient and simple synthesis of several 2-arylbenzimidazoles from the reaction of 4-methyl-1,2-phenylenediamine and aromatic carboxylic acids in the presence of zeolite catalyst is reported. The reactions are performed under microwave irradiation, and the catalyst could be recycled and used several times. The yields of products following recrystallization from benzene or dichloromethane are of the order of 26–97% [64].

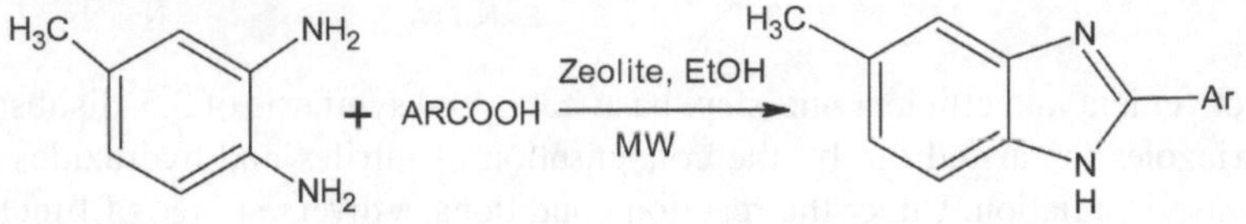

The reactions of 2-methyl benzimidazole or 2-methyl benzimidazolium iodide with aromatic aldehydes are accelerated under microwave irradiation by using $AC_2O$ or piperidine as a dehydrant or catalyst in the absence of any solvent. The approach provides an attractive and environmentally friendly pathway to several useful styryl dyes with benzimidazole nucleus [65].

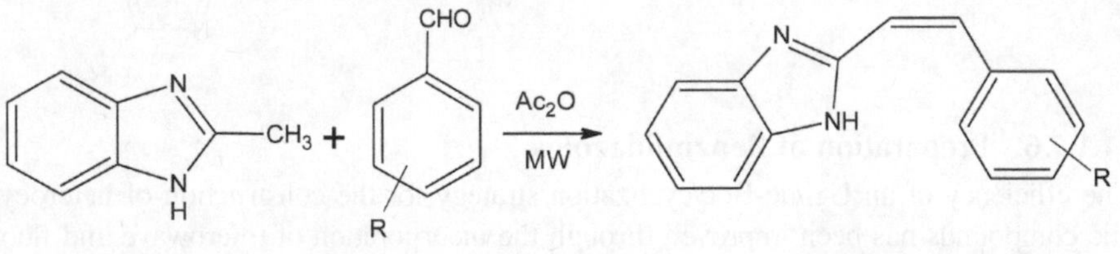

A microwave-assisted method for the synthesis of 2-substituted benzimidazoles in the presence of alumina-methanesulfonic acid (AMA) is reported. In addition, by this method some new bis-benzimidazoles from the direct reaction of phenylenediamine and dicarboxylic acid under microwave irradiation in good to excellent yields are described [66].

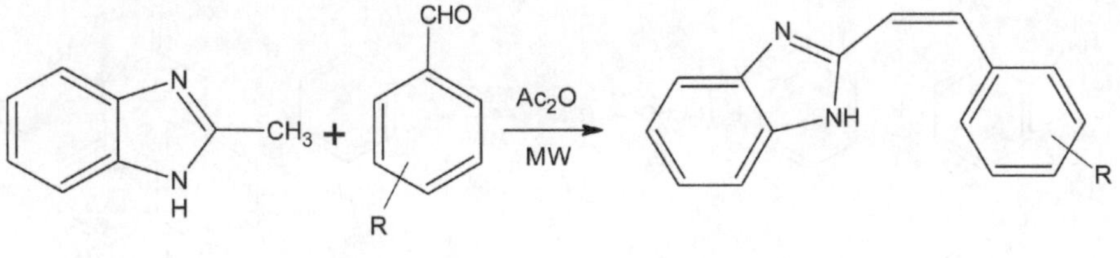

A series of 2-(substituted phenyl)-1H-benzimidazole derivatives with various 5- and 6-position substituents (-H, -CH$_3$, -CF$_3$) are synthesized via microwave irradiation using a short synthetic route and $Na_2S_2O_5$ as the oxidant. This simple, fast, and efficient preparation of benzimidazole derivatives are developed using readily available and inexpensive reagents (aldehydes and 1,2-phenylenediamines) under solvent-free conditions [67].

### 11.1.7.7 Preparation of Benzoxazoles

Benzoxazoles under MW are routinely prepared in a two-step sequence comprising base-catalyzed bis-acylation of ortho-aminophenols followed by a Lewis-acid-assisted cyclization-dehydration reaction. Microwave flash heating of readily available acid chlorides and ortho-aminophenols in sealed reaction vessels delivered benzoxazoles in a one-pot process without the aid of any additive, such as base or Lewis acid [68].

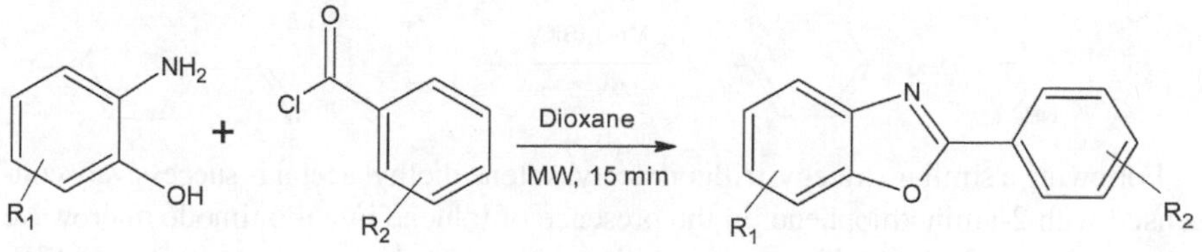

In a closely related publication, carboxylic acids were employed instead of acid chloride in a microwave-assisted direct synthesis of 2-substituted benzoxazoles. The reactions with 2-aminophenol were performed in a household microwave oven and worked well with aromatic, heteroaromatic, $\alpha$, $\beta$-unsaturated and aryl alkyl carboxylic acids [69].

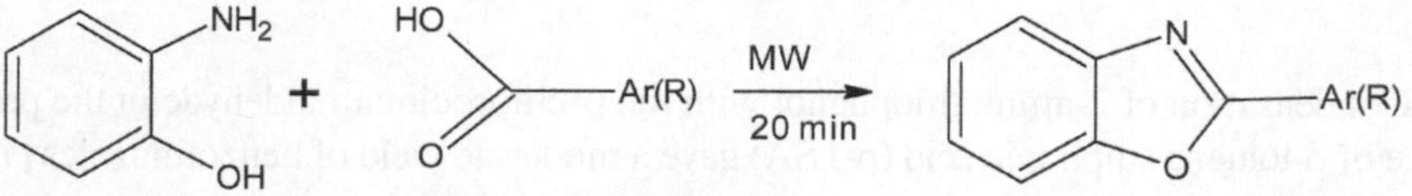

Martinez-Palou and co-workers have generated a 40-membered library of compounds from readily available aromatic 2-substituted amines and fatty acids by means of two efficient microwave-assisted protocols. A parallel synthesis using a monomode microwave (MOMW), an automated procedure using a multimode microwave (MUMW) and a robotic liquid handler are also described [70].

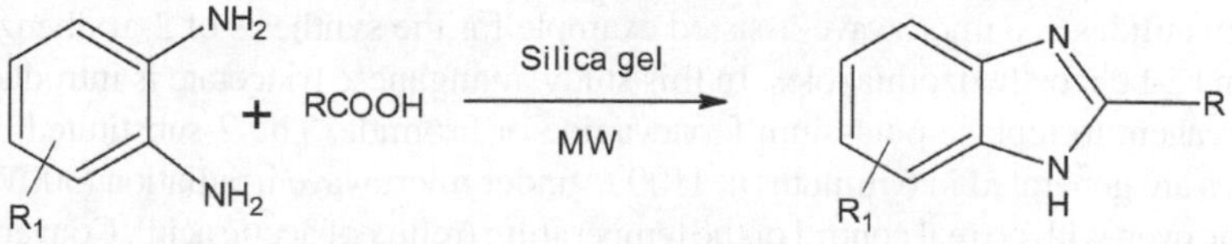

Seijas and co-workers have developed Lawesson's reagent (LW)-mediated microwave-assisted efficient access to benzoxazoles and benzothiazoles from carboxylic acids under solvent-free conditions [71].

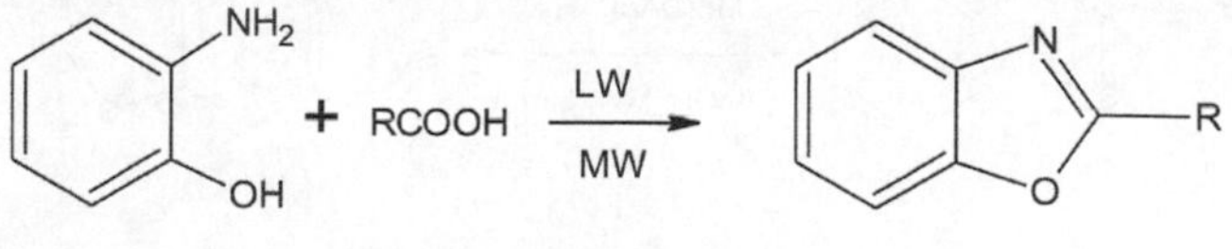

## 11.1.7.8  Preparation of Benzothiazoles

Microwave-assisted synthesis of benzothiazoles can be carried out by the condensation of a dinucleophile, such as 2-aminothiophenol, with an ortho-ester in the presence of KSF clay in a monomode microwave reactor operating at 60W under a nitrogen atmosphere [72].

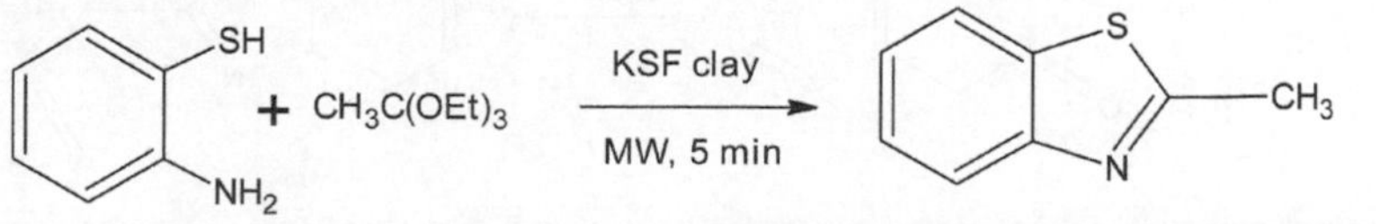

Solvent-free microwave-assisted synthesis of benzothiazoles is also described by an attack of the dinucleophiles on benzaldehydes and benzaldoximines [73].

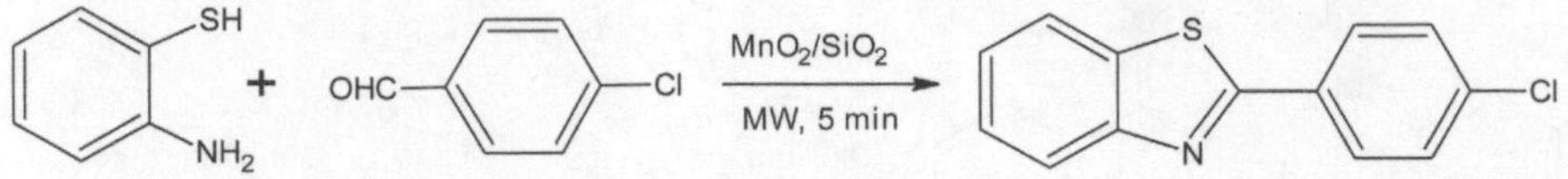

Following a similar strategy, trifluoroacetyl ketene diethyl acetal is successively condensed with 2-aminothiophenol in the presence of toluene in a multimode microwave oven to give the 2-(1,1,1-trifluoroacetonyl)benzothiazole ring in an excellent yield [74].

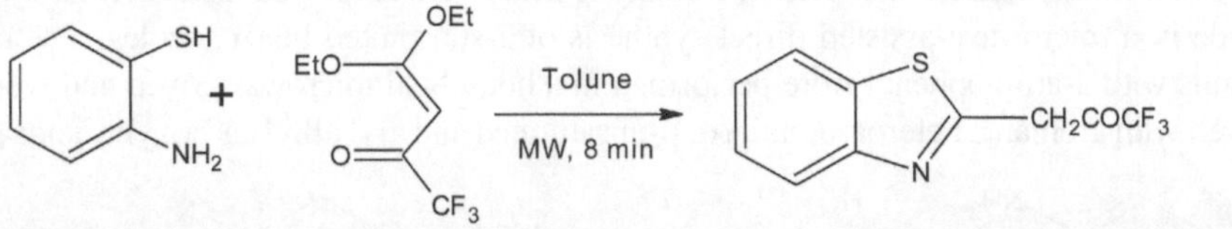

Condensation of 2-aminothiophenol with the β-chlorocinnamaldehyde in the presence of p-toluene sulphonic acid (p-TSA) gave a moderate yield of benzothiazoles [75].

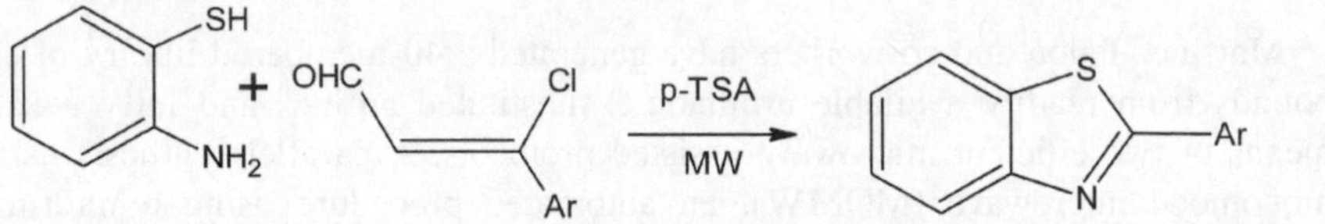

Manganese (III)-promoted radical cyclization of arylthioformanilides and α-benzoyl thioformanilides is a microwave-assisted example for the synthesis of 2-arylbenzothiazoles and 2-benzoylbenzothiazoles. In this study, manganese triacetate is introduced as a new reagent to replace potassium ferricyanide or bromide. The 2-substituted benzothiazoles are generated in 6 minutes at 1100°C under microwave irradiation (300W) in a domestic oven with no real control of the temperature (reflux of acetic acid). Conventional heating (oil bath) of the reaction at 1100°C for 6 hours gave similar yields [76].

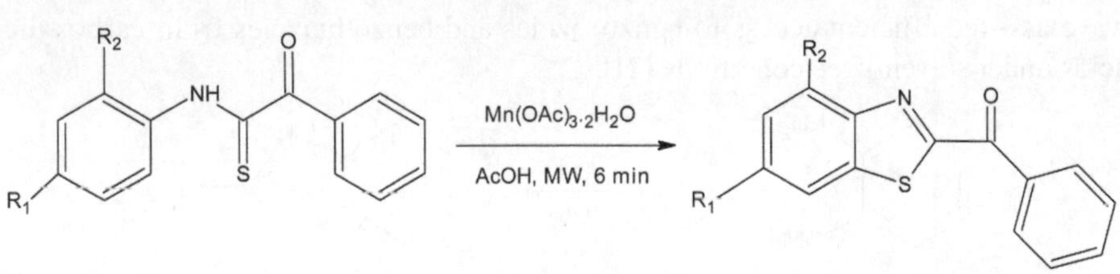

Microwave heating for the synthesis of 2-substituted benzothiazoles such as condensation of aromatic or aliphatic aldehydes with 2-aminothiophenol on $SiO_2$, aromatic aldehydes with 2-aminothiophenol in the presence of nitrobenzene/$SiO_2$, nitrobenzene/montmorillonite K10 or carboxylic acids [77, 78].

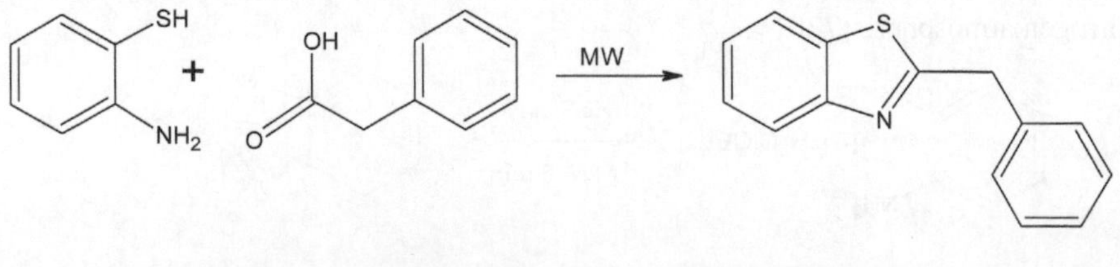

#### 11.1.7.9 Preparation of Imidazoles

The solvent-free microwave-assisted synthesis of 2,4,5-substituted and 1,2,4,5-substituted imidazoles is reported. Imidazoles are obtained as a result of the condensation of a 1,2-dicarbonyl compound with an aldehyde and an amine using acidic alumina impregnated with ammonium acetate as the solid support [79].

A simple, high-yielding synthesis of 2,4,5-trisubstituted imidazoles from 1,2-diketones and aldehydes in the presence of $NH_4OAc$ is described. Under microwave irradiation, alkyl-, aryl- and heteroaryl-substituted imidazoles are formed in yields ranging from 80 to 99%. Short syntheses of lepidiline B and trifenagrel illustrate the utility of this approach [80].

#### 11.1.7.10 Preparation of Indoles

The classical Fischer-indole synthesis from an aryl hydrazine and a ketone is sped-up by several 100-fold using microwave-assisted reactions. The synthesis is achieved in 30 seconds under microwave irradiation and in 2 hours under conventional conditions [81].

## 11.1.7.11    Preparation of 1,3,4-Oxadiazoles and 1,2,4-Oxadiazoles

4-[3-(aryl)-1,2,4-oxadiazol-5-yl]-butan-2-ones are synthesized from methyl levulinate and arylamidoximes. The reaction is carried out in a microwave oven without any solvent in a shorter time with high yields as compared to conventional heating [82].

2,5-disubstituted-1,3,4-oxadiazoles are synthesized from pyridinyl hydrazide and various substituted carboxylic acids under microwave irradiation [83].

1,3,4-Oxadiazoles can be rapidly and efficiently synthesized from a variety of carboxylic acids and acid hydrazides in a simple step. The use of commercially available PS-PPh3 resin combined with microwave heating delivered the product 1,3,4-oxadiazoles in high yields and purities [84].

An efficient, rapid, microwave-accelerated one-step synthesis of some 5-aryl-2-(2-hydroxy-phenyl)-1,3,4-oxadiazoles by reaction of salicylic hydrazide with carboxylic acids in the presence of thionyl chloride under neat conditions is described [85].

Microwave-assisted, as well as conventional, synthesis of 5-substituted-2-(2-methyl-4-nitro-1-imidazomethyl)-1,3,4-oxadiazoles containing the nitroimidazole moiety is carried out [86].

Amidoximes is reacted with isopropenyl acetate in the presence of KSF under microwave irradiation and produces 1,2,4-oxadiazoles. 1,2,4-Oxadiazoles can also be obtained by microwave irradiation from O-acylamidoximes adsorbed on alumina. 1,3,4-Oxadiazoles are obtained by irradiation of bis(acyl)hydrazines in thionyl chloride [87].

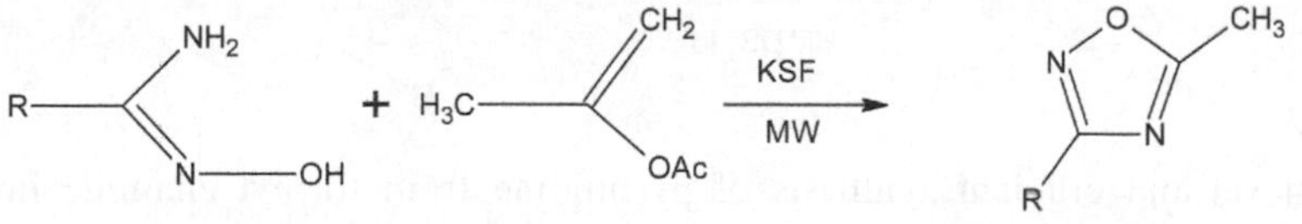

2,5-disubstituted-1,3,4-oxadiazoles are synthesized by the oxidation of 1-aroyl-2-arylidene hydrazines with potassium permanganate on the surface of a solid mineral support as well as in mixtures of acetone and water under microwave irradiation [88].

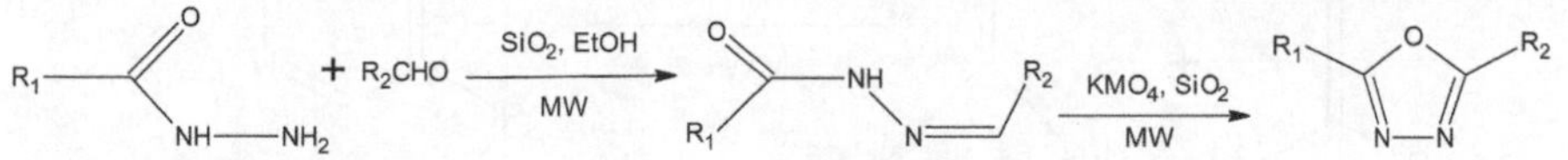

Microwave-assisted synthesis of 1,2,4-oxadiazoles of amidoximes under solvent-free conditions is found to be an efficient method for one-pot synthesis of 1,2,4-oxadiazole derivatives from amidoximes and acylchlorides. The method reported by Kaboudin and Navaee is an easy, rapid and high-yielding reaction for the synthesis of 1,2,4-oxadiazoles [89].

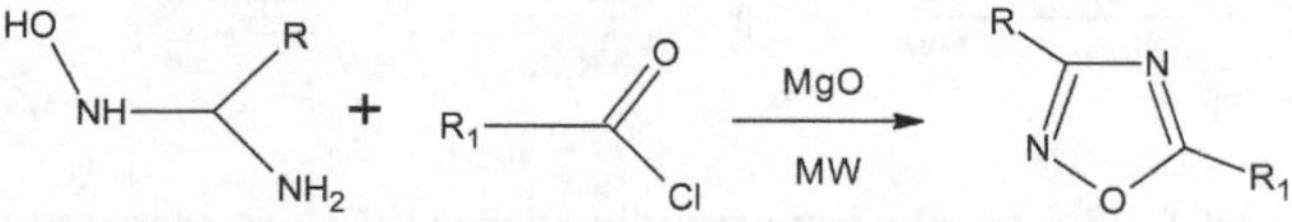

The one-pot three component condensation reaction between nitriles, hydroxylamine and aldehydes for the synthesis of 3,5-disubstituted-1,2,4-oxadiazoles under microwave irradiation and solvent-free conditions in excellent yields are reported [90].

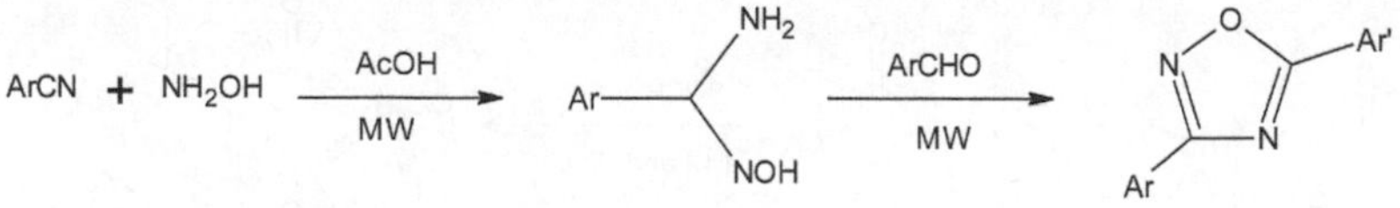

The one-pot microwave-assisted synthesis of substituted-1,2,4-oxadiazoles in solvent and under solvent-free conditions is performed exploring the importance of some coupling reagents. Good yields and shorter reaction times are the main aspects of this method [91].

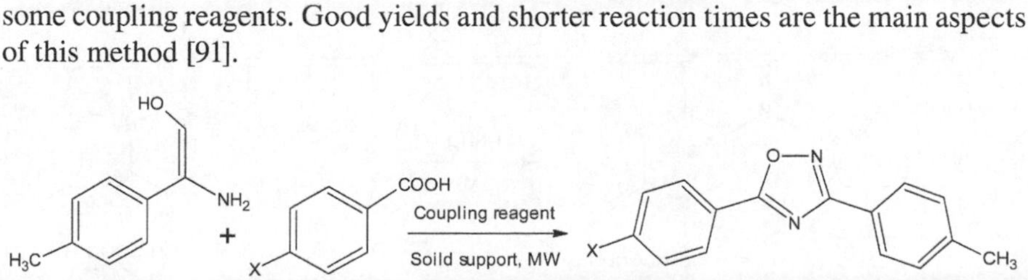

## 11.1.7.12 Preparation of Pyrimidine Derivatives

A simple, high yielding synthesis of pyrimidines from ketones in the presence of HMDS and formamide is reported under microwave irradiation [92].

A novel and efficient synthesis of pyrimidine from formyl enamide involves samarium chloride-catalyzed cyclisation of formyl enamides using urea as a source of ammonia under microwave irradiation [93].

A multicomponent microwave-assisted synthesis of pyrimidine derivatives is has been reported [94].

4-Substituted 1,2,3,4-tetrahydropyrimidine derivatives are prepared by Niharika et al. To a mixture of urea (0.1mole), substituted aldehydes (0.1mole) and ethyl aceto-acetate (0.1mole) in ethanol, 4 drops of concentrated hydrochloric acid is added and refluxed under microwave for 3 minutes at a power level of 840W [95].

Aromatic aldehydes on reaction with barbituric acid and urea/thiourea using dry conditions yielded corresponding pyrimido[4,5d]pyrimidines [96].

### 11.1.7.13 Preparation of Pyrrole Derivatives

Cyclisation of 1,4-diketones is carried out under microwave conditions to yield the corresponding pyrrole derivative. The reaction requires only two minutes under microwave procedure in contrast to conventional conditions that require 12 hours to achieve the conversion [97].

### 11.1.7.14 Preparation of Thiazole Derivatives

A series of substituted 2-amino thiazole derivatives are synthesized by the reaction of substitution of acetophenone with thiourea and iodine in a microwave oven [98].

Equimolar amounts of aryl ketones, thiosemicarbazide and substituted phenacyl bromide are mixed and subjected to microwave irradiation for 30–50 seconds at a heating of 300W to produce hydrazinyl thiazoles [99].

A novel one-pot three-component reaction is developed for the synthesis of thiazole derivatives from thioamides, α-haloketones and ammonium acetate at 110°C and/or under microwave irradiation under solvent-free conditions [100].

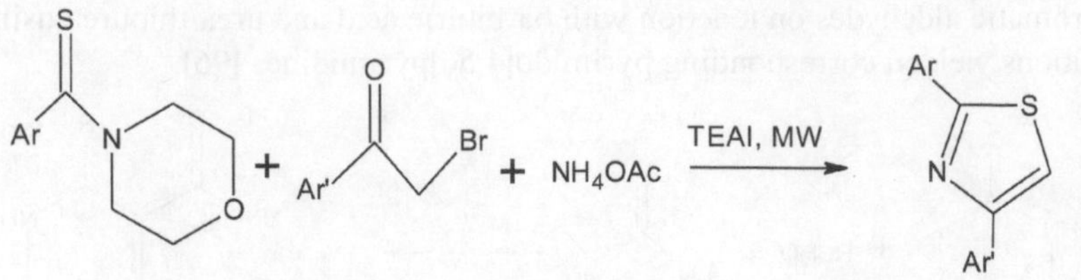

The synthesis of a series of 5-(2′-indolyl)-thiazoles, reported from the reaction of thioamides with 3-tosyloxypentane-2,4-dione, led to *in situ* formations of 5-acetyl-thiazole upon treatment with arylhydrazines in polyphosphoric acid [101].

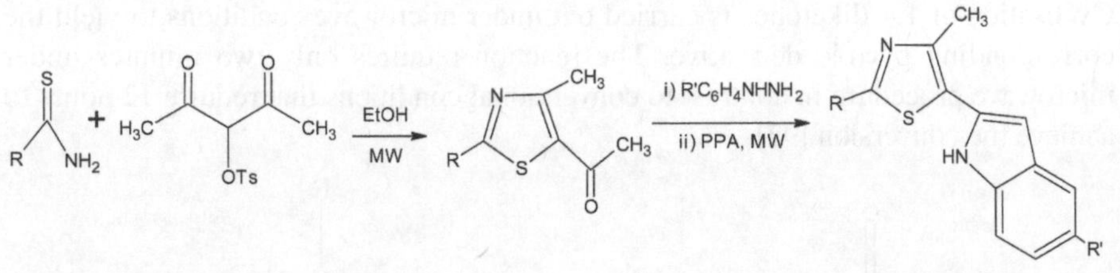

One-step synthesis of various substituted 2-cyanobenzothiazole is described by the condensation of the corresponding substituted ortho-aminothiophenol with eth-ylcyanoformate, employing an effective amount of Lawesson's reagent, under micro-wave irradiation and solvent-free conditions within a short period of time [102].

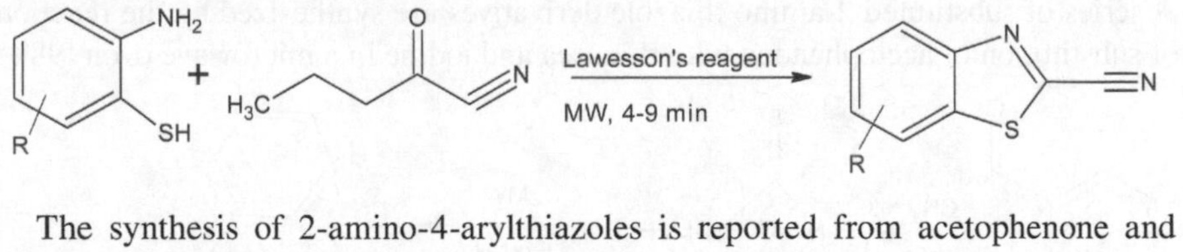

The synthesis of 2-amino-4-arylthiazoles is reported from acetophenone and thiourea under microwave irradiation [103].

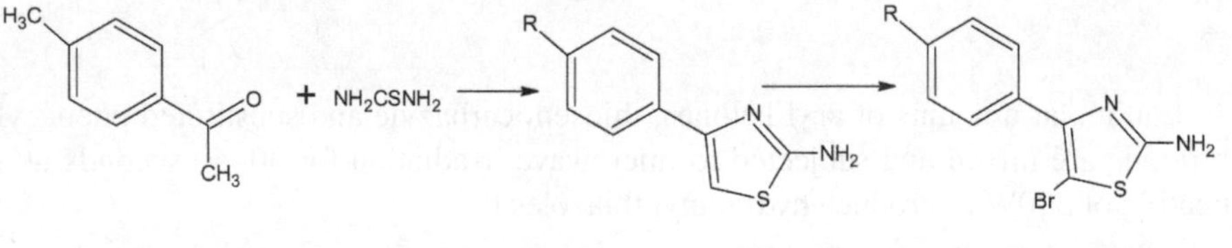

4-(o-Methoxyphenyl)-2-aminothiazole and 3,5-dichloro-salicylaldehyde are mixed with each other in a mortar andpestle and the reaction mixture is placed in a small conical flask at room temperature, then 1 ml of alcohol is added. The mixture is then exposed to microwave irradiation at 10% power for 10–20 seconds [104].

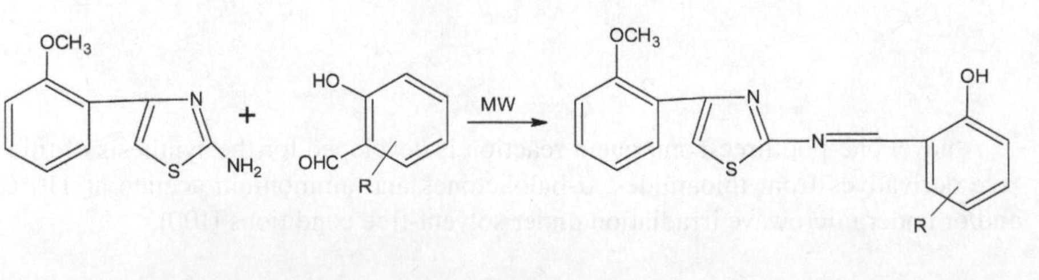

## 11.1.7.15 Preparation of 1,3,4-Thiadiazole Derivatives

Some novel compounds, namely, 3-(2-methyl-1H-indol-3-yl)-6-aryl-[1,2,4] triazolo[3,4-b][1,3,4]thiadiazoles via bromination of 2-methyl-3-[4-(arylideamino)-5-mercapto-4H- [1,2,4triazol-3-yl]-1H indoles are synthesized [105].

N-Phenyl thiosemicarbazide is synthesized from aromatic amine by refluxing with CS$_2$ and hydrazinehydrate in ethanol and from phenyl isothiocyanate by reacting with hydrazine hydrate in ethanol. The synthesized thiosemicarbazides where condensed with aromatic carboxylic acid in the presence of concentrated H$_2$SO$_4$ to form thiadiazole analogues [106].

Thiadiazole compounds are designed as anti-diabetic agents using docking studies. The designed thiadiazole derivatives are synthesized by cyclisation between aromatic acid and thiosemicarbazide using concentrated H$_2$SO$_4$ and condensing the product in the presence of aldehyde by microwave irradiation [107].

5-Phenyl-1,3,4-thiadiazol-2-amine is synthesized by reacting benzoic acid with thiosemicarbazide, which, on reaction with different aromatic aldehydes, afforded 5-phenyl-N-[(E)-phenylmethylidene]-1,3,4-thiadiazol-2-amine derivatives . The compounds on treatment with thioglycolic acid in the presence of ZnCl$_2$ gave 2-phenyl-3-(5-phenyl-1,3,4-thiadiazol-2-yl)-1,3- thiazolidin-4-one [108].

Thiadiazole is synthesized from aromatic acid and thiosemicarbazide, and Schiff bases are prepared by reacting with different aldehydes and isatin using glacial acetic acid as a catalyst [109].

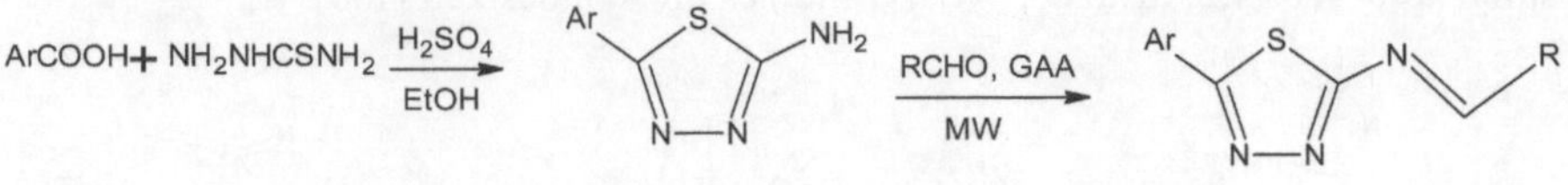

## 11.1.8  PHARMACOLOGICAL ACTIVITIES

### 11.1.8.1  Triazole Derivatives

1-[5-(Substituted aryl)-1H-pyrazol-3-yl]-3,5-diphenyl-1H–1,2,4-triazole derivatives were synthesized by the cyclisation of 1-(3,5-diphenyl-1H-1,2,4-triazol-1-yl)-3-(substituted aryl)prop-2-en-1-one and hydrazine hydrate in presence of a small amount of glacial acetic acid by Shantaram et al. The analgesic activity is evaluated using the acetic acid-induced writhing (abdominal constriction) test and hot plate method. The results of the study are given in Tables 11.2 and 11.3 [110].

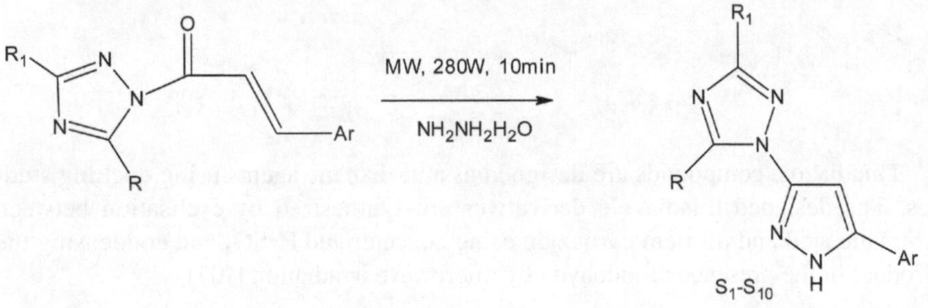

---

**TABLE 11.2**

**Evaluation of Analgesic Activity by Acetic Acid-induced Writhing Method**

| Treatment | Dose (mg/kg) | Percent Protection |
|---|---|---|
| Control | – | – |
| Ibuprofen | 10 | 71 |
| S1 | 100 | 66 |
| S2 | 100 | 61 |
| S3 | 100 | 59 |
| S4 | 100 | 67 |
| S5 | 100 | 53 |
| S6 | 100 | 62 |
| S7 | 100 | 42 |
| S8 | 100 | 48 |
| S9 | 100 | 57 |
| S10 | 100 | 63 |

---

## TABLE 11.3
### Evaluation of Analgesic Activity by Hot Plate Method

| Sr. No. | Treatment | Reaction Time in Seconds After Treatment (Mean ± S.E.M.) | | | %MPE |
| | | 15 min | 30 min | 60 min | |
|---|---|---|---|---|---|
| 1 | Control | 4.75±0.1683 | 4.75±0.2663 | 4.75±0.4541 | – |
| 2 | Pentazocine | 7.70±0.5611** | 9.67±0.4602** | 11.81±0.5254** | 69.02 |
| 3 | S1 | 7.65±0.6346** | 9.83±0.5645** | 11.67+0.4369** | 67.79 |
| 4 | S2 | 7.60±0.5234** | 9.87±0.4865** | 11.54±0.4356** | 66.44 |
| 5 | S3 | 7.52±0.6532** | 9.92±0.5433** | 11.79±0.4156** | 68.92 |
| 6 | S4 | 7.55±0.5585** | 9.72±0.5658** | 11.47±0.5541** | 66.02 |
| 7 | S5 | 6.45±0.5418** | 8.36±0.6256** | 9.27±0.5454** | 44.42 |
| 8 | S6 | 7.42±0.4769** | 9.79±0.5169** | 11.71±0.5075** | 68.21 |
| 9 | S7 | 6.70±0.4559** | 8.49±0.5474** | 9.37±0.5525** | 46.07 |
| 10 | S8 | 7.52±0.5585** | 9.77±0.5456** | 10.82±0.6084** | 60.65 |
| 11 | S9 | 7.39±0.4646** | 9.53±0.5136** | 10.53±0.6359** | 57.18 |
| 12 | S10 | 7.56±0.6636** | 9.59±0.5552** | 10.37±0.4427** | 55.13 |

*Note:* ** $p < 0.01$ Significant Difference

Some triazole derivatives are prepared by using a microwave oven. Antibacterial activities are determined by filter paper disc method against *Bacillus cereus* and *Klebsiella pneumoniae* bacteria. The investigation of antibacterial screening data revealed that the synthesized heterocycles exhibited moderate to promising antimicrobial activity against a moderate range of bacterial stains. The results are given in Table 11.4 [111].

## TABLE 11.4
### Antibacterial Activity of the Synthesized Compounds

| Compound No. | Zone of Inhibition (mm) | |
| | BC | KN |
|---|---|---|
| 2a | 14.8 | 13.7 |
| 2b | 14.1 | 13.1 |
| 2c | 13.2 | 11.3 |
| 2d | 8.7 | 7.1 |
| 2e | 8.2 | 6.9 |
| 3a | 7.8 | 7.1 |
| 3b | 13.9 | 13.5 |
| 3c | 12.7 | 12.1 |
| 3d | 8.2 | 7.9 |
| 3e | 7.9 | 7.3 |
| Gatifloxacin | 18.3 | 15.4 |

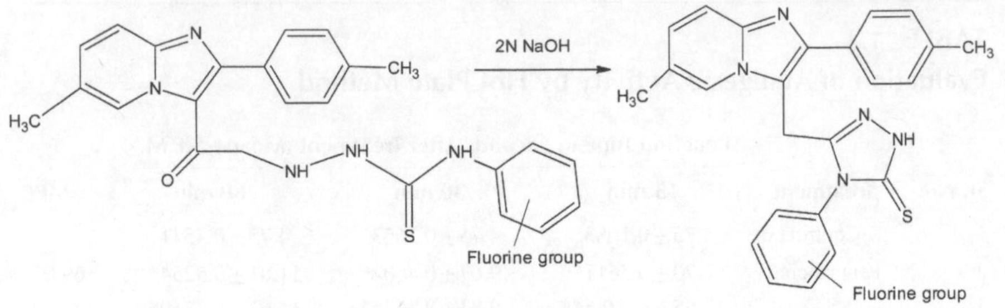

A series of 3-(4-methylcoumarinyl-7-oxymethyl)-6-substitutedphenyl-5,6-dihydr o-s-triazolo (3,4-b)(1,3,4)-thiadiazoles are prepared by reacting 5-(4-methyl couma rinyl-7-oxymethyl)-4-amino-3-mercapto(4H)-1,2,4-triazole with various aromatic aldehydes by microwave-assisted organic synthesis . Antimicrobial activity is deter-mined using the disc diffusion method by measuring the inhibition zone in mm. The compounds are screened *in vitro* for antibacterial activity against *Staphylococcus aureus* and *Escherichia coli* and antifungal activity against *Candida albicans*. The compounds are also evaluated for antioxidant activity using the DPPH method. The results of the study are given in Table 11.5 [112].

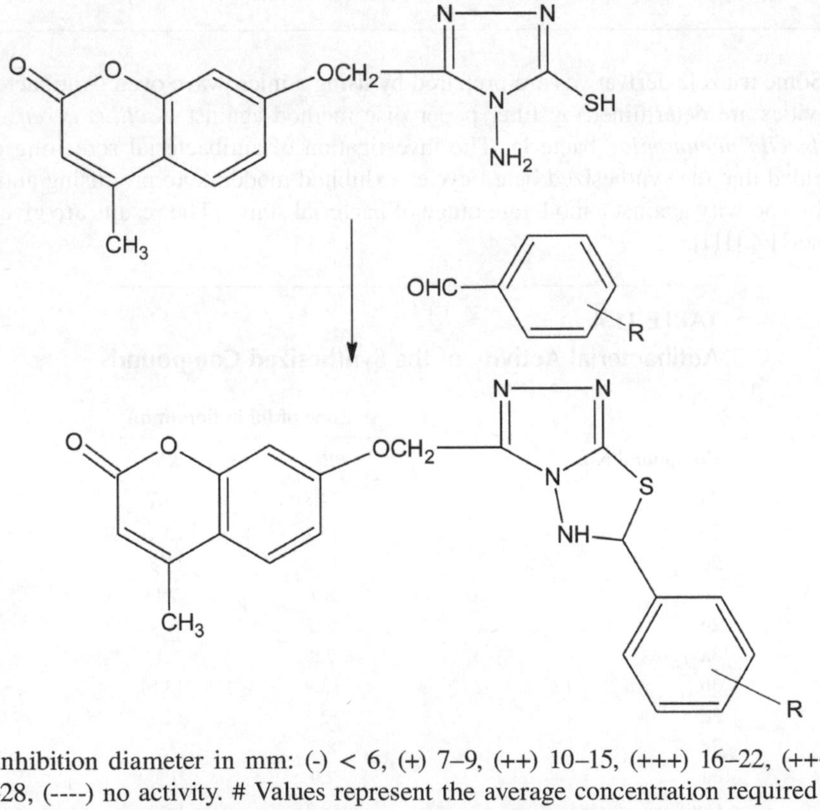

Inhibition diameter in mm: (-) < 6, (+) 7–9, (++) 10–15, (+++) 16–22, (++++) 23–28, (----) no activity. # Values represent the average concentration required for exerting 50% of antioxidant activity from three separate tests.

**TABLE 11.5**

**Antimicrobial and Antioxidant Activity Data of the Synthesized Compounds**

| Compd. | Antibacterial Activity | | Antifungal Activity | Antioxidant Activity |
|---|---|---|---|---|
| | SA | EC | CA | IC50 µg/ml# |
| 2a | − | − | − | 7.6 |
| 2b | ++ | +++ | ++ | 7.7 |
| 2c | − | ++ | − | − |
| 2d | ++ | ++ | − | 5.7 |
| 2e | − | ++ | − | 9.2 |
| 2f | ++ | ++ | − | 8.2 |
| 2g | − | − | − | 7.5 |
| 2h | − | − | ++ | 5.6 |
| 2i | − | − | − | 9.6 |
| 2j | − | − | ++ | − |
| Ciprofloxacin | ++++ | ++++ | ---- | ---- |
| Fluconazole | ---- | ---- | ++++ | ---- |
| Ascorbic acid | ---- | ---- | ---- | 5.4 |

Some triazolidine derivatives are prepared by using the microwave technique for their antimicrobial activity against the following microorganisms: *Escherichia coli, Pseudomonas putida, Bacillus subtilis, Streptococcus lactis, Aspergillus niger* and *Candida albicans*. Anti-oxidant activity screening assay is carried out by the ABTS method. Some compounds showed moderate or weak antimicrobial activity and promising antioxidant activity. The results of the study are given in Tables 11.6 and 11.7 [113].

**TABLE 11.6**

**Anti-Oxidant Assays by ABTS Method**

| Compounds | Inhibition (%) |
|---|---|
| 4a | 66.7 |
| 4b | 63.0 |
| 5a | 25.9 |
| 5b | 22.2 |
| 6a | 44.4 |
| 6b | 40.7 |
| 7 | 79.6 |
| 8 | 81.5 |
| ABTS control | 0 |
| Ascorbic acid | 88.9 |

**TABLE 11.7**

**Antimicrobial Activities of the Newly Synthesized Compounds**

| | Inhibition Zone (mm) | | | | | | |
|---|---|---|---|---|---|---|---|
| | Gram-negative | | Gram-positive | | Fungi | Yeast | |
| Comp. No. | EC | PP | BS | SL | AN | PS | CA |
| 4a | 8 | 4 | 4 | 6 | 3 | 2 | 0 |
| 4b | 0 | 0 | 0 | 0 | 0 | 0 | 0 |
| 5a | 5 | 3 | 5 | 5 | 4 | 3 | 0 |
| 5b | 10 | 9 | 10 | 8 | 6 | 5 | 0 |
| 6a | 0 | 0 | 0 | 0 | 0 | 0 | 0 |
| 6b | 0 | 0 | 0 | 0 | 0 | 0 | 0 |
| 7 | 0 | 0 | 0 | 0 | 0 | 0 | 0 |
| 8 | 10 | 8 | 8 | 8 | 5 | 5 | 0 |
| Chloram phenicol | 22 | 21 | 18 | 20 | 12 | 12 | 0 |
| Ampicilin | 24 | 20 | 19 | 24 | 15 | 14 | 14 |

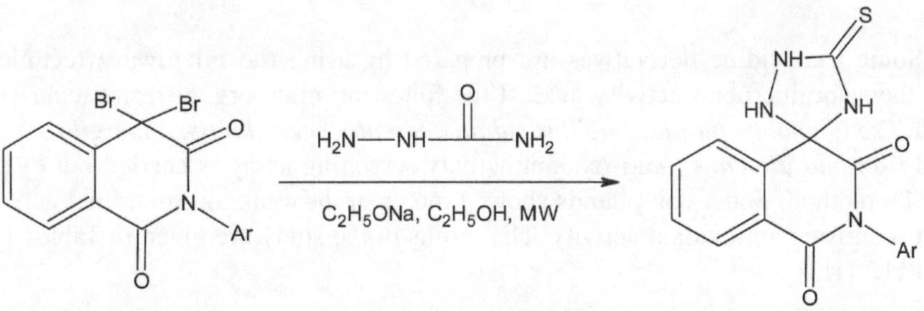

### 11.1.8.2 Thiazole Derivatives

A new series of thiazolyl-coumarin derivatives were synthesized by Moustafa et al. The new hybrid compounds are tested for *in vitro* antitumor efficacy over cervical (Hela) and kidney fibroblast (COS-7) cancer cells. Compounds 5f, 5h, 5m and 5r displayed promising efficacy toward the Hela cell line. In addition, 5h and 5r were found to be the most active candidates toward the COS-7 cell line. The four active analogs, 5f, 5h, 5m and 5r, are screened for *in vivo* antitumor activity over EAC cells in mice, as well as *in vitro* cytotoxicity toward W138 normal cells. Results illustrated that 5r has the highest *in vivo* activity and that the four analogs are less cytotoxic than 5-FU toward W138 normal cells. In this study, 3D pharmacophore analysis is performed to investigate the matching pharmacophoric features of the synthesized compounds with trichostatin A. In silico studies showed that the investigated compounds meet the optimal needs for good oral absorption with no expected toxicity hazards. The results of the study are given in Table 11.8 [114].

**TABLE 11.8**
**In Vitro Antitumor Activity of 5a-t toward Hela and COS-7 Cancer Cell Lines**

| Comp. No. | IC50 (μM) | | Comp. No. | IC50 (μM) | |
|-----------|-----------|-----------|-----------|-----------|-----------|
| | Hela | COS-7 | | Hela | COS-7 |
| 5a | >50 | >50 | 5l | >50 | >50 |
| 5b | >50 | >50 | 5m | 6.25 | 12.50 |
| 5c | >50 | >50 | 5n | >50 | >50 |
| 5d | >50 | >50 | 5o | >50 | >50 |
| 5e | >50 | >50 | 5p | >50 | >50 |
| 5f | 1.90 | >50 | 5q | >50 | >50 |
| 5g | >50 | >50 | 5r | 1.29 | 1.66 |
| 5h | 1.42 | 1.96 | 5s | >50 | >50 |
| 5i | >50 | >50 | 5t | >50 | >50 |
| 5j | >50 | >50 | Doxorubicin | 2.05 | 3.04 |
| 5k | >50 | >50 | – | – | – |

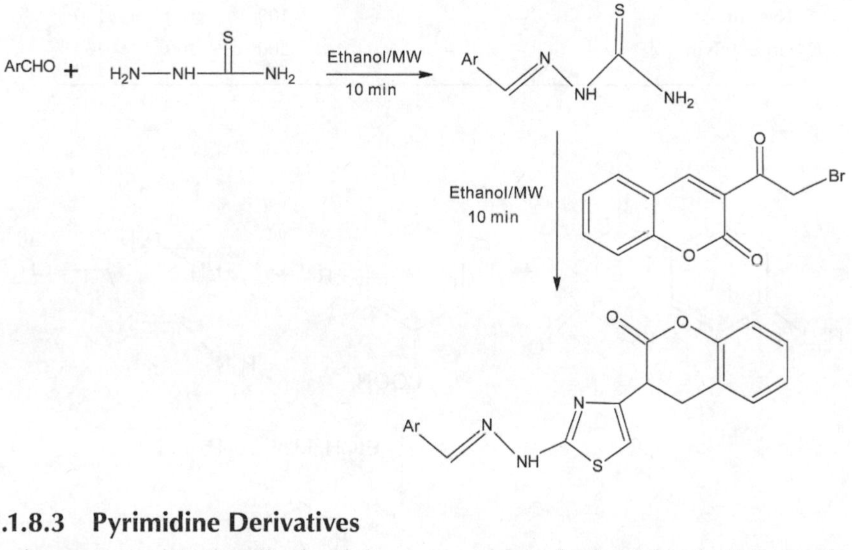

### 11.1.8.3 Pyrimidine Derivatives

A microwave-assisted synthesis of three new series of 1,2,4-triazolo [1,5-a]pyrimidines (PK-101 to PK-110) is synthesized by the mixture of 5-(methylthio)-2H-1,2,4-triazol-3-amine, 4,4,4-trifluoro-1-(4-methoxyphenyl)butane-1,3-dione and an appropriate aromatic aldehyde in ethanol under microwave conditions at 120°C for 10–15 minutes by Parthiv KC. The newly synthesized compounds are subjected to various biological activities viz., antimicrobial, antimycobacterial, anticancer and antiviral. All of the synthesized compounds (PK-101 to 110) are tested for their antibacterial and antifungal activity (MIC) in vitro by the broth dilution method. The results of the study are given in Table 11.9 [115].

**TABLE 11.9**

**Minimum Inhibition Concentration of the Test Substances**

| Compounds | Minimum Inhibition Concentration ($\mu$g mL$^{-1}$) | | | | | | |
|---|---|---|---|---|---|---|---|
| | SA | SP | EC | PA | CA | AN | AC |
| PK 101 | 500 | 500 | 1000 | >1000 | 1000 | 500 | 125 |
| PK 102 | 125 | 100 | 50 | 250 | 500 | 1000 | 500 |
| PK 103 | 100 | 1000 | 250 | 1000 | 1000 | 500 | 1000 |
| PK 104 | 50 | 500 | 250 | 250 | >1000 | 1000 | >1000 |
| PK 105 | 500 | 1000 | 500 | 1000 | 500 | 500 | 100 |
| PK 106 | 125 | 25 | 100 | 100 | 500 | >1000 | 500 |
| PK 107 | 250 | 1000 | 1000 | 250 | 500 | 1000 | 1000 |
| PK 108 | 100 | 250 | 100 | 500 | 500 | 1000 | 1000 |
| PK 109 | 250 | 125 | 250 | 250 | 100 | 1000 | 250 |
| PK 110 | 62.5 | 100 | 100 | 250 | 250 | 500 | 250 |
| Ampicillin | 250 | 100 | 100 | 100 | – | – | – |
| Chloramphenicol | 50 | 50 | 50 | 50 | – | – | – |
| Ciprofloxacin | 50 | 50 | 25 | 25 | – | – | – |
| Norfloxacin | 10 | 10 | 10 | 10 | – | – | – |
| Nystatin | – | – | – | – | 100 | 100 | 100 |
| Griseofulvin | – | – | – | – | 500 | 100 | 100 |

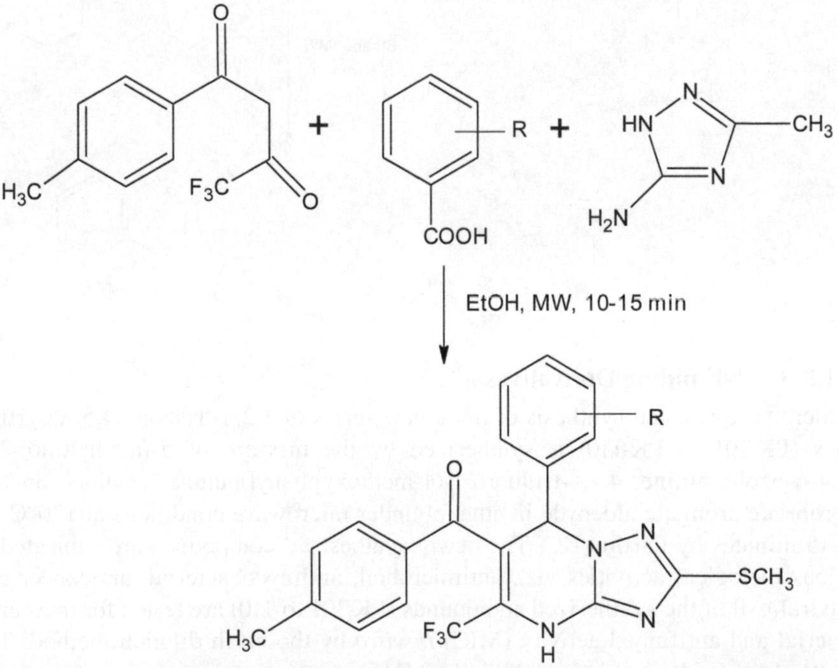

A simple protocol for the efficient preparation of aryl- and heteroaryl-substituted dihydropyrimidinone was achieved via initial Knoevenagel, subsequent addition and final cyclization of aldehyde, ethylcyanoacetate and guanidine nitrate in the presence of piperidine as a catalyst in solvent-free under microwave irradiation by Anjna et al. The anti-inflammatory activity is determined *in vivo* using the carrageenan-induced rat paw edema test. Antibacterial activity of the prepared compounds is tested by the disk diffusion method. An antifungal susceptibility test is done by the disk diffusion method using Sabouraud's dextrose agar medium. The synthesized compounds showed good anti-inflammatory, antibacterial and antifungal activity. The results of the study are presented in Table 11.10 [116].

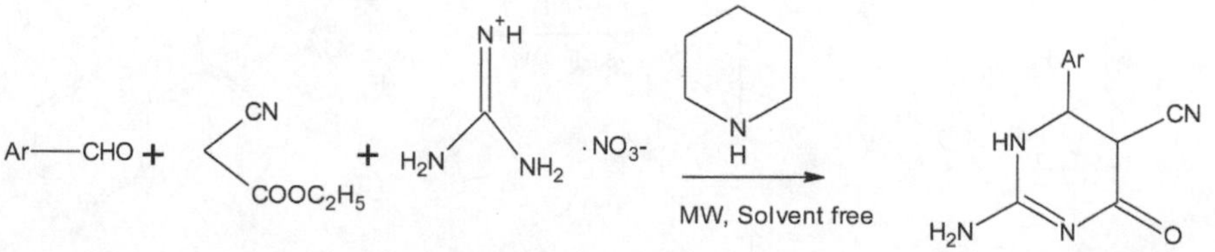

Synthesis of pyrimidino-4,6-(2,4-diazepine) derivatives are achieved by Green and one-pot synthesis using 1,2-diketones and 4,6-diaminopyrimidines as the precursor of condensation reaction and CTAB as the cationic surfactant. For the purpose of determining the efficiency of the substituent on 1,2-diketones, three different forms of diketones were used viz., Benzil, 4,4'-dimethyl Benzil and 4,4'-difluoro Benzil. The results showed that the use of 4,4'-difluoro Benzil, was better among the three Benzils, considering its yield and the time consumption. All three types of benzodiazepine derivatives obtained are independently used for evaluating their anxiolytic, sedative and hypnotic activity using albino mice as the subject. When the resultant pyrimidino-benzodiazepines is compared with standard diazepam, the following results are obtained: sedative and hypnotic activity > standard diazepam, but anxiolytic activity < standard diazepam. In most cases, fluoro substituted diazepine

**TABLE 11.10**
**Antimicrobial Evaluation of Synthesized Compounds**

| Compounds | Diameter of Zone of Inhibition in mm | | | | | | |
|---|---|---|---|---|---|---|---|
| | SA | BS | EC | PA | CA | AF | AN |
| 4d | 18 | 17 | 24 | 20 | 21 | 20 | 19 |
| 4e | 10 | 12 | 10 | 13 | 16 | 17 | 16 |
| 4f | 16 | 15 | 20 | 20 | 18 | 17 | 20 |
| 4g | 15 | 16 | 16 | 18 | 17 | 18 | 16 |
| 4h | 18 | 17 | 15 | 17 | 17 | 15 | 16 |
| Norfloxacin | 20 | 22 | 22 | 23 | – | – | – |
| Clotrimazole | – | – | – | – | 20 | 22 | 21 |

show deviation in activity. The results in locomotor activity suggest the appreciable alertness of the mice when exposed to pyrimidine-based diazepine drugs. To assess the anti-anxiety activity of test compounds, elevated plus maze (EPM) is used. To evaluate the sedative and hypnotic activity of the synthesized compounds, the Hole-Board Test is used. To evaluate the spontaneous locomotor activity of the synthesized compounds, an actophotometer is used. The improvement in locomotor activity when compared with diazepam shows that this drug is better than diazepam. The results of these studies are given in Table 11.11 [117].

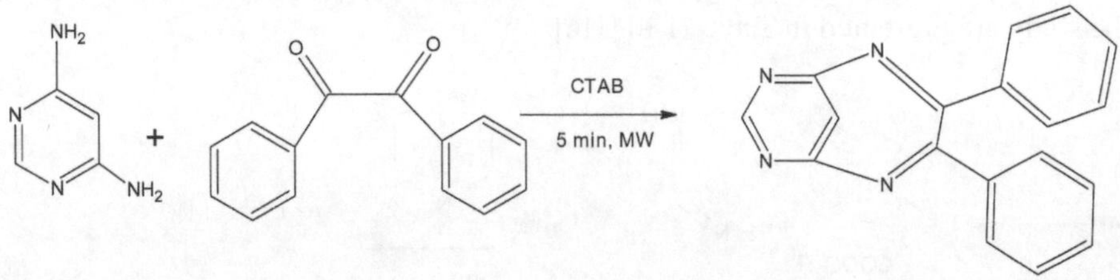

Spiro-pyrimidinethiones/spiro-pyrimidinones-barbituric acid derivatives were synthesized in a simple and efficient method using the one-pot three-component reaction of a cyclic 1,3-dicarbonyl compound (barbituric acid), an aromatic aldehyde and urea or thiourea in the presence of nanoporous silica SBA-Pr-SO$_3$H under solvent-free conditions by Ghodsi et al. Urease inhibitory activity of spiro compounds were tested against Jack bean urease using Berthelot alkaline phenol–hypochlorite method. Five of the thirteen compounds were inhibitors and two of them were enzyme activators. Analysis of the docking results showed that, in

**TABLE 11.11**
**Urease Inhibitory Activity**

| Product | Concentration (mM) | Inhibition (%) |
|---------|--------------------|-----------------|
| 4a | 1.7 | 48±0.026 |
| 4b | 1.0 | 2.8±0.046 |
| 4c | 1.0 | – |
| 4d | 1.7 | 51±0.052 |
| 4e | 1.0 | – |
| 4f | 1.0 | – |
| 4g | 1.0 | – |
| 4h | 1.7 | 59±0.019 |
| 4i | 1.0 | 11±0.042 |
| 4j | 1.7 | −42±0.061 |
| 4k | 1.0 | −13±0.039 |
| 4l | 1.0 | – |
| 4m | 1.0 | – |
| Standard | 0.38 | 90±0.023 |

most of the spiro molecules, one of the carbonyl groups was coordinated with both nickel atoms, while the other one was involved in the formation of hydrogen bonds with important active-site residues. The effect of inserting two methyl groups on –N-atoms of barbiturate ring, S-substituted compounds and ortho-, meta- and para-substituted compounds are investigated too. The results of the study are given in Table 11.11 [118].

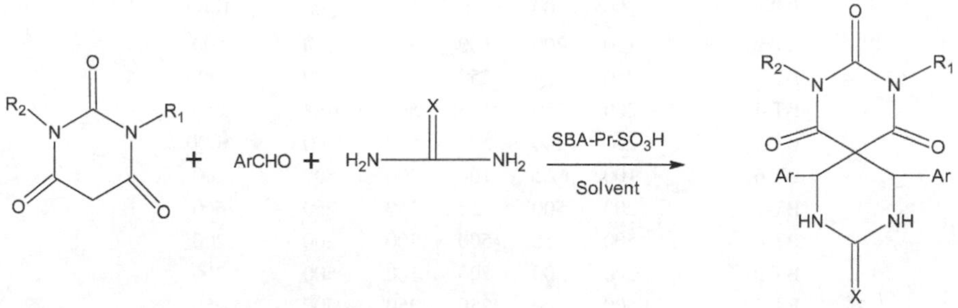

### 11.1.8.4 Benzimidazole Derivatives

Some new 2-substituted benzimidazole derivatives were synthesized from the microwave irradiation method by condensation of 2-nitroaniline with different carboxylic acids (aliphatic, aromatic and hetrocyclic) by Kuldeep et al. The compounds synthesized are identified by 1HNMR and FT-IR (Fourier-transform infrared spectroscopy) spectroscopic techniques. All compounds studied in this work are screened for their *in vitro* antimicrobial activities against the standard strains: *Escherichia coli*, *Pseudomonas aeruginosa*, *Bacillus subtilis*, *Bacillus pumilus*, *Candida albicans* and *Aspergillus niger* using agar well diffusion method. The diameter (in mm) of zone of inhibition is determined by the agar well diffusion method. Compound 2-pyridin-3-yl-1Hbenzimidazole (1f) is found to be the most active antimicrobial compound amongst the series. Compounds 2-(2-chloro-4-nitro-phenyl)-1H-benzimidazole (1a), 2-(1H-benzimidazol-2-yl)-6-nitro-benzoic acid (1e) also showed good antimicrobial activity [119].

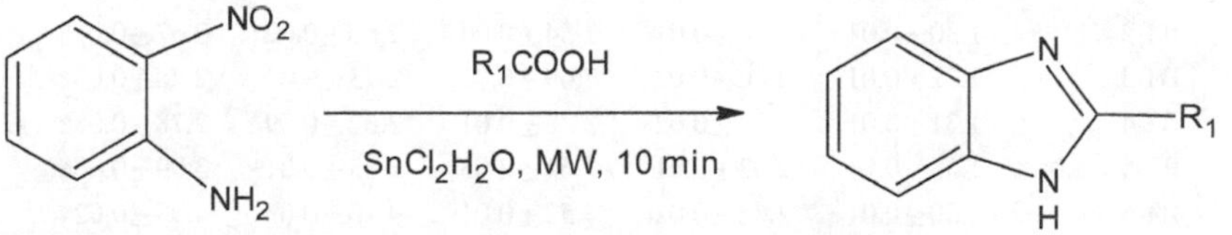

### 11.1.0.5 Benzothiazole Derivatives

Synthesis of benzothiazole derivatives (BT-1 to BT-10) are carried out from 2-aminothiophenol, aromatic aldehyde and phenyl iodonium bis-trifluoroacetate (PIFA) in ethanol. The synthesized compounds are subjected to antimicrobial screening by estimating the minimum inhibitory concentration (MIC) by adopting the serial dilution technique. All the synthesized newer benzothiazole derivatives (BT-1 to BT-10)

## TABLE 11.12
### Antimicrobial Study Benzothiazole Derivatives

| | Minimal Inhibitory Concentration (μg/ml) | | | | | |
| | Antibacterial Activity | | | | Antifungal Activity | |
| Compound | SA | SP | EC | PA | CA | AN |
|---|---|---|---|---|---|---|
| BT-1 | 100 | 100 | 250 | 250 | 200 | 1000 |
| BT-2 | 100 | 200 | 100 | 500 | 250 | 500 |
| BT-3 | 250 | 250 | 250 | 100 | 500 | 1000 |
| BT-4 | 200 | 250 | 125 | 500 | 1000 | 250 |
| BT-5 | 250 | 200 | 500 | 500 | 500 | 1000 |
| BT-6 | 100 | 62.5 | 100 | 200 | 500 | 200 |
| BT-7 | 200 | 500 | 62.5 | 250 | 250 | 500 |
| BT-8 | 500 | 250 | 500 | 200 | 200 | 200 |
| BT-9 | 200 | 500 | 200 | 100 | 500 | 200 |
| BT-10 | 500 | 200 | 250 | 250 | 500 | 250 |
| Streptomycin | 250 | 100 | 100 | 100 | NT | NT |
| Griseofulvin | NT | NT | NT | NT | 500 | 100 |

are evaluated in terms of analgesic activity by tail immersion test. All the tested compounds displayed varying degrees of analgesic activity. Benzothiazole derivatives bearing pyrazolyl system exhibited comparable or slightly less potent activity than the standard pentazocine. The results of the study are given in Tables 11.12 and 11.13 [120].

## TABLE 11.13
### Analgesic Activities of Benzothiazole Derivatives

| | Comparative Analgesic Potency to Pentazocine After Time in Minutes | | | | |
| Compound | 10 min. | 15 min. | 30 min. | 60 min. | 90 min |
|---|---|---|---|---|---|
| BT-1 | $1.30 \pm 0.01$ | $2.46 \pm 0.02$ | $2.50 \pm 0.02$ | $2.55 \pm 0.02$ | $2.61 \pm 0.02*$ |
| BT-2 | $1.30 \pm 0.01$ | $2.50 \pm 0.01$ | $2.54 \pm 0.04$ | $2.60 \pm 0.02$ | $2.67 \pm 0.01*$ |
| BT-3 | $1.32 \pm 0.01$ | $2.66 \pm 0.02$ | $2.69 \pm 0.02$ | $2.73 \pm 0.03$ | $2.78 \pm 0.03*$ |
| BT-4 | $1.31 \pm 0.01$ | $2.33 \pm 0.02$ | $2.38 \pm 0.01$ | $2.62 \pm 0.09$ | $2.78 \pm 0.08*$ |
| BT-5 | $1.30 \pm 0.01$ | $2.63 \pm 0.01$ | $2.66 \pm 0.03$ | $2.83 \pm 0.01$ | $3.09 \pm 0.08*$ |
| BT-6 | $1.30 \pm 0.01$ | $2.68 \pm 0.02$ | $3.52 \pm 0.01$ | $4.10 \pm 0.03$ | $4.98 \pm 0.02*$ |
| BT-7 | $1.31 \pm 0.01$ | $2.69 \pm 0.01$ | $3.62 \pm 0.01$ | $4.37 \pm 0.04$ | $5.00 \pm 0.01*$ |
| BT-8 | $1.30 \pm 0.01$ | $2.55 \pm 0.02$ | $3.49 \pm 0.01$ | $3.99 \pm 0.03$ | $4.70 \pm 0.02*$ |
| BT-9 | $1.31 \pm 0.01$ | $2.59 \pm 0.02$ | $3.59 \pm 0.02$ | $4.00 \pm 0.02$ | $4.64 \pm 0.03*$ |
| BT-10 | $1.31 \pm 0.01$ | $2.46 \pm 0.02$ | $2.52 \pm 0.02$ | $2.56 \pm 0.02$ | $2.61 \pm 0.01*$ |
| Gum acacia | $1.30 \pm 0.01$ | $1.24 \pm 0.01$ | $1.12 \pm 0.01$ | $1.15 \pm 0.01$ | $1.30 \pm 0.01*$ |
| Pentazocine | $1.30 \pm 0.01$ | $6.31 \pm 0.03$ | $6.39 \pm 0.04$ | $6.54 \pm 0.03$ | $6.72 \pm 0.02*$ |

Note: * $p < 0.05$ Non Significant

## 11.2 CONCLUSIONS

Microwave chemistry has already been combined with other enabling technologies and strategies, such as multi-component reactions, solid-phase organic synthesis or combinatorial chemistry. The combination of multidisciplinary approaches with microwave heating encourages scientists to initiate new and unexplored areas of complex organic synthesis. To date, microwave irradiation still needs to find promising combinations with more techniques to satisfy the increased synthetic demands in industry and academia. All in all, microwave heating is undoubtedly a bonanza for organic chemistry researchers. In the future, with lower costs, microwave synthesizers will become an integral part of and a standard technology in most synthetic laboratories and will continually make a valuable impact on organic synthesis.

## ABBREVIATIONS

| | |
|---|---|
| **%MPE** | Percent maximum possible effect |
| **°C** | Degree centigrade |
| **5-FU** | 5-Fluoro uracil |
| **ABTS** | 2,2′-azino-bis(3-ethylbenzothiazoline-6-sulphonic acid) |
| *AC* | *Aspergillus clavatus* |
| **Ac₂O** | Acetic anhydride |
| *AF* | *Aspergillus flavus* |
| **AMA** | Alumina-methanesulfonic acid |
| *AN* | *Aspergillus niger* |
| **BC** | *Bacillus cereus* |
| **BS** | *Bacillus subtilis* |
| **BT** | Benzothiazole |
| **BT** | Bulk temperature |
| *CA* | *Candida albicans* |
| **CMS** | Conventional microwave synthesis |
| **CS₂** | Carbon disulfide |
| **CTAB** | Cetrimonium bromide |
| **DNA** | Deoxyribonucleic acid |
| **DPPH** | 2,2-diphenyl-1-picrylhydrazyl |
| *EC* | *Escherichia coli* |
| **EPM** | Elevated plus maze |
| **FT-IR** | Fourier-transform infrared spectroscopy |
| **GHz** | Gigahertz |
| **H₂SO₄** | Sulfuric acid |

| | |
|---|---|
| **HMDS** | Hexamethyl disilazane |
| **IC50** | Half maximal inhibitory concentration |
| **KMnO₄** | Potassium permanganate |
| *KN* | *Klebsiella pneumoniae* |
| **KOH** | Potassium hydroxide |
| **LW** | Lawesson's reagent |
| **MASFOS** | Microwave-assisted solvent-free organic synthesis |
| **MHz** | Megahertz |
| **MIC** | Minimum inhibition concentration |
| **ml** | Milliliter |
| **mm** | Millimeter |
| **MPE** | Maximal possible effect |
| **MUMW** | Multimode microwave |
| **MW** | Microwave |
| **MWI** | Microwave Irradiation |
| **NaOH** | Sodium hydroxide |
| **NH₄OAc** | Ammonium acetate |
| **NMR** | Nuclear magnetic resonance |
| *PA* | *Pseudomonas aeruginosa* |
| **PCC** | Pyridinium chlorochromate |
| **PIFA** | Phenyliodonium bis-trifluoroacetate |
| *PP* | *Pseudomonas putida* |
| **PS** | Penicillium species |
| **PTC** | Phase transfer catalysis |
| **p-TSA** | p-toluene sulphonic acid |
| **S.E.M** | Standard error of mean |
| *SA* | *Staphylococcus aureus* |
| *SL* | *Streptococcus lactis* |
| *SP* | *Streptococcus pyogenes* |
| **SPE** | Solid-phase extraction |
| **TBAB** | Tetrabutyl ammonium bromide |
| **TLC** | Thin layer chromatography |
| **W** | Watt |

## REFERENCES

1. Ajmer SG, Karunesh K, Sonika R, Shashikant B. 2013. Microwave-assisted synthesis: A green chemistry approach. *Int. Res. J. Pharm. App. Sci.* 3(5): 278–285.
2. Somani R, Pawar S, Nikam S. 2010. Microwave-assisted synthesis and antimicrobial activity of Schiff bases. *Int. J. ChemTech Res.* 2: 860–864.
3. Rajak H, Mishra P. 2004. Microwave-assisted combinatorial chemistry: The potential approach for acceleration of drug discovery. *J. Sci. Ind Res.* 63(8): 641–654.
4. Wathey B, Tierney J, Lidstrom P, Westman J. 2002. The impact of microwave-assisted organic chemistry on drug discovery. *Drug Discov. Today* 7(6): 373–380.
5. Lidstrom P, Tierney J, Wathey B, Westman J. 2001. Microwave-assisted organic synthesis—A review. *Tetrahedron* 57(45): 9225–9283.
6. Gabriel C, Gabriel S, Grant EH, Grant EH, Halstead BSJ, Mingos DMP. 1998. Dielectric parameters relevant to microwave dielectric heating. *Chem. Soc. Rev.* 27(3): 213–224.

7. Strauss CR, Trainor RW. 1995. Developments in microwave-assisted organic chemistry. *Aust. J. Chem.* 48(10): 1665–1692.
8. Langa F, Cruz PD, Hoz AD, Diaz-Ortiz A, Diez-Barra E. 1997. Microwave irradiation: more than just a method for accelerating reactions. *Contemp. Org. Synth.* 4(5): 373–386.
9. Gaba M, Dhingra N. 2011. Microwave chemistry: General features and applications. *Indian J. Pharm. Educ. Res.* 45(2): 175–183
10. Montes I, Sanabria D, Garcia M, Castro J, Fajardo J. 2006. A greener approach to aspirin synthesis using microwave irradiation. *J. Chem. Educ.* 83(4): 628–631.
11. Loupy A, Petit A, Hamelin J, Texier-Boullet F, Jacquault P, Mathe D. 1998. New solvent free organic synthesis using focused microwaves. *Synthesis* 9: 1213–1234.
12. Krstenansky JL, Cotterill I. 2000. Recent advances in microwave-assisted organic syntheses. *Curr. Opin. Drug. Discov. Dev.* 3(4): 454–461.
13. Wilson NS, Sarko CR, Roth GP. 2004. Development and applications of a practical continuous flow microwave cell. *Org. Proc. Res. Dev.* 8(3): 535–538.
14. Charde MS, Shukla A, Bukhariya V, Chakole RD. 2012. A review on: A significance of microwave assist technique in green chemistry. *Int. J. Phytopharm.* 2(2): 39–50.
15. Sekhon BS. 2010. Microwave-assisted pharmaceutical synthesis: An overview. *Int. J. PharmTech Res.* 2(1): 827–833.
16. Larhed M, Hallberg A. 2001. Microwave-assisted high-speed chemistry: A new technique in drug discovery. *Drug Discov. Today* 6(8): 406–416.
17. Lew A, Krutzik PO, Hart ME, Chamberlin AR. 2002. Increasing rates of reaction: Microwave-assisted organic synthesis for combinatorial chemistry. *J. Comb. Chem.* 4(2): 95–105.
18. Ley SV, Baxendale IR. 2002. New tools and concepts for modern organic synthesis. *Nat. Rev. Drug Discov.* 1(8): 573–586.
19. Hayes BL. 2004. Recent advances in microwave-assisted synthesis *Aldrichimica Acta* 37(2): 66–76.
20. Chemat-Djenni Z, Hamada B, Chemat F. 2007. Atmospheric pressure microwave assisted heterogeneous catalytic reactions. *Molecules* 12(7): 1399–1409.
21. Liu Y, Lu Y, Liu P, Gao R, Yin Y. 1998. Effects of microwaves on selective oxidation of toluene to benzoic acid over a $V_2O_5/TiO_2$ system. *Appl. Catal. A. Gen.* 170(2): 207–214.
22. Gupta M, Paul S, Gupta R. 2009. General characteristics and applications of microwaves in organic synthesis. *Acta Chimica Slovenica* 56: 749–764.
23. Breslow R. 1991. *Acc. Chem. Res.* 24: 159–164.
24. Bose AK, Manhas MS, Banik BK, Robb EW. 1994. *Res. Chem. Intermed* 20: 1–11.
25. Martelanc M, Kranjc K, Polanc S, Koevar M. 2005. *Green Chem.* 7: 737–741.
26. Gedye RN, Rank W, Westaway KC. 1991. *Can. J. Chem.* 69: 700.
27. Hren J, Kranjc K, Polanc S, Koevar M. 2008. *Synthesis* 452–458.
28. Herrero MA, Kremsner JM, Kappe CO. 2008. *J. Org. Chem.* 73: 36–47.
29. Gedye RN, Smith E, Westaway K, Ali Baldisera H, Laberge L, Rousell J. 1986. *Tetrahedron Lett.* 27: 279.
30. Gedye MN, Smith FE, Westaway KC. 1988. *Can. J. Chem.* 66: 17.
31. Jones GB, Chapman BJ. 1993. *J. Org. Chem.* 58: 5558.
32. Katritzky AR, Zhang Y, Singh SK, Steel PJ. 2003. *Arkivoc* xv: 47–64.
33. Vass A, Dudas J, Varma RS. 1999. *Tetrahedron Lett.* 40: 4951–4954.
34. Perreux L, Loupy A, Volatron F. 2002. *Tetrahedron* 58: 2155–2162.
35. Paul S, Gupta M, Gupta R, Loupy A. 2001. *Tetrahedron Lett.* 42: 3827–3829.
36. Kabza KG, Chapados BR, Gestwicki JE, McGrath JL. 2000. *J. Org. Chem.* 65: 1210–1214.
37. Kappe CO. 2004. *Angew. Chem. Int. Ed.* 43: 6250–6255.
38. Hayes BL. 2004. *Aldrichim. Acta* 37: 66–69.

39. Varma RS. 2006. *Indian J. Chem. Sec. B* 45B: 2305–2307.
40. Seijas A, Vazquez-Tato MP. 2007. *Chim. Oggi.* 25: 20–26.
41. Seijas JA, Vazquez-Tato MP, Martinez MM, Corredoira GN. 1999. *J. Chem. Res.* 420–425.
42. Dahmani Z, Rahmouni M, Brugidou R, Bazureau JP, Hamelin J. 1998. *Tetrahedron Lett.* 39: 8453–8456.
43. Scharn D, Wenschuh H, Reineke U, Schneider-Mergener J, Germeroth L. 2000. *J. Comb. Chem.* 2: 361–369.
44. Bogda D, Pielichowski J, Borona A. 1996. *Synlett* 873–874.
45. Chakraborty D, Bordoloi M. 1999. *J. Chem. Res.* 118–122.
46. Gupta M, Wakhloo BP. 2007. *Arkivoc* (i): 94–98.
47. Heravi MM, Farhangi N, Beheshtiha YS, Ghassenizadeh M, Tabar-Hydar K. 2004. *Indian J. Chem.* 43B: 430–431.
48. Xu Q, Chao B, Wang YD, Dittmer C. 1997. *Tetrahedron* 53: 2131–2134.
49. Gutierrez E, Loupy A, Bram G, Ruiz-Hitzky E. 1989. Inorganic solids in "dry media" an efficient way for developing microwave irradiation activated organic reactions. *Tetrahedron Lett.* 30(8): 945.
50. Bosch AI, Cruez PD, Barra ED, Loupy A, Langa F. 1995. *Synlett* 1259.
51. Varma RS, Dahiya R, Kumar S. 1997. Clay catalyzed synthesis of imines and enamines under solvent-free conditions using microwave irradiation. *Tetrahedron Lett.* 38(12): 2039.
52. Villemin D, Martin B. 1995. Dry condensation of creatinine with aldehydes under focused microwave irradiation. *Synth. Commun.* 25: 3135.
53. Singh V, Singh J, Kaur P, Kad GL. 1997. Acceleration of the Pechmann reaction by microwave irradiation: Application to the preparation of coumarins. *J. Chem. Res.* (S): 58.
54. Furniss BS, Hanford AJ. 1989. *Vogel's Textbook of Practical Chemistry*, 5th ed. Pearson Education, Essex, UK, 1161–1162.
55. Aspinall SR. 1940. A synthesis of tetrahydropyrimidines, *J. Am. Chem. Soc.* 2160–2162.
56. Azarifar D, Ghasemnejad H. 2003. Microwave-assisted synthesis of some 3, 5-arylated 2-pyrazolines. *Molecules* 8(8): 642–648.
57. Niementowski SV. 1894. Niementowski quinoline synthesis. *Ber* 27: 1394.
58. Rostamizadeh S, Mollahoseini K, Moghadasi S. 2006. *Phosphorus Sulfur Silicon* 181: 1839.
59. Zamani K, Bagheri S. 2006. *Phosphorus, Sulfur Silicon* 181: 1913.
60. Bentiss F, Lagrenee M, Barby D. 2000. *Tetrahedron Lett.* 41: 1539.
61. Rostamizadeh S, Tajik H, Yazdanfarahi S. 2003. *Synth. Commun.* 33: 113.
62. Yeung KS, Farkas ME, Kadow JF, Meanwell NA. 2005. *Tetrahedron Lett.* 46: 3429.
63. Zhang W, Tempest P. 2004. *Tetrahedron Lett.* 45: 6757.
64. Mobinikhaledi A, Zendehdel M, Jamshidi FH. 2007. Synthesis and reactivity in inorganic. *Metal-Org. Nano-Metal Chem.* 37: 175.
65. Wang LY, Zhang XG, Qi JY, Zhang ZX. 2003. *Chem. Lett.* 14: 1116.
66. Niknam K, Raviz AF. 2007. *J. Iran Chem. Soc.* 4: 438.
67. Vazquez GN, Diaz HM, Soto SE, Torres-Piedra M. 2007. *Synth. Commun.* 37: 2815.
68. Pottorf RS, Chadha NR, Katkevics M, Ozola V, Suna E, Gadi H, Regberg T, Player MR. 2003. *Tetrahedron Lett.* 44: 175.
69. Kumar R, Selvam C, Kaur G, Chakraborti AK. 2005. *Synlett* 1401.
70. Martinez-Palou R, Zepeda LG, Hopfl H, Montoya A, Guzman-Lucero DJ, Guzman J. 2005. *Mol. Divers.* 9: 361.
71. Seijas JA, Vazquez-Tato MP, Carballido-Reboredo MR, Crecente-Campo J, Romar-Lopez L. 2007. *Synlett* 313.
72. Villemin D, Hammedi M, Benoit R. 1996. *Synth. Commun.* 26: 2895.
73. Bougrin K, Loupy A, Soufiaoui M. 1998. *Tetrahedron* 54: 8055.
74. Chandrashekher RA, Shanthan RP, Venkataranam RV. 1997. *Tetrahedron* 53: 5847.

75. Paul S, Gupta M, Gupta R. 2002. *Synth. Commun.* 32: 3541.
76. Mu XJ, Zou JP, Zeng RS, Wu JC. 2002. *Tetrahedron Lett.* 46: 4345.
77. Ben-Alloum A, BakkasS, Soufiaoui M. 1997. *Tetrahedron Lett.* 38: 6395.
78. Chakraborti AK, Selvam C, Kaur G, Bhagat S. 2004. *Synlett* 5: 851.
79. Usyatinsky AY, Khmelnitsky YL. 2000. *Tetrahedron Lett.* 41: 5031.
80. Samanta SK, Kylanlahti I, Kauhaluoma JY. 2005. *Bioorg. Med. Chem. Lett.* 15: 3717.
81. Sridar V. 1997. *Indian J. Chem. B* 36: 86.
82. De Freitas JJR, De Freitas JCR, Da Silva LP, Filho JRF, Kimura GYV, Srivastava RM. 2007. *Tetrahedron Lett.* 48: 6195.
83. Khan KM, Ullah Z, Rani M, Perveen S, Haider SM, Choudhary MI, Rahman A, Voelter W. 2004. *Lett. Org. Chem.* 1: 50.
84. Wang Y, Sauer DR, Djuric SW. 2006. *Tetrahedron Lett.* 47: 105.
85. Aamer S. 2007. *Chem. Heterocycl. Compd.* 8: 43.
86. Frank PV, Girish KS, Kalluraya B. 2007. *J. Chem. Sci.* 119: 41.
87. Oussaid B, Moeini L, Martin B, Villemin D, Garrigues B. 1995. *Synth. Commun.* 25: 1451.
88. Rostamizadeh S, Housaini SAG. 2004. *Tetrahedron Lett.* 45: 8753.
89. Kaboudin B, Navaee K. 2003. *Heterocycles* 60: 2287.
90. Adib M, Jahromi AM, Tavoosi N, Mahdavi M, Bijanzadeh HR. 2006. *Tetrahedron Lett.* 47: 2965.
91. Santagada V, Frecentese F, Perissutti E, Cirillo D, Terracciano S, Caliendo G. 2004. *Bioorg. Med. Chem. Lett.* 14: 4491.
92. Sriram T, Prasun KC. 2005. *Tetrahedron Lett.* 46(46): 7889–7891.
93. Karpov S, Muller TJJ. 2003. *Synthesis* 2815–2826.
94. Xavier AL, Alfredo MS, Emerson PSF, Janaina VA. 2013. Antinociceptive pyrimidine derivatives: Aqueous multicomponent microwave assisted synthesis. *Tetrahedron Lett.* 54(26): 3462–3465.
95. Niharika IS, Sandip SK, Hemlata MN, Praful SC, Jayendrasing PB, Rajesh JO. 2011. Microwave assisted synthesis of 4-substituted-1,2,3,4-tetrahydropyrimidine derivatives. *Int. J. Pharm. Pharmaceut. Sci.* 3(1): 109–111.
96. Shingare MS, Kategaonkar AH, Sadaphal SA, Shelke KF, Shingate BB. 2009. Microwave assisted synthesis of pyrimido[4,5d]pyrimidine derivatives in dry media. *Ukrainica Bioorg. Acta* 1: 3–7.
97. Danks TN. 1999. Microwave assisted synthesis of pyrroles. *Tetrahedron Lett.* 40(20): 3957.
98. Zeki ANA, Hanan AA. 2015. Microwave assisted synthesis, characterizations and antibacterial activity of some of thiazole derivatives. *Res. J. Pharm. Bio. Chem. Sci.* 6(2): 718–725.
99. Rajalakshmi R, Chinnaraja D. 2015. A facile, solvent and catalyst free, microwave assisted one pot synthesis of hydrazinyl thiazole derivatives. *J. Saudi Chem. Soc.* 19: 200–206.
100. Zali-Bocini H, Mansouri SG. 2016. One-step three-component and solvent-free synthesis of thiazoles from tertiary thioamides. *J. Iranian Chem. Soc.* 1–7.
101. Vaddula BR, Reddy B, Tantak M, Sadana R, Gonzalez M, Kumar D. 2016. One-pot synthesis and in-vitro anticancer evaluation of 5-(2′-indolyl)-thiazoles. *Sci. Rep.* 6: 23401.
102. Prajapati N, Vekariya R, Patel H. 2015. Microwave induced facile one-pot access to diverse 2-cyanobenzothiazole-a key intermediate for the synthesis of firefly luciferin. *Int. Lett. Chem. Phys. Astron.* 5(2): 81–89.
103. Castell RM, Villanueva-Novelo C, Caceres-Castillo D, Carballo RM, Quijano-Quinones R, Quesadas-Rojas M, Cantillo-Ciau Z, Cedillo-Rivera R. 2015. 2-Amino-4-arylthiazole derivatives as anti-giardial agents: Synthesis, biological evaluation and QSAR studies. *Open Chem.* 13: 1127–1136.

104. Netaji NK. Antibiofilm activity of thiazole schiff bases. 2016. *Int. J. Chem. Sci.* 14(4): 2535–2545.

105. Gomha SM. 2011. Synthesis under microwave irradiation of [1,2,4]triazolo[3,4-b] [1,3,4] thiadiazoles and other diazoles bearing indole moieties and their antimicrobial evaluation. *Molecules* 16: 8244–8256.

106. Verma AK, Martin A. 2014. Synthesis, characterization and antiinflammatory activity of analogues of 1, 3, 4-thiadiazole. *Int. J. Pharm. Arch.* 3(9): 1–5.

107. Datar PA, Deokule TA. 2014. Design and synthesis of thiadiazole derivatives as anti-diabetic agents. *Med. Chem.* 4(4): 390–399.

108. Harika MS, Nagasudha B, Raghavendra HG. 2014. Synergistic activity of thiadiazole and thiadiazolidinone derivatives against alloxan induced diabetes in rats. *Scholars Acad. J. Pharm.* 3(3): 301–305.

109. Chitale SK, Ramesh B, Bhalgat CM, Jaishree V, Puttaraj C, Bharathi DR. 2011. Synthesis and antioxidant screening of some novel 1, 3, 4-thiadiazole derivatives. *Res. J. Pharm. Technol.* 4(10): 1540–1544.

110. Shantaram GH, Popat BP, Ramdas BP, Appala R. 2014. Microwave assisted synthesis of 1-[5-(substituted aryl)-1H-pyrazol-3-yl]-3,5-diphenyl-1H–1,2,4-triazole as antinoci-ceptive and antimicrobial agents. *Adv. Pharmaceut. Bull.* 4(2): 105–112.

111. Nilesh GS. 2012. Green synthesis, characterization and biological evaluation of some triazole and thiadiazole. *J. Curr. Chem. Pharm. Sci.* 2(2): 100–106.

112. Manoj PK, Ravi TK, Chawla R, Bhuvana S, Sonia G, Gopalakrishnan S. 2010. Microwave assisted aynthesis and biological activity of novel coumarinyltriazolothia-diazoles. *Indian J. Pharmaceut. Sci.* 72(3): 357–360.

113. Mohamed MY, Mahmoud AA. 2010. Microwave assisted synthesis of some new heterocyclic spiro-derivatives with potential antimicrobial and antioxidant Activity. *Molecules* 15: 8827–8840.

114. Moustafa TG, Nadia SE, Eman RE, Mohamed ME, Nanting N. 2017. Microwave-assisted synthesis and antitumor evaluation of a new series of thiazolylcoumarin deriv-atives. *Excli. J.* 16:1114–1131.

115. Parthiv KC. 2016. Microwave assisted synthesis and biological evaluation of 1,2,4-tri-azolo [1,5-A]pyrimidines. *Int. J. Res. Innovation Appl. Sci.* 1(8): 1–5.

116. Anjna B, Srinivasa RJ, Tanuja K, Pradeep P, Shubha J. 2013. Microwave-assisted syn-thesis and biological evaluation of dihydropyrimidinone derivatives as anti-inflamma-tory, antibacterial and antifungal agents. *Int. J. Med. Chem.* Article ID 197612, 5 pages doi: 10.1155/2013/197612.

117. Govindaraj C, Kullagounder S, Venkatesan S. 2017. Green synthesis, characterization and anxiolytic, sedative and hypnotic activity of pyrimidine based diazepine deriva-tives. *Biomed. Res.* 28(2): 525–531.

118. Ghodsi MZ, Shima A, Sakineh F, Massoud A. 2015. Green synthesis and urease inhibi-tory activity of spiro-pyrimidinethiones/spiro-pyrimidinones-barbituric acid deriva-tives. *Iranian J. Pharmaceut. Res.* 14 (4): 1105–114.

119. Kuldeep K, Pathak DP. 2012. Synthesis, characterization and evaluation for antimi-crobial activity of 2-substituted benzimidazole derivatives. *Pharma Innovation* 1(9): 44–50.

120. Kondeti TN, Karumudi BS, Kota C. 2016. Green synthesis, biological evaluation of newer b enzothiazole derivatives. *J. Basic Appl. Res* 2(3): 226–231.

# 12 Microwave-Assisted Green Chemistry Approach
## A Potential Tool for Drug Synthesis in Medicinal Chemistry

*Biswa Mohan Sahoo, Bimal Krishna Banik,
and Jnyanaranjan Panda*

## CONTENTS

## 12.1  INTRODUCTION

Microwave-accelerated drug synthesis is a synthetic green technology used to speed up drug discovery and development processes. A characteristic of microwave-assisted synthesis includes a reduction in the time period of a chemical reaction. This process is uniform and instantaneous and, on many occasions, reactions can be performed under solvent-free conditions with higher product yield [1]. Microwaves are electromagnetic waves with wavelengths ranging from 0.1 to 100cm and frequencies between 0.3 and 300GHz. Figure 12.1 shows the electromagnetic spectrum in which the microwave region lies between infrared (IR) and radio wave frequencies. Microwave energy consists of both an electric and magnetic field [2].

Microwaves have sufficient momentum to activate chemical reactions. The basic mechanisms involved in microwave synthesis are dipolar polarization and ionic conduction. The heating effect in microwave-assisted synthesis is generally due to a change in the dipole moment of the reacting molecules [3]. Microwaves have the ability to convert electromagnetic energy into thermal energy, depending on the dielectric constant of the reactants and solvents. Solvents with high dielectric constants, such as water, methanol, ethyl acetate, acetone, acetic acid, formic acid, dimethyl formamide (DMF) and dimethyl sulfoxide (DMSO), are heated rapidly by microwave radiation, whereas solvents with low dielectric constants, such as chloroform, tetrahydrofuran, toluene, hexane and dichloromethane (DCM), are heated slowly [4]. Due to high polarity and stability at high temperatures, the above solvents are more commonly used as solvents in the microwave synthesis of medicinally important organic compounds [5].

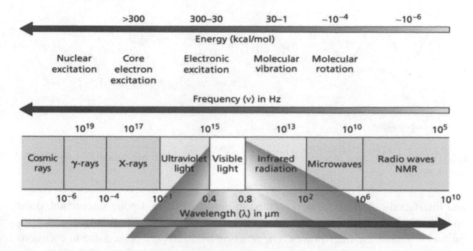

**FIGURE 12.1** Frequencies range of electromagnetic radiation.

Microwave-assisted synthesis is considered as an expedient green tool for the synthesis of new drug molecules. Green Chemistry technology provides environmentally friendly reactions without causing harm to the environment due to less hazardous chemical syntheses and reduced by-product formation [6]. During the 1990s, the Environmental Protection Agency (EPA) of US coined Green Chemistry, which describes the elimination or reduction in use or generation of hazardous chemicals to design and manufacture products with improved efficiency [7].

### 12.1.1 History and Development of Microwave Synthesis

The concept of microwave technology was developed in 1946 when Dr. Percy Le Baron Spencer had conducted laboratory tests on a magnetron, which generates an electromagnetic field. Accidentally, he observed that a candy bar in his pocket had melted on exposure to microwave radiation [8]. So, he applied this microwave technology as a heating method for chemical reactions. In 1947, Dr. Percy devised the first microwave oven for domestic use. Subsequently, the use of microwave irradiation (MWI) to carry out chemical reactions was reported by the groups of Gedye and Giguere/Majetich in 1986. A continuous microwave reactor (CMR) was invented in 1988 for flow through reactions [9]. In 1990, various research groups started experiments based on solvent-free reactions under microwave irradiation which eliminates the risk of any chemical hazards [10]. A laboratory scale microwave batch reactor (MBR) with pressure-resistant vessels had been assembled in 1995. Since 2000, microwave reactors (single-mode and multimode microwaves) are available for the synthesis of various organic compounds of medicinal significance [11].

### 12.1.2 Heating Mechanism of Microwave Synthesis

As mentioned before, the reacting materials of chemical reactions under microwave irradiation are heated by two mechanisms, such as dipolar interaction, ionic conduction

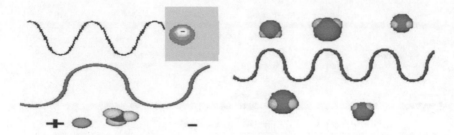

**FIGURE 12.2** Heating mechanism of microwave synthesis. (Ionic conduction and dipolar interaction).

and interfacial polarization, as presented in Figure 12.2. In dipolar interaction, polar ends of the reacting molecules align together and undergo oscillation by the electromagnetic field of the microwaves. The reacting materials are heated due to collisions and friction between the moving molecules in a chemical reaction [12]. But in the case of ionic conduction, free ions or ionic species present in the reaction medium are heated. The microwaves' electric field generates ionic motion due to the reacting molecules orientating themselves to the rapidly changing field, which leads to the instantaneous heating of the molecules [13]. The interfacial polarization method involves the combination of both dipolar polarization and ionic conduction mechanisms [14].

It is essential for the heating of a conducting material dispersed in a non-conducting material, for example, the dispersion of metal particles in sulfur where sulfur does not respond to microwave radiation but metal reflects most of the microwave energy [15]. Based on the reaction of materials towards microwave radiation, the general categories are as follows.

1. Materials are transparent to microwaves, e.g. sulfur
2. Materials reflect microwaves, e.g. copper
3. Materials absorb microwaves, e.g. water

### 12.1.3 MICROWAVE SYNTHESIS APPARATUS

The microwave-assisted synthesis apparatus comprises of two types of ovens, such as single mode and multi-mode microwave ovens. Microwave ovens consist of four components: (a) high power source, (b) waveguide feed, (c) oven cavity and (d) reaction vessel. The high power source consists of a magnetron, which is a thermo-ionic diode. An anode that heats a cathode generates microwaves [16]. The waveguide feed is a rectangular channel which causes the transmission of microwaves from the magnetron to the microwave cavity. The reflective walls prevent the leakage of radiation by increasing the efficiency of the oven. The oven cavity is designed in such a way that it receives large quantities of energy in the form of electric energy, as shown in Figure 12.3. Teflon and polystyrene are used for preparing the reaction vessels because these materials are transparent to microwave radiations [17].

There are various types of reactions involved in microwave synthesis, such as (a) microwave-assisted reactions using solvents, (b) microwave-assisted reactions

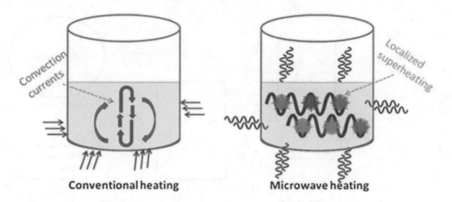

**FIGURE 12.3** Conventional and microwave heating method.

under solvent-free conditions, (c) microwave-assisted reactions using solid phase and (d) microwave-assisted reactions on mineral supports in dry media [18].

### 12.1.4 WORKING PRINCIPLE OF THE MICROWAVE OVEN

In a microwave oven, microwaves are generated by a magnetron. A magnetron is a thermo-ionic diode having an anode and a directly heated cathode. As the cathode is heated, electrons are released and are attracted towards the anode, as shown in Figure 12.4. The anode is made up of an even number of small cavities, each of which acts as a tuned circuit [19]. The anode is, therefore, a series of circuits, which are tuned to oscillate at a specific frequency or at its overtones. A very strong magnetic field is induced axially through the anode assembly and has the ability to bend the path of electrons as they travel from the cathode to the anode [20]. As the deflected electrons pass through the cavity gaps, they induce a small charge into the tuned circuit, resulting in the oscillation of the cavity, as

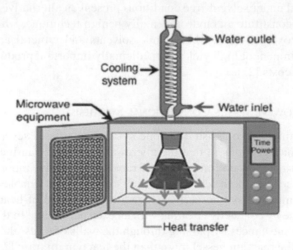

**FIGURE 12.4** Microwave Synthesis Apparatus.

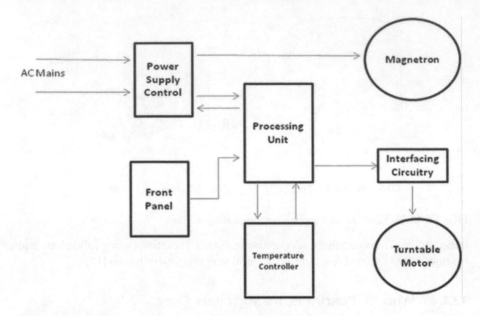

**FIGURE 12.5**  Block diagram of microwave processing unit.

presented Figure 12.5. Alternate cavities are linked by two small wire straps, which ensure the correct phase relationship. This process of oscillation continues until the oscillation has achieved sufficiently high amplitude. It is then taken off of the anode via an antenna. The variable power available in domestic ovens is produced by switching the magnetron on and off according to the duty cycle [21]. Microwave dielectric heating is effective when the matrix has a sufficiently large dielectric loss tangent (i.e. contains molecules possessing a dipole moment). The use of a solvent is not always mandatory for the transport of heat. Therefore, reactions performed under solvent-free conditions present an alternative in microwave chemistry and constitute an environmentally benign technique, which avoids the generation of toxic residues, like organic solvents and mineral acids, and thus allows the attainment of high yields of medicinally important products at reduced environmental costs [22].

### 12.1.5  MICROWAVE VERSUS CONVENTIONAL SYNTHESIS

In Table 12.1, microwave-assisted organic synthesis provides several advantages over conventional reactions in that the microwave allows for an increase in reaction rate, rapid reaction optimization and rapid medicinally important analogue synthesis. It also uses both less energy and solvent and it enables difficult and challenging compound synthesis. In the traditional method of heating, chemical synthesis is done by conductive heating with an external source. In this way, heat is passed into the substances by passing through the walls of a vessel. This results in the heating of the reaction vessel more than the reaction mixture [23]. Microwave synthesis, being a more innovative process of heating, instantaneously heats up the

**TABLE 12.1**

**Comparisons Between Conventional and Microwave Synthesis**

| Sl. No. | Conventional Synthesis | Microwave Synthesis |
|---|---|---|
| 1. | Reaction vessel gets heated first by a heat source (Oil bath, water bath, heating mantle, etc.) and then heat is transferred to the reaction medium. | The reaction mixture is heated directly because the reaction vessel is transparent to microwave radiation. |
| 2. | Physical contact between the reaction vessel and reacting materials is required. | No need for physical contact between the reaction vessel and reacting materials. |
| 3. | Thermal or electric source of heating takes place. | Electromagnetic wave heating takes place. |
| 4. | The heating mechanism involves conduction. | The heating mechanism involves ionic conduction and dipolar interaction. |
| 5. | The rate of heating is lower. | The rate of heating is significantly higher. |
| 6. | The highest temperature of the reaction is limited to the boiling point of the solvents or reaction mixture. | The temperature of the reaction can be raised above the boiling point of the solvents or reaction mixture (Super heating). |
| 7. | The reaction mixture is heated equally. | A specific component of the mixture is heated specifically. |
| 8. | Loss of heat is higher. | Loss of heat is lower. |

molecules present in the reaction mixture as this process does not depend on the thermal conductivity of the vessel. Microwave heating results in the instantaneous localized heating of the reaction mixture. This is due to the fact that microwaves directly couples up with the molecules which are present in the reaction mixtures. Thus, this process of heating is not dependent on the thermal conductivity of the reaction vessels [24].

## 12.1.5.1 Advantages

Uniform heating occurs throughout the material
Selective and high efficiency of heating
Process speed is increased
Desirable chemical and physical effects are produced
Floor space requirements are decreased
Better and rapid process control is achieved
Low operating cost
Activation of catalysts
Reduction in unwanted side reaction
Purity in the final product
Improved reproducibility

Less by-product formation
Environmental heat loss can be avoided
Low energy input
Rapid energy transfer to reacting materials
Use of green or eco-friendly solvents
Less solvent used and more solvent-free reactions achieved
Reduction in heat loss from the reaction vessel
Regio- and chemo-selective
Molecular diversity
Evaporation of solvent is reduced
Versatility of applied reaction conditions

### 12.1.5.2  Disadvantages

1. Heat force control is difficult
2. The closed container is dangerous because it may explode
3. *In situ* monitoring
4. Water evaporation
5. Expensive setup
6. More maintenance cost

### 12.1.5.3  Selection of Solvents in Microwave Synthesis

Table 12.2 provides information regarding the selection of solvents, which is an important factor for the synthesis of medicinally active compounds as most of the reactions take place in the solution phase of the reactant mixtures. The polarity of the solvent plays a major role as polar reaction mixtures have a better ability to couple up with microwave energy [25].

Various factors, such as dielectric constant, dipole moment, tangential delta, dielectric loss and dielectric correlation, affect the polarity of the solvents and in turn alter the absorbing power of the solvent. Dielectric constant is also known as the relative permittivity of the solvent (ability to store its electric charges). Loss tangent is a measure of the ability to absorb microwave energy and convert it into thermal energy or heat [26].

As per Maxwell's equation, $\tan\delta = \varepsilon''/\varepsilon'$ where $\delta$ = dissipation factor

$\varepsilon''$ = loss factor

$\varepsilon'$ = dielectric constant

**TABLE 12.2**
**List of Solvents Used in Microwave Synthesis**

| Name of Solvent | Boiling Point (°C) | Dielectric Constant($\varepsilon'$) | Loss Factors (tan$\delta$) |
|---|---|---|---|
| Water | 100 | 80.4 | 0.123 |
| Formic acid | 100.8 | 58 | 0.722 |
| DMSO | 189 | 47 | 0.825 |
| DMF | 153 | 36.7 | 0.161 |
| Acetonitrile | 82 | 36 | 0.062 |
| Methanol | 64.7 | 32.7 | 0.659 |
| Ethanol | 78.4 | 24.6 | 0.941 |
| Acetone | 56.6 | 20.6 | 0.054 |
| THF | 66 | 7.6 | 0.047 |
| Ethyl acetate | 77.1 | 6.2 | 0.059 |
| Acetic acid | 118.1 | 6.1 | 0.174 |
| Chloroform | 61.7 | 4.8 | 0.091 |
| Toluene | 110.6 | 2.4 | 0.040 |
| Benzene | 80.1 | 2.3 | – |
| Carbon tetrachloride | 76.7 | 2.2 | – |
| Hexane | 68.7 | 1.9 | 0.020 |

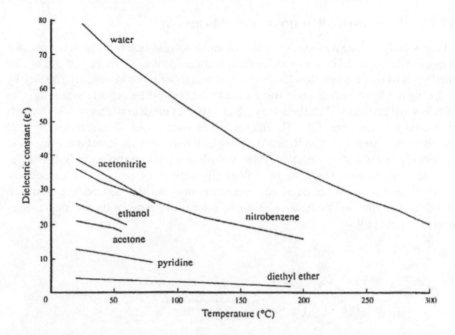

**FIGURE 12.6** Plots of dielectric constant against temperature for various solvents.

So, the reaction medium containing solvent with high tan δ value has efficient absorption of microwave radiation and causes rapid heating.

There are about 26 organic solvents available, out of which alcohols (methanol and ethanol) and alkanes (heptane and hexane) are environmentally friendly solvents, whereas solvents, like dioxane, acetonitrile, acids, formaldehyde and tetrahydrofuran (THF), are not suitable for the environment, as given in Figure 12.6 [27]. It is observed that methanol-water or ethanol-water mixtures are suitable for the environment as compared to pure alcohol. Table 12.3 presents types of solvents which can be classified as high (tanδ>0.5), medium (tanδ 0.1–0.5) and low (tanδ<0.1) microwave absorbing.

**TABLE 12.3**
**Types of Solvents Based on Their Absorbance Level**

| Sl. No. | Name of Solvents | Absorbance Level |
|---------|------------------|------------------|
| 1. | DMSO, Ethanol, Methanol, Propanol, Nitrobenzene, Formic acid, Ethylene glycol | High (tanδ>0.5) |
| 2. | DMF, water, Butanol, Acetonitrile, Acetone, Acetic acid | Medium (tanδ 0.1–0.5) |
| 3. | DCM, Chloroform, Carbon tetrachloride, Ethyl acetate, Pyridine, Toluene, Benzene, Chlorobenzene | Low (tanδ<0.1) |

### 12.1.7  SOLVENT-FREE REACTIONS UNDER MICROWAVE

The synthesis of drug molecules carried out under solvent-free conditions has become a more efficient technique in various chemical reactions. The main purpose of this method is to make a reaction less hazardous and more environmentally friendly by reducing the formation of toxic waste products. Solvent-free organic synthesis is an efficient and economical methodology which provides clean products with high yields in shorter reaction times [28]. The microwave-assisted solvent-free reactions are categorised into three types: (a) Reactions using neat reactants, (b) Reactions using solid mineral supports and (c) Reactions using solid-liquid phase transfer catalysis (PTC).

The heterogenous reactions of biologically active compounds are carried out between neat reactants in quasi equivalent amounts without any adduct. Mannich reaction is performed under a solvent-free microwave irradiation technique using neat reactants [29].

Alumina, silica, clays or zeolite are used as acidic or basic supports based on the type of organic reactions involved. Fluorine undergoes a solvent-free MWI reaction in the presence of $KMnO_4$-alumina to produce 9H-fluoren-9-one [30].

9H-fluorene                                                                9H-fluoren-9-one

Diphenyl-methane undergoes a solvent-free MWI reaction in the presence of $KMnO_4$-alumina to produce benzophenone [31].

Diphenylmethane                                                          Benzophenone

1,3-diphenylthiourea reacts with 2-chloroacetyl chloride in the presence of $K_2CO_3$-alumina and produces 1,3-diphenyl-2-thioxo-imidazolidin-4-one [32].

1,3-diphenylthiourea        2-chloroacetyl chloride        1,3-diphenyl-2-thioxoimidazolidin-4-one

Kidwai and co-workers carried out chemical reactions without using any solvents. Their work involved the synthesis of N-acetylated antibiotic cephalosporin without the use of solvent under microwave irradiation. The 7-amino-cephalosporanic acid adsorbs on the basic alumina and then brought into contact with microwave radiation for two minutes to generate the corresponding product [33].

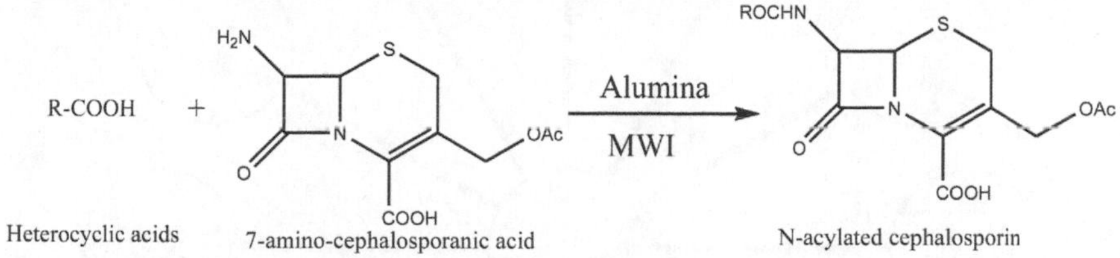

Heterocyclic acids    7-amino-cephalosporanic acid      N-acylated cephalosporin

PTC is specifically for anionic reactions as it involves anionic activation. A catalytic amount of tetra-alkyl ammonium salt or a cation complexing agent is added to the mixture of pure reactants. Reactions occurring in the liquid organic phase consist of electrophiles only and produce simple but medicinally important compounds [34].

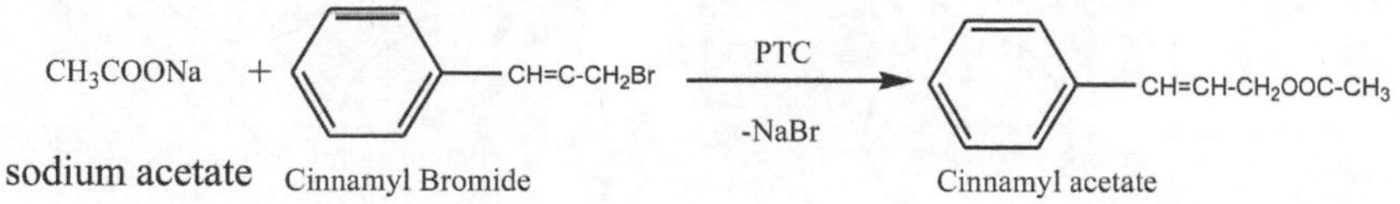

sodium acetate    Cinnamyl Bromide      Cinnamyl acetate

## 12.1.8 GREEN CHEMISTRY APPROACH

Green Chemistry, also called sustainable chemistry, involves a very efficient approach to prevent environmental pollution. The term "Green Chemistry" refers to the invention, design and application of chemical products and processes to reduce or to eliminate the use and generation of hazardous substances. There are 12 principles of Green Chemistry that provide a path for researchers to employ green technology [35]. Microwave-accelerated synthesis is also an important tool which follows a Green Chemistry approach by utilizing eco-friendly, non-hazardous, efficient solvents and catalysts in the production of various drug molecules, as given in Figure 12.7.

An organic synthesis is perfectly efficient or atom-economical if it generates waste which is not visible in percentage yield calculation.

%Yield= (Practical yield/theoretical yield) ×100

%Atom Economy= (Mass of atoms in desired products/Mass of atoms in all reactants) ×100

## 12.1.9 MICROWAVE-ASSISTED CHEMICAL REACTIONS

Figure 12.8 shows the various types of chemical reactions carried out under microwave irradiation, including oxidation reductions, hydrolysis, cycloaddition reactions, rearrangements, condensation reactions, cyclisation reactions, Diels-Alder reaction, dehydration, esterification, epoxidation, Buchwald–Hartwig reactions, Sonogashira reaction, glycosylation reactions, Mitsunobu reactions, Negishi and Kumada

**FIGURE 12.7**    Twelve principles of Green Chemistry Approach.

cross-coupling reactions, Heck reaction, Mannich reaction, Suzuki reaction, protection and deprotection, etc. [36]. These reactions are extensively used in medicinal chemistry.

For example, the Suzuki coupling of aryl chlorides with boronic acids is performed in an aqueous media using a palladium catalyst under microwave irradiation [37].

Mannich reaction is performed by the microwave-induced synthesis of $\beta$-amino ketones (yield of 97%) via the three-component condensation of a substituted methyl ketone, an aldehyde and an amine [38].

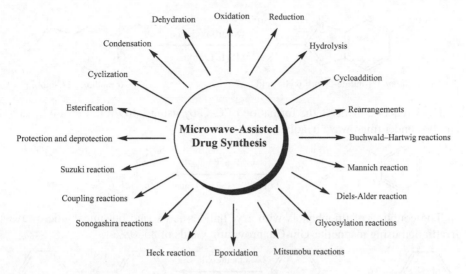

**FIGURE 12.8**   Microwave-assisted chemical reactions.

Hydrolysis of benzyl chloride is carried out under microwave irradiation to produce benzyl alcohol (yield of 98%).

$$\text{Benzyl chloride} \xrightarrow[\text{MWI/ 3 min}]{\text{H}_2\text{O}} \text{Benzyl alcohol}$$

Benzyl chloride —CH$_2$Cl  →  —CH$_2$OH Benzyl alcohol

The oxidation of toluene in the presence of KMnO$_4$ and aqueous KOH under microwave irradiation produces benzoic acid.

Toluene —CH$_3$ $\xrightarrow[\text{MWI/ 5 min}]{\text{KMnO}_4/\text{KOH}}$ —COOH Benzoic acid

Benzoic acid undergoes esterification with alcohol in presence of sulfuric acid under microwave irradiation.

Benzoic acid —COOH + C$_2$H$_5$OH $\xrightarrow[\text{MWI/ 6 min}]{\text{Conc. H}_2\text{SO}_4}$ —COOC$_2$H$_5$ Ethyl benzoate

6-Methoxy-indol-2-carboxylic acid undergoes decarboxylation under microwave irradiation to produce a corresponding product.

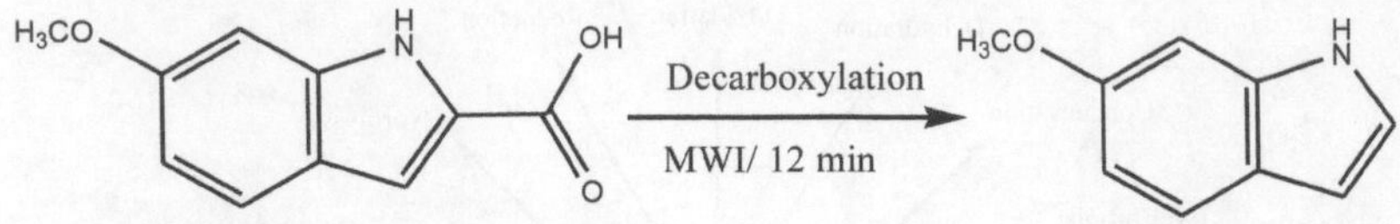

6-methoxy-1*H*-indole-2-carboxylic acid                                6-methoxy-1*H*-indole

Heck reaction involves the formation of C-C by using Pd(OAc)$_2$, P(o-tolyl)$_3$ as a catalyst under microwave irradiation.

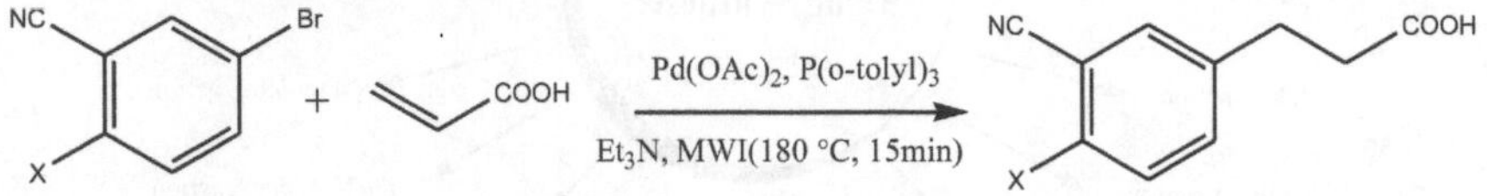

The coupling of thiophenols with aryl halides is carried out under microwave irradiation using magnetite-Glu-Cu, providing yields of 85–98%.

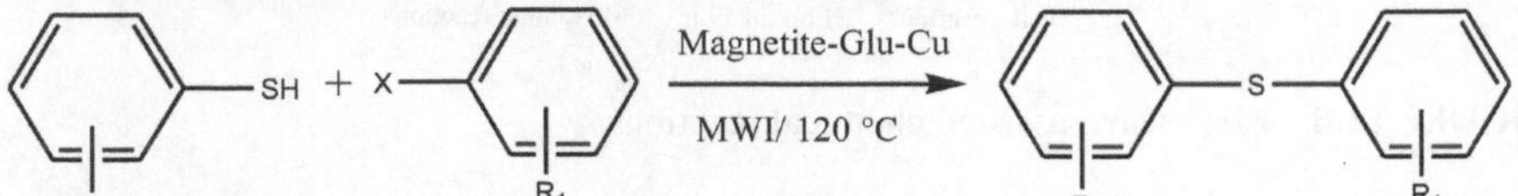

Nitration of a selected heterocyclic compound is carried out using Cu(NO$_3$)$_2$, Pb(NO$_3$)$_2$, KNO$_3$ and NaNO$_3$ in glacial acetic acid under MWI.

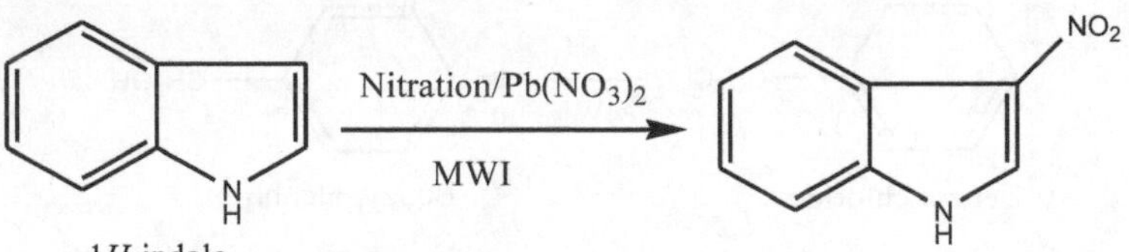

1*H*-indole                                                            3-nitro-1*H*-indole

## 12.2   APPLICATIONS OF MICROWAVE-ASSISTED SYNTHESIS IN MEDICINAL CHEMISTRY

### 12.2.1   SYNTHESIS OF PYRIMIDINE DERIVATIVES

The synthesis of tetrahydro pyrimidine, an important group of medicinally active agents, is performed based on Bignelli condensation in which an equimolar mixture of an aromatic aldehyde, ethyl acetoacetate and urea are taken. To this, a few drops of concentrated hydrochloric acid are added as a catalyst [39]. The reaction is subjected to microwave at 180W to get the product with a yield of 85.71% in one minute.

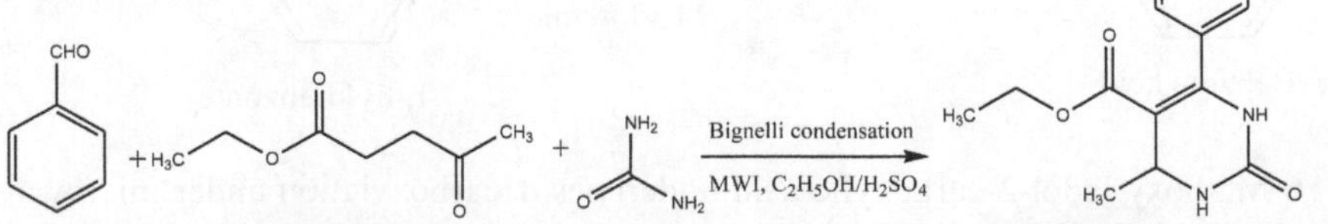

Benzaldehyde          Ethyl acetonacetate          Urea                    Tetrahydropyrimidine-5-carboxylate

**TABLE 12.4**
**Antimicrobial Activity of Compounds**

| Compound Number | Gram Positive Bacteria | | Gram Negative Bacteria |
| --- | --- | --- | --- |
| | SA | BS | EC |
| 2 | 21 | 19 | 23 |
| 4a | 18 | 20 | 25 |
| 4b | 17 | 18 | 18 |
| 4c | 22 | 20 | 17 |
| 5 | 19 | 12 | 17 |
| 8a | 12 | 18 | 11 |
| 8b | 21 | 18 | 12 |
| 8c | 18 | 15 | 23 |
| 8d | 22 | 12 | 17 |
| Chloramphenicol | 30 | 24 | 29 |
| DMSO | N.A. | N.A. | N.A. |

The synthesis of some novel azoles and azolo-pyrimidines are carried out under microwave. The antimicrobial assay is investigated using pathogenic microorganisms, such as gram positive (*SA*, *BS*) and gram negative (*EC*) bacteria, respectively. It is found that gram positive bacteria are more sensitive to the tested compounds, more so *SA* rather than *BS*. The activity is investigated by measuring the diameter of the inhibition zone (IZD) in mm, the results of which are presented in Table 12.4. The solvent DMSO does not exhibit any effect on bacteria [40].

The synthesis of modified pyrimidine and purine pyrano-nucleosides is carried out under microwave irradiation. These are screened as novel cytotoxic, antiviral agents and glycogen phosphorylase-B inhibitors [41].

## 12.2.2  SYNTHESIS OF 3H-QUINAZOLIN-4-ONE (NIMENTOWSKI REACTION)

The synthesis of 3H-quinazolin-4-one is carried out by the Nimentowski reaction in which methyl anthranilate and formamide are mixed and subjected to MWI at 350W for 40 minutes [42]. The reaction mixture is allowed to cool at room temperature and then poured into ice water to get the solid product with a 91% yield.

Methyl anthranilate          Formamide          Nimentowski reaction MWI 350W, 40 min          3H-quinazolin-4-one

The synthesis of the anticancer drug Gefinitib is carried out by using a microwave heating procedure via intermediate 6-benzyloxy-7-methoxyquinazolin-4(3H)-one. It acts as a tyrosine kinase inhibitor of the epidermal growth factor receptor and is used for the treatment of lung cancer [43].

Gefinitib

## 12.2.3  SYNTHESIS OF PURINE

The Suzuki-Miyaura reaction of 6-chloropurines is carried out with an arylation reagent, like sodium tetra-arylborate ($Ph_4BNa$), in water [44].

6-chloropurines          Suzuki-Miyaura reaction MWI, $Ph_4BNa$, $H_2O$          6-phenyl-5H-purine

## 12.2.4  SYNTHESIS OF QUINOLINE

The microwave synthesis of quinoline is carried out by a reaction between aniline and cinnamaldehyde using the acidic catalyst montmorillonite K10 [45].

Aniline            Cinnamaldehyde                              Quinoline

## 12.2.5  SYNTHESIS OF PYRAZOLES AND DIAZEPINS

Pyrazoles and Diazepins are produced under solvent-free conditions using MWI at 120°C in 5–15 minutes [46].

Benzene-1,2-diamine                                        Diazepin

## 12.2.6  SYNTHESIS OF PYRIDO-CARBAZOLES

The reactions of 2-(3-oxo-1,3-diarylpropyl)-1-cyclohexanones with phenylhydrazine hydrochloride in aqueous media under microwave results in the formation of pyrido-carbazoles [47].

phenylhydrazine hydrochloride    2-(3-oxo-1,3-di-phenylpropyl)-1-cyclohexanones        Pyridocarbazoles

### 12.2.6.1  Synthesis of Tetrahydro-Carbazole (Borsche–Drechsel Cyclization)

Tetrahydro-carbazole is synthesised based on Borsche–Drechsel cyclization in which equimolar quantities of cyclohexanone and phenyl hydrazine are taken along with few drops of glacial acetic acid (GAA) and then subjected to microwave at power level-2 (320W) for 10 minutes to produce the product with a yield of 95.6%.

Cyclohexanone        Phenylhydrazine                                          Carbazole

### 12.2.7   SYNTHESIS OF FLAVANONE

The synthesis of flavanone involves the reaction between 4-hydroxy-4-substituted acetophenone and hetero-aldehyde in the presence of ethanolic KOH under microwave for 3–5 minutes [48].

4-hydroxy-4-substituted acetophenone   Hetero-aldehyde                        Flavonone

### 12.2.8   SYNTHESIS OF ISONIAZID ANALOGUES

The synthesis of isoniazid (INH) analogues involves a reaction between benzaldehydes and dimedone in water with a catalytic amount of dodecylbenzenesulfonic acid (DBSA) under microwave. DBSA is a Bronsted acid which acts as a catalyst [49].

Benzaldehydes   5,5-dimethylcyclohexane-1,3-dione  Isonicotinohydrazide        Isoniazid (INH) Analogues

### 12.2.9   SYNTHESIS OF DIHYDRO-PYRIDINES (HANTZSCH SYNTHESIS)

The synthesis of Hantzsch dihydro-pyridines involves the reaction between aromatic aldehyde, ethyl acetoacetate and ammonium formate under solvent-free microwave irradiation.

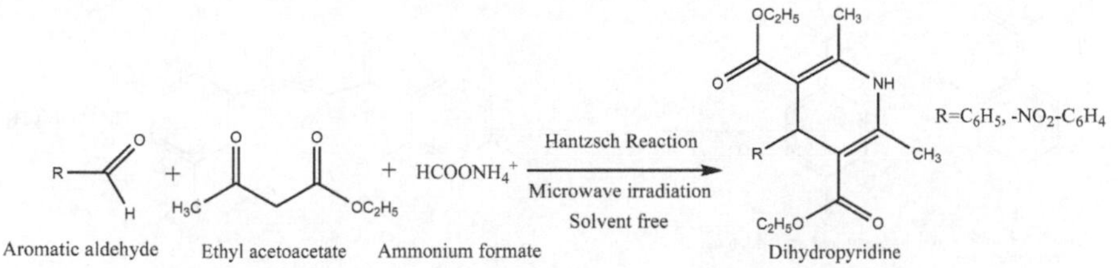

Aromatic aldehyde   Ethyl acetoacetate   Ammonium formate

Hantzsch Reaction
Microwave irradiation
Solvent free

R=C₆H₅, -NO₂-C₆H₄

Dihydropyridine

The design, synthesis and biological evaluation of novel 5H-chromeno-pyridines are carried out as potential anticancer agents. For an anticancer activity study, human A375, WM164 and MDA-MB-435 melanoma cells were cultured in 10% fetal bovine serum supplemented with a DMEM medium with 1% antibiotics at 37°C. The cell viability was determined by a MTS assay method. The MTS agent was added to detect the cell density in each well by reading the optical absorbance at 490nm, as given in Table 12.5 [50].

**TABLE 12.5**
**Anti-Proliferative Activity of Titled Chromeno Pyridine and Chromene Analogues**

| Compounds | IC50±SEM (µM) | | |
| --- | --- | --- | --- |
| | A375 | WM164 | MDA-MB-435 |
| 1a | 6.4±0.8 | 7.5±1.2 | 7.4±0.8 |
| 1b | 6.7±1.5 | 3.5±1.2 | 3.7±0.8 |
| 1c | 6.3±0.7 | 3.6±0.6 | 4.1±0.4 |
| 1d | 7.0±0.9 | 6.6±1.0 | 6.5±0.6 |
| 1e | 5.3±0.7 | 5.7±1.4 | 6.0±0.8 |
| 1f | 5.7±0.4 | 5.6±0.6 | 7.1±0.5 |
| Colchicine | 0.02±0.01 | 0.03±0.02 | – |

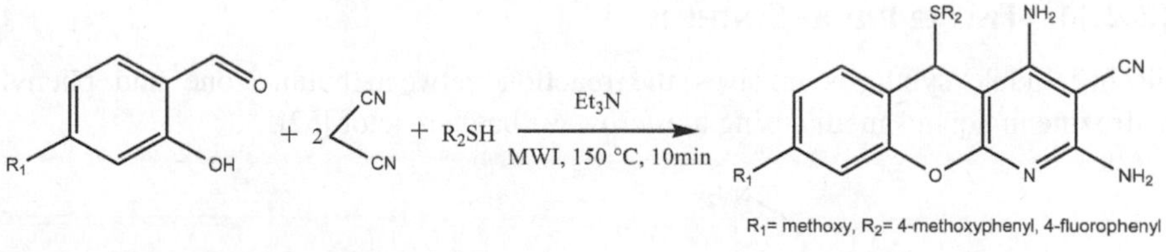

Et₃N
MWI, 150 °C, 10min

R₁= methoxy, R₂= 4-methoxyphenyl, 4-fluorophenyl

## 12.2.10 SYNTHESIS OF CHALCONE

The synthesis of chalcone is carried out using 2-hydroxy-4-substituted acetophenone and substituted aryl aldehyde in the presence ethanolic KOH under microwave for 5–7 minutes.

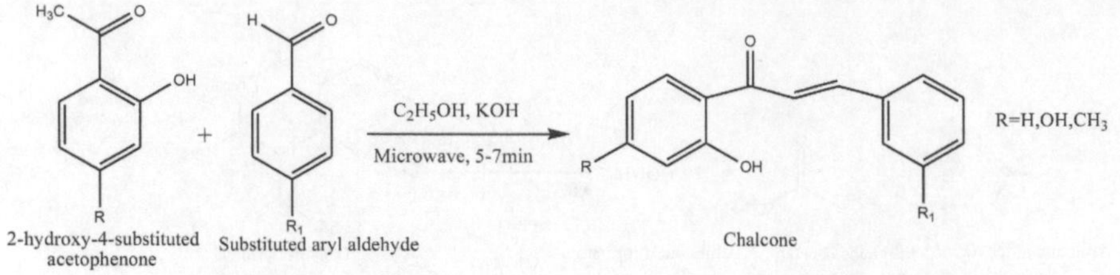

2-hydroxy-4-substituted    Substituted aryl aldehyde                    Chalcone
      acetophenone

Chalcones are also synthesized by reacting equimolar quantities of 2-acetyl hetero cyclic derivatives and respective aldehyde in the presence of alcoholic potassium hydroxide solution. The reaction mixture is subjected to microwave irradiation at 180W for 2–6 minutes. The synthesized compounds are screened for their antioxidant activity [51]. Table 12.6 presents the antioxidant activity of chalcones by using the DPPH method.

**TABLE 12.6**
**Antioxidant Activity of Chalcones by Using DPPH Method**

| Compound | Percentage Inhibition | | | |
| | 25 µg/ml | 50 µg/ml | 100 µg/ml | $IC_{50}$ |
| --- | --- | --- | --- | --- |
| 2a | 7.24 | 12.31 | 16.03 | 76.12 |
| 2b | 4.05 | 7.12 | 11.04 | 65.04 |
| 2c | 9.11 | 10.03 | 12.04 | 81.15 |
| 2d | 8.35 | 9.24 | 10.47 | 75.20 |
| 2e | 10.14 | 11.85 | 13.69 | 49.18 |
| Ascorbic acid | 16.13 | 38.11 | 62.34 | 0.61 |

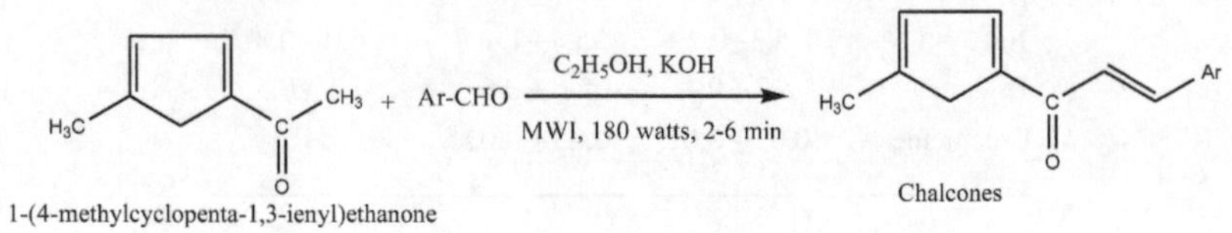

1-(4-methylcyclopenta-1,3-ienyl)ethanone                              Chalcones

## 12.2.11  Fischer Indole Synthesis

Fischer Indole synthesis involves the reaction between butan-2-one and phenyl hydrazine in aquous media using a microwave batch reactor [52].

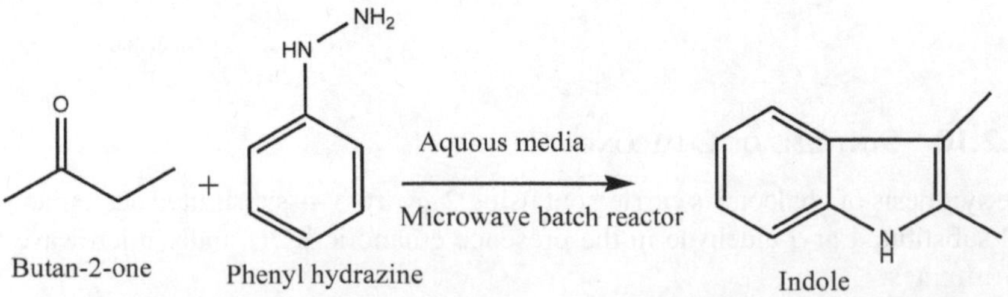

Butan-2-one    Phenyl hydrazine                                        Indole

## 12.2.12 Formation of β-Lactam

The formation of β-lactam involves the reaction between 2-(benzyloxy)acetyl chloride and N-benzylidenemethanamine in the presence of N-methyl morpholine and chlorobenzene under microwave irradiation [53].

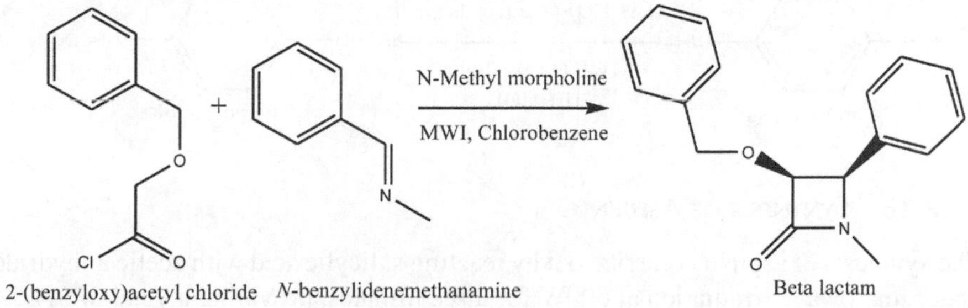

N-Methyl morpholine

MWI, Chlorobenzene

2-(benzyloxy)acetyl chloride   *N*-benzylidenemethanamine                    Beta lactam

## 12.2.13 Synthesis of Imatinib

Imatinib is a blockbuster anticancer drug which is synthesized by microwave-assisted solid phase synthesis using acid-sensitive resin, providing a high yield [54].

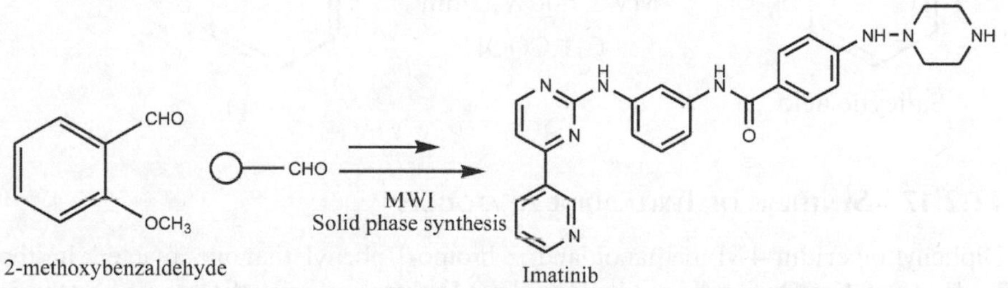

MWI
Solid phase synthesis

2-methoxybenzaldehyde                             Imatinib

## 12.2.14 Synthesis of Phenytoin (Benzillic Acid Rearrangement)

The synthesis of phenytoin is based on benzillic acid rearrangement in which benzil and urea are mixed in an ethanolic NaOH solution and subjected to microwave at 160W for 30 minutes to produce the product with a yield of 79% [55].

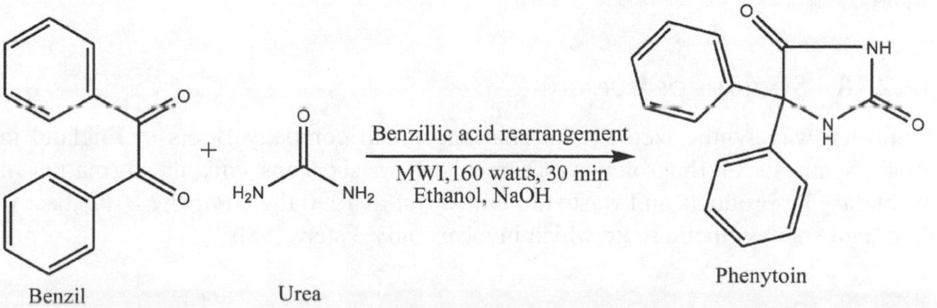

Benzillic acid rearrangement

MWI,160 watts, 30 min
Ethanol, NaOH

Benzil                        Urea                                              Phenytoin

## 12.2.15  SYNTHESIS OF PARACETAMOL

Synthesis of paracetamol is carried out by reacting 4-aminophenol with acetic anhydride in aqueous media and subjecting it to microwave irradiation at 600W for two minutes, providing a yield of 92% [56].

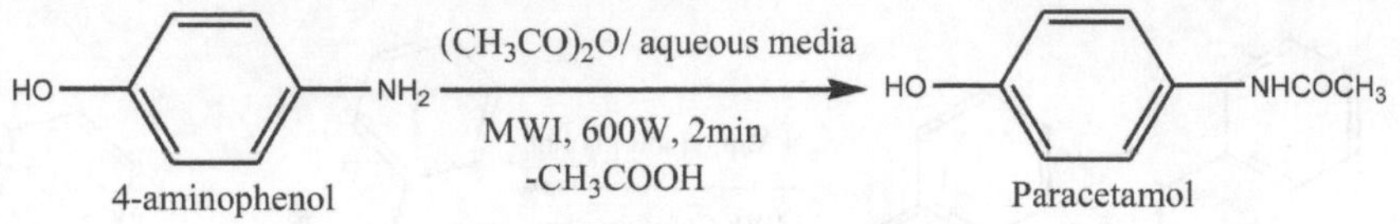

4-aminophenol

$(CH_3CO)_2O$/ aqueous media

MWI, 600W, 2min

$-CH_3COOH$

Paracetamol

## 12.2.16  SYNTHESIS OF ASPIRIN

The synthesis of aspirin is carried out by reacting salicylic acid with acetic anhydride under microwave irradiation at 600W for three minutes, providing a yield of 81%.

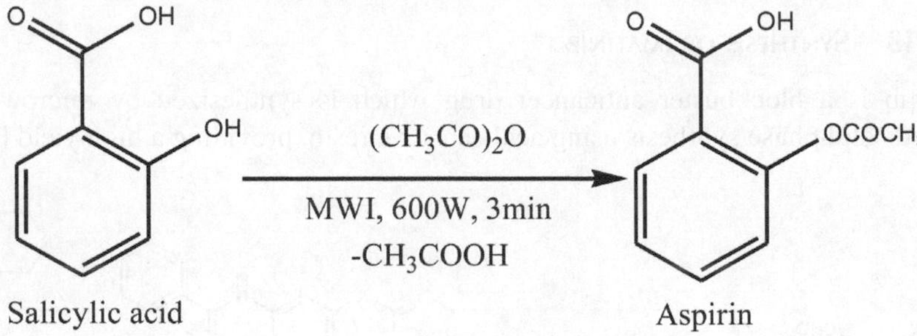

Salicylic acid

$(CH_3CO)_2O$

MWI, 600W, 3min

$-CH_3COOH$

Aspirin

## 12.2.17  SYNTHESIS OF TERFENADINE ANALOGUES

Diphenyl(piperidin-4-yl)methanol and 2-bromo-1-phenylethanone reacted in the presence of $K_2CO_3$ and acetonitrile under Microwave-Assisted Organic Synthesis (MAOS) for 5–45 minutes to produce terfenadine analogues [57].

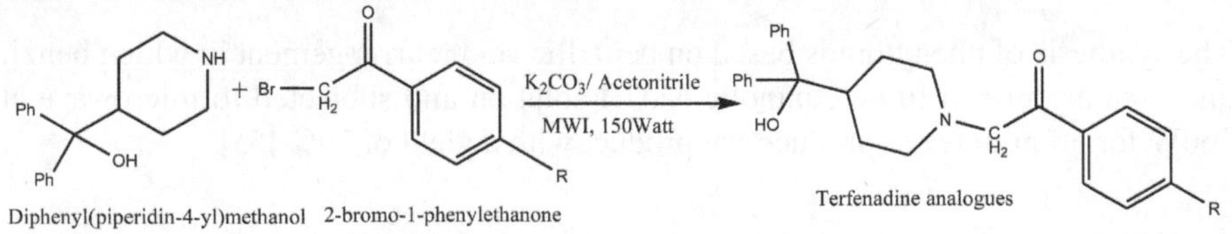

Diphenyl(piperidin-4-yl)methanol   2-bromo-1-phenylethanone

$K_2CO_3$/ Acetonitrile

MWI, 150Watt

Terfenadine analogues

## 12.2.18  SYNTHESIS OF IBUPROFEN

Ibuprofen was synthesized by the pharmaceutical company Boots in England in 1960. Synthesis of Ibuprofen was carried out in six steps with the formation of secondary by-products and waste materials. But in 1990 the company BHC discovered a greener synthetic route which involves only 3 steps [58].

Isobutylbenzene   1-(4-isobutylphenyl)ethanone   1-(4-isobutylphenyl)ethanol   Ibuprofen

## 12.2.19 Synthesis of 4-Bromostilbene

The reaction between 4-bromoiodobenzene and styrene in the presence of palladium acetate [Pd(OAC)$_2$] as a catalyst forms 4-bromostilbene under microwave irradiation at 60Wfor 4–8 minutes [59].

4-bromo-iodobenzene        Styrene        4-bromostilbene

## 12.2.20 Synthesis of Barbituric Acid

Barbituric acid is produced by the reaction between a malonic ester, urea and acetic anhydride under microwave irradiation at 60°C for seven minutes.

5-Arylidene barbituric acid is also prepared by the reaction between an aromatic aldehyde and barbituric acid in the presence of sodium acetate under solvent-free conditions [60].

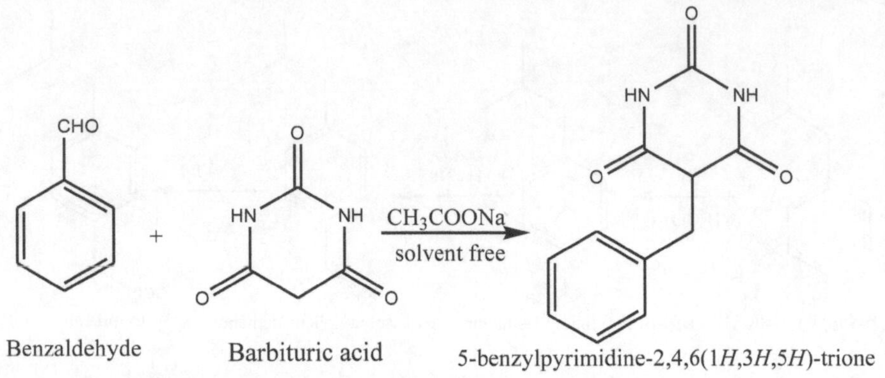

CHO
O
HN
NH
O
O
+
$CH_3COONa$
solvent free
O
HN
NH
O
O
Benzaldehyde
Barbituric acid
5-benzylpyrimidine-2,4,6(1*H*,3*H*,5*H*)-trione

## 12.2.22 Synthesis of Sulfonamides

The synthesis of sulfonamides involves the reaction between 4-toluenesulphonyl chlorides and aniline in presence of silica gel under solvent-free microwave irradiation, providing a yield of 95% in 40 minutes [61].

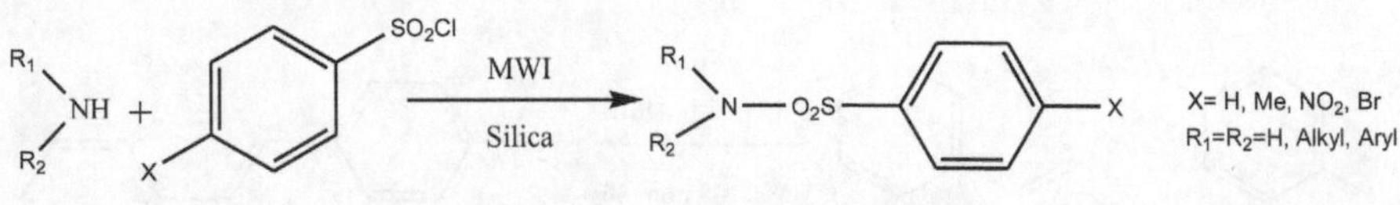

$R_1$
NH +
$R_2$
$SO_2Cl$
X
MWI
Silica
$R_1$
N
$O_2S$
$R_2$
X
X= H, Me, $NO_2$, Br
$R_1$=$R_2$=H, Alkyl, Aryl

## 12.2.23 Synthesis of Thiazolidine-Dione

The microwave-assisted synthesis of novel 5-[4-(substituted)benzylidene]thiazolidi ne-2,4-dione was carried out and evaluated for their anti-diabetic activity. *In vivo* antidiabetic activity is based on the oral glucose tolerance test (OGTT). Statistical comparisons are performed by one-way analysis of variance (ANOVA) followed by Dunnett's Multiple Comparison test and the values are considered statistically significant when P<0.05, as given Table 12.7 [62].

**TABLE 12.7**
**Anti-Diabetic Activity of Titled Compounds**

| Groups | AUC (Area Under the Curve) (0–120min) | Statistically Significant |
|---|---|---|
| Control | 14380 ± 485.0773 | Not significant |
| Pioglitazone | 12755 ± 132.5707 | P < 0.05 |
| 7a | 14365 ± 388.6837 | Not significant |
| 7b | 13185 ± 836.3761 | Not significant |
| 7c | 14450 ± 454.1751 | Not significant |
| 7d | 14175 ± 420.357 | Not significant |
| 7e | 12315 ± 277.5338 | P < 0.05 |

## 12.2.24 Synthesis of Quinoline

Microwave-assisted synthesis of 2-styryl-quinoline-4-carboxylic acids as anti-tubercular agents is developed through an acid-catalyzed condensation reaction.

## 12.2.25 Synthesis of Procaine

The microwave-assisted synthesis of the local anesthetic procaine from benzocaine was carried out [63].

Ethyl 4-aminobenzoate
(Benzocaine)

2-(diethylamino)ethyl 4-aminobenzoate
(Proocaine)

A novel series of 1,2,4-triazolo[4,3-a]pyridines was designed and synthesized under microwave irradiation conditions. These derivatives were evaluated for antifungal activity at a dosage of 100 parts per million (ppm) against *Stemphylium lycopersici* (*SL*), *Fusarium oxysporum* (*FO*) and *Botrytis cinerea* (*BC*), as described in Table 12.8 [64].

**TABLE 12.8**
**The Antifungal Activities of Title Compounds**

| Test Compounds | SL | FO | BC |
|---|---|---|---|
| 2a | 26.34 | 62.73 | 12.64 |
| 2b | 20.24 | 51.11 | 18.89 |
| 2c | 80.65 | 40.56 | 21.11 |
| 2d | 74.85 | 83.33 | 24.26 |
| 2e | 56.85 | 84.26 | 18.89 |
| 2f | 16.67 | 72.22 | 24.26 |
| 2g | 34.52 | 71.11 | 28.89 |
| 2h | 53.87 | 48.89 | 0.00 |
| 2i | 56.25 | 49.44 | 5.56 |
| 2j | 4.17 | 45.28 | 6.67 |
| 2k | 51.19 | 62.02 | 15.56 |
| Zhongshengmycin | 59.58 | – | – |
| Cyprodynil | – | – | 45.56 |

## 12.2.26 SYNTHESIS OF BENZIMIDAZOLE DERIVATIVES

Microwave-assisted synthesis of new benzimidazole derivatives was carried out with lipase inhibition activity [65].

## 12.2.27 SYNTHESIS OF ISATIN DERIVATIVES

The synthesis of some novel isatin derivatives were carried out under microwave irradiation and antimicrobial activity was performed [66].

## 12.2.28 SYNTHESIS OF QUINOXALINES

Quinoxalines were successfully synthesized in a few minutes by the condensation reactions of o-phenylene diamine (OPDA) with α-dicarbonyl compounds in ethanol under MWI [67].

o-phenylene diamine   Dicarbonyl compounds        Quinoxaline

The synthesis of aminopyrimidines bearing benzofuran was carried out under solvent-free microwave irradiation. The *in vitro* antimicrobial activity study was carried out against bacterial and fungal strains by using the cup-plate method. In the case of DMF, the +ve sign indicates the growth of microbes. Chloramphenicol and fluconazole are used as standard drugs for antibacterial and antifungal activity, respectively. Table 12.9 presents antimicrobial activity, which is determined by measuring the zone of inhibition in mm [68].

Benzocaine is synthesized from 4-amino benzoic acid and ethanol in the presence of dry hydrogen chloride under microwave irradiation at a power of 280W for 25 minutes [69].

4-aminobenzoic acid                             Benzocaine

**TABLE 12.9**
**The Antimicrobial Activity of Title Compounds**

| Compd. | Antibacterial Activity | | Antifungal Activity | |
|--------|-------------|-----------|----------|-----------|
| | *P. aerogenosa* | *S. aureus* | *A. niger* | *Curvularia* |
| 3a | 12 | 14 | 15 | 10 |
| 3b | 10 | 11 | 12 | 13 |
| 3c | 18 | 17 | 16 | 18 |
| 3d | 17 | 16 | 18 | 16 |
| 3d | 15 | 16 | 17 | 15 |
| Std. | 24 | 26 | 22 | 24 |
| DMF | +ve | +ve | +ve | +ve |

The synthesis of benzocaine derivatives was carried out by the condensation of urea, thiourea, semicarbazide and thiosemicarbazide with ethyl-4[(chloroacetyl) amino]benzoate under microwave irradiation in the presence of ethanol. The antimicrobial activities of the synthesized compounds are studied by the disc diffusion method, as given in Table 12.10. The zone of inhibition is measured in mm to estimate the potency of the test compounds [70].

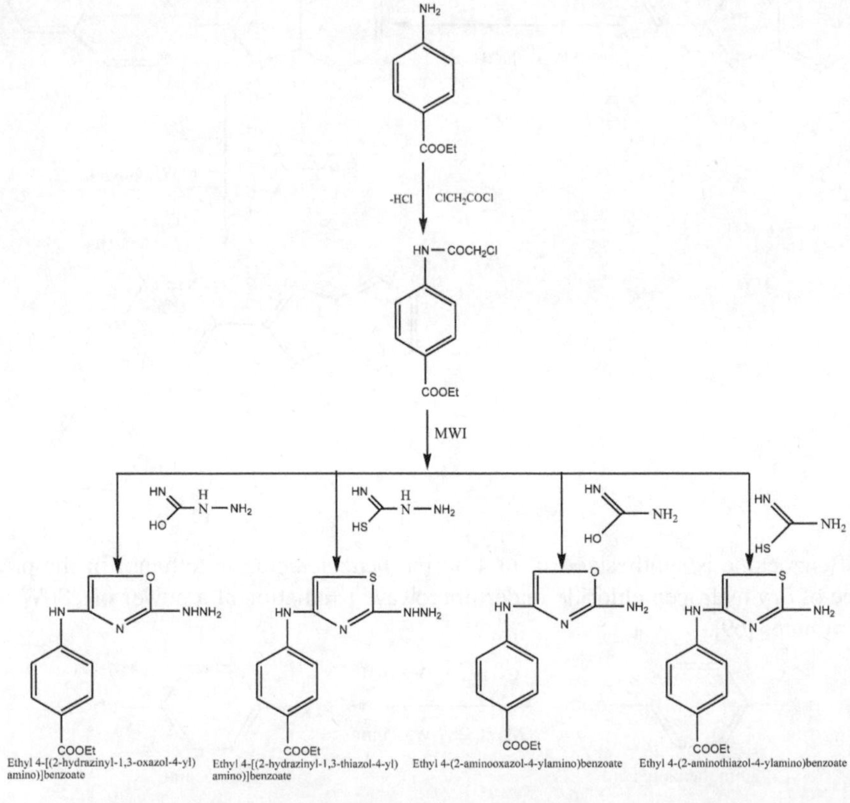

Ethyl 4-[(2-hydrazinyl-1,3-oxazol-4-yl) amino)]benzoate    Ethyl 4-[(2-hydrazinyl-1,3-thiazol-4-yl) amino)]benzoate    Ethyl 4-(2-aminooxazol-4-ylamino)benzoate    Ethyl 4-(2-aminothiazol-4-ylamino)benzoate

**TABLE 12.10**

**The Antimicrobial Activity of Title Compounds**

| Organisms | SA (µg/ml) | | | | BC (µg/ml) | | | | EC (µg/ml) | | | | PA (µg/ml) | | | |
|---|---|---|---|---|---|---|---|---|---|---|---|---|---|---|---|---|
| Compounds | Std | 10 | 20 | 30 | Std | 10 | 20 | 30 | Std | 10 | 20 | 30 | Std | 10 | 20 | 30 |
| A | 38 | 15 | 23 | 26 | 39 | 16 | 20 | 26 | 40 | 19 | 25 | 29 | 40 | 16 | 26 | 29 |
| B | 38 | 14 | 21 | 24 | 39 | 15 | 22 | 25 | 40 | 18 | 23 | 28 | 40 | 16 | 25 | 27 |
| C | 38 | 16 | 19 | 25 | 38 | 17 | 21 | 24 | 40 | 14 | 19 | 24 | 40 | 19 | 24 | 28 |
| D | 38 | 15 | 22 | 28 | 39 | 19 | 24 | 28 | 40 | 16 | 22 | 26 | 40 | 16 | 25 | 29 |

## 12.3 CONCLUSIONS

The microwave irradiation method is more advantageous as compared to the conventional method due to its rapid uniform heating, eco-friendly nature and greater product yield with less or no by-product formation. Therefore, microwave synthesis is one of the most important tools in medicinal chemistry and drug discovery processes. This Green Chemistry approach not only reduces the synthesis time but also the side reactions which improve the yield and reproducibility of the products. The use of polar solvents or solvent-free conditions is essential for microwave activation, which promotes the rate of chemical reaction and minimizes environmental pollution. Perhaps, the most important part of microwave-assisted medicinal chemistry is to identify the molecules that have medicinal activities. Once the structures of these molecules are available, it seems that these can be synthesized by microwave through the chemical manipulation of reactants, solvents and conditions. The most important limitation of microwave-induced medicinal chemistry is the non-availability of efficient routes for complex drugs that require multi-step processes. Nevertheless, microwave-assisted synthesis will be an efficient technology for the production of new chemical entities as drug candidates, which will cause an increase in lead generation and optimization.

## ABBREVIATIONS

| | |
|---|---|
| **5-HT** | 5-hydroxytryptamine |
| **ANOVA** | Analysis of variance |
| **ATP** | Adenosine triphosphate |
| **AUC** | Area under the curve |
| *BC* | *Botrytis cinerea* |
| *BS* | *Bacillus subtilis* |
| **cAMP** | Cyclic adenosine monophosphate |
| **CMR** | Continuous microwave reactor |
| **COX** | Cyclooxygenase |
| **DBSA** | Dodecylbenzene-sulfonic acid |
| **DCM** | Dichloromethane |
| **dGTP** | Deoxyguanosine triphosphate |

| | |
|---|---|
| **DHF** | Dihydrofolic acid |
| **DHFR** | Dihydro-folate reductase |
| **DHPS** | Dihydropteroate synthetase |
| **DMF** | Dimethyl formamide |
| **DMSO** | Dimethyl sulfoxide |
| **DNA** | Deoxyribonucleic acid |
| **DPPH** | 1,1–diphenyl-2-picrylhydrazyl |
| *EC* | *Escherichia coli* |
| **EPA** | Environmental Protection Agency |
| *FO* | *Fusarium oxysporum* |
| **FU** | Fluorouracil |
| **GAA** | Glacial acetic acid |
| **GABA** | Gamma-aminobutyric acid |
| **GHz** | Gigahertz |
| **H1** | Histamine |
| **HAD** | Hydrogen acceptor/donor |
| **HIV** | Human immunodeficiency virus |
| **$H_2O_2$** | Hydrogen peroxide |
| **$IC_{50}$** | Inhibitory concentration |
| **INH** | Isoniazid |
| **IR** | Infrared |
| **IZD** | Diameter of inhibition zone |
| **$K_2CO_3$** | Potassium carbonate |
| **$KMnO_4$** | Potassium permanganate |
| **KOH** | Potassium hydroxide |
| **MABA** | Microplate Alamar Blue Assay |
| **MAOS** | Microwave-Assisted Organic Synthesis |
| **MBR** | Microwave batch reactor |
| **MT** | Mycobacterium tuberculosis |
| **MTS** | (3-(4,5-dimethylthiazol-2-yl)-5-(3-carboxymethoxyphenyl)-2-(4-sulfophenyl)-2H-tetrazolium) |
| **MWI** | Microwave irradiation |
| **nACh** | Nicotinic acetylcholine |
| **NMR** | Nuclear magnetic resonance |
| **NRTI** | Nucleoside reverse-transcriptase inhibitor |
| **OGTT** | Oral glucose tolerance test |
| **OPDA** | o-phenylene diamine |
| *PA* | *Pseudomonas aeruginosa* |
| **$Ph_4BNa$** | Sodium tetra-arylborate |
| **pKa** | Acid dissociation constant |
| **ppm** | parts per million |
| **PTC** | Phase Transfer Catalysis |
| **PZA** | Pyrazinamide |
| **RNA** | Ribonucleic acid |
| **RNR** | Ribonucleotide reductase |
| *SA* | *Staphylococcus aureus* |

| SAR | Structure activity relationships |
|---|---|
| *SL* | *Stemphylium lycopersici* |
| TFA | Trifluoroacetic acid |
| THF | Tetrahydrofolic acid |
| THF | Tetrahydrofuran |
| ThMP | Thymidine monophosphate |
| TMPK | Thymidine monophosphate kinase |
| UV | Ultraviolet |

# REFERENCES

1. Loupy, A. 2006. *Microwaves in Organic Synthesis*, 2nd ed., Wiley-VCH, Weinheim, Germany, ISBN 978-3-527-31452-2.
2. Santagada, V., Frecentese, F., Perissutti, E., Fiorino, F., Severino. B., Caliendo, G. 2009. Microwave assisted synthesis: A new technology in drug discovery. *Mini. Rev. Med. Chem.* 9(3):340–358.
3. Santagada, V., Perissutti, E., Caliendo, G. 2002. The application of microwave irradiation as new convenient synthetic procedure in drug discovery. *Curr. Med. Chem.* 9(13):1251–1283.
4. Somani, R., Shirodkar, P., Dandekar, R., Gide, P., Tanushree, P., Kadam, V. 2010. Optimization of microwave assisted synthesis of some Schiff's bases. *Int. J. ChemTech. Res.* 2(1):172–179.
5. Madhvi, S., Smita, J., Desai, K.R. 2012. A brief review: Microwave assisted organic reactions. *Arch. Appl. Sci. Res.* 4(1):645–661.
6. Ravichandran, S., Karthikeyan E. 2011. Microwave synthesis–A potential tool for green chemistry. *Int. J. ChemTech. Res.* 3(1):466–470.
7. Polshettiwar, V., Varma, R.S. 2008. Microwave-assisted organic synthesis and transformations using benign reaction media. *Acc. Chem. Res.* 41(5):629–639.
8. Kappe, C.O. & Dallinger, D. 2006. The impact of microwave synthesis on drug discovery. *Nat. Rev. Drug Discov.* 5(1): 51–63.
9. Wathey, B., Tierney, J., Lidström, P., Westman, J., 2002. The impact of microwave-assisted organic chemistry on drug discovery. *Drug Discov. Today* 7(6):373–380.
10. Lidstrom, P., Tierney, J., Wathey, B., Westman, J. 2001. Microwave-assisted organic synthesis—A review. *Tetrahedron* 57(45):9225–9283.
11. Dubey, R., Dwivedi, S., Mehta, K., Joshi H. 2008. New era in the field of synthetic chemistry: Microwave assisted synthesis. *Pharmainfo.net* 6(3):6–12.
12. Mingos, D.M.P., Baghurst, D.R. 1991. Applications of microwave dielectric heating effect to synthetic problems in chemistry. *Chem. Soc. Rev.* 20(1):1–47.
13. Colombo, M., Peretto, I. 2008. Chemistry strategies in early drug discovery: An overview of recent trends. *Drug Discov. Today* 13(15–16):677–684.
14. Larhed, M., Hallberg, A. 2001. Microwave-assisted high speed chemistry: A new technique in drug discovery. *Drug Discov. Today* 6(8):406–416.
15. Mavandadi, F., Pilotti, A. 2006. The impact of microwave-assisted organic synthesis in drug discovery. *Drug Discov. Today* 11(3–4):165–174.
16. Shipe, W.D., Wolkenberg, S.E., Lindsley C.W. 2005. Accelerating lead development by microwave enhanced medicinal chemistry. *Drug Discov. Today Technol.* 2(2):155–161.
17. Polshettiwar, V., Nadagouda, M.N., Varma, R.S. 2009. Microwave-assisted chemistry: A rapid and sustainable route to synthesis of organics and nanomaterials. *Aust. J. Chem.* 62:16–26.

18. Polshettiwar, V., Varma, R.S. 2008. Greener and expeditious synthesis of bio-active heterocycles using microwave irradiation. *Pure Appl. Chem.* 80(4):777–790.

19. Ersmark K., Larhed M., Wannberg J. 2004. Microwave-enhanced medicinal chemistry: A high-speed opportunity for convenient preparation of protease inhibitors. *Curr. Opin. Drug Discov. Devel.* 7(4):417–427.

20. James, H.C. 2001. Catalysis for green chemistry. *Pure Appl. Chem.* 73(1):103–111

21. Badami, S., Minimathew, A., Thomas, S., Suresh, B. 2003. Use of microwave technique in pharmaceutical organic chemistry practical. *Indian J. Pharma. Educ.* 37(4): 199–203.

22. Caddick, S., Fitzmaurice, R. 2009. Microwave enhanced synthesis. *Tetrahedron* 65(17):3325–3355.

23. Gedye, R., Smith, F., Westaway, K. 1986. The use of microwave ovens for rapid organic synthesis. *Tetrahedron Lett.* 27(3):279–282.

24. Hayes, B.L. 2004. Recent advances in microwave-assisted synthesis. *Aldrichimica Acta* 37:66–69.

25. Tierney, J.P., Lidström, P. 2005. *Microwave Assisted Organic Synthesis.* Blackwell, USA and Canada, ISBN 1–4051-1560–2.

26. Wathey, B., Tierney, J., Lidström, P., Westman, 2002. The Impact of microwave-assisted organic chemistry on drug discovery. *Drug Discov. Today* 7(6):373–380.

27. Sheldon, R.A. 2005. Green solvents for sustainable organic synthesis. State of the art. *Green Chem.* 7:267–268.

28. Nuchter, M., Ondruschka, B., Bonrath, W., & Gum, A. 2004. Microwave assisted synthesis – A critical technology overview. *Green Chem.* 6:128. doi: 10.1039/b310502d.

29. Strauss, C.R., Trainor, R.W. 1995. Developments in microwave-assisted organic chemistry. *Aust. J. Chem* 48:1665–1692. dot:10.1071/CH9951665.

30. Sheldon, R.A., Arends, I., Hanefeld, U. 2007. *Green Chemistry and Catalysis.* Wiley, Wienheim, Germany, 1–2.

31. Krstenansky, J.L., Cotterill I. 2000. Recent advances in microwave-assisted organic syntheses. *Curr. Opin. Drug Discov. Devel.* 3(4):454–461.

32. Sekhon, B.S. 2010. Microwave-assisted pharmaceutical synthesis: An overview. *Int. J. PharmTech. Res.* 2(1):827–833.

33. Hugel, H.M. 2009. Microwave multicomponent synthesis. *Molecules* 14:4936–4972.

34. Bougrin, K., Loupy, A., Soufiaoui, M. 2005. Microwave-assisted solvent-free heterocyclic synthesis. *J. Photochem. Photobiol. C: Photochem. Rev.* 6:139–167.

35. Wardencki, W., Curylo, J., Namisni K. 2005. Green chemistry – Current and future issues. *Pol. J. Environ. Stud.* 14(4):389–395.

36. Gedye, R.N., Smith, F.E., Westaway, K.C. 1988. The rapid synthesis of organic compounds in microwave ovens. *Can. J. Chem.* 66(1):17–26.

37. Miao, G., Ye, P., Yu, L., Baldino, C.M. 2005. Microwave promoted Suzuki reactions of aryl chlorides in aqueous media. *J. Org. Chem.* 70(6):2332–2334.

38. Patel, D., Patel, B. 2011. Microwave assisted organic synthesis: An overview. *J. Pharm. Res.* 4(7):2090–2092.

39. Mobinikhaledi, A., Forughifar, N. 2006. Microwave assisted synthesis of some pyrimidine derivatives. Phosphorus Sulfur Silicon Relat. Elem. 181(11): 2653–2658. doi: 10.1080/10426500600862977.

40. Gomha, S.M., Farghaly, T.A., Mabkhot, Y.N. 2017. Microwave-assisted synthesis of some novel azoles and azolopyrimidines as antimicrobial agents. *Molecules* 22(3):346. doi: 10.3390/molecules22030346.

41. Manta S., Dimopoulou A., Kollatos N., Komiotis, D. 2017. Rapid microwave-assisted synthesis of modified pyrimidine and purine pyrano nucleosides as novel cytotoxic, antiviral agents and glycogen phosphorylase B inhibitors. *Med. Chem.* 7(5): 865–868.

42. Alexandre, F., Besson, T. 2002. Microwave-assisted Niementowski reaction. Back to the roots. *Tetrahedron Lett.* 43(21):3911–3913.
43. Feng Li, Y.F., Qingqing, M. 2007. An efficient construction of quinazolin-4(3H)-ones under microwave irradiation. *ARKIVOC* (1):40–50.
44. Qu, G.R., Xin, P.Y., Niu, H.Y. Guo, H.M. 2011. Microwave promoted palladium-catalyzed Suzuki–Miyaura cross-coupling reactions of 6-chloropurines with sodium tetraarylborate in water. *Tetrahedron* 67(47):9099–9103.
45. Adam D. 2003. Microwave chemistry: Out of the kitchen. *Nature* 421(6923): 571–572.
46. Manmohan S. 2012. Green chemistry potential for past, present and future perspectives. *IRJP.* 3(4):31–36.
47. Nerkar, A.G., Pawale, D., Ghante, M.R. and Sanjay, D. 2013. Microwave assisted organic synthesis of some traditional and named reactions: A practical approach of green chemistry. *Int. J. Pharm. Pharm. Sci.* 5(3):564–566.
48. Sagrera, G.J., Gustavo, A. 2005. Microwave accelerated solvent-free synthesis of flavanones. *J. Braz. Chem. Soc.* 16(4):851–856.
49. Manjashetty, T.H., Yogeeswari, P., Sriram, D. 2011. Microwave assisted one-pot synthesis of highly potent novel isoniazid analogues. *Bioorg. Med. Chem. Lett.* 21(7): 2125–2128.
50. Banerjee, S., Pfeffer, S., Pfeffer, L.M. 2015. Design, synthesis and biological evaluation of novel 5H-chromenopyridines as potential anti-cancer agents. *Molecules* 20(9): 17152–17165; doi:10.3390/molecules200917152.
51. Ahmad, M.R., Sastry, V.G. 2016. Synthesis of novel chalcone derivatives by conventional and microwave irradiation methods and their pharmacological activities. *Arab. J. Chem.* 9(1):S931–S935.
52. Creencia, E.C., Tsukamoto, M., Horaguchi, T. 2011. One-pot-one-step, microwave-assisted Fischer indole synthesis. *J. Heterocycl. Chem.* 48(5):1095–1102.
53. Bose, A.K., Banik, B.K. Manhas, M.S. 1995. Stereocontrol of β-lactam formation using microwave irradiation. *Tetrahedron Lett.* 36(2):213–216.
54. Leonetti, F., Capaldi, C., Carotti, A. 2007. Microwave-assisted solid phase synthesis of imatinib, a blockbuster anticancer drug. *Tetrahedron Lett.* 48(19):3455–3458.
55. Fernand, A. Gbaguidi, S. D. 2011. A high yield synthesis of phenytoin and related compounds using microwave activation. *Afr. J. Pure Appl. Chem.* 5(7):168–175.
56. Olariu, T., Suta, L.M., Popoiu, C. 2014. Alternative synthesis of paracetamol and aspirin under non-conventional conditions. *Rev. Chim.* 65(6):633–635.
57. Holzer, M., Ziegler, S., Albrecht, B. 2008. Identification of terfenadine as an inhibitor of human CD81-receptor HCV-E2 interaction: Synthesis and structure optimization. *Molecules* 13(5):1081–1110. doi: 10.3390/molecules13051081.
58. Baran, A.U. 2013. Comparative study of microwave-assisted and conventional synthesis of ibuprofen-based acyl hydrazone derivatives. *Turkish J. Chem.* 37(6):927–935.
59. Negishi, E. 2002. *Handbook of Organopalladium Chemistry for Organic Synthesis.* Volume 2, John Wiley & Sons, New York, 1167.
60. Uttam, B.M. 2016. A solvent free green protocol for synthesis of 5-arylidine barbituric acid derivatives. *Org. Chem.: Indian J.* 12(3):1–6.
61. Maasoumeh, J., Abdolreza, R., Marzieh, A. 2009. Catalytic activity of silica gel in the synthesis of sulphonamides under mild and solvent-free conditions. *Appl. Catal., A.* 358:49–53.
62. Patel, K.D., Chhaganbhai, N.P., Patel, G.M. 2016. Microwave assisted synthesis and antidiabetic activity of novel 5-[4-(Substituted) Benzylidine]thiazolidine-2,4-dione. *Med. Chem.* 6(10): 647–651.
63. Nodiţi, G. 2014. Microwave-assisted synthesis of local anaesthetic procaine from benzocaine. *Rev. Chim.* 65(1): 65–67.

64. Zhai, Z., Shi, Y., Yang, M. 2016. Microwave assisted synthesis and antifungal activity of some novel thioethers containing 1,2,4-triazolo[4,3-a]pyridine moiety. *Lett. Drug Des. Discov.* 13(6):521–525.

65. Mentese, E., Ozil, M., Serdar, U. 2013. An efficient synthesis of benzimidazoles via a microwave technique and evaluation of their biological activities, *Monatsh Chem.* 144(7): 993–1001.

66. El-Faham, A., Hozzein, W. N., Mohammad, A.M. 2015. Microwave synthesis, characterization, and antimicrobial activity of some novel isatin derivatives. *J. Chem.,* Article ID716987.

67. Costa, C.F., Souza, M.V. 2017. Microwave-assisted synthesis of quinoxalines—A review. *Curr. Microwave Chem.* 4(4):277–286.

68. Kumar, D.B., Prakash, G.K., Kumaraswamy, M.N. 2006. Microwave assisted facile synthesis of amino pyrimidines bearing benzofuran and investigation of their antimicrobial activity. *Indian J. Chem.* 45B:1699–1703.

69. Megha, M., Bodhe, L. G., Rathi, U.N., Mahajan, V.V.R. 2017. Development of microwave assisted synthesis for common molecules. *Int. Res. J. Pharm.* 8 (1):25–27.

70. Prasad, S.R., Saraswathy, T., Niraimathi, V. 2012. Synthesis, characterization and antimicrobial activity of some hetero benzocaine derivatives. *Int. J. Pharm. Pharm. Sci.* 4(5):285–287.

# Index